Lecture Notes in Earth Sciences 40

Sunday W. Petters

Regional Geology of Africa

Springer-Verlag
Berlin Heidelberg New York
London Paris Tokyo
Hong Kong Barcelona
Budapest

Author

Sunday W. Petters
Department of Geology
University of Calabar
Calabar, Nigeria

"For all Lecture Notes in Earth Sciences published till now please see final page of the book"

ISBN 3-540-54528-X Springer-Verlag Berlin Heidelberg New York
ISBN 0-387-54528-X Springer-Verlag New York Berlin Heidelberg

Printed in Germany

Typesetting: Camera ready by author
Printing and binding: Druckhaus Beltz, Hemsbach/Bergstr.
32/3140-543210 - Printed on acid-free paper

Dedicated to:

Wissenschaftskolleg zu Berlin

- Institute for Advanced Study -

Preface

This book represents the first attempt in three decades to marshall out available information on the regional geology of Africa for advanced undergraduates and beginning graduate students. Geologic education in African universities is severely hampered by the lack of a textbook on African regional geology. This situation is greatly exacerbated by the inability of most African universities to purchase reference books and maintain journal subscriptions. Besides, geologic information about Africa is so widely dispersed that a balanced and comprehensive course content on Africa is beyond the routine preparation of lecture notes by university teachers. Since geology is a universal subject and Africa is one of the largest landmasses on Earth with one of the longest continuous records of Earth history, there is no doubt that geologic education in other parts of the world will benefit from a comprehensive presentation of African geologic case histories. The scope of this text also addresses the need of the professional geologist, who may require some general or background information about an unfamiliar African geologic region or age interval.

Africa occupies a central position in the world's mineral raw materials trade. Because of its enormous extent and great geologic age, the diversity and size of Africa's mineral endowment is unparalleled. Africa is the leading source of gold, diamond, uranium, and dominates the world's supply of strategic minerals such as chromium, manganese, cobalt, and platinum. Consequently, African nations from Algeria to Zimbabwe depend solely on mineral exports for economic survival. The geologic factors which govern economic mineral deposits are stressed in this text.

The geological history of Africa spans 3.8 billion years, a record that is unique both in duration and continuity. Few other parts of our planet match the plethora of geologic phenomena and processes that are displayed in the African continent. From the various stages of crustal evolution decipherable from the Archean of southern Africa, through the plate tectonics scenarios in the ubiquitous Late Proterozoic-Early Paleozoic Pan-African mobile belts and in the Hercynian and Alpine orogenies of northwest Africa, to the East African Rift Valley, Africa is replete with excellent examples and problems for a course on regional tectonics.

Teachers of igneous and metamorphic petrology can hardly ignore Africa's anorogenic magmatism (e.g.. layered ultramafic intrusives such as the Great Dyke and the Bushveld Complex; the Tete gabbro-anorthosite pluton; alkaline complexes; basaltic volcanism), or tantalizing high-grade metamorphic terranes such as the Limpopo belt, the Namaqua mobile belt, and the Mozambique belt.

From the extensive Precambrian supracrustal sequences throughout the continent with enormous thicknesses of sedimentary rocks that have hardly been deformed or metamorphosed, to the stratigraphic evolution of Africa's present-day passive continental margin, there is a complete spectrum of facies models upon which to base a course on basin analysis and stratigraphy.

To maintain its integrity a course on historical geology anywhere in the world must address the theory of Continental Drift beyond invoking past continuities between West Africa and South America. Past connections between West Africa and eastern North America must equally be explored, so also connections between northeast Africa and Arabia, and the paleogeography of southern Gondwana where Africa occupied centre stage. The Precambrian fossil record, the transitions from reptiles to the earliest mammals and dinosaurs, and the evolution of Man are among Africa's unique contributions to the history of life and the story of organic evolution. Although it lies today in the tropics Africa was the theatre of the Earth's most-spectacular glaciations. Even after the scene of continental glaciation had shifted to the northern continents only lately during the Pleistocene, Africa still witnessed spectacular climatic fluctuations during the Quaternary. Certainly students of archeology and paleoanthropology cannot overlook the Quaternary paleoenvironmental record of the Olduvai Gorge in Tanzania, the Lake Turkana basin in Kenya, the Nile valley, the Sahara, and southern Africa.

But since African examples have already been cited in standard geologic textbook, I have often been asked why it has become necessary to revive the idea of a full-length textbook on African geology, 30 years after this idea was abandoned by the geologic community. My simple answer, as already stated, is that the wealth of available geologic information about Africa is so enormous and fascinating, but so diffuse, that an attempt must be made to assemble and pass on this knowledge.

Berlin, May 1991 Sunday W. Petters

Acknowledgements

I would like to acknowledge the unique opportunity afforded me by two German institutions to write this text. This text was prepared during eleven months of residence at the Wissenschaftskolleg zu Berlin (Institute for Advanced Study Berlin). I thank this institution for fellowship support, and excellent bibliographic and secretarial assistance through which I enjoyed limitless access to geologic literature on Africa, and equally important, had the manuscript typed. Secondly, I wish to thank Prof. Dr. Eberhard Klitzsch of the Technical University Berlin who is the project leader of Special Research Project "Geoscientific Problems in Arid and Semiarid Areas" (Sonderforschungsbereich 69) funded by the German Research Foundation (DFG). Special Project 69 is devoted to geoscientific research in northeast Africa (Egypt, Sudan, Somalia, Ethiopia) and has graduate students from various parts of Africa. My visit to the Wissenschaftskolleg was through the recommendation of Professor Klitzsch. During the preparation of the manuscript I benefited enormously from discussions with and suggestions from the geologists in Special Project 69.

The idea of writing a textbook on the regional geology of Africa was conceived during my 15 years of teaching various geology courses at five Nigerian universities. During this period I sought to enrich my course contents by visiting several European libraries and museums. In this respect I wish to thank Dr. M.C. Daly and his wife and Dr. C.S. Orereke for their hospitality during my visit to the University of Leeds in 1984. Dr. M. Oden was my host in London that year during my visit to the British Museum of Natural History, the Geological Museum, and the Imperial College library. I thank Prof. P. Bowden and Dr. J.A. Kinnaird for their hospitality during the colloquium on African geology at St. Andrews University in 1985. Professor H.P. Luterbacher was very helpful during my visit to the University of Tübingen in 1987.

I am generally greatly indebted to all geologists who have worked in Africa, from whose publications I have drawn the material for this text. I would also like to thank especially all those who sent very important books, reprints, and pre-prints of their publications on Africa. These include Profs. J.B Wright, S.J. Culver, R. Caby, B.-D. Erdtmann, L.B. Halstead, V. Jacobshagen, J.A. Peterson, P. Wilde, C.O. Ofoegbu, J.D. Fairhead, I. Valeton, J.R. Vail, and N.J. Jackson. Professors E. Buffetaut and L.L. Jacobs supplied illustrations of African vertebrates.

I am greatly indebted to Prof. Rushdi Said who was also in residence at the Wissenschaftskolleg during the 1989/90 session. Professor Said's constant advice and encouragement kept up my spirits. I thank Profs. R.K. Olsson, R.C. Murray and B.W. Andah for encouraging me to pursue this project over the years.

I am very grateful to Drs. M.C. Daly and G. Matheis who read through the Precambrian chapters and offered very useful suggestions. I also thank Drs. S. Muhongo and H. Schandelmeier for their comments on some of the Precambrian chapters. Dr. Muhongo greatly improved my coverage of East Africa. Professor N. Rutter kindly reviewed the Quaternary chapter.

My special thanks go to Ms R. Plaar for the preparation of the manuscript, and all the secretarial staff of the Institute for their invaluable help. Special appreciation goes to Mrs Maria A. Gowans and Ms Linda O'Riordan who prepared the final camera-ready manuscript, for their patience and hard work. I would also like to acknowledge Mr Reinhard Prasser, who, in addition to his great hospitality, served at the final moment as the liaison with the publisher. Excellent bibliographic assistance was provided by Mrs Gesine Bottomley and her staff at the Wissenschaftskolleg and by Mrs Evelyn Kubig of the Geology Library of the Technical University, Berlin. Messrs Umo Harrison, E. Umo, Joe Sams, and Richard Ingwe and his colleagues rendered cartographic assistance.

I thank Professor Charles Effiong, Vice-Chancellor of the University of Calabar for moral and material support. Dr. Alfred Koch, chairman and managing director of Mobil Producing Nigeria and Mr. Wande Sawyerr, exploration manager of Mobil, also encouraged this project.

Finally, on behalf of my wife Janet, and my children, Mfon, Emem, Ekanga and Unwana, who were with me in Berlin, I wish to express profound gratitude to the Rektor of the Wissenschaftskolleg, Prof. Dr. Wolf Lepenies and his wife for their hospitality. All the staff of the Wissenschaftskolleg were very friendly to us, and for this we are very grateful.

TABLE OF CONTENTS

CHAPTER 7 PRECAMBRIAN GLACIATION AND FOSSIL RECORD

CHAPTER 8 PALEOZOIC SEDIMENTARY BASINS IN AFRICA

Chapter 1 Introduction

1.1 The Physical Setting of Africa

Africa is the second largest continent, occupying one-fifth of the land surface of the Earth. Surrounded on all sides by oceans, the African continent is like a huge island. The boundaries of the African plate (Fig.1.1), except on the northern side, lie along mid-oceanic ridges. The African plate is growing in size as new material is accreting along these spreading centres. But what Africa gains is lost elsewhere by subduction in the global system of moving plates. World-wide estimates of the rates of plate motion indicate that the African plate is moving slowly towards the northeast at the rate of about 2 cm/yr.

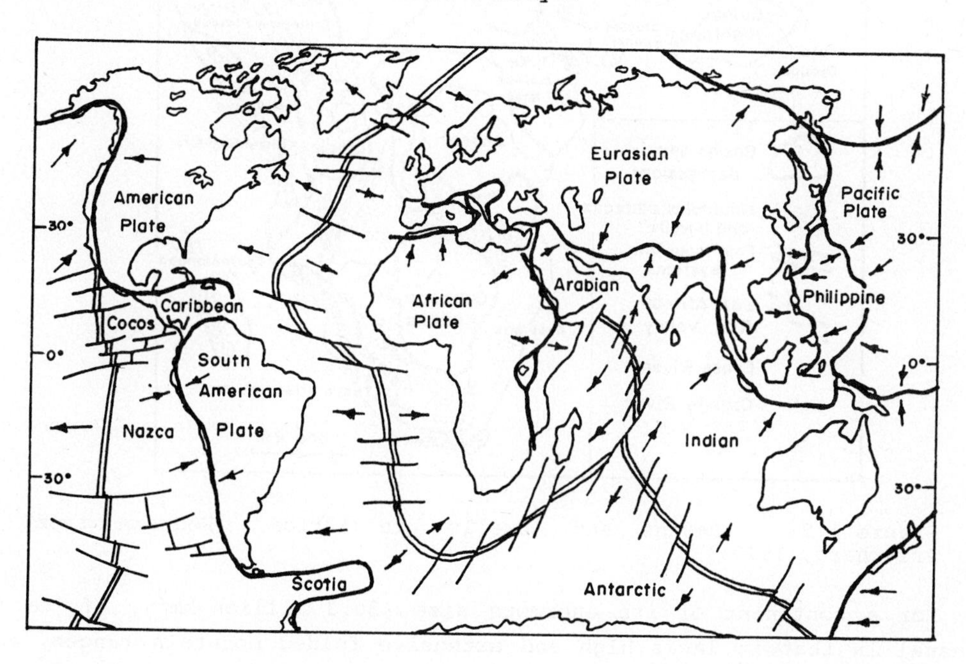

Figure 1.1: Major plates of the Earth; spreading directions are shown with arrows. (Redrawn from Braithwaite, 1987.)

Africa is the most tropical of all the continents, for it lies almost evenly astride the equator, and extends from 37°51'N to 37°51'S. However, the African climate and vegetation are quite extreme. They range from extremely hot and arid in the Sahara in the north and the Kalahari and

Namib deserts in the southwest (Fig.1.2), through tropical rain forests to tundra on the highest snow-capped mountain peaks located right on the equator. A Mediterranean type of climate and vegetation prevails in the northern and southern extremities of the continent with low shrubs, evergreen bushes, and forests. The climate and vegetation of Africa are discussed in greater detail in the final chapter in relation to the environmental changes in Africa over the last 2.5 to 1.8 million years.

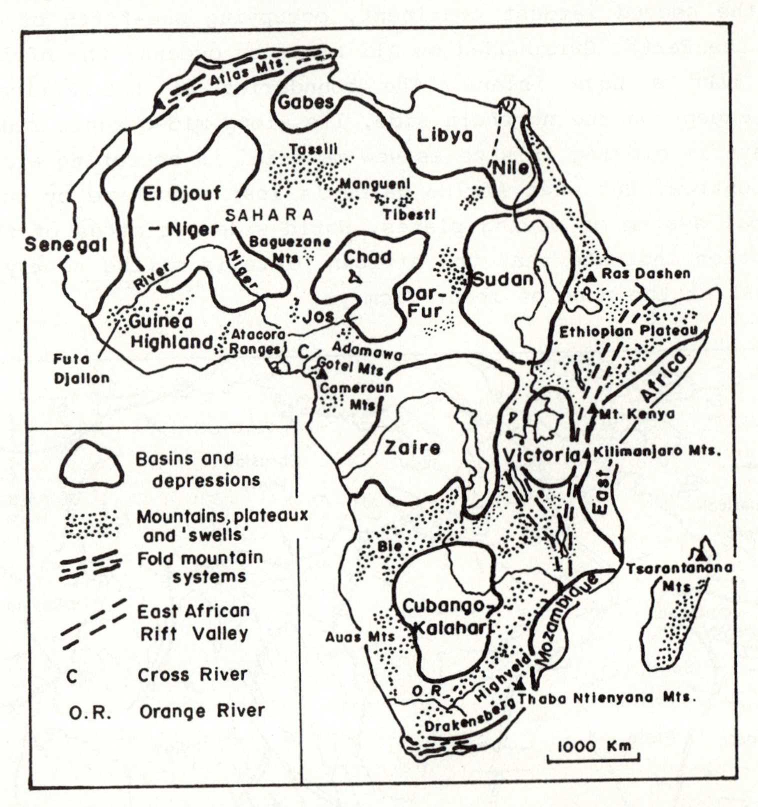

Figure 1.2: Basins and "Swells" in Africa. (Redrawn from Pritchard, 1979.)

For a continent of its enormous size (30.3 million km^2), Africa is unusual in that it lacks high and extensive folded mountain ranges, except the Atlas ranges (2,100 m high) in the northwest and the Cape ranges in South Africa (1,800 m high). This morphology, however, belies the fact that Africa has the largest area of basement terrain with ancient mountain belts which have been completely bevelled and exposed at their deep roots. Steady uplift, deep weathering, and erosion are the dominant surface processes that have shaped the African continent over the last 450 million years.

The topography of Africa is characterized by basins and swells (Fig.1.2). Basement upwarps form large domes or shields while extensive interior basins lie in broad basement downwarps. The swells are highest where capped by volcanic flows as in East Africa and central West Africa. Generally the continent can be described as a large uneven plateau that is higher in the eastern and southern parts and lower in the west and north. The continent thus appears to be tilted to the northwest.

Rising abruptly above sea level to a rolling upland 2,000-2,400 m high, the Ethiopian swell (Fig.1.2) is part of an eastern African swell which continues through Kenya where it is 3,000-4,000 m high, and extends with interruptions into South Africa. The East African Rift Valley has ruptured through these swells and created some of the most spectacular horst and graben landscapes on Earth. West of Lake Victoria, on the western arm of the rift valley system, towers the Ruwenzori Mountain, a basement horst 5,000 m high clad with snow. From the Ruwenzori the rift valley drops down a fault scarp to the rift floor, 4,000 m below. Such spectacular fault scarps are not unusual along the rift valley in Kenya and Ethiopia. On the basement upwarps which form the shoulders of the rifts in Kenya and Tanzania, stand two snow-clad mountains, Mt. Kenya (5,199 m high) right on the equator, and Mt. Kilimanjaro (5,895 m) in Tanzania to the south. Both mountains are actually composite volcanoes, and part of the vast volcanic fields associated with the East African Rift Valley. Lakes, great and small, fresh and saline, occur in the rifts, the deepest being Lake Tanganyika (1,470 m).

Prolonged crustal stability punctuated by uplifts have sustained cycles of scarp retreat and erosion of extensive surfaces, especially in eastern Africa where regional planation surfaces were first recognized. Uplifts have also created great escarpments in the southeastern part of the continent. But by far the most profound and widespread geomorphic process that characterizes Africa is the development of erosion surfaces and resultant heavily leached residual soils. Rich in secondary oxides of iron (laterite), aluminium (bauxite) or both, and deprived of nutrients, these soils are inimical to agriculture. They are, however, sometimes conducive to the concentration of mineral deposits such as bauxite, manganese, iron ore, and gold.

Another characteristic African product of prolonged deep tropical weathering, scarp retreat, and stream incision are inselbergs. These are small isolated steep-sided residual hills made of resistant rock. Inselbergs break the monotony of the African great plains. They are best developed in open woodlands and grasslands on the plateau country of Africa where they have created a distinctive scenery, for example in the savanna region of West Africa, the Masai steppe of Kenya, and in the Great Karoo of South Africa.

African drainage systems also bear the imprints of uplift in that they are frequently interrupted by waterfalls and rapids. This has endowed most parts of the continent with almost limitless hydro-electricity potential. While some of the principal rivers such as the Nile, the Niger and the Orange discharge their sediment load into large deltas across their mouths, others such as the Zaire River empty through submarine canyons into deep-sea fans on the ocean floor. The major deltas including the Cross River delta in southeastern Nigeria also construct deep-sea fans. For the most part the African continental shelf is narrow and sandy, but along the eastern African coasts and shelves there are areas of carbonate sedimentation, where coral reefs thrive.

With huge reserves of petroleum in the Niger delta, and diamonds strewn in alluvial terraces along the Orange River in South Africa, and diamonds on the beaches of Namibia and on the shallow shelf of the Orange delta, the economic potentialities of African rivers sometimes sound like fairy tale.

1.2 Geological History and Mineral Deposits of Africa

The geological record of Africa spans at least 3.8 billion years of Earth history. Few other continents, notably West Greenland, North America, and the USSR match this antiquity and continuum of geological history.

Whilst only North America exceeds Africa in the overall spatial extent of the rocks that formed between 3.8 and 2.5 billion years, in South Africa alone the rocks of this age have supplied over half of the world's gold. Most of the world's chrome reserves lie in the Great Dyke of Zimbabwe (Fig.1.3) which is about 2.5 billion years old. Apart from mineral production, the wealth of information that has accrued from the highly diverse and very peculiar rocks of this age in southern Africa has inspired classical geological models and treatises about this period. Widespread occurrences of these early rocks in the African basement, either as ancient nuclei or as relicts, attest to the consolidation of what is now Africa, so long ago. Although models of this phase of Earth history indicate a hotter planet with smaller, thinner and more mobile plates than exist today, enough evidence is emerging from Africa to suggest that tectonic processes 3.8-2.5 billion years ago were generally compatible with plate tectonic processes. These rocks, minerals, and models are discussed in Chapter 3.

Figure 1.3: Outline map of Africa showing some major mineral deposits.

The period between 2.5 and 1.75 billion years ago was crucial in Earth history. Large parts of southern Africa had by that time attained sufficient stability and rigidity to form the sites of extensive intracontinental sedimentary basins. Shallow water sandstones and limestones with algal mats accumulated profusely for the first time in these primordial seas. But the global importance of these early South African basins lies in their gold, uranium, manganese, iron ore, fluorite, copper and zinc deposits. Chapter 4 deals with this phase of African geological history.

In central and western Africa mountain-building processes quite similar to those of later geological periods created major mountain chains 2.5-1.75 billion years ago. Gold, diamond and manganese in Ghana, and rich uranium and manganese deposits in Gabon (Fig.1.3) are among the major mineral deposits of this period.

Studies on paleomagnetism which have led to reconstructions of past continents show that one supercontinent emerged from the above episode of mountain-building. From the matching of the rocks dated between 1.75 billion years and 950 million years it has been established that Africa and South America formed one continent this long ago. Africa, like other parts of this supercontinent, was mostly quiescent, except for rifting and mountain-building in the eastern part of central Africa (Chapter 5). Paleomagnetic reconstructions of past continental positions and assemblies during the period between 950 and 450 million years, and distinctive rock types reveal that mountain-building processes operated in accordance with modern plate tectonics. Chapter 6 is replete with many fine examples of ancient mountain chains in Africa that formed during this period. Now exposed at their deep roots in linear belts throughout western, central and eastern Africa, these mountain chains, like the Alps and the Himalayas, formed by the opening and closing of oceans involving the collision of ancient continents. The rifting and volcanism which preceded the opening of one of Africa's oceans between 950 and 450 million years ago created one of the world's largest deposits of cobalt and copper in the Zambian-Zairean copperbelt (Fig.1.3).

It is pertinent to mention at this juncture a major Africa-inspired contribution to the geological sciences--the theory of Continental Drift. From the original ideas of Alexander von Humboldt in the 19th Century, through the theoretical formulations of Alfred Wegener in 1912, and the practical demonstrations of Alex du Toit in 1937, the connection between Africa and South America has been the focus of the Continental Drift theory. Whilst the early workers derived their restorations of the unity between both continents from the matching of the present coastlines, and on the evidence of Permo-Carboniferous glaciations and other geological similarities, modern workers base their reconstructions on radiometric ages and paleomagnetism. Reconstructions of Gondwana, the southern continent to which Africa and South America belonged at the end of the

950-450 year interval (Late Proterozoic-Early Paleozoic), have furnished the framework for understanding the subsequent geological history of the African continent. The history of life in the Precambrian and the record of Africa's early glaciations are reviewed in Chapter 7.

African sedimentary basins, the subjects of Chapters 8 and 9, record essentially the history of marine transgressions and regressions, except along the Atlas and Cape fold belts where mountain-building processes in other parts of the world marginally affected Africa. Paleozoic marine transgressions climaxed in the Early Silurian, the Mid-Devonian, and in the Early Carboniferous. Paleomagnetic reconstructions of the shifting positions of Gondwana reveal that the South Pole was located in northwest Africa in the Late Ordovician. This caused widespread continental glaciation in Africa, followed by the extensive Early Silurian transgression after the melting of the polar ice caps. During the Late Carboniferous-Permian southern Gondwana moved near the South Pole, thus triggering another widespread glaciation which affected all of southern Gondwana. This marked the beginning of a distinctive phase of continental sedimentation known as the Karoo cycle. Referred to as Gondwana formations in India, South America, Australia, Antarctica, the deposits of the Karoo cycle accumulated mostly in continental rifts. They contain extensive coal measures and uranium, the distinctive southern *Glossopteris* flora, and unusually abundant reptiles with mammal-like features showing transitions towards the earliest mammals and dinosaurs.

The Mesozoic-Cenozoic history of Africa was dominated by the fragmentation of Gondwana and the formation of the present continental margins and marginal basins along the Atlantic, Indian Ocean, and the Red Sea and the Gulf of Aden. Major intracontinental rift basins in Africa formed during the break-up of Gondwana. The igneous activities that attended this continent-wide phase of rifting, from the end of Karoo sedimentation to the initiation of the East African Rift Systems, are reviewed in Chapter 10.

Chapter 2 The Precambrian of Africa: An Introduction

2.1 Tectonic Framework

A means of appreciating the vastness of Precambrian crust in Africa relative to other continents is to glance at the tectonic or geological map of the world. The tectonic map of the world recently compiled by Condie (1989) shows that Africa has the largest area of Precambrian crust, followed by North America and Antarctica.

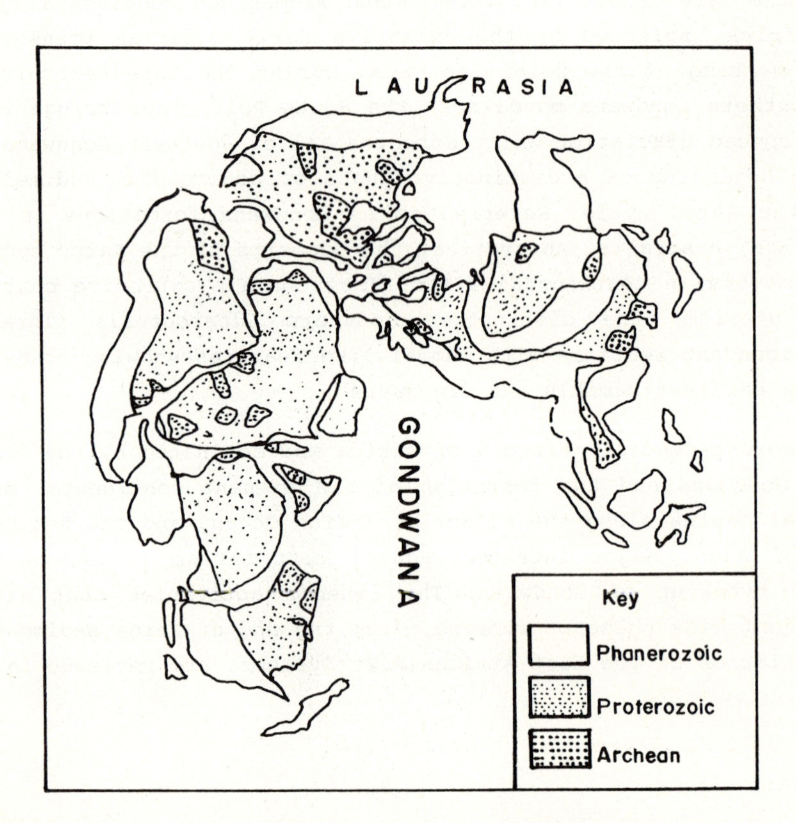

Figure 2.1: Pre-Mesozoic drift reconstruction of the Earth showing approximate extent of the Precambrian. (Redrawn from Windley, 1984.)

But in lieu of a global geological or tectonic map which cannot be conveniently reproduced here, the relative extent of the African Precambrian can still be appraised from a highly schematic pre-Mesozoic drift reconstruction of the continents which simply depicts the Precambrian and Phanerozoic regions of the world (Fig.2.1). This map clearly shows that Africa is almost entirely made up of Precambrian rocks, except along the northwestern and southern margins of the continent where narrow Phanerozoic mountain belts abut the Precambrian landmass. A study on African geology is therefore essentially a study on the Precambrian, especially in the many African countries that are completely underlain by Precambrian rocks. (Fig.2.2).

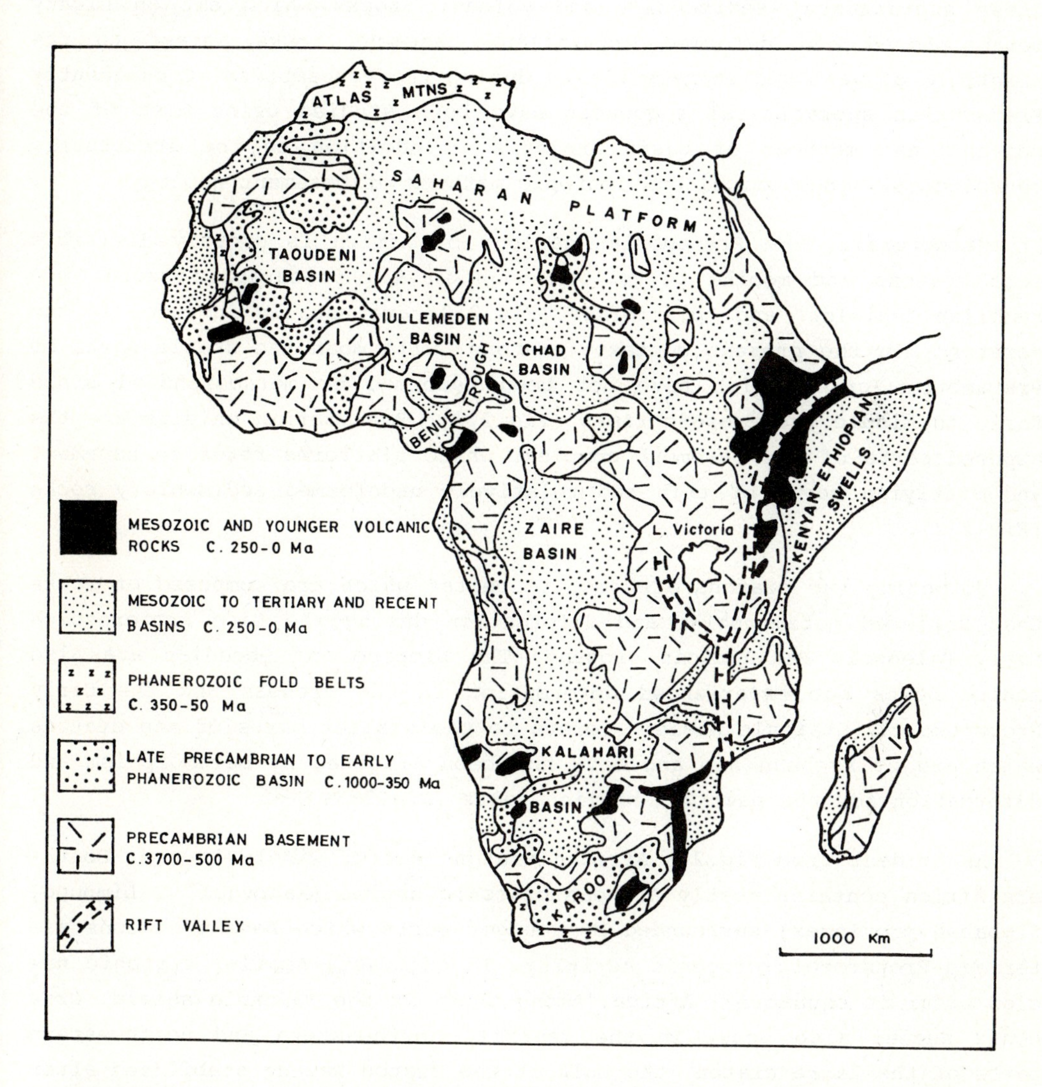

Figure 2.2: Geological outline map of Africa showing basement outcrops and basins. (Redrawn from Wright et al., 1985.)

The unparalleled diversity of African Precambrian rocks and mineral wealth and the complete span of the Precambrian age represented on the continent reinforce the preeminence of the Precambrian in Africa.

The term "basement complex" is commonly loosely used in African countries to refer to Precambrian rocks even though "basement" rocks range into the Early Paleozoic and contain significant amounts of unmetamorphosed and many undeformed Late Proterozoic-Early Paleozoic sequences. These supracrustal sedimentary and volcanic rocks which sit on highly metamorphosed and deformed crystalline basement rocks attest to the existence of vast sedimentary basins during the Precambrian. Consequently Precambrian supracrustal sequences have been studied using most of the conventional methods of basin analysis in addition to the structural, petrological, geochemical and isotopic methods of basement geology.

Structurally the Precambrian geology of Africa is grossly divisible into cratons and mobile belts. Although in a physiographic sense Precambrian "shields" and "platforms" (Fig.2.2) are usually included in the "craton", in the present context "cratons" will mean the stable parts of Precambrian crust which have not been deformed or metamorphosed since Early to Middle Proterozoic times (Fig.2.3). Precambrian shields are the exposed parts of the basement complex, while platforms refer to basement and overlying cover of thin and relatively undeformed sedimentary rocks (Fig.2.2).

Bordering the cratons are "mobile belts" which are composed of rocks that suffered metamorphism and deformation during the Late Proterozoic-Early Paleozoic Pan-African orogeny. The Limpopo and Ubendian are also mobile belts but experienced deformation in the Archean and the Early Proterozoic. "Cratonic nuclei" refers to the smaller parts of the cratons which are of Archean age and have not been affected by metamorphism and deformation for the past 2.5 billion years (Fig.2.3).

As evident from Fig.2.3 African cratons differ widely in age. Southern Africa contains mostly Archean cratonic nuclei (Kaapvaal , Limpopo, Zimbabwe provinces) surrounded by younger parts which became cratons after Mid-Proterozoic orogenic activity. In contrast, smaller cratonic nuclei occur in equatorial Africa. Among these is the Tanzania shield. Cratonic nuclei also occur in the central, northeastern and northwestern parts of the Zaire craton, the bulk of the craton having stabilized after an Early Proterozoic orogeny, like the West African craton. The Bangweulu block in central Africa is entirely of Early Proterozoic age and has only locally been involved in major orogenic activity since then. A poorly exposed and poorly defined cratonic area of considerable tectonic significance seems to stretch north of the Zaire craton as far as Jebel Uweinat, where Archean and Early Proterozoic rocks outcrop in the northern part of what is regarded as the East Saharan craton.

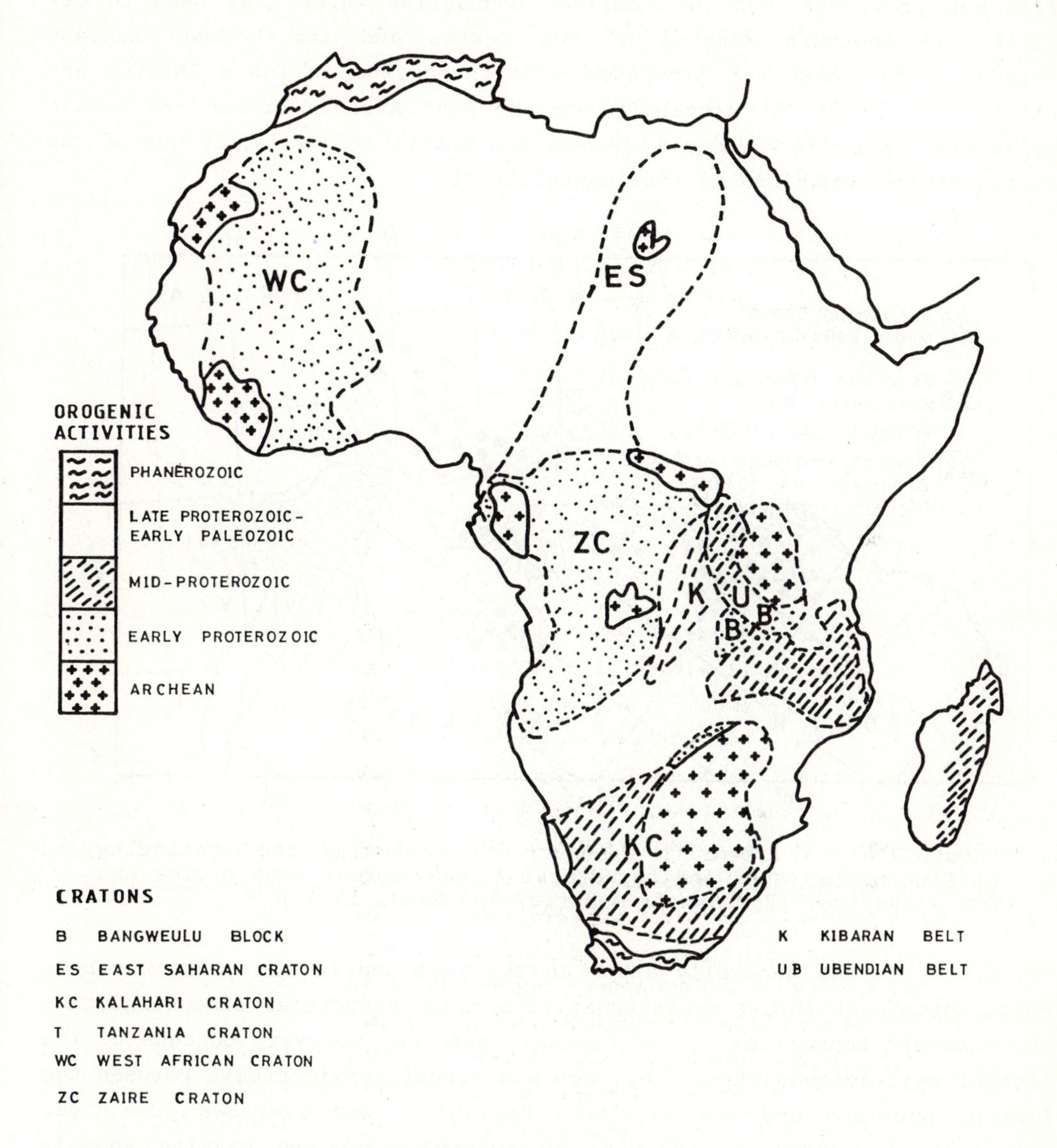

Figure 2.3: Cratons and mobile belts in Africa.

The boundaries between cratons and mobile belts are sometimes clearly defined based on structural, geophysical, radiometric, and metamorphic discontinuities. Thus, the limits of the West African craton have been clearly defined by the so-called circum-West African craton belt of gravity highs (Briden et al., 1981; Roussel and Lécorche, 1989). In one of the earliest systematic regional radiometric age surveys across the Precambrian of West Africa and South America, Hurley and Rand (1973) identified age provinces with well-defined boundaries which they used to delimit the southern margins of the craton and its Archean nucleus (Fig.2.4). The same age provinces were recognized in South America and utilized to prove the continuity of the West African craton and mobile belts with those of Venezuela, Guyana and Brazil (Fig.2.4), in one of the strongest confirmations of continental drift.

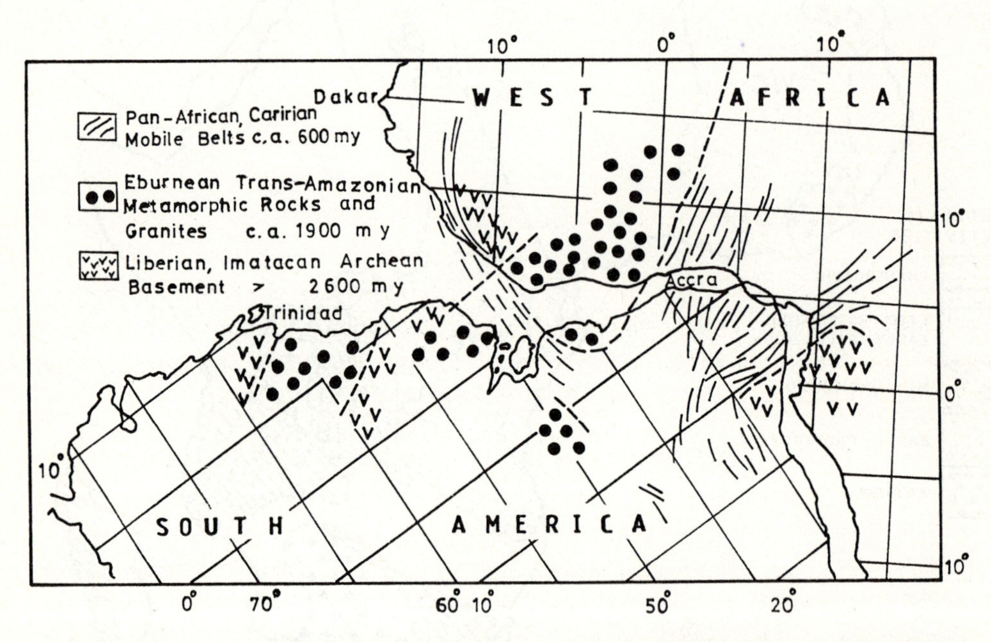

Figure 2.4: Pre-drift reconstruction showing the continuity of Precambrian ages and structural trends across West Africa and South America. (Redrawn from Hurley and Rand, 1973.)

Since cratons generally acted as the foreland to the younger mobile belts, prominent thrust zones constitute major structural discontinuities and tectonic boundaries around cratonic margins. However, because of the lack of well-defined structural, age and stratigraphic breaks between the Limpopo province and the adjoining Kapvaal and Zimbabwe provinces, metamorphic isograds in addition to seismological and gravity anomaly discontinuities, have been adopted as the northern and southern boundaries of the Limpopo province. Furthermore, in southern Africa Fairhead and Henderson (1977) showed that the gross distribution of earthquakes clearly outlines the major cratonic areas, seismicity being confined to the mobile belts. Fundamental differences also exist in the thermal structure between southern African cratons and mobile belts, with the latter exhibiting greater heat flow than the former. These differences reflect the cold and stable nature of the cratons which have thick lithosphere in contrast to the surrounding mobile belts which are often characterized by thicker crust but thinner lithosphere, especially along segments that are intruded by abundant granitoids and sliced by numerous shear zones (Black, 1984). Further attesting to the fundamental differences between African cratons and mobile belts is the fact that the zones of Mesozoic rifting which led to the break-up of Gondwana (Fig.2.1) were located along the all-encircling Late Proterozoic-Early Paleozoic Pan-African mobile belts.

2.2 The Precambrian Time-Scale

Cahen et al. (1984) presented a benchmark compilation and interpretation of available radiometric ages in Africa upon which they based their interpretation of the tectonic evolution of the continent. Their synthesis has provided the most cogent and comprehensive geochronological framework for describing the Precambrian regional geology of Africa and correlating it with those of other world regions.

Various Precambrian time-scales have been proposed for various regions of the world, but neither a review nor a critique of these geochronological scales will be attempted here. Let it suffice for our present purpose to simply highlight the principal age boundaries and broad subdivisions which are tenable for Africa. Among the available comprehensive and authoritative discussions of the Precambrian time-scale are those of Harland et al. (1982), James (1978), Salop (1983) and Sims (1980).

According to Cahen et al. (1984) the most significant "chronological milestones" for Africa occurred at 2.5 Ga (G stands for giga, meaning one billion years) for the Archean-Proterozoic boundary; 1.75 Ga for the Early-Middle Proterozoic boundary; and also Porada (1984) proposed 950 Ma (M denotes mega, meaning one million years) for the Middle-Late Proterozoic boundary. The subdivisions of the Archean used here (Fig.2.5A) are those recommended by the International Union of Geological Sciences (IUGS) Subcommission on Stratigraphy (Sims, 1980) which were also adopted by Tankard et al. (1982) for southern Africa. The Precambrian geochronological scale adopted for our present purpose is almost identical with the one used by Tankard et al. (1982).

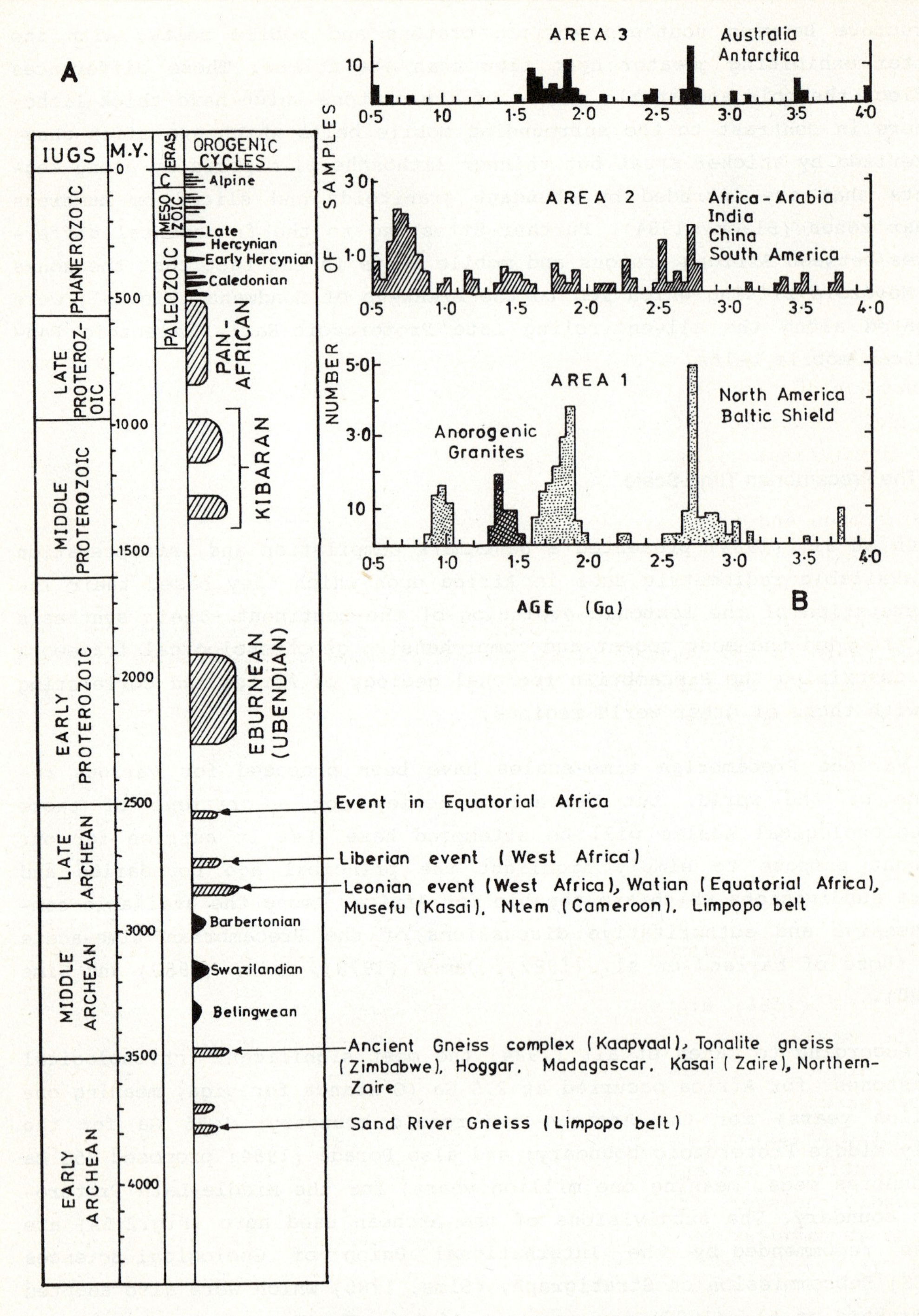

Figure 2.5: Distribution of African orogenies (A); B: Frequency distribution of U-Pb zircon ages from different Precambrian regions. (Partly redrawn from Condie, 1989.)

It is considered relevant here to outline the various arguments that have been advanced in support of some of the above boundaries, especially for the Proterozoic. Windley (1984) favoured the adoption of 2.5 Ga as the world-wide Archean-Proterozoic boundary. But he added that if the geological features that are associated with this major turning point in earth history are to be considered then a better Archean-Proterozoic boundary for Africa could perhaps be at about 3.0 Ga. In essence the Archean-Proterozoic boundary is diachronous, if the criterion for it is crustal evolution. The distinguishing features of the Archean-Proterozoic boundary which mark a major phase in the evolution of the earth's crust are: 1) the presence of a major unconformity separating highly deformed Archean rocks from slightly or undeformed and unmetamorphosed Early Proterozoic cratonic platform sequences; 2) Early Proterozoic mobile belts are stratigraphically and structurally very similar to modern orogenic belts formed by plate tectonics; and 3) there is greater abundance of greenstones and granulite-gneiss belts in the Archean. Condie (1989) related these and other geodynamic changes across the Archean-Early Proterozoic boundary to the cooling of the earth and the rapid growth of stable continental crust during the Late Archean. Since the oldest cratonic basin is the Pongola basin of the Kaapvaal province which is dated at between 3.1 Ga and 2.9 Ga, it means that crustal stability and rigidity had been attained earlier in southern Africa. But as Cahen et al. (1984) emphasized it was not until about 2.5 Ga that larger areas of the African continental crust became sufficiently stable and rigid (cratonic) to form the sites of platform basins. Also, the Great Dyke of Zimbabwe is dated at about 2.5 Ga. This dyke which is intruded into an abortive rift structure suggests, like other Archean-Early Proterozoic rifts in the cratonic basins of the Kapvaal province, that widespread stable continental crust had existed in southern Africa by 2.5 Ga.

Kröner (1984) argued that the Middle-Late Proterozoic boundary in Africa should be placed at the beginning of the last major and thoroughgoing orogenic phase, the Pan-African orogeny. During this orogenic phase which started about 950 Ma ago, Africa was differentiated into the presently observed pattern of cratons and mobile belts. Pan-African orogenic activity was episodic and lasted until about 450 Ma in the Early Paleozoic. However, for the upper boundary of the Precambrian, the age of 570 Ma recommended by the IUGS, is adopted here. This boundary is debatable. It has been placed at 590 Ma by Harland et al. (1982), and even at 650 Ma by Salop (1983). As emphasized by Windley (1984) the problem centres around where to assign the earliest known assemblage of soft-bodied organisms (Ediacaran fauna) which appeared towards the close of the Precambrian and heralded the beginning of the Phanerozoic eon. Just as the geochronological terms that have been proposed for the age interval containing Ediacaran faunas and rocks have varied (e.g. "Eocambrian", "Vendian", "Ediacarian", and "Infracambrian" in Africa), opinion is divided on whether to assign the Ediacarian interval to the latest Precambrian, the earliest Phanerozoic, or even to a separate "Ediacarian" or transitional age interval.

2.3 Orogenic Cycles in Africa

Orogenic cycles or orogeny can be defined as the tectonic processes which produced crustal deformation, metamorphism and magmatism in mountain belts. An orogenic or mobile belt is the crustal region where orogenic activity has taken place. Generally an orogeny involves a progressive sequence of tectonic events which in plate tectonics terms often implies the opening and closing of an ocean. Since orogenic processes are now better understood within the framework of plate tectonics it is better to consider African orogenic cycles in the context of plate tectonics even though some of the familiar and entrenched terminologies of the classical concept of the "geosynclinal cycle" will be used in this text for ease of expression. Furthermore, the successful applications of plate tectonics to the interpretations of African Precambrian terranes (e.g. Burke and Dewey, 1973; Burke et al., 1976; Caby, 1987, 1989; Daly, 1988; Lécorche et al., 1989; Kröner, 1984, 1989; Porada, 1989; Schandelmeier et al., 1989; Shackleton, 1986; Tankard et al., 1982; Wright et al., 1985) have greatly enhanced our understanding of African orogenic cycles and tectonic processes. The Late Proterozoic orogenic cycle in Africa has been most successfully interpreted within the framework of the Wilson Cycle (Burke et al., 1977) which attributes orogenic cycles and the formation of mountain belts to the opening and closing of oceans.

The Wilson Cycle incorporates the stages of the classical geosynclinal cycle. It starts with doming, rifting and the accumulation of nonmarine rift sediments and volcanics (Fig.2.6A), followed by crustal subsidence and the opening of an ocean basin during which thick piles of geosynclinal deposits accumulate along both the passive continental margins and on the adjacent seafloor. Subduction starts at some point and the ocean begins to close; and concomitantly, syntectonic magmatism is generated above the subduction zone. The ultimate result of ocean closure is continent-continent collision which causes intense deformation and plutonism; and finally differential uplift, volcanism and molasse deposition.

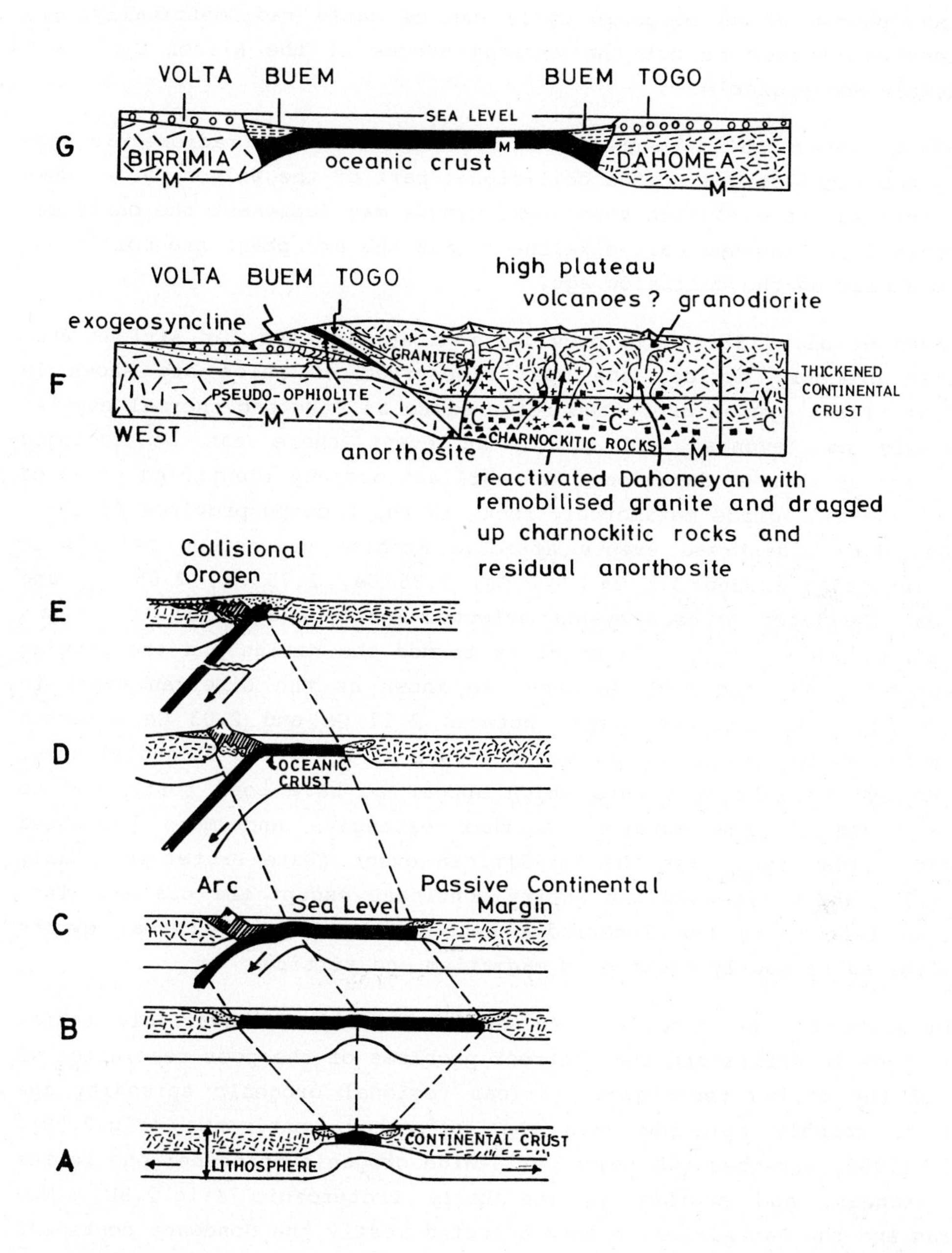

Figure 2.6: Idealized stages of the Wilson Cycle compared with a Pan-African collision suture in the southern Trans-Saharan mobile belt of West Africa. (Redrawn from Burke and Dewey, 1973; Condie, 1989.)

Whether formed by Wilson Cycle processes or according to some other geodynamic model, the important point that is stressed here is that the terminal phases of an orogenic cycle can be dated radiometrically, regardless of whether or not the various stages of the Wilson Cycle are completely decipherable.

Due to metamorphism, magmatism and deformation most radiometric ages record the final stages of the collisional part of the Wilson Cycle. However, this is not exclusively so--dyke events may represent the early extension and rifting and calc-alkaline magmas the arc phase pre-collision--hence a blur on the collision age.

Cahen et al. (1984) have provided radiometric ages for African Precambrian orogenic cycles, the most widespread of which are shown in Fig.2.5A. These cycles are often referred to as "tectono-thermal events" or simply as "events" and within an event there can be tectonic "episodes" of shorter duration . The earliest orogeny identified produced some of the high-grade metamorphic rocks in the Limpopo province at about 3.8 Ga. More widespread events affected Archean greenstone belts with ages clustering around 3.5 Ga, 3.2 Ga, 2.95 Ga, 2.75 Ga, 2.65 Ga, and 2.55 Ga. The later Archean events affected mostly equatorial Africa and West Africa where the 2.9 Ga event is termed the Watian and the Leonian respectively; and the 2.75 Ga event is known as the Liberian event in West Africa. The Eburnean event, between 2.27 Ga and 2.03 Ga affected nearly the whole continent, and was followed at 1.4 - 1.3 Ga and at about 1.10 Ga by the Kibaran events which appear to have been restricted to Africa south of the equator. Another extensive and more prolonged tectono-thermal cycle was the Pan-African event (Late Proterozoic-Early Paleozoic) which affected the entire continent except the cratons. Tectonic activities in the interludes between the tectono-thermal events were limited to mostly anorogenic magmatism and rifting.

In spite of the uncertainties which surround the available radiometric ages in Africa and the inherent problems of the poor resolution of some of the dating techniques, African regional orogenic episodity appears to roughly coincide with that of other continents (Fig.2.5B). Condie (1989) stressed two major world-wide orogenic episodes: one in the Late Archean, and another in the Early Proterozoic (Fig.2.5B). The Kibaran and the Pan-African events affected mostly the Gondwana continent (Fig.2.1) and did not seem to have strong counterparts in other continents.

Two notable features of African orogenic cycles deserve mention. First, outside the cratons, orogenies repeatedly affected the same zones of crustal weakness which ofttimes were the sites of earlier mobile belts. This is profoundly true of the Pan-African belts of East Africa where the Pan-African orogeny was superposed on the Kibaran and Ubendian (Eburnean) mobile belts (Fig.2.3). The consequence of this reworking or reactivation of older terranes is the preservation of older basement rocks, structures and radiometric ages as relicts in the younger rocks, a problem that has bedeviled Precambrian tectonic interpretations. Secondly, it is evident from Fig.2.3 that the distribution of orogenic cycles indicates the progressive growth of the continent with time. Most of the continental masses of the world are believed to have formed during the Late Archean-Early Proterozoic; and in Africa this is evident from the widespread Archean and Early Proterozoic relict ages found in the younger mobile belts. A natural consequence of the Wilson Cycle or orogenic cycle is the addition of new continent-type crust to the volume of the continents, due to oceanic subduction and calc-alkaline magmatism. This process of cratonization resulted from plate motions and collision. Cratonization is indeed evident in the Precambrian crustal evolution of Africa.

2.4 Dominant Rock Types

Before traversing the vast Precambrian terranes of Africa in an odyssey of immense time span, it is useful to distil out of the medley of Precambrian rocks a few salient characteristics of their composition and structure. In this regard the synthesis of Wright et al. (1985) is germane. Wright et al. stressed the point that regardless of geological age Precambrian rocks can be grouped into a basic stratigraphy of the basement complex, supracrustals, and granitic intrusions.

In addition to grouping them into basement, supracrustals and granitic intrusions, Precambrian rocks can also be broadly categorized according to their ages. As already pointed out the Archean-Proterozoic boundary separates Archean rocks with different characteristics from Proterozoic rocks, most of which exhibit features that are similar to those of modern orogenic belts formed as a result of plate tectonics. Condie (1989) enumerated the key and contrasting features of Archean and Proterozoic rocks.

Archean crustal provinces are dominated by two major rock types: high-grade rocks and granite-greenstone belts. Proterozoic rocks, in contrast, are highly varied. Seven major rock associations have been recognized in them. These are: a quartz-pelite-carbonate association which was characteristic of platform basins; bimodal volcanic-arkose-conglomerate and mafic dykes of continental rift or aulacogen tectonic setting; small amounts of greenstones which are similar to modern volcanic arc rocks; ophiolites like those of modern ocean ridges or back-arc basins; anorogenic granite-anorthosite complexes which were restricted to the Middle Proterozoic; mafic dyke swarms; and layered igneous intrusions. While arc and cratonic lithological assemblages have been recognized back to the Early Archean, continental rift and ophiolite assemblages are rare or non-existent prior to about 2.0 Ga. In Africa ophiolites became widespread during the Pan-African orogeny as a result of the operation of the Wilson Cycle in most parts of the continent.

Because of post-orogenic isostatic uplift and consequent erosion many African Precambrian orogenic belts are exposed at very deep crustal levels (Fig.2.6 F). Consequently most of the characteristic stratigraphic assemblages of convergent plate margins and plate collision sutures are lost (Burke and Dewey, 1973). Collision zones in deeply eroded mobile belts such as the Limpopo, Mozambique and Benin-Nigeria provinces are represented by cryptic sutures and high-grade metamorphic rocks.

The above outline of some of the parameters that will be used in subsequent chapters to discuss the Precambrian geology of Africa leans heavily on plate tectonics, even though this model has hardly been presented here in any comprehensive or systematic manner. Plate tectonics also aids our understanding of the processes that controlled the distribution of mineral deposits during Precambrian times (Sawkins, 1990). Hitherto, Precambrian metallogeny in Africa has simply been viewed as a function of age: older cratons contain important gold, iron, manganese, chromium, asbestos and diamond deposits; while younger mobile belts are characterized by major deposits of copper, lead, zinc, cobalt, tin, beryllium, and niobium-tantalum (Clifford, 1966).

Chapter 3 The Archean

3.1 Introduction

At the very beginning of geological time the Archean eon is very significant. A complete range of Archean rocks is represented in Africa, some of which, for example komatiites and greenstone belts, were first described from this continent. Being largely underlain by stable cratons, Africa perhaps stood the best chance of preserving the Archean geologic record either in isolated cratonic nuclei completely removed from later orogenic activities, or as relicts that had survived in the younger polycyclic mobile belts (Fig.2.3). A nearly complete span of Archean times, about 3.9 Ga to 2.5 Ga, is represented in Africa, where like in West Greenland, the oldest rocks on Earth are found.

In terms of overall spatial extent, the Archean rocks of Africa come second after those of North America. However, in the Republic of South Africa and Zimbabwe alone, the diversity of Archean rocks, their enormous mineral wealth (gold, diamond, chromite, cobalt, uranium, etc.), and their paleontological record are so far unmatched anywhere else in the world. Furthermore, the oldest well preserved cratonic sedimentary basins are found in South Africa which have furnished the earliest reliable record of the paleoenvironmental conditions that prevailed on the primordial Earth. It is hardly surprising therefore that southern Africa has been the cornerstone to our understanding of the early history of our planet, and consequently this region has inspired classical geological models and treatises on the Archean eon (e.g. Condie 1981; Nisbet, 1987).

In Africa, like elsewhere, several peculiar problems confront the study of the Archean. First, the Archean has been assigned a very long time duration, lasting for about 1.3 billion years, which is nearly a half of the remaining span of geological time. Since Archean rocks contain only algal stromatolites and doubtful bacteria and no index fossils by means of which stratigraphic subdivisions and correlation can be established, Archean regional stratigraphy is therefore very imprecise and uncertain, especially in a continent like Africa where vast geographical areas and thick rock sequences belong to this interval. Radiometric dating, field mapping, structural analysis, petrology, and geochemistry, and stratigraphic analysis are the primary tools for unravelling the Archean record. But further compounding the problems of interpretation, are the structural complexities found in Archean terranes, which are usually the products of multiple episodes of deformation, metamorphism and magmatism.

Some very peculiar rock types also occur in the Archean which in the absence of modern analogues have evoked a wealth of speculations about their origin. These include komatiites, banded iron-formations, and Archean greenstone belts. Conversely, the absence or rarity in the Archean of rocks such as carbonates and evaporites which are good paleoenvironmental indicators, attest to rather unusual conditions, which to say the least were very much unlike modern times. Archean oceans and atmosphere had different compositions from modern times; the biosphere did not exist; and since there was no vegetation cover, the rates of weathering and erosion must have been profoundly greater.

The Archean physical surrounding is believed to have suffered greater meteoric impacts; there were more volcanic eruptions; and since higher amounts of heat emanated from the mantle, conditions in and around the Archean ocean must have been simmering, even below the nascent crust, down to the lithosphere. Archean plate tectonic processes though compatible with the subduction of ocean crust, differed from the later Proterozoic and Phanerozoic ones as evident from the extensive occurrence of komatiites, tonalites and trondhjemites.

A striking feature of Archean rocks in all parts of the world (Fig.3.1) is the remarkable similarity of their gross lithologies. Two major lithological assemblages today characterize the Archean: greenstone belts and high-grade metamorphic terranes. Greenstone belts consist of thick and deeply infolded compact dark-green altered basic to ultrabasic predominantly volcanics and associated sediments which have suffered low-grade metamorphism and intensive granitic intrusions. Sharply contrasting are the high-grade terranes comprising various granitic gneisses, amphibolites and metasediments which have been subjected to high-grade metamorphism, often at the granulite facies. The structural relationships of granite-greenstone and high-grade terranes are often uncertain so that their relative ages are often debatable.

The regional pattern of the crustal evolution in Africa during the Archean is one in which the granite-greenstone terranes of the Kaapvaal province of South Africa were the earliest to stabilize, between 3.2 Ga and 3.1 Ga; followed by the Zimbabwe province at about 2.5 Ga; while those of the Zaire-Tanzania craton and the West African craton also stabilized at the end of the Archean (Fig.3.2). Regionally the greenstone belts show a northward decrease in Africa in their state of preservation and in their lithofacies development. Repeated metamorphism during and after the Archean reduced the preservation of greenstone belts outside the Kalahari craton and caused the preponderance of high-grade gneiss and granitoid terranes in the northern cratons of Africa. Below, the Archean provinces of Africa are considered from the south to the north, beginning with the Kalahari craton where they are best preserved and better known.

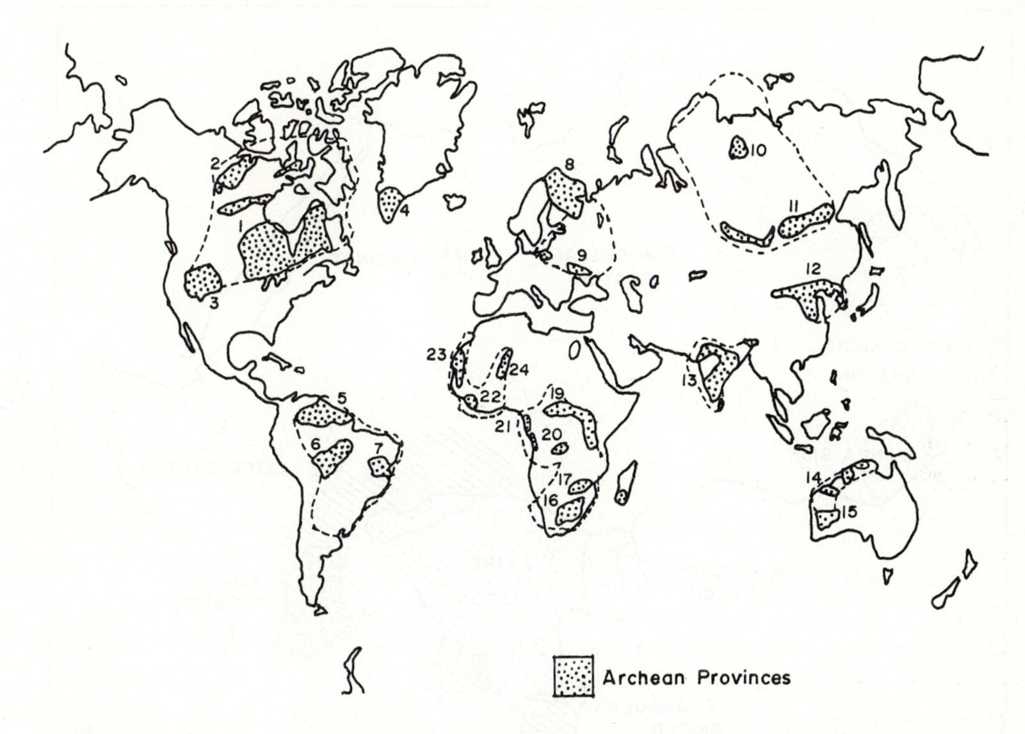

Figure 3.1: Archean provinces of the Earth: 1, Superior; 2, Slave; 3, Wyoming; 4, North Atlantic; 5, Guyana; 6, Guapore; 7, Sao Francisco; 8, Kola; 9, Ukrainian; 10, Anabar; 11, Aldan; 12, Chinese; 13, Indian; 14, Pilbara; 15, Yilgarn 16, Kaapvaal; 17, Zimbabwe; 19, NE Zaire Craton; 20, Kasai; 21, NW Zaire Craton; 22, Liberian; 23, Mauritanian; 24, Ouzzalian. (Redrawn from Condie, 1981.)

3.2 Kalahari Craton

The Kalahari craton comprises the Kaapvaal craton to the south and the Zimbabwe craton to the north, separated in the middle, around the tri-state (Botswana, Zimbabwe, South Africa) border, by the Limpopo orogenic belt (Fig.3.3). But to avoid the redundancy of the term craton, and in order to emphasize the lithologic, structural, metamorphic and radiometric age similarities and distinctiveness of each Precambrian region, the terms tectonic province (Kröner and Blignault, 1976) or domain are used here. The Limpopo province is included in the Archean part of the Kalahari craton because it was stabilized during the Late Archean tectonothermal events which also affected the Zimbabwe province to the north.

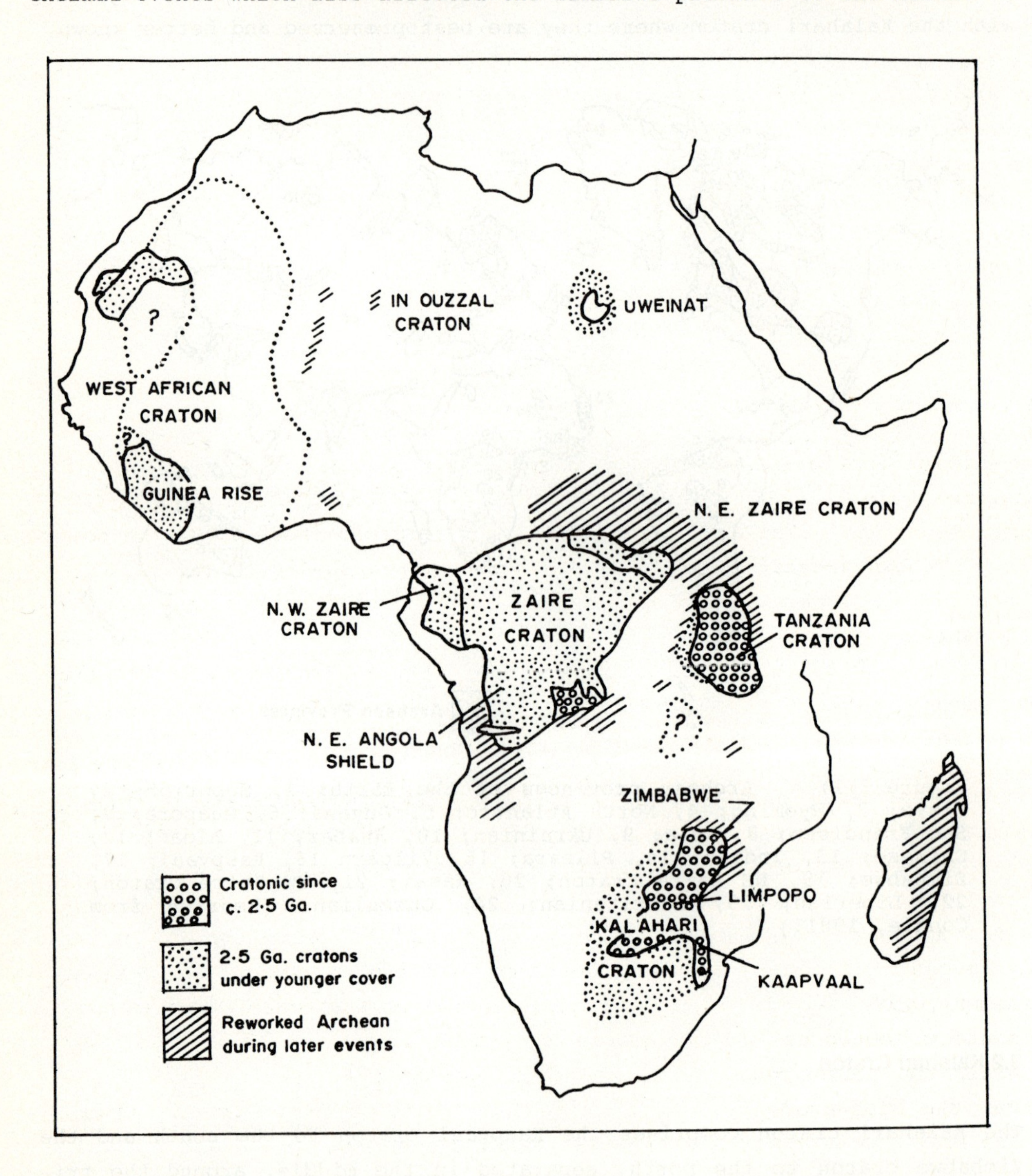

Figure 3.2: Distribution of Archaean cratonic nuclei in Africa. (Redrawn from Cahen et al., 1984.)

Unlike the Kaapvaal and Zimbabwe provinces, rocks of granulite facies metamorphism with stronger deformation predominate in the Limpopo province which is believed to represent the root zones of the Archean geosuture along which there was repeated shearing and overthrusting of the Kaapvaal province over the Zimbabwe province (Coward et al., 1976; Burke et al., 1977). The tectonic link between these three provinces has manifested in the progressive increase in metamorphic grade from the borders of the Zimbabwe and Kaapvaal provinces towards the central zone of the Limpopo province. As shown by Tankard et al. (1982) the Kalahari craton is surrounded on all sides by younger Proterozoic Pan-African mobile belts. The geology of the Kaapvaal, Zimbabwe and Limpopo domains, are summarized below according to the three principal Archean rock assemblages - the high-grade gneissic basement, the greenstone or schist belts, and the intrusive granitoids. The earliest cratonic sedimentary basin in the Kaapvaal province is also discussed.

3.2.1 Kaapvaal Province

Detailed investigations of the Archean greenstone belts of the Kaapvaal province (Fig.3.2) by several workers (e.g. Viljoen and Viljoen, 1969; Anhaeusser, 1971; Tankard et al., 1982) has rendered these among the best known Archean greenstone belts in the world. However, the global significance of the Barberton Mountain Land, the principal greenstone belt of the Kaapvaal province, lies in its excellent geologic exposures, the location of some of the earliest evidence of life, and in the fact that the Barberton Belt is the type locality of komatiites, the unique Archean magnesian ultramafic lavas.

Together with the more northerly greenstone belts of the Kaapvaal province, the Kaapvaal supracrustals comprise only about 10 % of the exposed Archean rocks in the Kaapvaal province, the vast remainder being granulites and granitoids which engulf the narrow keel-shaped greenstone belts. Although their structural relationships are very complex, it has been suggested that the gneissic terranes were the contemporaneous sialic basement which existed during the accumulation of the oceanic volcano-sedimentary sequences of the greenstone belts (Paris, 1987). Since they contain the oldest rocks in the Kalahari craton and are also more extensive, the high-grade rocks are presented first, followed by the greenstone belts, and the late-or-post-tectonic granitoids.

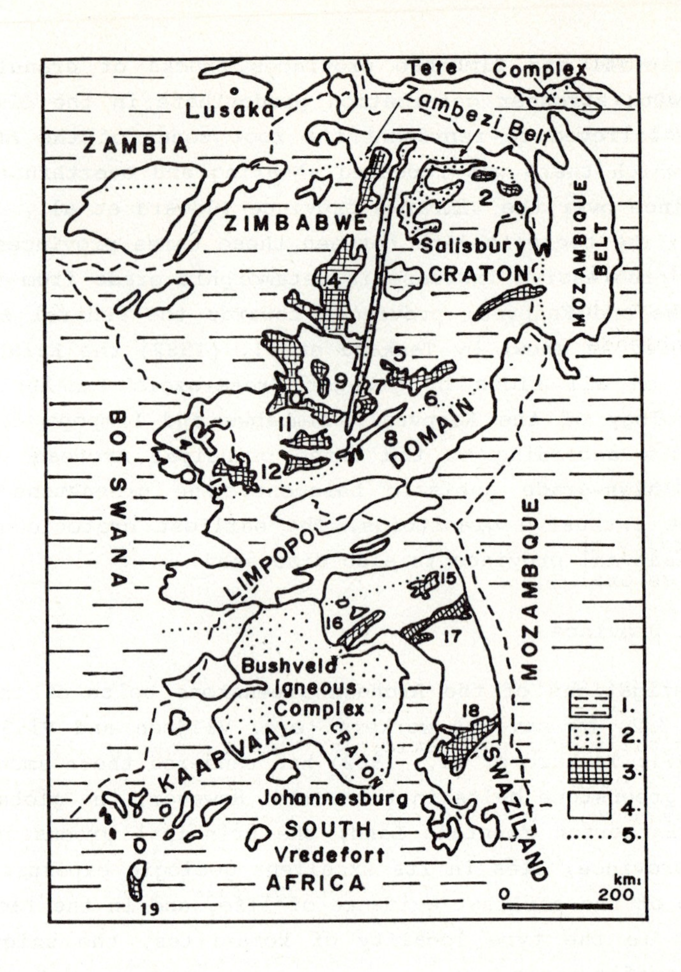

Figure 3.3: Exposed part of the Kalahari Craton. 1, Cover rocks; 2, Igneous complexes; 3, Greenstone belts; 4, Granites and gneisses; 5, Margins of mobile belts. The numbered greenstone belts are: 1, Salisbury-Shamva; 2, Makaha; 3, Gwelo; 4, Midlands; 5, Mashaba; 6, Victoria; 7, Belingwe; 8, Buchwa; 9, Shangani; 10, Bulawayo; 11, Gwanda; 12, Antelope; 13, Tati; 14, Matsitama; 15, Sutherland; 16, Pietersburg; 17, Murchison; 18, Barberton; 19, Amalia. (Redrawn from Cahen et al., 1984.)

Ancient Gneiss Complex

This is a collective term for the basement gneisses of the central Swaziland area south of the Barberton Mountain greenstone belt (Tankard et al., 1982). Similar gneissic terranes which are less well known, occur to the north of the Barberton Mountain Land (Fig.3.3). The Ancient Gneiss

Complex, as summarized by Tankard et al. (1982), comprises (in order of decreasing age), the Bimodal Gneiss Suite, migmatite gneisses of unknown age, the Dwalile Metamorphic Suite, the Tsawela Gneiss, the Mponono Intrusive Suite, lenses of homogeneous medium-grained quartz monzonite, and the Mkhondo Valley Metamorphic Suite. The gross structural relationship between these gneisses is one in which the 3.5 Ga Bimodal Gneiss Suite of interlayered siliceous low-potassium leucocratic tonalites, and the amphibolites of the Dwalile Metamorphic Suite are intruded by the Tsawela biotite-hornblende tonalite gneiss which has been dated at about 3.3 Ga. The Mkhondo Valley Metamorphic Suite of unknown age, consists of layered amphibolites, while the migmatite gneisses appear to grade into the Bimodal Gneiss Suite within which the Mponono Intrusive Suite occurs as sheet-like intrusions of hornblende anorthosite.

Structurally the Ancient Gneiss Complex shows a very complex superposition of several generations of strong deformation which produced isoclinally folded gneissic layers and quartz veins and in which the axial-planar schistocity in the Bimodal Gneisses are cross-cut by the intrusive contacts of the Tsawela tonalite gneiss.

Petrologically and geochemically, the various rocks of the Ancient Gneiss Complex of Swaziland, like their counterparts in Archean gneissic terranes elsewhere, are tonalitic in composition. Although they represent the high-grade metamorphic end-products of a variety of magmatic and supracrustal parent materials, the preponderance of tonalitic and andesitic magmas evokes comparisons with modern tectonic regimes where similar magmas are generated (Nisbet, 1987). Since these tonalitic gneisses are so voluminous in Archean terranes and will be encountered in all the African provinces, it is important to mention the salient geochemical characteristics of the Kaapvaal gneisses which relate to their possible origin. The Bimodal Suite and the Tsawela tonalite gneiss show low initial $^{87}Sr/^{86}Sr$ ratios, low δ^{18} values, and low K_2O contents which suggest the derivation of their parent magmas from mantle sources (Tankard et al., 1982), possibly from the partial melting of sinking basaltic crust in a manner that evokes analogy with the generation of tonalitic batholiths above modern subduction zones (Nisbet, 1987). The absence of intermediate rocks in the Bimodal Gneiss Suite rules out its derivation from the fractionation of basaltic parent magmas. However, the high Rb/Sr and K/Na ratios, enrichment in light REE_s, slight depletion of heavy REE_s and the prominent negative Eu anomalies in the Mkhondo Valley Metamorphic Suite suggest that these could have originated later by the partial melting of pre-existing trondhjemitic-tonalitic gneisses (Tankard et al., 1982).

The Barberton Greenstone Belt

Of the six greenstone belts in the Kaapvaal province (Fig.3.4), the Barberton belt is the largest and the best preserved (Fig.3.5). The Barberton belt extends as a wedge-shaped mountain chain for over 140 km between the Drakensberg escarpment in the west and the Lebombo Range in the east. The greenstones of the Barberton belt are termed the Swaziland Supergroup. This group consists of a thick volcano-sedimentary pile with predominantly ultramafic to mafic volcanics at the base, followed upward by a cyclical sequence of graywackes, shales and chert, which passes upward into another cyclical sequence of conglomerates, quartzites with minor shale interbeds. Because of slight metamorphism, only to lower greenschist facies, these supracrustals have retained their original sedimentary structures by means of which their paleoenvironments have been precisely determined.

The Swaziland Supergroup underwent several episodes of intensive deformation in which the entire sequence was repeatedly folded, and thrust, to the extent that stratigraphic successions are repeated, thus complicating enormously the determination of the true thickness of the Swaziland Supergroup. Therefore, the usual description of the Barberton greenstone belt as synclinorial keels set in a "sea" of granitoids merely gives an overall regional structural setting for this and other greenstone belts. The age of the Swaziland Supergroup ranges from 3.5 Ga to about 3.2 Ga (Cahen et al., 1984) based on the ages of its basal volcanics and those of the surrounding granitoids respectively. The occurrence of granitoid gneisses similar to those of the Ancient Gneiss Complex in one of the basal tectonic slivers in the Swaziland Supergroup suggests that the Ancient Gneiss Complex or its equivalent was probably the sialic basement upon which the volcano-sedimentary sequence was deposited (Paris, 1987).

Three major lithostratigraphic sequences make up the Swaziland Supergroup (Fig.3.6). These are, the lower ultramafic-mafic Onverwacht Group; the middle predominantly graywackes Fig Tree Group; and the upper alluvial-deltaic Moodies Group. The entire supergroup is thickest in the northern part of the Barberton belt (Fig.3.7) where the depositional basin was deepest, and thinner in the south which apparently was undergoing uplift and thrusting at the time the northern part of the basin was filling. The South African Committee for Stratigraphy (1980) assigned a thickness of 24 km to the Swaziland Supergroup which was highly improbable even on stratigraphic grounds (Burke et al., 1976), and on geophysical evidence, upon which Darracott (1975) had earlier based an estimate of 8 km. Also, Paris (1987) gave an estimate of 8 km after removing the effects of nappes and polyphase deformation which had repeated many stratigraphic sections, a factor which had not been taken into account in previous estimates. The revised stratigraphy of the Swaziland Supergroup proposed by Paris (1987) is shown in Table 3.1.

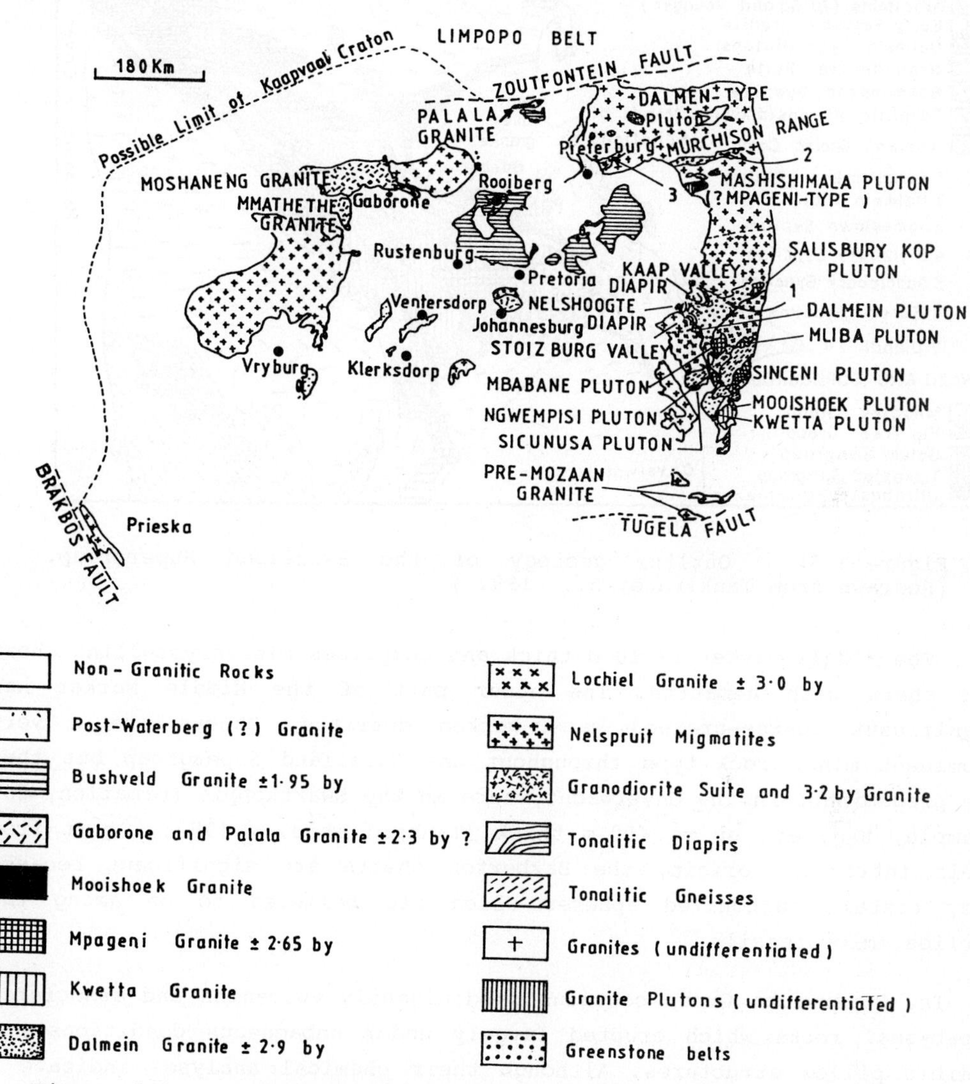

Figure 3.4: Outline geologic map of Kaapvaal province. (Redrawn from Condie, 1981.)

Table 3.1 shows that the Onverwacht Group comprises six formations, the lower three of which belong to the Tjakastad Subgroup -- a sequence of ultramafics and mafics (Fig.3.6). The upper three formations are mainly calc-alkaline volcanics belonging to the Geluk Subgroup. A regionally persistent unit, the Middle Marker occurs at the base of the Geluk.

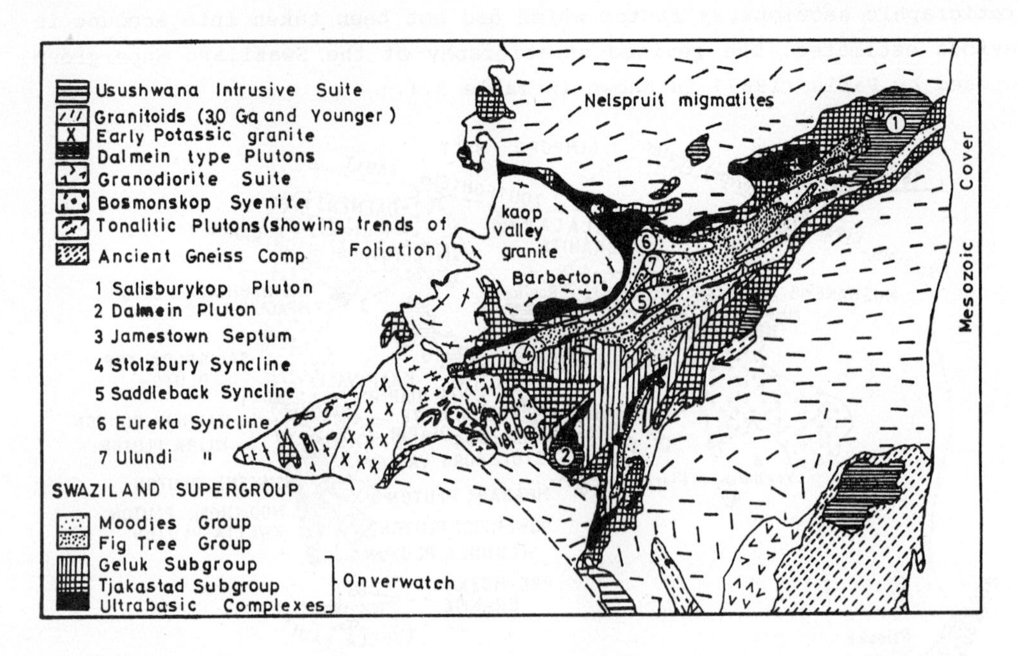

Figure 3.5: Outline geology of the Swaziland Supergroup. (Redrawn from Tankard et al., 1982.)

The Middle Marker is 10 m thick and comprises microcrystalline chert and chert with hematite. The upper part of the Middle Marker has significant coarse-grained water-worked detritus. Cherts are a very prominent minor rock type throughout the Swaziland Supergroup but they are predominant in the Onverwacht where in the Swartkoppie Formation, for example, they are up to 400 m thick (Tankard et al., 1982). Apart from their intriguing origin, the Barberton cherts are significant because they contain carbonized spheres which are believed to be among the earliest microfossils.

The Onverwacht Group contains predominantly volcanics and associated hypabyssal rocks which erupted largely under subaqueous conditions and exhibit pillow structures. Although their chemical analyses indicate a wide range of composition from ultramafic to felsic, by far the most notable are the highly magnesian lavas known as komatiites, of which the Komati Formation of the Onverwacht Supergroup is the type sequence. Komatiites are ultramafic rocks with an MgO content of about 18 %, while komatiitic basalts are those with MgO in the range of 10-18 % (Nisbet, 1987). They commonly exhibit spinifex or quench textures. Chemically, ko-

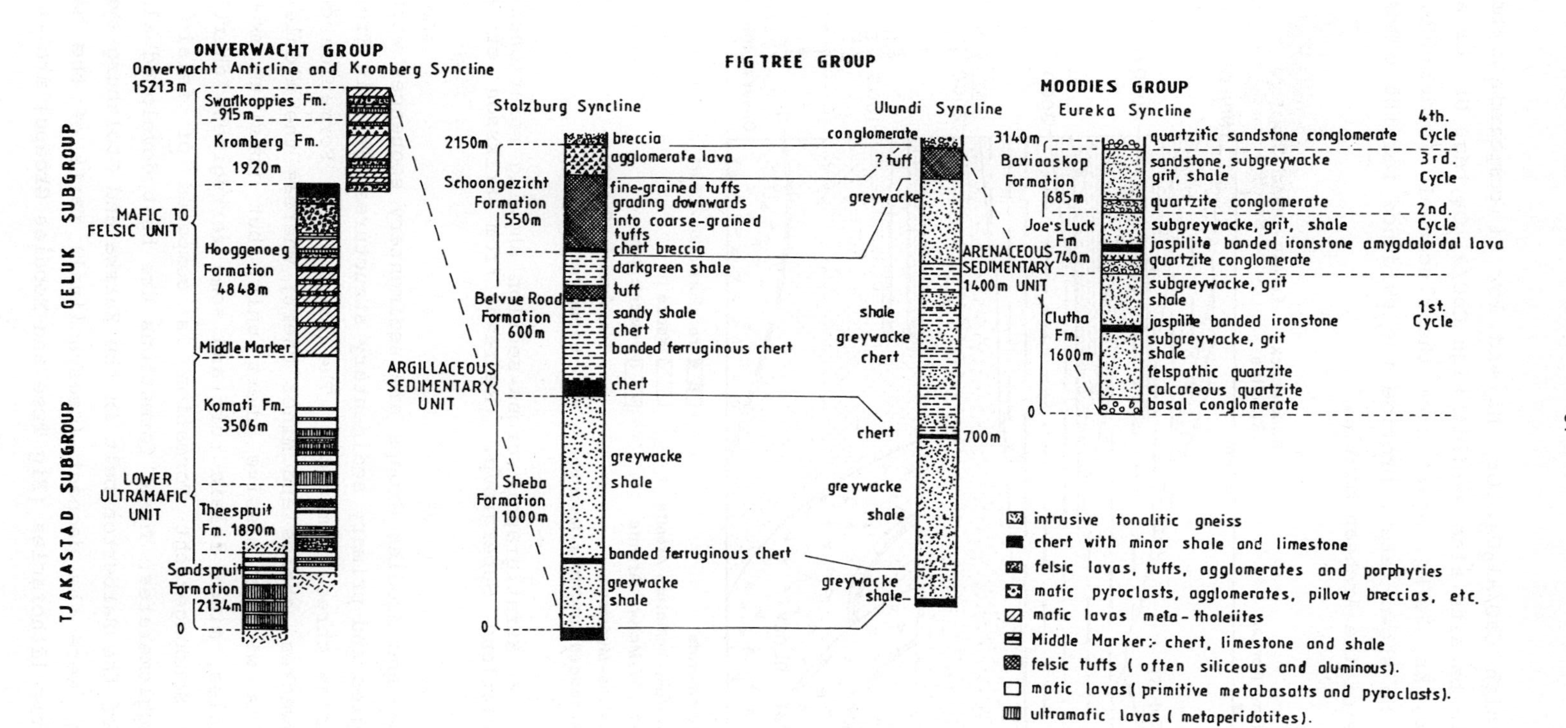

Figure 3.6: Stratigraphic columns for the Swaziland Supergroup. (Redrawn from Condie, 1981.)

matiites have high CaO/Al_2O_3, Cr, Ni and low Ti compared to basalts, while komatiitic basalts also exhibit high CaO/Al_2O_3, Mg, Ni, Cr and low alkalis, Ti, Nb, Zr, Fe/Mg. Apart from their economic importance, these rocks, as will be shown later, provide the evidence for the composition and temperature of the Archean mantle.

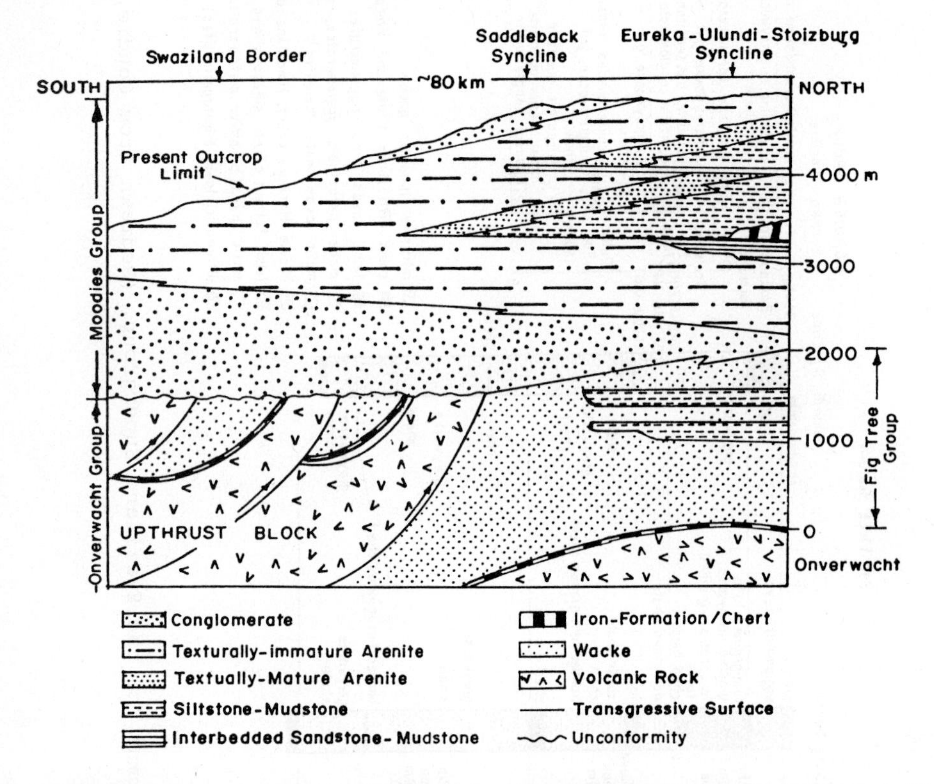

Figure 3.7: Stratigraphic cross-section showing relationships in the Swaziland Supergroup. (Redrawn from Eriksson et al. (1988.)

The Fig Tree and Moodies Groups are sedimentary sequences with diagnostic lithologies and primary sedimentary structures. The Fig Tree Group (Fig.3.6) contains three formations. The lower Sheba Formation consists mainly of graywackes, shales and minor chert; whereas the middle Belvue Formation begins with a massive chert unit, but consists mostly of graywackes, shales, minor felsic tuff and some ferruginous chert bands. The overlying Schoongezicht Formation is composed of felsic tuffs, breccias and agglomerates. These formations are best developed along the northern part of the Barberton belt in the Eureka and Stolzburg synclines (Fig.3.5) which were the deepest (geosynclinal) parts of the basin in which the clastic lithofacies (Fig Tree and Moodies Groups) are thickest.

Table 3.1: Stratigraphy of the Barberton greenstone belt based on the South African Committee for Stratigraphy (A), and as revised (B) by eg. Paris (1987).

A.

MOODIES GROUP about 5km thick			3 sedimentary cycles (conglomerates, quartzites, shales, greywackes, jaspilites, magnetic shales)
FIG TREE GROUP about 2km thick		Schoongezicht Formation Belvue Road Formation Sheba Formation	cherts, shales, greywackes banded ferruginous cherts
ONVERWACHT GROUP about 17km thick	Geluk Subgroup	Swartkoppie Formation Kromberg Formation Hooggenoeg Formation	mafic to felsic volcanic cycles, cherts
		Middle Marker (chert)	
	Tjakastad Subgroup	Komati Formation Theespruit Formation Sandspruit Formation	ultramafic to mafic volcanic cycles, cherts

B.

MALOLOTSHA GROUP ~2 km	F5 quartz-arenite, siltstone F4 conglomerate in matrix of both chert and single crystal quartz grains, chert-quartz arenite, conformable to unconformable	CONTINENTAL ALLUVIAL FAN
DIEPGEZET GROUP ~2 km	F3 chert-arenite, conglomerate in matrix of chert grains F2 ferruginous and tuffaceous siltstone, ferruginous chert-arenite F1 jaspilites, ferruginous chert, ferruginous tuff, shale and siltstone, conformable (?)	OCEANIC PROGRADING SUBMARINE FAN
ONVERWACHT GROUP ~3 km	volcaniclastic unit (distal and proximal turbidites facies and subaerial facies), mafic and ultramafic unit	OPHIOLITE ARCHEAN OCEANIC CRUST
	Unconformity or tectonic contact	
GRANITOID		SIALIC CRUST

(Predominantly silicified: spans from F3 through the Onverwacht Group)

The deepest part of this basin is represented by the Sheba Formation where the graywackes and shales display typical Bouma turbidite facies. Also, in the overlying Belvue units there are prograding fine-grained clastic deposits with intercalated banded iron-formations and chert, which Eriksson et al. (1988) interpreted as the lower submarine fan and basin floor, and basin slope environments (Fig.3.8). The presence of soft sediment folding in the iron-formations suggest gravity displacement in a slope environment. An overall upward coarsening of the top part of the Fig Tree Group and the presence of conglomerates in the overlying Schoongezicht Formation suggest basin filling and shoaling due to the northward progradation of proximal more landward sediments from the south. The conformably overlying Moodies Group with its strongly conglomeratic lithofacies was deposited during this phase of basin filling when deltaic and alluvial conditions were established in what had been a deep turbidite basin (Fig.3.8).

Sedimentary structures and textural characteristics have been successfully utilized (Eriksson et al., 1988) for detailed paleoenvironmental interpretations of the Moodies Group lithofacies. In the northern Eureka syncline the contact of the Moodies Group with the Fig Tree Group is gradational, with conglomeratic beds which thicken upward. Although poorly sorted, these conglomerates are well stratified, displaying internal grading and weak imbrication, channelization and the intercalation of plane or cross-stratified sandstones which suggest deposition in environments similar to modern-day upper alluvial plains. The lower conglomeratic facies (lower part of the Clutha Formation) is similar to the modern conglomeratic longitudinal bar facies which contains pebbles, cobbles and boulders separated by channels. In the Clutha Formation these lithofacies are represented by cross-bedded sandstone beds (Fig.3.9). In general, conglomerates are more prominent in the southern source region of the Clutha Formation, while subarkose and quartz arenite become more abundant northward with shale and banded iron-formations being predominant in the north. Overlying the conglomeratic beds of the Clutha Formation are thinly interbedded plane-to cross-bedded sandstones and shales which show bimodal-bipolar paleocurrent patterns, indicating tidal current-induced reversals of flow directions (Fig.3.9) on tidal flats. The planar, cross-bedded sandstones indicate flood tidal deltaic facies, while the channelized sandstones with small-scale trough , planar and herringbone cross-beds and superimposed ripple structures reflect low-tide sand flats with shallow tidal channels. The plane-bedded sandstones probably formed as washover sand sheets while the mudstones within the thin sandstones represent tidal flats.

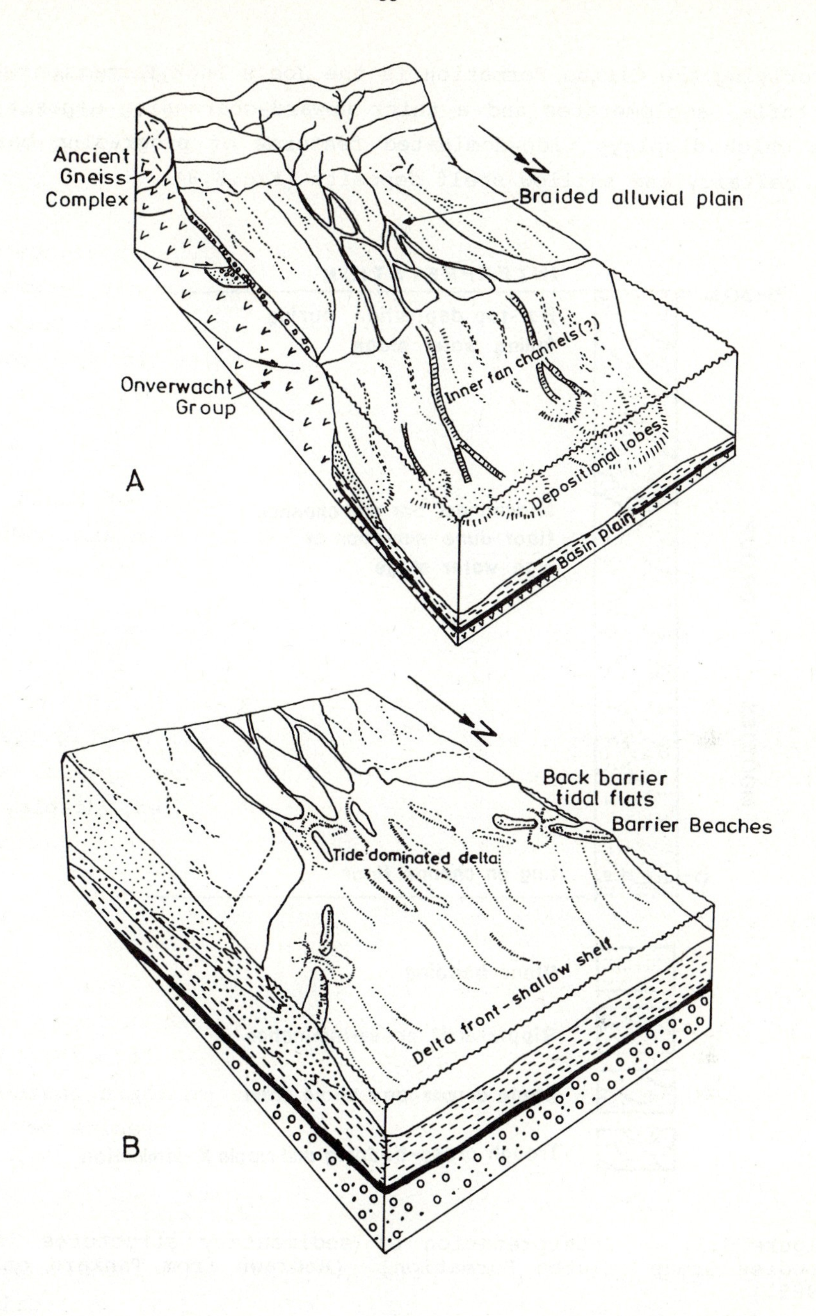

Figure 3.8: Depositional models for the Fig Tree Group (A), and Moodies Group (B). (Redrawn from Eriksson et al., 1988.)

Overlying the Clutha Formation is the Joe's Luck Formation which contains tuffs, agglomerates and a thick upward-coarsening depositional sequence which displays tide-dominated features of prograding barrier islands, deltaic, and shallow shelf deposits (Fig.3.9).

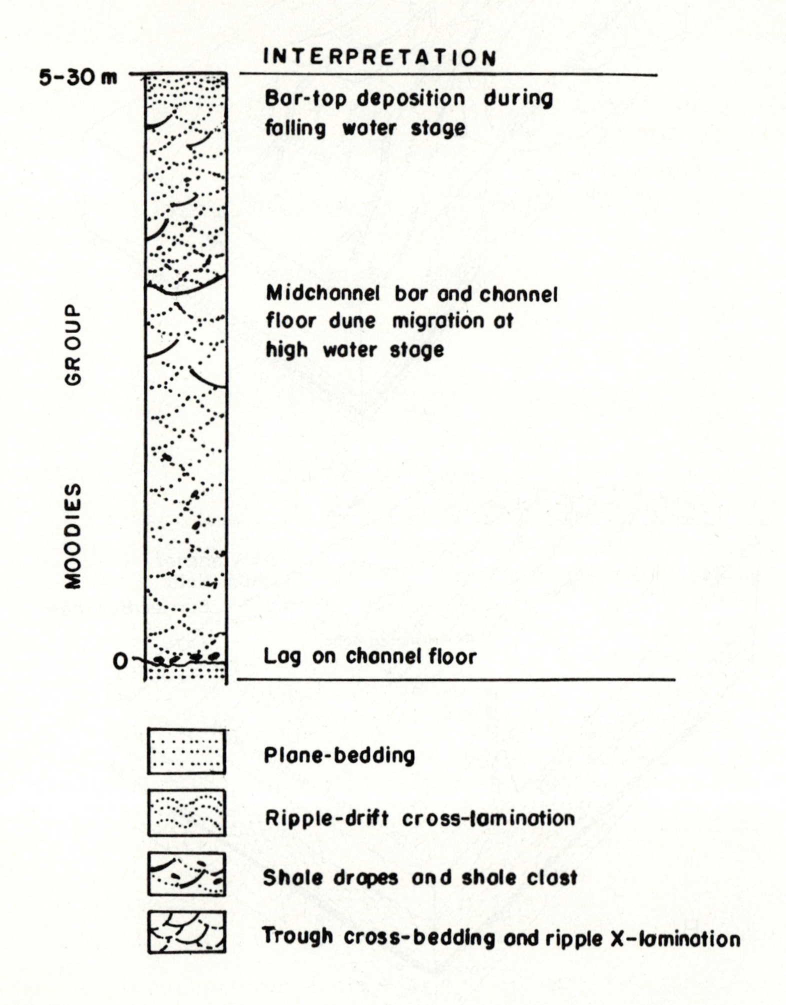

Figure 3.9: Interpretation of sedimentary structures in the Moodies Group (Clutha Formation). (Redrawn from Tankard et al., 1982.)

The banded iron-formations at the base of the sequence formed in the deeper part of the shelf under quiet conditions which favoured chemical and suspension sedimentation, far away from clastic influx. The top part of the Moodies Group was deposited during a regression when there was a return to tidal flat and alluvial plain environments.

According to Eriksson et al. (1988), the paleogeography of the Barberton belt during Fig Tree and Moodies sedimentation was one in which sediments were derived from a southern uplifted mixed sialic-volcanic-chert terrain and deposited along a northward-facing continental margin (Fig.3.8). The abrupt transition from submarine fan sedimentation to braided plain in the Fig Tree and the basal Moodies and the absence of shallow marine facies reflect sedimentation along a steep continental margin with a narrow shelf. Shallow marine and extensive coastal sediments higher up in the Moodies suggest the development of a wider shelf with braided alluvial plain and deltaic sedimentation in which coastal reworking formed barrier complexes with extensive back-barrier tidal flats.

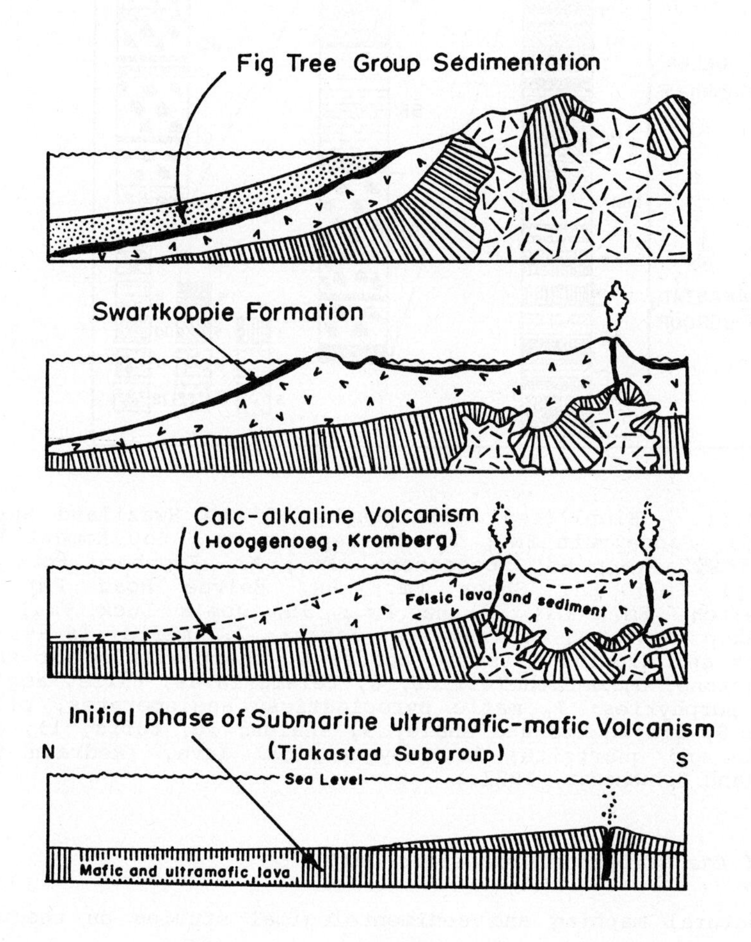

Figure 3.10: Model for the stratigraphic and tectonic evolution of the Barberton greenstone belt. (Redrawn from Lowe and Knauth, 1977.)

As postulated by Lowe and Knauth (1977), the tectonic and sedimentological evolution of the Barberton belt (Fig.3.10) involved an initial phase of submarine mafic-ultramafic volcanism during which the Tjakastad Subgroup was formed (Fig.3.11). This was followed by calc-alkaline volcanism represented by the Hooggenoeg and Kromberg Formations; and by the deposition of the Swartkoppie Formation; and the deposition of the Fig Tree Group, during uplift of the granitic southern sediment source areas.

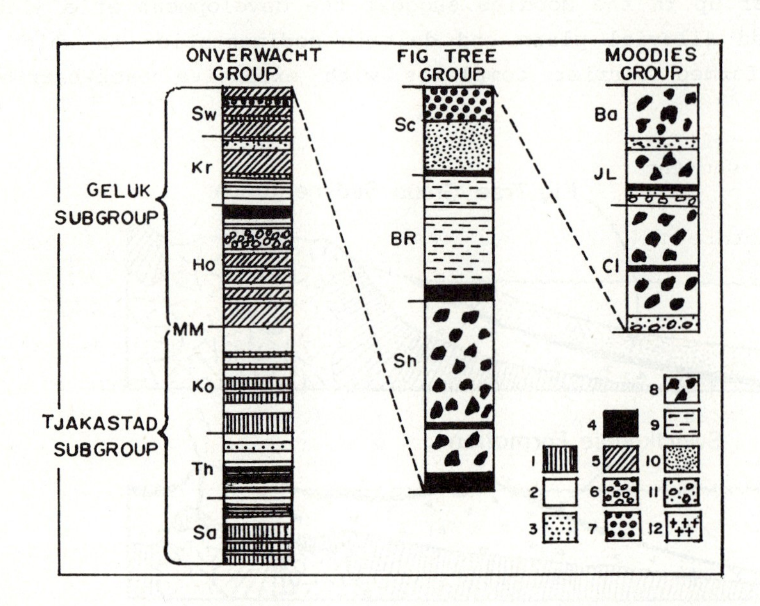

Figure 3.11: Simplified stratigraphy of the Swaziland Supergroup. Sa, Sandspruit Fm.; Th, Theespruit Fm.; Ko, Komati Fm.; MM, Middle Marker; Ho, Hooggenoeg. Fm.; Kr, Kromberg Fm.; Sw, Swartkoppi Fm.; Sh, Sheba Fm.; BR, Belvue Road Fm; Sc, Schoongezicht Fm.; Cl, Clutha Fm.; JL, Joe's Luck Fm.; Ba, Baviaanskop Fm.; 1, ultramafic lavas; 2, mafic lavas; 3, siliceous and aluminous felsic tuffs; 4, chert with minor shale and limestone; 5, metatholeiites; 6, felsic lavas, tuffs, agglomerates, porphyries; 7, mafic pyroclastics, agglomerates, pillow brecias; 8, graywackes and shale; 9, shales; 10, tuffs; 11, conglomerate and quartzite; 12, amygdaloidal lava. (Redrawn from Frazier and Schwimmer, 1987.)

Structure of the Barberton Greenstone Belt

In his structural mapping and sedimentological studies on the southern part of the Barberton belt, Paris (1987) identified four major episodes of deformation during which stratigraphic units had been dismembered and repeated, leading to the 24 km thickness which had been previously assigned to the Swaziland Supergroup. Although Paris applied new stratigraphic terminologies (Table 3.1), he also identified a prograding submarine facies overlain by a continental alluvial fan sequence. Undoubtedly these are the equivalents of the Fig Tree and Moodies respectively. The structural picture presented by Tankard et al. (1982) for the northern part of the Barberton belt also reveals polyphase deformation like in the south.

Regionally, the Barberton belt is a broad synclinorium trending NE-NNE and comprising tight to isoclinal synforms which are generally overturned to the west and separated by tectonic slices (Fig.3.12). Several deformation episodes have been recognized in the Eureka and Ulundi synclines. Both synclines were formed during major and minor folding episodes about NE- or NNE-striking and SE-dipping axial planes. This was followed by shortening along NNW axes which led to the development of transverse slaty cleavage and schistosity, and a concomitant downward-facing structure with steeply-plunging folds. These folds formed sporadically along the Eureka syncline. The next deformations produced a large northwestward-trending synform which buckled the major synclines and produced crescentic interference minor conjugate folds and faults. There is no doubt that compressive forces played a role in these deformations (Paris, 1987).

Granitoid Emplacement and Cratonization

As evident in Fig.3.4 granitoids are the most pervasive rocks in the eastern part of the Kaapvaal province. They were emplaced into the greenstone belts and adjacent gneisses from about 3.35 Ga to 2.6 Ga, after the volcanism and deposition of the Swaziland Supergroup. Tankard et al. (1982) gave the following sequence of events during the emplacement of the major granitoids. The Granodiorite Suite, a coarse-grained, slightly foliated rock ranging in composition from hornblendite to granodiorite and predominantly granodiorite and tonalite, was emplaced into the southern gneissic terrane of the Barberton belt at about 3.35 Ga. Between 3.4 Ga and 3.1 Ga, several leucotonalitic, trondhjemitic and tonalitic plutons intruded the lower members of the Onverwacht Group along the southwestern and northwestern margins of the Barberton belt, and thereby marked the end of the deposition of the Swaziland Supergroup and the approximate period for the stabilization of this part of the Kalahari craton.

One of the largest granitoid plutons is the Kaap Valley quartz diorite. Even more extensive are the Nelspruit porphyritic granite and migmatites which underlie a vast area north of the Swaziland Supergroup.

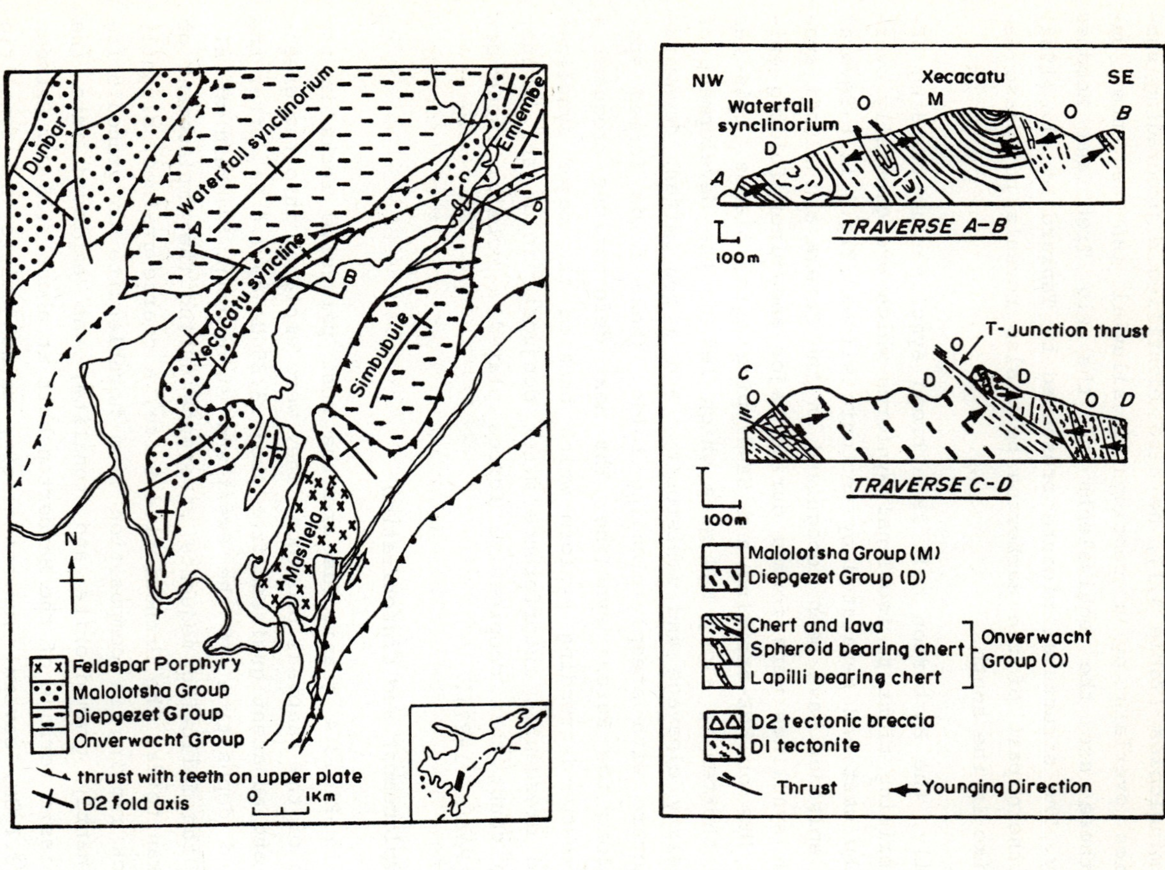

Figure 3.12: Structural relationships in the Swaziland Supergroup. (Redrawn from Paris, 1987.)

Within the Nelspruit porphyritic granite is the Hebron granodiorite, which together with the Nelspruit Suite was emplaced around 3.2 Ga. The most extensive intrusive suite, however, is the younger Lochiel sheet-like quartz monzonite which surrounds the southeastern margin of the Swaziland Supergroup. Mineralized pegmatites with cassiterite, beryl and monazite are common in the Lochiel complex. The Lochiel quartz monzonite was emplaced syntectonically at about 3.0 Ga.

The geochemical models for the origin of these tonalites are based on their characteristically low K/Na ratios (< 0.5); variable K/Rb ratios; limited range of Rb/Sr ratios (close to 0.1); enrichment in light REE and low $\delta^{18}O$ values; all of these suggest mantle derivation probably by partial melting of the Onverwacht Group basalts. However, the Lochiel quartz monzonite with its high initial $^{87}Sr/^{86}Sr$ ratios and high $\delta^{18}O$ values could have been generated from crustal sources.

Other Greenstone Belts in the Kaapvaal Province

The smaller greenstone belts in the Kaapvaal province are less well known than the Barberton belt. These outcrop mainly to the north of the Barberton belt (Fig.3.3) while two belts, the Muldersdrif and the Amalia are eastern inliers. The occurrence of these other belts far away from the Barberton belt suggests that the Archean greenstone belt of the Kaapvaal province was originally much more extensive than what now remains.

1. *Pietersburg Belt*. This belt extends in a northeasterly direction from beneath the cover of the Transvaal Supergroup, for about 100 km, with an average width of about 10 km, tapering out between two granitoids at its northeastern end. It has a lower sequence of ultramafic to mafic lavas which are chemically similar to the komatiitic lavas of the Barberton belt. There are minor chert bands and banded iron-formations, and an upper sequence of conglomerates, quartzites and shale interbeds, like the Moodies Group. There is no calc-alkaline suite in the Pietersburg belt. Three periods of deformation affected the belt during which the volcano-sedimentary pile was weakly metamorphosed to the greenschist facies. Amphibolite facies rocks occur adjacent to granitoid intrusions probably due to contact metamorphism as well as a regional increase in metamorphism northeastward towards the high-grade Limpopo province.

2. *Murchison Belt*. This is a narrow greenstone belt some 140 km long and up to 15 km wide. The sequence here starts with Onverwacht-type ultramafic-mafic succession and passes into quartzite metasediments and acid pyroclastic rocks, all of which have been subdivided into six litho-

stratigraphic units (Tankard et al., 1982). However, the stratigraphic relationship between these units, being uncertain, is based entirely on structural interpretations. The Murchison belt was intruded by the Rooi-water layered complex and granitoids.

3. *Sutherland Belt*. This occurs at the extreme northeastern part of the Kaapvaal province. It is about 60 km long and 15 km wide. The exposures here are poor comprising schistose and massive mafic-ultramafic rocks with minor intercalations of banded iron-formations, chert and quartz-sericite schist.

4. *Muldersdrif and Amalia Belts*. Both belts lie to the west of the Barberton belt. The Muldersdrif underlies a small area of about 150 km^2 on the edge of the Johannesburg Dome and contains predominantly basic and ultrabasic massive and schistose metavolcanics with remnants of layered ultrabasic intrusions (Tankard et al., 1982).

The Amalia belt of western Transvaal is preserved as an inlier surrounded by the Ventersdorp Supergroup. It consists mainly of graywackes, grits, shales, ironstones, lavas and tuffs.

3.2.2 Pongola Basin

In this basin the Pongola Supergroup unconformably overlies granitoids in the southern part of the Kaapvaal province (Figs. 3.4), in what was probably the earliest stabilized cratonic shelf environment, between 3.1 Ga and 2.9 Ga. The Pongola Supergroup is a sequence of tholeiitic volcanics and tidal flat sandstones and shales. It is divisible into a lower Nsuze Group, up to 8 km thick, comprising about 7.5 km of interfingering basalts, basaltic andesites, dacite and rhyolites, with tholeiitic affinity. A 30-m thick carbonate with stromatolitic structures occurs within the Nsuze volcanics. This carbonate body is laterally discontinuous with recrystallized dolomite and clastic-textured dolomites which display well-developed herringbone cross-bedding and contain large stomatolite-derived intraclasts, all of which suggest tidally influenced environments (Grotinger, 1989). This is believed to be the oldest known carbonate platform deposit which probably accumulated on a high-energy carbonate ramp, based on the evidence for tidal activity in both carbonates and siliciclastic sediments, and on the comparatively sparse occurrence of stromatolites.

The upper part of the Pongola Supergroup consists of an unconformable sequence (Fig.3.13), up to 1,800 m thick, in which there are mostly mature quartz arenites, shale and banded iron-formations. This is the

Mozaan Group which contains sedimentary structures that are suggestive of sedimentation in tidal flats and tide-dominated shelf environments (Tankard et al., 1982).

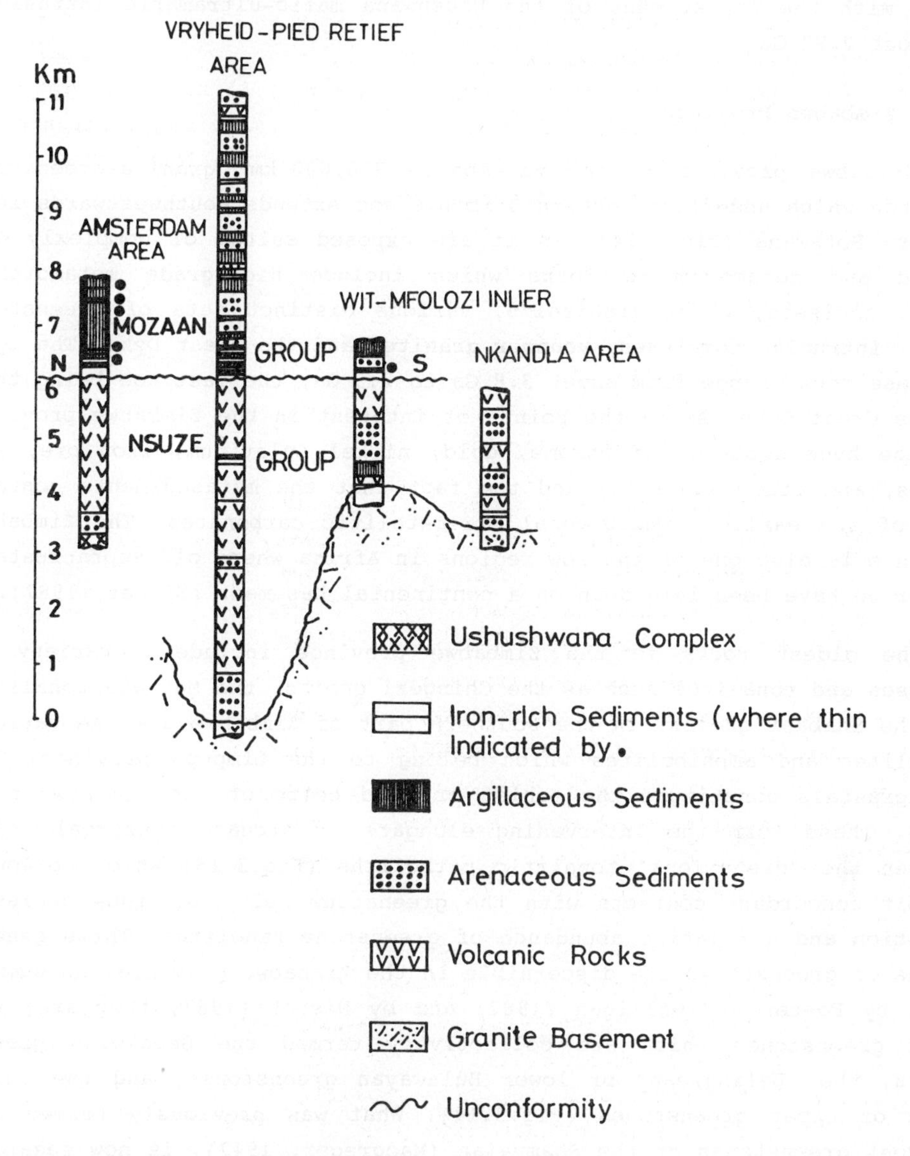

Figure 3.13: Stratigraphic columns for the Pongola Supergroup. (Redrawn from Nisbet, 1987.)

The shelf sediments in the Mozaan Group consist of ironstones with shales and siltstones which are arranged in upward-coarsening motifs that suggest beach sedimentation. The banded iron-formations at the base of the Mozaan contain interlaminated chert-jasper and magnetite and are believed to be chemical precipitates in distal shelf environments far away from detrital influx. Pongola Supergroup volcanism and sedimentation ended with the emplacement of the Usushwana mafic-ultramafic intrusives at about 2.87 Ga.

3.2.3 Zimbabwe Province

The Zimbabwe province is an oval-shaped 300,000 km^2 granite-greenstone province which underlies eastern Zimbabwe and extends southwestwards into eastern Botswana (Fig.3.14). In it are exposed suites of complexly deformed and metamorphosed rocks which include high-grade metamorphic rocks, gneisses, older granitoids, various distinct sets of greenstone belts, intrusive complexes, younger granites and the Great Dyke. The ages of these rocks range from about 3.8 Ga to 2.5 Ga, the last age being that of the Great Dyke. Among the points of interest in the Zimbabwe province are the huge deposits of chrome, gold, nickel, platinum, iron ore, asbestos, and other minerals; and the fact that the metasediments contain some of the earliest unequivocal stromatolitic carbonates. The Zimbabwe province is also one of the few regions in Africa where old supracrustals appear to have been laid down on a continental basement (Nisbet, 1987).

The oldest rocks in the Zimbabwe province include a variety of gneisses and tonalites such as the Chingezi gneiss, the Mashaba tonalite, and the Shabani gneiss. In the southern part of Zimbabwe are the various granulites and amphibolites which belong to the Limpopo province. The supracrustals constitute the schist or gold belts of the Zimbabwe province. These form the intervening elongate or arcuate synformal belts between the "gregarious" tonalitic batholiths (Fig.3.15) which commonly exhibit concordant contacts with the greenstone belts and show gneissic foliation and a relative abundance of greenstone xenoliths. Three generations of greenstones are discernible in the Zimbabwe province. As summarized by Foster and Gilligan (1987) and by Nisbet (1987) they are: the older greenstones which are collectively termed the Sebakwian greenstones, the "Belingwean" or lower Bulawayan greenstones, and the Bulawayan or upper greenstones (Fig.3.14). What was previously termed the youngest greenstones or the Shamvaian (Macgregor, 1947), is now regarded as the upper part of the Bulawayan Group, representing the terminal phase of greenstone volcanism and the accumulation of granitic detritus in probably local isolated basins (Wilson, 1972). Representative gneissic, greenstone and granitoid domains from the Zimbabwe province are described below in order to illustrate the stratigraphic, lithologic and structural characteristics of this province.

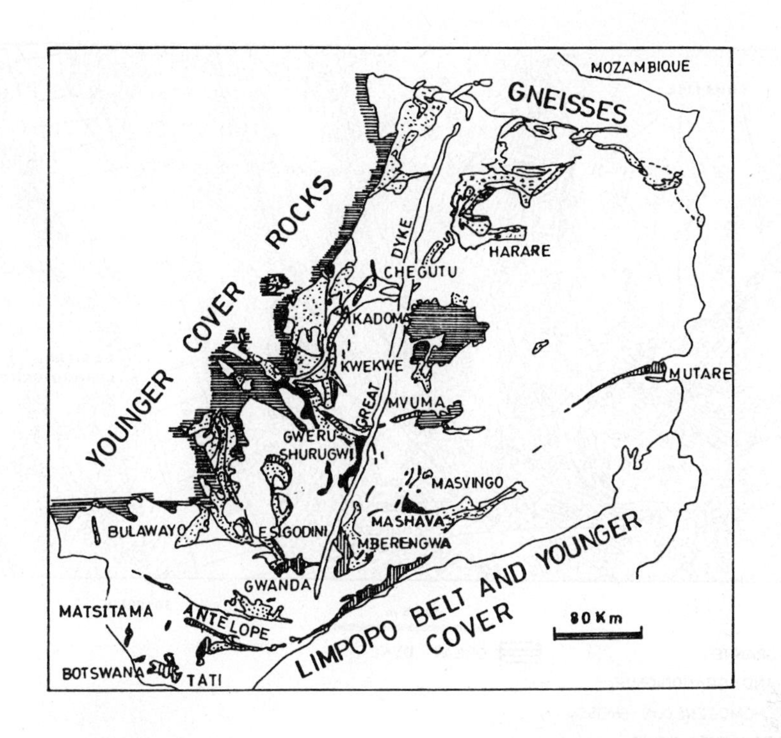

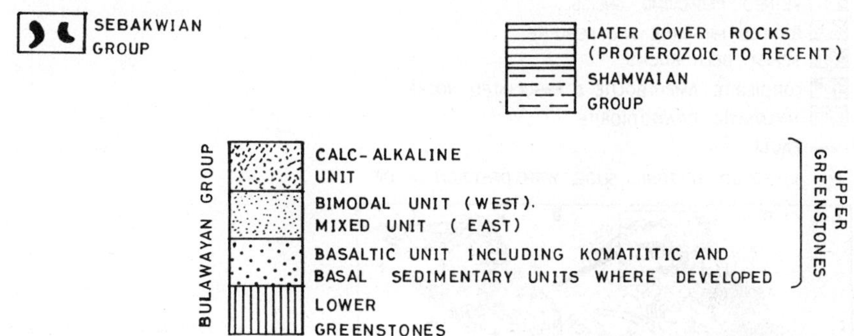

Figure 3.14: Archean greenstone belts of Zimbabwe. (Redrawn from Foster and Gilligan, 1987.)

Gwenoro Dam Basement Gneisses

This is a structurally complex basement region south of the Gwelo schist belt, between the towns of Gweru and Shurugwi. It illustrates the field

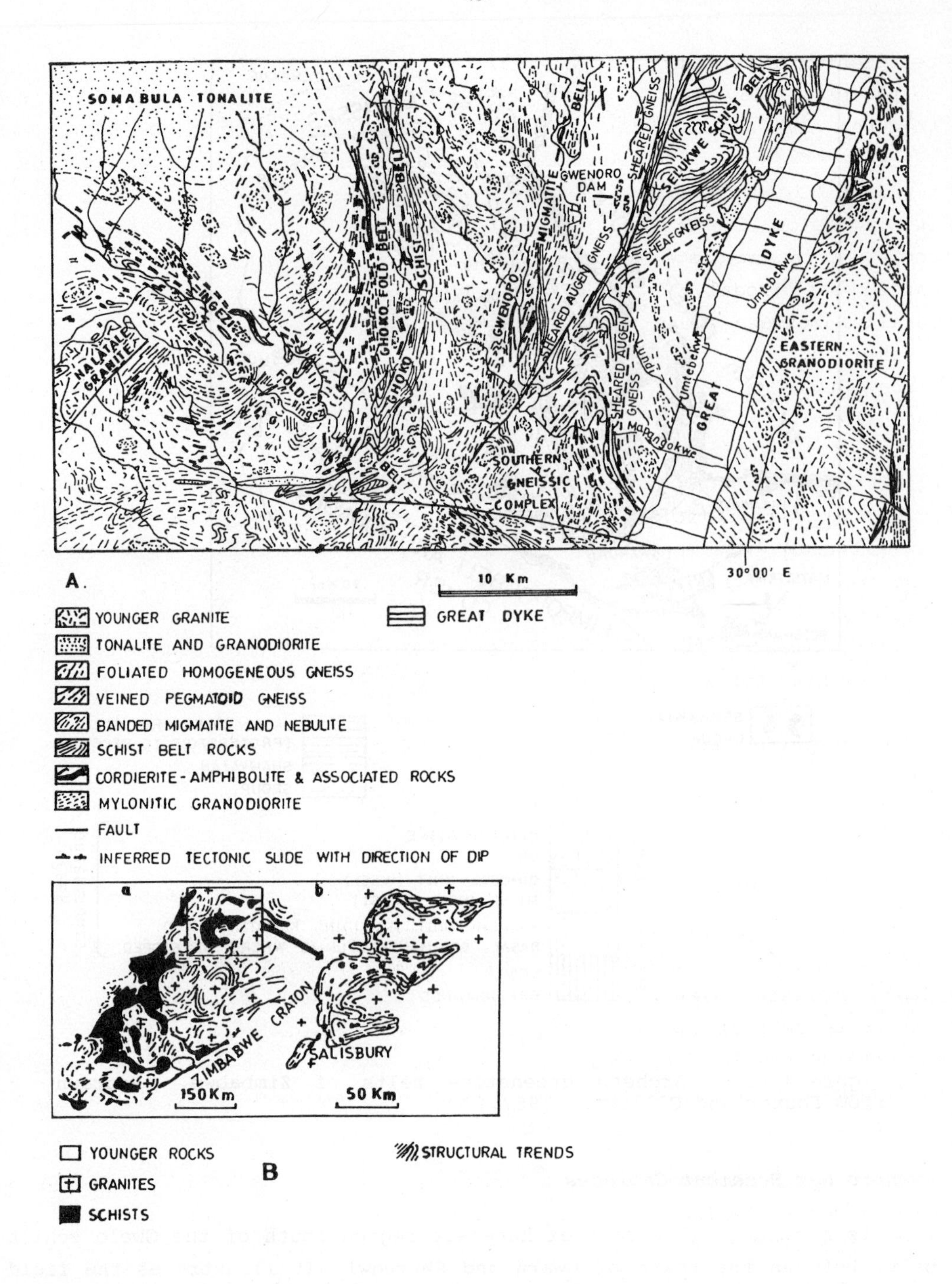

Figure 3.15: A, Outline geological map of the Gwenoro Dam area; B, Structural relationships in Zimbabwe greenstone belts. (Redrawn from Condie, 1981; Frazier and Schwimmer, 1987.)

relations between gneisses, associated granitic plutons and greenstone belts (Fig.3.15, b). The gneisses range in composition from banded gneisses with alternations of quartzo-feldspathic bands and biotite-rich layers, to faintly foliated, homogeneous gneisses. There are locally abundant migmatite-agmatite-nebulite terranes. Since the gneisses are mainly of tonalite or trondhjemite composition within mafic enclaves, and rocks of intermediate composition are rare, the entire terrane is regarded as bimodal. As depicted in Fig.3.15A, foliation is generally parallel between the gneisses and the greenstone belts especially close to their contacts; but foliation becomes variable in the intervening regions. Stronger foliation develops where there are abundant supracrustal inclusions in the gneisses. Some of the inclusions show progressive degrees of assimilation and fragmentation by the gneisses. Sometimes, trains of inclusions show a transition from the Ghoko greenstone belt into the adjacent foldbelt comprising banded migmatites and nebulites. These trends and transitions of inclusions are useful for tracing the connections between greenstone belts (Condie, 1981). The gneissic complex-granite pluton contacts range from sharp and discordant to gradational. Sheared contacts render it difficult to ascertain whether such contacts are unconformities or intrusive.

Older Greenstone Belt (Sebakwian Group)

The Sebakwian greenstone succession comprising lavas and sediments which accumulated around 3.4 Ga, are best developed in the south-central part of Zimbabwe, in the Shurugwi, Lower Gwelo and the Mashava regions. These are believed to be the oldest greenstones which have been mostly metamorphosed to the amphibolite facies. Outside the south-central occurrences, they are found scattered elsewhere, mostly as infolded remnants within gneisses. Near the town of Shurugwi the lower part of the Sebakwian sequence includes magnesian basalts and possible komatiites, and minor metapelites and banded ironstones. The lower Sebakwian in this region is intruded by a major suite of ultramafic bodies, some of which bear chromite.

Resting unconformably on the lower sequence is the sedimentary Wanderer Formation which has a very diverse assortment of sediments showing rapid lateral facies variation from conglomerates to pelites and banded iron-formations. Clasts of talc-carbonate rocks, chromite, metabasalt, jaspilite chert, granite and gneiss suggest that the conglomerates were eroded from a highly varied terrain. As shown in the cross section (Fig.3.16) through Shurugwi, the Sebakwian sequence has been deformed into a large nappe structure before the intrusion of the Mont d'Or granite about 3.35 Ga ago. In this nappe the greenstone succession is inverted. The nappe, which is about 10 km wide and can be traced in a northwesterly direction for about 60 km, appeared to have transported the supracrustal sequence over a distance of about 50 km (Stowe, 1984).

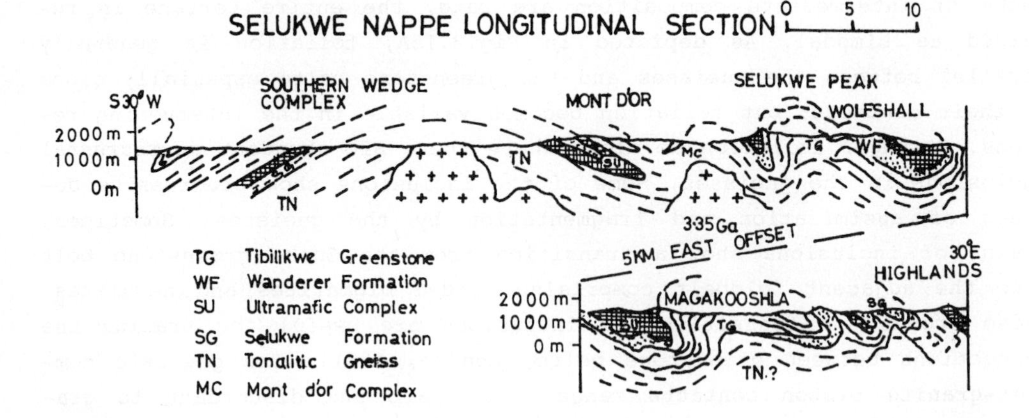

Figure 3.16: Cross-section through the Selukwe nappe. (Redrawn from Nisbet, 1987.)

Bulawayan Greenstones

These greenstones of which the Mberengwa belt is a microcosm, has been widely recognized and correlated across vast parts of the Zimbabwe province (Fig.3.14). The lower part of the succession is almost entirely unconformable upon an older basement comprising the Chingezi gneiss, the Shabani gneiss and the Mashaba tonalite which range in age from 3.5 Ga to 2.9 Ga. Two supracrustal sequences separated by an unconformity, make up the Bulawayan greenstones which formed between 2.7 Ga and 2.6 Ga. These are the "Belingwean" or the Lower Bulawayan greenstones and the Upper Bulawayan greenstones. In the Mberengwa belt, the lithostratigraphic separation between both greenstone successions is highlighted by the presence in the Upper Bulawayan sequence of a distinct basal marker bed, the Manjeri Formation (Fig.3.17).

The Manjeri Formation ranges in thickness from 0 to 100 m and comprises a coarse and persistent conglomerate bed with clasts of the underlying tonalite, overlain by a sequence of shallow-water siltstones and a thin dolomite which passes upward into chert and a thick sequence of graywacke, followed by banded iron-formations. Many Zimbabwe greenstones contain a Manjeri-type marker unit which can be used regionally to separate the lower from the upper Bulawayan greenstones. The upper and lower greenstones are characterized by lower sequences of bimodal volcanics comprising pillowed mafic and ultramafic and felsic volcanics. However, major regional lithofacies developments occur in the Upper Bulawayan succession.

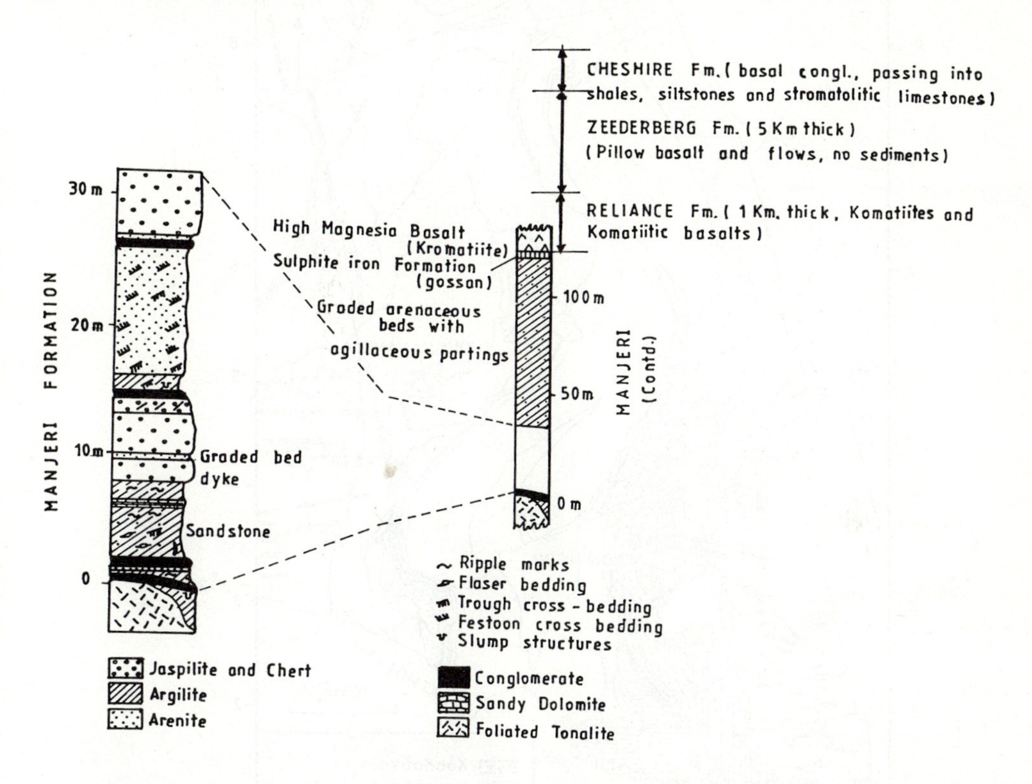

Figure 3.17: Stratigraphic column for the Bulawayan greenstone near Shabani. (Redrawn from Condie, 1981.)

First, the upper part of the western greenstone successions contain appreciable calc-alkaline volcanics (Fig.3.14) as found in the Maliyami and Felsic Formations, while the eastern greenstones remained bimodal. Calc-alkaline volcanic suites are not common in the Archean. Secondly, there is a major carbonate body in the Upper Bulawayan sequence, in the Cheshire Formation, from which sedimentological and isotope geochemical studies of some of the earliest stromatolites, have shed considerable light on Archean shallow marine environments (Abell and McClory, 1987).

1. *Mberengwa Belt*. In the Mberengwa greenstone belt the upper greenstones rest on the lower greenstones in a synform (Fig.3.18). The lower greenstones belong to the Mtshingwe Group which has been subdivided into four formations. The lower Hokonui Formation comprises 2-3 km of dacitic

pyroclastics and andesitic flows with a spectacular vent agglomerate which includes huge blocks of the underlying tonalitic country rock.

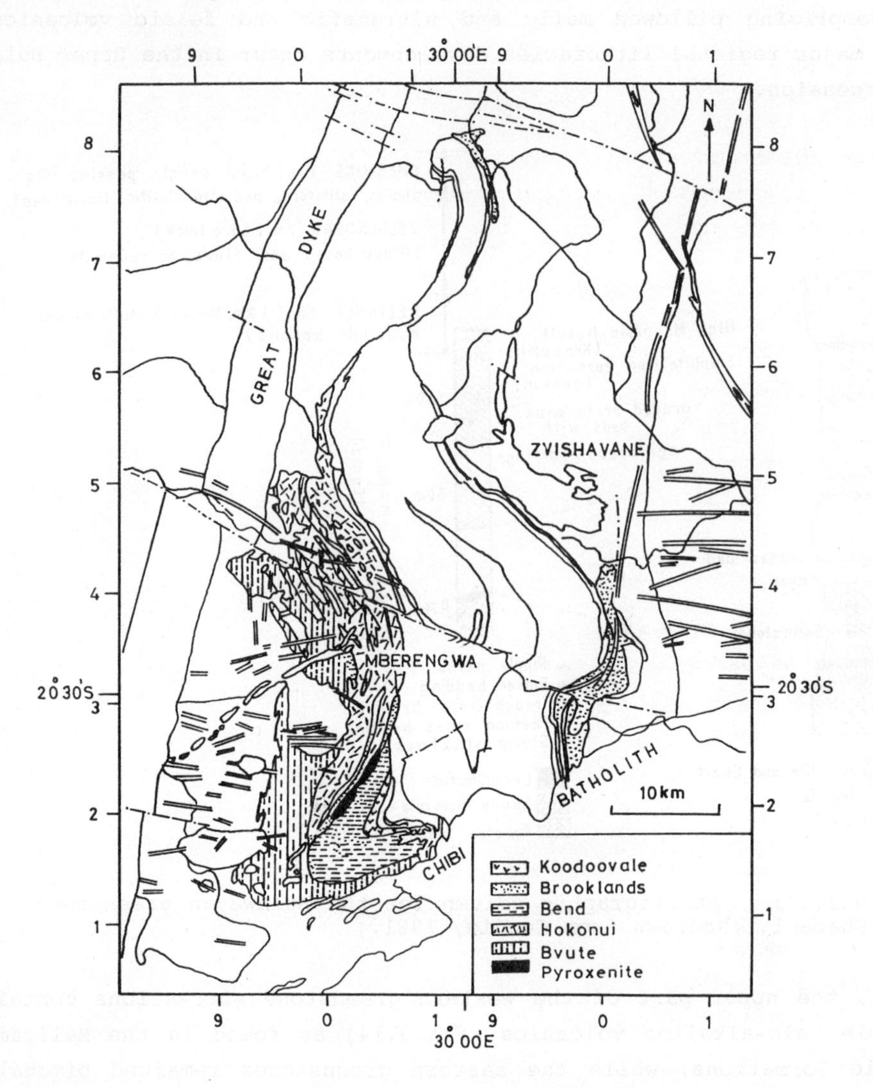

Figure 3.18: Belingwe greenstone belt. (Redrawn from Nisbet, 1987.)

This tonalite intrudes into the Hokonui in some places. In the southern part of the Mberengwa belt, the Bend Formation, a remarkable pile of komatiites, komatiitic basalt and banded ironstone, overlies the Hokonui unconformably, and is 2-5 km thick. The absence of clastic sediments in the Bend Formation suggests that volcanism possibly occurred well away from clastic detritus. The eastern part of the Mberengwa belt, in contrast, contains a thick sequence of coarse conglomerates and breccias which passes into an assortment of komatiite and komatiitic basalts, shales and iron-formations. These belong to the Brooklands Formation which is probably the lateral equivalent of the Bend mafic-ultramafics, thus suggesting a paleogeographic transition from trough edge proximal lithofacies into a basinal facies with banded iron-formations, shales and komatiites. Overlying the Bend Formation is the Koodoovale Formation which consists of 1,000 m of coarse conglomerates and felsic volcanics. The entire Mtshingwe Group was warped and partly eroded before the deposition of the upper Bulawayan greenstones in the synform.

In the Mberengwa belt the Ngezi Group constitutes the youngest greenstones which overlie all the older strata with a marked unconformity (Fig.3.18). Since the Ngezi clearly oversteps older greenstones and gneisses, this means that not all greenstones represent Archean oceanic material as previously believed, rather some greenstones like those in Mberengwa were laid down on pre-existing continental crust (Bickle et al., 1975; Nisbet, 1987).

From the intertidal and shallow-water sandstones and stromatolitic limestones of the Manjeri Formation, there is an upward change into deeper water deposits, overlain by the lavas of the Reliance Formation, about 1,000 m thick. These lavas include komatiitic basalts and komatiites, erupted as flows, pillow lavas and tuffs. The 5.5-km thick Zeederbergs Formation overlies the Reliance volcanics and comprise mostly pillowed and massive tholeiitic basalts with little or no interbedded sedimentary material. The Cheshire Formation, up to 2.5 km thick, occurs at the top of the Ngezi Group. It comprises a basal conglomerate (derived from the Zeederbergs Formation) which passes laterally into iron-formations, and upward into an assortment of shallow-water sediments, including very extensive limestones which, in places, are profusely stromatolitic. Figure 3.19 is the inferred paleogeographic setting for the upper part of the Ngezi Group (Nisbet, 1987).

2.*Midlands Greenstone Belt*. The Midlands greenstone belt, comprising the Kwekwe-Gweru and the Chegutu-Kadoma greenstone belts (Foster and Gilligan, 1987), contains some of the best exposures of the Upper Bulawayan greenstones in the western part of the Zimbabwe province (Fig.3.14). The western greenstones as already noted are peculiar in their calc-alkaline volcanic suites and in the significant development of the Shamvaian Group. At Kwekwe the lowest lithostratigraphic unit, the Mafic Formation, is a sequence of pillowed mafic flows interlayered with chert and minor felsic tuffs and conglomerate with granitic clasts which indicate the presence of earlier pre-greenstone sialic crust. The Mafic Formation is intruded by the Rhodesdale batholiths or ultramafic bodies, and is overlain by the Maliyami Formation (Fig.3.20), which consists of a thick sequence of intercalated mafic and andesitic flows and andesitic to dacitic pyroclastics (Condie, 1981). The lavas which are dated at 2.7 Ga, are augite andesites in which clinopyroxene is typically fresh within an altered groundmass. The presence of aggregates of serpentine in some lavas imply that olivine was present. Some fine-grained amygdaloidal lavas contain low-greenschist facies mineralogy and rare groundmass pyroxene. Porphyritic lavas include examples with altered feldspar phenocrysts, 1-3 mm long, and some chlorite pseudomorphs after orthopyroxene. Devitrified glass, rarely found in Archean terrane, is present (Harrison, 1970). Based on its petrography and chemical composition, Nisbet (1987) likened the Maliyami volcanics and its contemporaneous tonalitic intrusion, the Sesombi Balholith, to modern basaltic andesitic assemblages above subduction zones where, as in the Sierras and the Cascades of western North America, a young calc-alkaline belt of intrusives is surrounded by lavas of similar age. The Maliyami Formation is conformably overlain by the Felsic Formation comprising mainly andesitic to dacitic flows, breccias, and tuffs with lesser amounts of mafic volcanics. An erosional surface implying folding, uplift and stripping, marks the top of the volcanic pile and its unconformable contact with the overlying graywackes, phyllites and conglomerates of the Shamvaian Group.

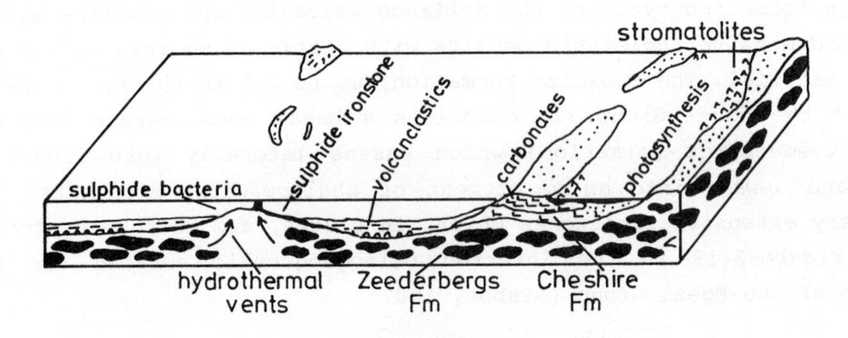

Figure 3.19: Paleogeographic setting for the Cheshire Formation. (Redrawn from Nisbet, 1987.)

3. *Tati Greenstone Belt*. In the Tati belt of northeastern Botswana, a thick sequence of volcanics and sediments unconformably overlie highly deformed meta-arkoses. The lower one-third of the Tati succession comprises a bimodal sequence, the Lady Mary Formation which is a sequence of mafic and ultramafic flows and sills with minor felsic tuffs, arkose, and

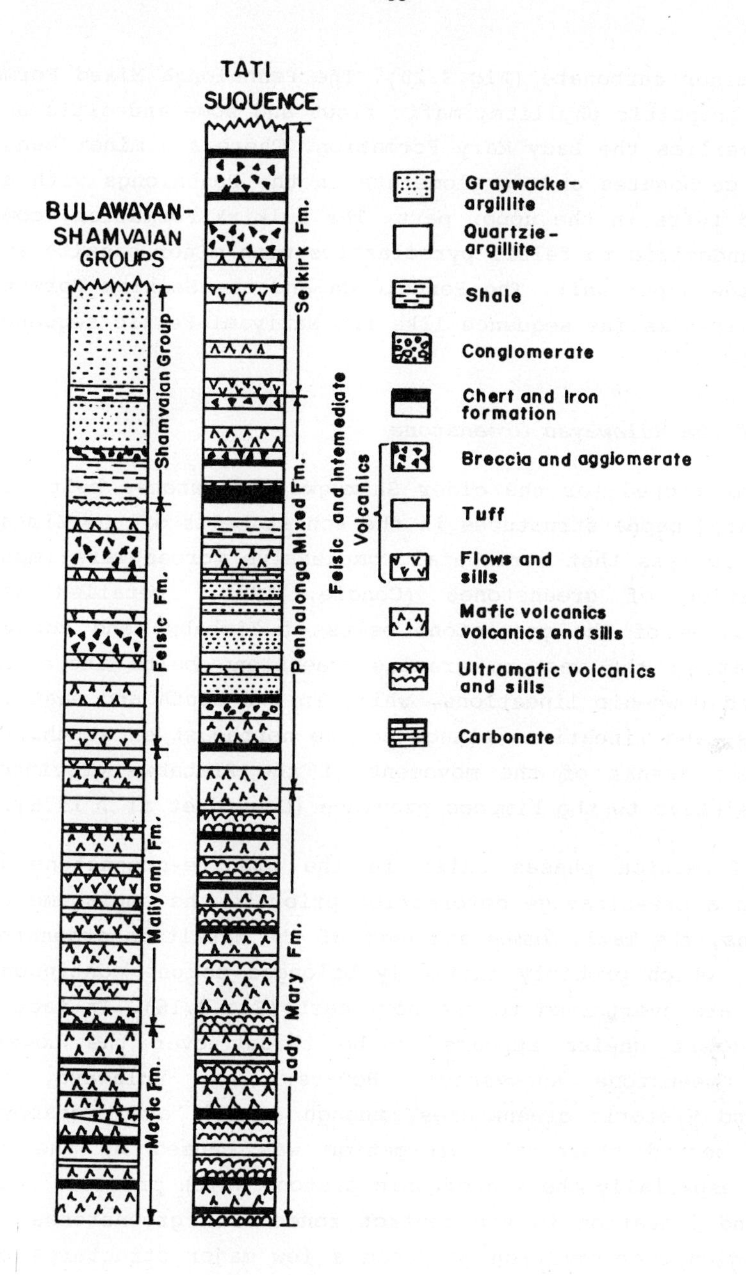

Figure 3.20: Stratigraphic columns for some Zimbabwe greenstone belts. (Redrawn from Condie, 1981.)

chert with minor carbonate (Fig.3.20). The Penhalonga Mixed Formation of graywackes, graphitic phyllite, mafic flows and some andesitic and felsic volcanics overlies the Lady Mary Formation. There are minor banded iron-formations, carbonates and conglomerate in the Penhalonga with andesitic breccias and tuffs in the upper part. The Selkirk Formation, composed of dominantly andesitic to felsic pyroclastics with minor basalts and chert, constitute the upper unit. The Penhalonga and the Selkirk Formations represent a calc-alkaline sequence like the Maliyami-Felsic sequence of the Midlands belt.

Structure of the Bulawayan Greenstone

As already mentioned for the older Shurugwi greenstones belt, there are well-documented nappe structures in the schist belts of the Zimbabwe province. This implies that horizontal compressive forces were important in the deformation of greenstones (Condie, 1981). Detailed structural (strain) studies of the greenstone belts of Zimbabwe and Botswana have revealed that in the western granite-greenstone belts there are steep foliation and down-dip lineations, while in the south and east the foliation curves, and lineations plunge to the northeast or south-southwest, probably as a result of the movement of the Zimbabwe province to the southwest relative to the Limpopo province (Coward et al., 1976).

Four deformation phases exist in the granite-greenstone belts of Zimbabwe. In a pre-cleavage deformation prior to the emplacement of diapiric plutons, the Tati, Vumba and part of the Matsitama greenstone belts of Botswana, which probably initially belonged to one continuous greenstone belt, are overturned to the northeast (Fig.3.16). In fact, the tonalitic basement gneiss appears to be thrust over the Lower Gwanda (Antelope) greenstone succession. However, the Bulawayo, Shangani, Mberengwa and Victoria greenstones, though folded, are probably autochthonous. A second phase of deformation was caused by the intrusive granitoids, especially the syntectonic plutons which produced local steep foliation and lineation in the contact zones with greenstones. The main phase of regional deformation produced a few major structures but widespread cleavage in both granite and greenstone terranes. During this time the greenstones were shortened by up to 65 %. Late deformation produced crenulations and tight folds which deform the earlier fabrics.

Igneous Intrusion and Cratonization

The terminal phase of deformation and metamorphism in the Zimbabwe province was accompanied by the emplacement of mafic and granitoid intrusives. The cratons, which had stabilized, became the sites of large cratonic basins in some regions such as the Kaapvaal province. But in some parts of the Kalahari craton, there was renewed magmatism during which large intrusive bodies rose to fill major fractures which had developed in the newly formed continental masses. It was during this renewed phase of magmatism between 2.7 Ga and about 2.5 Ga that large mineralized ultramafic intrusives, such as the Mashaba ultramafic suite and the Great Dyke, respectively, were emplaced, cutting across the pre-existing granite-greenstone terranes of Zimbabwe. The Great Dyke is, however, considered as marking the inception of the Proterozoic, hence it is discussed under the Early Proterozoic.

1. *Mashaba Ultramafic Suite*. Scattered around the perimeters of the Mberengwa greenstone belt are several major mafic to ultramafic intrusions which have been dated at about 2.7 Ga. These are collectively termed the Mashaba Ultramafic Suite (Fig.3.14), the most notable of which is the Shabani Complex. Mafic or komatiitic dykes which may not be related to the main intrusive cut across most of the granite-gneiss country rocks. The Shabani Complex is a large slab of mainly ultramafic rock which is exposed along the northeastern edge of the Upper Bulawayan greenstones. It consists of a simply differentiated sill-like igneous body in which layers of dunite pass upward through peridotite and pyroxenite into gabbro. It outcrops over an area of about 15 km by 2.5 km and is about 1,500 m thick, with 60° dip in the southward direction. Seventy per cent of the sill is dunite; 20 % consists of peridotite and pyroxenite, while gabbro constitutes up to 10 %. The Shabani Complex has one of the world's largest deposits of asbestos.

The Shabani Complex probably represents a stratified magma chamber which was fed from below by ultramafic liquids, in which doubly diffusive processes could have operated to produce the voluminous low-density eruptive basalt lavas of the Zeederbergs Formation (Nisbet, 1987).

2. *Younger Granitoids*. Granitoids such as the Sesombi tonatite and the more potassic Chilimanzi Suite (Fig.3.14) were intruded into the greenstone belts around 2.6 Ga. The Sesombi, with its low initial $^{87}Sr/^{86}Sr$ ration (0.701) may have been derived from the mantle or from deep crustal granulites. This renders the tonalites indistinguishable from the greenstones, hence they probably originated from the same melting process. With the initial $^{87}Sr/^{86}Sr$ ratio of 0.7025-0.7045, the Chilimanzi Suite may have had a more crustal component at its source.

3.2.4 Limpopo Province

The Limpopo belt, as it is commonly known, is a major zone of Archean high-grade metamorphic and igneous rocks located between the Kaapvaal and the Zimbabwe cratons (Fig.3.21). Mason (1973) divided the Limpopo province into three subdivisions. The following account on this province is culled mostly from Tankard et al. (1982). There are two marginal zones each adjacent to the Zimbabwe and Kaapvaal provinces. The marginal zones are characterized by highly sheared rocks striking parallel to the Limpopo belt and composed chiefly of deformed granitoids and subordinate sequences of greenstone affinity, all of which have been metamorphosed to the granulite facies. There is a central zone comprising basement (ca. 3.8 Ga old) and supracrustal rocks which have been metamorphosed to granulite and amphibolite facies. The marginal zones are separated from the central zone by shear belts. Although the timing of all the periods of regional metamorphism in the adjoining terranes is not completely known, a major tectono-thermal event occurred at about 2.7 Ga (Van Reenen et al., 1987). The overall progressive increase in metamorphic grade towards the Limpopo province from both adjoining provinces suggests a relationship between these three provinces. The increase in metamorphic grade outwards from the centres of the Zimbabwe and Kaapvaal provinces suggests either differential uplift in which the Limpopo belt represents deeper crustal levels of granite-greenstone terranes or there was increase in geothermal gradient towards the Limpopo belt.

Northern Marginal Zone (N.M.Z.)

This extends like a wedge from southern Zimbabwe and tapers westward dying out south of the Great Dyke (Fig.3.21). It is bounded to the south by the Tuli-Sabi shear zone, while to the west in Botswana, the boundary between the N.M.Z. and the central zone of the Limpopo province is not so well defined. In Botswana the Central Zone is believed to grade northwards directly into the Zimbabwe province north of the Tuli-Sabi shear belt.

The cover rocks of the N.M.Z. are represented by linear relicts of granulite-grade greenstone belts in which metasediments occur with compositions such as ferruginous (quartz-magnetite-pyroxene rocks), calcareous (quartz-bytownite-diopside rocks), and possibly pelitic rocks (cordierite-sillimanite-biotite-sapphirine rocks and sapphirine-hyperthene rocks). Low-pressure granulite metamorphism took place before 2.9 Ga producing two pyroxene assemblages throughout the N.M.Z., as far as Botswana. Between 2.7 Ga and 2.6 Ga the metamorphic rocks of the

N.M.Z. were strongly deformed into major upright folds. Charnockites and enderbites were produced after granulite metamorphism of the granulite gneisses. This was as a result of the segregation and injection of quartzo-feldspathic and granite veins and magmas crystallizing under conditions of low water pressure. Deformation during this phase caused the sinistral displacement of the Zimbabwe province for about 200 km in the eastern areas, while in the western region there was a displacement of about 50 km. This was followed by another shearing event, before the intrusion of the Chilimanzi Intrusive Suite of granitoid batholiths. These batholiths were intruded north of the N.M.Z. between 2.7 Ga and 2.6 Ga. They resulted from the partial melting of the granulite gneisses which formed at about 2.9 Ga.

Central Zone in the Limpopo Valley

The Limpopo valley falls within the Central Zone where the gneisses are believed (Tankard et al., 1982) to have undergone at least six periods of deformation between 3.8 Ga and 2.6 Ga. The basement to the Central Zone pre-cratonic cover is made up of the Sand River Gneiss (Fig.3.22) which are among the oldest rocks known in Africa (3.8 Ga). The Sand River Gneiss comprises nebulitic layered gray gneisses of granodioritic, tonalitic and quartz dioritic composition with metabasite dykes. These are overlain by the Beitbridge Sequence which has a carbonate facies in the Messina area (Fig.3.22). Other rocks of the Beitbridge Sequence include the quartzo-feldspathic gneisses of the Diti-Shanzi Metamorphic Suite which has intercalations of banded iron-formations, marble and metaquartzite, all of which suggest sedimentary parent rocks. Also in the Beitbridge Sequence, are the Messina and Nuli Metamorphic Suites which comprise metaquartzites of probably deep-water eugeosynclinal origin (Tankard et al., 1982). The pre-cratonic intrusions of the Central Zone include the Messina Intrusive Suite (metamorphosed anorthosite and leucogabbro) which was emplaced between 3.2 and 3.1 Ga, and the Bulai Granitoid Gneisses emplaced at about 2.7 Ga. The emplacement of the Bulai granitoid was followed by a widespread and intensive episode of crustal shortening which was coeval with similar tectonic events in the Zimbabwe province. Throughout the Limpopo valley the flat-lying gneisses were deformed by this thermo-tectonic event into tight isoclinally upright folds with NE axial trends, caused by shearing and strong axial plane schistocity.

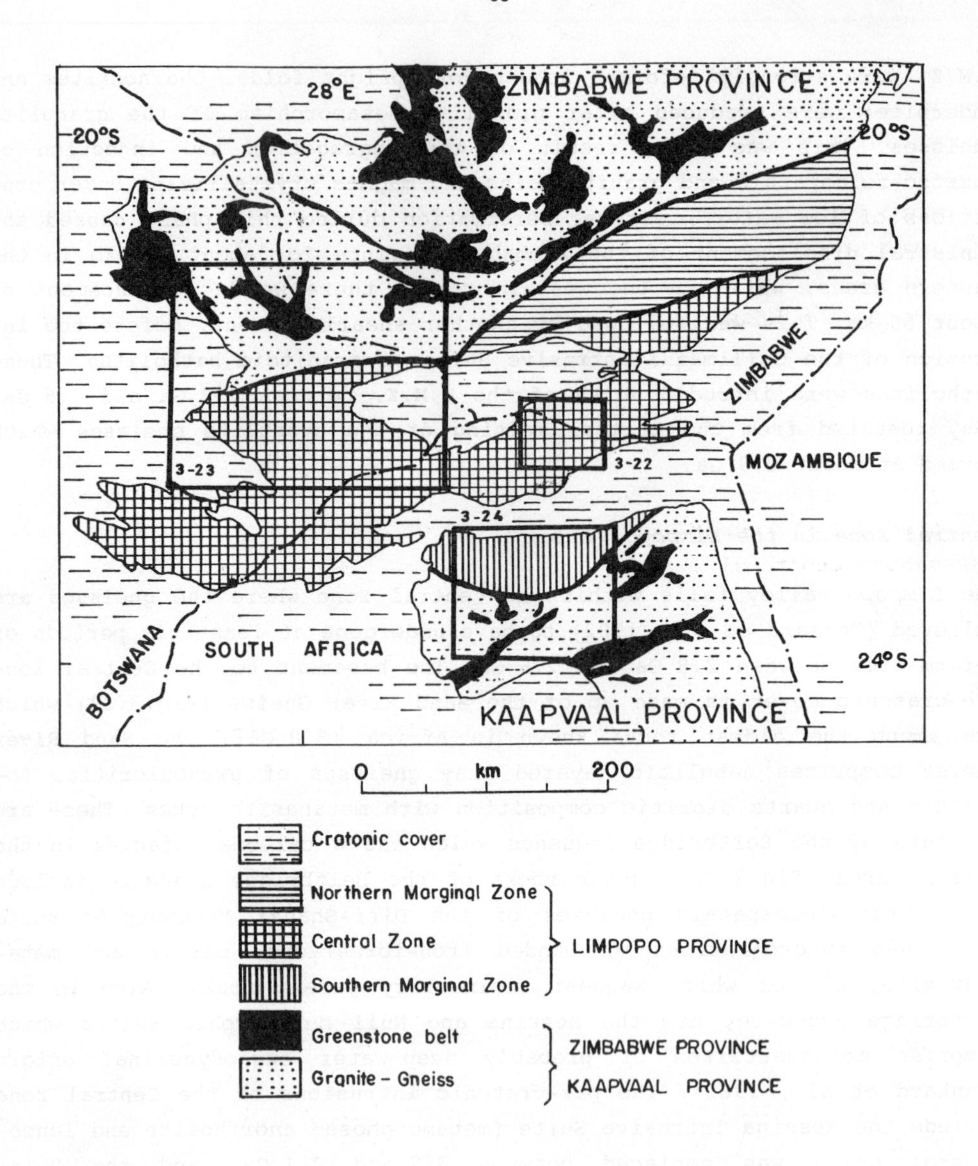

Figure 3.21: Tectonic Zones in the Limpopo province. (Redrawn from Tankard et al., 1982.)

Central Zone in Botswana

The northern margin of the Tuli-Sabi shear belt is believed to be the structural boundary between the Limpopo province and the Zimbabwe province (Fig.3.23); otherwise there is no precise boundary. At Baines Drift there are gray layered tonalitic gneisses of unknown age probably representing basement. The age of the overlying metasediments of the Baines Drift Formation is also uncertain although the entire Baines Drift Metamorphic Suite is generally believed to represent the high-grade equivalents of the shallow-water facies of the 2.7 Ga-greenstone belts such as the nearby Matsitama belt in Zimbabwe. Metamorphism is much less intense in the Matsitama belt than in the Central Zone; the structurally complex Matsitama sequence consists of current-bedded quartzites, marbles, and shales and basalts and dolerite sills. Near Baines Drift there are large sheets of layered metagabbro-anorthosites which were folded into nappes during the major deformation phase of 2.6 Ga, which appears everywhere in the Limpopo province. The strata-bound Ni-Cu sulphide deposit at Pikwe was formed during the intrusion of the layered metagabbro-anorthosite.

Southern Marginal Zone (S.M.Z.)

This zone which displays the metamorphic and deformational transition from a typical low-grade granite-greenstone terrane to the high-grade gneiss grade was described in detail by Van Reenen et al. (1987). As shown by these authors the Southern Marginal Zone represents a cross section through the Archean crust. In this zone (Fig.3.24) steep northward-dipping, typical granite-greenstone lithologies, comprising the mafic, ultramafic, felsic, and volcano-sedimentary assemblages of the Pietersburg, Sutherland, and Rhenosterkoppies greenstone belts of South Africa (Fig.3.24), are tectonically juxtaposed with and overlain by progressively higher lithologies from south to north.

The Pietersburg greenstones, at least 3.45 Ga old, is at greenschist-grade in the central and southwestern parts and is succeeded along shear zones by amphibolite-grade rocks in the northeast. The Rhenosterkoppies and Sutherland greenstone belts show similar arrangements of metamorphic grades relative to shear zones and are surrounded by the tonalitic and trondhjemitic Baviaanskloof Gneiss which is about 3.5 Ga old. The Baviaanskloof Gneiss and the greenstone assemblages can be followed uninterrupted across the transition from amphibolite grade to granulite grade. At this transition there is a significant change in deformational style in which high-grade greenstones are highly reduced compared to the more extensive outcrop of the lower-grade lithologies to the south (Fig.3.24). In the granulite terrane, metamorphosed greenstone assemblages and their intrusive granodioritic plutons have yielded ages around 2.65 Ga which reflects a widespread tectono-thermal event of this age.

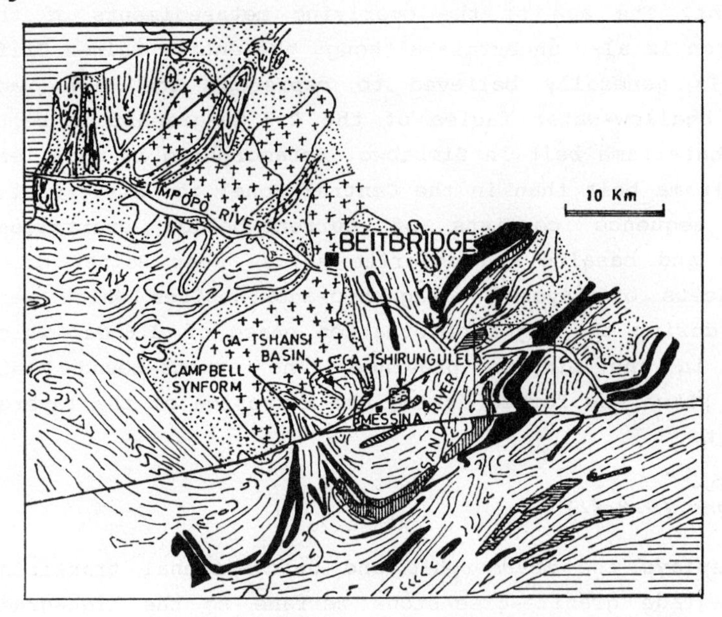

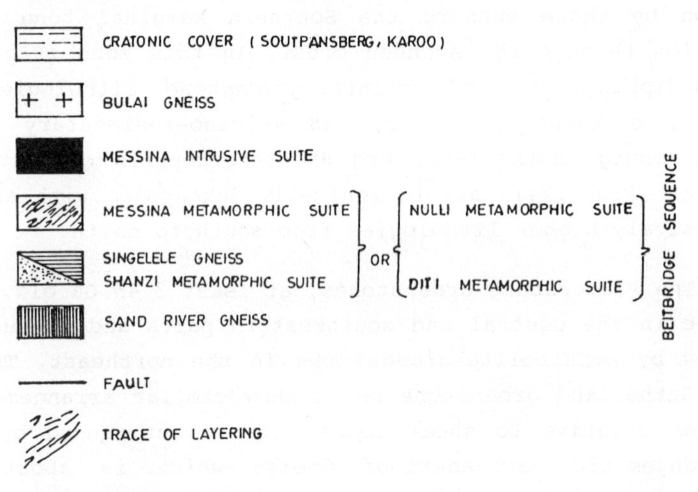

Figure 3.22: Type area of the Central Zone. (Redrawn from Tankard et al., 1982.)

Tectonic Models

Various models have been proposed for the origin of the Limpopo belt. The only points of agreement, as summed up by Shackleton (1986), are that the Limpopo belt shows evidence of drastic tectonic crustal thickening, and a complex deformation sequence of great intensity, suggesting continent-continent plate collision. The various collision models involve relative movements of the Kaapvaal and Zimbabwe cratons either as dextral motion on the Tuli-Sabi shear zone (Coward, 1976), or compressional motion as both cratons rotated towards one another (Barton and Key, 1981; Fripp, 1983; Light, 1982; Van Reenen et al., 1987).

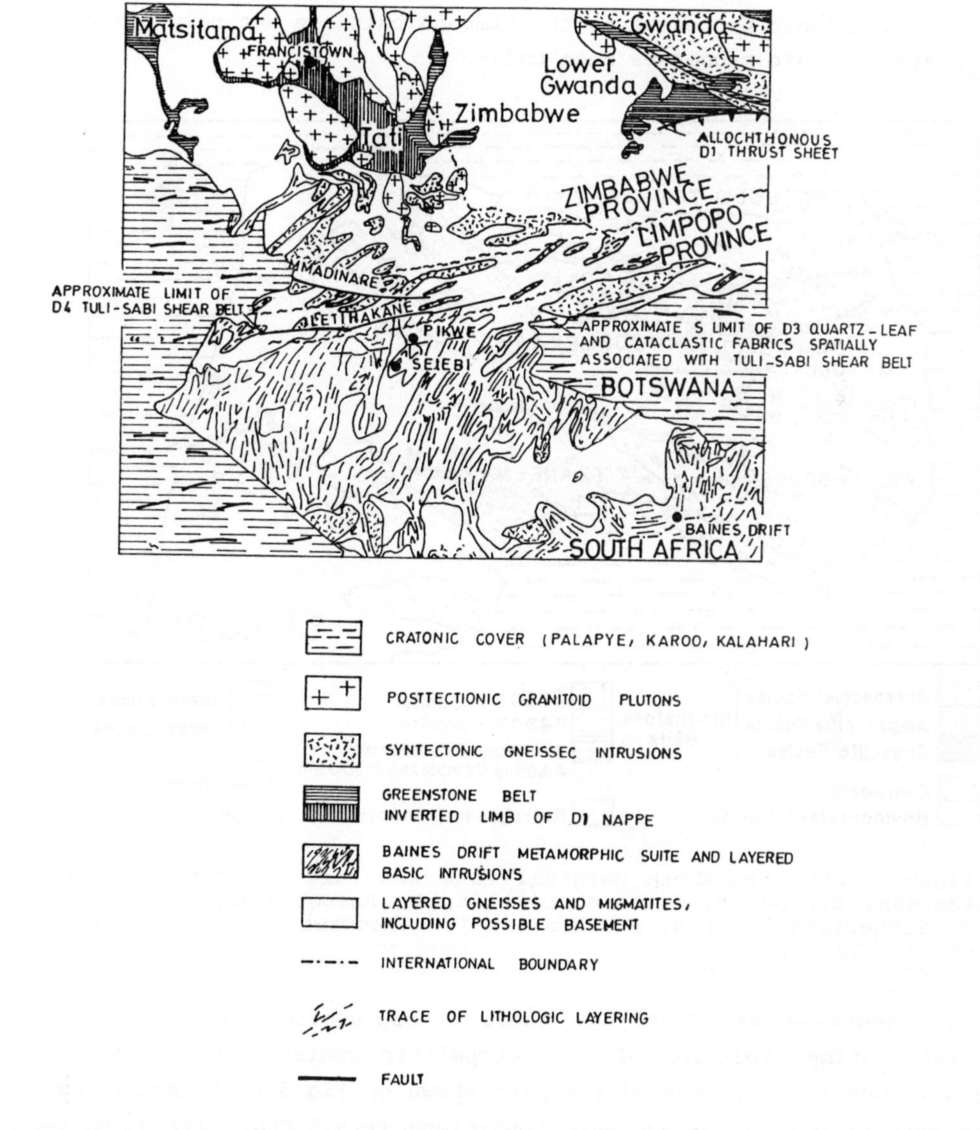

Figure 3.23: Central Zone in Botswana. (Redrawn from Tankard et al., 1982.)

The early tectonic settings during the evolution of the Limpopo belt are, however, quite uncertain in the absence of true remnants of oceanic crust (ophiolites) or calc-alkaline magmatism which could have been associated with the existence of an ocean that was subducted between the Zimbabwe and Kaapvaal cratons prior to collision. But as suggested by Burke et al. (1977) these indicators of geosutures are missing because the Limpopo belt has been sheared and uplifted to expose deeper crustal levels than in the adjoining granite-greenstone terranes. Thus, even at 2.7 Ga the tectonic processes in the Limpopo belt are entirely compatible with plate tectonic processes seen today.

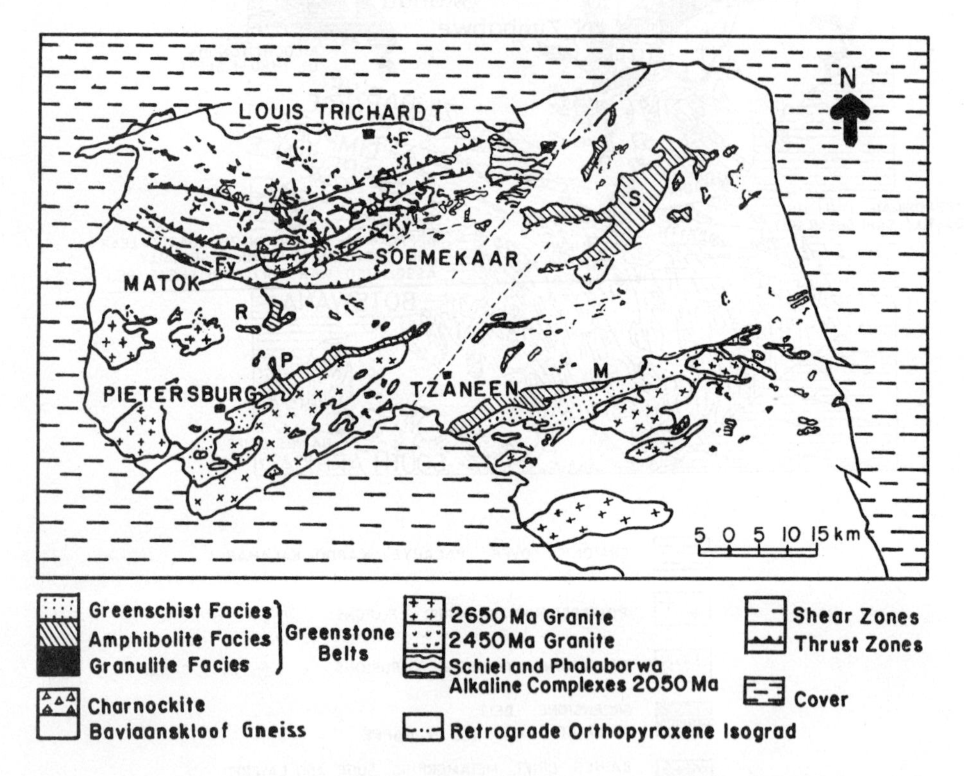

Figure 3.24: Southern Marginal Zone and northern part of the Kaapvaal craton. P, Pietersburg belt; R, Rhenosterkoppies belt; S, Sutherland belt; M, Murchison belt. (Redrawn from Van Reenen et al., 1987.)

Van Reenen et al. (1987) in their interpretation of the pressure-temperature-time evolution of the metapelitic gneisses of the Southern Marginal Zone (S.M.Z.) traced the path shown on Fig.3.25,A. According to these authors maximum metamorphic conditions (P>9.5 Kbar; T>800° C) were followed by rapid isothermal decompression of about 2.5 Kbar between about 2.7 Ga and 2.65 Ga, as evident in the decompression textures of cordierite and hypersthene after garnet.

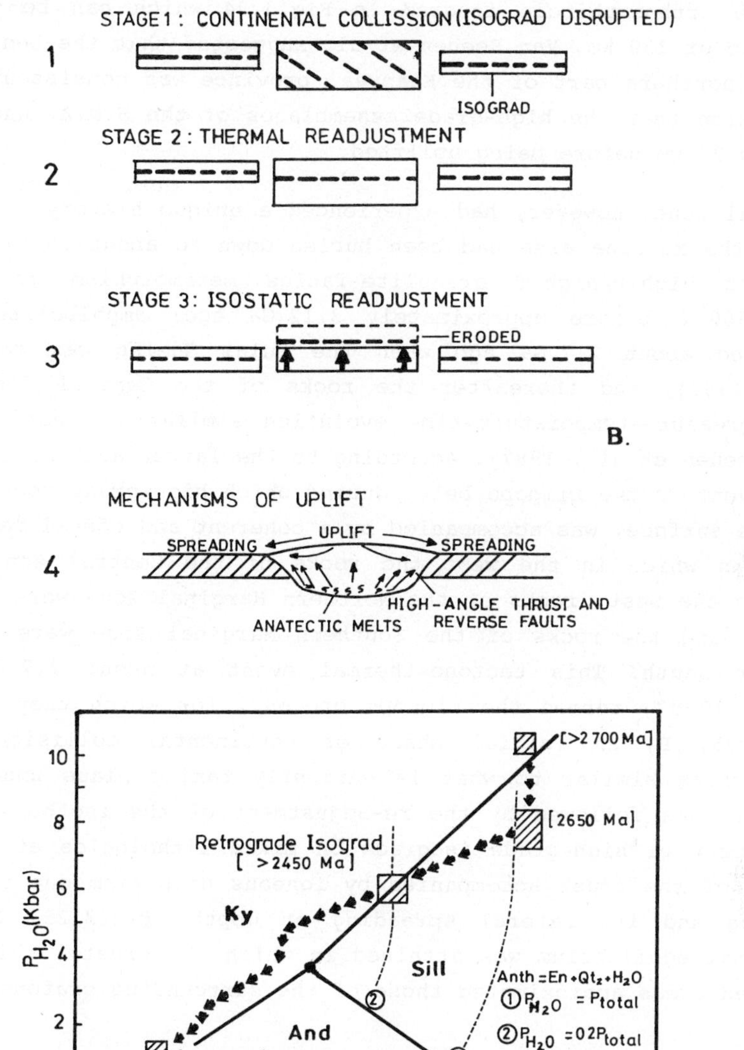

Figure 3.25: Pressure-temperature-time path for the Southern Marginal Zone. (Redrawn from Van Reenen et al., 1987.)

During this decompression vast volumes of granitic melts were produced in virtually all rock types; and the Matok pluton was emplaced. The granulite-grade rocks of the S.M.Z. were uplifted during this event and the granulite terrane was established in this zone at that time. The southern margin of this dehydrated granulite terrane was subjected to a regional encroachment of CO_2-rich fluids which, by rehydration, caused the retrograde orthoamphibole isograd in Fig.3.24 which can be traced over a distance of 150 km. Van Reenen et al. suggested that the behaviour of the entire northern part of the Kaapvaal province was consistent with their observation that the high-grade assemblages of the S.M.Z. had been buried down to 27 km before being uplifted.

The Central Zone, however, had experienced a unique history in which the rocks of the Messina area had been buried down to about 35 km where they underwent high P/high T granulite-facies metamorphism at about 10 Kbar and 800° C before approximately 3.12 Ga ago; amphibolite conditions existed about 2.7 Ga ago when the Bulai Gneiss was emplaced (Shackleton, 1986); and thereafter the rocks of the Central Zone experienced a pressure-temperature-time evolution similar to that of the S.M.Z. (Van Reenen et al., 1987). According to the latter authors the decompression event in the Limpopo belt, during which high-grade rocks were brought to the surface, was accompanied by a coherent and coeval regional deformation, in which in the west the rocks of the Central Zone were transported to the west; rocks of the Northern Marginal Zone were thrust to the north; and the rocks of the Southern Marginal Zone were transported to the south. This tectono-thermal event at about 2.7 Ga, Van Reenen et al. (1987) termed the Limpopo orogeny, for which they postulated (Fig.3.25, B) an initial phase of continental collision with crustal thickening similar to what is currently taking place under the Himalayas. This was followed by the re-adjustment of the isotherms, and the establishment of high-grade isograds on mixed lithologies at depth; rapid rebound of the crust accompanied by igneous diapirism due to anatectic melting and its lateral spreading at depth (Fig.3.25, B, 4). Finally, crustal equilibrium was attained in which the crustal thickness of the uplifted areas approximated those of the surrounding cratons.

3.2.5 Archean Mineralization on the Kalahari Craton

Archean high-grade amphibolite-granulite regions have not supplied many mineral deposits because of their high metamorphic grade and strong deformation. However, as already pointed out, an economic deposit of Ni-Cu occurs in a 50 m wide amphibolite layer in the Selebi-Pikwe area of Botswana in the central zone of the Limpopo province (Fig.3.23).

By far, the granite-greenstone belts have yielded the highest economic mineral potential which are ranked among the world's largest sources of Au, Ag, Cr, Ni, Cu, and Zn. Before reviewing southern Africa's enormous mineral deposits (Fig.3.26), the greenstone metallogenic framework will first be considered in general terms. A unique characteristic of Archean terranes all over the world is their remarkable similarity in the types and modes of mineral occurrences, hence Archean greenstone belts constitute a distinct metallogenic province in the different shield regions of the world. There is a close relationship between the mineral deposits and the major granite-greenstone rock types. Chromite, nickel, asbestos, magnesite and talc occur in ultramafic flows and intrusions; gold, silver, copper and zinc are found in the mafic-felsic volcanics; iron ore, manganese and barytes occur in sedimentary rocks; and granites and pegmatites are the sources of lithium, tantalum, beryllium, tin, molybdenum and bismuth. The primary source of most of the gold mineralization in southern Africa were the mafic-felsic volcanics. Gold mineralization occurs in four modes viz., as stratiform deposits, massive sulphide, quartz lode and as disseminations (Fig.3.27). Most of the quartz lodes occur within the margins of surrounding granitoid plutons (Anhaeusser, 1976). Stratiform-type deposits are found in banded iron-formations where they occur in the oxide facies or in the sulphide and carbonate facies as a result of leaching and precipitation of gold by circulating volcanic thermal brines at temperatures below 400° C (Fripp, 1976).

Gold

Gold occurs in the granite-greenstone belts of Zimbabwe and South Africa, but the largest deposits are concentrated in the Archean-Proterozoic Witwatersrand succession which will be treated later. In Zimbabwe where gold mining dates back to the Middle ages, there are many varieties of gold deposits. As summarized by Hutchison (1983), of the 100 larger mines in Zimbabwe, 30 are stratiform mineralizations associated mainly with the Sebakwian greenstones, seven occur in more massive but stratiform sulphide deposits without stratigraphic control, 56 are in quartz lodes mostly in the Bulawayan and Shamvaian successions but also in the Sebakwian, while seven are strata-bound disseminated deposits predominantly in the Bulawayan and Shamvaian successions.

In the Sebakwian where most of the stratiform gold deposits occur in Zimbabwe, mineralization is found in several thin beds of banded iron-formations that are interlayered with mafic and felsic water-deposited tuffs. The individual gold-bearing beds, generally less than 5 m thick, are confined to sulphide beds and mixed sulphide-carbonate facies in the iron-formations (Foster and Gilligan, 1987). Gold occurs as minute

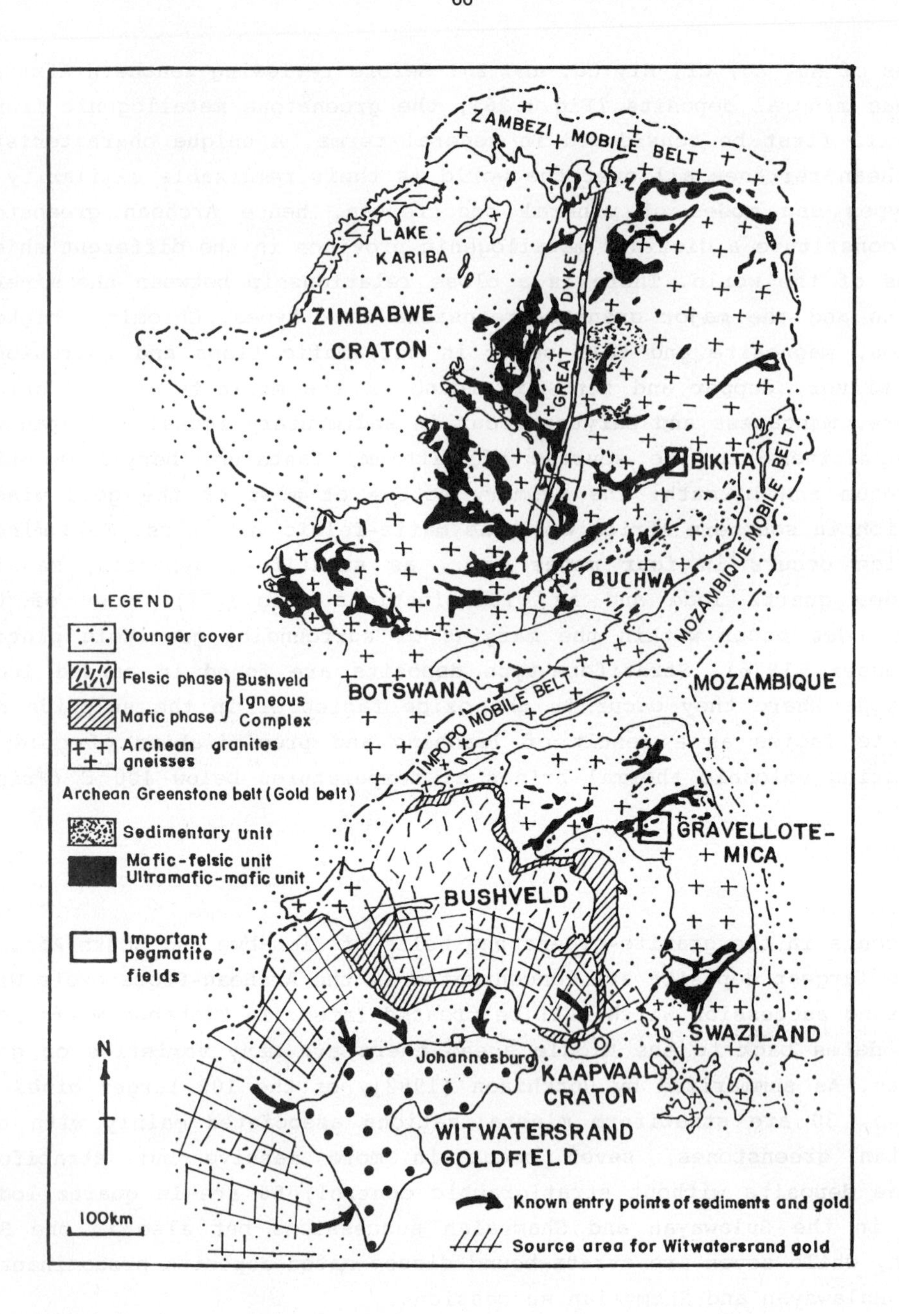

Figure 3.26: Some mineralizations on the Kaapvaal craton. (Redrawn from Hutchison, 1983.)

50-micron grains of native gold with arsenopyrite, in an ore grade which averages 11 ppm (gt^{-1}).

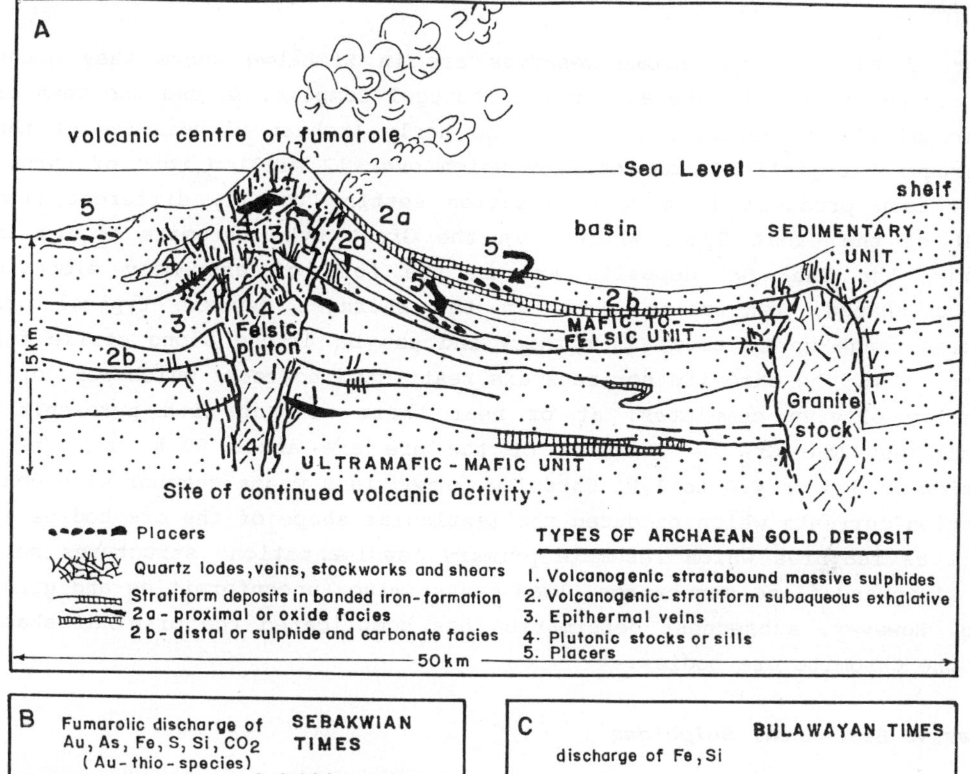

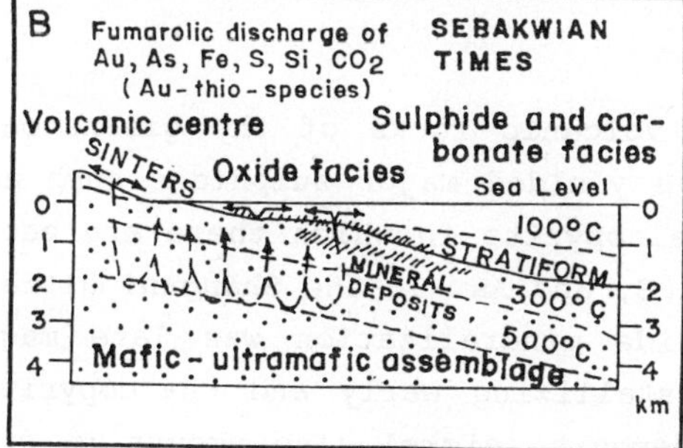

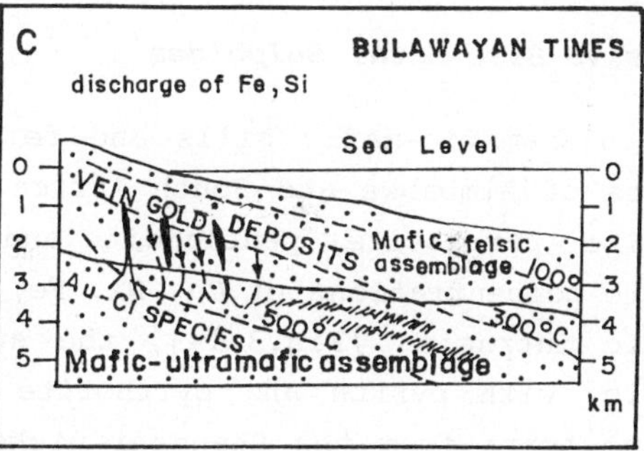

Figure 3.27: Relationship between Archean greenstone mineralization and volcanic-exhalative activity (A); B, model explaining the association of stratiform gold and sulphides with iron-formation; C, later lode or vein deposition in tectonic fractures. (Redrawn from Hutchison, 1983.)

Lode deposits of gold occur in the thick mafic-felsic volcanic rocks of the Bulawayan greenstones where the distribution of the veins are structurally controlled. Lode gold usually occurs in association with pyrite, chalcopyrite, stibnite and galena. Hydrothermal solutions were responsible for transporting the gold and sulphides form the mafic-felsic source rocks into the veins. Silver is recovered as a by-product of gold mining in Zimbabwe.

Chrome

Most of the world's chrome reserves are in Zimbabwe where they occur mainly in the Great Dyke and in the Shurugwi Complex. Around the town of Shurugwi chrome deposits occur in the Early Archean ultramafics of the Shurugwi Ultramafic Formation from which over 12 million tons of chrome have been produced in a mineralization setting totally different from that of the Great Dyke. Whereas in the Great Dyke chromite occurs in seams, the Shurugwi deposits are in the form of extremely elongate ribbon-shaped, lenticular pods. In the Shurugwi deposit typical ore bodies range up to about 15 m thick and can be traced along strike for over 300 m. The chromite horizons are restricted to rocks which were originally olivine cumulates, at or near their contact with pyroxenitic cumulates. The Cr_2O_3 contents of the ore are about 60 % with Cr/Fe ratios of 3.5 or 4.0 to 1.0. Chromite formed in a magma chamber with convection currents which produced the lenticular shape of the ore bodies as well as features which resemble primary "sedimentation" structures such as slump and flame structures, load casts, minor unconformities and grading. However, subsequent deformation has complicated the original shape of the chromite ore bodies.

Massive Base-Metal Sulphides

The ultramafic-mafic sills and felsic volcanic rocks of the greenstone belts of Zimbabwe and South Africa have yielded major sulphides such as pyrite, pyrrhotite, pentlandite and chalcopyrite in which there are economic concentrations of Cu, Ni, Fe, Co, S, and As. In the Shangani ultramafic intrusion (Fig.3.14), the sulphide mineralization was late magmatic, with pyrite and pyrrhotite crystallizing early and chalcopyrite late. Apart from the Shangani nickel deposit, nickel also occurs in extrusive ultramafics which are stratigraphically equivalent to the Bulawayan komatiites of the Mberengwa greenstone belt. This is because of the characteristic association of nickel with komatiitic fluids especially where such fluids are saturated with sulphur.

In the Murchison greenstone belt of South Africa mercury and stibnite are mined from the volcano-sedimentary sequence. The stibnite ore bodies at this locality occur as concordant lenses which are strung out along a strike distance of about 50 km. The host rock consists of altered volcanics, dolomites and ironstones. Mercury and antimony were concentrated in these sediments as a result of submarine hydrothermal and genetically related igneous activity which is represented by the associated quartz prophyry or basic lavas (Hutchison, 1983).

Iron Ore

Although the best development of banded iron-formation occurs around the Archean-Proterozoic boundary, the Buchwa iron ore body in Zimbabwe is a good example of an economic deposit of a mid-Archean banded iron-formation. This deposit lies in the Buchwa syncline in southern Zimbabwe (Fig.3.26). It occurs in a host rock of jaspilite which outcrops as a large 900 m high hill above the surrounding granitic terrain on account of its resistance to weathering. The ore body occurs near the top of the hill and thereby facilitates mining operations. This iron ore probably formed through the replacement of silica by iron oxide, either by circulating solutions during metamorphism, or through a process of supergene enrichment involving the fluctuation of the ground water table.

Pegmatite Mineralization

The best known Archean example in Southern Africa is in Bikita at the eastern extremity of the Victoria greenstone belt (Fig.3.26). Apart from economic deposits of petalite, lepidolite, spodumene, pollucite, beryl, eucryptite and amblygonite in zoned pegmatites (Fig.3.28), there was also tin mineralization at Bikita which although now exhausted, was the geologically oldest tin deposit. The tin occurred disseminated with tantalite and microlite in marginal pockets in quartz-rich zones in large masses of lepidolite greisen. As already pointed out, the Murchison greenstone belt is an important pegmatite region in South Africa, where there are important deposits of mica, beryllium, lithium, tantalum, columbium, corundum, feldspar, emerald, and beryl which are mined from the pegmatites which intrude biotite schists. Tungsten mineralization occurs widely in the Bulawayan greenstones of Zimbabwe. The above pegmatites occur within amphibolite facies metamorphic terranes and are apparently unrelated to later intrusive granitoids. They are the products of anatectic melts which were produced in situ or nearby during regional metamorphism.

Corundum

This is economically important in the Archean greenstone belts of Zimbabwe. Boulder corundum deposits are most important and are characterized by features which reflect the metamorphism of Al-rich sediments.

Most deposits occur near the granitoid-greenstone or mafic dyke-granitoid contacts. They appear conformable with the greenstone stratigraphy and the corundum occurs as lenses in Al-rich schist and contains one or more of andalusite, sillimanite, or kyanite. The associated rocks include ultramafic rocks (talc, schists, serpentinites), ironstones, argillaceous sediments or gneisses.

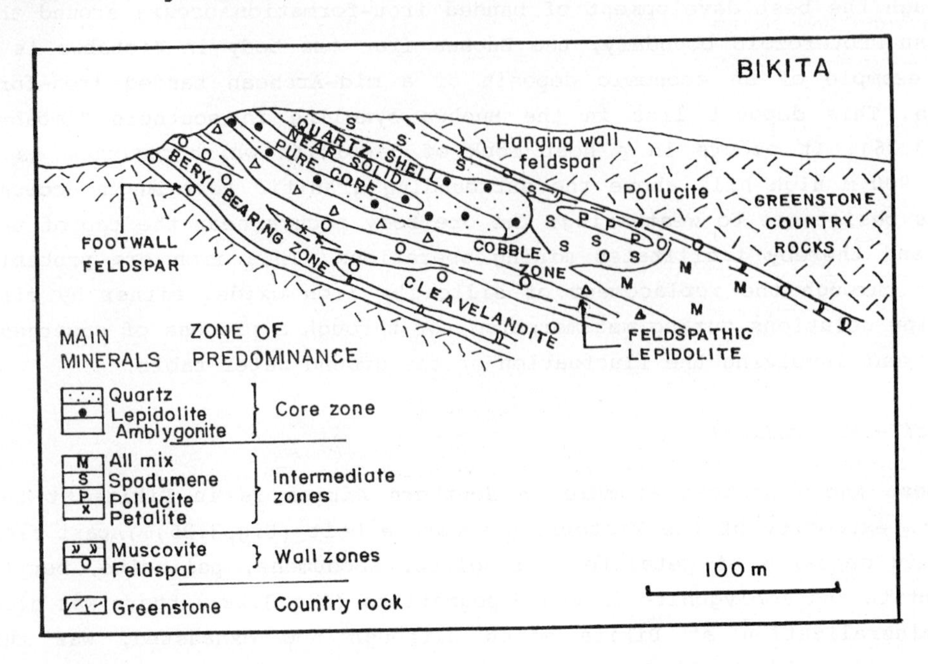

Figure 3.28: Cross-section through the main Bikita pegmatite in Zimbabwe. (Redrawn from Hutchison, 1983.)

Asbestos

Large asbestos deposit occurs in the Shabani ultramafic complex in Zimbabwe. Many Archean ultramafics in Zimbabwe and South Africa contain chrisotile fibres of good quality, and Southern Africa ranks third in the world production of chrisotile fibre (Anhaeusser, 1985).

Most occurrence are in serpentinized ultramafic sills although some are in volcanic rocks. There are three modes of asbestos mineralization in southern Africa, viz., (i) layered ultramafic bodies associated with the mafic-ultramafic portions of greenstones successions; (ii) layered ultramafic bodies associated with the mafic to felsic portions of greenstones; and (iii), intrusive ultramafic bodies which post-date volcanism but pre-date granitoid intrusion (e.g. Shabani). More than 3 million tons of asbestos fibre have been mined mostly from the Shabanie mine in Zimbabwe. Other deposits occur in other parts of the Mashaba ultramafic suite from Filabusi and Norma to Zvishavane and Mashava along a 150 km arc (de Kun, 1987). In the Shabanie mine the ore bodies occur in the central footwall dunite of the Shabani Complex with the main fibres concentrated in zones where extensive fractures and faulting aided hydrothermal alteration of dunite (Fig.3.29). Other mineralizations in the greenstone belts of southern Africa include magnesite and talc in the ultramafics and beryl in the volcanogenic deposits.

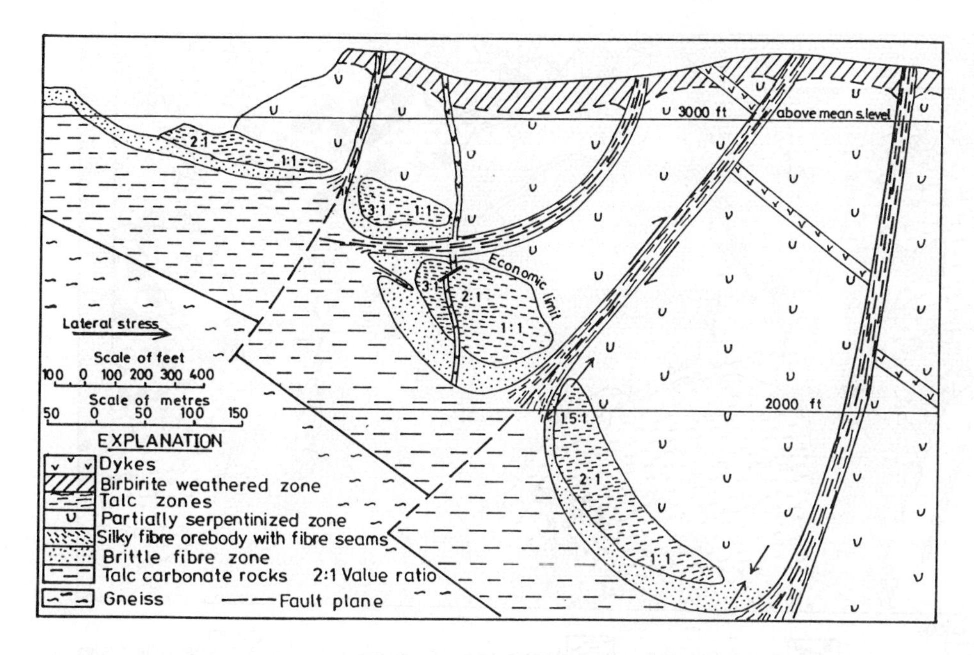

Figure 3.29: Composite cross-section through the Shabanie asbestos mine. (Redrawn from Nisbet, 1987.)

3.3 Zaire Craton

Archean terranes are exposed in three principal parts of the Zaire craton in equatorial Africa (Fig.3.2). In the southwestern part there is the Kasai-NE Angola shield which comprises high-grade charnockitic gneisses and migmatites. In the northwestern part of the craton lie the Ntem-du Chaillu charnockitic-granitoid massifs stretching from southern Cameroon to Congo Republic, as the foreland to a younger Precambrian mobile belt. In the northeastern part of the Zaire craton, however, is a vast and varied Archean gneissic and granite-greenstone terrane which spans the southern part of the Central African Republic, NE Zaire, southern Sudan and western Uganda (Fig.3.30). The predominantly high-grade terranes in the western parts of the craton will be described first before the northeastern parts where there is more information and a clearer picture of the Archean granite-greenstone assemblages and tectono-thermal events. Remnants of Archean rocks in the later Proterozoic mobile belts which completely surround the Zaire craton are indications of an initially more extensive Archean terrane in this region.

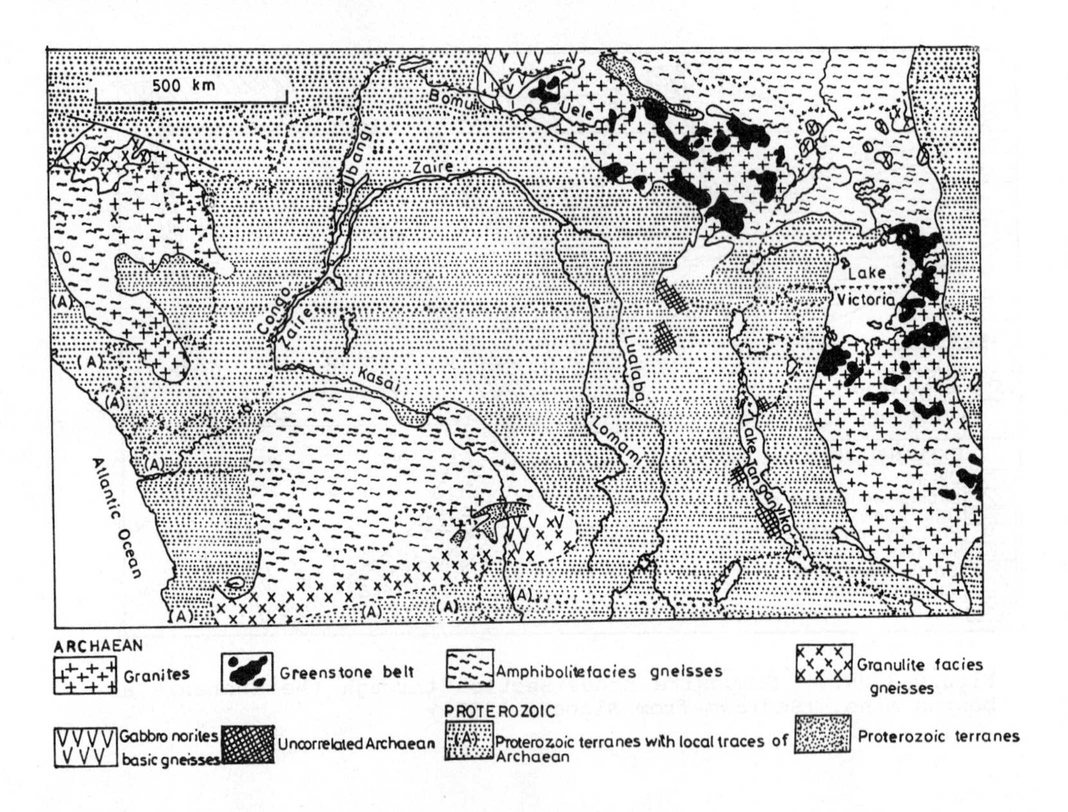

Figure 3.30: Archean cratonic nuclei on the Zaire craton. (Redrawn from Condie, 1981.)

3.3.1 Kasai-NE Angola Shield

This region includes the basement of western Kasai and southwestern Shaba provinces of Zaire and northeastern Luanda in Angola (Fig.3.31). An ancient metamorphic basement is exposed here on a shield which dips gently to the north where at about 4°S it is bounded by an important fault. In the east the Archean terrane is bounded by the Katangan system while Phanerozoic rocks cover the southern and western ends of the shield.

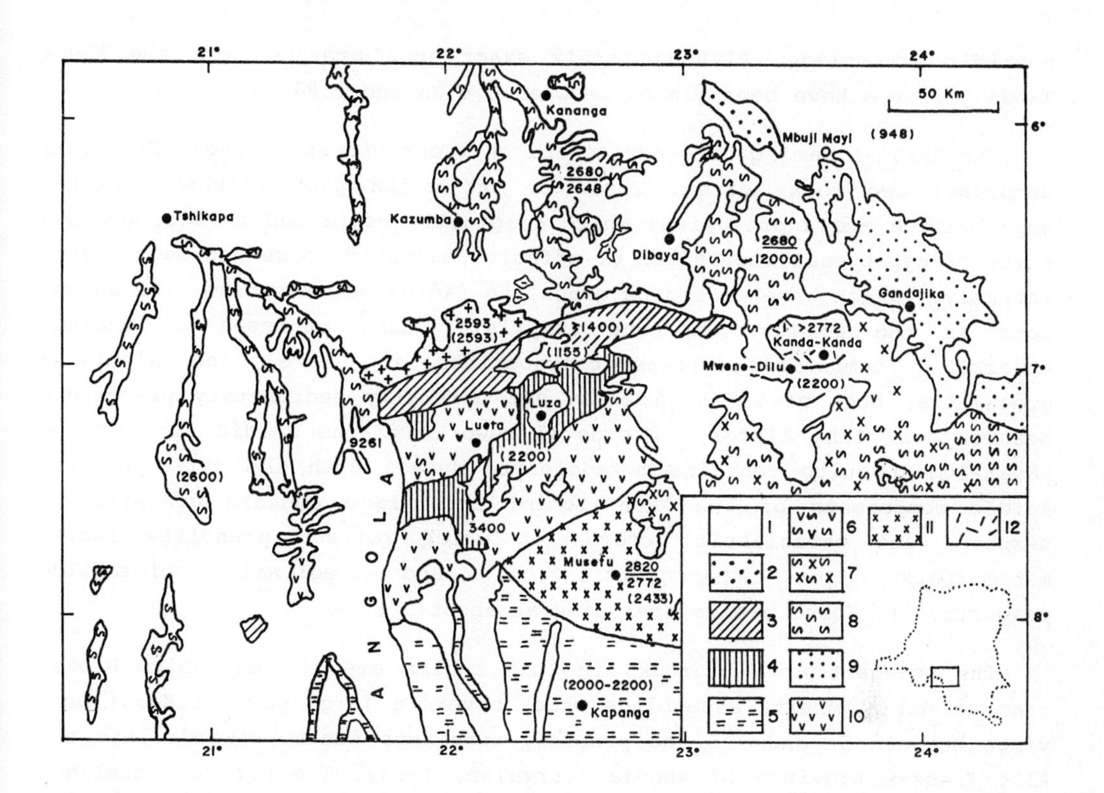

Figure 3.31: Central Kasai province, Zaire showing: 1, Karoo to Recent cover; 2, Mbuyi Mayi Supergroup; 3, Lulua Group; 4, Luiza Supergroup; 5, Gneisses, Dibaya Granite, and migmatite assemblage; 6, Amphibolites; 7, Gneisses; 8, Migmatites; 9, Granites; 10, Gabbro-norite; 11, Charnockites of the Kasai-Lomami gabbro-norite and charnockite assemblage; 12, Kanda Kanda tonalitic and granodioritic gneisses and Upper Luanyi gneisses. (Redrawn from Cahen et al., 1984.)

As shown by Cahen and Lepersonne (1967), most of the Kasai and NE Angola shield between lats. 7°S and 11°S and longs. 22°E and 24°30′E, inclusive of the Dibaya-Luiza-Kanda Kanda type areas, are underlain mainly by poorly exposed gneisses and migmatites (Fig.3.31). The oldest rocks, dated at about 3.4 Ga, are the Upper Luanyi granite gneisses with pegmatites of amphibolite facies. These are separated from the adjacent Kanda Kanda gray tonalites and granodiorite gneisses by inferred faults. The Kanda Kanda gneisses contain diffuse lenses of alaskite gneisses which are hololeucocratic pink rocks that probably formed near the limit of granulite facies metamorphism or as intrusions which originated from the nearby charnockitic rocks, the Kasai-Lomami gabbro-norite and charnockitic assemblage. Although their exact ages are unknown, the Kanda Kanda gneisses have been placed between 3.4 Ga and 2.82 Ga.

The Kasai-Lomami gabbro-norite and charnockitic assemblage (Fig.3.31) comprises two rock suites. There is the mafic part (gabbro, norite, amphibolites and anorthosite) which originally comprised a heterogeneous suite of hypabyssal intrusives or effusive magmatic rocks and deep-seated intrusives which have undergone granulite facies metamorphism. The second part is the acidic component comprising dark gneisses of charno-enderbitic composition (true charnockites are rare) and aluminous granulites, both of which probably had partially sedimentary precursors before their long history of crustal reworking. The acidic part of the Kasai-Lomami suite contains metadolerite dykes. Both the mafic and the acidic components of the Kasai-Lomami assemblage contain granoclastic textures and cataclastic deformation which reflect granulite facies metamorphism (or charnockitization) and regional deformation respectively. These events are dated at about 2.8 Ga.

The youngest Archean rocks in this region are the extensive Dibaya granite and migmatite assemblage which covers a large part of Kasai province between 5° and 7°S and probably continues southwestwards into the Alto Zambeze province of Angola (Carvalho, 1983). The Dibaya assemblage comprises granite-to-tonalite migmatic gneisses and calc-alkaline granites. These rocks have prophyroclastic to heteroclastic structure and are locally mylonitic. They have yielded radiometric ages which suggest the transformation of pre-existing Archean rock into the Dibaya granite and migmatite assemblage between 2.82 Ga and 2.56 Ga. South of Dibaya, deformation (cataclasis) decreases so that structurally undeformed granites appear. The Malafundi anatectic granites occur in the southern area linked to the migmatites.

Archean rocks are also exposed in several places in the basement of western Angola (Fig.3.32), for example at Malanje, Dondo, and south of the River Cuanza near Cariango where Archean assemblages show east-west and NE-SW trends and comprise enderbites, charnockites, kinzigites and granulite gneisses. South of Nova Lisboa in the area between 13° and $15^{\circ}40'$S and $14^{\circ}30'$ and $17^{\circ}30'$E, the volcano-sedimentary Jamba Group is of Archean age.

3.3.2 NW Zaire Craton

This part of the Zaire craton is a broad basement upwarp which constitutes the foreland of the younger West Congolian mobile belt (Pan-

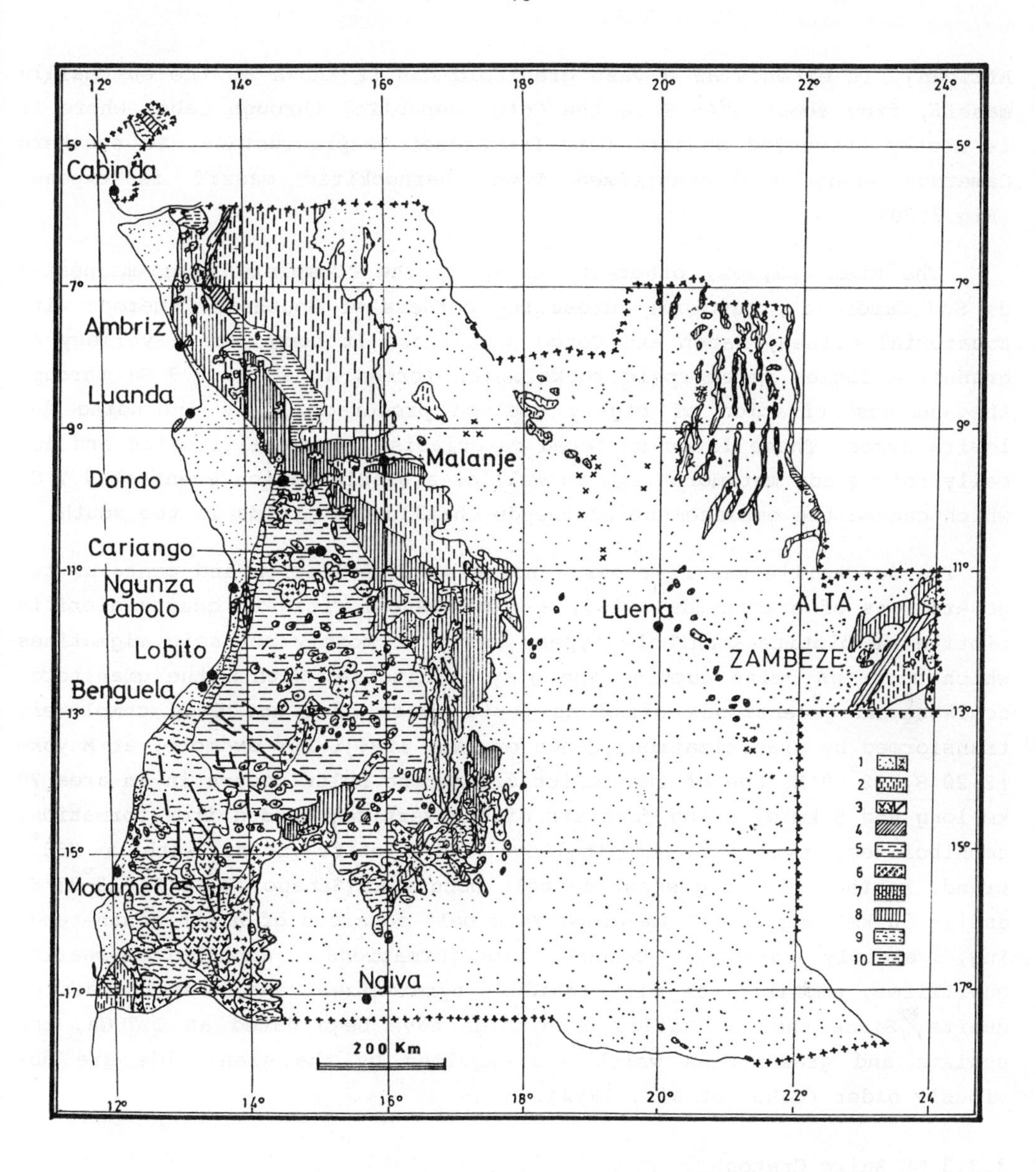

Figure 3.32: Tectonic map of Angola. 1, Phanerozoic cover; 2, Pan-African; 3, Kibaran; 4, Kibaran-Eburnean; 5, Eburnean; 6, Eburnean-Archean; 7, reactivated Archean; 8, Archean; 9, older Archean reactivated; 10, older Archean. (Redrawn from Carvalho, 1983.)

African). It extends as a vast granitoid massif known as the du Chaillu massif, from about 3°45'S in the Congo Republic, through Gabon where it is mostly concealed beneath Late Proterozoic supracrustals, to southern Cameroon where the granitized Ntem charnockitic massif is exposed (Fig.3.30).

The Ntem complex, otherwise known as the "Complexe calco-magnésien du Sud Cameroun", outcrops across the borders of Southern Cameroon with Equatorial Guinea, Gabon and Congo Republics. It comprises a variety of granulite facies metamorphic rocks which formed at about 2.9 Ga through the charnockitization of highly evolved precursor rocks including dolerite dykes. These later suffered cataclasis, recrystallization and locally retrograde metamorphism, as well as a granitization event at 2.7 Ga which caused the emplacement of the du Chaillu granitoids in the south.

The du Chaillu massif shows a north-south foliation and contains two generations of granitoids, viz., a gray granodioritic to quartz dioritic biotite or biotite-amphibole types, and pink mostly potassic migmatites which occur as veins cutting the gray granitoids. Within the granitoid, schists and greenstones exist as septa which have not been completely transformed by granitization (Cahen et. al; 1984). For example, at Mayoko (2°20'S, 12°50'E) one of the relict greenstone belts occurs in an area 20 km long and 5 km wide and consists of sub-vertical banded iron-formation, amphibolites, pyroxeno- amphibolites and biotite-gneiss with a N60°E trend. Another relict greenstone belt occurs at Zanago (2°45'S, 13°33'E) and is 30 km long and 25 km wide. This belt consists of north-south-tending, steeply dipping banded iron-formations, amphibolite-bearing quartzites, amphibolites with residual pyroxenites, and a small mass of dunite. Since the du Chaillu granitoids have been dated at 2.7 Ga, the schists and greenstones which are engulfed by the granitoids are obviously older (Cahen et al., 1984).

3.3.3 NE Zaire Craton

In the northwestern part of the Zaire craton (Fig.3.33) Archean gneisses and granite-greenstone terrane underlie most of the Haut Zaire administrative province and extend into the neighbouring territories of Central African Republic, western Uganda and southern Sudan. In this vast and geologically complex terrane, Archean rocks comprise the following assemblages: (i) old basement gneisses which date from about 3.5 Ga and are known as the Bomu and West Nile Gneissic Complexes; (ii) scattered greenstone belts known as the Ganguan greenstone belts in the west and as the Kibalian greenstone belts in the east, both of which represent two periods of greenstone emplacement between 3.2 and 2.6 Ga; and (iii), two main generations of granitoid emplaced at between 2.9 Ga and 2.7 Ga and at about 2.5 Ga. During these intrusive events the Upper Zaire Granitoid Massif broke through the greenstones belts and the basement granulite gneisses (Table 3.2).

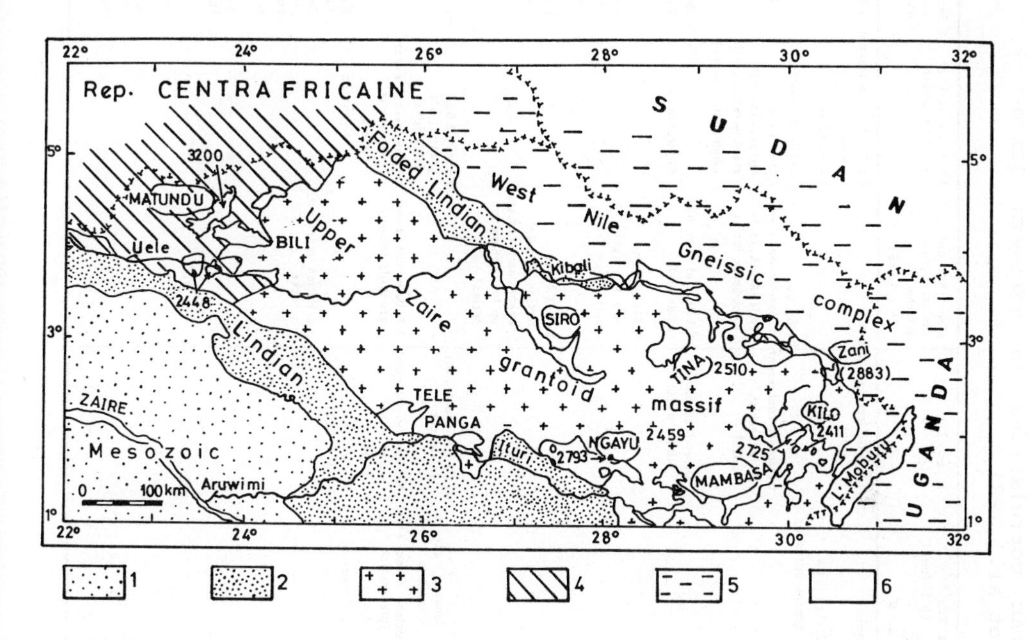

Figure 3.33: Granite-greenstone belts of NE Zaire: 1, Mesozoic cover; 2, Lindian Supergroup; 3, Upper Zaire granitic massif; 4, Bomu gneissic complex; 5, West Nile gneissic complex; 6, Greenstone belts. (Redrawn from Cahen et al, 1984.)

Bomu Gneiss Complex

This is exposed around the confluence of the Bomu (Mbomou) and Uele rivers where it underlies northern Zaire and the southern parts of Central African Republic (Fig.3.33). Several gneiss complexes, the Bomu, Béréme, Nzangi and Monga gneisses actually constitute the basement complex in this region (Cahen et al., 1984). The largest is the Bomu amphibole-pyroxene gneisses which form a large outcrop of about 50,000 km^2, and occupy a synformal structure in northern Zaire. The Bomu gneisses are schistose and garnetiferous and have undergone retrograde metamorphism. They also contain massive and banded tonalitic and monzonitic granitoids.

At the eastern border of the Bomu complex are the synformal Béréme gneisses, a suite of micaceous quartzites, mica schists, biotite gneisses and amphibole gneisses. Both the Bomu and Béréme gneisses have ENE and NW

Table 3.2: Early Precambrian correlations in northern Zaire craton. (Redrawn from Cahen et al., 1984.)

1. Bomu region (N. Zaire)	2. Moto (N. E. Zaire)	3. Southern Ituri (N. E. Zaire)	4. Kilo (N. E. Zaire)	5. Aru-Zani (N. E. Zaire)	6. West Nile and central Uganda extreme (N. E. Zaire)	7. Northern part of Tanzania shield	8. Sequence of assemblages and events	
GAGUAN	KIBALIAN							
		c. 2.05 Ga tectono-thermal event		c. 2.05 Ga tectono-thermal event		c. 2.05 Ga tectono-thermal event	c. 2.05 Ga tectono-thermal event	
2.45±0.03 Ga granite intrusions	2.51±0.06 Ga granite intrusions	2.46±0.03 Ga granite intrusions	2.41±0.13 Ga granite intrusions			2.47±0.06 Ga granite intrusions	2.46±0.05 Ga granite intrusions	
2.60±0.03 Ga tectono-thermal event (amphibolite facies)	tectono-thermal event (greenschist facies)			tectono-thermal event (greenschist facies)	2.55 Ga tectono-thermal event (late-Aruan, up to amphibolite facies)	2.54±0.04 Ga tectono-thermal event (greenschist to amphibolite facies)	2.57±0.04 Ga (late-Aruan tectono-thermal event)	A
	Upper Kibalian			Upper Kibalian (Kibali beds)	Kibalian of West Nile (Adida belt)	Kavirondian	deposition of Upper Kibalian, Kibalian of W. Nile and Kavirondian	R
					2.64±0.09 Ga late-to post-tectonic monzonite		2.64±0.09 Ga late-to post-tectonic monzonite	U
					2.68±0.06 Ga tectono-thermal event (early-Aruan, up to amphibolite facies)		2.68±0.06 Ga early-Aruan tectono-thermal event	A N
	2.89±0.07 Ga tonalite, etc. intrusions	2.79±0.07 Ga tonalite, etc. intrusions	2.72±0.08 Ga tonalite, etc. intrusions	(2.88 Ga) granite, etc. intrusions		2.85±0.08 Ga granite, etc. intrusions	2.84±0.05 Ga tonalite, etc. intrusions	
2.98±0.05 Ga tectono-thermal event	tectono-thermal event (greenschist facies)	tectono-thermal event (greenschist facies)	tectono-thermal event (greenschist facies)	tectono-thermal event (greenschist facies	2.91 Ga Watian event (granulite facies)	tectono-thermal event (greenschist facies)	2.95±0.04 Ga WATIAN tectono-thermal event (Watian)	
	Lower Kibalian	Lower Kibalian	Lower Kibalian	Lower Kibalian (Aru beds)		Nyanzian	Deposition of lower Kibalian and Nyanzian	
Ganguan					pre-Watian precursor rocks		Deposition of Ganguan pre-Watian precursor rocks (age unknown)	
tectono-thermal event ?		tectono-thermal event					tectono-thermal event	
?Bereme gneisses		Ituri paragneisses					?Bereme gneisses and Ituri paragneisses	
3.4 Ga tonalite precursors Nzangi gneisses							tonalite precursors Nzangi gneisses	
high grade tectono-thermal event							high grade tectono-thermal event	
mafic precursors Bomu gneisses							mafic precursors of Bomu gneisses	

structural trends, the later structures being older than the former. To the south of the Bérémé complex are the Nzangi gneisses which outcrop around Bondo. These are massive or banded basic-to-intermediate gneisses which are associated with quartzites and mica schists. In the Nzangi gneisses medium-to high-grade metamorphism was followed by retrograde metamorphism and by the intrusion of the Bondo granite. The Monga gneisses occur to the south of the Bomu gneisses and include mica schists with the gneisses.

The geological history of these gneisses began with the deposition of the probably oceanic precursors of the Bomu basic gneisses and the precursors of the Nzangi gray gneisses at about 3.5 (Cahen et al., 1984). This was followed by high-grade tectono-thermal activity which ended with the intrusion of tonalites at about 3.41 Ga. The emplacement of the Bérémé gneisses followed before the deposition of the Ganguan greenstone and schist belts about 3.2 Ga to 3.1 Ga ago.

West Nile Gneissic Complex

Included in this collective term are several basement gneisses which are poorly exposed from the West Nile Province of Uganda and NE Zaire, through southern Sudan, into Central African Republic. The descriptions presented below mainly apply to the better known parts in West Nile and NE Zaire (Hepworth, 1964; Lavreau, 1980). This region represents what were probably coeval continental areas during the deposition of the nearby Kibalian greenstones. This terrane attained granulite facies during the tectono-thermal episodes which subjected the greenstones to low-grade metamorphism. However, no continental basement is known which directly underlies the Kibalian greenstones.

Granulite rocks belonging to the so-called pre-Watian assemblage of the West Nile Province of Uganda and northeasternmost Zaire contain charnockitic dolerite dykes and are characterized by isoclinal folds with vertical axial planes trending E or ENE. Before they were deformed and intruded by dolerite dykes, the parent rocks to these granulites were probably of volcano-sedimentary origin (Cahen et al. 1984). These parent rocks were metamorphosed at greater crustal depths to granulite facies during the Watian tectono-thermal event at about 2.9 Ga. Charnockites developed at this stage. The Watian event was probably followed by the formation of volcano-sedimentary rocks which were later metamorphosed into the so-called western grey gneisses group comprising well-layered gneissic rocks which are predominantly upper amphibolite facies biotite-hornblende gneiss with microcline. The metasedimentary origin of the western gray gneisses is suggested by the conformable fuchsite and sillimanite-bearing quartzites. These gneisses extend from northeastern Zaire into the northern part of the West Nile Province. Their relationship with the pre-Watian granulite gneisses is, however, uncertain. The western gray gneisses exhibit NE-plunging folds with steep axial planes, and metamorphism which are dated 2.68 Ga and assigned to the Aruan tectono-thermal event.

In the southern part of the West Nile Province and in the adjoining NE Zaire there are low metamorphic grade rocks which are referred to as the eastern gray gneisses. These are probably equivalent to the western gray gneisses. The remaining parts of the West Nile Gneissic Complex are made up of other gneisses which are probably equivalent to the eastern and western gray gneisses.

Ganguan Greenstone and Schist Belt

This occurs in several exposures mainly east of the Bomu-Uele confluence completely detached from the eastern Kibalian belt (Fig.3.33). In the northern part, the Ganguan rests unconformably on the Bomu mafic gneisses and on a part of Bérémé gneisses, with which they share NE-trending folds. To the south, the Ganguan is unconformable on the Nzangi mafic and intermediate gneisses and have been folded with its Nzangi basement. Lithologically the Ganguan comprises (in ascending order): sericite quartzites and quartz phyllites, quartz-poor talc schist, sericite schist, chlorite schists, black schists, and phyllites. The phyllites are cut by gold-bearing quartz reefs. The Ganguan greenstones are older than the 2.98 Ga tectono-thermal event which affected them. However, at Matundu near the Central African Republic border, the Ganguan greenstones are older, having formed at about 3.2 Ga and suffered two deformation events, whereas only the 2.9 Ga-event affected other Ganguan belts.

Northwest of the Bomu gneisses are two other greenstone belts, the Bandas and the Dekoa belts, lying in Central African Republic. Both belts have been correlated with the Ganguan greenstones (Poldevin et al., 1981). The Bandas greenstone belt, about 250 km long, comprises a lower predominantly volcanic unit, and an upper metasedimentary unit (Fig.3.34). The lower unit consists of about 500 m of quartz-feldspathic schists; a middle unit of alternating tholeiitic basalts and itabirites, about 2,600 m thick; and an upper suite 1,000 m thick, composed of itabirites, basalts and andesites. The metasedimentary sequence in the Bandas belt consists of graywackes intercalated with acid volcanic tuffs, and is about 500 m thick. These rocks are intruded by granitic rocks. Deformation was polyphase resulting in isoclinal folds and NWW-WNW structural trends and steep plunges. The Dekoa greenstone belt extends for about 150 km and lies to the south of the Bandas belt. It consists of mostly gneissic granitoids with associated basic rocks and banded iron-formation which exhibit NW foliation.

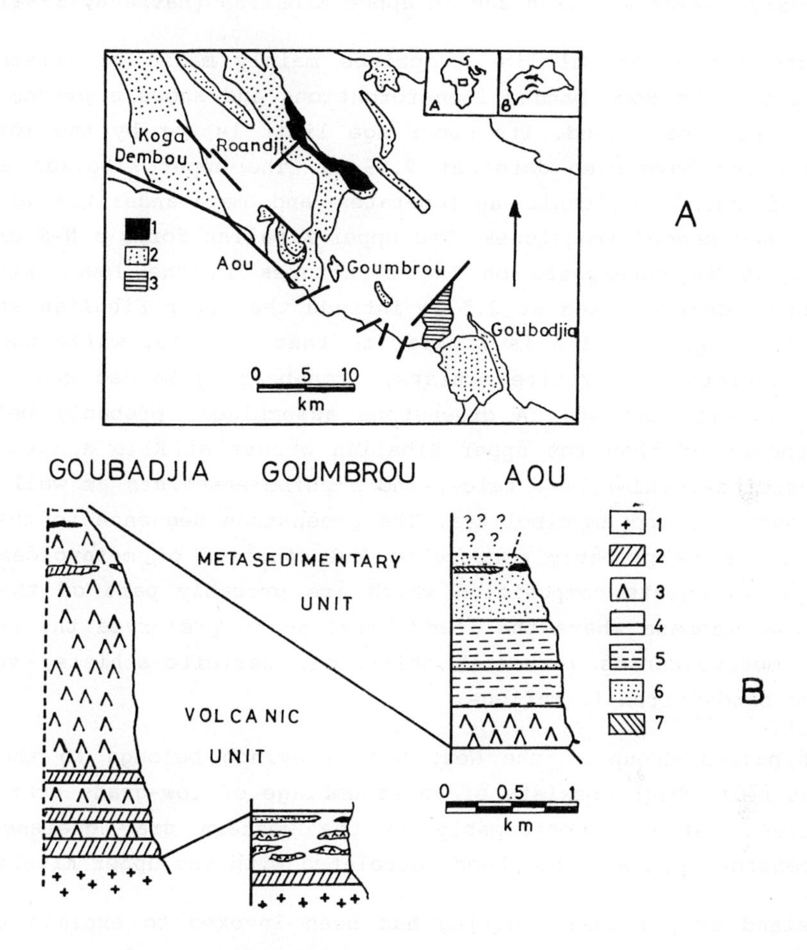

Figure 3.34: A, Bandas greenstone belt. B, Stratigraphic columns for Bandas greenstones; 1, granitoids; 2, tuffs; 3, basalt; 4, rhyolite sills; 5, andesites; 6, graywackes; 7, acid volcanics. (Redrawn from Poidevin et al., 1981.)

Kibalian Greenstone Belts

Although exposed in several isolated belts (Fig.3.33) separated by the Upper Zaire granitoid massif, the Kibalian greenstone belts probably represent one previously continuous greenstone terrane, which has now been separated into an eastern facies and a western facies. The eastern facies which outcrop at Moto, Zani, Kilo, Ngayu and Mambasa show a predominance of mafic to intermediate volcanics while the western facies which is exposed at Sili-Isiro and Tina contain mostly banded iron-formations and less mafic rocks. The Kibalian is further subdivided into a regionally more extensive lower Kibalian and an upper Kibalian (Lavreau, 1984).

At Moto the lower Kibalian comprises mainly mafic to intermediate metavolcanics with some banded iron-formations and shows a predominantly E to ESE structural trend. Its upper age limit is set by the intruding tonalites which have been dated at 2.89 Ga. The upper Kibalian at Moto consists of mainly volcanic agglomerates and meta-andesites and some quartzites and banded ironstones. The upper Kibalian forms a N-S oriented synform which is superposed on the structures in the lower Kibalian. Granitoids which are dated at 2.5 Ga intrude the upper Kibalian at Moto. At Zani the lower Kibalian is similar to that at Moto, while the upper sequence consists of sericite schists, iron-bearing banded quartz-phyllite and chlorite schists. A greenstone assemblage, probably belonging more to the lower than the upper Kibalian occurs at Kilo and comprises mostly sericite-, chlorite-, talc-, and amphibole-schists as well as andesites, dacites, and amphibolites. The greenstone sequence to the south at Ngayu river is probably monocyclic and consists of metavolcanics of mafic to intermediate composition which are probably part of the lower Kibalian. At Mambasa where the lower Kibalian is preserved the sequence comprises metavolcanics (greenish chlorite-, sericite-schists; volcanic tuffs) and banded iron-formations.

The Kibalian Group of the West Nile Province belongs to the Adida greenstone belt which consists of an assemblage of low-grade schists and amphibolites resting unconformably on the western gray gneisses. The Adida greenstone sequence has been correlated with the upper Kibalian.

An island arc tectonic setting has been invoked to explain the Kibalian greenstones, some of which are believed to be of oceanic origin (Cahen et al., 1984).

Granitoids

These are the most extensive rock types on the NE Zaire craton, covering about five or six times more surface area than the schist and greenstone belts. The granitoids are mostly orthogneisses which were derived from the reworking of monzonite granites and tonalites. There are two generations of granitoids (Table 3.2). The first generation is dated at about

2.84 Ga and belongs to the Upper Zaire Granitoid Massif which consists of tonalites with diorites and granodiorites. The second group which is dated at about 2.46 Ga is the most abundant and consists of medium-to coarse-grained quartz monzonites which intrude the first group.

Gold Mineralization

The granite-greenstone terrane of NE Zaire has been a major source of gold in Zaire since production first began in 1904, and by the early 1980's the total output from this region had reached 350 tonnes (Lavreau, 1984). Half of the production has come from placer deposits, the extents of which generally do not overstep the limits of the Kibalian greenstone belts. Lavreau (1984) presented an interpretation of the controls of gold mineralization in the NE Zaire greenstones (Fig.3.33). Eluvial gold deposits also occur as a result of the pronounced tropical weathering of the greenstones in which supergene processes has also enriched gold concentration. Such deposits have been mined near Kilo and at Subani north of the Moto greenstone belt (Fig.3.33).

Other geologic features which influenced the concentration of gold apart from placers, regoliths, and supergene enrichment, are linked to the development of shear zones in the Kabalian belts. Au-bearing quartz veins were emplaced into Early Proterozoic and Pan-African shear zones which had later cut across the greenstones or tonalites. Gold impregnations also occur along the foliation planes within some greenstone units. Quartz veins account for more than half of the primary gold extracted from NE Zaire. In the Kilo belt, for example, gold-bearing quartz veins are located along zones, three of which are subparallel with an ENE-WSW strike, while a fourth trends north-south; some of these are located on a major lineament which is observable on Landsat images (Fig.3.35). The north-south structure is a major shear zone (Fig.3.35, b) with several mylonitic zones along which quartz-veins, sometimes gold-bearing, are arranged in an "en échelon" pattern.

Gold also occurs as impregnations near the banded iron-formation horizons in the volcano-sedimentary parts of the greenstones, without being related to quartz veins. In such deposits the mineralization was probably deposited by syngenetic enrichment by the co-precipitation of gold together with iron and silica, or as a result of the supply of gold through fumaroles.

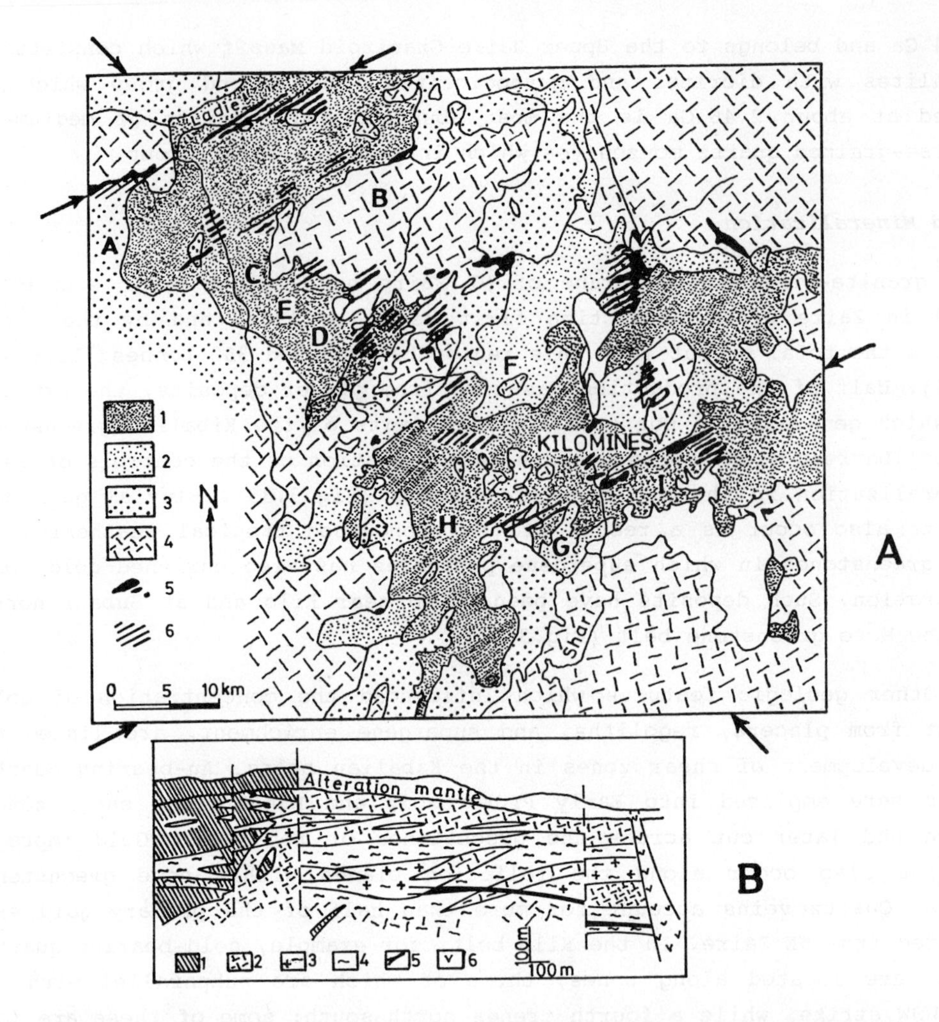

Figure 3.35: A, Kibalian of Kilo. 1, mafic; 2, andesitic; 3, amphibolites; 4, granitoids; 5, itabirites and gossans; 6, mineralized areas. B, N-S cross-section through Isuru-Kanga shear zone; 1, amphibolites; 2, granitoids; 3, mylonitized rocks; 4, cataclazed rocks; 5, lamprophyre; 6, dolerite. (Redrawn from Lavreau, 1984.)

3.4 Tanzania Craton

3.4.1 Geologic Framework

The Tanzanian craton or shield (Fig.3.36) largely coincides with the Central Plateau of Tanzania, and consists of Archean granitoids, gneisses, migmatites and irregularly shaped and scattered greenstones and schist belts which extend northwards into the eastern borders of Lake Victoria (Nyanza) and the surrounding region of southwestern Kenya and southeastern Uganda. Since the Late Archean this region has remained largely cratonic with subsequent tectonism being confined to the Late Proterozoic Pan-African Mozambique belt to the east; the Early Proterozoic Usagaran mobile belt to the south-south-east, the Ubendian belt to the west and southwest; and the mid-Proterozoic Kibaran belt to the northwest (Fig.3.36).

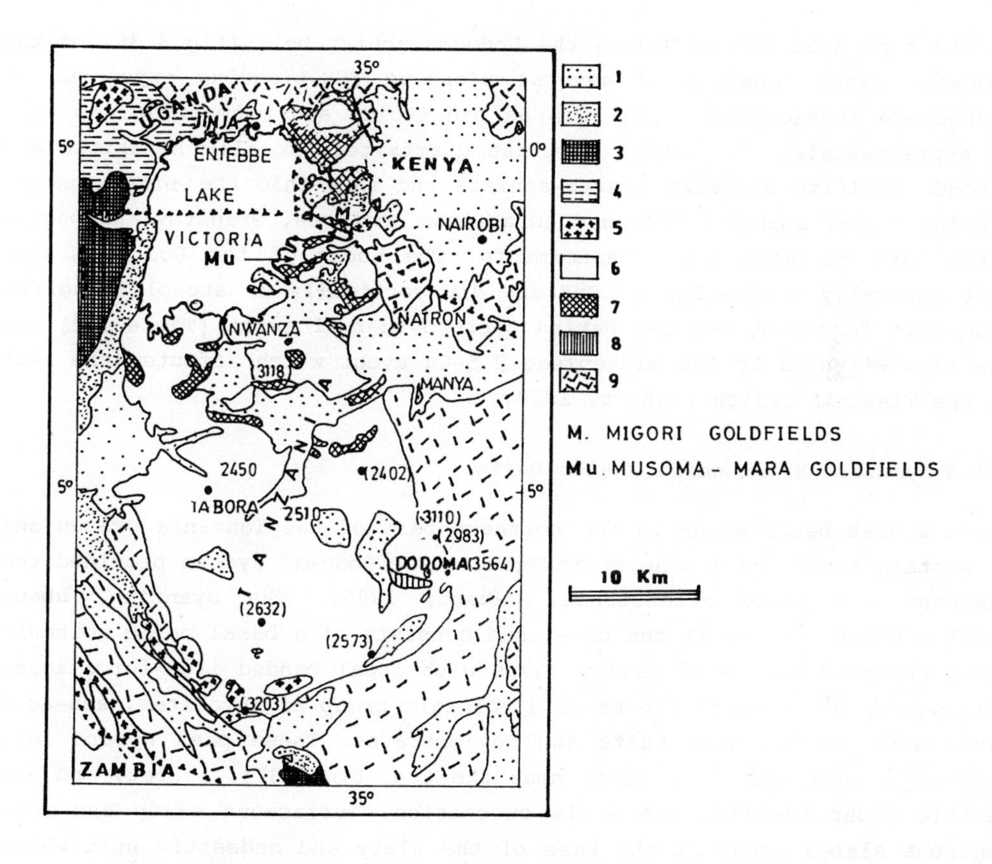

Figure 3.36: Geological sketch map of the Tanzania craton. 1, Mesozoic-Cenozoic cover; 2, Bukoban and equivalents; 3, Karagwe-Ankolean (Kibaran) and Ukingan (in the south); 4, Buganda-Toro; 5, Mubende granite; 6, Granites, migmatites and gneisses; 7, Kavirondian and Nyanzian; 8, Dodoman; 9, Gneisses, etc, (Redrawn from Cahen et al., 1984.)

The Archean terrane of the Tanzania shield, unlike those on most African cratons cannot be readily delimited into separate gneiss, greenstone and granitoid terranes. Rather, there are distinct schist belts surrounded by a vast granitoid-migmatite-gneiss terrane. As summarized by Bell and Dodson (1980), there are three schist belts, the Dodoma, the Nyanzian and the Kavirondian, separated by series of granitoids which constitute the bulk of the Archean terrane of the Tanzania craton. Widespread granitoid magmatism occurred at about 2.54 Ga on the Tanzania craton, and assimilated a lot of pre-existing high-grade rocks, supracrustal material, and granitic rocks (Cahen et al., 1984).

3.4.2 Dodoma Schist Belt

In its type area around Dodoma the Dododma schist belt (Fig.3.36) or the Dodoman System consists of several elongate ESE-trending exposures of high-grade metamorphic rocks such as granitoids and migmatites, as well as supracrustals. The latter comprise quartzites, locally hematitic and banded; sericite schists; quartz-schists and talc-chlorite and corundum-bearing rocks; amphibolites; and hornblende gneisses. Granites and pegmatites have intruded the metasediments. The rocks of the Dodoma schist belt generally strike ESE or ENE and dip vertically or steeply. The Dodoma belt formed during the Watian tectono-thermal event (Table 3.2) and was also affected by the widespread 2.5-Ga event which affected the rest of the Tanzania craton (Gabert, 1990).

3.4.3 Nyanzian-Kavirondian Schist Belts

These schist belts occur in the northern part of the Tanzania craton and in western Kenya and southeastern Uganda. The Dodoman System provided the basement for these greenstones (Gabert, 1990). The Nyanzian, about 7,500 m thick, occurs at the base, and consists of a basal mafic volcanic group composed mostly of pillow lavas with local banded iron-formations, followed by an intermediate-to-acid volcanic group of rhyolites, sub-acid lavas with intercalated tuffs and agglomerates. These pass upward into graywackes with andesitic tuffs near the top (Fig.3.37). A slaty and andesitic group overlies the whole succession. Tuffaceous silty and ferruginous slates occur at the base of the slaty and andesitic unit while banded iron-formation occurs near the top. The Nyanzian has simple folds and shear belts with local isoclinal folding and thrusts.

Resting unconformably on the Nyanzian is another volcano-sedimentary sequence, the Kavirondian System, about 1,500 to 3,000 m thick. At the base of the Kavirondian is a pelitic sequence of slates, mudstones and phyllites with some fine-grained sandstones and volcanics (Fig.3.37). Arkoses with pebble beds and agglomerates occur in the upper part of the Kavirondian. The Kavirondian rocks have structures which are subparallel to the Nyanzian within which they are infolded. Structurally, the Kavirondian is simpler than the Nyanzian. Granitoids, mostly post-Kavirondian, ranging in age from 2.7 Ga to 2.5 Ga, intrude the above succession, for example at Buteba and Masaba in Uganda, and at Mumias and Migori in Kenya.

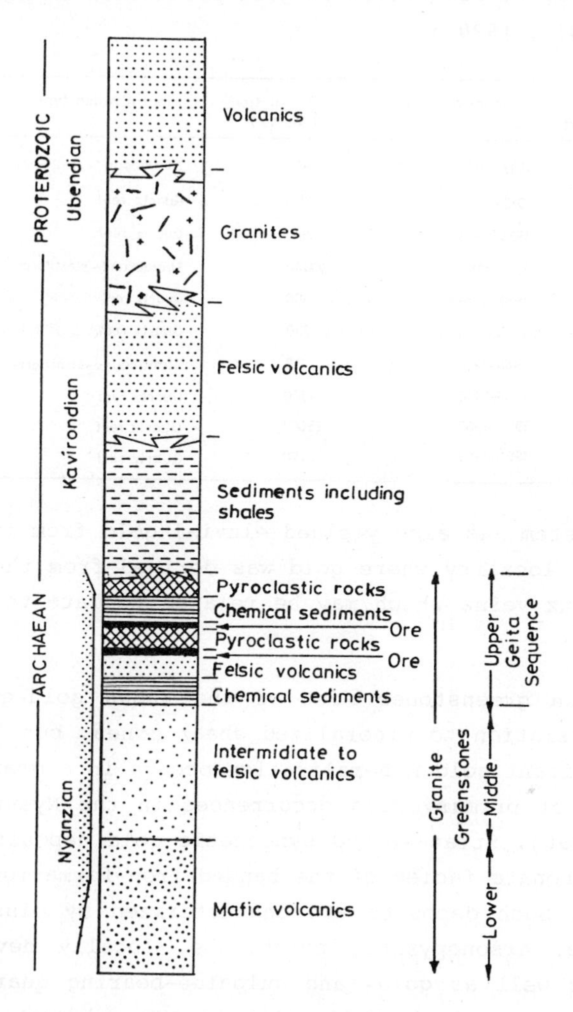

Figure 3.37: Schematic stratigraphy of the Precambrian of Geita area. (Redrawn from Kuehn et al., 1990.)

3.4.4 Gold Mineralization on the Tanzania Craton

The greenstone belts of the Nyanzian System are an important host rock for syngenetic and epigenetic gold mineralization (Fig.3.36). Gabert (1990) and Kuehn et al. (1990) provided an evaluation of the geological factors controlling these mineralizations. Small but steady production (Table 3.3) has been made from the Nyanzian greenstones in Tanzania, Kenya and Uganda over the years.

Table 3.3: Summary of statistics on gold production in East Africa. (From Kuehn et al., 1990.)

	Goldfield mine / prospect	Productive years	Production (kg Au)	Mineralization type	Host rock
NYANZIAN	Migori	1933-1966	950	Stratabound-stratiform	Basalts, BIF
	Buhemba	1913-1970	12170	Quartz reef	Mafic schist
	Kiabakari	1933-1966	8810	Quartz reef	Andinol rock
	Geita mines	1938-1966	27440	Stratabound-stratiform	BIF, tuffs
	Buck Reef	1982-present	100	Quartz reef	Basalts
	Canuck	1945-1953	230	Quartz reef	BIF
	Mahene	1946-1956	15	Stratabound-stratiform	BIF
	Sekenke	1909-1956	4300	Quartz reef	Diorite
UBENDIAN	Lupa	1935-1960	25000	Quartz reef	Basalt
	Mpanda	1950-1960	2170	Quartz reef	Gneisses, schist

The Dodoma System has also yielded eluvial gold from its supracrustal rocks at the type locality where gold was derived from the weathering of gold-bearing quartz veins which may be related to late-tectonic granitic activities.

In the Nyanzan greenstones most of the known gold-quartz vein deposits occur in relation to mineralized shear zones, but there is a rare case where significant gold mineralization occurs in a granite. There are three main types of primary gold occurrences in the Nyanzian greenstone belts (Gabert, 1990). Strata-bound syngenetic gold deposits occur in the sulphidic and carbonate facies of the banded iron-formations, and the accompanying tuffs. Such deposits are characterized by disseminated gold, auriferous pyrite, arsenopyrite, pyrrhotite, locally developed massive pyrite bodies, as well as gold- and sulphide-bearing quartz and calcite veinlets. Another type of mineralization is the epigenetic-hydrothermal-type in the form of tectonically controlled quartz veins or reefs, which occur preferentially in carbonized mafic metavolcanics of the greenstone belt. This type is characterized by the following paragenesis: native gold, pyrite, pyrrhotite, chalcopyrite and arsenopyrite. The third mode of gold mineralization is the epigenetic-metasomatic gold-sulphide impregnations of both the banded iron-formation host rocks and the quartz reef wall rocks, and quartz-sulphide replacements. This type of occurrence is limited to tectonic zones and occurs mainly in the banded iron-formation. The quartz-sulphide replacement is characterized by pyrite, pyrrhotite, chalcopyrite, arsenopyrite, galena and sphalerite paragenesis.

Ore bodies often occur preferentially along the contact between banded iron-formations and tuffs, or within tuffs near the banded iron-formation. They also exhibit structural control in which they are concentrated in the fold hinges of small-scale folds which occur in medium- and large-scale folds; and the positions of the reef orebodies, which often extend to considerable depths, are often controlled by zones of intense folding, fracturing and shearing in the greenstone belts especially at their contacts with granites. In strata-bound gold mineralization there was lithological and stratigraphical control in which the medium-laminated sulphide-facies of the banded iron-formation seemed to offer the best places. Gold-sulphide mineralization preferred the banded iron-formation and associated tuffs and the sulphide ores show a distinct preference for the iron-rich bands.

In the Migori goldfield in southwestern Kenya (Fig.3.36) the Nyanzian host rocks consist of mafic volcanics, banded iron-formations, graywackes, conglomerates, shales and andesitic volcanics. Gold occurs in the form of auriferous quartz reefs and impregnations of host rocks such as the banded iron-formations and mafic volcanics. The main sulphide ores are pyrite, pyrrhotite, arsenopyrite and chalcopyrite. As shown by Ogola (1987) the quartz reefs, up to 300 m long and 2 m wide, were controlled by shear zones and faults (Fig.3.38, B). Several small mines including the Maccalder mine, have produced gold from the Migori greenstone belt (Table 3.3).

In the Musoma-Mara goldfields in northern Tanzania (Fig.3.36) production came from quartz reefs, mineralized shear zones, and planar stratiform orebodies. The deposits are mostly associated with mafic volcanics, and also with felsic volcanics and banded iron-formation, and even granites; and in the Buhemba mine, which used to be the largest producer, shear-zone-hosted quartz reefs constitute most of the orebodies (Kuehn et al., 1990). In the nearby Geita-Kahama goldfields strata-bound stratiform gold deposits occur in the oxide and sulphide facies of the banded iron-formations in the form of Fe-sulphide-rich bodies (pyrite, pyrrhotite, arsenopyrite, with or without chalcopyrite). Mafic volcanic-hosted quartz reef mineralization also occurs near granite intrusions. The Nzega-Sekende goldfields have produced gold mainly from structurally controlled quartz reefs with strong associated wallrock alteration; gold is hosted here by dolerite or mafic volcanics.

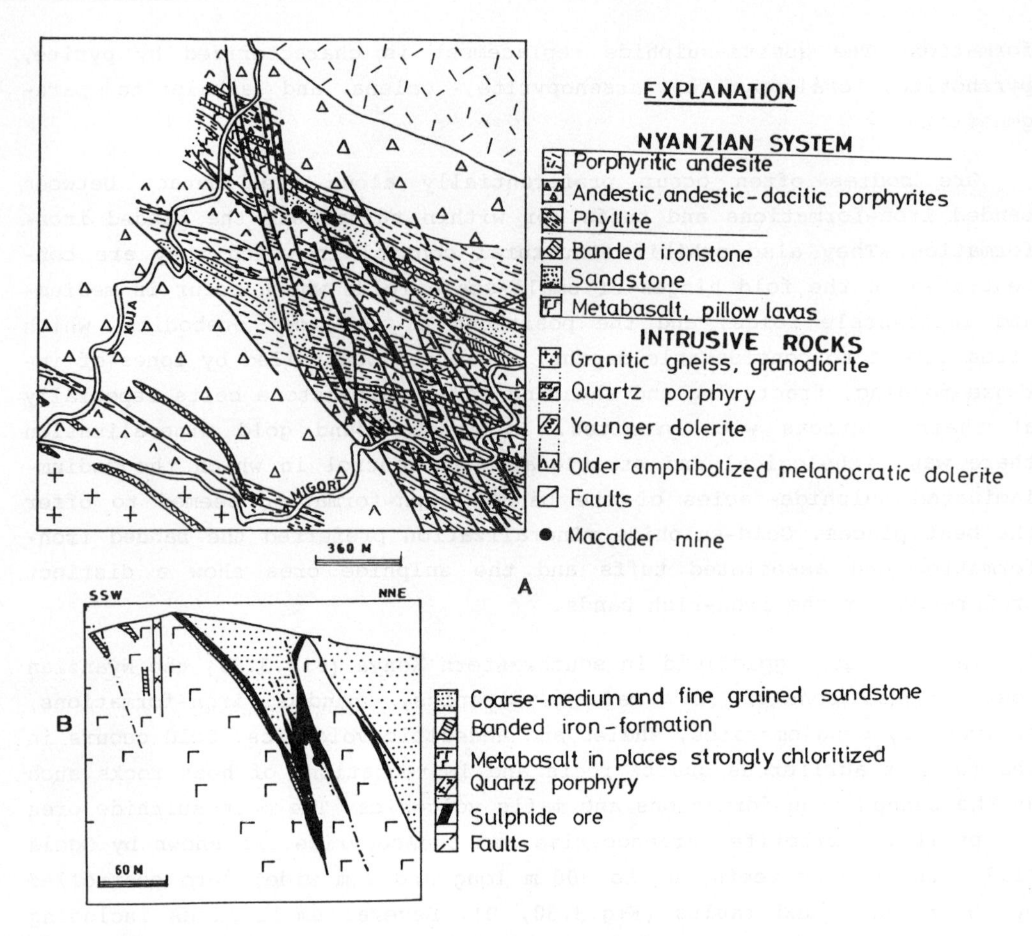

Figure 3.38: Geological sketch map (A) and section (B) of the area around the Macalder Mine. (Redrawn from Ogola, 1987.)

3.5 West African Craton

The West African craton (Fig.3.39) is exposed in two segments, the Guinea Rise to the south and the Reguibat Shield to the north, with a shallow downsag, the Taoudeni basin lying in the centre. The craton is bounded to the north, east and west by Pan-African mobile belts.

3.5.1 Guinea Rise

An Archean nucleus of the West African craton lies in a tectonic unit known as the Kanema-Man Domain, on the southwestern corner of the Guinea

Rise, otherwise known as the Dorsale de Man Shield (Fig.3.39). The Kanema-Man tectonic domain underlies an area of about 150,000 km^2, covering nearly all of Sierra Leone and Liberia, southeastern Guinea, and a small part of the western Ivory Coast (Fig.3.40). It is bounded to the southwest by the Pan-African reactivated Kasila belt, to the east by the Mt. Trou-Sanssandra fault zone which separates the Kanema-Man domain from the Early Proterozoic Eburnean mobile belt; and in the north, the Archean gneissic basement grades into undifferentiated basement and the Eburnean supracrustals, along the Guinea-Mali frontiers. While radiometric dates and field descriptions, as presented in Cahen et al. (1984) and Wright et al. (1985), have enabled the recognition of major tectono-thermal events at about 2.96 Ga and 2.75 Ga in most of the granitic gneiss basement and in the granite-greenstone terranes, the lack of such data in the northern parts of the Kanema-Man domain has hindered the delimitation of the precise northern extent of the Archean nucleus of the West African craton.

Granitic Gneiss Basement

The basement complex in Sierra Leone and Liberia has not been assigned local names. But the adjoining parts in Guinea have been subdivided into the Mahana Gneiss, east of the Simandou supracrustal belt, the Macenta Gneiss to the west, and the Guinea Gneiss to the northwest (Table 3.4). In Ivory Coast the basement complex, which is known as the Migmatite and Gneiss Formation has two higher-grade (granulite facies) areas referred to as the Mt. Douan Formation and the Man charnockite complex.

In general at least 85 % of the Kanema-Man domain comprises granitic gneisses and migmatites. According to Bessoles (1977) and Wright et al. (1985) these are mostly quartzo-feldspathic biotite and hornblende-bearing rocks which vary in metamorphic grade from amphibolite to granulite facies. The composition is predominantly granodiorite, but there is a range from diorite through tonalite to granite. Metamorphosed and deformed relicts of basaltic dykes occur as small lenses and sheets of amphibolite within the gneisses. The Man charnockite massif of southwest Ivory Coast includes a granite series of charnockites sensu stricto, a noritic series, an amphibole-pyroxenite series, magnetite quartzites, and lit-par-lit injection gneisses. Two phases of tectonism have been identified in the Man charnockites. During the first phase metamorphism reached granulite facies, while the second phase involved cataclasis and retrogressive metamorphism of the various units which contain the hypersthene gneisses and granites of the first phase. Retrogressive metamorphism was in the hornblende-granulite and amphibolite facies. The Man charnockites contain rocks which have been dated at 3.1 Ga. The main structural trend in the basement of the Kanema-Man domain is north to northeast, but towards the north it swings to the northwest (Fig.3.40).

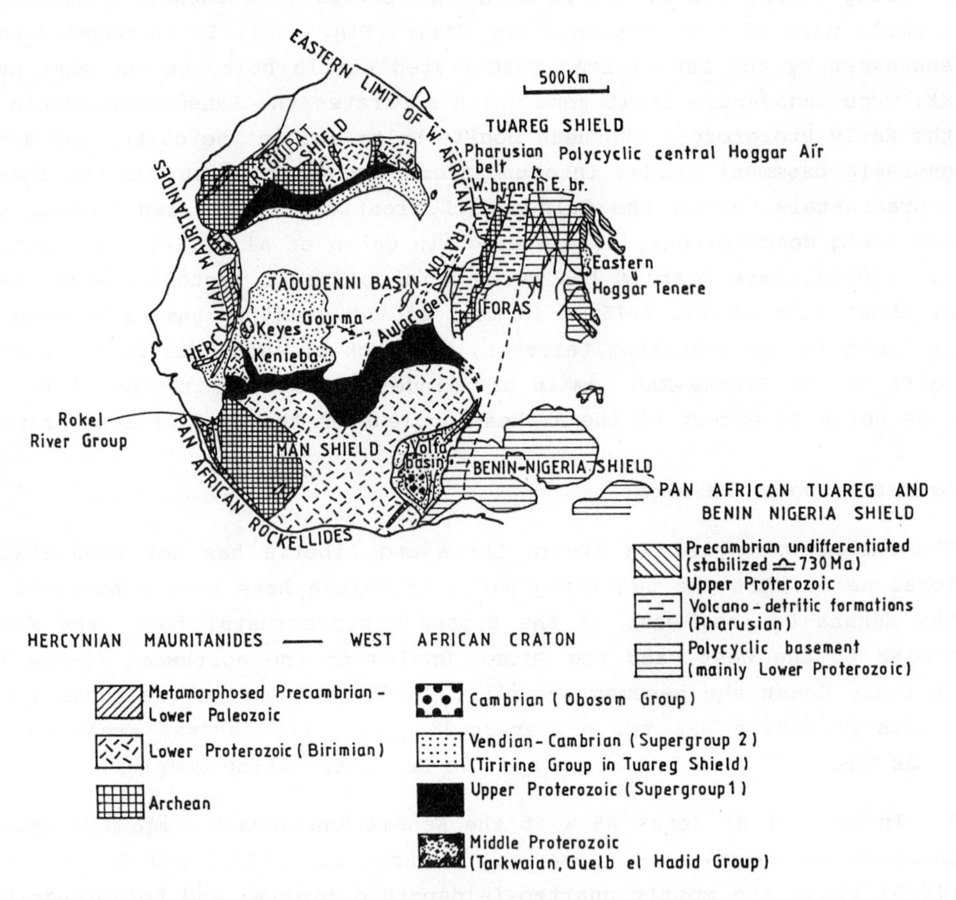

Figure 3.39: Schematic Precambrian geology of West Africa. (Redrawn from Black and Fabre, 1983.)

Greenstone Belts

In the Kanema-Man domain relicts of greenstone belts occur mainly in north to northeasterly trending synformal structures which are surrounded by basement granitic gneisses and autochthonous and paratochthonous granitoids. Based on their sizes, stratigraphic thicknesses, lithologies and metamorphic grade, the supracrustal sequences of the Kanema-Man domain are divisible regionally into a western part covering most of Sierra Leone, and an eastern part which occurs in southeastern Sierra Leone, Liberia and southwestern Ivory Coast (Rollinson, 1978). The western part is characterized by larger typical greenstone belts, up to

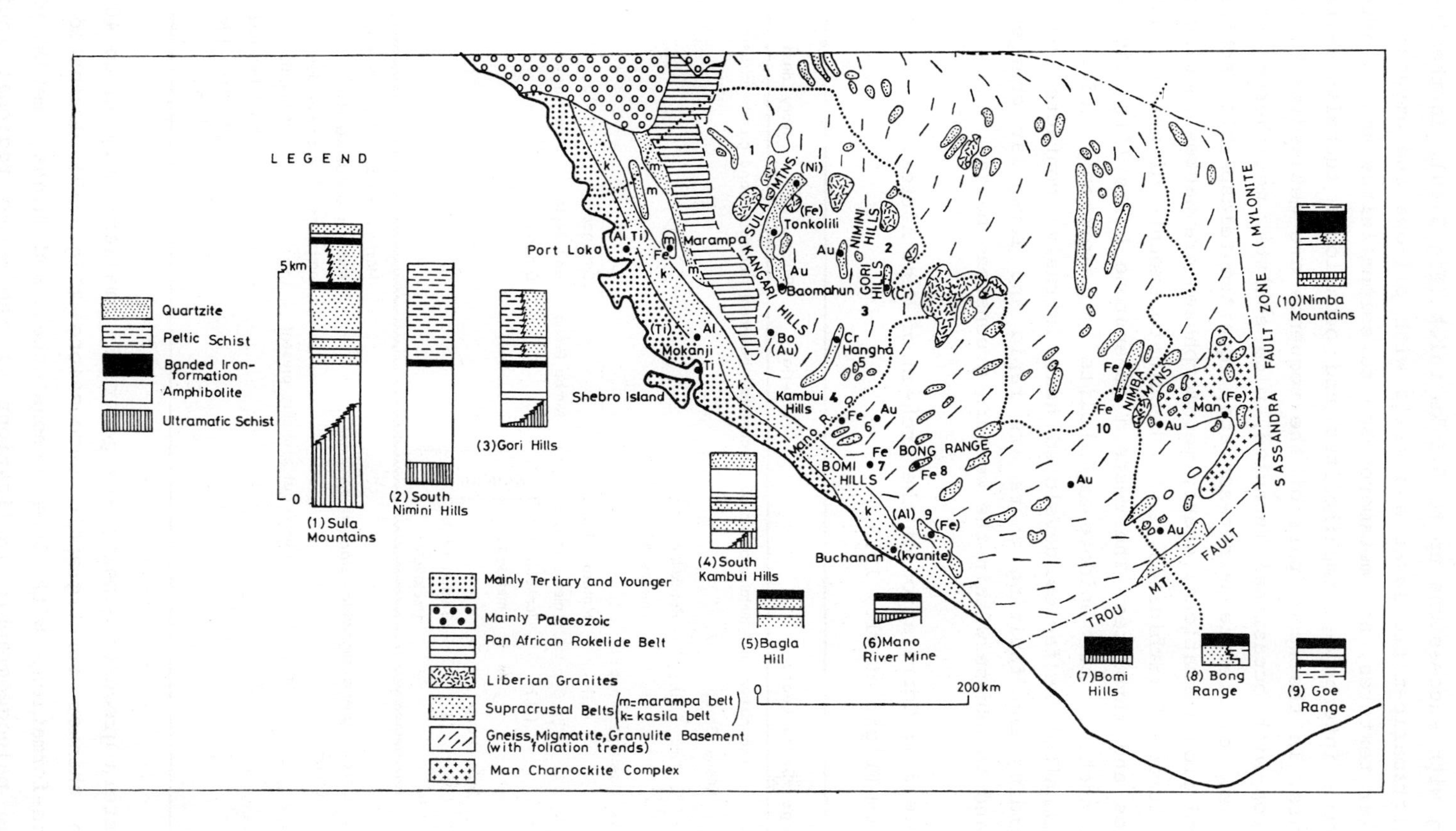

Figure 3.40: Schematic geologic and mineral deposits map of the Archean nucleus of West Africa. (Redrawn from Wright et al., 1985.)

130 km long with successions up to 6.5 km thick (Fig.3.40). In the lower part are ultramafic-mafic lavas and sills with pillows and amygdaloidal and vesicular textures, now metamorphosed to serpentinites and chloritic schists, with interbedded amphibolites and occasional fuchsite-bearing metasediments. In the upper part of the sequence are metasediments such as quartzites with occasional fuchsite, mica schists, metacherts, meta-conglomerates, cordierite-garnet-schists, quartzo-feldspathic schists, minor banded iron-formations, weakly metamorphosed graywackes showing the typical features of turbidites, and subordinate amounts of metavolcanics with dacites and rhyolites. The western greenstone belts (Fig.3.40) exhibit greenschist to epidote-amphibolite metamorphic facies with almandine-amphibolite facies developed near intrusive contacts. Due to intense folding and faulting there are rapid and irregular changes in lithology and thickness within the western greenstone belts.

Table 3.4: Correlation of the Archean of West Africa. (From Wright et al., 1985.)

Event and approximate age	Sierra Leone			Liberia	Guinea	Ivory Coast
Liberian, 2.75 Ga	deformation and metamorphism of supracrustals, basement reactivation and granite intrusion					
Broadly contemporaneous deposition of supracrustal sequences	Kasila Group			Nimba Group	Simandou and Beyla Groups	Mt. Gao Formation
	Marampa Group (U.)	Rotokolon Fm. (metased.)				
	Marampa Group (L.)	Matoto Fm. (metavolc.)				
	Sula Group (U.)	Tonkolili Fm. (metased.)	Kambui			
	Sula Group (L.)	Sonfon Fm. (metavolc.)	Kambui			
unconformity						
Leonian, 2.95 Ga	Loko Group	gneiss-migmatite-granulite basement			Macenta Gneiss, Mahana Gneiss and Guinea Gneiss	Migmatite and Gneiss Formation
pre-Leonian, 3.1 Ga and older		not identified but probably present				Mt Douan Formation and Man charnockite complex

The eastern greenstone belts are generally smaller, only up to 40 km long, with thinner stratigraphic successions (Fig.3.40) dominated by banded iron-formation, with less greenstones, and higher metamorphic grades. The paleogeographic implications of the above regional facies distribution is that the eastern greenstones probably developed in shallow ensialic shelf basins whereas the western facies were deposited in a deeper, less stable and ensimatic geosynclinal setting (Wright et al., 1985).

1. *Kanema Greenstone Belt*. This belt in central Sierra Leone is the type area of the Kanema-Man domain; it consists of a basement complex of granites, acid gneisses, granulite facies rocks and greenstone belts of schistose sediments and volcanics. Two separate suites of greenstones have been recognized in the area, namely, an older Loko Group comprising amphibolites with subordinate serpentinites, quartzites and banded iron-formations and a younger greenstone succession, the Kambui Supergroup comprising a lower metavolcanic formation, the Sonfon Formation and an upper sequence of metasediments, the Tonkolili Formation.

The Loko Group supracrustal greenstone belt was affected by a major tectono-thermal event dated 2.96 Ga. During this event which is known as the Leonean event, the granitoid basement on which the Loko Group was deposited was metamorphosed to the granulite facies while the Loko greenstones were metamorphosed into amphibolites and serpentinites. East-west structures and fabrics which formed during the Leonean event consist of tight minor folds with an axial planar schistocity. The Leonean event terminated with the formation of autochthonous granitoids with K-feldspar porphyroblasts.

The Kambui Supergroup supracrustals accumulated after the Leonean tectono-thermal event. The Sula Group, about 6 km thick, is the predominant unit in the Kambui Supergroup. It comprises basal ultramafic serpentinitic extrusives and a thick sequence of banded and massive amphibolites with some pillow horizons. In the overlying Tonkolili Formation tuffs occur near the base, but this unit is mostly clastic with iron-formation horizons near the top. The Kambui Supergroup was metamorphosed to low-pressure greenschist-to-amphibolite facies during the Liberian event at about 2.75 Ga. During this event the Kambui Supergroup was deformed into anticlines and synclines which are, however, more pronounced in the underlying Loko Group. Northerly structural orientations resulted from the Liberian event. The Liberian event terminated with the formation of ovoid-to-ellipsoid granitoids which are elongated parallel to the Liberian foliation trend.

2. *Kasila Domain*. A belt of granulite facies rocks, the Kasila Group, extends along the coast from Liberia to Guinea (Fig.3.40). Further inland are related discontinuous greenstone belts belonging to the Marampa Group

(Fig.3.41, inset). The lower part of the Marampa Group consists of metamorphosed basaltic and andesitic volcanics in which pillows and porphyritic and flow structures are preserved with largely serpentinized ultramafic intercalations (Wright et al., 1985). The upper part is metasedimentary with pebbly and cross-bedded quartzites, fuschsite-quartzite, manganiferous quartzite, quartz-schists, mica schists, garnet schists, gneisses and banded iron-formation. The Kasila Group consists of high-grade metabasic igneous rocks (now pyroxene-bearing basic granulites) with metamorphosed and dismembered remnants of large layered anorthositic and gabbroic rocks, now occurring as lenses up to 200 m thick and 50 km long. The Kasila also contains minor quartz-magnetite, quartz-diopside and aluminosilicate-bearing rocks which represent highly metamorphosed equivalents of banded iron-formations, marbles, and pelites respectively (Wright et al., 1985).

Although the Kasila Group is lithologically and stratigraphically similar to greenstones, Wright et al. (1985) interpreted it as the root zones of nappes which formed as a result of Alpine-Himalayan-type continental collision and suturing of two Archean continental plates, a West African cratonic plate, and a South American (Guyana) plate (Fig.3.41, B). The higher metamorphic grade of the Kasila Group, and the occurrences of layered anorthositic rocks (suggesting deeper crustal levels), are suggestive of deep crustal processes which perhaps took place during subduction and plate collision. Collision and thrusting of the Kasila Group (Fig.3.41) are also suggested by the predominance of NW-SE-trending foliation with consistently moderate dips to the southwest; and the presence of a northeastern boundary zone of sheared gneisses and mylonites which dip at moderate angles to the southwest, suggesting medium-angle thrusting onto the granitic basement. According to Wright et al. (1985) the Marampa Group probably represents nappes that were derived from the Kasila belt (Fig.3.41). This is suggested by the recumbent fold structures, low-angle thrust-faulted contacts of the Marampa with the underlying granitic basement and the lower metamorphic grade which increases from greenschist in the east to amphibolite in the west.

After they were amalgamated with the West African craton along the mylonite suture zone in Late Archean times (Figs. 3.41, B), this suture was reactivated as the internal zone of thrusting and ductile shearing during the Late Proterozoic - Early Paleozoic Pan-African tectono-thermal event (Lécorche et al., 1989; Williams and Culver, 1988).

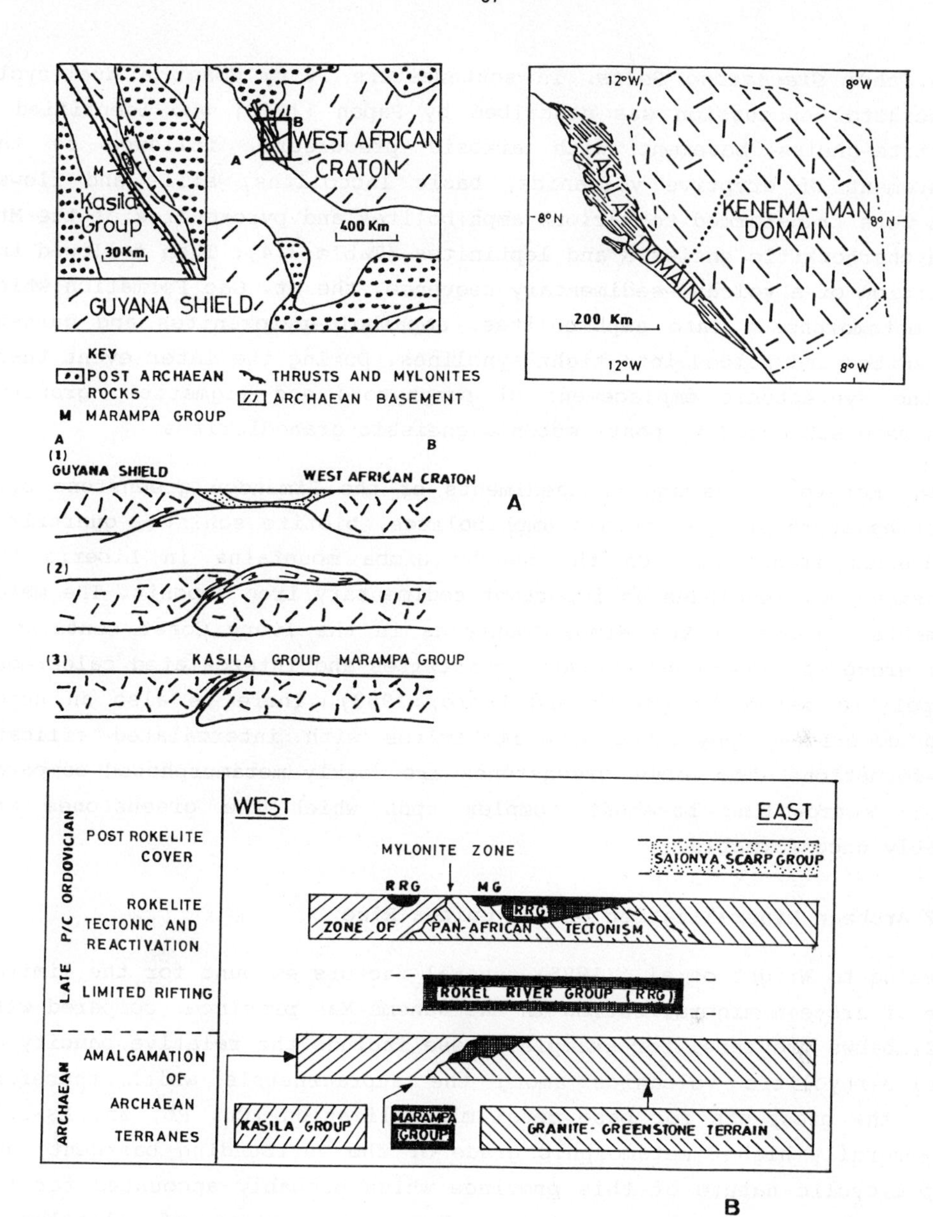

Figure 3.41: A, tectonic model for the Kasila and Marampa Groups; B, time-terrane evolution of Precambrian of Sierra Leone. (Redrawn from Wright et al., 1985; Williams and Culver, 1988.)

Under the Pan-African tectono-thermal regime the Kasila and the Marampa Groups suffered retrograde metamorphism and penetrative mylonitization.

3. *Other Greenstone Belts.* In southwestern Ivory Coast a dual-cycle supracrustal succession was described by Papon (1973) who identified a migmatite-gneiss basement with arkosic precursors, followed by the emplacement of eruptive volcanics, basic laccoliths, sills and flows, which were transformed to various amphibolites and pyroxenites of the Mt. Douan charnockitic gneisses and leptinites (Table 3.4). Then followed the deposition of a volcano-sedimentary sequence, the Mt. Gao Formation which were metamorphosed into amphibolites, amphibole-pyroxenites and garnet-pyroxenites and folded into tight synclines. During the later event there was the syntectonic emplacement of porphyroid and migmatitic granites which were succeeded by post-tectonic gneissic granodiorites.

The metavolcanics and metasediments of the Simandou greenstone belt of southeastern Guinea contain amphibolites, biotite schists, quartzites and banded ironstones. On the nearby Nimba mountains in Liberia the greenstone belt includes an important sedimentary iron deposit. The meta-sedimentary rocks of the Nimba Mountains in the Ivory Coast contains a lower group of basal conglomeratic quartzites and intercalated calcareous and pelitic sediments (Black and Fabre, 1983). There is also an upper group comprising phyllites and itabirites with intercalated silicate iron-formations. The Nimba greenstones are weakly metamorphosed compared to the surrounding basement complex upon which the greenstones are probably unconformable.

3.5.2 Archean Mineralization on the Guinea Rise

According to Wright et al. (1985) several factors account for the limited range of Archean mineralization in the Kanema-Man province, compared with the Zimbabwe province, for example. These include the relative paucity of andesite-rhyolitic volcanics among the supracrustals which therefore limits the prospects for base metal mineralization (Cu, Pb, Zn, As-Bi); the generally higher metamorphic grade of the surrounding basement, and the polycyclic nature of this province which probably accounted for the loss of potential ore-forming elements during the passage of volatiles.

Gold has been mined from placer deposits around the greenstone belts and from gold-quartz veins in Sierra Leone and Liberia (Fig.3.40). With dwindling production from alluvial mining of river gravels, commercial interest has shifted to the primary source in amphibolites and ferruginous schists (metamorphosed banded iron-formations), where gold-quartz veins occur sometimes containing pyrite and arsenopyrite, or tourmaline. Iron deposits which lie mainly in the eastern greenstone belts are the principal Archean mineral resource of West Africa, coming from

Liberia, Guinea and Côte d'Ivoire. Liberia, a leading producer, holds a reserve of over 4.4 billion tonnes at Nimba, Wologisi-Bea Mts., Mano Riverhills, Bomi Hills and Bong-Zaweah (de Kun, 1987). The average ore grade is about 65 % Fe and ranges from 35 to 70 % Fe. In the largest mine at Nimba, the banded iron-formation is up to 500 m thick (Fig.3.42). Magnetite and hematite are the main ore minerals. Deep weathering during the formation of peneplains at about 1,300 m above sea level in the Cretaceous and earlier at 1,600 m above sea level, caused supergene up-grading of the Nimba itabirite from 38 % Fe to 40 - 68 % Fe in the weathered ore zones (Fig.3.42, B). In the nearby Republic of Guinea the reserve is also vast, over 2.5 billion tonnes, especially in the Simandou-Nimba region where the grade is also about 65 % Fe. Here the banded ironstone is 350 - 1,000 m thick with a strike length of about 80 km. Other deposits occur in Côte d'Ivoire and Sierra Leone where both the quality and reserves are economic. Archean itabirite-type iron deposits also occur in the Imataca complex in Venezuela (Fig.3.42, A), thus supporting the link between the Guyana and West African cratons in Archean times.

3.5.3 Reguibat Shield

The Reguibat shield is the northern part of the West African craton. It is bounded to the south by the Taoudeni intracratonic basin, to the north by the Tindouf basin and to the west by the West African polyorogenic mobile belt (Fig.3.43). Like the Kanema-Man shield to the south, the Reguibat shield is composed of Archean rocks in the west and centre, with Lower Proterozoic rocks predominating in the east. The Archean basement of the Reguibat shield, known as the Amsaga Group in the west and the Ghallaman Group to the east, has been divided into a migmatitic, and a metamorphic unit comprising at the base typical granulites. The metamorphic unit consists of (from bottom to top): charnockitic pyroxeno-amphibolites; granulites; pyroxene-amphibolite rocks, sillimanite-gneisses, granulites, biotite gneisses; amphibolites, marbles, and ferruginous quartzites. These assemblages are intruded by metagabbros, anorthosites and serpentines and are cut by small massifs of biotite granite and by beryl-tourmaline pegmatites. The Amsaga assemblage is about 10 km thick. In the central part of the Reguibat shield the Archean includes abundant leptynites, biotite and muscovite gneisses, amphibolites, marbles and quartzites. The ferruginous quartzites which are typical of the western part of the shield are absent from the central part. In the eastern Reguibat shield Archean rocks are represented by leptynites, amphibolites, garnet gneisses, and migmatites.

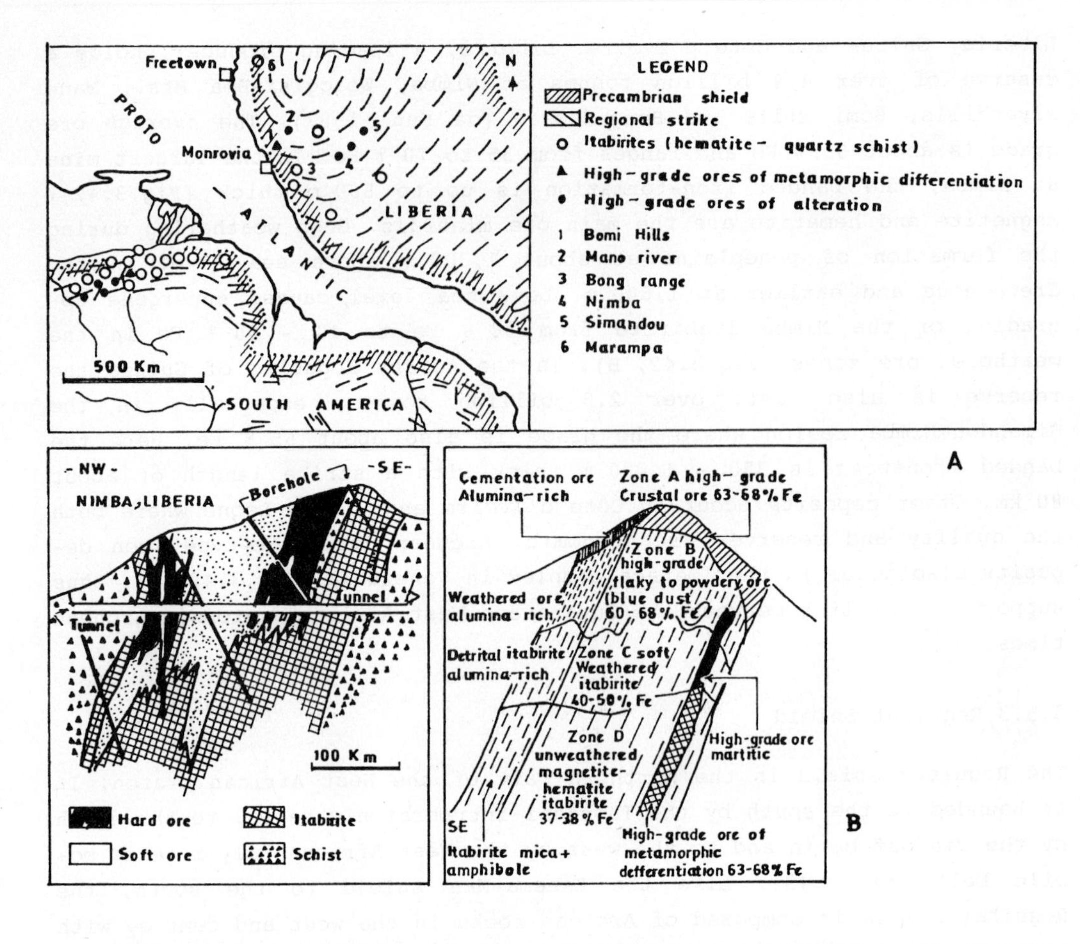

Figure 3.42: Archean iron mineralization (itabirite) on the West African craton and the Imataca Complex in Venezuela (A). B, supergene enrichment of Nimba itabirite (Liberia). (Redrawn from Hutchison, 1983.)

The metamorphic grade varies from low pressure granulite facies to amphibolite facies and decreases towards the east. Syntectonic granites are abundant in the eastern zone and include two-mica granites, porphyritic biotite granites, granodiorites and diorites. The gneisses have yielded ages indicating a major tectono-thermal event around 2.7 Ga, whereas the granites are about 2.4 Ga old (Black and Fabre, 1983). Eburnean reactivation produced biotite ages falling in the range of 2.0 - 1.5 Ga.

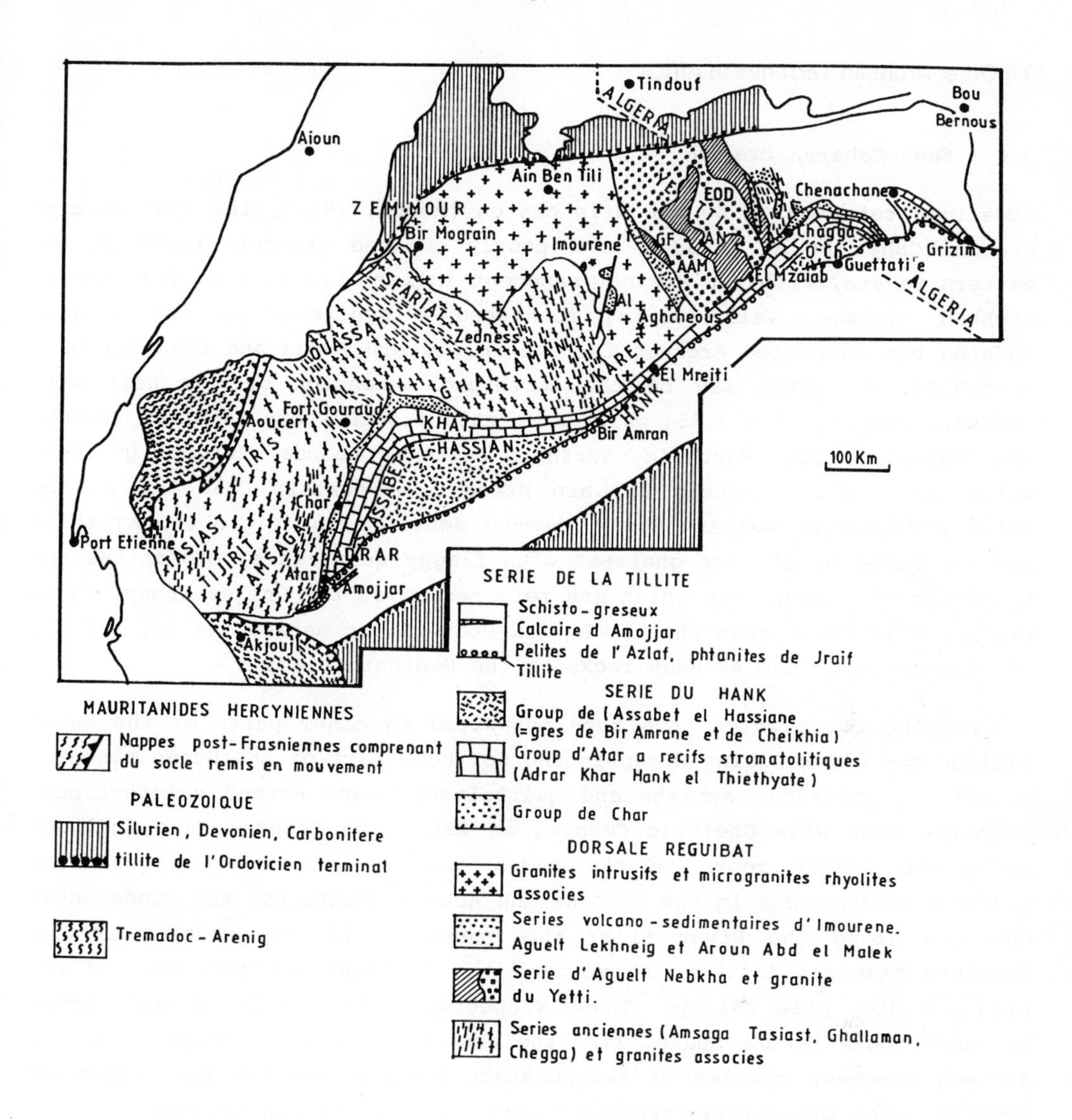

Figure 3.43: Geologic sketch map of the Reguibat shield. (Redrawn from Bessoles, 1983.)

3.6 Other Archean Terranes in Africa

3.6.1 East Saharan Craton

Sometimes referred to as the Nile craton (Rocci, 1965), the East Saharan craton (Cahen et al., 1984) is a poorly exposed cratonic block in the eastern Sahara, with the Uweinat basement inlier (Fig.3.2) as the exposed Archean nucleus. Vail (1988a) furnished a synthesis on the largely undated but suspected Archean terrane in western Sudan and Chad Republic (Fig.3.44, A) which may be the northward extension of the West Nile Gneissic Complex (Fig.3.33) which stretches from eastern Cameroon through the Central African Republic, northern Zaire to Uganda. Ancient gneisses which may include Archean age are present in eastern Chad and in the Dafur province of western Sudan. Around Jebel Marra (Fig.3.44, A) there are exposures of biotite gneisses with flaggy quartzites, migmatites and mylonites of unknown age which are referred to as the Gneiss Group; these are probably older than the nearby Mid-Proterozoic metasediments, and may be comparable to the Archean rocks in the Uweinat inlier.

Archean rocks which have been suspected in other parts of the Sudan include the Older Plains Group in the northern part of the Sudan basement comprising gneisses, schists and quartzites; these extend uninterrupted into the West Nile Gneissic Complex of Zaire and Uganda. Also included among the oldest rocks in the Sudan are the charnockitic granulite gneisses which occur in the Imatong and Acholi Mountains; the Fundamental Gneisses along the River Nile; the granulite facies gneisses in the Sabaloka basement inlier; and most of the basement gneisses west of the Nile and Blue Nile valleys. These suspected Archean - Early Proterozoic basement rocks in the Sudan, like the West Nile Gneissic Complex and the Archean basement complex of southwestern Nigeria and the Bur region of Somalia, were thoroughly reworked during the Pan-African orogeny.

Jebel Uweinat

Archean charnockitic granulites are exposed on the Jebel Uweinat (Oweynat) basement inlier in northeast Africa (Fig.3.44) at the junction of the Libyan, Egyptian and Sudanese frontiers. The Uweinat basement comprises biotite and amphibolite gneisses, migmatitic biotite gneisses and charnockitic gneisses (Cahen et al., 1984). Paleozoic - Mesozoic sedimentary rocks unconformably overlie the Uweinat basement, while a variety of syenites, granites and other igneous rocks intrude both basement and sediments. Among the basement rocks is the Karkur Murr series, a group of pyroxene granulites containing charnockitic, noritic, and diopsidic

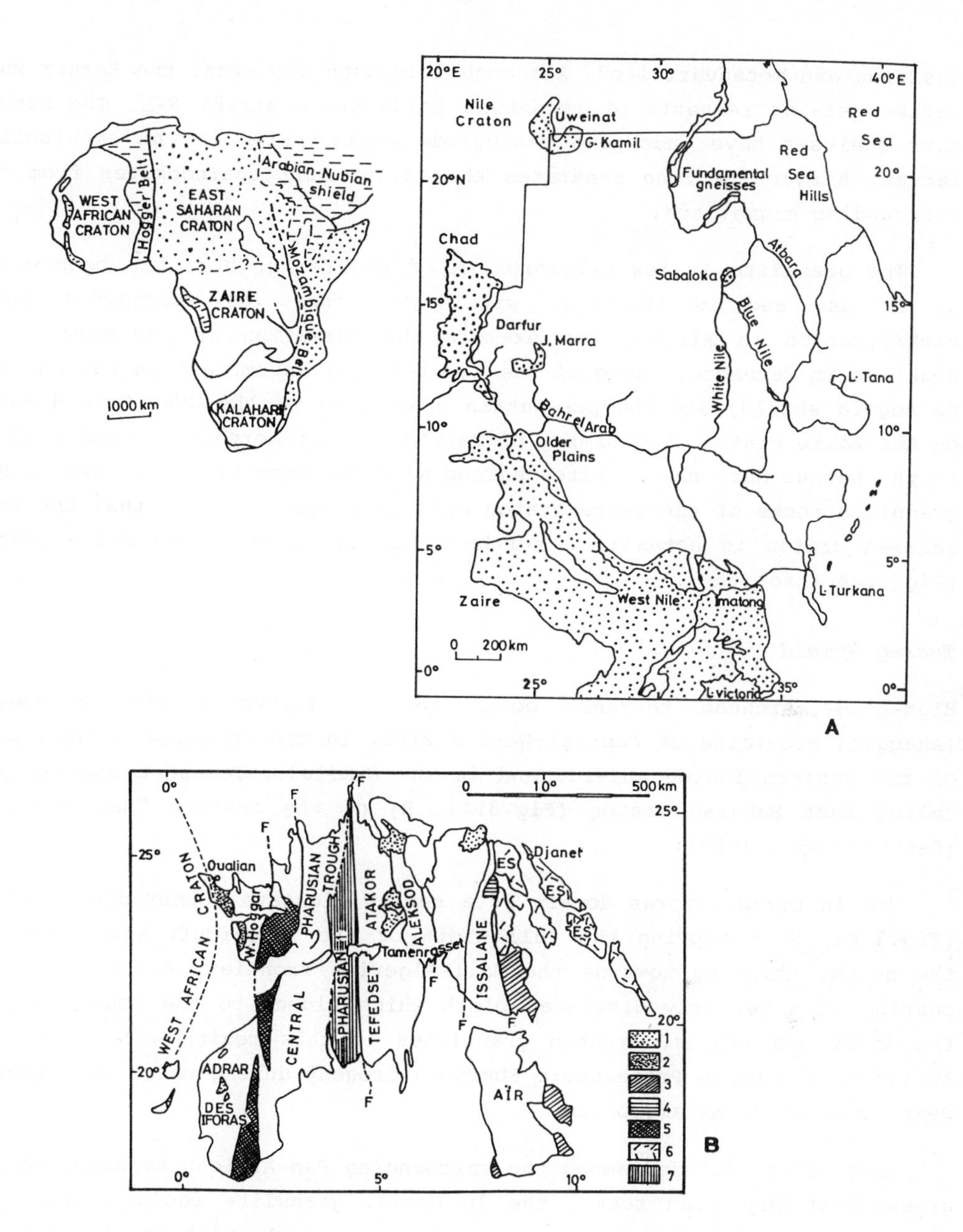

Figure 3.44: Archean components of the so-called East Saharan craton. A, Sudan; B, Tuareg Shield. (Redrawn from Cahen et al., 1984; Vail, 1988a.)

gneisses and metaquartzites. Although folded on E-W axes, the Karkur Murr series retains remnants of recumbent folds which strike N-S. The Karkur Murr gneisses have undergone retrograde metamorphism in the amphibolite facies. A mylonite zone separates the Karkur Murr charnockites from the surrounding migmatites.

The granulite facies metamorphism of the Karkur Murr has been dated at 2.9 Ga, and is therefore within the phase of widespread charnockitization in Africa, for example the formation of the Ntem charnockites in Cameroon; those of the Kasai-Lomami assemblage in the Kasai - NE Angola shield; and the pre-Watian granulites of the West Nile Complex on the Zaire craton. The timing of retrograde metamorphism around 2.63 Ga in the Karkur Murr charnockites agrees with the same event in these other granulite rocks of the Zaire craton and reinforces the view that the East Saharan craton is actually a northern continuation of the Zaire craton (Fig.3.44 inset).

Tuareg Shield

High-grade Archean terranes occur in the Tuareg shield or Hoggar (Ahaggar) Mountains of central West Africa, in the In Ouzzal-Iforas part of the western Tuareg shield, and in the Oumelalen-Temasent and the so-called East Saharan craton (Fig.3.44, B) of the eastern Tuareg shield (Cahen et al., 1984).

The In Ouzzal-Iforas domain is a narrow elongate submeridional block (Fig.3.44, B) occupying the full width of the Tanezrouft-Adrar zone in the north, thinning towards the Mali-Algerian frontier, and then re-appearing in a westerly displaced block which widens to the south through the Adrar des Iforas. Archean granulites in this region were later affected by the Early Proterozoic Eburnean orogeny during which they underwent lower grade metamorphism.

Originally included among the surrounding Pan-African amphibolite and greenschist Suggarian rocks, the In Ouzzal granulite facies rocks were later found to have N-S shear zone tectonic contacts with the Pan-African rocks. These Archean granulites occur as high level nappes which were emplaced from the SSE to NNW direction before the deposition of Late Proterozoic sediments (Boullier et al., 1978). The In Ouzzal granulite complex consists of a range of aluminous metapelites, banded magnetite quartzites, leptynites and marbles commonly associated with norite and lenses of pyrigarnite and lherzolite. Gray gneisses and associated potassium augen-gneisses are the oldest rocks in the region, dated

3.48 Ga. Interstratified within the metasediments of the In Ouzzal granulite complex are lenticular bodies of orthypyroxene-sillimanite granulites varying in thickness from 100 cm to 150 cm. These Al-Mg-rich rocks, according to Keinast and Ouzegane (1987), have undergone widespread high pressure (10 Kbar), high temperature (750° C - 805° C), metamorphism at crustal depths of up to 35 km, between 3.3 Ga and 2.9 Ga. These rocks were later affected by lower pressure metamorphism during the Eburnean orogeny (2.1 Ga) during which other granulites formed in the region.

3.6.2 Madagascar

The Antongilian of Madagascar (Fig.3.45) comprises Archean terranes which are older than 3.0 Ga. It consists of gray gneisses and the more potassic gneisses which are exposed on the Antongil massif north of Tamatave, the Masora Group to the south, the granite-migmatitic facies on the anticlinal ridges in the interior of Madagascar and the cores which outcrop in the younger belts (Cahen et al., 1984). The Antongilian gneisses were metamorphosed at about 3.48 Ga (Cahen et al., 1984). However, these rocks comprise mostly granitoids (Antongil granites) and migmatites (Ambodiriana migmatites). The granitoids range between granodiorites and quartz oligoclasolites on the one hand and calc-alkaline, sometimes potassic leucocratic granites on the other. Granodiorites predominate followed by tonalites, diorites and amphibolites.

3.7 Archean Tectonic Models

3.7.1 Classical Models

Since some of the earliest tectonic models for the origin of Archean greenstone belts were actually developed from investigations of the Archean of southern Africa, it is pertinent to briefly review some of these models and thence cursorily examine some of the prevailing views about the origin and development of Archean terranes especially in the light of plate tectonics. Discussions on Archean tectonic models can be found in Condie (1981), Nisbet (1987), Kröner (1989), and in many other Precambrian research papers which are devoted to this controversial topic.

According to the classical or downsagging basin model (Anhaeusser et al., 1969; Viljoen and Viljoen, 1969; Glikson, 1972), Archean greenstone belts were initiated along fractures in downwarps at the boundary between thin continental crust and oceanic crust or in parallel fault-bounded troughs or rifts in the primitive sialic crust. Basin development began with the accumulation of volcanic material which as it thickened caused the basin to downsag beneath the heavy volcanic pile. As the basin sank into the crust diapiric tonalitic granites rose and invaded the greenstones in and around its margin. The intrusion of tonalitic diapirs caused the compressional deformation of the greenstone basin into the characteristic synformal shape of Archean greenstone belts. Among the major deficiencies of the classical or downsagging model are its failure to explain the origin of the extensive high-grade gneissic terranes which surround the granite-greenstone belts and the fact that the high-grade terranes were not considered as the possible basement to the greenstones. As shown in the southern part of the Barberton belt the gneissic terrane was the sialic basement to the greenstones (Anhaeusser et al., 1969; Viljoen and Viljoen, 1969; Glikson, 1972), as was the case also in the Zimbabwe province, and on the Tanzania shield.

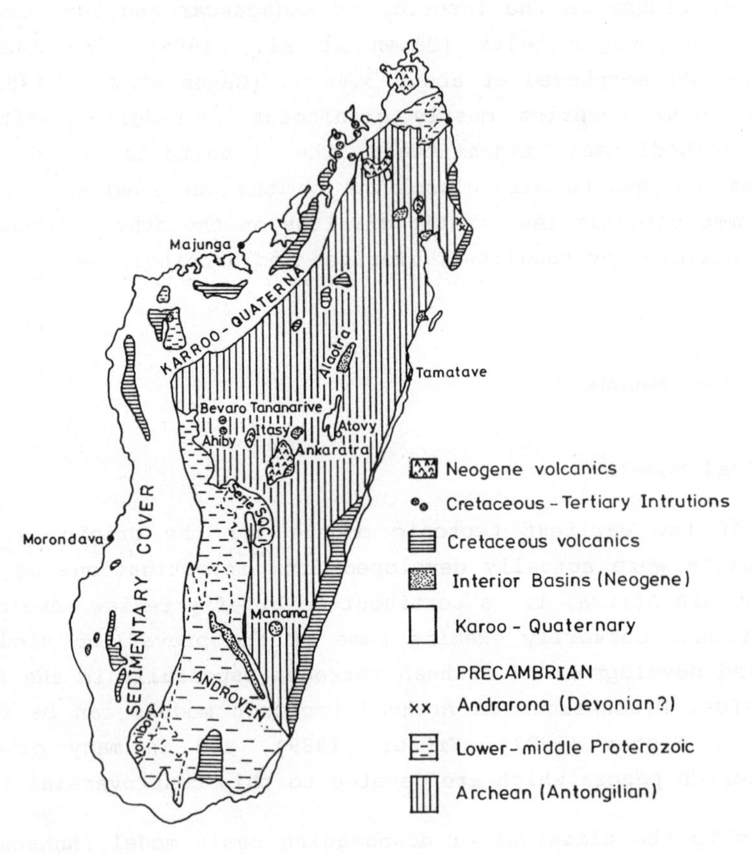

Figure 3.45: Geologic sketch map of Madagascar. (Redrawn from Hottin, 1976.)

3.7.2 Back-arc-Marginal Basin Models

Of the many subduction-related plate tectonic models that have been proposed to explain Archean granulite-gneiss and granite-greenstone belts, the island-arc subduction model has been the most popular (Condie, 1989; Nisbet, 1987; Windley, 1984). Archean granite-greenstone belts are similar to modern island-arcs in their basal low-K tholeiitic volcanics, calc-alkaline volcanics and in the chemistry of their tonalitic plutons. Archean greenstone belts are similar to modern marginal back-arc basins (Fig.3.46) where the volcano-sedimentary sequences are usually best preserved, unlike other subduction settings where the oceanic lithosphere is often destroyed by subduction and later by erosion after uplift. The back-arc basin setting has therefore been the preferred tectonic setting for Archean greenstone belts. The Tertiary Rocas Verdes complex of southern Chile has been cited as showing features which are remarkably similar to Archean greenstone belts (Tarney et al., 1976). Similarities include the synformal structure of the Rocas Verdes basin, its size, greenschist facies metamorphism, structural style, the stratigraphy of the volcano-sedimentary pile, the geochemistry of the volcanics, and the younger intrusive tonalite-granodiorite. However, the main difference lies in the absence of undisputed ophiolites in Archean greenstone.

The marginal back-arc basin model offers a broad spectrum of depositional settings which would accommodate the variety of volcanogenic, continental and marine deposits found in African greenstone belts. This ranges from the deep flysch turbidite basin environments represented in graywackes such as the Fig Tree Group in the Barberton belt, the Manjeri Formation in Zimbabwe, and the Tonkolili Formation in Sierra Leone, to proximal deltaic and intertidal environments like those of the Kalahari craton (e.g. Moodies Group, Cheshire Formation, Shamvaian Group). Chemical sediments such as chert and banded iron-formation accumulated in parts of the shelf and deep basin which were far away from detrital sedimentation, while stromatolites like those of the Cheshire Formation flourished on intertidal flats, lagoons and bays. The volcanic centres in the back-arc basins were not only sediment sourcelands, but also sources of mineralizing fluids rich in gold and sulphides.

3.7.3 Archean Plate Tectonics

A full consideration of the tectonic implications of the high Archean geothermal gradient and the properties of the Archean asthenosphere, lithosphere, and crust is not intended here since very authoritative syntheses are available in Condie (1981), Nisbet (1987), and Frazier and

Schwimmer (1987), to mention a few. However, since the crustal evolution of the African continent is fundamental to its regional geology and metallogeny, it is pertinent to consider cursorily the prevailing views on Archean crustal evolution. The overview presented below is culled from Burke and Dewey (1973); and from Frazier and Schwimmer (1987), who although writing about North America, found the Archean of South Africa and its global tectonic implications worthy of a full-blown account.

Burke and Dewey (1973) and Frazier and Schwimmer (1987) considered Archean tectonic processes to have evolved through three main phases, namely: the primitive phase being the earliest Archean tectonic regime; the permobile stage which characterized most of Archean history; and the Archean-Proterozoic transitional phase.

The model for the primitive phase (Frazier and Schwimmer, 1987) assumes a thick convecting asthenosphere, a thin lithosphere and an ultramafic-mafic crust which initially contained little sialic material. Small-scale convection dominated in the asthenosphere so that heat and mantle-derived magmas ascended at randomly oriented "hot spots", causing the development of three-armed rifts in the thin lithosphere and the spreading of the resulting three plates radially away from the three-armed rifts. The plates probably moved across the surface until they met plates with opposing motion in which case underthrusting of one plate under the other could have resulted (Fig.3.47). Since the geothermal gradient was too high in the Archean for eclogite to form from basalt as is the case in the Phanerozoic where eclogite forms along a down-going plate and pulls the plate steeply down into the mantle, the down-going Archean plate probably simply slipped beneath the overriding plate at a very low angle as shown in Fig.3.47. Instead of eclogite, garnet granulite, a lower density material, could have formed, which did not have sufficient density to pull the down-going slab steeply into the mantle. Calc-alkaline material which increased in quantity with time in the Archean, was probably generated due to the partial melting of the down-going slab (Fig.3.47 A); or as a result of the thickening of volcanic material at hot spots (Fig.3.47 B); or still, the lithosphere could have buckled at zones of opposing plate motion (Fig.3.47 C).

Burke and Dewey (1973) highlighted the characteristics of the permobile phase (Table 3.5) to include: the absence of continents or large cratonic areas; individual dominantly granodioritic areas which were bounded by and generally intruded the curved greenstone belts which included calc-alkaline volcanics and volcaniclastics; greenstone belt outcrops which were wider but shorter than modern plate sutures on account of the smaller sizes of Archean plates; and the general absence of large volumes of sedimentary units in the Archean because of the absence of stable cratonic areas where they could have been deposited.

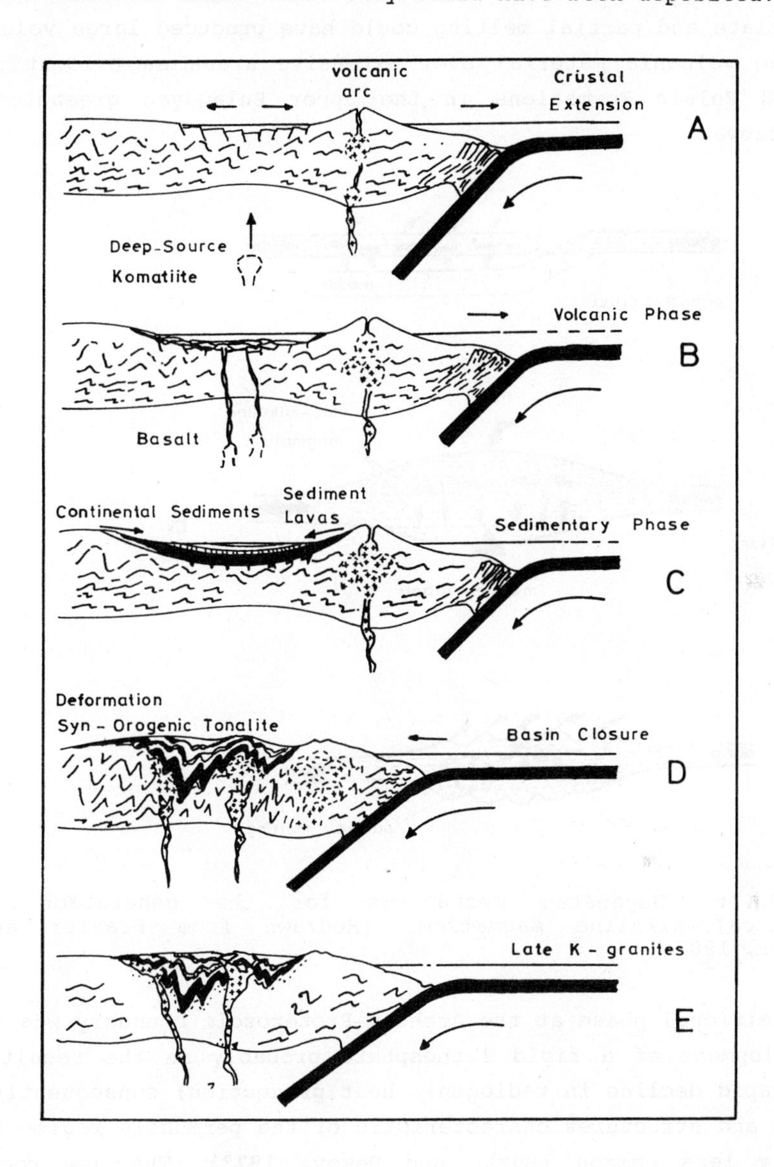

Figure 3.46: Back-arc marginal basin model for the development of Archean greenstone belts. (Redrawn from Windley, 1984.)

Subduction during the permobile regime (Fig.3.48, B) could have occurred at the boundary between simatic and sialic crustal elements as a result of the buckling of the lithosphere. Low-angle subduction of the down-going plate and partial melting could have produced large volumes of calc-alkaline volcanic material over extensive areas as evident in the Maliyami and Felsic Formations in the upper Bulawayan greenstones of western Zimbabwe.

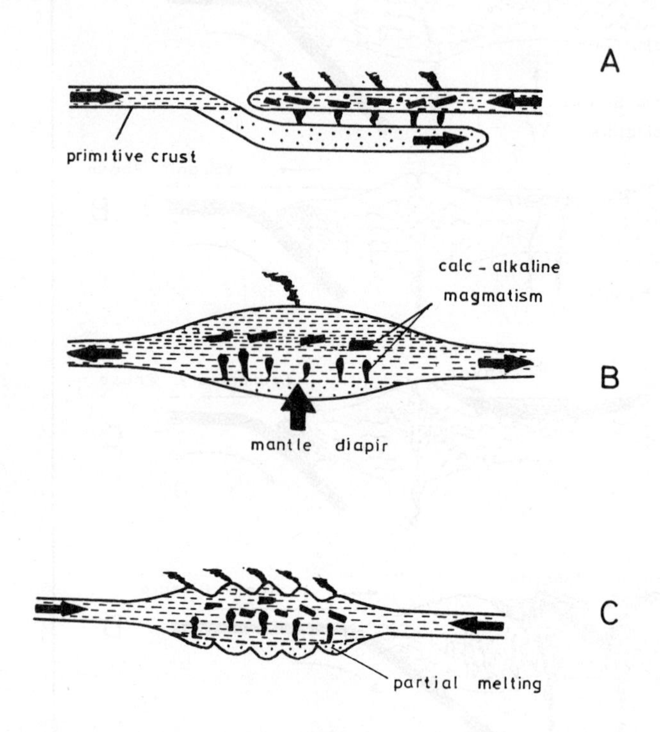

Figure 3.47: Suggested mechanisms for the generation of Archean calc-alkaline magmatism. (Redrawn from Frazier and Schwimmer, 1987.)

The transitional phase at the Archean-Proterozoic boundary was marked by the development of a rigid lithosphere probably as the result of a relatively rapid decline in radiogenic heat production; consequently rock associations and structures characteristic of the permobile regime became progressively less common (Burke and Dewey, 1973). The new rock associations which developed during the transition phase as a consequence of the availability of a rigid lithosphere and extensive stable cratonic areas were dyke swarms, the Great Dyke of Zimbabwe, and the Witwatersrand Triad, the oldest thick extensive group of rocks deposited on continental crust. The Great Dyke was emplaced around 2.5 Ga, at the Archean-Proterozoic boundary.

Table 3.5: Lithospheric evolution through the permobile, transitional, and plate tectonic stages. (From Burke and Dewey, 1973.)

PERMOBILE PHASE	TRANSITIONAL PHASE	PLATE TECTONICS REGIME		
			600–200	200–
		semiquantitative approach from paleomagnetic data qualitative approach from paleoclimatic indicators	semiquantitative approach from paleomagnetic data qualitative approach from paleoclimatic indicators; qualitative approach from founal provinces	quantitative approach from magnatic anomalies
pervasive permobile conditions no craton or platforms	localisation of permobile regions / progressive stabilisation of cratons-local semi stable platforms and troughs	continents oceans orogenic belts		
	aulacogen; first major dike swarms	extensive cratonic instability-clastic basins and troughs-basaltic dike swarms-vulcanicity	stable cratons - major marine transgressions flood continents	
non-linear orogenic patterns-greenstone belt tectonics		narrow well-defined linear/arcuate orogenic belt		
		miogeoclines [continental shelves] sharp boundaries with eugeosynclinal facies		
		strongly zonal sedimentary igneous and metamorphic facies relationships		
		ophiolite lithologies - pseudo-ophiolite sequences	ophiolite sequences	
		?	blueschists low rank	high rank
? andalusite silimanite	? cordierite - - -	kyanite		
calc-alkaline assemblages				
sediment supply from rising granodiolite terranes and volcanic complexes	sediment supply from cratons	sedimentary polarity reversals from cratonic to tectonic land supply		
		exogeosynclines - marginal thrust zones		
pervasive reworking of basement in permobile regions		basement reactivation by Tibetan Plateau-type mechanism-cryptic sutures		
high heatflow - thin lithosphere growth of permobile crust by convective scum mechanism or vertical growth of granodiorite-terranes	sharp decline in heatflow -rapid thickening of lithosphere; cratonic stabilisation completed - lithosphere suffiently strong to crack on a large scale	lithosphere thickens-plate margins become more narrowly-defined		
		increasing stresses of subduction zones and higher pressure assemblages		
		increasingly-narrow and better-defined accreting plate margins leads by 600 my to generation of ophiolite sequences [= oceanic crust and mantle]		
		increasing proportion of alkalic to tholelitic basalt in dike swarms		
		progressive sharpening of facies changes in orogenic belts		
		progressive thinning of asthenosphere by basal accumulation of refractory lithosphere slabs		
		continental accretion in volcanic arcs		
SWAZI. SUPERIOR	HURONIAN	LABRADOR KIBARA	DAHOMEY. APPAL.	URALS ALPS
2700	2000	1750	600	200

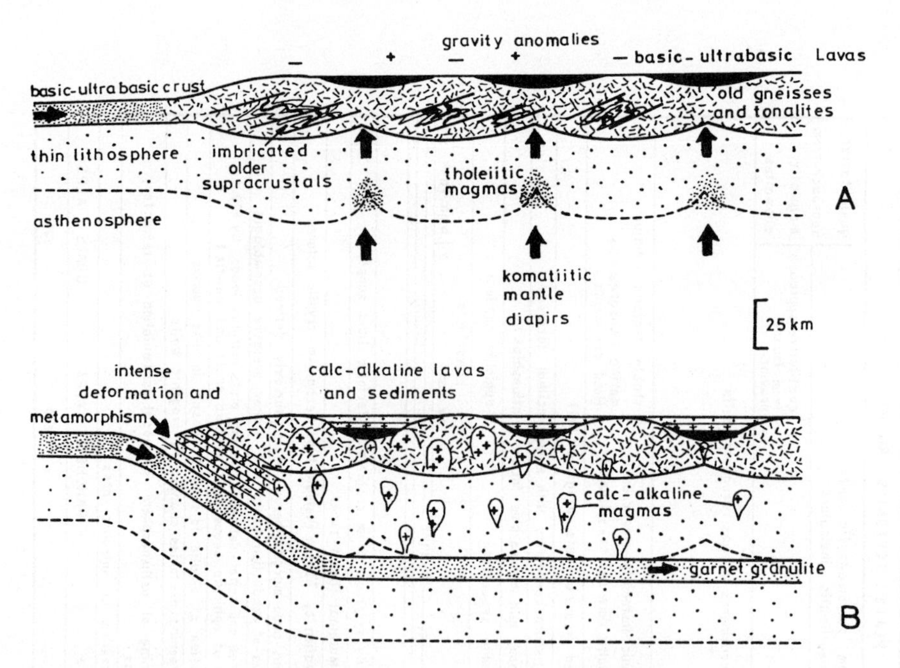

Figure 3.48: Actualistic model for permobile Archean tectonics. (Redrawn from Frazier and Schwimmer, 1987.)

After the falling near-surface geothermal gradient reached 15° C/km, and eclogite formed instead of granulite, steeply dipping subduction which ever since has characterized convergent plate-boundaries, was probably initiated (Frazier and Schwimmer, 1987). This final dramatic change in the tectonic behaviour of the Earth, according to Frazier and Schwimmer, was geologically "sudden" because only a small change in the geothermal gradient would have been needed to cross the phase boundary from granulite to eclogite.

Chapter 4 Early Proterozoic Cratonic Basins and Mobile Belts

4.1 Introduction

Although the Early Proterozoic, by most conventional definitions started from about 2.5 Ga and lasted in Africa until about 1.75 Ga (Cahen et al., 1984), the crustal evolution of Africa precludes strict adherence to this time framework in several respects. First, the assembly of large Archean continental blocks out of the mêlée of colliding island-arcs established stable crustal areas in the Kaapvaal province as early as around 3.0 Ga. Between 2.8 Ga and 1.8 Ga, vast portions of this province had become extensive stable cratons on which intracontinental basins developed (Kröner, 1989) successively, as sites of Pongola, Witwatersrand, Ventersdorp, and Transvaal-Griqualand West and the Waterberg-Soutpansberg-Umkondo-Matsap (Fig.4.1) epicontinental sedimentation (Tankard et al., 1982). The long continuum of cratonic sedimentation dating from the Late Archean in southern Africa therefore prohibits any strict adherence to the geochronological limits of the Early Proterozoic in this region. Secondly, outside the cratonic areas in southern Africa crustal instability prevailed from Early to Middle Proterozoic, giving rise to the Namaqua and Natal mobile belts (Fig.4.1). Although two distinct orogenic cycles are distinguishable in the Namaqua belt, the Early Proterozoic Orange River orogeny (2.0-1.7 Ga) and the Namaqua orogeny (1.2-1.0 Ga), the latter being of mid-Proterozoic age, recurrent and protracted orogenic activity in the same region renders it again impracticable to enforce the arbitrary Early Proterozoic age limits. Equally significant is the fact that after the Early to Middle Proterozoic tectogenesis the Namaqua mobile belt stabilized and became part of the Kalahari craton.

Thus, although this chapter dwells principally on the Early Proterozoic in Africa, the continuum of Late Archean and Middle Proterozoic geologic processes in southern Africa has necessitated the inclusion of these other age intervals in this chapter. Also considered here are the Eburnean orogenic belts of the West African and Zaire cratons, the Ubendian mobile belts of central Africa, and the areas affected by Early Proterozoic anorogenic magmatism.

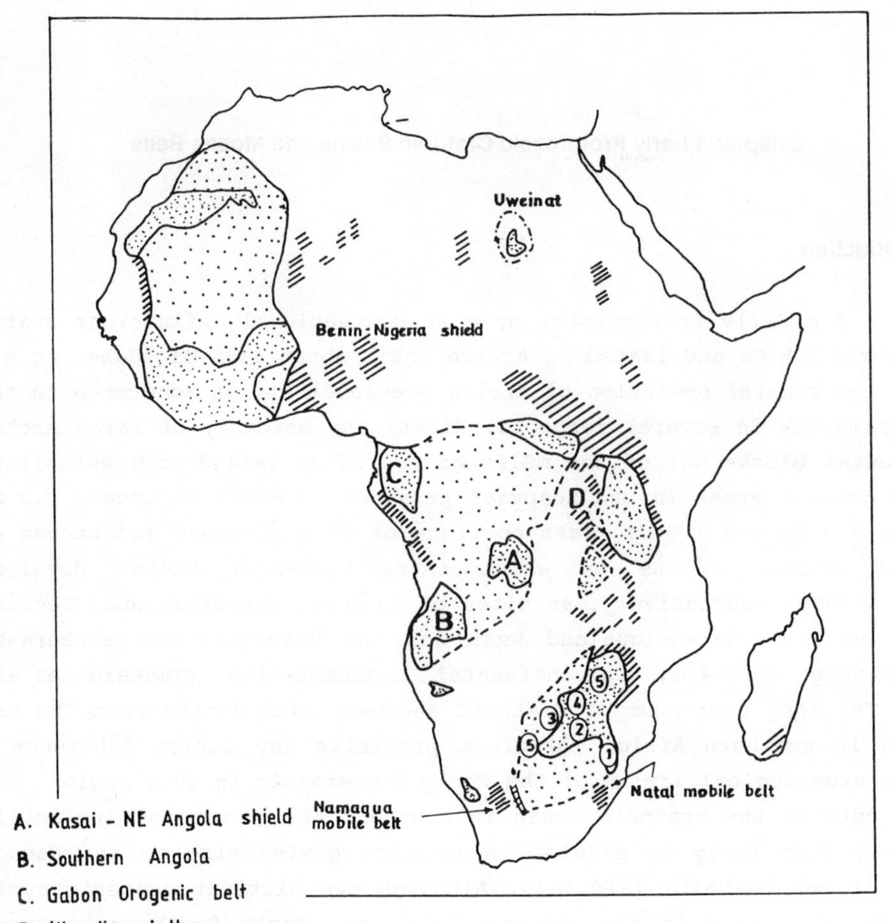

Figure 4.1: Distribution of Early Proterozoic rocks in Africa. 1, Pongola basin; 2, Witwatersrand basin; 3, Ventersdorp basin; 4, Transvaal-Griqualand West basin; 5, Waterberg-Soutpansberg-Umkondo-Matsap basins. (Redrawn from Cahen et al., 1984.)

4.2 Kalahari Cratonic Basins

4.2.1 Introduction

The Late Archean-Early Proterozoic in the Kalahari craton was marked by the formation of extensive intracratonic basins in areas that had attained crustal stability (Fig.4.2,A). Tectono-thermal activity was re-

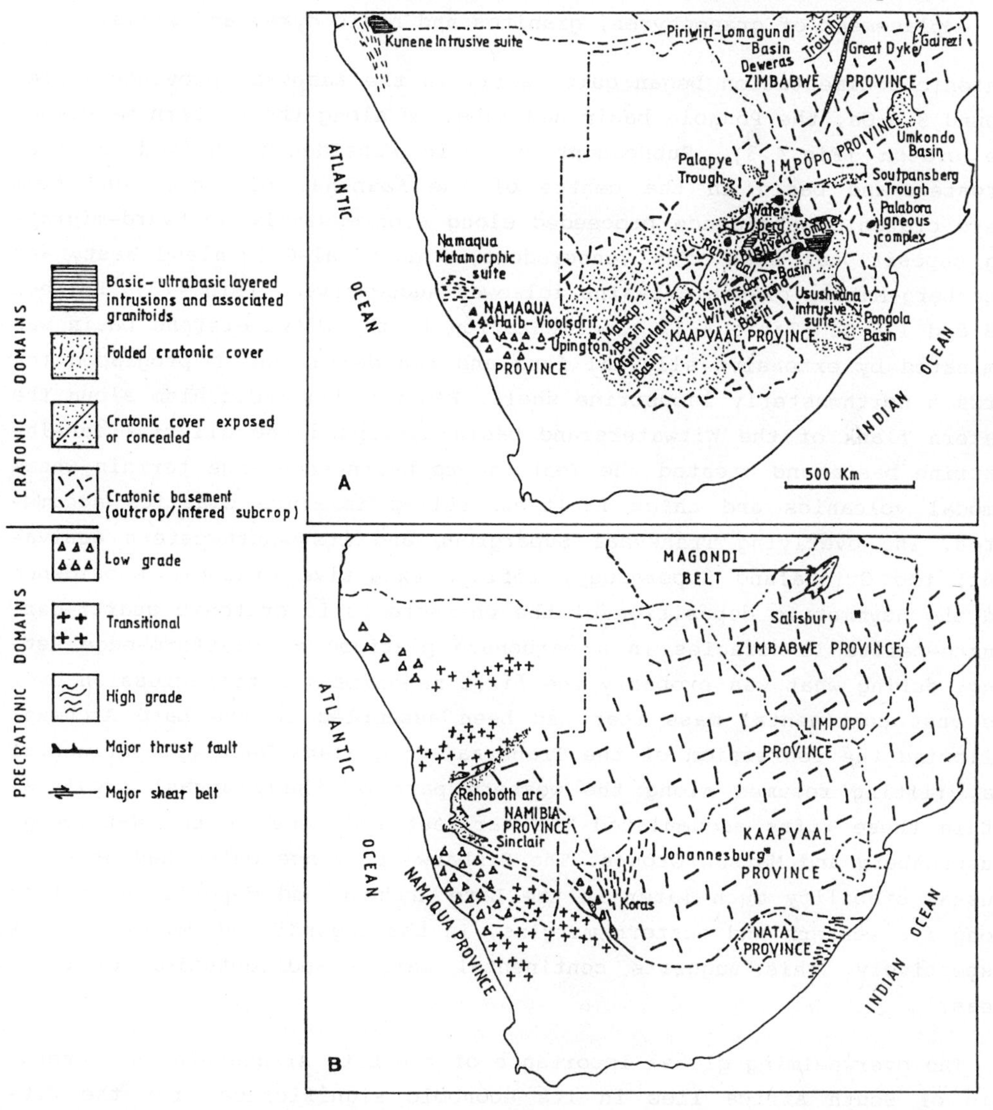

Figure 4.2: Tectonic outline map of the Kalahari craton showing A, supracrustal development from 2.9 Ga to 1,8 Ga; B, Early to Middle Proterozoic orogenic belts from 2.1 to 1.0 Ga. (Redrawn from Tankard et al., 1982.)

stricted to the Namaqua and Natal mobile belts along the southern parts of the craton, and to the Magondi mobile belt on the northwestern margin (Fig.4.2,B). Elsewhere, igneous activity periodically broke out through the stable continental crusts of the Zimbabwe and Kaapvaal cratonic nuclei. These Early Proterozoic magmatic suites include layered igneous intrusions such as the Great Dyke of Zimbabwe and the Bushveld Complex of the Kaapvaal province. The major magmatic events were followed closely by the emplacement of granophyres, granites and basic dykes and sills.

Cratonic sedimentation began quite early in the Kaapvaal province where, around 3.0 Ga, the Pongola basin had subsided along the eastern margin of the craton (Fig.4.3). Subsequent cratonic subsidence shifted to the Witwatersrand basin in the centre of the Kaapvaal province, and from there cratonic subsidence proceeded along progressively westward-migrating depositional axes. The Ventersdorp, Traansvaal-Griqualand West, and Waterberg-Matsap cratonic basins followed successively (Fig.4.3), between 2.8 and 1.8 Ga (Table 4.1). Sedimentation in the Witwatersrand basin was dominated by extensive alluvial fans and fan deltas which prograded towards a northwesterly lacustrine shelf. Faulting and volcanism along the western flank of the Witwatersrand basin disrupted the alluvial fan-lacustrine basin and created the Ventersdorp basin-and-range terrain where bimodal volcanics and thick fault-controlled immature clastics accumulated. The overlying Transvaal Supergroup and its southwestern equivalent, the Griqualand Supergroup, reflect extensive cratonic subsidence and the asymmetric deposition of the characteristic cratonic quartz-carbonate-pelite lithofacies in a carbonate platform to platform-edge setting, during what was probably the first major marine transgression over the vast continental mass that had been assembled in the Late Archean. Following the regression of the Transvaal-Griqualand West epicontinental sea, rifting resumed along the western part of the Kaapvaal province. Within these rifts accumulated the earliest red beds of the Waterberg, Soutpansberg and Matsap Groups. The Zimbabwe province which had attained crustal stability much later in the Late Archean had deposition centred along its western and eastern margins, in the Magondi and Umkondo basins respectively. This suggests continental margin sedimentation in these areas.

The overwhelming global importance of the Late Archean-Early Proterozoic of South Africa lies in its economic significance, for the Witwatersrand basin dominates the world's gold and uranium production. The Transvaal basins and the Namaqua Metamorphic Complex are repositories of vast strata-bound mineralizations such as manganese, iron ore, fluorite, copper and zinc. Furthermore, layered intrusives such as the Bushveld Complex, and the Great Dyke of Zimbabwe host some of the most spectacu-

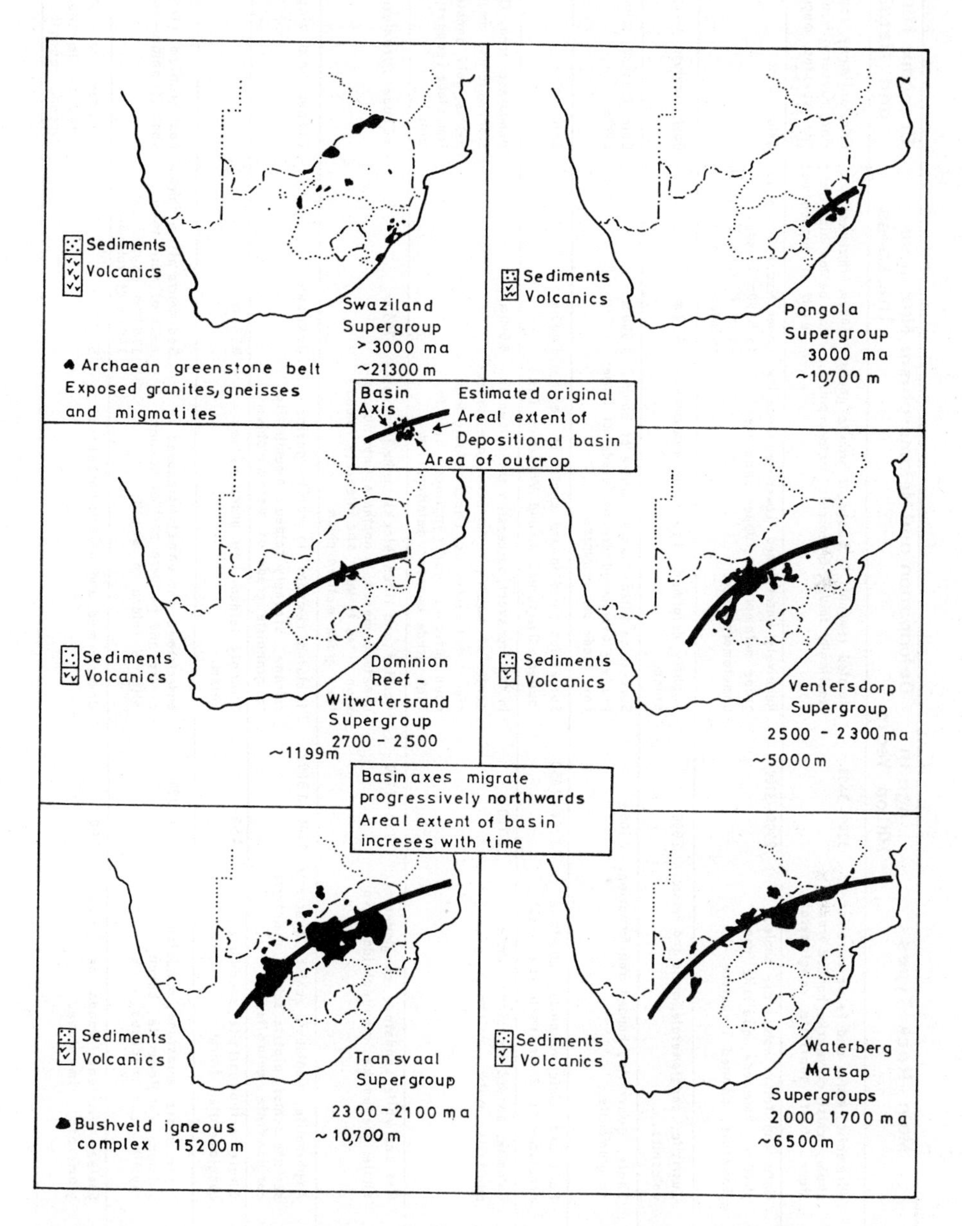

Figure 4.3: Progressive development of Archean to Early Proterozoic cratonic basins on the kalahari craton. (Redrawn from Anderson and Biljon, 1979.)

lar magmatic ores known, viz., chromium, platinum, nickel, and vanadiferous and titaniferous magnetite.

Table 4.1: Main features of South African supracrustal sequences. (From Anderson and Biuljon, 1979.)

Sequences	Main Rock Types	Age in Million Years	Deformation and Metamorphism	Maximum Thickness	Economic Minerals and Metals
SWAZILAND SUPERGROUP	Ultramafic, mafic and Felsic volcanics, shale, schist, greywacke, banded iron-stone, chert, quartzite, conglomerate.	3200-3400	Isolated remnants in Basement granite-gneiss, Isoclinally folded, Greenschist metamorphism.	15000m (thickness may be too high as result of duplication)	Gold, antimony, copper, lead, zinc, mercury, iron, chrysotile asbestos, magnesite, talc (nickel)
PONGOLA GROUP	Mafic and felsic volcanics, shale, phyllite, limestone, quartzite conglomerate, banded ironstone.	2900-3100	Relatively underformed, Lower sequence low grade metamorphism, Upper sequence unmetamorphosed.	Lower Nsuzi Group up to 5500m in south. Upper Mozaan Group up to 5000m in north.	Gold.
DOMINION	Quartzite, conglomerate, mafic and Felsic volcanics.	2800	Slightly deformed, Low grade metamorphism.	2650 m	Gold, uranium, pyrophyllite.
WITWATERSRAND SUPERGROUP	Shale, quartzite, banded iron-formation, conglomerate.	2700±	Succession tilted towards centre of basin, faulted; Overturned around Vredefort Dome; Low grade metamorphism.	7500 m	Gold, uranium, silver, pyrite.
VENTERSDORP SUPERGROUP	Mafic and felsic volcanics, quartzite conglomerate, shale, carbonate rock.	2300-2650	Succession tilted toward centre of basin, faulted; Overturned around Vredefort Dome.	5000 m	Gold.
TRANSVAAL SEQUENCE GRIQUALAND WEST SEQUENCE	Dolomite, banded ironstone, shale, quartzite, mafic volcanics	2200	In the Transvaal, succession tilted towards centre of basin - in northern Cape succession tilted westwards; Local folding into open anticlines and synclines; Low to high grade contact metamorphism	8500m	Manganese, iron, fluorite, crocidolite and amosite asbestos, gold, andalusite, limestone (copper, lead, zinc).
WATERBERG GROUP (OLIFANTSHOEK SEQUENCE)	Red and white sandstone, conglomerate, shale, greywacke, amygdaloidal lava	1750-2100	Faulted and tilted blocks in the Transvaal-unmetamorphosed; In northern Cape succession tilted to west and intensely folded - low grade metamorphism.	7700 m	(Copper, uranium ?)
KHEIS GROUP	Amphibolite, amphibole schist, quartzite sericite schist, biotite schist quartzite, conglomerate, dolomite.	1100-1300?	Isolated remnant in granite-gneiss terrane, Strongly folded; Amphibolite to granulite grade of metamorphism.	Unknown.	Copper, lead, zinc, silver.
CAPE SUPERGROUP	Shale, orthoquartzite, sandstone, conglomerate, tillite.	450	Strongly folded; Low grade metamorphism.	8500 m	—
KAROO SEQUENCE	Tillite, shale, mudstone, sandstone, siltstone, conglomerate, basalt, rhyolite, carbonaceous rock.	160-300	Underformed and unmetamorphosed in central and northern part of basin; Slightly folded in south.	Sediments attain thickness of more than 10,000 m capped by 1400 m of volcanics.	Coal, uranium, clays, oil-shale (titanium-sands).
MESOZOIC AND YOUNGER	Shales, siltstone, sandstone, conglomerate, limestone.	160	Underformed and unmetamorphosed.	5700m±.	(Oil), gas (coal). alluvial diamonds, ilmenite, rutile zircon, monazite, salt.

4.2.2 Witwatersrand Basin

The Witwatersrand basin, in its tectonic setting (Fig.4.4), represents an intracratonic trough or yoked basin (Pretorius, 1975) that developed from the rifting of the foreland of the Kaapvaal cratonic nucleus (Burke et

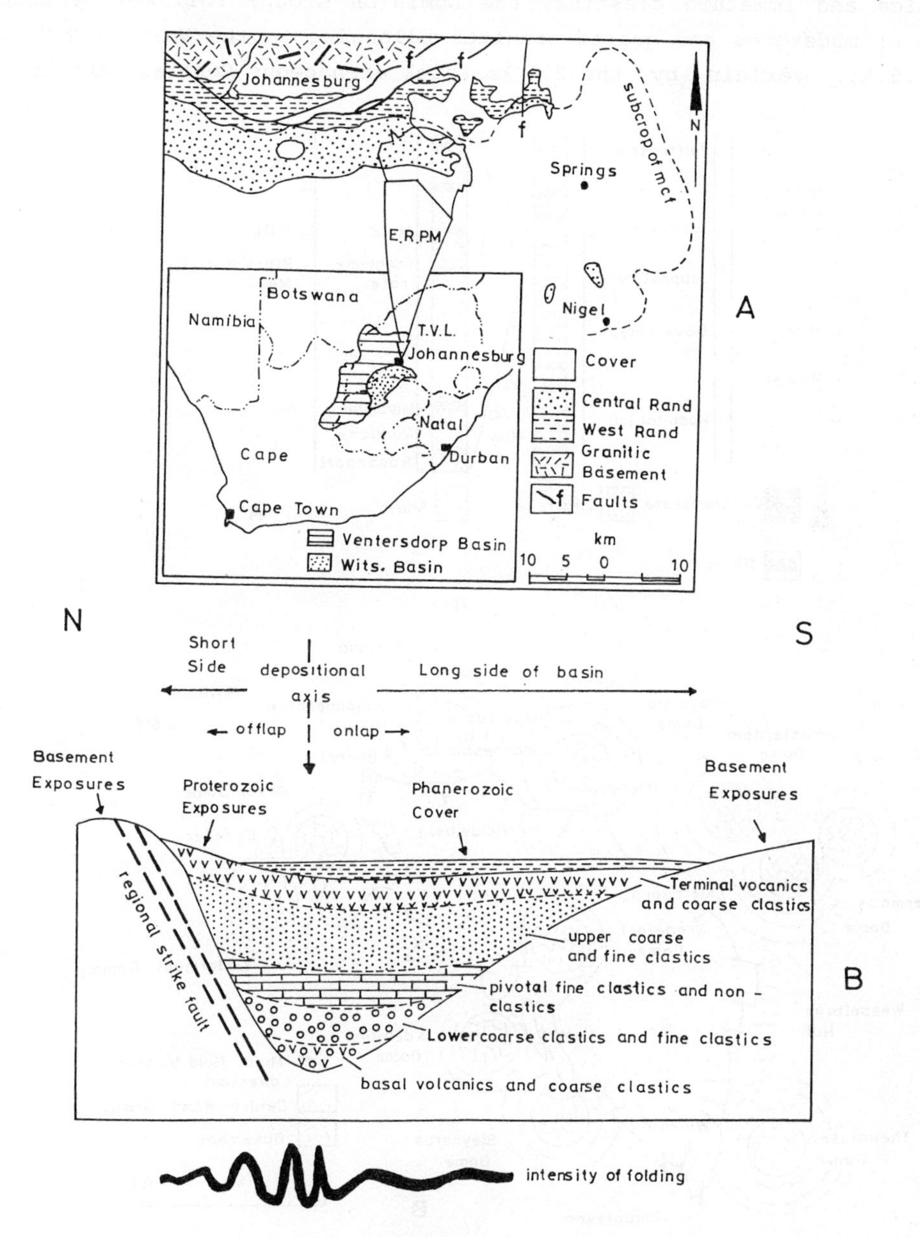

Figure 4.4: Geologic outline map (A) and cross-section (B) of the Witwatersrand basin. (Redrawn partly from Martin et al., 1989.)

al., 1986). Up to 16 km of clastic sediments, and felsic and mafic volcanics representing the Witwatersrand and the Ventersdorp Supergroups respectively accumulated in what was a one-sided rift basin (Fig.4.4,B). The Witwatersrand basin was initially filled with about 2.7 km of mixed volcanics and immature clastics, the Dominion Group; followed by about 4.5 km of mudstones and quartz arenites belonging to the West Rand Group (Fig.4.5,A), overlain by the 2.4 km thick arenaceous Central Rand Group

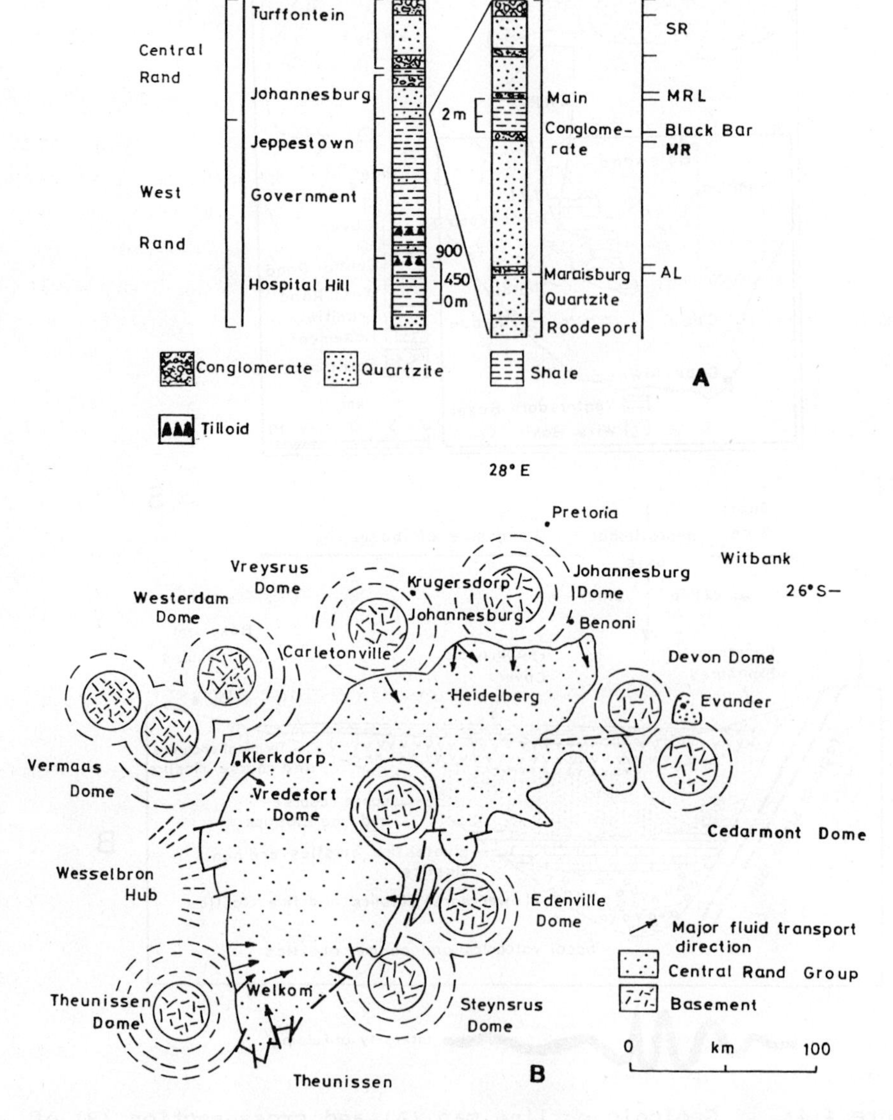

Figure 4.5: A, Stratigraphic succession in the Witwatersrand Supergroup; B, Distribution of the Central Rand Group. (Redrawn from Tankard et al., 1982

(Tankard et al., 1982). The conglomerates of the West Rand Group host placer gold and uranium mineralization.

Since the Witwatersrand Supergroup has been metamorphosed only to the lower greenschist facies, the unusually well preserved sedimentary fabric and primary sedimentary structures have allowed detailed paleoenvironmental reconstructions (Tankard et al., 1982). According to Tankard et al. the depositional model of the Witwatersrand Supergroup was essentially one in which alluvial fans and fan deltas prograded across finer-grained lacustrine deposits on a humid and semi-arid landscape that was devoid of vegetation. Along the fan-delta shorelines there was tidal and wave reworking of sediments while finer grained clastics and chemical sediments accumulated offshore. Epeirogenic tilting and warping controlled the geometry of the sedimentary units, unconformities, and the paleocurrent patterns in the basin. Diapiric granite doming accelerated fluvial sedimentation and the formation of rich gold placers in the Central Rand Group (Fig.4.5,B). Exploration for gold and uranium has furnished the detailed geologic information that is available on this ancient basin.

Stratigraphy

The Dominion Group is preserved in an area of about 15,000 km^2, in western Orange Free State and southwestern Transvaal, where up to 2.7 km of Dominion strata rest non-conformably on Archean basement granites 3.0 Ga to 2.7 Ga old. Intrusive and extrusive volcanics in the Dominion Group have been dated at 2.8 Ga, which implies the initiation of the Witwatersrand basin around that time. An immature, cross-bedded fluvial sandstone with placer concentrations of detrital pyrite, monazite, gold and uraninite occurs at the base of the Dominion Group. A mixed volcano-sedimentary interval follows comprising basaltic andesite and tuff, acid and andesitic lavas, volcanic breccia and quartzo-feldspathic lava.

Comformably overlying the Dominion Group is the West Rand Group (Fig.4.5) which occurs over an area of about 42,000 km^2 with an average thickness of 4,650 m, and a maximum thickness of about 7,500 m in the northwest. A basal sandstone of tidal inlet and beach origin comprises quartz pebble lag deposits overlain by trough cross-bedded, fine grained and plane-bedded mature quartz arenite with bimodal-bipolar paleocurrent directions. Beach and tidal sandstones and siltstones with intercalated offshore shales and a thin contorted iron-formation (Hospital Hill Subgroup) grade upward into the fine-medium-grained, upward-fining, planar- and trough-bedded subarkosic fan-deltaic arenaceous deposits of the Government Subgroup. The occurrence of diamictites within the Government Subgroup has been interpreted as the evidence of the oldest known glaciation, representing ice-rafted moraines which were displaced into a distal shelf environment by submarine flow (Tankard et al., 1982). The upper part of the West Rand Group comprises fluvial, fine-grained, trough and planar cross-bedded subarkose with bipolar paleocurrent directions and proximal shelf siltstones and shales which pass gradationally into the overlying Central Rand Group (Fig.4.5,A).

The Central Rand Group is of tremendous economic importance being the world's largest source of gold. Its sedimentology has been unravelled in great detail in the course of gold mining; this has shed considerable light on the paleogeography of these unmetamorphosed ancient sediments. The Central Rand Group is exposed over an area of about 9,750 km^2; its maximum thickness is about 2,880 m near the Vredefort Dome. Lithologically this group consists of predominantly coarse-grained, cross-bedded subgraywackes which were deposited in alluvial fan fluvial packages with an average thickness of 250 m. The alluvials fans formed mostly along the northern and western fault-bounded margins of the basin (Figs.4.4,B; 4.5,B;4.6) and not on the more gentle southeastern margin of the basin. At the centre of the basin was a shallow lake or an enclosed inland sea. With time the centre of the basin became unstable and the various goldfields developed in the down-warps between the basement uplifts. From the surrounding highlands (Fig.4.6) short braided streams emerged and flowed over the piedmont plain. Bimodal cross-stratification in the sandstones show a down-slope component which was produced in the stream channels while the other component is related to distribution by longshore currents in fan deltas along the lake margin (Fig.4.6).

Two horizons of diamictites occur in the Central Rand Group (Fig.4.5,A). These are characterized by clasts supported in an argillaceous matrix; they probably originated as high-density muddy suspension which accumulated downslope in arid or semi-arid alluvial fans. In such an environment with sparse vegetation cover, weathered rock debris could have been wetted and displaced by rainstorms or melting snow.

Several generations of dykes and sills interlace the Witwatersrand Supergroup which were emplaced during the overlying Ventersdorp volcanism; during the intrusion of the Bushveld Complex; during the emplacement of Karoo dolerites; and during the Cretaceous when kimberlite dykes were intruded.

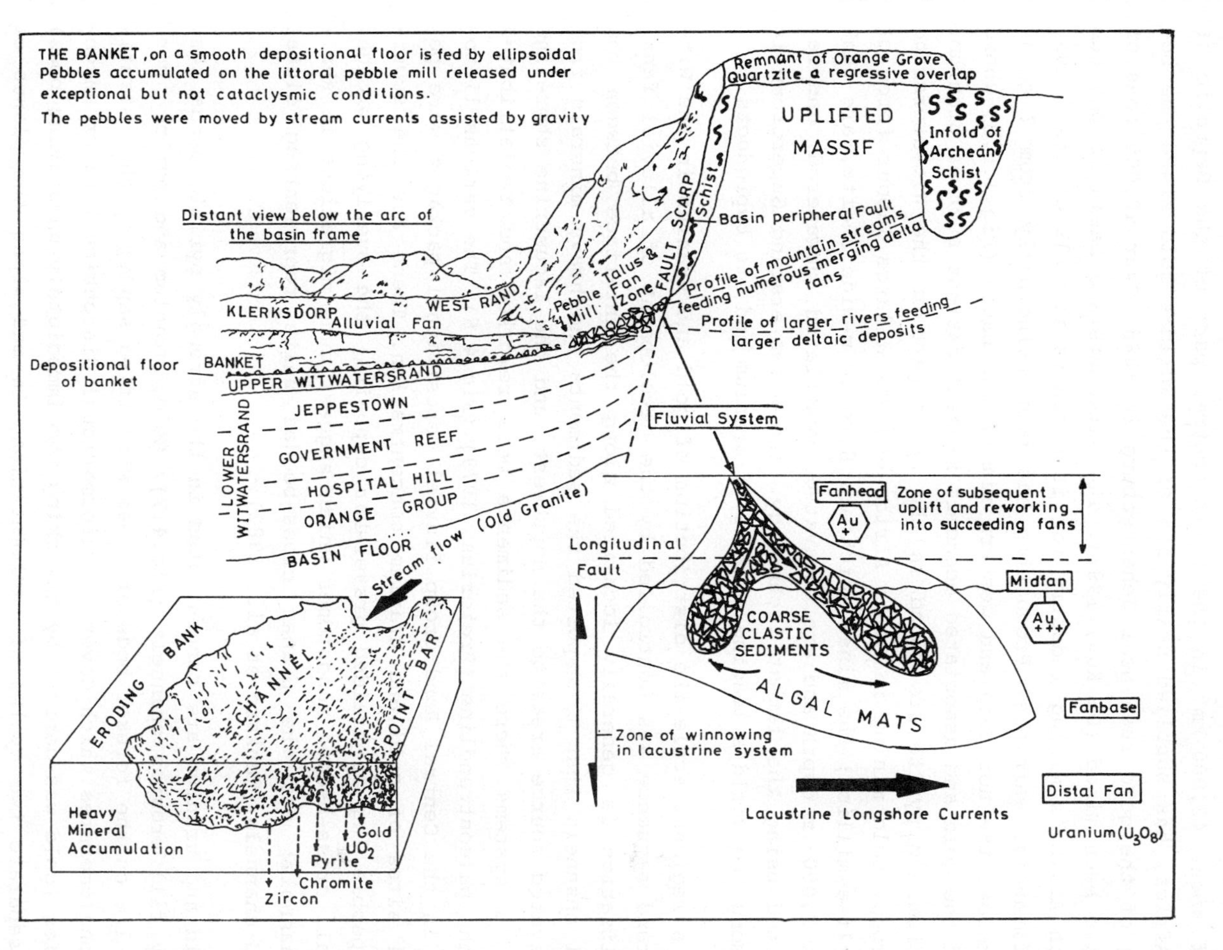

Figure 4.6: Geologic section and facies model for the Witwatersrand alluvial goldfield. (Redrawn from Hutchison, 1983.)

Mineralization

The Witwatersrand Supergroup, deposited between 2.8 and 2.3 Ga, over an area of about 42,000 km^2 in the north-central part of the Republic of South Africa, has supplied nearly 55% of the world's gold since the beginning of the gold rush near Johannesburg in 1886. Over 40,000 tons of gold have been mined (De Kun, 1987). The Witwatersrand basin is the hub of South Africa's mining industry for it accounts for 70% of the industry's financial earnings. Production has been principally from 7 goldfields along the northern and western rim of the basin (Fig.4.5,B) where most of the gold and associated economic mineralizations such as uranium, osmiridium and pyrite are found in the upper part of the Witwatersrand Supergroup. Gold mining in South Africa engages a large labour force of over half-a-million; the mines which are 8 to 10 m-wide shafts, are sunk to over 2,000 m depth and are excavated, ventilated, dewatered, and refrigerated using the latest technology. The Witwatersrand Supergroup is also among the world's leading sources of uranium and its byproducts.

On a regional scale the distribution of gold and uranium in the Witwatersrand sequence was controlled by the development of alluvial fans. Mineralization is generally localized along the interface between the fluvial channels that transported the sediments and heavy minerals from the elevated source areas to the northwest, and the lacustrine shoreline fan delta systems where the sediments were reworked and redistributed along the paleostrandline (Pretorius, 1979) (Fig.4.6). The various lithofacies in the Central Rand Group fluvial depositional packages were potential sites of gold and uranium concentration. These are the scour-based pebble lag or braided gravel-bar deposits; the overlying trough, unimodal, cross-bedded sandstones which suggest dune migrations in shallow channels; and the planar cross-bedded units representing distal braided channels (Tankard et al., 1982).

Gold and uranium are concentrated in the strongly pyritic sands that usually fill erosion channels (Fig.4.7). Gold, uranium and pyrite particles lie on the foreset beds of cross-stratified sands; in the stratiform conglomerates which cover intraformational discontinuities or unconformities (gold was derived by reworking the immediately-underlying beds) which separate two cycles of sedimentation; in mud along the planes of unconformity that separate short depositional cycles; and in carbon seams which are developed on or immediately adjacent to planes of unconformity. However, the greatest concentration of gold and uranium occurs in or immediately adjacent to bands of conglomerate (locally known as bankets). These are preferentially developed at or near the base of each cycle of sedimentation. Within the Dominion Group and the Central Rand Group in the Johannesburg area, these conglomerates occupy about 8% of the total thickness whereas on the southern margin of the basin they constitute only 1% of these lithostratigraphic units (Hutchison, 1983). The bankets or conglomerate reefs (Fig.4.7) consist of well-rounded pebbles which form 70% of the rock volume; the pebbles range in diameter from 20 mm to 50 mm . They are mostly oval to subangular vein quartz, accompanied by chert, jasper, quartzite, and other minerals. The matrix is compact consisting of secondary quartz, phyllosilicates, sericite, pyrophyllite, muscovite, chlorite, and chloritoid.

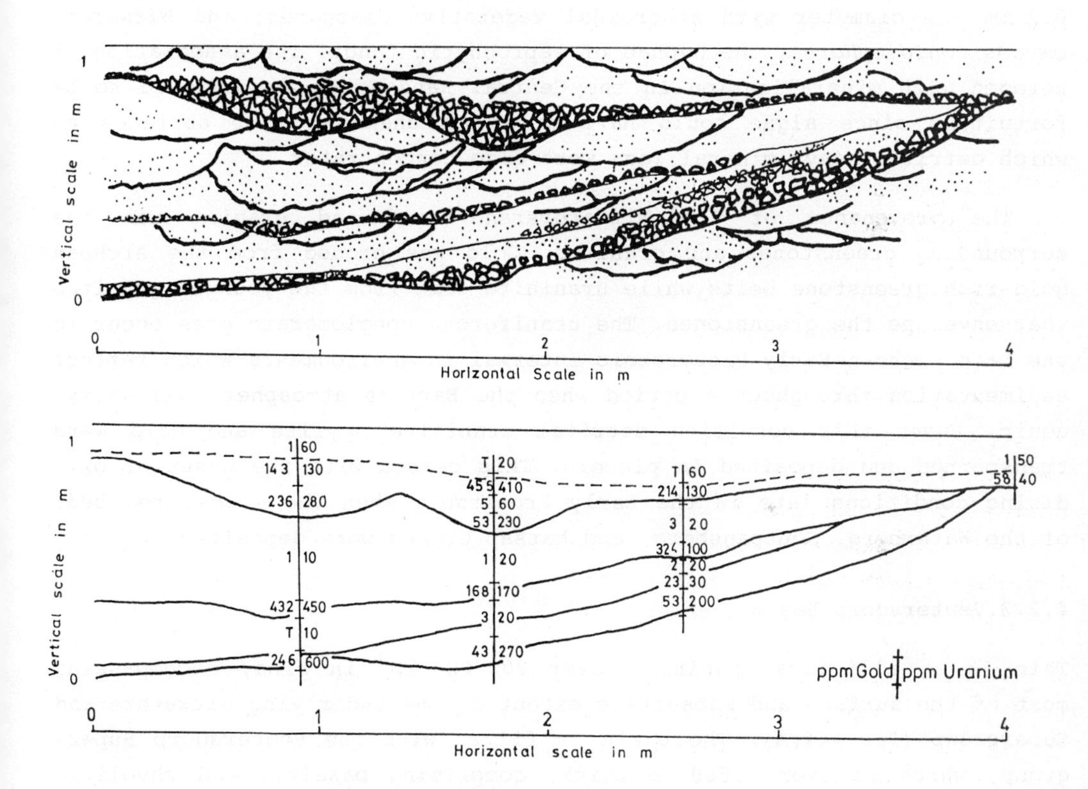

Figure 4.7: Concentration of gold and uranium in pyritic fluviatile erosion channels of the Steyn placer. (Redrawn from Tankard et al., 1982.)

In general gold particles may be distributed in a dispersed manner throughout the banket matrix; as isolated clusters; or as thin streaks on either the hanging wall or the footwall contact. Since gold and uranium were concentrated hydraulically, good correlation exists on a local scale, between their distribution both vertically (Fig.4.7) and horizontally (Fig.4.6). In the Steyn and Basal placers of the Central Rand Group, gold was concentrated in the upper and lower fan plains whereas uraninite, zircon, and chromite were transported further basinward into the distal fans (Fig.4.6).

Kerogen seams occur in association with the placer deposits of the Witwatersrand Supergroup (Tankard et al., 1982). The seams, consisting of hydrocarbons with organic sulfur and oxygen, originated by polymerization of biochemical compounds produced by decaying primitive micro-organisms. At least two types of organisms have been identified in the kerogen. These are *Thuchomyces lichenoides*, a fibrous form 0.5-5.0 mm long and 0.2 mm in diameter with spheroidal vegetative diaspores; and *Witwateromyces conidiophorus*, the hyphae of saprophytic fungi. The association of kerogen with mineralization in the Central Rand Group is believed to be fortuitous since algae could have grown on channel scour surfaces on which detrital heavy mineral lags were also concentrated.

The provenance of the Witwatersrand gold and uranium was the surrounding greenstones and granites. Gold was eroded from the Archean gold-rich greenstone belts while uraninite came from the younger granites that envelope the greenstones. The uraniferous conglomerate ores occur in the Late Archean-Early Proterozoic interval in environments which reflect sedimentation throughout a period when the Earth's atmosphere was anoxygenic. Under this condition detrital uraninite, pyrite and gold were transported and deposited in placers. This ceased with the onset of oxidizing conditions late in the Early Proterzoic when the oldest red beds of the Waterberg, Soutpansberg, and Matsap Groups were deposited.

4.2.3 Ventersdorp Basin

This is an elliptical basin of over 200,000 km^2 in area, encompassing most of the surface and subsurface extent of the underlying Witwatersrand Supergroup (Fig.4.4,A). The basin is filled with the Ventersdorp Supergroup, which is over 7,860 m thick, comprising basaltic and rhyolitic lavas at the base, overlain by conglomeratic subgraywackes and subordinate shale and limestone at the top. The stratigraphy of the Ventersdorp Supergroup is complicated by local unconformities, lenticular and wedgelike depositional geometries, repeated lithologies, overlaps, and folding and faulting which occured during deposition.

At the base of the succession is the Klipriviersberg Group which was deposited during the phase of stress and faulting that followed subsidence in the Witwatersrand basin. Faults acted as conduits for voluminous (1,830 m thick) continental tholeiitic basalts which are porphyritic, amygdaloidal, tuffaceous, and with breccia. This volcanic sequence is conformable upon the Witwatersrand Supergroup in the northeastern part, but on the northwestern margin of the basin the contact is unconformable with a basal conglomerate that is mineralized with gold and uranium that were reworked from the underlying truncated Central Rand Group.

The middle Platberg Group, about 5,000 m thick, succeeds unconformably. This is a sequence of immature clastics with coarse scree deposits, sandy graywackes and debris flows. Graben margin boulder conglomerates which were shed from surrounding horsts change basinward into multiple alluvial fan systems and graben lake stromatolitic calcareous shales. Basin-and-range topography prevailed during Platberg times and persisted throughout the deposition of the overlying Pinel Group alluvial fan deposits.

4.2.4 Transvaal-Griqualand West Basins

The Transvaal-Griqualand West basin system (Figs.4.2.A;4.8) are the remnants of a once extensive epeiric basin which covered a large part of the Kaapvaal cratonic nucleus, up to an area of about 500,000 km^2, from about 2.3 Ga to 2.1 Ga. The Lobatse basement arch (Fig.4.8) separated the Transvaal basin to the northeast from the Griqualand West basin to the southwest. The thickest parts of both basins, with about 12 km of sediments, lay in the Transvaal basin. The lithostratigraphic units in both basins are, however, correlatable across the 130-km gap occupied by the Lobatse arch (Fig.4.9). The Transvaal and Griqualand West Supergroups which occupy the northeastern and southwestern basins respectively are lithologically varied epicontinental deposits which host a great variety of strata-bound mineral deposits such as iron ore, manganese, gold, fluorite, lead, zinc, vanadium, asbestos, aluminous minerals, and limestone deposits.

Stratigraphy

At the base of the Transvaal Supergroup are the cross-bedded, conglomeratic and arkosic deltaic sandstones and shales of the Wolkberg Group. This group sits unconformably upon the Ventersdorp Supergroup and is equivalent to the much thinner Vryburg Formation in the Griqualand West

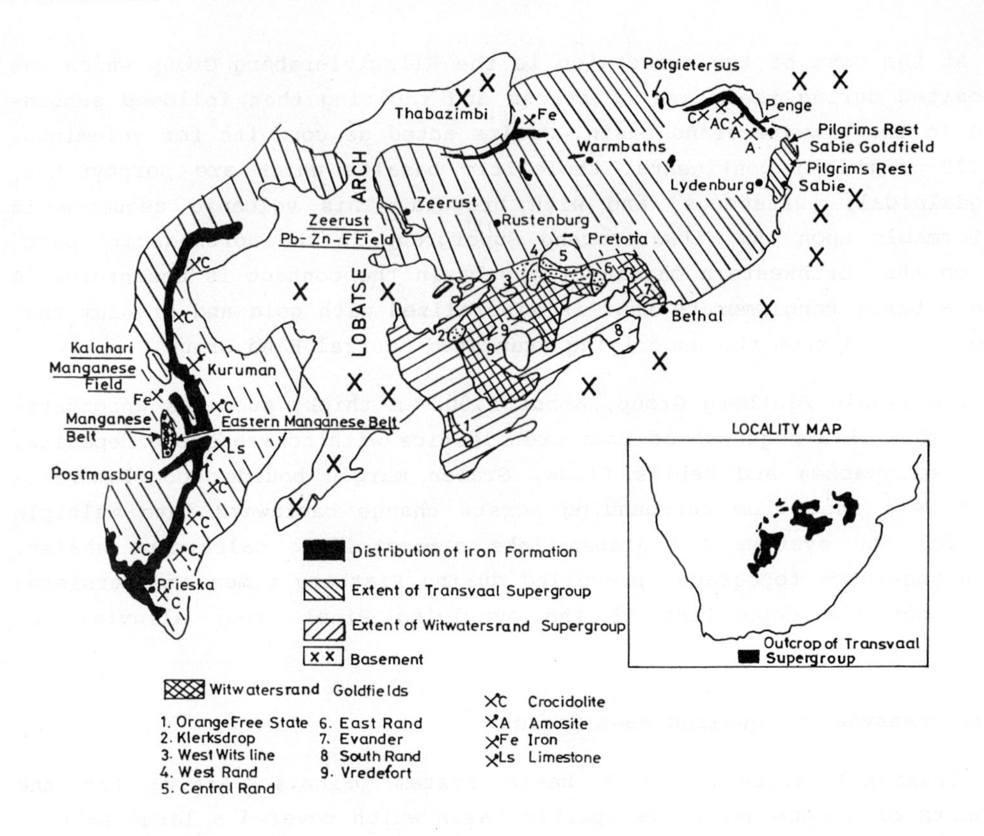

Figure 4.8: Distribution of the Witwatersrand, and Transvaal-Griqualand West Supergroups and their mineralization. (Redrawn from Anhaeusser and Button, 1976.)

basin. The overlying Chuniespoort and Ghaap Groups (Fig.4.9,B) are a thick succession of mostly dolomites and dolomitic limestones with extensive banded iron-formations. Like similar Precambrian sequences in Australia (Hamersley Group), these carbonates represent the oldest extensive carbonate platforms in the geological record. Tidal-flats, subtidal environments, platform edge and basinal carbonate facies are well displayed in the Chuniesport and Ghaap sequences (Tankard et al., 1982).

Tidal flat sedimentary structures include flat-laminated, low-relief and domical algal stromatolites; fibrous, bladed dolomites which pseudomorph after supratidal gypsum; tepee structures which suggest evaporitive tidal flats; and quartz oolitic dolomites. Subtidal paleoenvironmental indicators include dark, Fe-poor dolomites characterized by large, elongate stromatolitic mounds with no exposure-related features. These stromatolites grew below wave base. As shown in Fig.4.10 the carbonate plat-

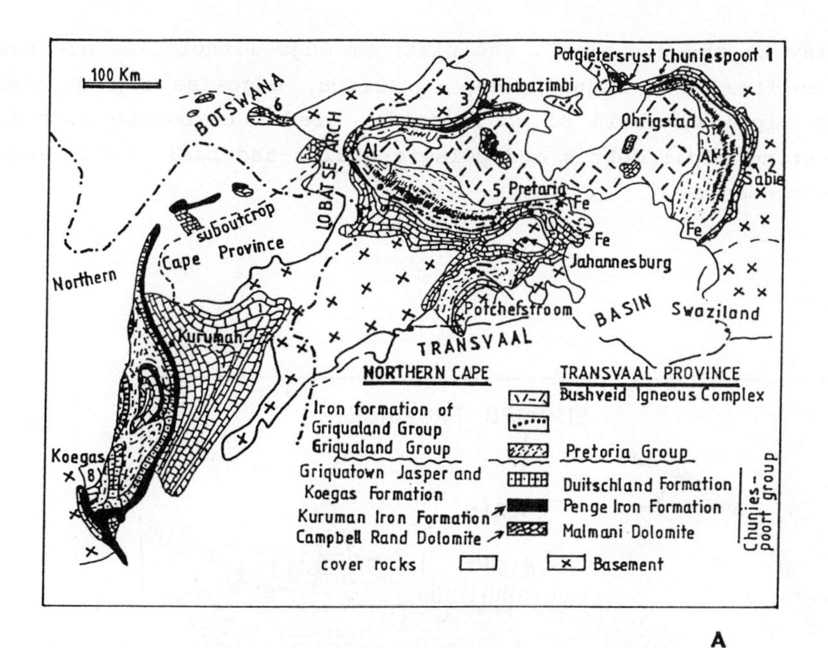

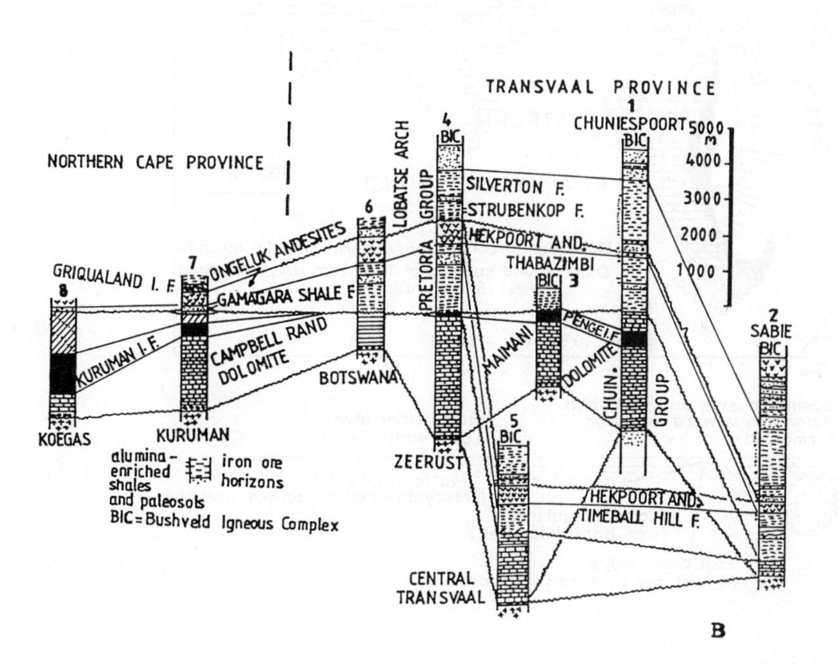

Figure 4.9: Geologic sketch map (A) and stratigraphic columns (B) for the Transvaal-Griqualand West Supergroups. (Redrawn from Reimer, 1987.)

form edge lay to the southwest. The platform edge lithofacies are represented by oolites with megaripples, oncolites, edgewise rip-up breccia and current ripples; while slump breccias, graded carbonate turbidites with alternating algal debris and pelagic shales and basic tuffs suggest basinal environments.

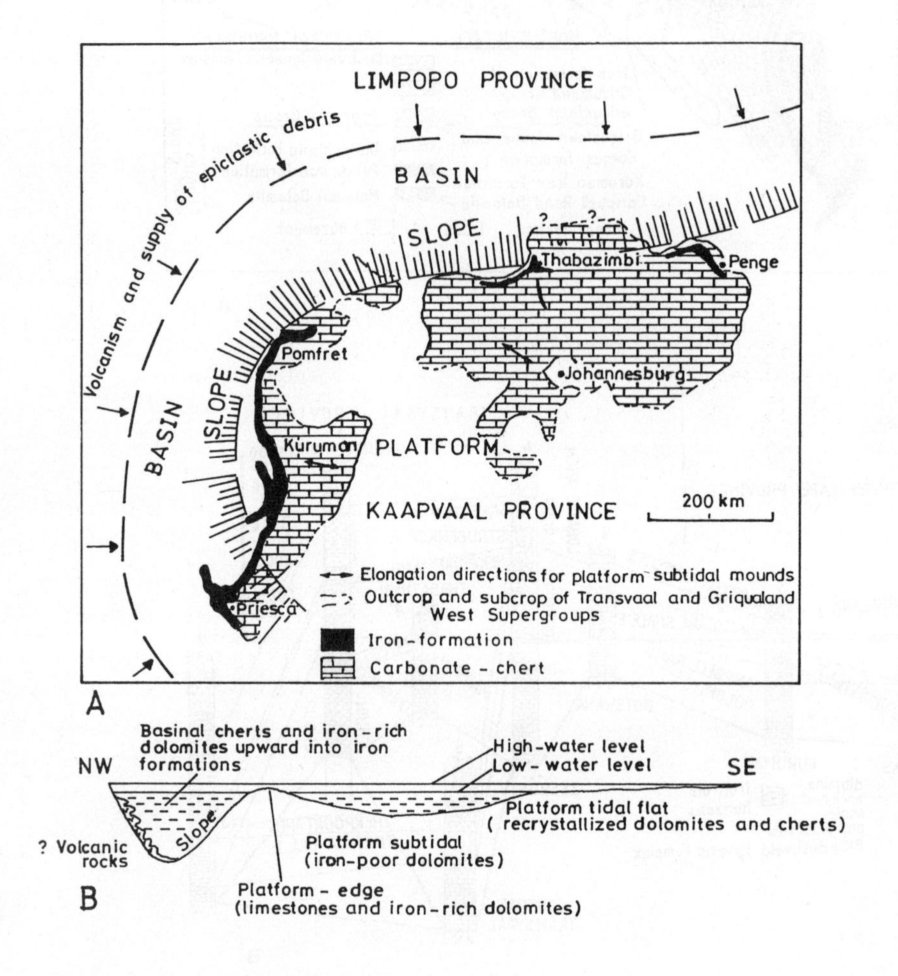

Figure 4.10: Tectonic setting for the Ghaap and Chuniespoort carbonates. (Redrawn from Tankard et al., 1982.)

Basinal iron-formations, the Asbeshuewels Subgroup, and its metamorphosed equivalent, the Penge iron-formation of the Transvaal area (Fig.4.9), contain lithologic units which are laterally persistent over very long distance. These are chemical deposits with stacked vertical cycles of volcanic material, chert, siderite, magnetite and microbanded hematite.

The unconformably overlying Pretoria and Postmasburg Groups (Fig.4.9) are primarily regressive, with fluvial arkoses, tidal-flat arenites and shales, shoal water quartz arenites and oolitic ironstones and volcanics, all of which signal a dramatic change in the pattern and source of sedimentation. The emergence of most parts of the Transvaal basin as a vast land surface with little relief and little erosion, following the regression of the Chuniespoort-Ghaap epeiric sea, favoured pronounced weathering and paleosol development (Reimer, 1987). Products of intensive weathering occur in the Pretoria Group as alumina-enriched sediments, and as "minette"-type ironstones, which are probably the oldest ironstones of this type in the geologic record. The more basinward Postmasburg Group, however, contains chemical sediments such as banded iron-formation and manganese, in the Griqualand West basin. Pronounced weathering caused the large-scale mobilization of silica into the sea where it contributed to the formation of chert whereas iron was mobilized into deltaic coastal complexes where it formed the oolitic and pisolitic "minette"-type ironstones of the Pretoria group. The paleosols with over 20% Al_2O_3 was the source of alumina enrichment in the Pretoria Group shales.

Another interesting aspect of the stratigraphy of the Transvaal and Griqualand West Supergroup relates to the implications of large-scale carbonate sedimentation to the evolution of the primitive atmosphere. The carbonate platforms of the Chuniespoort and Ghaap Groups, apart from being among the oldest in the geological record, are stromatolitic virtually throughout their 1,500-m thickness. Since there were no metazoans in the Early Proterozoic seas to graze upon them, the stromatolite communities were able to colonize several environments. Well-preserved blue-green algal filaments have been recovered from these stromatolites proving that large amounts of oxygen were being generated by photosynthesis some 2.2 Ga ago. Whereas there were limited oxidizing conditions below the regional unconformity separating the Chuniespoort-Ghaap carbonates from the Pretoria-Postmasburg Groups, there is evidence above this unconformity, that following the deposition of the carbonates, oxidizing conditions developed in the hydrosphere and probably also in the atmosphere (Tankard et al., 1982). The presence of manganese oxide in the chert breccia which defines the unconformity suggests that manganese was not taken into solution during subaerial weathering of the dolomite but rather was precipitated upon oxidation to its higher valency. With more oxygen available the ironstones in the Pretoria Group and the iron-formations in the upper Postmasburg Group are deficient in ferrous iron. The presence of significant quantities of manganese in quadrivalent-state minerals in the calcareous shales which are interlayered within the iron-formation are further evidence suggesting the prevalence of oxidizing conditions.

4.2.5 Mineralization in the Transvaal-Griqualand West Supergroups

Iron and Manganese

As aforementioned "minette"-type ironstones and banded iron-formations are present in the Transvaal and Griqualand West Supergroups. The "Minette" type is mined around Pretoria, with reserves of about $6x10^9$ tons at 45% Fe.

Manganese occurs in calcareous sediments interbedded in the iron-formation immediately above the Ongeluk lava in the Griqualand West Supergroup (Fig.4.11). Known as the Kalahari Manganese Field, this is among the world's largest reserves of manganese, with over 800 million tons. Up to three layers of conformable manganese ore, with strike lengths of over 50 km, are developed within the basal 100 m of the banded iron-formation. The lowest manganese band is up to 25 m thick. Near-surface supergene-enriched ores and the primary ore are mined.

The primary control of manganese mineralization was the original sedimentary environment. Discrete bands of chemically deposited manganese are interbedded within the banded iron-formation (Fig.4.12). Iron and manganese oxide facies occurred in the nearshore part of the depositional basin, while limestones, iron-formation and manganese carbonates represent the distal basinal facies. Manganese was mostly held in solution by seawater with small addition from hydrothermal sources (Beukes, 1989; Hutchison, 1983). With the onset of more oxidizing conditions, chemical precipitation of manganese took place in deep marine water far away from clastic depositional sites.

Gold

Gold was first mined on a large-scale in South Africa from the Pilgrims Rest-Sabie gold deposits (Figs.4.8;4.11) in eastern Transvaal in 1872. Mining was by the washing of elluvial and alluvial sediments and by working oxidized and sulphide ores. After nearly a century of mining production ceased (Anhaeusser and Button, 1976). Quartz-pyrite reefs, a few centimetres to tens of centimetres thick, cross-cutting veins, sausage-

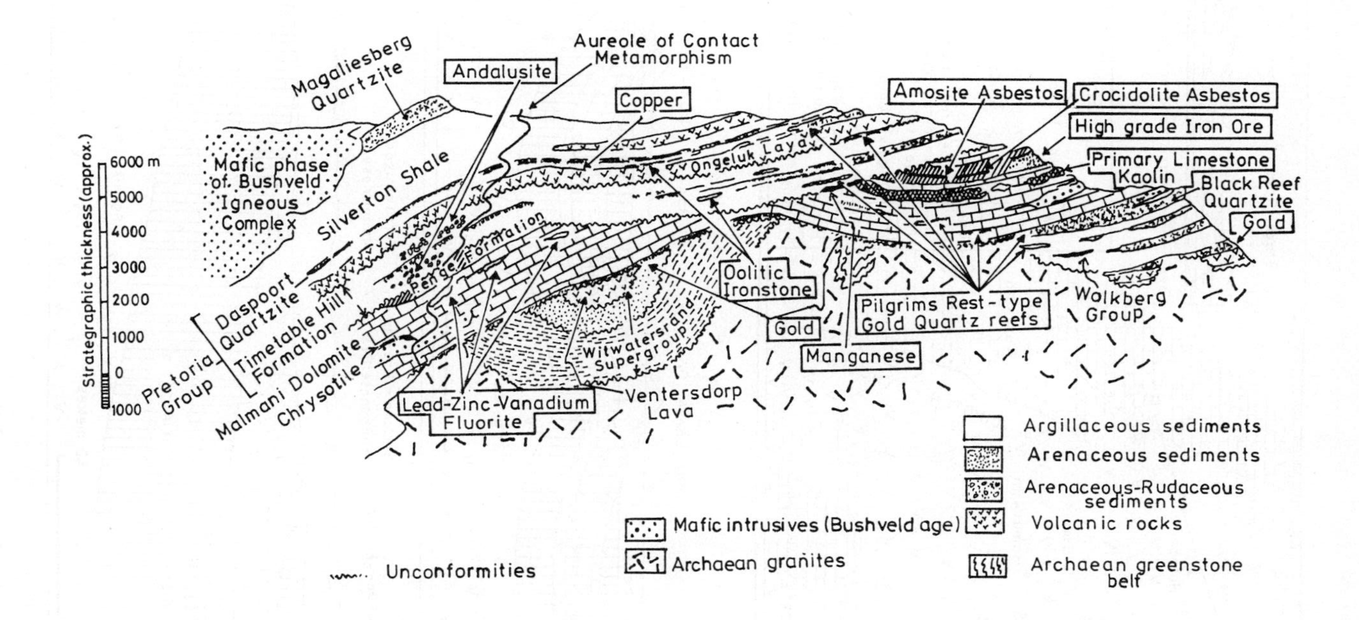

Figure 4.11: Schematic section depicting the stratigraphic setting for stratiform ore deposits in the Transvaal basin. (Redrawn from Anhaeusser and Button, 1976.)

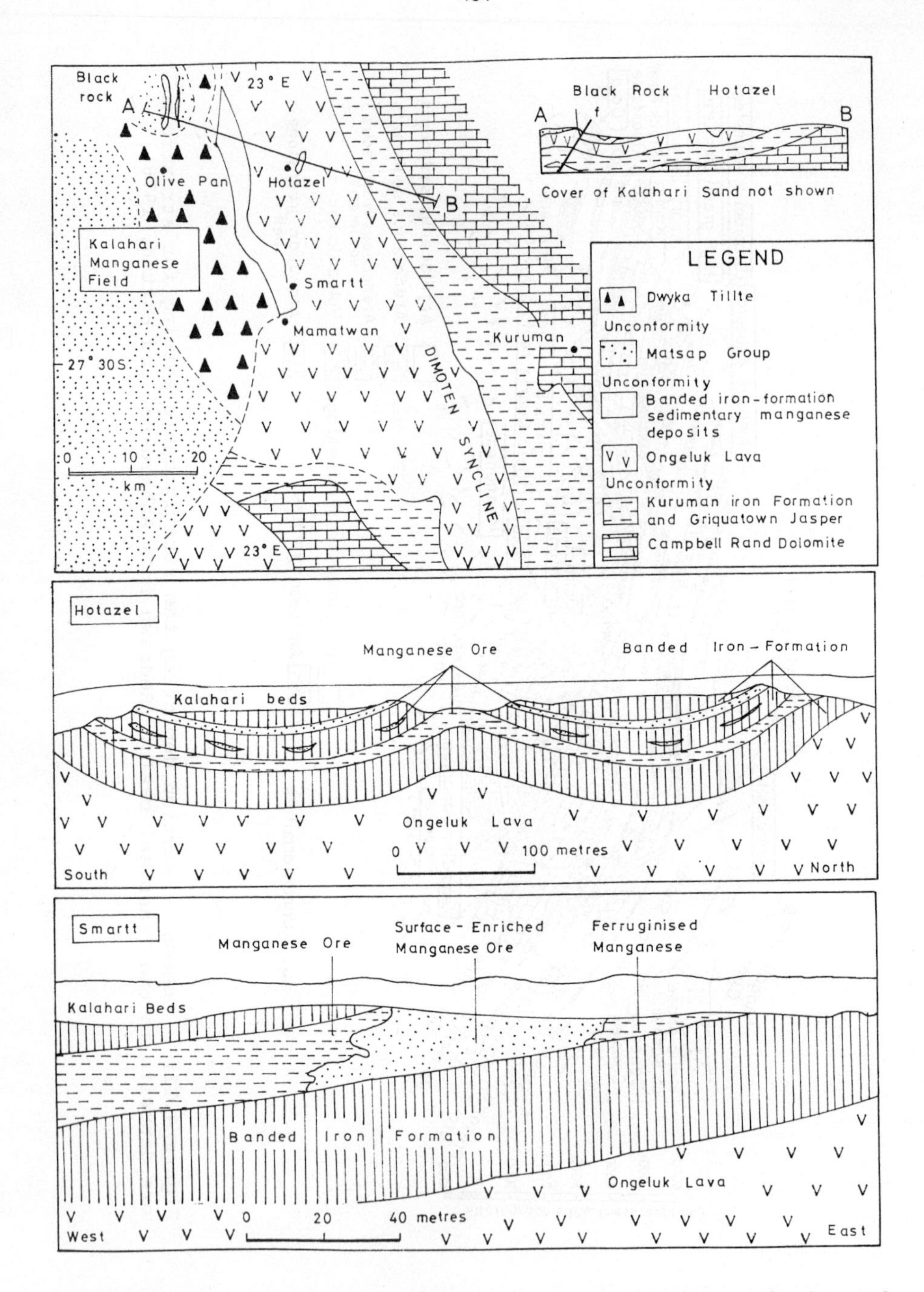

Figure 4.12: Outline geology and cross-sections of the Kalahari manganese ores and banded iron-formation. (Redrawn from Hutchison, 1983.)

like swellings in stratiform quartz veins and gold impregnations were the primary types of mineralization.

Stratified gold veins are hosted by the Malami Dolomite, the upper Wolkberg Group, the Black Reef Quartzite and the basal 1,700 m of the Pretoria Group (Fig.4.11). Mineralizing fluids concentrated the auriferous ores by migrating through conformable passageways created by intrastratal tectonic movements which also produced slickensided surfaces around the reefs, offset dykes, as well as non-penetrative cleavage in the shaly rocks. The reefs, which are banded, dip gently to the west and extend up to 10 km along strike. Mineralogically, the reefs consist of early-phase quartz; carbonates and pyrite with some scheelite, arsenopyrite, pyrrhotite, sphalerite and galena; with gold and chalcopyrite among later mineralization. In addition to gold, byproducts such as silver, copper, arsenic, bismuth and pyrite were mined. The origin of the Pilgrims Rest-Sabie gold deposits has been attributed to hydrothermal fluids sourced by the Bushveld Igneous Complex, some 80 km west of the goldfield. However, there is a strong possibility that the mineralizing fluids could have been generated from the Transvaal Supergroup, especially the shales in the dolomite which contain about 0.1 ppm Au (Anhaeusser and Button, 1976).

Base Metals

Lead, zinc, vanadium and fluorite mineralization in the Malami Dolomite near Zeerust in the western Bushveld (Figs.4.8;4.11) is perhaps the oldest known Mississippi Valley-type mineral province in the world. The production of lead and zinc has ceased, but fluorite, with 100-150 million tons of total reserve and a tenor of 15% calcium fluorite, is one of the largest deposit in the world (Hutchison, 1983).

The Malami Dolomite fluorite deposit has many features in common with the Mississippi Valley-type ore, including the fact that both are telethermal stratiform desposits in carbonate rocks which are capped by an overlying impervious layer which is chert in the case of the Malami Dolomite (Fig.4.11). The Zeerust fluorite deposit is located on a paleohigh where mineralization was controlled by the paleo-porosity of the dolomite such as vuggy horizons and paleo-karsts (Fig.4.13). The ore bodies occur at and near the top of the shallow-dipping Malami Dolomite of the Chuniespoort Group, beneath the unconformity which separates the overlying Pretoria group. The ore bodies which include breccia pipes, veins, disseminations and stratiform replacements, occur over a strike-length of about 60 km. The ore is of several varieties including algal ore, which occurs as irregular nodular masses and less commonly as laminae of fluorite in stromatolites where the effects of telethermal solutions was weak. Black spar occurs as finely crystallized black fluorite

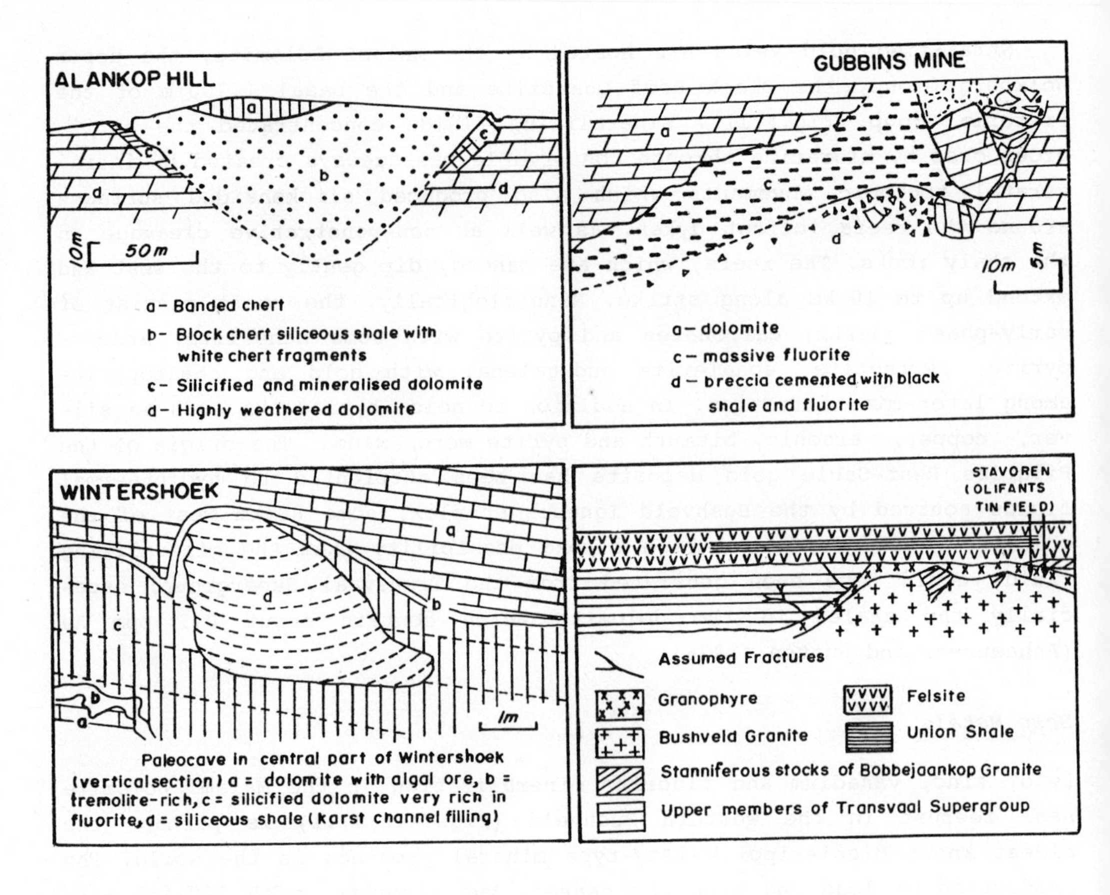

Figure 4.13: Zeerust fluorite mineralization and tin lodes. (Redrawn from Hutchison, 1983.)

replacing dolomite (the black colour is due to carbon inclusions inherited from the dolomite). There is irregularly banded spar possibly deposited by void-filling along the main pathway of fluid movement in cavernous dolomite; and also breccia spar, which forms a cement of breccia bodies in the dolomite, creating pipe-like bodies which are locally associated with lead-zinc mineralization. Fluorite also occurs around paleo-sink holes in association with impure black chert (Hutchison, 1983). Major east-northeast and north-northwest tectonic zones in the Kaapvaal province which controlled the emplacement of the Bushveld Igneous Complex and numerous younger alkaline intrusions such as the Pilanesberg Complex, controlled the introduction of fluorine from the alkaline intrusives into the Malami Dolomite (Hutchison, 1983).

Industrial Minerals

The Transvaal Supergroup is also a rich source of asbestos, aluminous minerals and limestone. Chrysotile asbestos occurs around the Transvaal basin (Fig.4.11) where Bushveld-age sills have intruded into and thermally altered the Malami Dolomite. During the formation of the deposits Mg and Si were supplied by dolomite and chert respectively while the sill was the sources of the water and the heat needed for the reactions. An alteration zone, a metre or two above the upper chilled contacts of the sills, shows dedolomitization into calcite, and serpentine pseudomorphs after chert in which the nodular and algal laminations in the chert as well as ripple marks are inherited. A variety of aluminous minerals including andalusite, staurolite and kyanite occur in the thermal aureole of the Bushveld Complex with the aluminous shales of the Pretoria Group. Andalusite is recovered from the Marico district in western Transvaal (Fig.4.11). Primary limestone is quarried at several locations in the Chuniespoort Group.

4.2.6 Waterberg, Soutpansberg, and Matsap Basins

These are coeval Early- to mid-Proterozoic intracratonic basins which are strung out along the western margin of the Kalahari craton, from the Limpopo province to the Kaapvaal province (Fig.4.14). They are characterized by ubiquitous red beds which were deposited between 2.1 Ga and 1.1 Ga. These are the earliest red beds in the geological record, which imply that oxidizing conditions must have existed in the Earth's atmosphere either during sedimentation or shortly afterwards (Tankard et al., 1982).

Waterberg Basin

This basin is filled with about 5 km of continental beds, the Waterberg Group. The Waterberg Group rests unconformably on the Kaapvaal craton and comprises red conglomeratic and medium-grained to granular arkosic sandstones, planar and trough cross-bedded fluvial sandstones, with intercalations of trachyte and quartz porphyritic lavas (Fig.4.14) in the southern part of the basin.

Soutpansberg Trough

This is a yoked basin or faulted trough located within the previously uplifted and eroded high-grade gneiss terrane of the Limpopo province. It is filled by the Soutpansberg Group which, in the lower part comprises 3,600 m of lavas with subordinate sandstone intercalations (Fig.4.14). In the upper part of the group are 800 m of litharenites, lithic wackes, siltstones, mudstones, pyroclastics, and thin conglomerates. Planar cross-beds and upward-fining depositional cycles suggest fluvial sedimentation.

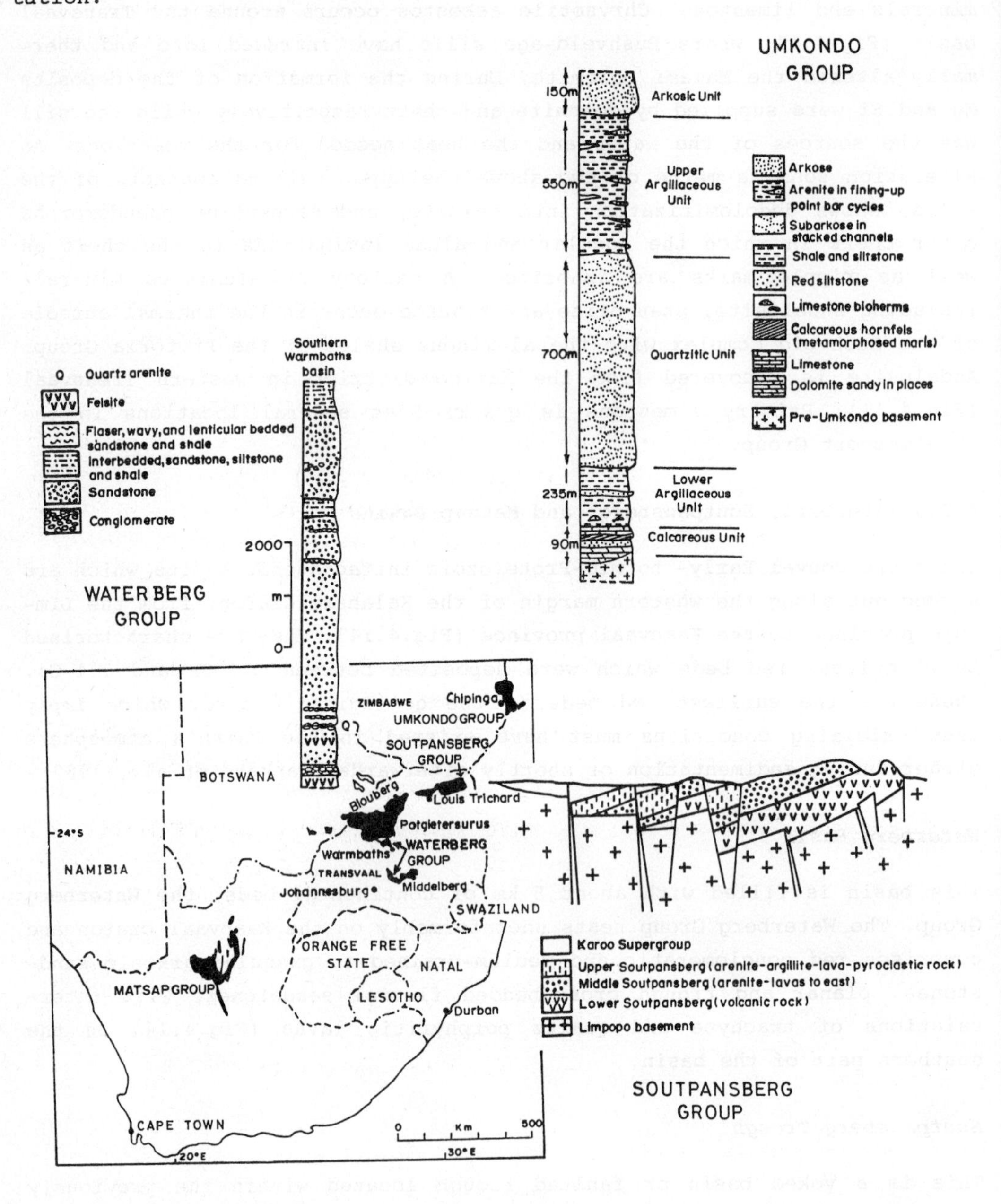

Figure 4.14: Distribution and representative geologic columns for the Waterberg, Soutpansberg, and Matsap basins. (Redrawn from Tankard et al., 1982.)

Matsap Basin

This basin lies on the southwestern margin of the Kaapvaal province where a thick sequence of continental beds and lavas, the Matsap Group, sits unconformably on the Griqualand West Supergroup. The lower unit, the Lucknow Formation, about 1,000 m thick, consists of sandstone, shale, and subordinate conglomerates, dolomitic limestones and volcanics. The overlying Hartley formation has an average thickness of 1,300 m and comprises amygdaloidal lavas with intercalated tuffs, quartzites and breccia quartzites. In the Volop Formation at the top arkosic cross-bedded sandstones predominate with minor chert and jasper; this unit is about 1,500 m thick.

While the older stratigraphic sequences, such as the Transvaal and Griqualand West Supergroups are abruptly terminated on the edge of the Kaapvaal province, the Matsap Group which accumulated on the southwestern margin of the craton at about 2.1-1.8 Ga, shows lithologic continuity with the adjoining Namaqua Mobile belt.

4.2.7 Umkondo Epeiric Basin

Stratigraphy

The basement complex of the Zimbabwe craton is nonconformably overlain in the northeast by relatively unmetamorphosed cratonic sequences of predominantly clastic sediments which show an increase in metamorphism and deformation away from the craton. Although the stratigraphic sequence in the Umkondo Basin is termed the Umkondo Group in the south and the Gairezi Group in the north (Fig.4.2,B), these sequences are almost in outcrop continuity. The Umkondo Group is about 1,250 m thick comprising limestones overlain by sandstones, shales, and altered basic lavas.

The calcareous unit at the base of the Umkondo Group (Fig.4.14) is underlain by reworked basal conglomerates which are succeeded by supratidal dolomites and lacustrine or lagoonal siltstones with greenish gray shales and feldspathic arenites. Supratidal environment for the dolomite was inferred from diagnostic features such as delicate, crinkly algal laminations, tepee-like structures, ghost bird's eye structures, ripples, cross-bedding and intraclast breccia which are indicative of supratidal flat or sabkhha (Tankard et al. 1982). The overlying lower argillaceous unit is of shallow shelf origin. This is succeeded by a thick sequence of deltaic cross-bedded arkosic subgraywackes and wackes. Near the top of the Umkondo Group is the upper argillaceous unit which contains point-bar sequences, overlain by arkosic sandstones and lavas. The facies progression in the Umkondo Group is one in which continental lakes and sabkhas were drowned by a marine transgression coming from the east, after which fan deltas prograded across the basin, and with reduced terrigenous influx meandering fluvial systems deposited alluvial plain clays over abandoned deltaic lobes. In the Chipinga area (Fig.4.14) a higher gradient braided river system prograded eastward across the meandering river belt and deposited the upper arkosic unit.

From west to east the Umkondo and Gairezi sequences show a progressive increase in metamorphic grade, from greenschist to high amphibolite facies. There is a corresponding west to east increase in structural complexity with folding and thrusting along north-south to NE-SW axes. Lithologic units lose their identity within the adjoining Mozambique belt. The age of the Umkondo Group falls between the age of the underlying basement (2.5 Ga) and that of the intruding Umkondo dolerites and lavas which were emplaced about 1.78 Ga, after the deposition of the sediments (Cahen et al., 1984).

Mineralization

The Umkondo lavas which overlie the upper arkosic unit are copper-bearing. They were probably the source of the copper-in-sandstone epigenetic mineralization in several parts of the Umkondo Group (Anhaeusser and Button, 1976). Copper-bearing solutions formed ores in the Umkondo quartzites and shales. At the Umkondo Mine bornite occurs as nodules in shale and as disseminations in quartzite. There are sporadic occurrences of kyanite in pelitic gneisses which correlate with the Umkondo Group. Kyanite, which sometimes occurs with sillimanite, is found here because these rocks have been involved in the high-grade regional metamorphism and deformation of the Zambezi mobile belt. Kyanite is mined, and there are prospects where the host sedimentary succession consits of interdigitating pelitic and semi-pelitic Umkondo gneisses.

4.3 Anorogenic Magmatism on the Kalahari Craton

Apart from the deposition of thick cratonic sequences another consequence of cratonization in the Kaapvaal and Zimbabwe provinces in the Late Archean was the initiation of a mega-fracture system along a NE-SW trend running from the Zimbabwe province to the Kaapvaal province. The Great Dyke and the Bushveld Complex were emplaced along these major fractures or incipient intracontinental rift lineaments during the Early-mid

Proterozoic (McConnell, 1972). Swarms of mafic dykes and extrusives such as the Moshonaland and Umkondo dolerites were emplaced during this phase, as well as the Palabora carbonatite complex, though not along the same structural lineament (Fig.4.2,A).

The Great Dyke was emplaced into the Zimbabwe granite-greenstone terrane at about 2.46 Ga, the Bushveld Complex was intruded into the Kaapvaal basement and overlying cratonic sediments from 2.05 Ga (the age of the layered mafic phase) to 1.4 Ga (the age of the last associated granites), while the Mashonaland and Umkondo dolerite dyke swarms intruded the high-grade Archean gneisses and greenstone belts of Zimbabwe at 1.9 Ga and 1.78 Ga (Cahen et al., 1984). The Great Dyke and the Bushveld Complex contain classic examples of magmatic mineralization while the mafic bodies, as already noted, are associated with stratiform copper deposits.

The Early-mid-Proterozoic phase of ultramafic intrusions was worldwide. Other intrusives are the Stillwater Complex in Montana, U.S.A., the Sudbury Irruptive in Canada, the giant dyke suites in West Australia and the dolerite dyke swarms in West Greenland. Dyke intrusion was associated with regional stress systems that developed during early abortive attempts to break up the newly assembled continental plates (Windley, 1984).

4.3.1 The Great Dyke

Occurrence, Composition, and Origin

The Great Dyke is a spectacular layered ultramafic intrusion, 480 km long; and with an average width of 8 km. It consists of four distinct but continuous elongate, gently inward-dipping layered lopolithic subcomplexes known as the Musengezi, Hartley, Selukwe and Wedza (Fig.4.15). For most of its length, the Great Dyke, its two parallel satellite dykes on either side, known as the Umvimeela and East Dykes, and the gabbro-filled Popoteke fracture zone further east, were intruded into the basement complex greenstone belts and batholitic granite. Dolerites of the Mashonaland suite and granitic veins cut the dyke in places. At its southern end the Dyke splits into a number of smaller bodies, the Southern Satellites which cut the high-grade metamorphic rocks of the Limpopo belt. East-northeast faults have disrupted and displaced the dyke at its northern end.

Each subcomplex has a broadly gentle synclinal structure in which the layering dips inwards towards the dyke centre at angles as steep as 25°

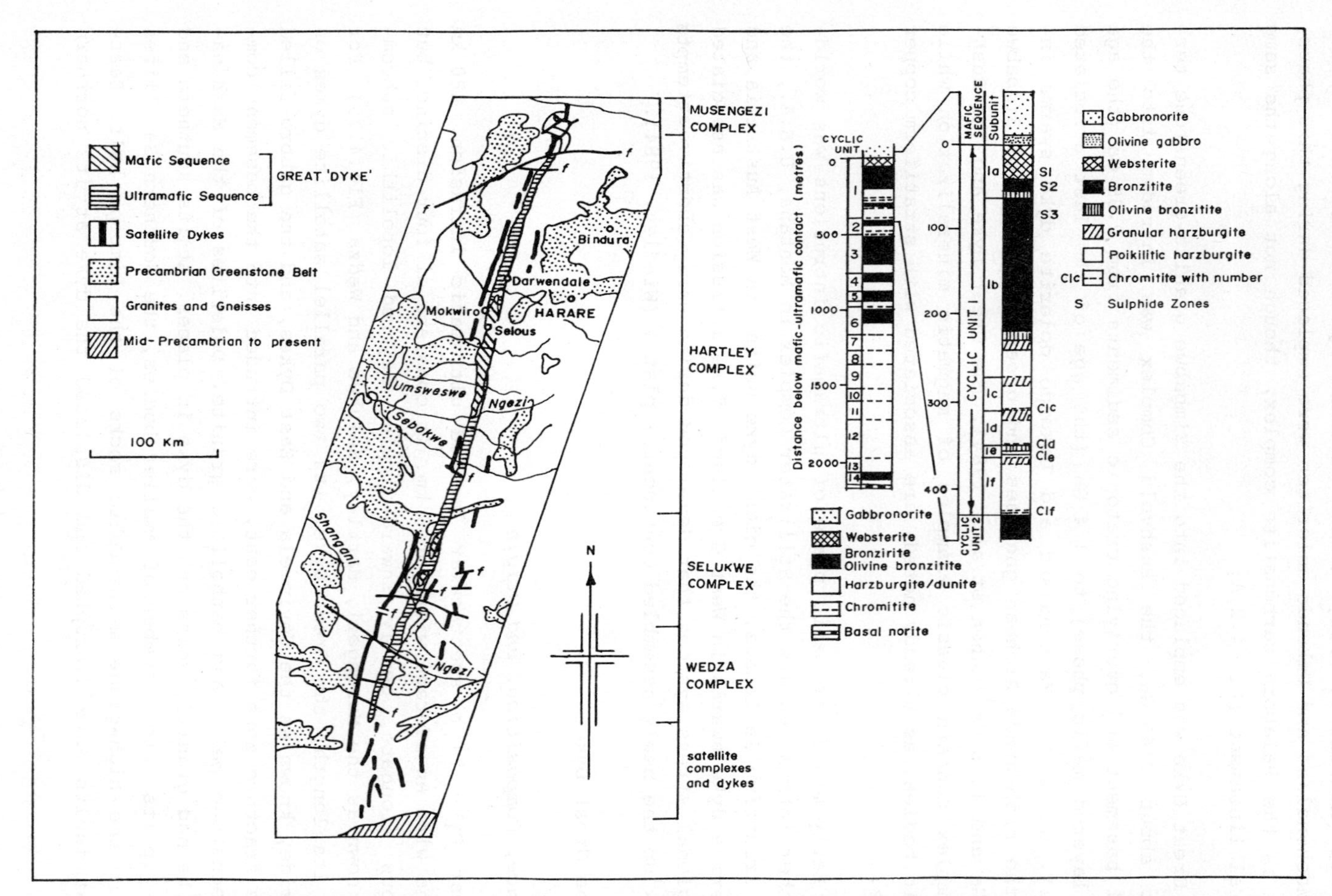

Figure 4.15: The Great Dyke of Zimbabwe and the cyclic units of its Ultramafic Sequence. (Redrawn from Wilson, 1989.)

near the outer contacts and as shallow as 5° in the central areas. Gravity studies across the dyke indicate that at depth the cross-section is V-shaped with the possibility of deep central feeder dykes. The contact between the dyke and the surroundig country rocks is steep, sheared and desilicified but not chilled. Xenoliths of country rock have been incorporated into the gabbroic parts of the dyke.

In his description of the geology and mineralization of the Great Dyke Wilson (1987, 1989) showed that the layered succession of the dyke consists of two parts, a lower Ultramafic Sequence, 2,000 m thick; overlain by the Mafic Sequence of gabbroic rocks, about 1000 m thick (Fig.4.15). The Ultramafic Sequence was further divided into a lower Dunite Succession and an upper Bronzitite Succession, the latter comprising cyclic units. Each cyclic unit comprises a lower chromite overlain by a dunite which in turn passes gradationally through a complex dunite-harzburgite-olivine bronzitite to a bronzitite layer. The olivine bronzitites or harzburgites exhibit small-scale layering on the scale of a few centimeters. This resulted from the rhythmic fluctuation in the proportions of olivine and orthopyroxene. On the weathered surface harzburgite has a very characteristic appearance with the pyroxenes being resistant to weathering and producing sharp protuberances. Websterite marks the top of the ultramafic succession in all four subcomplexes and is immediately overlain by the gabbroic rocks.

Wilson's (1987) study of the mineral chemistry and geochemical modelling suggests that the Great Dyke originated from tholeiitic magma which was highly magnesian, with 15% MgO (a komatiitic basalt liquid). The parent liquid was repeatedly injected into a stratified magma chamber about 1 km high and with a doubly diffusive chamber. In explaining the origin of Cyclic Unit I (Fig.4.15), for example, Wilson (1989) invoked a series of pulses of hot primitive magma which were injected into cooler and more evolved resident magma in the chamber. The pulses were of sufficient volume and turbulence to drive the resulting hybrid mixture into the primary field of chromite, to give the thick chromitites at the base of Subunits Ic and Id. However, Subunit Ib which was initiated by a smaller pulse of magma failed to produce chromite, but caused olivine to replace orthopyroxene as the dominant cumulus phase. Continued fractionation during Subunit Ib caused olivine to once again give way to cumulus orthopyroxene with bronzitite forming the dominant lithology of the cyclic unit. The last phase of magma injection during the emplacement of the Ultramafic Sequence, caused olivine to appear only temporarily as a narrow olivine bronzitite in Subunit Ia. It has been suggested that magma was injected as fountains. The hybrid liquid produced by the entrainment of the more evolved resident magma by the turbulence in the fountains, would have broken down into a series of double diffusive layers that created a compositionally stratified magma column. The double diffusive convection system could have been sustained by successive pulses of new magma. Since Cyclic Unit I maintains similar stratigraphy and lithologies throughout the Great Dyke, the above genetic mechanism generally applies to the lower Ultramafic Sequence in the dyke.

Mineralization

The Great Dyke contains one of the largest deposits of chromium in the world. Chromium reserves in the dyke, computed to a depth of 150 m, holds 190 million tons of metallurgical chromium with 48% Cr_2O_3 and a chromium/iron ratio of 2.8; 350 million tons of chemical/refractory ore at 49% Cr_2O_3 and a chromium/iron ratio of 2.3; and 60 million tons of eluvial ore (Hutchinson 1983). Cyclic Unit I, the uppermost unit of the Ultramafic Sequence is important because it is host to several sulphide layers bearing platinum group elements.

4.3.2 Bushveld Igneous Complex

Occurrence

This is the world's most extensive layered ultrabasic-basic complex, occupying an area of about 65.000 km^2 in the centre of the Transvaal cratonic basin into which it was intruded. It consists of four lobes, a western lobe, southeastern and eastern lobes which are largely covered by Mesozoic rocks, and a northern lobe (Fig.4.16). Domical areas of deformed sedimentary rocks of the Transvaal Supergroup and older formations which represent upfolded portions of the country rock, occur within the cores of the western and eastern lobes (Fig.4.16). The central part of the complex comprises granite, microgranite, felsite, and granophyre. As depicted in Fig.4.16, the Transvaal Supergroup also forms the floor of the complex. The layered sequence of the ultrabasic and basic rocks in each lobe dips, usually at low angles of between 10° and 25°, towards the centre. Gravity studies reveal that each lobe could be regarded as sill-like in form, thinning laterally from a feeder dyke near the centre.

Igneous Stratigraphy

The Bushveld Complex consists of the Rustenburg Layered Suite which is about 7.6 km thick and consists of mafic and ultramafic rocks that have been correlated regionally based on persistent marker horizons such as the main Chromite Seam; the Merensky Reef, and the main Magnetic Seam, all of which are easily mappable. The complex also contains granites, granophyres and felsites, and various satellite intrusions which are located near the Vredefort Dome. The Rustenburg Suite has been subdivided into five zones (Fig.4.16), which from the base upwards are: the Chill Zone, the Basal Zone, the Critical Zone, the Main Zone and the Upper Zone (Tankard et al., 1982; Hutchinson, 1983; Sawkins, 1990).

A zone of norites which are highly contaminated with sedimentary rocks, notably quartzite, intrudes transversely into the Pretoria Group of the Transvaal Supergroup. This is the Chill Zone which may be missing in some places. The lowest unit of the complex is often the Basal Zone in

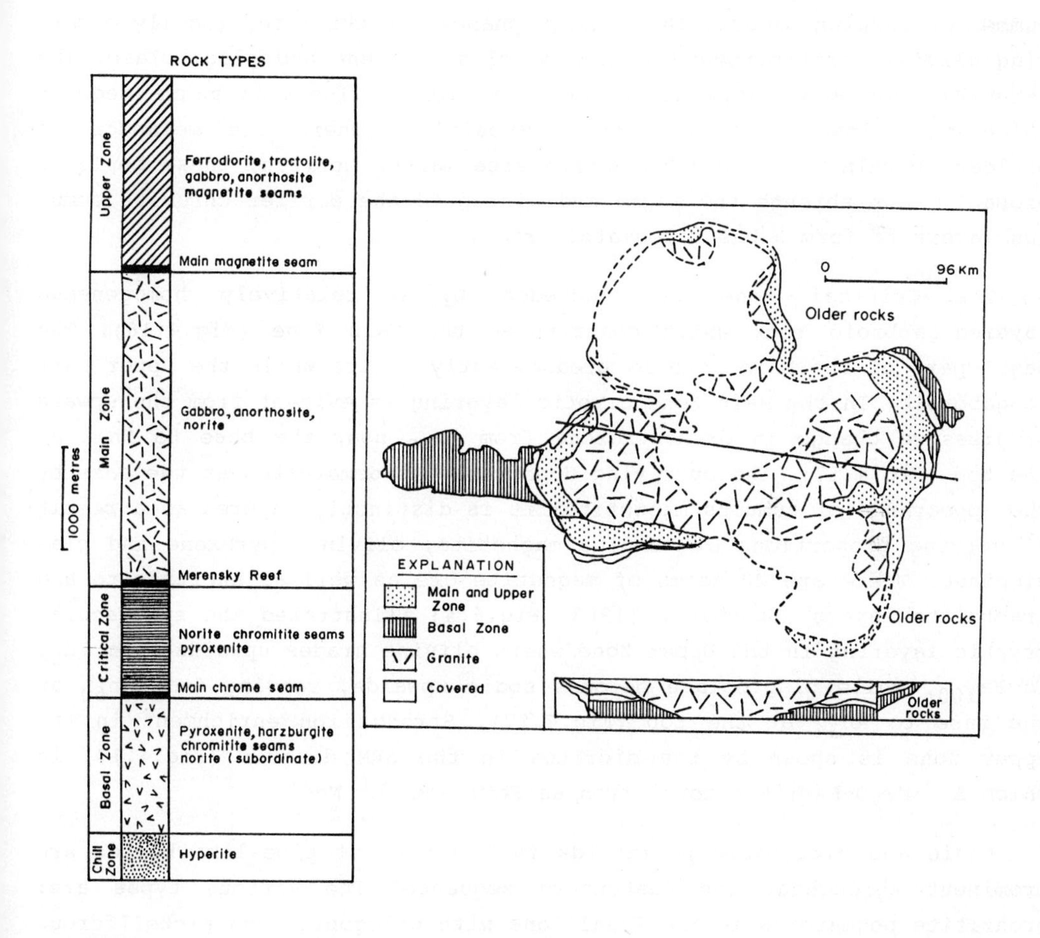

Figure 4.16: Geologic sketch map and schematic sections through the Bushveld Igneous Complex. (Redrawn from Sawkins, 1990.)

which the rocks are pyroxenite and harzburgite containing chromite seams which are richer in aluminium and poorer in iron than in the higher stratiform series. The Basal Zone displays marked layering with thick layers. In the overlying Critical Zone layering is even on a larger scale and ranges from banded norite to variable successions of pyroxenite, norite, anorthosite and chromite. The Critical Zone contains most of the chromitite seams; chromite is a cumulus phase within the anorthosite suite of this zone. The Merensky Reef and the overlying Bastard Reef are the two most complete cyclic units which terminate this zone. After the formation of both units, chromite and olivine disappear as cumulus phases. The Merensky Reef is unique because it is probably the most persistent magmatic-sedimentary layer of all layered complexes in the world, and because it represents a complete mineral grading with the maximum number of cumulus phases. The cumulus phases are chromite, locally occurring olivine, orthopyroxene with some clinopyroxene and plagioclase. The Merensky Reef apparently differentiated from an isolated magma segment which was close to a basaltic composition. There are melanocratic boulders within an underlying anorthosite marker horizon; these boulders probably sank through the magma and disrupted the earlier chromite cumulus layers to form a basal pegmatoid reef.

The Critical Zone is succeeded by a relatively homogeneous layered gabbroic rock which constitutes the Main Zone (Fig.4.16). The basal part of the Main zone is predominantly norite while the upper part is gabbroic. In the Main Zone cryptic layering is evident from the upward progressive change in orthopyroxene from En_{75} near the base to En_{40} at the top (Fig.4.17). The overlying Upper Zone is demarcated at the base by the appearance of cumulus magnetite; it is distinctly layered as a result of varying proportions of cumulus magnetite, olivine, pyroxene and plagioclase. There are 20 seams of magnetite ore as well as anorthosite and tractolite layers. Hutchison (1983, Fig.6.4) illustrated the spectacular cryptic layering in the Upper Zone where olivine grades upwards from Fa_{54} to Fa_{100}, while plagioclase becomes sodic upwards, ranging from An_{60} at the base to An_{34} at the top (Fig.4.17). Strong iron enrichment in the Upper Zone is shown by the diorites in the AFM diagram (Fig.4.17) in which A = Na_2O+K_2O; F = total iron as FeO; and M = MgO.

Basic and ultrabasic pegmatoids in the form of pipe-like bodies are prominent throughout the Rustenburg sequence. The various types are: bronzitite pegmatoids in the Basal Zone with phlogopite and nickeliferous sulphides; dunite pipes in the Basal and Critical Zones; diallagite pegmatoids which form irregular masses, anastomosing veins and pipe-like bodies in Critical, Main and Upper Zones; pegmatoids in the Critical Zone

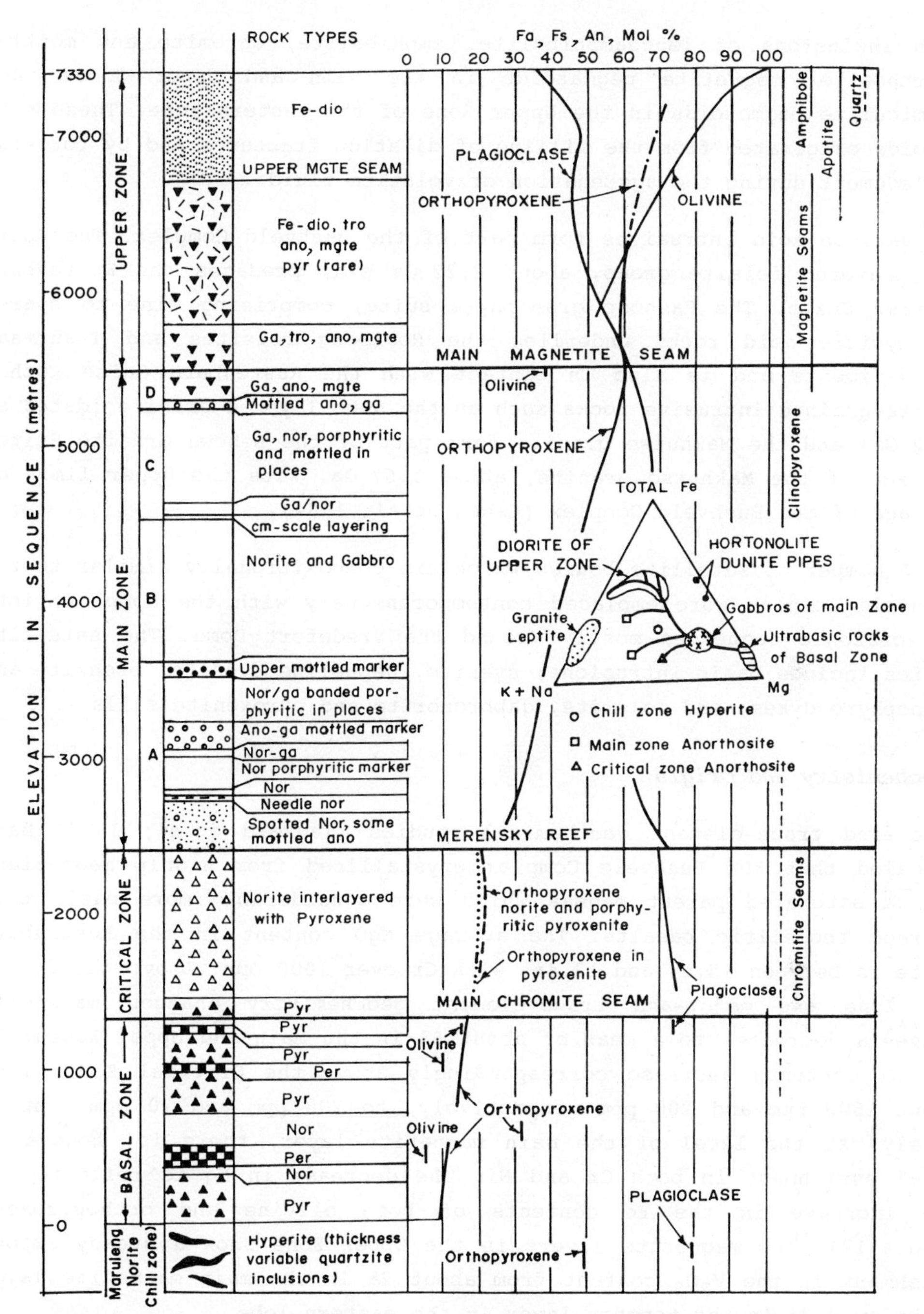

Figure 4.17: Bushveld Complex showing mineral trends in the cryptic layering and differentiation trends (AFM diagram). An, anorthite; ano, anorthosite; dio, diorite; fe-dio, ferrodiorite; fa, fayalite; fs, ferrosilite; ga, gabbro; mgte, magnetite; nor, norite; per, peridotite; pyr, pyroxenite; tro, troctolite. (Redrawn from Hutchison, 1983.)

with inclusions of leucoamphibolite, amphibolite, chromite and mottled anorthosite; magnetite pegmatoids in the Main and Upper Zones; and vermiculite pegmatoids in the Upper Zone of the eastern lobe. These pegmatoids originated from the filling of dilation fractures and by forceful emplacement during the aggregation of volatite fluids.

Various acid intrusives form part of the Bushveld Complex. The Rooiberg layered felsite group, about 2.22 Ga old, predates the Rustenburg Layered Suite. The Rashoop granophyre suite, comprising fine to coarse porphyritic acid rocks underlies the Rooiberg felsites and Transvaal metasediments and is also in contact with the Rustenburg Suite. Other coarse-grained intrusive rocks such as the Nebo layered granite (dated at 1.92 Ga) and the Makhutso granite form part of the Lebowa Granite Suite. The age of the Makhutso granite, about 1.67 Ga, sets the upper limit of the age of the Bushveld Complex (Cahen et al. 1984).

A number of satellite bodies which are compositionally similar to the Bushveld Complex were emplaced contemporaneously with the complex, into the cratonic sequences mostly around the Vredefort Dome. The satellite bodies include basic intrusions; syenite, nepheline syenite, bronzite and granophyre dykes; and dolerite, gabbronorite and pyroxenite sills.

Geochemistry and Origin

Major-and trace-element geochemical studies (Tankard et al; 1982) have revealed that the Bushveld Complex crystallized from highly magnesian, silica saturated parent magmas which were considerably more basic than average tholeiitic basalts. The average MgO content in the Rustenburg Suite is between 13.0% and 13.4%, with Cr over 1000 ppm. Above the Critical Zone the major-and trace-element geochemistry changes markedly. Magnesia decreases to a mean of about 6% in the Main and Upper Zones; Cr and Ni contents decrease correspondingly above the Critical Zone, from about 1500 ppm and 200 ppm respectively, to 200 ppm and 70 ppm respectively. At the level of the main magnetite layer, there is, however a local enrichment in both Cr and Ni. The decrease in Mg is reflected in the increase in the Fe contents of both olivine and orthopyroxene (Fig.4.17). The magnetite layers in the Upper Zone show a steady impoverishment in the V_2O_5 content from about 2% in the main magnetite layer to below 0.3% in the topmost layer in the eastern lobe.

A mantle-derived magmatic source for the Rustenburg Suite is well established although the mechanism for the repetitive and persistent igneous layering, up to 100 km along strike, has remained uncertain. The two prevailing hypotheses for their origin are that magma differentiation took place at depth, with subsequent intrusion of the separate fractions with or without further differentiation; or that magma differentiation occurred in situ with or without additions of magma to the differentiating body. The latter hypothesis is supported by the trends in major-element geochemistry. However, it is generally accepted that the Bushveld Complex represents the emplacement of olivine tholeiite magma through five main vents (western, eastern, central, northern, southeastern). Shallow melting at a mantle magma source generated the magma which induced the partial melting of the sialic crust, to produce liquids from which the Nebo granite and its variants crystallized. The felsites could have been derived from a crustal source. As already noted the emplacement of the Bushveld Complex caused extensive contact metamorphism and deformation in the surrounding country rocks.

Mineralization

The Bushveld Complex is very rich in sulphide ores with very extensive deposits in the Merensky Reef, the UG 2 Chromite Layer, and the Platreef. Disseminated sulphides are up to 2 volume percent in the Rustenburg Layered Suite and in the Merensky Reef. Pyrrhotite is the most abundant sulphide, followed by pentlandite, chalcopyrite, pyrite, cubanite and mackinawite.

Chromite occurs in commercial quantity in the Critical Zone with chrome-ore reserves of about $2,300 \times 10^6$ tons down to a vertical mining depth of only 300 m, with ten times this reserve estimate below (Hutchison, 1983). The chromite layers extend for distances, up to 65 km; and are believed to be the products of cumulate magmatic sedimentation in which an increase in oxygen fugacity was responsible for the formation of the massive chromite layers.

Vanadiferous magnetite occurs as plug-like pegmatite bodies and as cumulate layers; it is mined at the Kennedy's Vale plug-like mass where there is about 2% vanadium pentoxide. The main magnetite layers are in the Main Zone and in the Upper Zone where the content of vanadium pentoxide varies from 2% in the lowest layer to 0.3% in the uppermost layer. The ore minerals include magnetite and ilmenite, with up to 14% titanium dioxide. The vanadium pentoxide resources of the main magnetite layer, 1.8 m thick, are estimated at 17×10^9 kg in 2×10^9 tons of ore (Hutchison, 1983).

Cassiterite occurs in the late-stage granite intrusions of the Lebowa Granite Suite. The Lebowa granites were derived by anatexis of the basement rocks of the Bushveld Complex, during which fluorine, which is highly enriched in these rocks, acted as the flux. Since the fluxed magmas were highly volatile they became enriched in tin and other trace metals by volatile stripping or scavenging. Fluorine-fluxing of sialic crustal rocks in fluorine-rich zones within an old, deep-seated and persistent tectonic trend in the Kaapvaal province, accounts for the occurrence of late-stage tin-bearing granites in the mega-fracture zone, along which the Bushveld Complex and the Great Dyke were emplaced.

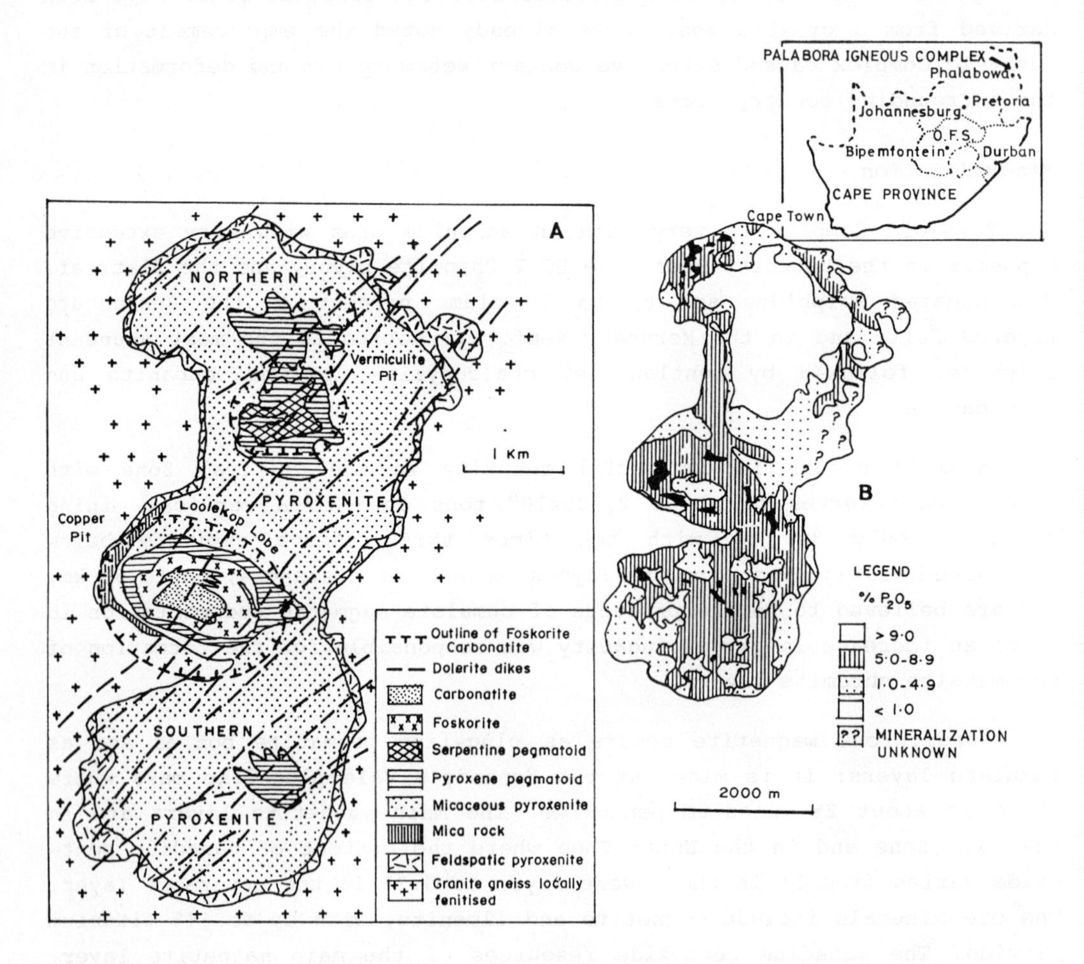

Figure 4.18: A, outline geologic map of the Palabora Igneous Complex where disseminated copper ores are mined by open-pit techniques from the Loolekop carbonatite. B, distribution of phosphate in the complex. (Redrawn from De Jager, 1988; Sawkins, 1990.)

4.3.3 Palabora Igneous Complex

The Palabora Igneous Complex, comprising alkaline rocks and carbonatites, is one of the oldest ring complexes and perhaps the most atypical complex of this type in Africa (Vail, 1989b), with an age of about 2.05 Ga. Located in the Archean gneiss terrane of the eastern Transvaal (Fig.4.18, inset), the Palabora (Phalaborwa) complex consists of three centres of intrusion (Northern Pyroxenite, Loolekop Lobe, Southern Pyroxenite) around which various rock types are arranged in a zonal pattern with a north-south alignment (Fig.4.18,A). The quartzo-feldspathic gneissic basement is intruded by micaceous pyroxenites and syenites with an intervening zone of fenetization, the youngest member, a søvite carbonatite, lying in the centre in the form of arcuate concentric bands.

The presence of copper sulphides, vermiculite, apatite, and magnetite renders Palabora one of most highly mineralized complexes of this type in the world (Vail, 1989b). Sawkins (1990) attributed the 300 million tons of copper reserve (at 0.69% Cu) to disseminated copper mineralization associated with the final phase of the irregular dike-like carbonate intrusion, and to the presence of numerous copper-bearing veinlets within the surrounding carbonatites and pyroxentites. The major sulphide minerals in the orebody include chalcopyrite and bornite, with small amounts of cubanite, pyrrhotite and various nickel, cobalt, copper, lead and zinc sulphides. Figure 4.18,B shows the distribution of phosphate (fluorapatite) with reserves of about 13×10^9 tons (de Jager 1988).

4.4 Vredefort Dome

Located on the southern part of the regional lineament on which the Great Dyke and the Bushveld Igneous Complex are situated, is the Vredefort dome (Fig.4.19,A). The Vredefort dome is an almost circular structure with a collar of steeply dipping and overturned strata of the Witwatersrand, Ventersdorp and Transvaal Supergroups (Fig.4.19,B) surrounding a core of Archean deep-level granulites and granite gneiss, thus revealing a cross-section through the crust of the Kaapvaal province. The radius of the basement core is 18 km and the widths of the collar and rim synclinorium are 17 km and 20-30 km, respectively (Tankard et al. 1982). The core of the Vredefort Dome comprises a ring of Archean granite gneiss surrounding a central part of granulite facies rocks. The Vredefort Dome experienced four metamorphic events.

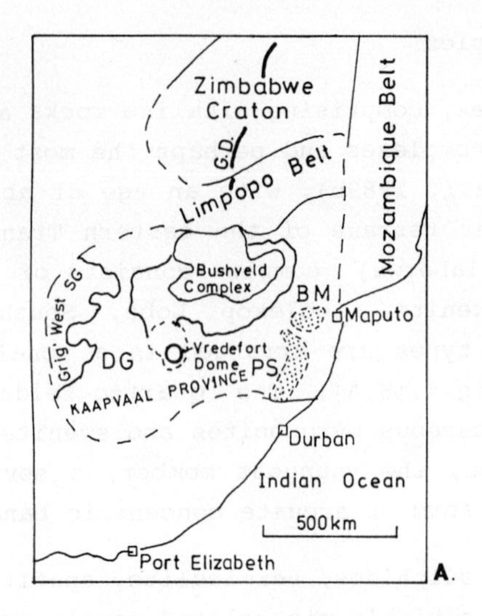

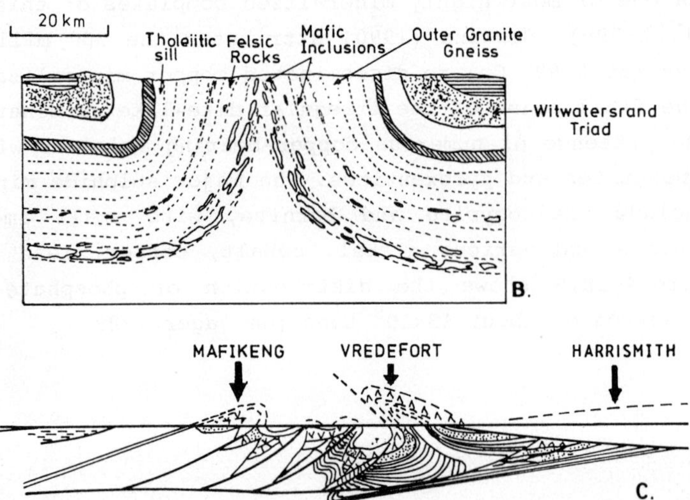

Figure 4.19: Regional tectonic setting for the Vredefort dome (A) and schematic cross profiles (B,C). (Redrawn from de la Winter, 1989; Marsh et al., 1989; Nisbet, 1987.)

The latest event that has been identified is shock metamorphism at about 2.0 Ga, represented by shatter cones and pseudotachylite veinlets which were caused by fracture-filling with mobile silicate material produced at high temperature and strain during the formation of the dome. This was preceded by thermal metamorphism of the Witwatersrand Supergroup which produced albite-epidote-hornfels and hornblende - hornfels facies rocks. An earlier metamorphic event had resulted in low-grade metamorphism of the Witwatersrand rocks, and this in turn was preceded by the earliest metamorphic event which had produced granulite-facies rocks. Although the presence of coesite and stishovite in the Vredefort rocks suggests that the dome probably originated from an extra-terrestrial impact event, another view is that metamorphism and uplift of the dome probably took place during an explosive intrusion at about 2.0 Ga (Nisbet, 1987), at the time of the Bushveld Complex was emplaced. Both models for the origin of the Vredefort dome are depicted in Fig.4.19, where the explosive intrusion model is shown in Fig.4.19, B whereas Fig.4.19, C implies the extra-terrestrial impact origin of this enigmatic feature.

4.5 Namaqua Mobile Belt

Wrapped round the Kaapvaal-Zimbabwe Archean cratonic nucleus is a curvilinear network of Proterozoic mobile belts (Fig.4.20, inset), of which

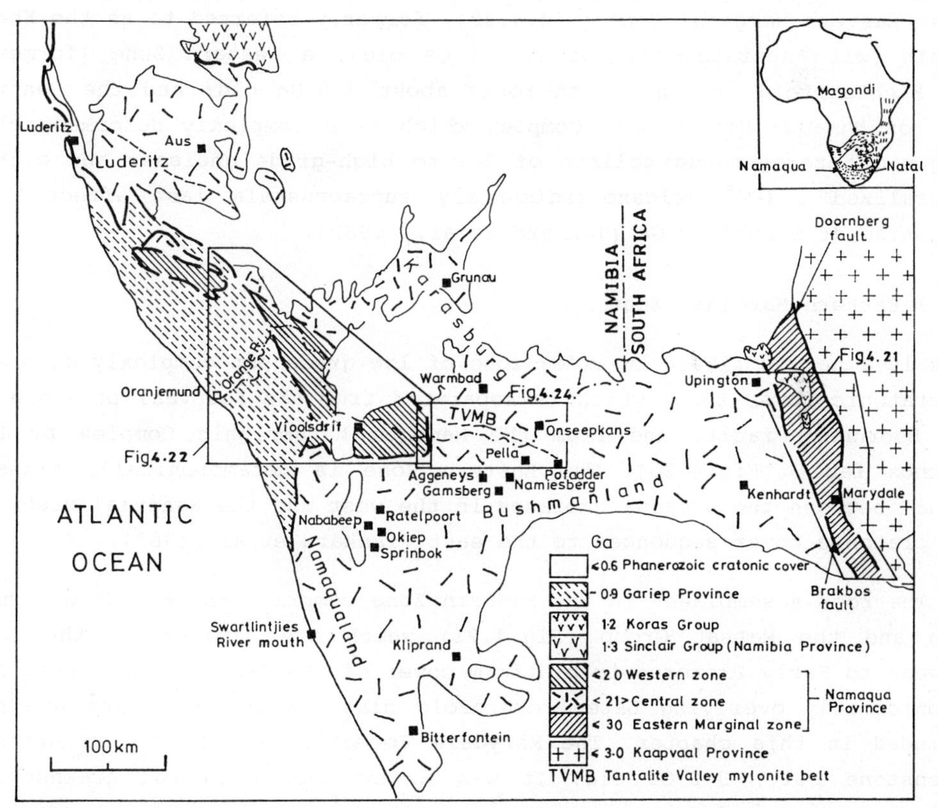

Figure 4.20: Tectonic zones of the Namaqua province and geographic subdivisions of the Central Zone. (Redrawn from Tankard et al., 1982.)

the Early-mid Proterozoic Namaqua and Natal mobile belts are located along the southern margins of the Kaapvaal province, whilst the Magondi belt is situated on the northwestern part of the Zimbabwe province. At the end of prolonged and intermittent tectonic activities the Namaqua and Magondi belts stabilized and became parts of the Kalahari craton (Clifford, 1966).

The Namaqua belt is a highly complex high-grade metamorphic polyorogenic mobile belt comprising several terranes of varying ages (Fig.4.20), ranging from about 2.0 Ga to 1.0 Ga. Prolonged tectonism involving high-grade metamorphism, crustal reworking and shearing renders the Namaqua belt tectonically comparable with the Limpopo belt. The Namaqua belt forms most of the crystalline basement complex of southwest Africa stretching from southern Namibia to the southwestern part of South Africa. The Namaqua basement is concealed beneath the Nama and Karoo sediments and volcanics to the north and south and by deformed metasediments of the Late Proterozoic Gariep Group in the west. The Namaqua province consists of an Eastern Marginal Zone (Fig.4.20), formerly referred to as the Kheis domain (with supracrustals about 3.0 Ga old); a Western Zone (formerly the Richtersveld domain), with rocks about 2.0 Ga old; and the Central Zone or Namaqua Metamorphic Complex which is a complexly deformed polyorogenic heterogeneous collage of low-to high-grade gneisses and highly mineralized 1.3-Ga volcano-sedimentary supracrustals last affected by tectonism at about 1.0 Ga (Tankard et al., 1982).

4.5.1 Eastern Marginal Zone

This is a narrow (15-30 km wide) zone of low-grade but complexly deformed precratonic cover rocks which is separated from the Kaapvaal province by the Doornberg fault, and from the Namaqua Metamorphic Complex by the Brakbos fault (Fig.4.20). The Eastern Zone is metamorphically transitional between the Namaqua gneisses in the west and the Kaapvaal basement and cratonic cover sequences to the east (Tankard et al. 1982).

The rock assemblages in the Eastern Zone comprise the Marydale Formation and the Matsap Group (Fig.4.21) which are similar to the Late Archean to Early Proterozoic platform cover of the Kaapvaal province. The unconformably overlying Late Proterozoic clastics and volcanics are not included in this chapter. The Marydale Formation represents an Archean greenstone belt, 3,0 Ga old. It was metamorphosed to the greenschist facies at about 1.9 Ga. Late Archean granitoids of the Kaapvaal province intruded the Marydale Formation at about 2.9-2.5 Ga. The Matsap Group is the only Kaapvaal cratonic sequences that has been directly traced into the Eastern Zone where it is represented by metasediments such as quartz-sericite schists, hematite quartzites, metaconglomerates and schistose basic lavas.

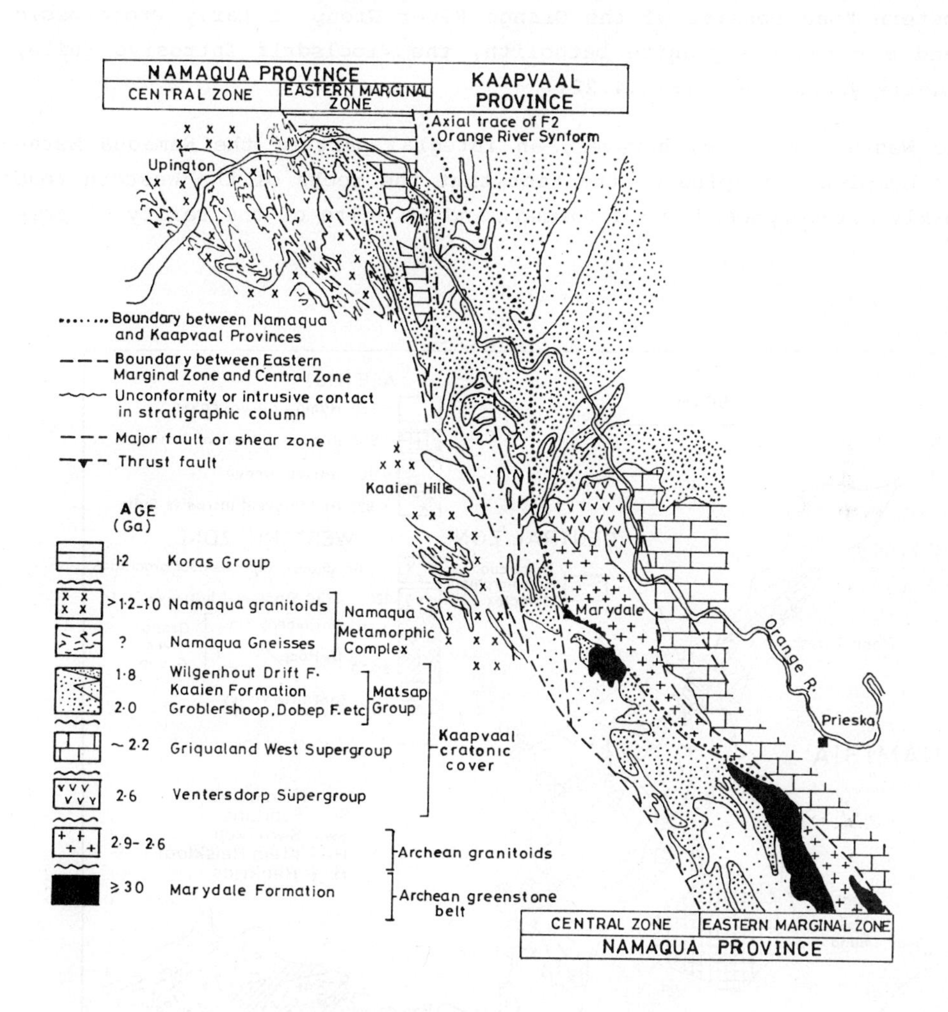

Figure 4.21: Eastern Marginal Zone and adjacent areas of the Central Zone and the Kaapvaal province. (Redrawn from Tankard et al., 1982.)

Structurally, the most characteristic features of the Eastern Zone include Archean isoclinal folds of variable sizes; Early-Late Proterozoic refolding which resulted in complex fold interference patterns; and mid-Proterozoic NNW faults and shear belts. A late metamorphic event at about 1.35 Ga produced greenschists to granulite facies rocks.

4.5.2 Western Zone

The vast Central Zone or Namaqua Metamorphic Complex is occupied in its west-central part by a small wedge-shaped belt of low-grade supracrustal rocks and high-level intrusions, known as the Western Zone. The rocks of the Western Zone consist of the Orange River Group of Early Proterozoic age, and a composite granite batholith, the Vioolsdrif Intrusive Suite, of slightly younger age (Fig.4.22).

The Western Zone is, however, an integral part of the Namaqua Metamorphic Complex, in spite of the fact that the rocks of the Western Zone are weakly metamorphosed and deformed, and in spite of the rarity of pre-

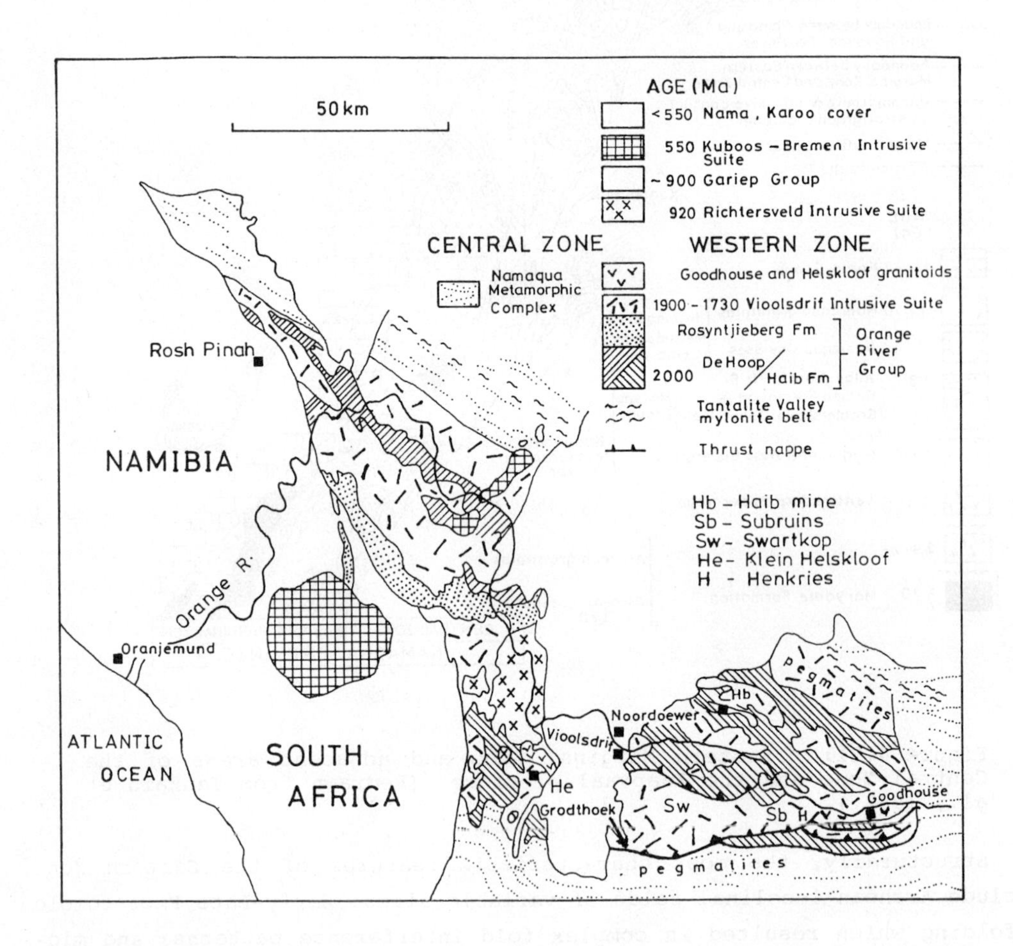

Figure 4.22: Western Zone of the Namaqua province. (Redrawn from Tankard et al., 1982.)

served basement rocks. The Orange River Group comprises the De Hoop Subgroup of intermediate and acid volcanics; the coeval predominantly acid and basic volcanic Haib Subgroup which is dated at 2.0 Ga; and the conformably overlying Rosyntjieberg Formation which comprises metaquartzites with ripple marks and cross-bedding and intercalations of magnetite iron-formations, chlorite schist and metapelite. Largely cataclastic regional foliation accompanied by the low-to medium-grade metamorphism of the Orange River Group and the Vioolsdrif Intrusive Suite, trends east-west along the nose of the Western Zone, but swings from a NW-SE to N-S direction in the northwestern area.

The Vioolsdrif Intrusive Suite was emplaced between 2.0 Ga and 1.87 Ga, structurally below the Orange River Group. It comprises a small basal basic—ultrabasic layered suite, extensive tonalites, and granodiorites with minor diorite. The intermediate rocks in the Vioolsdrif suite generally exhibit low to moderate initial $^{87}Sr/^{86}Sr$ ratios, averaging 0.7031, suggesting major additions of mantle-derived calc-alkaline volcanics to the crust during the emplacement of the Vioolsdrif Suite in the Early Proterozoic. Porphyry-type copper and molybdenum sulphides are found in the porphyritic granites of the Vioolsdrif Suite; these are the orebodies in the Haib mine.

4.5.3 Central Zone (Namaqua Metamorphic Complex)

This vast medium-to high-grade metamorphic terrane consists of a heterogeneous basement and an overlapping sequence of supracrustal volcano-sedimentary rocks which witnessed tectono-thermal events between 1.9 Ga and 1.75 Ga in the Early Proterozoic and again at 1.1 Ga in the mid-Proterozoic (Moore et al., 1989). The Central Zone (Fig.4.20) has been subdivided into the Namibian part, the Namaqualand sector and the Bushmanland section (Tankard et al., 1982), which are here described in that order. Since primary mineral assemblages and sedimentary features have been considerably obliterated in the Namaqua Metamorphic Complex by metamorphism and deformation, the following account stresses the nature of the parent rocks which were referred to by their bulk composition, and assigned to tectonic environments on the basis of their geochemistry (Tankard et al., 1982). Because of structural complexities involving large-scale discontinuities and thrusting, the rock types are not presented in their inferred order of superposition, but rather in a structural sequence.

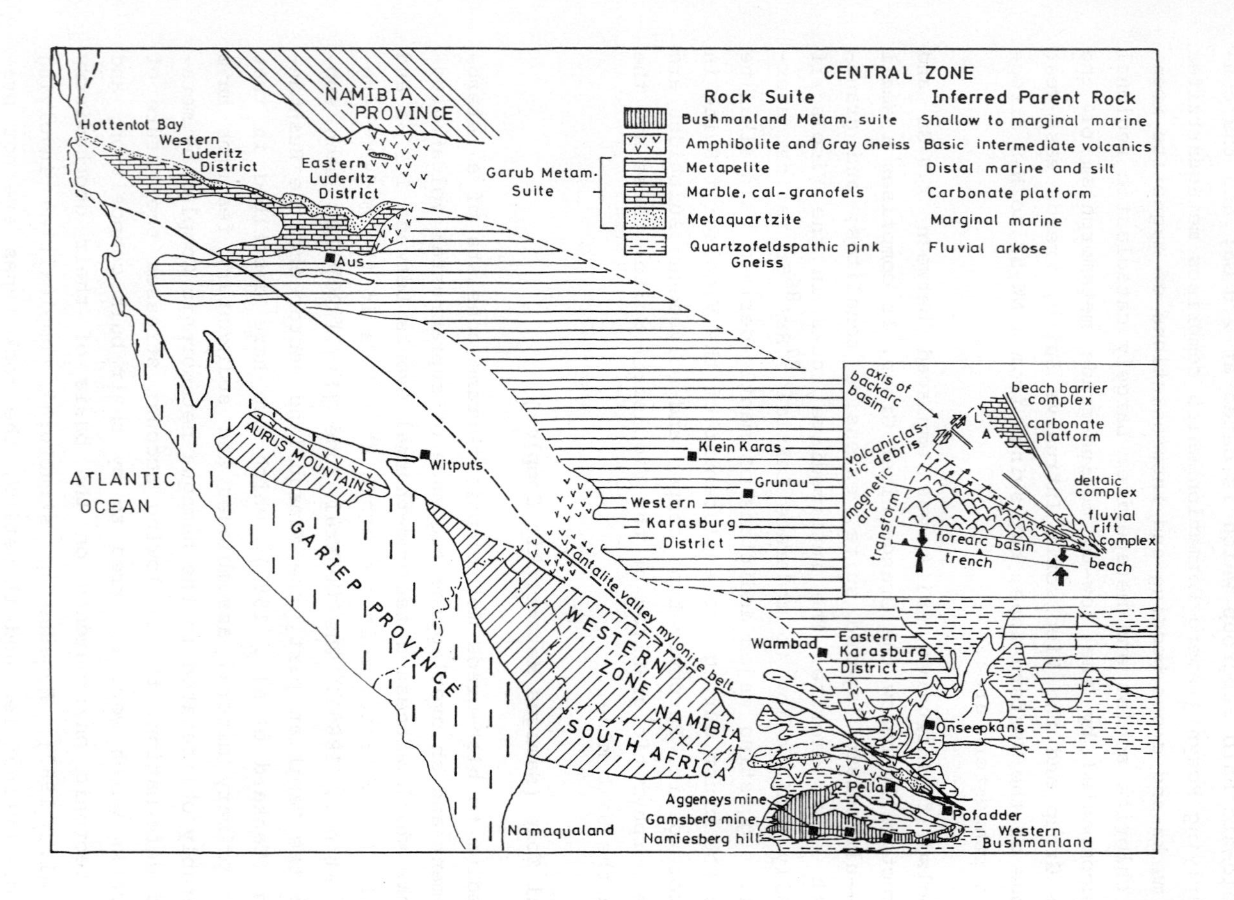

Figure 4.23: Suggested lithofacies map of the pre-tectonic cover sequence in the Namibia Central Zone (Namaqua province). Inset depicts tectonic setting for the Namibian supracrustals in a back-arc basin and intracontinental rift. (Redrawn from Tankard et al., 1982.)

Central Zone in Namibia

The Namaqua Metamorphic Complex in Namibia (Fig.4.23) consists of an igneous and metamorphic basement and a heterogeneous sequence of layered gneisses which represents the pretectonic cover. This cover in the western part (Luderitz district) is referred to as the Garub Metamorphic Suite comprising metapelite gneisses with intercalations of metaquartzite, graphite, and dolomite. The dolomite represents the most extensive carbonate rocks in the Namaqua province and has been interpreted as a carbonate shelf association (Fig.4.23) which broadens westward. A wide tract of gray biotite-hornblende gneisses of volcanic and plutonic origin separates the igneous rocks of the Vioolsdrif Suite in the Western Zone from the broad metapelite region to the north. In the southeastern part of the Namibian sector, there is a broad zone of pink quartzo-feldspathic gneisses with relict cross-bedding and arkosic geochemical signatures which point to an ancient continental rift. According to Tankard et al. (1982) the Early-mid-Proterozoic paleo-tectonic setting of this part of the Namaqua Metamorphic Complex (Fig.4.23, inset) was one in which a continental rift trended northwest through a delta, into a paleo-gulf which was filled with terrigenous muds and silt, with a carbonate platform situated in the northwest. Along the southern margin of the paleo-gulf lay the calc-alkaline magmatic chain of the Vioolsdrif Suite and the volcanic and plutonic rocks now represented by the biotite-hornblende gneisses. A back-arc basin and intracontinental rift model was postulated as in Fig.4.23 (inset) by Tankard et al. (1982) who also argued that the carbonate-quartz shelf association of the Garub Metamorphic Suite (dated 2.2-2.1 Ga) are probably equivalent to the carbonates of the Griqualand West Supergroup. This implies that the Garub dolomites accumulated on a western shelf at the same time with those of the epeiric Griqualand West Supergroup in the Kaapvaal cratonic basin (Figs.4.23, inset; 4.21).

Namaqualand

In Namaqualand (Fig.4.20) basement has not been identified. The pre-tectonic rock assemblages belong to the Namaqualand Metamorphic Suite (Tankard et al., 1982), also known as the Okiep or Nababeep sequence. At the base of the Namaqualand Suite there is a wedge of pink-weathering quartzo-feldspathic biotite gneiss of probable continental arkosic origin. This is overlain by metasediments such as metapelites, gneisses, metaquartzites, minor metaconglomerates, calc-silicate rocks and marble. Part of the Namaqualand Metamorphic Suite shows geochemical characteristics of tholeiitic lavas, suggesting the latter as the protoliths. The Namaqualand Suite is famous for copper mineralization near Springbok, where there are garnetiferous quartzo-feldspathic gneisses and metaquartzites as well as syntectonic granitoids and granitoid gneisses.

Bushmanland

In the Central Zone of Bushmanland (Fig.4.20) a poorly exposed basement comprising layered biotite gneisses with folded schistocity is overlain by a thin but extensive metamorphic cover sequence. The supracrustal rocks in this terrane merge with those of the Namibian and Namaqualand regions nearby, and comprise a variety of gneisses, schists and metaquartzite which are metamorphosed mainly at the lower amphibolite facies.

In eastern Bushmanland pink gneisses of pelitic origin are overlain by metamorphosed basic and calcareous rocks or by a metamorphosed clastic sequence in which trough cross-bedding and pebble sheets are well preserved. Based on its major element geochemistry the pink gneisses are believed to be derived from shallow-water pelitic sediments that were deposited possibly in a fore-arc or back-arc basin between a volcanic arc and its trench or on the continental side of an arc (Tankard et al., 1982). Quartz arenites overlie these pelitic rocks and probably originated from sediment reworking. In some parts of Bushmanland this unit was followed by the accumulation of low-energy predominantly siliceous clastic sediments and by Kuroko-type mineralization of Fe, Mn, Zn, Pb, Cu, and S, which are capped by predominantly calcareous metasediments. At the top of the Bushmanland structural sequence are layered gneisses of granitoid or basic composition which may represent intermediate to basic metalavas. Since the Bushmanland supracrustals seem to be younger (1.7 Ga), correlations with the Eastern province and the Matsap Group are unlikely, because the latter is about 1.8 Ga old (Tankard et al., 1982).

Igneous Intrusions in the Central Zone

The Namaqua Metamorphic Complex was intruded by early syntectonic augen granitoid gneisses, charnockites, gabbroids and mid-syntectonic granitoids in the Namibian sector (Fig.4.24), while in Namaqualand early syntectonic basic intrusions are represented by metabasites with tholeiitic metalavas, highly deformed basic dykes and sheets, and layered intrusions. Syntectonic intrusive augen granitoid gneisses and basic intrusives occur in Bushmanland.

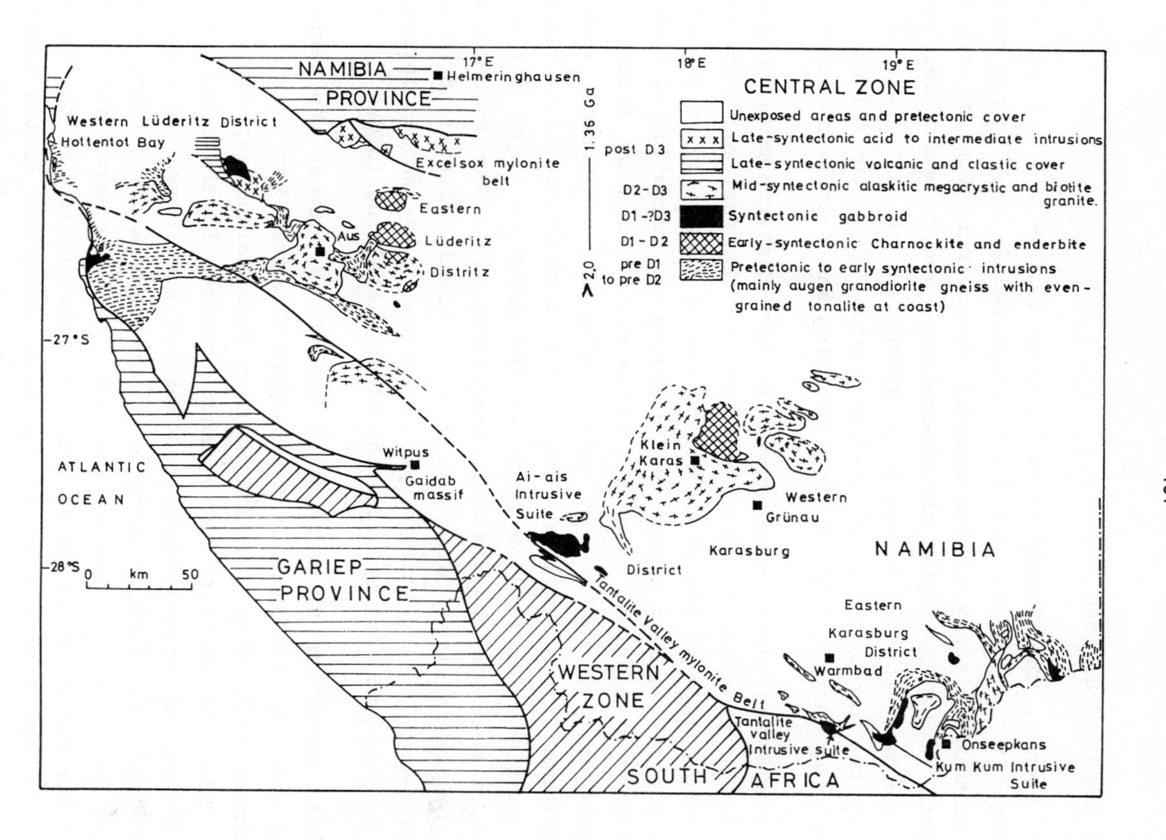

Figure 4.24: Central Zone (Namaqua province) syntectonic intrusions and cover sequence. (Redrawn from Tankard et al., 1982.)

Tectonics of the Central Zone

Two orogenic events, the Orange River orogeny (2.0 - 1.7 Ga) and the Namaqua orogeny (1.2 - 1.0 Ga) have been recognized in the Central Zone. Thus the structural history of the Namaqua Metamorphic Complex can be simplified into early low-angled thrust-dominated events which were accompanied by the most intense degree of metamorphism, up to the granulite facies, and by the intrusion of sheet-like bodies. This was followed by an event involving upright folding, late buckling and ductile shearing.

Major thrusts and shear zones occur in the Eastern Marginal Zone which separate it from the Kaapvaal province to the east and from the Bushmanland subprovince to the west. In the latter region up to four deformation episodes have been recognized involving the formation of thrust sheets, imbricate structures, ramps and large scale recumbent folds; the thrusting episodes were followed by open folding and subvertical shearing. Early deformation in Namqualand resulted in a regional finite fabric and folding ranging from shallow folds in the north to tight and isoclinal folds in southern and eastern Namaqualand. This was followed by periods of steeply inclined folding with axial trends of east-northeast and north-northwest, during which major periclinal folds formed throughout eastern Namaqualand and re-folded the older flat-lying folds.

Major shear zones define most of the boundaries between the various terranes and zones in the Namaqua Metamorphic Complex. For example the Groothoek thrust belt (Fig.4.22) in western Namaqualand separates the Western Zone from the rocks of the Namaqua province in the south. As a result of the Namaqua tectogenesis, the Western Zone moved in a southwesterly direction (Fig.4.25,B) along a thrust plane which strikes east-west and dips at about 30° towards the north (van der Merwe and Botha, 1989). Thrusting was accompanied by prograde metamorphism in the granitoids of the Western Zone and by retrograde metamorphism in the Namaqua province; with the metamorphic grade generally increasing towards the south.

The Tantalite Valley mylonite belt separating the Central Zone in Namibia from the Western Zone (Fig.4.23) trend roughly northwest for about 510 km and marks a zone of mostly dextral displacement for at least 85 km under greenschist to upper amphibolite metamorphism (Tankard et al., 1982).

Watkeys (1989) observed that although the Namaqua mobile belt is of Proterozoic age, it shows similarities with the Limpopo belt of Archean age. Both high-grade terranes record an initial crust-forming event fol-

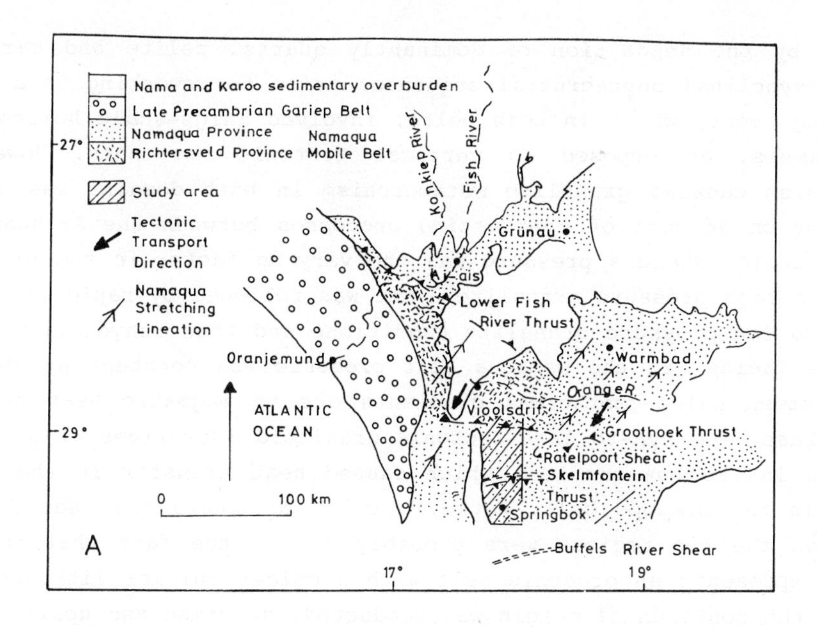

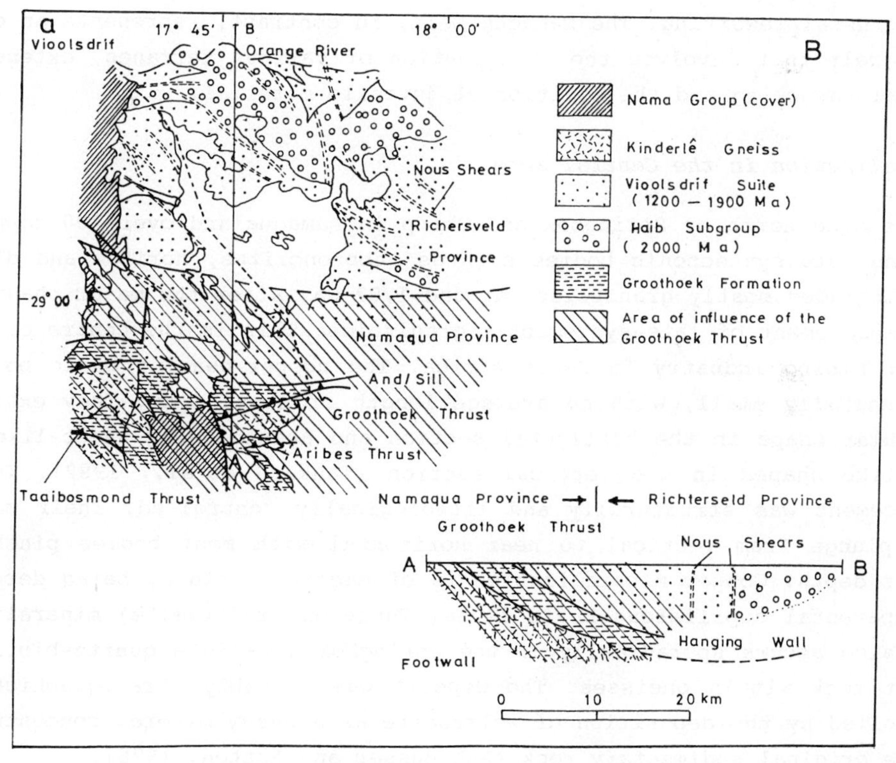

Figure 4.25: Geology and tectonics of western Namaqualand. (Redrawn from van der Merve and Botha, 1989.)

lowed by the deposition of dominantly quartz, pelite and carbonate or miogeosynclinal supracrustal sequences prior to reworking in a crust-deforming event, which in both belts, involved large-scale horizontal displacements, as opposed to vertical tectonic movements. However, the mechanism causing granulite metamorphism in both domains was different, perhaps on account of contrasting processes between the Archean and the Proterozoic. Whereas pressure was the varying factor in the Limpopo belt where a high pressure granulite event was followed by rapid depressurisation to low pressure granulite conditions and then rehydration to amphibolite facies, in the Namaqua belt pressure was constant at about 5 kb, with steep paleo-geothermal gradients due to magmatic heat transfer in the crust. The syn-to late-tectonic granitoid intrusives at about 1.2 - 1.0 Ga in the Namaqua belt which caused heat transfer in the crust are rare in the Limpopo belt. Watkeys (1989) concluded that the differences between the two regions were probably due to the fact that the Limpopo belt represents an orogenic belt with a colder thicker lithospheric root where the continental margin was subducted, deformed and uplifted without much crustal reworking. The Namaqua belt, in contrast, represents an orogenic belt that involved the amalgamation of various terranes, extensive crustal reworking and the addition of juvenile material.

Mineralization in the Central Zone

In the area north of Springbok and Okiep in Namaqualand over 700 copper-bearing late-syntectonic bodies such as gabbronorites, norites and diorites intruded mostly granitoids, around 1.07 Ga, after the major phase of the emplacement of late-syntectonic granitoids. This is the centre of the copper mining industry in South Africa. The copper-bearing basic bodies are generally small, with an average length of about 1 km. They exhibit irregular shape in the horizontal section and anastomosing sheet-like or pipe-like shapes in the vertical section (Tankard et al., 1982). Their emplacement was structurally and lithologically controlled. Their major axes plunge from vertical to near horizontal with most bodies pinching out at depth. Copper mineralization is of magmatic origin, being derived from parental cupriferous basic lavas. Tungsten (wolframite) mineralization also occurs sporadically in the Springbok area in a quartz-biotite-garnet rock within gneisses. The deposit was probably stratigraphically controlled by the deposition of wolframite as a heavy mineral concentrate in the original sedimentary rock (Anhaeusser and Button, 1976).

At Copperton in southeastern Bushmanland, strata-bound Cu-Zn mineralization occurs between quartzo-feldspathic gneisses and an overlying siliceous to pelitic gneiss. The large ore reserves in this deposit are the second largest source of zinc in South Africa, after those of western Bushmanland which are described below.

The Aggenys-Gambsberg-Swartberg area in western Bushmanland (Fig.4.23) has the largest zinc deposit in Africa with over 150 million tones of ore at 7% Zn and 0.5% Pb (Goossens, 1983; Tankard et al., 1982). This deposit is part of a huge base metal and Algoma-type iron mineralization. Base metal mineralization occurs at three distinct stratigraphic horizons in the Aggenys-Gamberg sequence, in intimate association with banded iron-formation and quartzites containing Mn and sulphides. The ore minerals include sphalerite, argentiferous galena and chalcopyrite with barite horizons occurring as lateral equivalents of the sulphide ores. The following are the evidence for the depositional environments and the source of the metals.

Fine banding in the ores suggest precipitation under quiet sedimentary conditions. Lead, strontium and sulfur isotopic studies suggest that the metals were derived from upper crustal sources. The iron oxide facies with barite and sulphides suggest a submarine exhalative origin and precipitation from seawater under reducing conditions in a proximal or distal position from the volcanic vent (Strydom et al., 1987). Erosion and renewed supply of clastics after the chemogenic phase is suggested by the overlying metamorphosed arenite and conglomerate, in a sequence of mainly mafic metavolcanic rocks. The mafic volcanics probably represent extrusive tholeiitic lava flows in a shallow basin which was part of an arc to back-arc tectonic setting.

The genetic model for the base metal mineralization and Algoma-type iron-formation in western Bushmanland postulated by Watson et al. (1989) is very instructive in that it places the mineralization within a lucid regional tectonic framework (Fig.4.23). An Early Proterozoic plate collision event between 1.9. Ga and 1.75 Ga, involving the basement components in the area, was followed by the extrusion of subaerial acid volcanic rocks to the north in the Vioolsdrif area (Fig.4.23) of the Western Zone. Immature and reworked metal-rich subaerial volcanic detritus were deposited in a closed sedimentary basin near the original suture. The accumulating deposits were affected by compressional events during which metal-rich brines were expelled. These compressional events which increased in intensity with time, finally ended at the onset of intensive tectonic activity as evidenced by the reduction of the size of the basin and by the presence of intraformational conglomerates and eruptive tholeiitic volcanic rocks at about 1.65 Ga. In the anoxic portions of the basin metal sulphide deposits accumulated from brines expelled along incipient thrust faults near the southern boundary of the basin. This was followed by low-angle thrust faulting which preserved the supracrustal rocks in thrust slices and associated synformal folds. A long history of tectonic activity, granite intrusion and regional metamorphism then followed which apparently lasted over the ensuing 500 million years (Watson, 1989).

According to von Knorring and Condliffe (1987) some of the best known tantalum pegmatites in Namibia are those of the Tantalite Valley which are located about 30 km south of Warmbad (Fig.4.24). These pegmatites belong to the much larger pegmatite region in the Orange River belt within the Namaqua Metamorphic Complex. Three of the larger lithium pegmatites (Homestead, White City, Lepidolite) in the Tantalite Valley, contain abundant tantalum mineralization which occur in gabbroic rocks. Tantalite-microlite, montebrasite, spodumene, caesian beryl, common beryl, and bismuth minerals have been mined from these pegmatites.

4.6 Natal Province

This Early to Middle Proterozoic orogeny, located along the eastern coast of South Africa covering a distance of about 280 km in the Natal Province (Fig.4.26,A), is a gneissic terrane which probably extends westward underneath the cratonic cover along the southern margin of the Archean Kaapvaal province, to link up with the Namaqua mobile belt (Fig.4.26, inset). Tankard et al. (1982) grouped the orogenic belts of southern Africa into the Early-mid-Proterozoic Namaqua and Natal belts, and the Late Proterozoic to Early Paleozoic Pan-African belts (Damara, Gariep, Saldanian origins). The main differences between both groups of orogenic belts, according to Tankard et. al., are the extreme rarity of recognizable basement beneath the older orogens. The supracrustal successions in the Natal and Namaqua belts have been correlated with the Early Proterozoic sequences in the Kaapvaal province, such as the Griqualand West and Matsap sequences; and these older cratonic cover often terminate against the orogens with a tectonic contact (Figs.4.21; 4.26,A). All the Pan-African orogens, in contrast are underlain by rigid crystalline basements, and contain thick miogeosynclinal and eugeosynclinal sequences.

Only the northern part of the Natal mobile belt is exposed, where allochthonous rocks (Fig.4.26,B,C), some of which resemble typical Archean greenstones, have been extensively thrust onto the Archean cratonic foreland of the Kaapvaal province (Tankard et al., 1982). The Natal belt

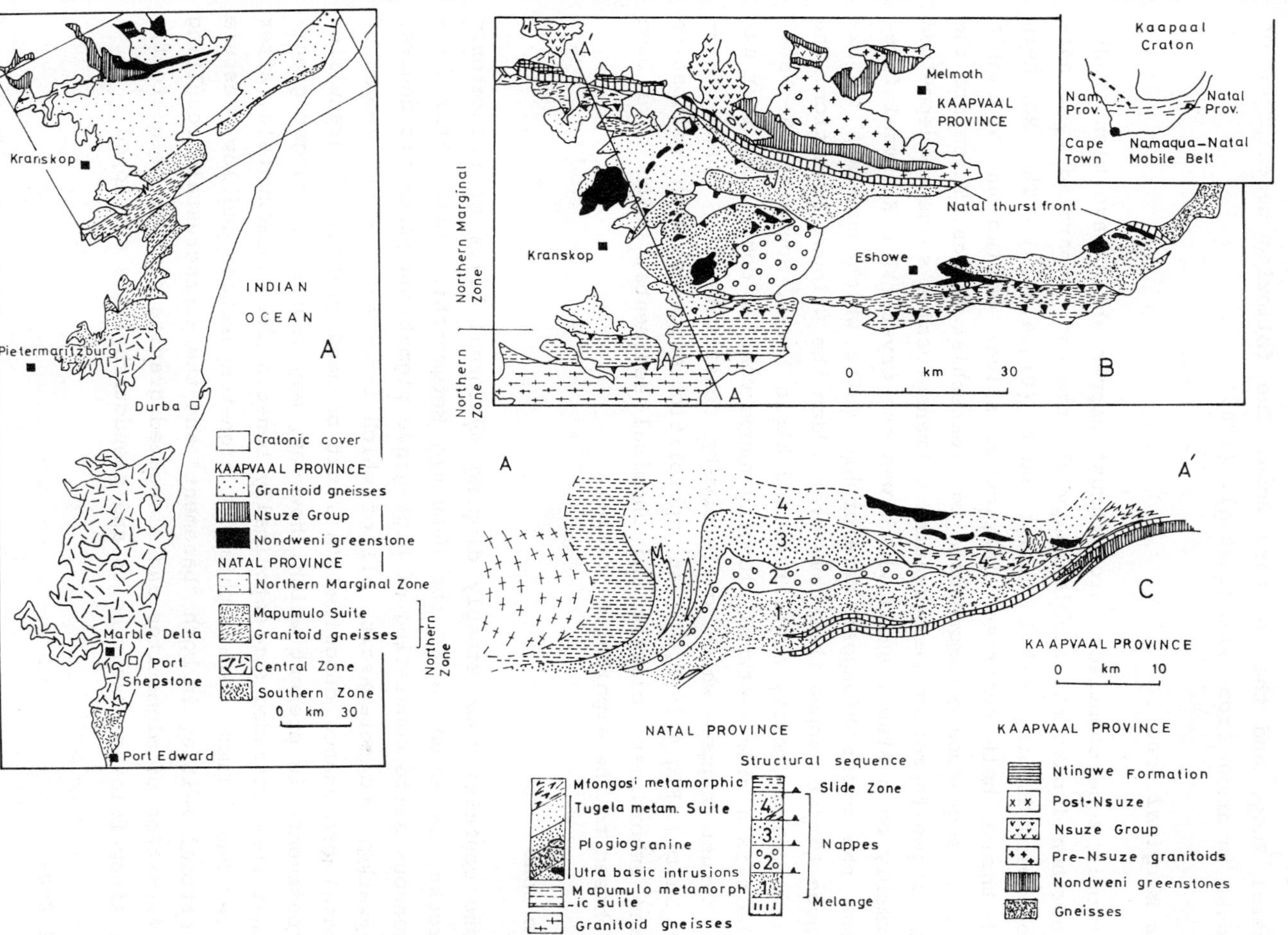

Figure 4.26: Geologic map (A) and cross-sections through the Natal province. (Redrawn from Tankard et al., 1982.)

has been subdivided into the Northern Marginal Zone, the Northern Zone, the Central Zone, and the Southern Zone. The following description of these zones is drawn from Tankard et al. (1982).

Northern Marginal Zone

This tectonic zone comprises large thrust nappes which have been transported northward onto the stable Kaapvaal foreland, separated by a narrow, southward-dipping Natal thrust belt (Fig.4.26,B). The Northern Marginal thrust belt comprises imbricated slices of two unconformable supracrustal sequences metamorphosed to greenschist facies. These are the younger Ntingwe formation representing a transgressive sequence deposited unconformably on a stable shelf which was the crystalline Kaapvaal foreland; and the older Mfongosi Metamorphic Suite which consists of allochthonous thrust nappes that have overridden the Ntingwe Formation. The Mfongosi comprises mostly argillites and basic lavas. South of the Natal thrust belt is a wide westward-plunging nappe complex comprising four extensive thrust nappes which collectively constitute the Tugela Metamorphic Suite (Fig.4.26,B,C), an ophiolitic sequence with metabasite lavas and subordinate clastic and chemical sediments, which have been metamorphosed to the amphibolite facies.

Northern Zone

This zone contains four steeply dipping synformal belts of precratonic cover rocks, referred to as the Mapumulo Metamorphic Suite. This is a heterogeneous suite consisting of high-grade migmatized psammitic gneiss, metagraywacke; and metabasalt, all of which represent clastic wedges of continental provenance deposited in fault-bounded grabens. A granulite-grade "basement" is present which, however, may represent younger granitoids that were intruded syntectonically beneath the supracrustals, after the latter had formed. Metamorphism at high-temperature and low-pressure caused partial melting in both "basement" and the supracrustals resulting in the formation of migmatites and elongated granite plutons of anatectic origin. After folding, shear belts and pseudotachylites developed.

Central Zone

This is a large zone characterized by extensive batholithic bodies of massive or foliated biotite or hornblende granitoids. The batholithic masses are separated by septa of augen gneisses, charnockitic intrusions, and metamorphic suites that formed under high-temperature low-pressure metamorphism which locally attained the granulite facies. These rocks probably represent relicts of a formerly much more extensive granulite terrane which has largely been retrogressed to upper amphibolite facies.

Southern Zone

This narrow zone is characterized by a more extensive development of the charnockite-granolite assemblage than is found in the northern parts. Here the precratonic cover sequence has been intruded by granite, folded isoclinally, metamorphosed to the granulite facies, and intruded by an extensive charnockitic suite.

Tectonic Model

The Northern Marginal Zone of the Natal mobile belt contains the tectonic imprints of a major collision zone. Apart from its northerly verging thrusts and fold nappes which override thrust sheets on the Kaapvaal foreland, its supracrustals, the Mfongosi lavas and argillites and the Tugela basic-ultrabasic extrusives with subordinate sediments, are believed to represent parts of an ophiolite suite that was obducted northward. The latter view is supported by the association of pillowed metalavas, subordinate metachert and carbonate rocks, with layered basic-ultrabasic intrusions of the Tugela Suite, in which case the latter and the Mfongosi lavas may represent metamorphosed oceanic crust. The rocks of the Natal province have not been extensively dated, and the few ages suggest tectonism around 1.0 Ga. With Proterozoic paleomagnetic reconstructions of past continental positions suggesting that the eastern extension of the Natal province lies in the high-grade terrane of Dronning Maud Land in Antarctica, it is conceivable that plate movements and collision probably involved an eastern cratonic nucleus in Antarctica and the Kaapvaal province being driven against each other.

4.7 Magondi Mobile Belt

Stratigraphy and Structure

The Magondi belt is located on the northwestern margin of the Zimbabwe cratonic nucleus, where it extends in the north-northeasterly direction over about 250 km, with a width of about 50 km (Fig.4.27,A). The Magondi mobile belt contains a thick supracrustal sequence that underwent polyphase deformation and metamorphism from greenschist to amphibolite facies during the Magondi orogeny between 2.15 Ga and 2.0 Ga; and was later in-

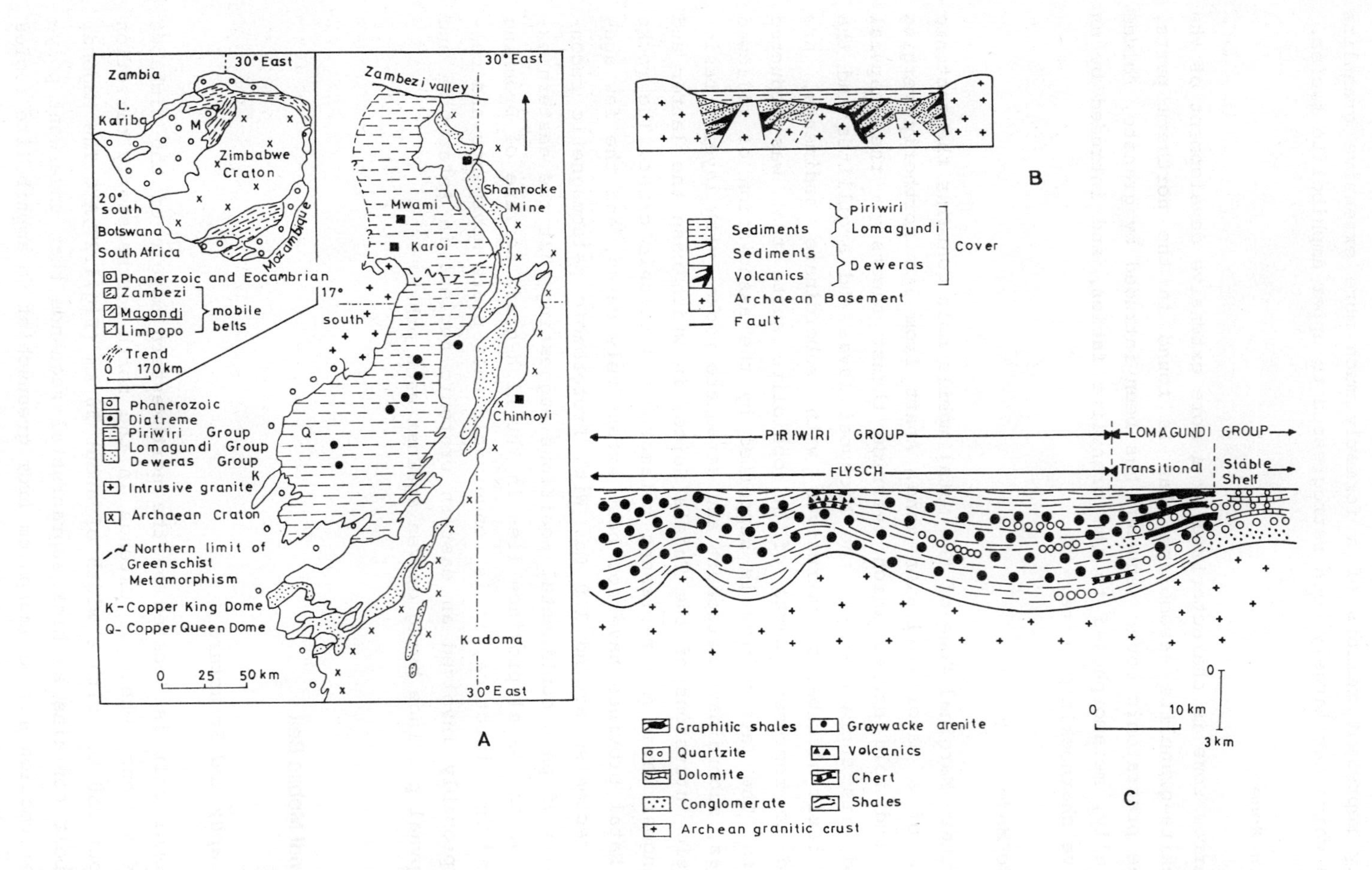

Figure 4.27: Geologic map (A) and stratigraphic sections (B, C) across the Magondi mobile belt. (Redrawn from Leyshon et al., 1988.)

truded with veins during a late stage of the Pan-African Zambezi orogeny along its northern margin. The metasediments and metavolcanics which constitute the Magondi Supergroup are divisible into the Deweras, Lomagundi and Piriwiri Groups (Fig.4.27,A).

The Deweras Group comprises a lower sequence of continental red beds of the alluvial fan facies with conglomerates, lithic wackes and arkosic arenites resting unconformably on the cratonic basement of Zimbabwe (Fig.4.27,B); intercalated volcanics with alkali metabasalts, massive, vesicular and amygdaloidal lavas; and an upper arenaceous formation with argillites and subordinate arenites which represent the shallow water deposits of an extensional basin. Masters (1989a,b) discussed the sedimentology of the copper-bearing lower sequence of the Dewaras Group. The lower sequence consists of the initial rift alluvial fan deposits, the Mangula Formation and rocks of a playa environment, the Norah formation, both of which have been studied in detail in the Norah copper mine. The alluvial fan and braided stream deposits of the Mangula Formation consists of trough cross-bedded, braided-stream conglomerates and arkosic arenites of the mid-fan environment which are vertically and laterally interfingered with lithic wackes, siltites and carbonate-bearing meta-argillites of the distal fan environment. The Norah formation is a sequence of a rapidly alternating succession of shallow water plane-bedded or massive arkosic arenites, interbedded with thinly bedded anhydrite-bearing dolomites, dolomitic argillites and argillites. The dolomitic units, which are very thinly bedded and with flaser and lenticular bedding, represent arid evaporitic playa flat deposits on the shores of an ephemeral playa lake which were subjected to rapidly alternating wet and dry spells. Since it represents an ephemeral playa environment, the Norah Formation is impersistent so that when the playa lake shrank or disappeared during renewed tectonism along the rift margins, the alluvial fan lithologies of the Mangula Formation prograded over the Norah playa facies.

The Lomagundi Group which succeeds the Dewaras Group is predominantly a clastic sequence with pyritiferous argillites and stromatolitic carbonates, while the Piriwiri Group consists essentially of phyllites, slates and graywackes, with minor bands of chert and quartzite. Leyshon and Tennick (1988) interpreted the Lomagundi Group as stable shelf or miogeosynclinal tidal flat, subtidal, marginal orthoquartzite deposits with carbonate platform development, and the Piriwiri Group as the coeval flysch deposits of the eugeosynclinal basin (Fig.27,C).

The Magondi belt is characterized by SE-directed thrust tectonics. The bulk of the deformation in the central and southern parts of the Magondi belt can be explained by a thin-skinned tectonic model, whereas in the extreme south both thin-and thick-skinned models apply (Fig.4.28,B). The central and southern parts of the belt attained greenschist facies, but towards the north regional metamorphism increased to the amphiobolite and in some places granulite facies. Apparently some form of horizontal tectonic processes operated in the Magondi belt between 2.15 Ga and 2.0 Ga, although the details are not yet clear.

Mineralization

Strata-bound copper mineralization occurs in the Deweras Group where copper mines are situated within the clastic sediments at Molly and Norah (Fig.4.27,A). In these mines disseminated copper occurs in a zone up to 200 m thick, with a strike length of 2 km in host rocks of argillites, quartzites, arkoses, grits and conglomerates. The major ore minerals are bornite and chalcopyrite (Leyshon and Tennick, 1988). The by-products which are produced include gold, silver, platinum, palladium and selenium. Mineralization was related to the evaporites of the Norah formation. The copper-silver ore bodies were formed diagenetically by oxidizing chlorine-rich basin brines which reacted with the evaporites, picked up metals absorbed onto alteration products of labile minerals in the red beds, moved along permeable horizons and the basal unconformity of the Deweras Group,and precipitated sulphides under reducing conditions in the Mangula and Norah Formations (Masters, 1989). During subsequent orogenic events, the ores were recrystallized and mobilized into quartz-microcline veins injected into surrounding rocks and into local cleavages and cleavage-parallel veinlets.

A variety of mineralizations occur in the Piriwiri and Lomagundi Groups (Leyshon and Tennick, 1988). Over 50 gold deposits are known in the Piriwiri Group where gold mineralization occurs in an unusual red earth with boulders of Lomagundi dolomite; in dark brown vein quartz; and along the northeast-trending Piriwiri Mineral Belt usually within or adjacent to small diatremes (Fig.4.27,A). Around the Copper Queen and Copper King domes are sulphide deposits (pyrrhotite, chalcopyrite, galena, arsenopyrite, pyrite with magnetite, cubanite, marcasite, etc.) which occur in discontinuous lenses in a skarn which is part of the Piriwiri Group. Although the intrusive granites of the domes could be considered as sources of the metals, Leyshon and Tennick (1988) considered the Cu-Pb-Zn ores as syngenetic strata-bound mineralizations. Pegmatites and quartz veins are the sources of mica, beryl, tantalo-columbite,

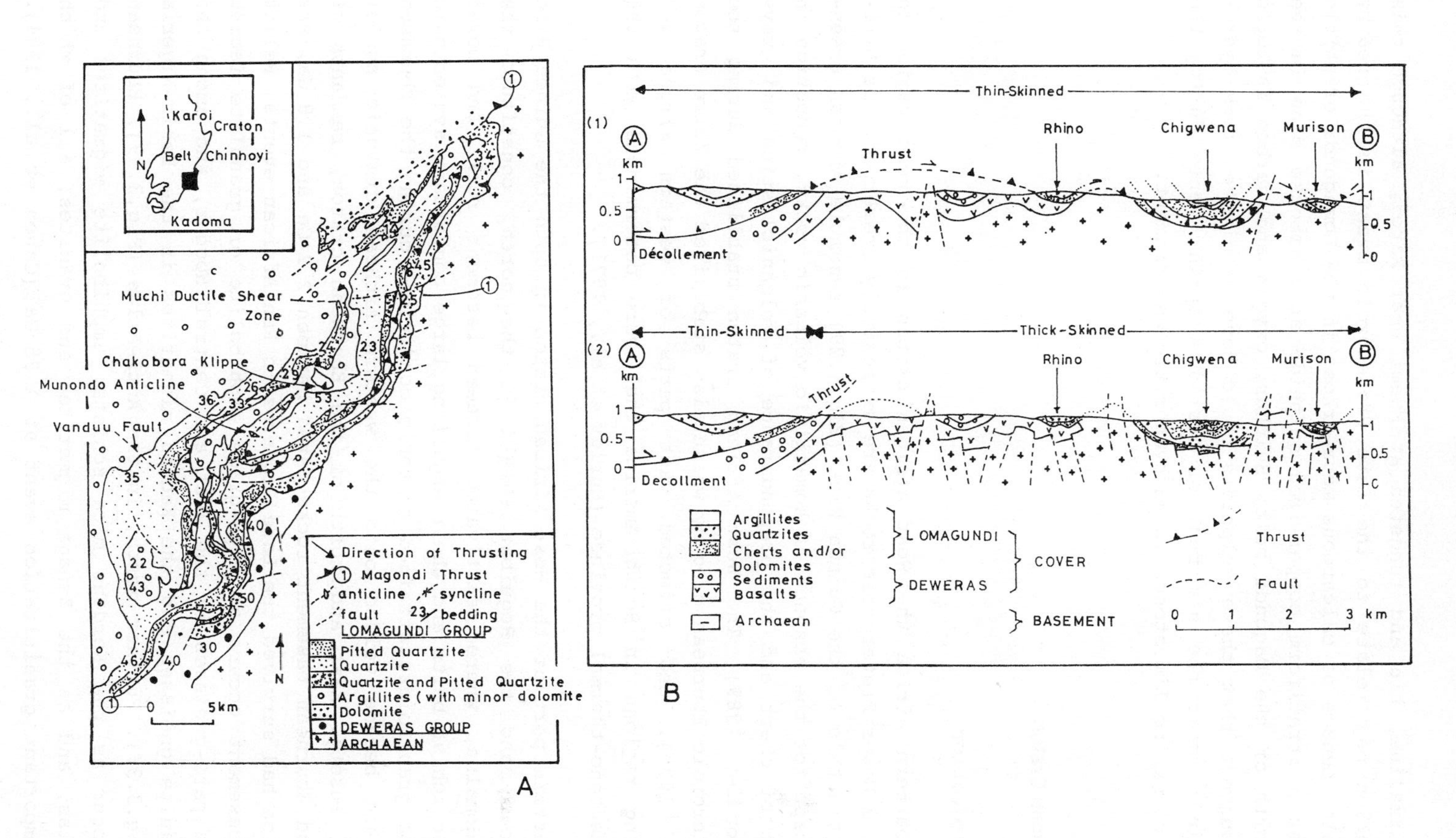

Figure 4.28: Tectonic map (A) and structural sections (B) showing suggested thin- and thick-skinned tectonic models for the Magondi belt. (Redrawn from Leyshon et al., 1988.)

topaz, tourmaline, tin and tungsten at Mwami near Karoi, although this mineralization may relate to the Late Proterozoic event in the nearby Zambezi belt. Lenses of calcareous meta-arkose in the Lomagundi graphitic schist host a stratiform copper-iron sulphide at Shamrocke Mine in the extreme north of the Magondi belt. Sulphur, oxygen and carbon isotopic evidence suggest that these sulphides could have originated from hydrothermal fluids near the sea bed, or near hot spring vents during the formation of marine limestones in an evaporitic environment.

4.8 West African Craton

4.8.1 Introduction

Outside southern Africa the West African craton is the next region in Africa where Lower Proterozoic rocks are extensively preserved. The Birimian supracrustals of the Guinea Rise (Fig.4.29) resemble Archean greenstones except for the absence of komatiitic volcanic rocks, reduction in the amount of chert and the preponderance of volcaniclastics and graywackes (Condie, 1989). The West African craton stabilized during the Early Proterozoic Eburnean orogeny which also stabilized the Zaire craton (Clifford, 1970), and affected vast parts of western Africa and neighbouring regions in South America that were coterminous with the Eburnean tectono-thermal province (Hurley et al., 1971).

The eastern part of the West African craton in both the Guinea Rise in the south and the Reguibat shield in the north, constitute the Eburnean domains, where reactivated Archean basement rocks and Lower Proterozoic schist belts and their engulfing large concordant syntectonic batholithic granitoids, attained cratonic conditions during the Eburnean orogeny, and became welded onto the western Archean cratonic nuclei (Fig.4.29, inset). East and north of the Eburnean craton, remnants of Archean and Eburnean basement rocks dated between 2.2 Ga and 1.8 Ga, are known, which had survived the Late Proterozoic Pan-African events. Relict Eburnean basement occurs as high-grade amphibolite or granulite facies ortho- and para-gneisses in the West and Central Hoggar, and among the Ibadan granite gneisses and the Kaduna migmatites in the Benin-Nigeria shield (Fig.3.39). Further north, in the Anti-Atlas (Fig.3.39), Eburnean inliers occur as the Kerdous mica schists, amphibolite migmatites and augen gneiss, and as the Zenaga migmatites and granites, all of which show an important granitization event at 1.95 Ga (Cahen et al., 1984).

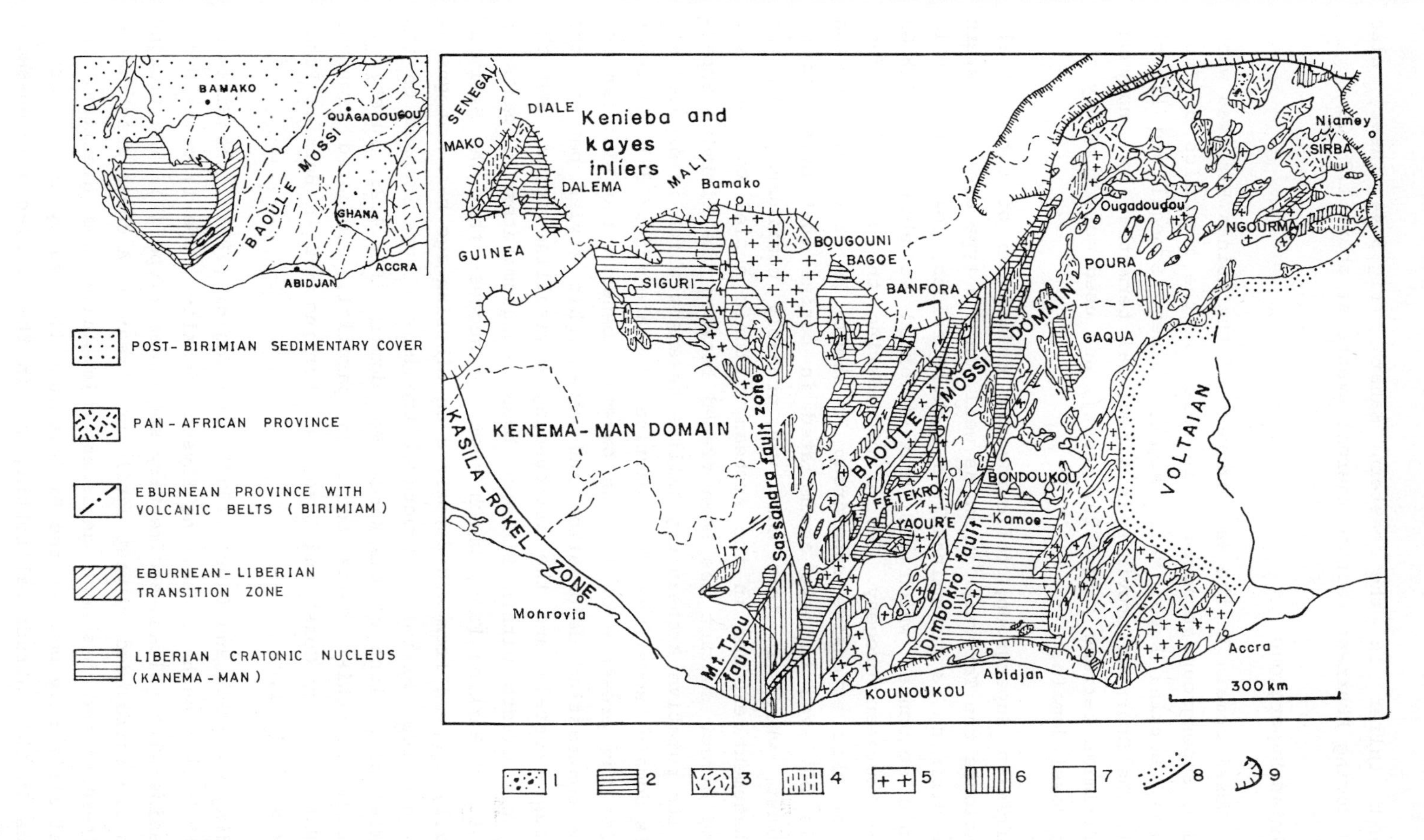

Figure 4.29: Sketch map of the Birimian of West Africa. 1, Tarkwaian facies; 2, Sedimentary flysch; 3. Volcano-clastic facies; 4, Greenstone; 5, Eburnean granitoids; 6, ? pre-Birimian; 7, Undifferentiated "basement" of the Baoule-Mossi domain; 8, Voltaian Supergroup; 9, Limit of Phanerozoic cover; 10, Recent. (Redrawn from Cahen et al., 1984).

The Uweinat inlier in the eastern Sahara (Fig.3.44, inset) also cratonized during Eburnean tectono-thermal events at about 1.8. Ga.

4.8.2 Birimian Supergroup

The Baoulé-Mossi domain in the eastern part of the Guinea Rise was where the Birimian Supergroup accumulated in parts of the republics of Ivory Coast, Ghana, Burkina Faso, Mali, Niger, Guinea, Senegal and Liberia (Fig.4.29). The Eburnean province is a regional term that is synonymous with the Birimian terrane (Fig.4.29, inset), and used more commonly than the Baoulé-Mossi domain.

The Birimian Supergroup, following the definition of Cahen et al. (1984), includes the Tarkwaian Group as well. In its type area in western Ghana, the Birimian is predominantly a tightly folded, low-grade clastic and volcanic sequence, which was last deformed at about 2.13-1.8 Ga during the Eburnean orogeny. Before considering the regional characteristics of this extensive supergroup it should be observed that the Birimian is believed to have accumulated in basins that developed on Archean crust, which in this case was probably continuous with the Liberian basement exposed in the Archean cratonic nuclei to the west. However, no direct contact has been traced between the Birimian Supergroup and its putative Archean crystalline basement which is exposed all around this domain, sometimes in a reactivated state (Fig.3.39). Indirect evidence for the presence of Archean basement beneath the Birimian includes the contrasting deformation and metamorphic styles between the Birimian supracrustals and the surrounding crystalline rocks, and the nature of the clasts within Birimian basal conglomerates in parts of Ivory Coast and Burkina Faso, which seems to suggest that the Birimian was partially derived from and deposited on Liberian-age basement gneisses and migmatites. Birimian structural trends are superimposed on the Archean axial directions of the Kanema-Man domain (Fig.4.29, inset). In general Birimian schist belts occur as parallel, elongated troughs, trending NNE in Ivory Coast and Ghana but turning ENE in Burkina Faso, and to the NW in Mali.

According to Black and Fabre (1983), and Tagini (1971), the Birimian occurs either in troughs containing flysch deposits (Type I) or as broad shallow basins with volcano-sedimentary sequences (Type II). The troughs and basins are surrounded by vast antiformal areas of Archean reactivated and undifferentiated basement gneisses, migmatites and granites. The superficial similarity between the Birimian and the nearby Archean greenstones lies in this structural setting and in the Birmian volcano-sedimentary lithology which includes basic lavas and intrusives, and phyllites and graywackes. As summarized by Black and Fabre (1983) Type I Birimian schist belts are predominant in Ivory Coast and in western Burkina Faso, while Type II is extensively developed in Ghana, eastern Burkina Faso and Niger. In southern and central Ivory Coast Type I Birimian belts are very similar, and were believed to start with basal greenstones, and metabasalts including pillow lavas; to be disconformably overlain by volcano-sedimentary rocks composed of metarhyolites, rhyolitic and dacitic tuffs, quartzites, phyllites and a persistent gondite horizon; which were unconformably succeeded by flysch-type deposits (Fig.4.30,A). Type II successions are mostly shallow water subcontinental deposits composed of volcano-detrital sequences with much acid volcanism and minor ultrabasic bodies and basic to intermediate lavas occurring with subvolcanic intrusions. Type II sedimentary formations are represented by sericite-schists, graywackes, quartzites, metaconglomerates and calc-chlorite schists. On its westernmost outcrop in

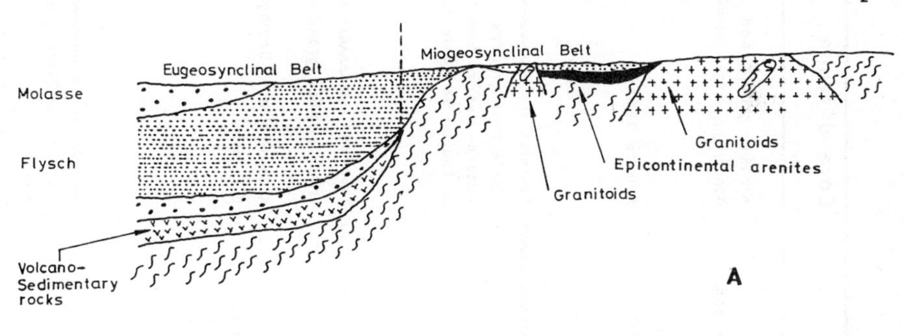

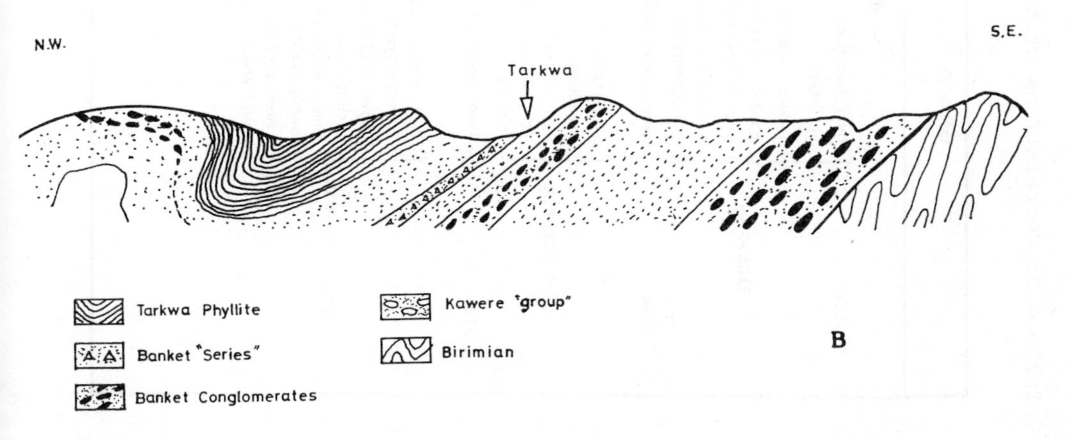

Figure 4.30: A, possible tectonic model for the Birimian; B, Tarkwaian Group. (Redrawn from Bessoles, 1977.)

Table 4.2: Correlations of the Birimian Supergroup and Eburnean events. (From Cahen et al., 1984)

Age	Group	Ghana	Côte d'Ivoire	Liptako, NE Haute Volta, and W Niger
2·03 Ga	TARKWAIAN	Huni formation (quartzites and phyllites) Tarkwa formation (phyllites) Banket formation (quartzites and conglomerates) Kewese formation (conglomerates)	Windéné granite and Bondoukou type granites Kinkéné series	Epizonal sediments of the Amarasinde and Bellekoiné formations —— the Liptakoian
2·13 Ga		~~~~~~~~~~	Eburnian II ~~~~~~~~~~ ?	~~~~~~~~~~
	UPPER BIRRIMIAN	Syntectonic and intrusive granitoids Basic volcanic formation Acid volcanic formation Volcanic arenaceous formation	Baoule type granites Volcano-clastic formation de Louga = Séries de Inahiri	Granitoids Mesozonal metamorphites of the Dibirshi and Tambao formations
2·27 Ga		~~local unconformity~~~~~~~~~~	Eburnian I ~~~~~~~~~~	
	LOWER BIRRIMIAN	Upper arenaceous formation (sandy flysch) Upper argillaceous formation (pelitic flysch) Middle arenaceous formation (sandy-pelitic flysch) Lower argillaceous formation Lower arenaceous formation	Orthogneissified granitoid Flyschoids Kounoukou and Doulayeko paragneiss	
2·6 Ga			~~~~~~~~~~ ?	~~~~~~~~~~
		?	Niega-Pauli Plage gneissified granite Monogaga paragneiss	Pre-Birrimian crystalline basement

the Kenieba and Kayes inliers (Fig.4.29) in eastern Senegal and Guinea, the Birimian consists essentially of low-grade volcano-detritic sequences with flysch, intruded by a variety of Eburnean granites.

A major discrepancy existed between the Birimian lithostratigraphic succession erected in Ghana and in the francophone parts of West Africa (e.g. Wright et al., 1985; Kesse, 1985). Whereas in Ghana a thick flysch clastic sequence underlies the volcano-clastic group, the opposite was believed to be the case in the francophone region (Fig.4.30,A), until Ledru et al. (1989) confirmed the regional occurrence of the metasedimentary sequence at the base of the Birimian, thus supporting the correlations of Cahen et al. (1982), shown in Table 4.2, a correlation which has recently been modified by Leube et al. (1990).

The Birimian in Ghana

As shown by Kesse (1985) the Birimian Supergroup in Ghana was previously subdivided into a lower metasedimentary Birimian and an unconformable predominantly volcanic upper Birimian, which is also unconformably overlain by a gold-bearing molasse sequence, the Trakwaian Group (Table 4.2). The Lower Birimian is weakly metamorphosed and changes upward from mainly phyllites and subgraywackes to phyllites and slightly metamorphosed tuffs, graywackes and feldspathic sandstones. The conglomerate horizons in the arenaceous units were believed to have been derived by the erosion of basement. Some of the phyllites are silicified in the upper part and also contain pyrite and fine carbonaceous matter. The Upper Birimian was assigned mostly basaltic and andesitic lavas and tuffs often spilitic, interstratified with graywackes and graphitic phyllites; to which was added overlying manganiferous phyllites and gondites. A previous depositional model for the Birimian (Fig.4.30,A) considered it as eugeosynclinal flysch (Bessoles, 1977), a view which has been modified in Ghana.

Leube et al. (1990) presented a significantly different stratigraphic interpretation for the Birimian Supergroup in Ghana, in which they stressed lateral lithofacies relationships within this group. The most notable aspect of their re-interpretation was the recognition of the fact that the so-called Upper Birimian, comprising essentially metalavas, was the lithofacies equivalent of the Lower Birimian metasedimentary rocks. Leube et al. (1990) drew attention to the parallel regional disposition of evenly spaced belts of folded Birimian metalavas which alternate with terranes consisting of isoclinally folded metasedimentary rocks and granitoids; the latter constituting the Birimian basins shown in Fig.4.31. The volcanic belts, 15-40 km wide and about 90 km apart from

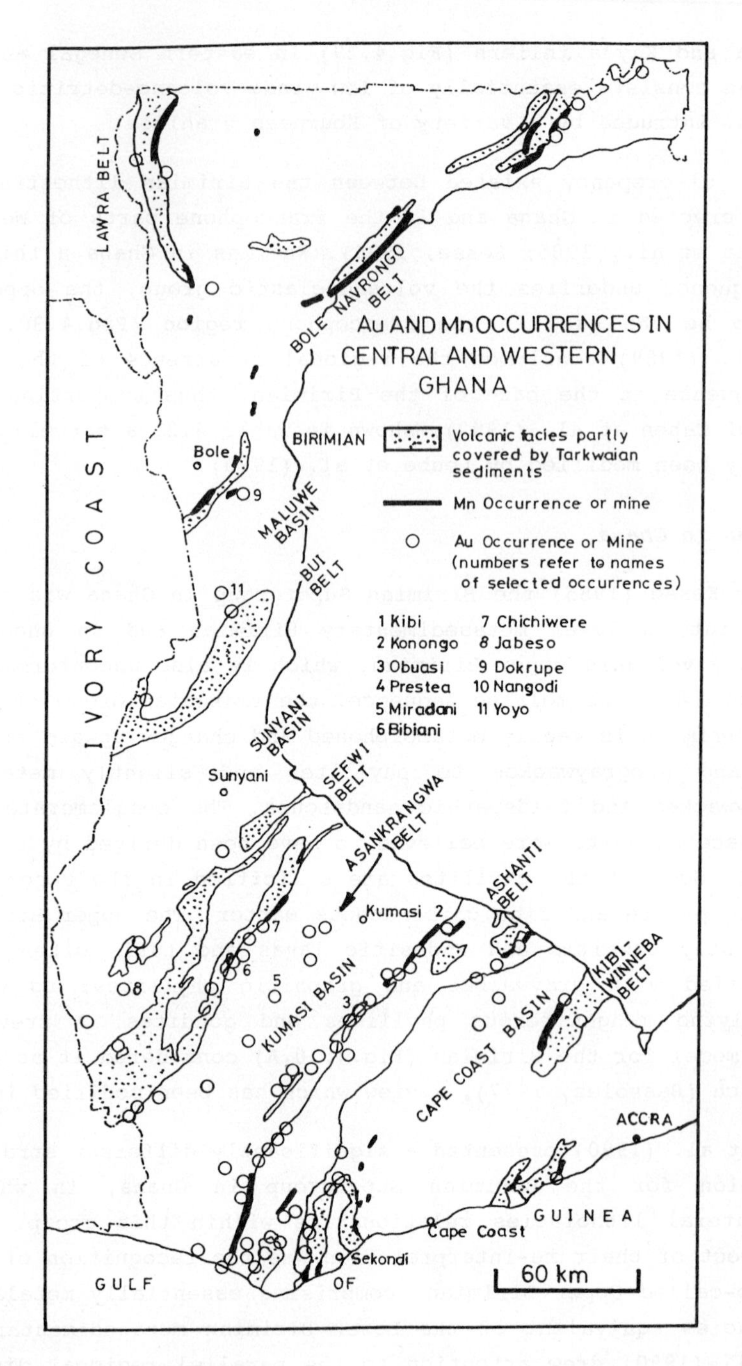

Figure 4.31: Simplified geology of the Birimian in western Ghana showing gold and manganese mineralization. (Redrawn from Leube et al., 1990.)

one another, have been traced by gravity surveys eastward beneath the flat-lying Late Proterozoic-Early Paleozoic Voltaian sediments. In southern Ghana the Birimian belts are less eroded, whereas in northern Ghana and in Burkina Faso where the level of erosion is deeper due to the uplift of the West African shield, the volcanic belts are narrower and granitoids dominate on account of the exposure of deeper levels. Most of the Birimian metalavas are tholeiitic with chemical characteristics similar to mid-oceanic ridge basalts (MORB), and with greater resemblance to Archean tholeiites than to Phanerozoic varieties (Leube et al., 1990). They show chemical affinity to rift-related settings (MORB and continental rifts); and virtually no relationship with collision-related tectonic environments,whether arc or calc-alkaline. However, Leube et al. (1990) pointed out that the Birimian metalavas did not originate in a setting identical with or similar to a modern mid-oceanic or back-arc basin ridge; rather, it seemed as if the Birimian tholeiites formed in a tension-related tectonic domain. Lithologically the metalavas and their interbedded pyroclastics are now between pumpellyite-prehnite facies and almandine-amphibolite subfacies.

The Birimian metasediments, according to Leube et al. (1990), are divisible into volcaniclastic rocks; turbidite-related wackes; argillitic rocks; and chemical sediments; these lithofacies show gradational boundaries between them. Figure 4.32,A shows the facies model and lithologic characteristics of the various lithofacies recognized by Leube et al. The volcaniclastic facies represents volcanic islands or chains where tholeiitic lavas and pyroclastics erupted; and the turbidite-related wackes were the pyroclastic and epiclastic sediments transported down slopes at basin edges by turbidity currents. Argillites, a widespread lithology in the Birimian basins of Ghana, represent basinal muds and silts, while the carbonaceous schist occurring in the transition zones between the volcanic chains and basin sediments suggests the growth of some form of plant-like organism. Chemical sediments such as chert, carbonates, manganese and sulphides occurred in the basin in relation to volcanic and hydrothermal emanations which were also the primary centres of gold formation.

Leube et al. (1990), in their depositional model for the Birimian Supergroup of Ghana, postulated that the deposition of the tholeiitic flows (metalavas) occurred simultaneously with that of the metasediments (Fig.4.32,A) and that this is evident in the transition from volcanic rocks into sedimentary rocks along strike, and in the interfingering between the two major rock types (Fig.4.32,B). Also, Sm-Nd isotopic analyses of samples from the Birimian sedimentary basins and from volcan-

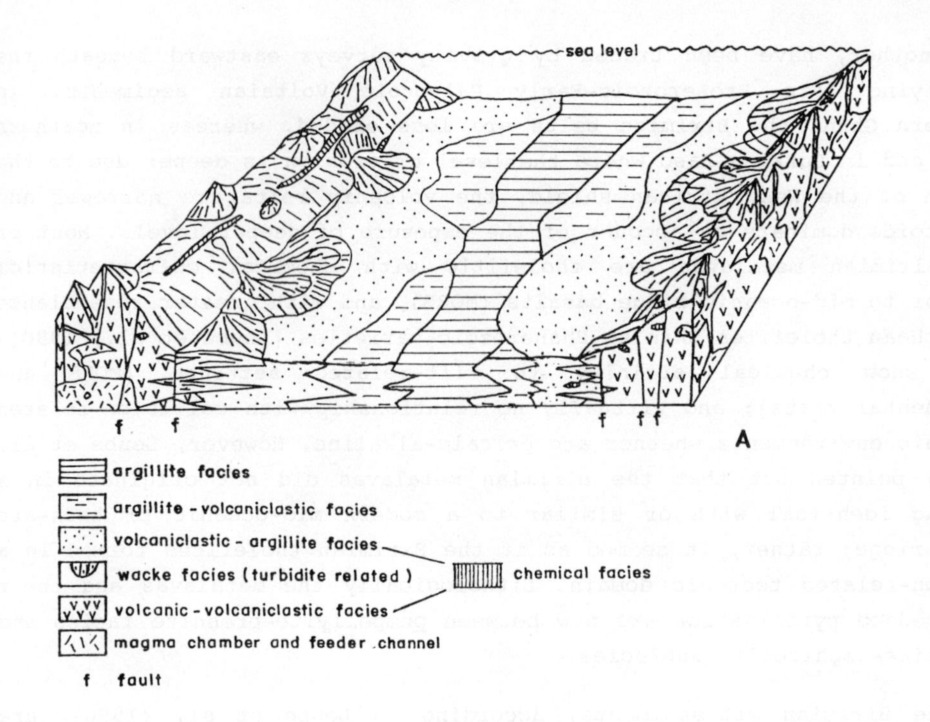

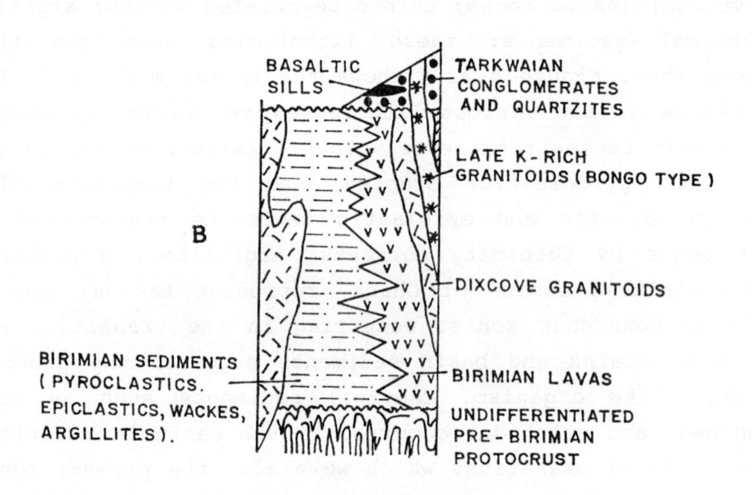

Figure 4.32: A, suggested facies relationship in the Birimian of Ghana; B, schematic relationship of major rock units, excluding the Tarkwaian. (Redrawn from Leube et al., 1990.)

ic flows in the volcanic belts confirm their contemporaneity and their accumulation between 2.31 Ga and 2.01 Ga. Birimian sediments in Ghana were therefore not of Archean provenance, but were derived from contemporaneous volcanism along adjoining volcanic belts.

About the unconformably overlying Tarkwaian Group, the traditional view (e.g. Kesse, 1985) has been that this is a gently folded molasse sequence (Fig.4.30,B) that was deformed during the waning phase of Eburnean tectonism. The Tarkwaian is believed to be 2,500 m thick in the type area and up to 6.7 km thick in other parts of Ghana.

At the base of the Tarkwaian is a conglomeratic member known as the Kawere "group"; this is succeeded by quartzites, grits and conglomerates of the Banket "series"; which in turn pass upward into the Tarkwa phyllite (Table 4.2; Fig.4.30,B), over which is the Huni sandstone. The Tarkwaian consists of coarse, poorly sorted, immature sediments which were deposited as piedmont alluvial fans in elongated intracratonic intermontane rift basins bounded by Birimian granitoids and supracrustals, from which the Tarkwaian sediments were derived.

In their account of the economically important Tarkwaian Sequence, Leube et al. (1990) emphasized the following sedimentological characteristics. The Tarkwaian consists of conglomerates, quartzites, arkosic rocks and minor phyllites; and shows significant vertical and lateral variability representing alluvial fan deposits with braided stream channel patterns. Pebbles in the stratigraphically lower conglomerates are poorly rounded and locally of the size of boulders; they are also poorly sorted and polymictic as a result of short distances of transportation. Higher up in the stratigraphic sequence the quartz-pebble conglomerates are "clean" quartzites having been more reworked. Pebble composition includes granitoids, basalts, rhyolites, tuffs, black pyritic chert, carbonaceous schist, quartz, Mn-rich rock, and siliceous and hematite-bearing hornfels. The sandstone facies reveals cross-bedding and channel scouring, which are indicative of shallow water conditions. The Tarkwaian sediments are only weakly metamorphosed.

Finally, of relevance to any paleogeographic reconstructions and tectonic models for the Birimian of Ghana, is the observation of Leube et al. (1990) that the volcaniclastic belts cannot be regarded as remnants of a previously more widespread stratigraphic sequence, hence each volcanic belt has to be considered as a well-defined independent structural unit.

The Birimian in Other Parts of the Guinea Rise

No attempt was, however, made by Leube et al. (1990) to extend their Birimian stratigraphic model to areas outside Ghana. Thus, in spite of the efforts of Ledru et al. (1989) and Milesi et al. (1989) to redress the conflicting stratigraphic interpretations between Ghana and her neighbours, major differences still persist. However, no attempt is made here either, to resolve these conflicts. As was previously the case in Ghana, the francophone geologists have retained the term Lower Birimian to refer to the basal metasedimentary succession (Table 4.2) comprising flysch-type sediments which are assumed to lie unconformably on the Archean craton. They have also retained the Upper Birimian for the volcanogenic unit comprising tholeiitic submarine flows at the base and calc-alkaline pyroclastics at the top, separated by a sequence of cherts, black sediments and acid pyroclastics.

Milesi et al. (1989) considered the Lower Birimian succession to have evolved through three main stages. Flysch-type deposits initially accumulated in a steady subsiding basin, followed by strong sedimentary instability leading to the deposition of coarser, turbidite-type clastic sediments. Sedimentation was accompanied by hydrothermal activity which resulted in the appearance of cryptocrystalline tourmaline with disseminated sulphides in breccia pipes and in the matrix of the immediately surrounding sandstones and conglomerates. With a decrease in ferrigenous input clay-carbonate sedimentation followed, with intervals of acidic volcanic activity represented by pyroclastic and epiclastic deposits, dykes, and sills. The tourmaline-bearing turbidites also contain gold deposits which are associated with volcanism and hydrothermal activity which appeared late during the deposition of the Lower Birimian.

The stratigraphic position of the Lower Birimian metasediments at the base of the Supergroup has been upheld to be valid for Ivory Coast, Burkina Faso, Mali, and Senegal, by Ledru et al. (1989) and by Milesi et al. (1989). These authors also retained the previously held view (Cahen et al., 1984) that a major unconformity separates the Lower from the Upper Birimian (Table 4.2), an opinion not shared by Leube et al. (1990) for Ghana, where Wright et al. (1985), had also reported a conformable relationship between the Lower and Upper Birimian.

Granitoids and Structure of the Birimian

Abundant granitoids are closely associated with the Birimian schist belts (Fig.4.32,B). These comprise pre-tectonic granites from which some of the basal pebbles of the Birimian were derived; and large concordant, foliated syntectonic batholithic granitoids (Baoulé type in Ivory Coast, Cape Coast type in Ghana), with biotite-hornblende and 2-mica varieties which are either emplaced in pre-Birimian migmatites and gneissic basement or produced by granitization of the folded Birimian geosynclinal formations. Two-mica granites are subordinate in the eastern part of the Baoulé-Mossi domain where Type II Birimian belts predominate; the majority of the syntectonic and post-tectonic granites contain biotite and hornblende. The post-tectonic intrusives are small and subcircular discordant granodiorites with localized thermal aureoles. They are referred to as the Bondoukou type in Ivory Coast and as the Dixcove type in Ghana. Basic to intermediate intrusives are known in Ghana among the Tarkwaian Group, including a norite body about 20 km long near Tumu in northern Ghana.

In Ghana the structure of the Birimian rocks is characterized by isoclinal folds with near vertical axial planes; locally developed open symmetrical folds in the volcanic belts with slight vergence to the southeast; an axial plane cleavage developed parallel to bedding throughout the steeply inclined sediments; and by a weak second cleavage striking oblique or perpendicular to the first cleavage (Leube et al., 1990). Cleavage is more pronounced and more regular around the granitoid intrusions. Three phases of fold deformation occur in Ghana. In the Ashanti belt (Fig.4.31) high-angle reverse faults or upthrusts are found in mines.

The Tarkwaian Group which rests unconformably on Birimian strata shows different structural styles in each of the five volcanic belts (Leube et al., 1990). In the Bui belt (Fig.4.31) the Tarkwaian Group forms a large symmetrical synclinal trough, 110 km long and up to 15 km wide. In addition to symmetrical structures in the Ashanti belt, there are locally developed tight, vertical and isoclinal folds; and from the centre of the Ashanti belt towards the west and the east, the open folding turns into thrusts directed towards the centre, in which strata are repeated and the Birimian is overthrust by the Tarkwaian. Leube et al. (1990) attributed Tarkwaian structures to tension-related folding by gravity during the deposition of Tarkwaian sediments rather than to a post-Tarkwaian tectonic event. Tarkwaian style and intensity of deformation are significantly different from those in the Birimian strata.

Outside Ghana, the Birimian Supergroup also exhibits three phases of deformation and metamorphism, D_1, D_2 and D_3 (Table 4.2) in which D_1 is generally penetrative, coaxial, and affects only the Lower Birimian Group (Milesi et al., 1989). During D_2 and D_3 the tectonic setting was one of transcurrent movement, in which D_2 (sinistral) and D_3 (dextral) transverse dislocations caused folding and also controlled the emplacement of granitoids. The large D_3 transcurrent faults which are hardly seen in Ghana are best developed in the northern Birimian belts and have imposed NE-SW structural trends on the northern terranes.

Tectonic Models for the Birimian Supergroup

According to Milesi et al., (1989), an early orogenic phase (Eburnean I of Cahen et al.(1984)), which they termed the Burkinian orogeny (Table, 4.2), had affected the Lower Birimian at deep crustal levels in the Ivory Coast. Judging from the extensive area affected by a first schistocity, Ledru et al.(1989) concluded that the Burkinian orogeny probably resulted from several plate collisions at different times during a phase of continental accretion. The Upper Birimian metavolcanics (Table 4.2), on the other hand, accumulated in discontinuous and independent basins. Ledru et al. (1989) considered them as in-situ zones of rifting, for example the Mako belt in eastern Senegal, the Gaoua belt in Burkina Faso, and the Yaouré belt in Ivory Coast. One could, perhaps add the Birimian belts of Ghana, and consequently relate their origin not to plate collision, but rather to a different geodynamic process -- continental rift development -- as was postulated by Leube et al. (1990) in their geodynamic model for the Birimian.

Leube et al. (1990) proposed a tectonic model for the Birimian of Ghana which involved small-scale, equidimensional, parallel and contemporaneously operating convection cells in the upper mantle which caused the rifting of a highly thinned protocrust as well as linear eruptions of tholeiitic magmas (Fig.4.33). With the growth of the volcanoes above sea level, volcanic-derived sediments accumulated in the slowly subsiding parallel basins. After volcanic activity ceased and lateral compression was generated by two adjacent contracting sialic blocks, crustal shortening and folding resulted, leading to the emplacement of the Baoulé type granitoids. The final stage in the evolution of the Birimian belt was caused by the reactivation of rifts and the formation of intermontane basins in the volcanic belts. The Tarkwaian molasse represents the infilling of these later basins with clastic sediments that were derived from Birimian rocks eroded nearby. This was followed by gravity deformation of Tarkwaian sediments, and by the intrusion of mafic sills and post-Eburnean Bondoukou or Dixcove type K-rich granitoids.

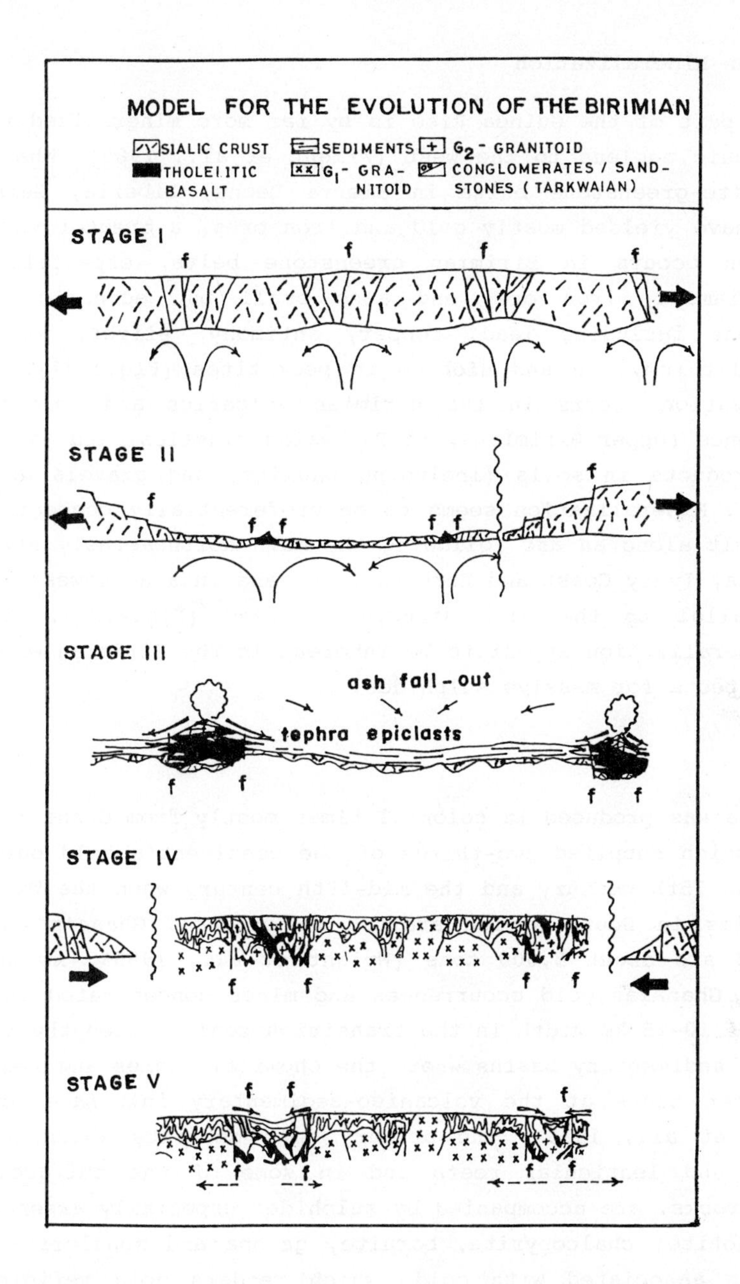

Figure 4.33: Tectonic model for the Birimian. (Redrawn from Leube et al., 1990.)

4.8.3 Birimian Mineralization

The Eburnean part of the Guinea Rise is by far more mineralized than the Archean cratonic nucleus to the west (Wright et al., 1985). Whereas the Archean granite-greenstone belts in Sierra Leone, Liberia, Guinea and Ivory Coast have yielded mostly gold and iron ores, a greater variety of mineralization occurs in Birimian greenstone belts, especially gold, manganese, diamond, iron ore, and also small but economic sulphide mineralizations including lead, copper, antimony, silver, nickel and cobalt, and lithium, tin and niobium in pegmatites (Fig.4.34). Most of the mineralization occurs in the Birimian volcanics and volcano-sedimentary sequence (Upper Birimian), in Tarkwaian clastics, and as residual weathering products in soils (including bauxite) and gravels above the supracrustals. Mineralization seems to be preferentially concentrated in an eastern belt along an arc following the main northeasterly structural trend in Ghana, Ivory Coast and Burkina Faso, and in a northwesterly belt in Mali parallel to the main structural trend (Fig.4.34). Gold and manganese mineralization appear to be inherent in the Birimian which also has good prospects for massive sulphides.

Gold

Gold in Africa was produced in colonial times mostly from Ghana (formerly Gold Coast) which supplied two-thirds of the continent's gold output between the late 15th century and the mid-19th century when the Witwatersrand goldfields in South Africa started to produce. Ghana's estimated gold reserves are about 5,000 tons (Wright et al., 1985). As shown in Fig.4.31 most Ghanaian gold occurrences and mines concentrated in narrow "corridors" of 10-15 km width in the transition zone between the volcanic belts and the sedimentary basins where the chemical facies and regionally extensive shear zones at the volcanico-sedimentary interface are also found (Leube et al., 1990). The primary gold deposits which occur in quartz veins and lenticular reefs and in some of the tuffaceous and argillaceous rocks, are accompanied by sulphides especially arsenopyrite, pyrite, pyrrhotite, chalcopyrite, bornite, galena and sphalerite. Up to 10% silver is associated with gold, which renders gold refining more profitable.

Gold lodes are associated with persistent and deep-seated shear zones and are also located along the transition zones between the Birimian phyllites and the volcaniclastics, and generally in country rocks comprising metamorphosed carbonaceous and manganiferous argillites, tuffs, graywackes and basic and intermediate magmatic rocks. In Ghana

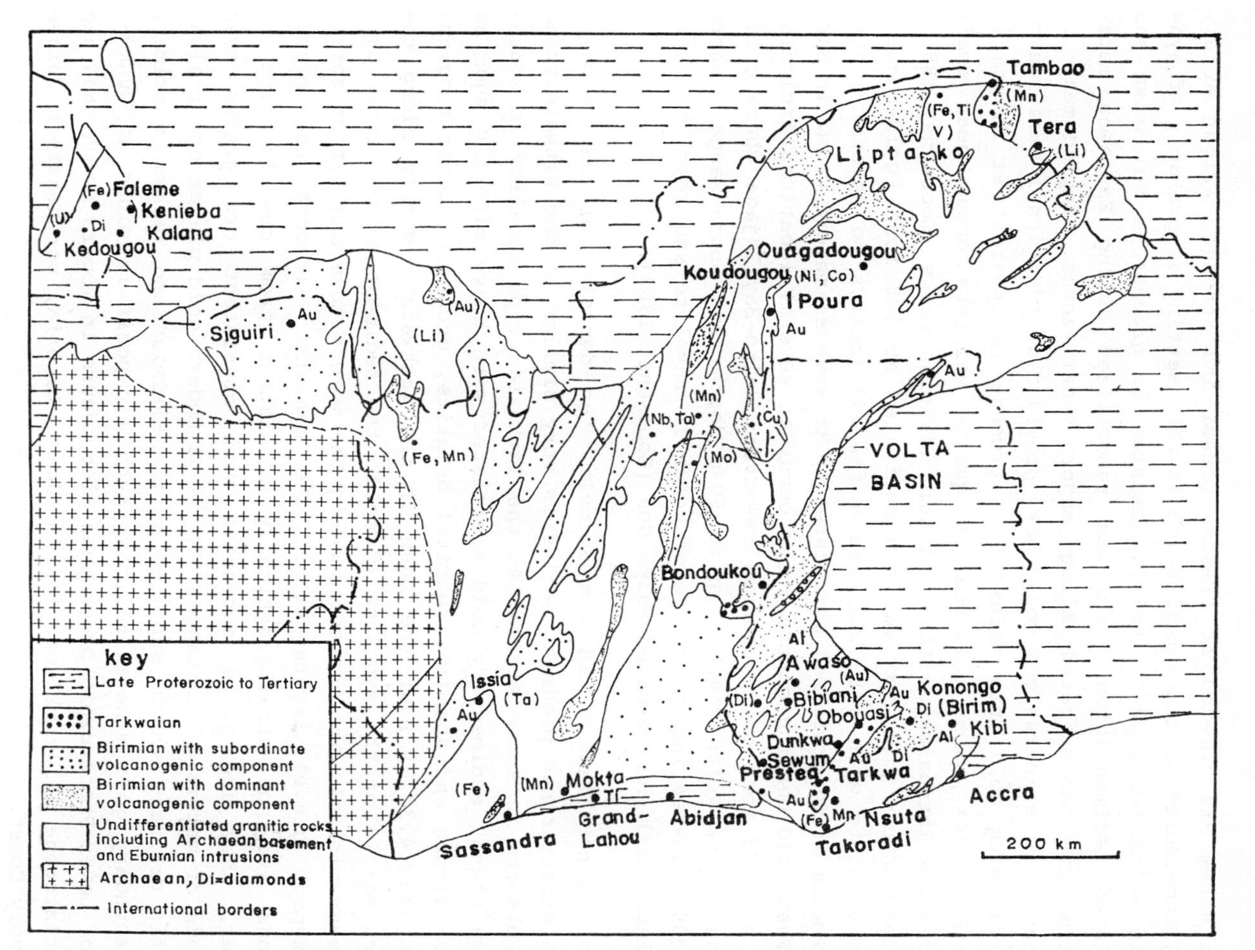

Figure 4.34: Mineralization of Lower Proterozoic rocks in West Africa. (Redrawn from Wright et al., 1985).

active mines are located at Presta, Obuasi and Konongo. Wright et al. (1985) considered gold mineralization in this setting as probably of syngenetic volcano-exhalative origin, related to greenstone volcanism and associated sedimentation. Gold was remobilized during Eburnean metamorphism and concentrated along major shear zones. Emplacement of Eburnean granite intrusions could have provided an additional heat source for mobilizing gold into veins. This process is apparent at Kalana in Mali (Fig.4.34) where gold-bearing quartz veinlets occur near small granodiorite intrusions within volcanic Birimian rocks (Goossens, 1983). At Kalana where gold reserves are estimated at 100 tons, gold-bearing quartz veinlets with sulphides are a few millimeters to several centimeters wide along preferred structures up to several meters wide. In Burkina Faso gold-bearing quartz veins occur at several locations including Poura with 23 tons of recognized reserves; Guiro-Bayildjaga where gold-bearing quartz veins reach down to 50-m depth; and Boura-Gangaol where over 300 gold-bearing quartz veins occur in rhyolite and rhyolitic tuff associated with disseminated sulphides (Goossens, 1983).

Gold mineralization in the alluvial fan deposits of the Takwaian Group intracratonic piedmont basins is very similar to the Witwatersrand paleo-placers in South Africa. At the Tarkwaian type locality in southwestern Ghana, sedimentary gold associated with heavy minerals such as rutile, zircon and abundant detrital hematite, occur in banket conglomerates (Fig.4.30,B) near the base of the Tarkwaian, mostly along the eastern margin of the syncline close to the sediment source area. Gold and heavy minerals are concentrated in the more mature and better sorted matrix-poor gravel layers within the bankets. Although the gravel horizons are not stratigraphically persistent from one location to another, they are quite extensive with consistent ore grades over hundreds of meters of distances. In Ghana modern gold places occur along the Ofin drainage system supplied by the Birimian and Tarkwaian primary and secondary gold mineralizations; such a deposit is mined at Dunkwa. Gold of Lower Proterozoic provenance also occurs in modern beach sand and in Pleistocene alluvial deposits in Ghana, Liberia, northeastern Guinea and Ivory Coast.

Manganese

Manganese oxides in the Birimian Supergroup are concentrated by the weathering of protores such as manganiferous phyllites, gondites (highly metamorphosed manganese sediments comprising spessartite and quartz), Mn-carbonates and silicate horizons in the Birimian volcaniclastics or sometimes near the transition between the metalavas and the metasediments.

Manganese ores occur mostly in Ghana, southern Ivory Coast, Burkina Faso, and in southwestern Niger. But by far the largest manganese mineralization is in the Birimian metalava sequence at Nsuta in southern Ghana where, in addition to metallurgical-grade ore, there used to be large amounts of an unusual battery-grade pure MnO_2, to which the locally-derived name nsutite was given. The nsutite ore, now almost depleted, formed as a result of deep weathering on a Tertiary peneplain during which pure manganese oxides were produced by supergene enrichment processes. Nsutite ore bodies occur in a sequence of manganiferous phyllites and gondites 450-600 m thick below the top of the Birimian Supergroup in the upper part of the phyllite group. Wright et al. (1985) believed that manganese in the phyllites and gondites could have had a submarine volcano-exhalative origin, and that with the common association of gold with manganese in the Birimian, manganese could be used as a pathfinder element in the geochemical exploration for gold.

Diamonds

Diamonds have been produced from Birimian rocks in Ghana and the Ivory Coast. In the Birim diamond field in southern Ghana, the largest single diamond-producing area in West Africa, alluvial diamonds come from a band of Birimian conglomerates about 200 m wide and 50 km long. The Birimian diamond sources are in the Lower Proterozoic rocks and are unrelated to the much later West African Mesozoic kimberlites which are also an important source of diamond.

Iron

Iron ores in the Lower Proterozoic of West Africa are not extensive. Whereas the Early Proterozoic was the climax for banded iron-formations in other parts of the world including South Africa, in West Africa the climax was attained in the Late Archean. As observed by Wright et al. (1985) manganiferous sediments seem to have superseded iron-formation in the Birimian unlike other regions where Early Proterozoic iron-formations are generally thicker and more extensive than those of Archean times. However, in easternmost Senegal there is a major Early Proterozoic iron deposit at Falémé with estimated reserves of 400 million tons of magnetite ore at 45-50% Fe, and 100 million tons at 62-65% Fe. Other iron deposits, of probably magmatic origin, occur in norites and gabbros in Burkina Faso and near Takoradi in Ghana.

Base Metal Deposits

According to Goossens (1983) a massive sulphide deposit rich in Zn and Ag occurs in the Birimian belt of Burkina Faso at Perkoa (Fig.4.34) in ores containing pyrrhotite, pyrite and sphalerite, and 20 oz Ag/ton. The grade of zinc is over 4% for a minimum of 30-m width, and exceeds 20% in drill intersections ranging from 3 to 15 m. Perkoa was the first volcanogenic massive sulphide deposit discovered in the Birimian. Goossens (1983) also reported deposits with porphyry copper affinities in the northern Birimian belt in Niger and Burkina Faso. These mineralizations occur at Kourki in Niger where 0.06% Mo and 0.07% Cu were found; and Gorem and Gaoua in Burkina Faso with 40 million tons at 0.15% Cu and 0.05% Mo, and 22 million tons at 0.08% Cu respectively. Although these deposits were considered sub-economic they contain rich copper veins that have been mined from time to time.

4.8.4 The Reguibat Shield

As in the Guinea Rise to the south, Archean basement rocks in the western part of the Reguibat shield give way to Lower Proterozoic igneous rocks and volcano-sedimentary assemblages in the east (Fig.3.39). However, in the eastern domain of the Reguibat shield Archean amphibolite facies and retrogressive greenschist facies migmatites (Chegga assemblage) occur in the eastern extremities where they had survived the Eburnean orogeny (Fig.4.35). From available geochronological data (Cahen et al., 1984) it is apparent that the Eburnean orogeny in the eastern Reguibat shield was slightly later than in the Guinea Rise, since it occurred in the former region between 2.0 Ga and 1.87 Ga, unlike the latter where the Eburnean events occurred between 2.4 Ga and 1.8 Ga (Table 4.2). The little disturbed volcano-clastic sequence and extensive rhyolitic volcanism of the eastern Requibat shield also contrasts with the Birimian Supergroup.

The Yetti Group and its equivalent the Aguelt Abd el Malek Group, are the oldest Proterozoic assemblages in the Reguibat shield (Table 4.3). These are preserved in NW-SE-trending belts, where they comprise thick masses of detrital and pyroclastic material with sills, dykes and volcanic flows, all of which are weakly metamorphosed at greenschist facies and intruded by the Yetti and Aï Ben Tili granites. The Yetti Group suffered two phases of deformation before the intrusion of granites at about 2.03 Ga; the first deformation being characterized by recumbent isoclinal folds, and the second by open folds of great amplitude with vertical or inclined axial planes.

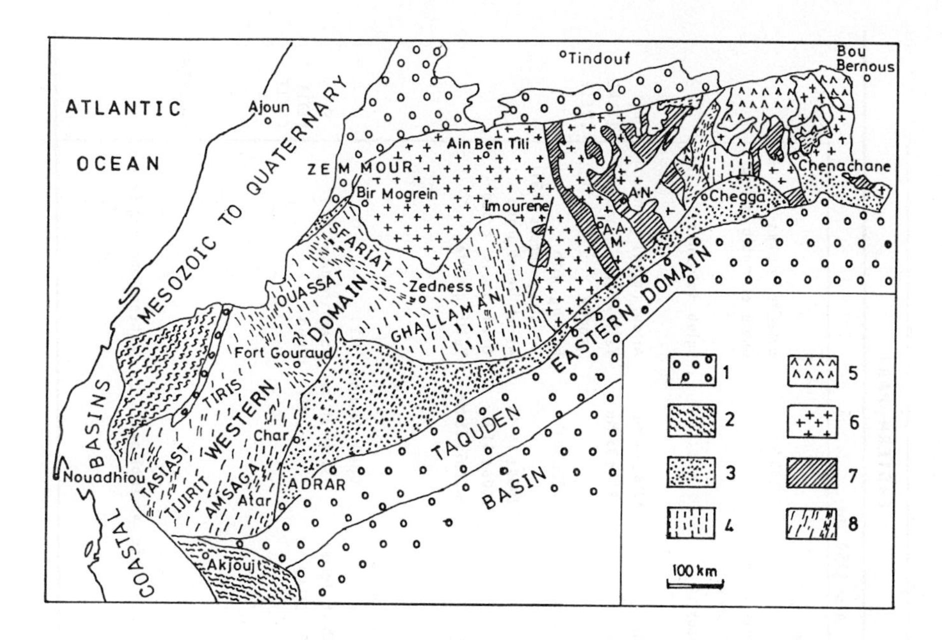

Figure 4.35: Geologic sketch map of the Reguibat Rise showing: 1, Late Proterozoic - Paleozoic; 2, Mauritanide (Hercynian) nappes; 3, Late Proterozoic, 4, Guelb El Habib Group; 5, Eglab volcanics; 6, Intrusive granites and associated rhyolites; 7, Oued Souss Group and Yetti Group; 8, Basement groups (Amsaga, Tasiast, Ghallaman, Chegga Groups and associated granites) (Redrawn from Cahen et al, 1984).

Early phases of Eburnean tectonism were followed by the deposition of volcano-detrital groups (e.g. The Oued Souss Group) which are preserved in NNW-SSE structures and were folded and slightly metamorphosed at about 2.02 Ga before the onset of extensive anorogenic magmatism from 1.97 Ga to 1.87 Ga. Among The Eburnean magmatic rocks in the eastern Reguibat domain are abundant high-level granitoids including porphyritic biotite adamellites, coarse pink biotite granites, sub-volcanic gabbros, diorites and granite porphyries, which have invaded the thick and vast sequence belonging to the Aftout rhyolites and ignimbrites with subordinate dacite and andesites near the base. There are also sub-volcanic alkaline granites and a nepheline syenite complex. Resting unconformably on this volcano-detrital and intrusive assemblages, is the Guelb El Habib Group, comprising unfolded and unmetamorphosed coarse-grained clastic rocks which are considered as the Eburnean molasse (Table 4.3). Dolerite intrusions into the Guelb El Habib, dated around 1.6 Ga, probably also belong to the regional post-Eburnean anorogenic magmatic phase.

Table 4.3: Tectonic events in the Reguibat Rise. (Redrawn from Cahen et al., 1984).

1 SOUTH-WESTERN PROVINCE		2 EASTERN PROVINCE: WESTERN PART		2 EASTERN PROVINCE: EASTERN PART		
Supergroup 1 of the cover of the Taoudenni basin Base at > 1050 Ma						
1546 ± 32 Ma		1563 ± 28 Ma	Tabatanat syenite			
	biotite closure ages; ? 1600 Ma dolerites			? 1600 Ma	dolerite dykes	
		1755 ± 65 Ma	Bir Moghrein granite (and closure of its biotites)		Guelb el Habib group	
1811 ± 56 Ma 1872 ± 52 Ma		1877 ± 35 Ma	El Archeouat rapakivi granite		Eglab volcanics	
	biotite closure ages	1912 ± 47 Ma	Tiguesmat granite	1918 ± 14 Ma	Attout (pink) granite	C
		1970 ± 46 Ma	Aïn Ben Tili granite	1946 ± 33 Ma	Yetti-Eglab junction granite	EGLAB CYCLE
		2022 ± 50 Ma	Folding with thrusting nappes	2022 ± 50 Ma	Folding with thrusting nappes (P3)	
				2039 ± 49 Ma	Akilet Deilel group	
				2008 ± 41 Ma	Oued Souss group	
				> 2057 ± 65 Ma	Yetti granite	
		2039 ± 49 Ma	Yetti granite			
2050 ± 119 Ma		2057 ± 66 Ma	Imourene group			
		2039 ± 49 Ma	Aioun Abd el Malek group			
				Open folding (P2)		
	Polyphase deformation	Polyphase deformation, greenschist facies	Aguelt Nebkha group	Isoclinal recumbent folding (P1)	Yetti group	B YETTI CYCLE
SW ← - - - - - -	- - - - - - → E					
	2539 ± 54 Ma Ghallaman granites					
Amsaga assemblage:						A
c. 2710 Ma migmatitic complex		? Hassi el Fogra syenite				BASEMENT
Saouda group	3270 ± 347 Ma Ghallaman gneisses	Hassi el Fogra 'Series' ← - - - - - -	Chegga Assemblage	- - - - - →	Chenachane-Erg Cheich groups	
(granulite to amphibolite facies)						

4.9 Zaire Craton

4.9.1 Introduction

Unlike the Kaapvaal craton which had mostly attained crustal stability by the end of the Archean, the Zaire and West African cratons are predominantly the products of the Early Proterozoic Eburnean orogeny (Clifford, 1970). In the Zaire craton there is unambiguous evidence for the Eburnean orogenic cycle in the Kasai-NE Angola shield; southern Angola; and in the Gabon orogenic belt, where Eburnean rocks provide the basement and foreland to the younger Pan-African West Congolian mobile belt. As in the West African craton the Eburnean orogeny was apparently quite prolonged in the Zaire craton where orogenic episodes are known at about 2.4 Ga and 2.2-2.0 Ga in the Kasai-NE Angola shield; at about 2.15 Ga in southern Angola; and at about 2.0 in the internal zone and foreland of the West Congolian belt, and in the Gabon orogenic belt (Fig.4.1).

4.9.2 Kasai - NE Angola Shield

On the uplifted southern part of the Zaire craton patches of Archean to Proterozoic basement gneisses, migmatites and metasediments are exposed. A 2.2-2.0-Ga tectono-thermal event was the last orogeny which affected this part of the craton (Cahen et al., 1984). However, the earliest Eburnean event, locally termed the Mubinji orogeny (Cahen et al., 1984), had recrystallized and rehomogenized localized Archean basement gneisses and also deformed and metamorphosed the Luiza metasedimentary cover at about 2.42 Ga (Table 4.4). The Luiza Supergroup is a metasedimentary sequence of quartzites, mica-schists and banded iron-formations lying unconformably on the Archean Kanda-Kanda tonalitic and granodioritic gneisses (Fig.3.31). Similar metasedimentary rocks occur near Mufo in NE Angola as outliers of the Luiza Supergroup. A later Eburnean tectonothermal event at 2.2-2.0 Ga caused more widespread metamorphism of basement rocks and the emplacement of anorogenic granites and pegmatites some of which cut the younger Lukoshi metasedimentary formations.

The Lulua Group which either post-dates the Luiza Supergroup or is the lateral foreland equivalent of it, is a metasedimentary and metavolcanic assemblage about 6 km thick, lying to the north of the Luiza Supergroup (Fig.3.31) in a belt about 170 km long and 20 km wide. The metasedimentary components of the Lulua Group are slates and quartzites, which are interstratified with greenstones comprising spilitic basalts, lavas (including unmodified pillow lavas) and post-tectonic granodiorite (Cahen et al., 1984).

Table 4.4: Major tectonic events in Kasai and adjacent parts of the Zaire craton. (Redrawn from Cahen et al, 1984).

Event		
6. LOMAMIAN OROGENY	c. 975 - 948± 20 or 937± 20 Ma.	
	— Mbuji Mayi supergroup	
5. POST-LULUA FOLDING	Pre-1155 ± 15 Ma (age of a post-tectonic syenodiorite sill)	
	—? Lulua group (interstratified spilitic lavas 1468 ± 30 Ma.	
4. c. 2200 - 2000 Ma orogeny	pegmatites:	c. 1920 Ma
	post-tectonic granites:	2037 ± 30 Ma
	syntectonic events:	2200 - 2050 Ma
	— perhaps Lukoshi formations	
3. MUBINDJI OROGENY	metamorphism at	2423 ± 48 Ma
	— Luiza metasedimentary group	
2. MOYO-MUSEFU EVENT		
b. Moyo episode:	closure of biotite:	2560 Ma
	Malafudi granite:	2593 ± 92 Ma
	migmatization and cataclasis:	2680 ± 5 Ma
	— Dibaya migmatite and granite assemblage	
a. Musefu episode:	charnockitization and granulite facies metamorphism:	2820 Ma
	— Kasai-Lomami gabbro-norite and charnockite assemblage	
1. PRE-MOYO-MUSEFU 'CYCLE'		
	— Kanda Kanda tonalite and granodiorite gneiss: undated	
	— Upper Luanyi granite gneiss:	c. 3400 Ma

In outcrop the Luiza is mostly bounded by longitudinal faults and was fragmented by late NNE-SSE faults which produced horizontal displacements. Lulua beds are folded along a WSW-ENE trend with a northerly vergence and a northerly attenuation of folding and metamorphism. In the western outcrop some weakly folded and unmetamorphosed beds of the Lulua Group rest unconformably with a basal conglomerate on crystalline basement. Based on the age of an interstratified lava, the Lulua Group is not younger than 1.46 Ga and may be older than 2.0 Ga (Cahen et al., 1984).

4.9.3 Eburnean Basement of Southern Angola

After consolidation during the Eburnean orogeny the basement complex in southwestern Angola remained as a stable crustal block between two Late Proterozoic orogens, the West Congolian orogen to the north and the Damara-Kaokoveld orogen of Namibia to the south (Figs.4.1;3.32). In addition to Archean gneisses (Fig.3.32), the basement complex of southwestern Angola contains a large terrane of Eburnean gneisses and migmatites formed of older Archean protoliths, separating schist belts of Eburnean-deformed low-grade metasediments. These metasediments occur as rafts and synclinal keels among rhyolitic to andesitic volcanics and related syntectonic granitoids and post-tectonic porphyroblastic granites (Unrug, 1989). Their volcano-sedimentary lithologies include basal conglomerates; meta-arenites; metagraywackes with crystalline limestone lenses; ortho-amphibolites; talc schists, chlorite schists, migmatitic gneisses and hypabyssal and volcanic rocks. As shown by Carvalho et al. (1987) and Cahen et al. (1984), lithostratigraphic subdivisions and correlations of the Eburnean assemblages of Angola are still very tentative.

Summing up the array of available geochronological data from the region, Cahen et al. (1984) concluded that an orogeny affected most of this region at about 2.15 Ga, during which the main metamorphism, granitization and deformation took place, followed by extensive late- and post-tectonic, and anorogenic granitic intrusions and volcanic activity between 2.05 Ga and 1.75 Ga or 1.65 Ga. Low initial Sr isotope ratios in the "homogeneous regional granites" of southern Angola suggest the addition of juvenile crustal material (Unrug, 1989).

4.9.4 Eburnean Basement in the Internal and Foreland Zones of the West Congolian Orogen

This polyorogenic domain is part of the Atlantic Rise or basement swell along the equatorial Atlantic margin of central Africa (Fig.4.36). On this basement rise are exposed a wide range of basement rocks ranging from Archean granitoid and charnockitic massifs of southern Cameroon, Gabon and Congo Republic to Eburnean rocks deformed during the Late Proterozoic West Congolian deformation. Eburnean rocks outcrop extensively as the basement to the West Congolian mobile belt and on the cratonic foreland of this orogen. Table 4.5 shows the correlations of the Eburnean lithostratigraphic sequences in this region, from southern Cameroon to northern Angola (Cahen et al., 1984).

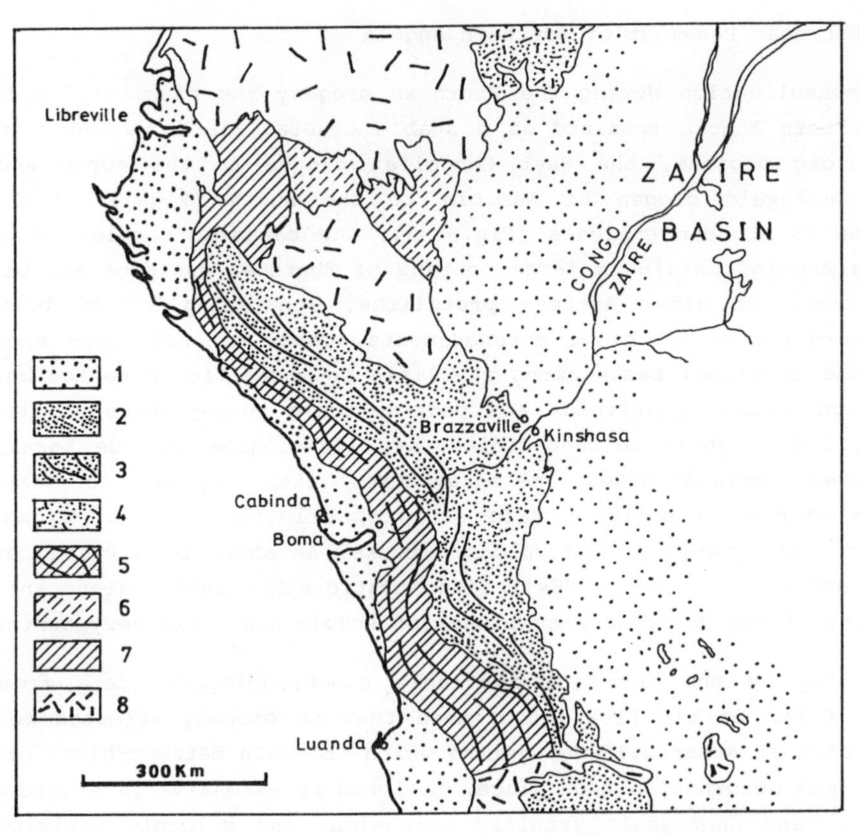

Figure 4.36: Geologic sketch map of the Atlantic Rise showing: 1, Phaerozoic cover; 2, Tabular West Congolian sequence; 3, Folded West Congolian; 4, Sembe-Ouesso Group; 5, pre-West Congolian formations in the West Congolian orogen; 6, Francevillian; 7, Ogooué-schists; 8, Archean basement (du Chaillu). (Redrawn from Cahen et al., 1984.)

In the internal zone of the West Congolian orogen the earliest Eburnean imprints occur as reactivated charnockitic basement in northern Gabon and southern Cameroon as well as in northern Angola. Basement reactivation occurred simultaneously with the deformation and metamorphism of cover rocks known as the Kimezian Supergroup in northern Angola and its correlatives, the "Série de la Loémé" in Congo and the "Série de la Doussa" in Gabon. The Kimezian Supergroup consisted of schists, quartzite and limestones which are now migmatites and gneisses. These had originally been metamorphosed in the amphibolite facies. They are folded along mainly NE, NNE and N-S trends. The "Série de la Loémé" in Congo consists mainly of mica schists and two-mica gneisses with garnets, and para-amphibolites. The deformation and metamorphism of these cover rocks occurred at about 2.08 Ga during what is locally known as the Tadilian orogenic event (Cahen et al., 1984).

Table 4.5: Stratigraphic subdivisions of the supracrustals in the internal zone of the West Congolian orogen. (Redrawn from Cahen et al., 1984.)

1. AGE	2. BAS-ZAIRE	3. CONGO	4. GABON-MAYOMBE SOUTH	5. GABON-MAYOMBE NORTH
V. 0.73-0·6 Ga	West Congolian supergroup	Système du Congo occidental Système des Monts Bamba	Système du Congo occidental Système des Monts Bamba	Système de la Noya
	disconformity to unconformity		unformity	presumed unconformity
IV. 1.0 Ga	Mayumbian supergroup	Séries de la Loukoula	Séries de la Loukoula	Groupe des chlorito-schistes
	unconformity		unconformity	unconformity
III	Zadinian supergroup Upper Zadinian or Tshela-Vangu group	Série de la Bikossi Groupe supérieur	Série de la Douigni	
	disconformity at least			Groupe des micaschistes
	Lower Zadinian or Matadi-Palaba group	Groupe inférieur	Série des Monts Kouboula	
TADILIAN OROGENY (2.08 Ga)	unconformity	? unconformity		
II. 2.1 Ga	Kimezian supergroup	Série de la Loémé	Série de la Doussa	Groupe des gneiss à plagioclase acides
				unconformity
I. 2.9-3.0 Ga				Groupe des gneiss à plagioclase basiques

In the Bas-Zaire and northern Angola the Kimezian is unconformably overlain by the Zadinian Supergroup, comprising two distinct groups: a lower group with micaceous quartzites, mica schists and acidic lavas; and a disconformable upper group which begins with abundant mafic lavas and is followed by chlorite and sericite-schist, quartzites, and some acidic pyroclastics in Bas-Zaire. In the latter area tholeiitic lavas which are andesitic towards the top, and over 500 m thick, occur at the base of the upper group. Zadinian sedimentation, metamorphism and deformation all fall between 2.08 Ga (the age of the underlying Kimezian) and 1.02 Ga, the age of the upper part of the overlying Mayumbian Supergroup (Table 4.5). Deformation, which was not intense in Bas-Zaire and Congo, was initially along ENE to NNE trending axes with increasing intensity towards the southern part in Angola. Metamorphism was at a lower grade than during the Late Proterozoic West Congolian orogeny.

4.9.5 Gabon Orogenic Belt

The Tadilian orogeny (Cahen et al., 1984) which produced the Eburnean basement in the internal and foreland zones of the West Congolian orogen (Fig.4.36) is better known further north in central Gabon, where the Gabon orogenic belt (Fig.4.37) comprises a major Eburnean tectono-thermal event interpreted as a collision zone (Ledru et al., 1989). Ledru et al. demonstrated that the metamorphic rocks of the Gabon orogenic belt display horizontal deformation and petrological characteristics which attest to Eburnean plate collision that involved major crustal shortening.

Stratigraphy of the Gabon Orogenic Belt

On the eastern slopes of the Atlantic Rise, the Archean du Chaillu medium-to high-grade granite-gneiss terrane (3.15-2.6 Ga old) is unconformably overlain in eastern Gabon by the Early Proterozoic Francevillian Supergroup in the Franceville and Booué basins. The undeformed Francevillian, dated between 2.3 Ga and 2.0 Ga, is overthrust on its western side by the Ogooué Metamorphics (Fig.4.37) which are also considered to be of Early Proterozoic age. The Ogooué Metamorphics are exposed in a broad synclinorium and consist essentially of paragneisses, with orthogneisses occurring locally at the base (Ledru et al., 1989). Both the Francevillian Supergroup and the Ogooué Metamorphics constitute the Gabon orogenic belt.

As summarized by Bonhomme et al. (1982), the Francevillian Supergroup comprises five subdivisions or formations. At the base is the coarse feldspathic conglomeratic Mabinga sandstone formation, about 1,000 m

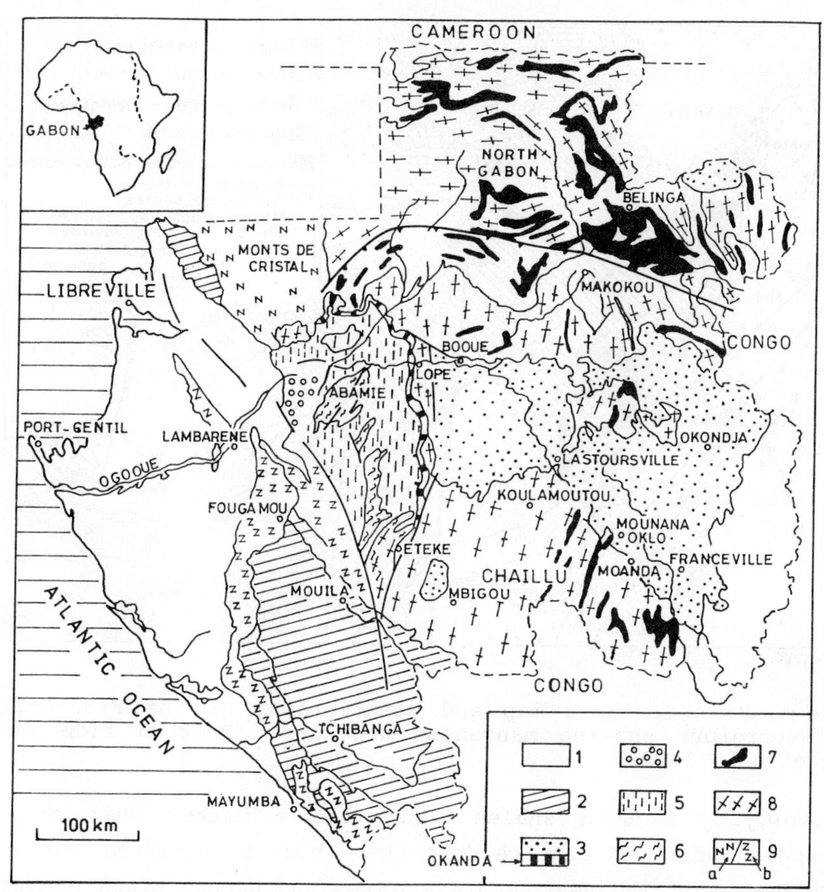

Figure 4.37: Geology of the Gabon orogenic belt. 1, Mesozoic cover; 2, Upper Proterozoic of the Nyanga Syncline; 3, Lower Proterozoic (Francevillian); 4, N'Djole Basin; 5, Ogooué-Metamorphics; 6, Migmatitic domes; 7, Iron formation; 8, Chaillu and North Gabon granitic domains; 9a, medium- to high-grade gneisses of the Monts de Cristal; 9b, Lambaréné Migmatities 3.09 Ga old. (Redrawn from Ledru et al., 1989.)

thick. The sandstones comprise a red facies with hematite, dolomite, anhydrite and gypsum, while a discordantly overlying non-oxidized facies contains pyrite and asphaltic material. Rich uranium deposits are mined from the top of the Mabinga conglomerate sandstone formation at Oklo and Mounana (Fig.4.38). Slightly discordant over the Mabinga sandstones are about 600 m of calcareous black shales (Bangombé shales) and lenticular channel sandstones, the Poubara sandstones. The rich and productive manganese deposits at Moanda formed from the weathering of a top horizon rich in manganese carbonate which occurs with banded iron-formation containing siderite and greenalite. With an annual production of about 1.6 million tons and reserves of up to 100 million tons, supergene processes had enormously upgraded the manganese content from 15 % in the black shales to about 45 % in the ores (Hutchison, 1983).

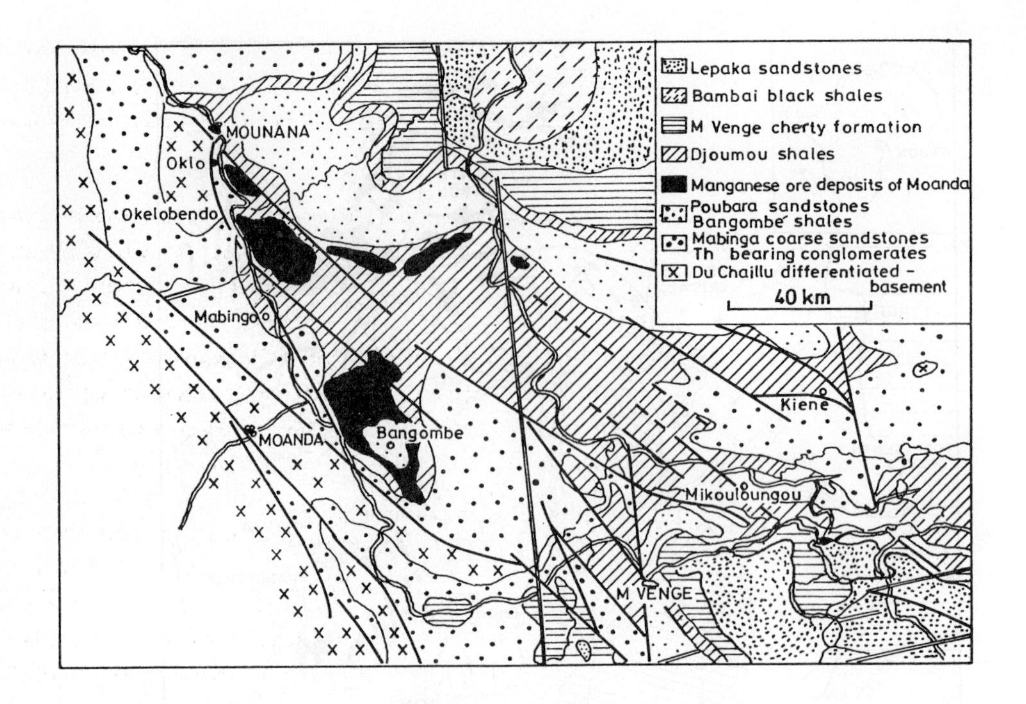

Figure 4.38: Geologic map and stratigraphy of the Francevillian Supergroup showing manganese deposits. (Redrawn from Bonhomme et al., 1982.)

The overlying Djoumou shales constitute a marker unit of mostly thickly banded chert and massive dolomite, about 50 m thick, which oversteps onto basement. The next unit comprises the Bambai black shales with associated ignimbritic tuffs at the top. The Lepaka sandstones constitute the uppermost unit and comprise alternations of sandstones and shales.

In the Franceville basin the Francevillian Supergroup is unmetamorphosed and undeformed except in the zones of early fracturing which had favoured the concentration of uranium. But further west in the Booué basin (Fig.4.37) where the Francevillian rests on the Archean basement of the Lopé horst with a basal conglomerate and sandstone unit, the Supergroup had undergone a phase of deformation accompanied by low-grade metamorphism (Ledru et al., 1989). Further to the west, the Francevillian clastic rocks terminate, having been overthrust by the Ogooué Metamorphics (metasandstones and metapelites) and by the migmatitic and granitic rocks of the Abamié and Diany Miyolé domes. Ledru et al. (1989) discussed the structure and petrology of the Francevillian Supergroup and the Ogooué Metamorphics in detail, thus furnishing the basis for the present account. Ogooué rocks at the base contain mainly basic to acid metavolcanics which wrap around the migmatitic domes. The Okanda or Intermediate series occurs at the western contact of the Lopé horst (Fig.4.37) and has a lithology that is identical with the basal part of the Francevillian, but shows structural and metamorphic characteristics which appear to be continuous with the Ogooué series.

Structure and Metamorphism

The Francevillian Supergroup in the Booué basin suffered two main superposed phases of deformation. An earlier phase of folding produced recumbent isoclinal folds with an axial plane schistocity which is parallel to bedding in pelitic beds and resulted in the repetition of resistant sandstone intervals. Deformation resulted in eastward overturned folds which were accompanied by greenschist-facies metamorphism. A later phase of folding produced some tight folds with N140°E axes. However, the tabular Francevillian to the east did not show such polyphase deformation, whereas to the west, the Okanda series exhibits two deformation phases of greater intensity, with intermediate-grade metamorphism in which kyanite appeared. Metamorphism in the Okanda series suggest burial to a depth of about 15 km. But since the maximum thickness of the Francevillian did not exceed about 2,000 m, this depth of burial has been attributed to tectonic thickening of the crust. The eastward overturned recumbent folds, flat-lying schistocity, décollment at the Francevillian-Archean contact, and the presence of late metamorphic shear zones (Fig.4.39) all suggest that the western part of the Francevillian was involved in horizontal deformation and crustal thickening.

In contrast to the Francevillian, the Oguooé Metamorphics show three phases of deformation. An early phase, which was overprinted by the main phase during which folds were produced which are overturned away from the axis of the Abamié migmatitic dome (Fig.4.39). The folds of the second phase diverge away from the core of the Abamié anticlinorium and while they are overturned here westward, they are nearly horizontal on the eastern margin. The third phase of deformation is represented by open folds with subvertical or near-horizontal axial planes.

Tectonic Model for the Gabon Orogenic Belt

The salient structural features of the Gabon mobile belt which Ledru et al. (1989) in their model have sought to explain are the eastward-directed thrusts, the superposition of progressively deeper structures, and the inverted metamorphic isograds. An east to west examination of this mobile belt revealed the following general tectonic features (Ledru et al., 1989): (1) peak metamorphism was reached in the Abamié anticlinorium where the structures are centered and from which they diverge; (2) the eastern flank of the Ogooué synclinorium has been affected by eastward-directed horizontal tectonics with the thrusting of the medium-grade Ogooué Metamorphics onto the low-grade Francevillian; (3) the Francevillian on either side of the Lopé horst is parautochthonous, having been detached from its Archean basement and affected by the same tangential tectonics, whereas in the Lastoursville basin, more than 100 km to the east, the Francevillian is truly autochthonous; and (4) the Archean has been involved to varying degrees, with Archean basement tectonic slices being incorporated in places. The Abamié gneissic dome has been interpreted as a mature, mushroom-shaped diapir.

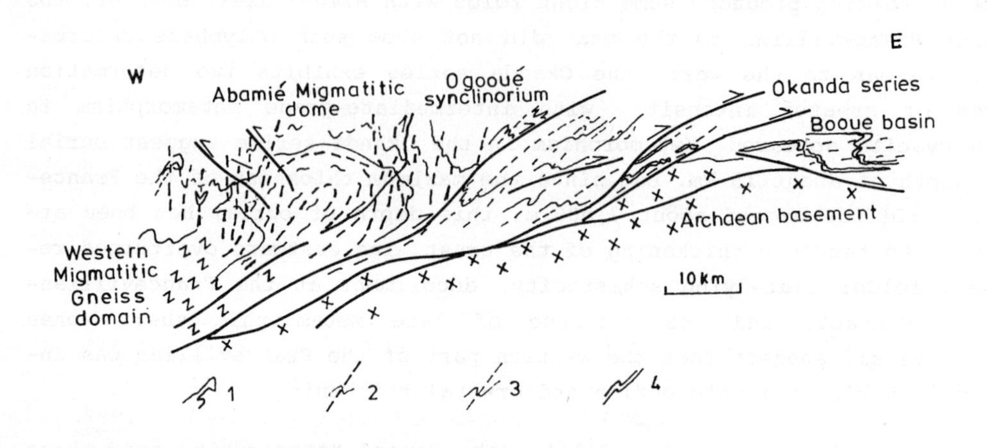

Figure 4.39: Structural section across the Gabon orogenic belt. 1, S1; 2, S2 in the sillimanite zone correlative with the foliation in the Abamié migmatites; 3, S2 in the staurolite-muscovite-biotite-garnet zone correlative with the first schistocity in the Francevillian; 4, stratification in the Francevillian. (Redrawn from Ledru et al., 1989.)

Ledru et al. (1989) likened the tectonic features in the western part of the Gabon mobile belt to those in continent-continent collision belts. Such belts contain broad areas of medium-to high-grade metamorphism ; nappes emplaced during Barrovian metamorphism upon a prograde metamorphic foreland; and migmatitic domes with granitic intrusive suites. Ledru et al. placed the tectonic events recorded in the rocks of the Gabon orogenic belt in the Eburnean orogeny (about 2.0 Ga). This suggest an important regional convergence during the Early Proterozoic leading to the crustal shortening seen in the Ogooué series and the Francevillian. These features were probably generated in a collisional setting.

Ledru et al. (1989) drew attention to the presence of komatiitic flows in a volcanic sequence in the Eteke area (Fig.4.37) probably in the basal Okanda series, and to a substantial volcanic-subvolcanic belt with ultrabasic rocks and tholeiitic metabasalts, located near the base of the Francevillian Supergroup in the southern region. Also, the base of the Ogooué series contains acid and basic volcanics. These magmatic suites point to an extensional regime during the deposition of the basal Proterozoic rocks in the Gabon mobile belt, and suggest the possibility of a full Wilson Cycle operating in the Gabon belt during the Eburnean orogenic cycle. Needless to say, the orogeny of the Gabon belt did not terminate at the southern frontier of Gabon Republic, but affected a belt which continued southward into what we considered earlier (Ch.4.9.4) as the Eburnean basement of the internal and foreland zones of the West Congolian orogen (Fig.4.37).

4.10 The Ubendian Belt of Central Africa

4.10.1 Introduction

The Ubendian orogenic belt (Fig.4.1) of central Africa was affected by the Early Proterozoic Eburnean orogeny, locally termed the Ubendian orogeny (McConnell, 1972), and dated between 2.05 Ga and 1.85 Ga (Cahen et al., 1984). The Ubendian belt is exposed along the Western Rift Rise, a north-south highland region which encompasses the Western Rift Valley, the Ruwenzori Mountains, and the Zaire-Nile watershed (upland separating the Zaire basin from the Tanzania plateau) (Fig.4.40). Whereas the Western Rift Rise is topographically the product of Mesozoic-Cenozoic regional uplift, geologically it is a polyorogenic belt bearing the imprints of Early, Middle and Late Proterozoic tectono-thermal activities. Furthermore, within the Ubendian belt are scattered remnants of Archean rocks which were once continuous with the Tanzania craton to the east and with the Zaire craton to the west.

Between the Late Archean and about 1.85 Ga in the Early Proterozoic, thick supracrustal assemblages accumulated and were deformed and metamorphosed during the Ubendian event. The Ubendian belt extends from northern Malawi and northeastern Zambia, western and southern Tanzania,

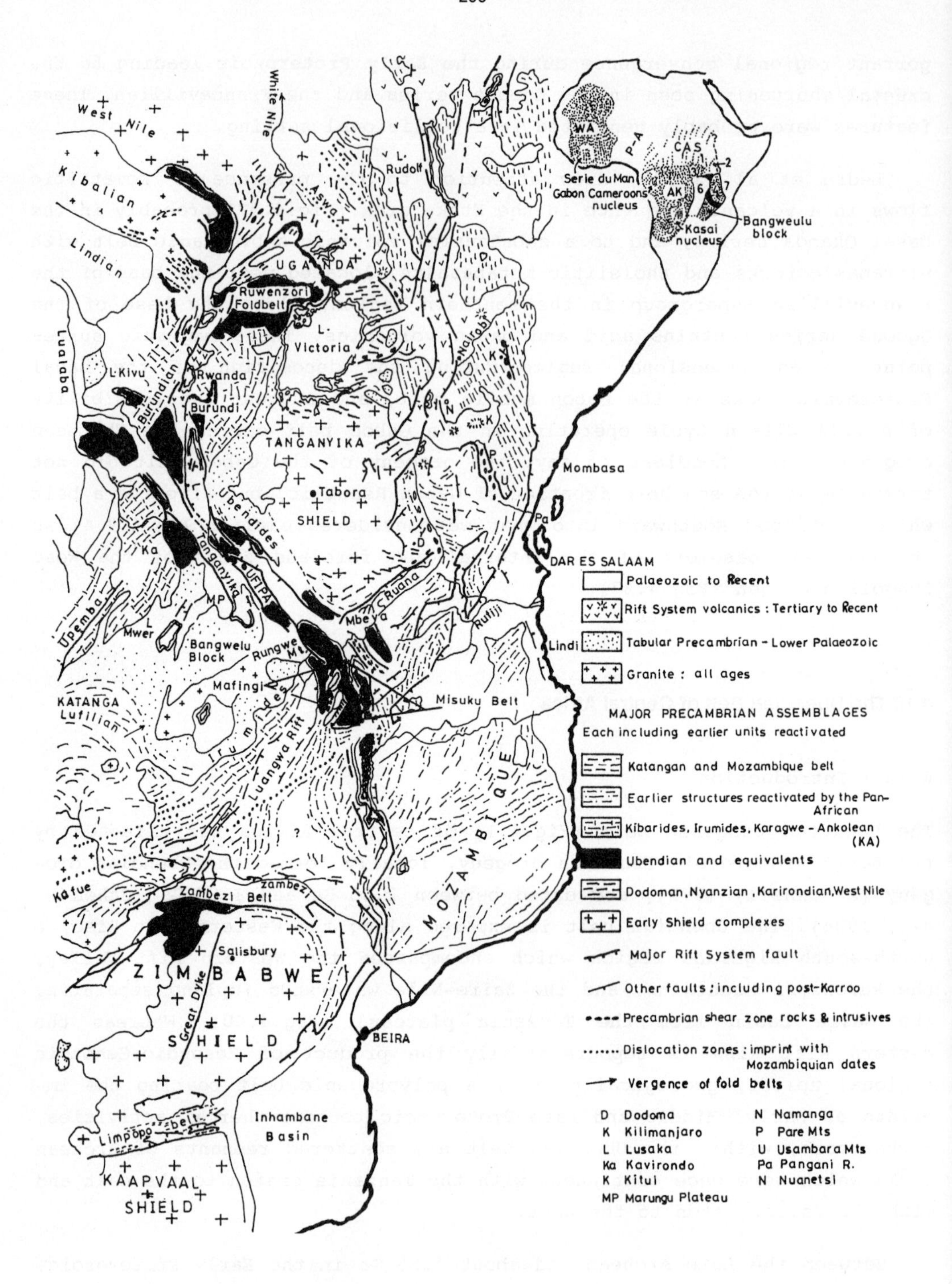

Figure 4.40: Outline Precambrian tectonic map of East Africa. (Redrawn from McConnell, 1972.)

Burundi, Rwanda and neighbouring parts of Zaire, to the Ruwenzori Mountains and the adjoining parts of western and central Uganda and northeastern Zaire.

4.10.2 Ubendian Rock Assemblages and Tectonism

Malawi and NE Zambia

In its southernmost parts in northern Malawi and northeastern Zambia the Ubendian belt is referred to by the local name, the Misuku belt. The Misuku belt (Fig.4.40) is a structural unit of gneisses which has been subdivided into the Songwe gneisses, the Chambo gneisses and the cordierite gneisses or Jembia River granulites (Cahen et al., 1984). Along the Malawi-Tanzania border the Songwe gneisses form a southeast-trending group of micaceous and amphibolite gneisses, schists and ferruginous quartzite separated by a shear zone from the pelitic and semi-pelitic Chambo garnet-sillimanite-mica gneisses and schists, with widespread migmatites and some augen-gneisses and ferruginous quartzite. The cordierite-sillimanite gneisses with east-southeast foliation constitute the southernmost group of gneissic rocks; these have tectonic contacts with the Chambo gneisses within which they are sometimes interlayered.

In northern Malawi and northeastern Zambia the Ubendian tectonothermal events show varying metamorphic grades, from greenschist facies in the south, through lower-to-middle amphibolite facies, to middle and upper amphibolite facies in the north where cordierite is completely destroyed. Two episodes of tectonic movements affected the Misuku gneisses which resulted in a regional foliation and fold pattern with isoclinal large-scale folds with southeast-trending axial planar schistocity, followed by boudinage and irregular folds in the Chambo gneisses. The Nyika granites and their associated migmatites and satellite granite bodies were emplaced at 2.05 Ga during the peak of the Ubendian orogeny.

Ubendian Terranes along the Southwestern Margin of the Tanzania Craton

This region is the type area of the Ubendian belt where a variety of high-grade metamorphic rocks of both sedimentary and igneous origin lie in several discrete blocks or terranes (Fig.4.41,A). The Ubendian terranes extend from Lake Tanganyika through the Ufipa plateau and margins of Lake Rukwa to the northern margins of Lake Malawi. The Ufipa gneiss complex corresponds to the Chambo gneisses in the south (northern

Malawi and NE Zambia), while the Ubende "Series" corresponds to the Songwe gneisses.

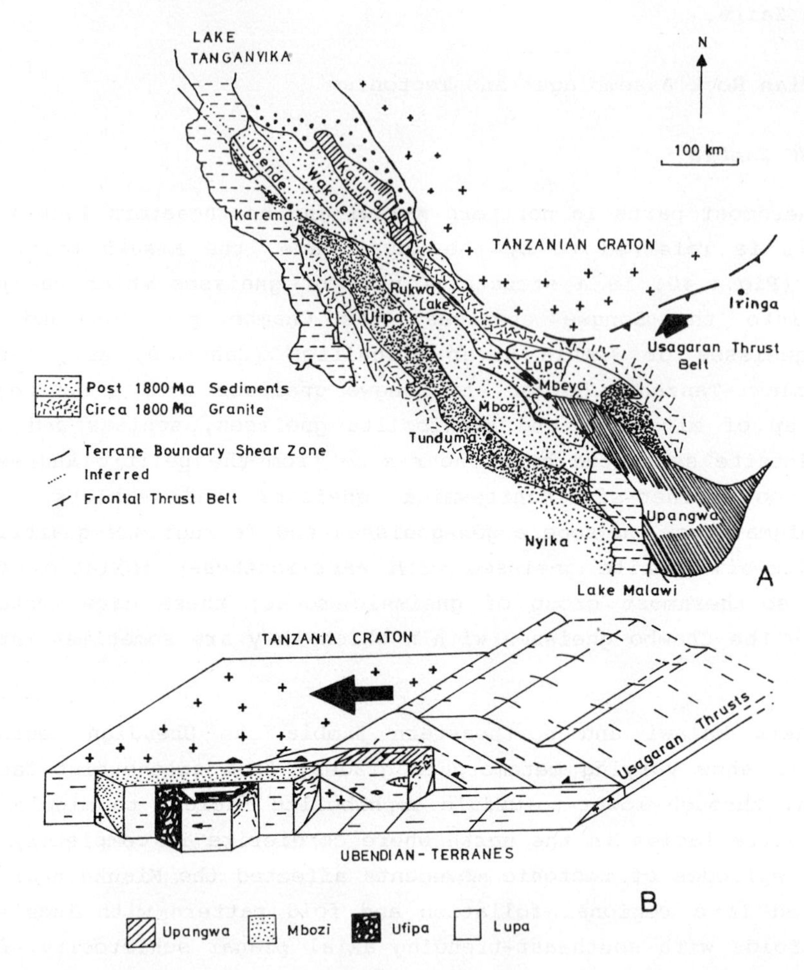

Figure 4.41: A, Ubendian terranes, each bounded by shear zones; B, tectonic model for the evolution of Ubendian belt terranes. (Redrawn from Daly, 1988.)

Figure 4.41 A shows the terranes that have been recognized on lithostratigraphic and structural criteria (Daly, 1988) in the Ubendian type area. These comprise, from south to north: (1) the Upangwa (meta-anorthosite) with NW-SE stretching lineation trend; (2) the Mbozi (meta-basites and intermediate granulites with quartzites) with NE-SW stretching lineation trend; (3) the Lupa (meta-volcanics); (4) the Ufipa (gneissic granite) with stretching lineations trending NW-SE; (5) the Nyika (cordierite granulites) with east-west lineation; (6) the Ubende (meta-basites) with ENE-WSW lineation; and (7) the Wakole terrane comprising alumino-silicate schists. Together these terranes occupy a NW-

SE-trending belt which extends over a distance of about 400 km, with major shear zones defining the individual terrane boundaries.

Daly (1988) presented a structural analysis and tectonic model for the Ubendian terranes. The terranes are internally heterogeneously deformed with extensive tracts of foliated mylonitic and ultra-mylonitic gneisses in which the main foliation is folded generally about axes which are sub-parallel to the elongation of the terranes. Variations in finite strain orientation and the lithological differences among the terranes noted above, emphasize the discrete nature of each terrane. The terranes are bounded by major steep ductile and brittle shear zones which persist for up to 600 km and had long and complex structural history. Daly's (1988) tectonic model (Fig.4.41,B) for the Ubendian terranes suggests that the terranes were accreted adjacent to the Tanzania craton during the Early Proterozoic orogeny which he termed the Usagaran orogeny. The Usagaran event caused a series of NW-directed thrust sheets to be overthrust onto the Tanzania craton as shown diagrammatically in Fig.4.41 B. During the late stage of the Usagaran thrusting the Ubendian terranes probably developed as a series of tectonic slivers which accreted laterally onto the Tanzania craton.

Along the southeastern margin of the Tanzania craton is the ENE to NNE-striking Usagaran belt (Fig.4.41,A), wherein lies the Usagaran Supergroup which was folded during the Early Proterozoic orogeny against the Tanzania craton, its foreland. Later, the Usagaran belt in turn acted as the foreland to the Late Proterozoic Mozambique belt. The Usagaran Supergroup unconformably overlies the Archean basement of the Tanzania craton. It is a sequence of psammitic and pelitic metasediments comprising the basal Konse Series of quartzites, amphibolites, hornblende gneisses and marbles, conformably overlain by the middle and upper biotite and hornblende gneisses of psammitic and pelitic origin. Thus, the Usagaran is predominantly of the amphibolite facies with pyroxene granulites (showing retrogressive metamorphism to the amphibolite facies), and some charnockites, migmatites, interlayered quartzites, various intrusive rocks, and eclogite. The Usagaran is unconformably overlain by the Ndembara Series which is up to 3,500 m thick consisting of intermediate-to-acidic volcanic rocks and subordinate phyllites and quartzites (Cahen et al., 1984).

The occurrence of eclogites within the Usagaran amphibolites and in garnet-bearing gneisses showing the oval-shaped foliation pattern of a precursor volcanic plug, suggests that the eclogites originated during high-pressure metamorphism of probably mantle-derived precursor rocks

(Muhongo, 1989). The eclogites were probably emplaced when the Usagaran gneissic complexes were thrust in the southeast-northwest direction over the Tanzania craton, whereas the charnockites may represent a cryptic suture (Dewey and Burke, 1973). The Usagaran rocks are isoclinally folded with easterly dips and are thrust over the craton. The Ndembara was folded and metamorphosed at greenschist facies in the north and at the amphibolite facies in the south, and was also intruded by granites at about 1.86 Ga and 1.77 Ga.

To the north and northwest of Kalemie (Zaire) along the western part of Lake Tanganyika are poorly-known gneisses, garnet-bearing mica schist and whitish-to greenish interstratified quartzites which are intruded by various mafic rocks and granites dated at 1.8 Ga. These gneisses were also deformed by the Ubendian events and comprise the northern extension of the Irumide belt. In this region the metamorphic grade ranges from low-grade to the amphibolite and hornblende granulite facies. Near the southeastern tip of Lake Tanganyika and Zambia is the Chocha Group, a sequence of greenschist-to-amphibolite facies metasediments and meta-volcanics which rest unconformably on the Ubendian rocks. The Chocha Group is younger than 2.05 Ga but is older than its accompanying granites such as the Kate and Luongo granites which are dated at about 1.8 Ga.

The Ubendian in Burundi, Rwanda and Zaire

Archean basement rocks in this region (Fig.4.42) were subjected to retro-gressive metamorphism in the greenschist facies and deformation during the Ubendian orogeny. The Kazigwe amphibolite facies complex resulted from the mylonitization and retrogressive metamorphism of the Archean Kikuka gneiss complex in southwestern Burundi. On the Itombwe plateau in south Kivu, Zaire (Fig.4.42), the Archean basement has superposed upon it tightly folded mica schists, mica-quartzites and gneisses with north-south strikes. These deformed supracrustals are the products of the Ubendian orogeny in view of their structural continuity with the more southerly Ubendian type area. Around Uvira in Zaire, similarly deformed gneisses and metasediments with north-south trends appear to be the pro-ducts of the Ubendian deformation, especially as they are cut by pegmatites which are 2.03 Ga old.

The Ruwenzori Fold Belt

The Ubendian belt terminates northward as the Ruwenzori fold belt (Fig.4.40). The Ruwenzori belt is an east-trending, strongly tectonized orogenic belt which is situated south of the Archean basement complex of

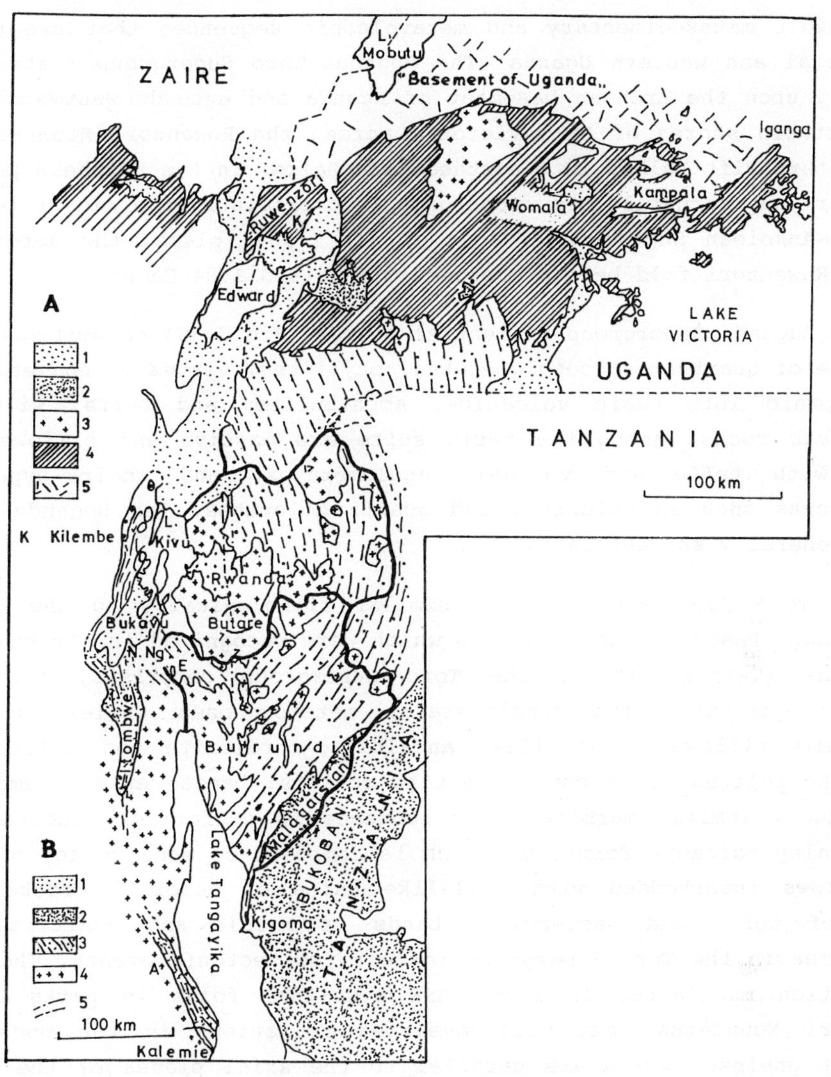

Figure 4.42: A, Ruwenzori fold belt showing: 1, Phanerozoic cover; 2, Karagwe-Ankolean (Kibaran); 3, Main granites of the Ruwenzori belt; 4, Buganda-Toro Supergroups; 5, Basement of Uganda, north of the Ruwenzori belt. B, Western Rift rise showing: 1, Cenozoic cover; 2, Bukoban and Malagarasian tabular sequences, folded formations with trends in the Itombwe Supergroup, and Upper Rumbu alkaline complex; 3, Burundian and Karagwe-Ankolean Supergroups (Kibaran belt); 4, Basement affected by Ubendian orogeny (C. 2.05 Ga). (Redrawn from Cahen et al., 1984.)

Uganda and east of the Archean Kibalian granite-greenstone belt of northeastern Zaire. The rock assemblages of the Ruwenzori fold belt belong to the Buganda-Toro Supergroup (Fig.4.42) which are the Early

Proterozoic metasedimentary and metavolcanic sequences that are exposed in central and western Uganda. The Buganda-Toro Supergroup rests unconformably upon the Archean basement of Uganda and extends westwards along the northern shores of Lake Victoria across the Ruwenzori Mountains and the Western Rift Valley into northeastern Zaire. In the southern part the Ruwenzori,supracrustals are unconformably overlain by the mid-Proterozoic Karagwe-Ankolean Supergroup. Cahen et al. (1984) placed the deformation of the Ruwenzori fold belt at between 2.5 Ga and 1.84 Ga ago.

The Buganda Supergroup of central Uganda (Fig.3.36) comprises a basal sequence of quartzites, conglomerates, phyllites, slates or shales, which pass upward into basic volcanics, amphibolites and tuffs, with some ultrabasic rocks. Among the basic suite are massive and pillowed lava flows with tuffs and volcanic agglomerates, and their hypabyssal equivalents such as dolerites and quartz dolerites. The Buganda Supergroup generally strikes east-west.

The Toro Supergroup is very similar lithologically to the Buganda Supergroup. Based on the sequences which were recognized in the Ruwenzori Mountains (Tanner, 1973), the Toro Supergroup comprises, from base upwards: quartzite and conglomerate; metamorphosed tholeiitic lavas (sometimes pillowed) and sills; andalusite-cordierite and sillimanite-muscovite pelites; less common biotite schists; banded epidote-amphibole rocks and dolomitic marbles. This sequence is conformably succeeded by the Stanley Volcanic Formation which is composed of massive and pillowed lava flows interbedded with sill-like units of doleritic rocks, with bands of tuffs and less-common bands of marble and quartzite. The structures in the Toro Supergroup reflect two tectonic events. The first deformation manifested in major ENE synclinal folds in parts of the Ruwenzori Mountains, and well-developed foliations in the underlying basement gneisses which are parallel to the axial planes of the folds. These major folds were affected by a younger event with axial traces extending north-south.

The Buganda and Toro Supergroups are equivalent lithostratigraphic units, not only because of their lithologic similarities but also because the base of the combined Buganda-Toro Supergroups extends almost continuously from the Ruwenzori Mountains to the Jinja area in central Uganda (Tanner, 1973). There are also structural similarities in the folding, schistocity and cleavage, with two folding episodes and schistosities.

The Luhule-Mobisio Group in the adjoining parts of northeast Zaire is lithologically similar to the Buganda-Toro Supergroups. The Luhule-Mobisio Group trends east-west and consists of (from bottom to top): slates and phyllites with small pebbles; quartzites with conglomerates and locally preserved ripple marks; shales, phyllites; quartzites; and a basic complex comprising intrusions and lavas. In the eastern outcrops this sequence appears to form a syncline with steep flanks, while the western outcrops are isoclinally folded with steep axial planes. The intense deformation of the Luhule-Mibisio Group during the Ubendian tectono-thermal event at about 2.0 Ga also affected the surrounding granite-greenstone belts of northeast Zaire. However, metamorphism in the Luhule-Mobisio Group was weak.

When traced southward, the north-south Ubendian structural trends in the southern outcrops of the Toro Supergroup appear to link up with the Ubendian belt of Rwanda through an Ubendian terrane of anatectic granite gneisses and migmatites. Cahen et al. (1984) placed the Buganda-Toro assemblage within the 2.05-1.85 Ga interval during which there was either one tectono-thermal event or two. An earlier event at about 2.1 Ga could have metamorphosed the lower part of the Toro Supergroup into gneisses and schists before the deposition and tectonism of the upper part of the Toro Supergroup. Regionally, the basic-to-ultrabasic lavas of the Buganda-Toro Supergroup, and possible intrusives which occur from northeastern Zaire to the area east of Jinja in central Uganda, have been considered as ophiolites which occur over a distance of about 550 km. The lavas and their related intrusives exhibit the same degree of high temperature-low pressure metamorphism as the surrounding metasediments and range from an unmetamorphosed state, through the greenschist facies, to the amphibolite facies, with the latter mainly developed on the Ruwenzori Mountains. Except on the Ruwenzori Mountains where there are the imprints of the mid-Proterozoic Kibaran event, other parts of the Buganda-Toro terrane show no effects of later tectono-thermal events.

Mineralization

The Ubendian belt is generally poorly mineralized. A stratiform syngenetic copper-cobalt mineralization occurs at Kilembe in the Toro Supergroup (Fig.4.42). The stratiform sulphide deposits at Kilembe occurs in a banded amphibolite horizon within a sequence of schists and amphibolites which are believed to be of sedimentary origin (Tanner and Bailey, 1971). Disseminated pyrite and chalcopyrite mineralization characterizes all the basic rocks of the Buganda Supergroup and the Stanley Volcanic Formation of the Toro Supergroup.

In Tanzania shear zones and fissures in the Ubendian rocks were invaded by mineralizing hydrothermal fluids during late-orogenic igneous activity, resulting in epigenetic lead, copper and gold mineralization at Mpanda (Fig.3.36), and gold mineralization at Lupa near the Rukwa trough (Harris, 1981). The Mpanda mineralization occurs as fissure veins and as disseminations and replacements in shear zones where there are galena and chalcopyrite, with some silver and gold in a quartz-siderite gangue. Gold mineralization in the Lupa goldfield is associated with granodiorites which, with related diorites, albitic rocks, and potash-rich alaskites, intrude Ubendian migmatitic gneisses. The gold occurs at Lupa in well-defined veins in fissures and shear zones in association with pyrite and occasional minor chalcopyrite; in association with varying proportions of pyrite, chalcopyrite and galena; or without sulphides in a siliceous gangue, sometimes in association with chloritic material.

4.11 The Bangweulu Block

Previously referred to as the Zambia craton (Kröner, 1977), or the Zambia nucleus (Clifford, 1970), the Bangweulu Block (Drysdall et al., 1972) is a small Eburnean craton situated in northeastern Zambia (Fig.4.40, inset). The Bangweulu Block is completely surrounded and isolated from the nearby Zaire and Tanzania cratons by the Ubendian mobile belt to the east, the Middle Proterozoic Kibaran foldbelt and the coeval Irumide foldbelt to the northwest and southeast, respectively, and by the Late Proterozoic Lufilian arc in the southwest (Fig.4.40). Anderson and Unrug (1984) showed that the Bangweulu Block comprises a crystalline basement of schists, granitoids, metavolcanics and a weakly deformed sedimentary cover which formed during the Eburnean event, between 2.0 Ga and 1.8 Ga, as a continuation of the Ubendian mobile belt.

4.11.1. Geological Evolution

Anderson and Unrug (1984) did not identify Archean rocks in the Bangweulu Block, hence they attributed it solely to the Eburnean. The basement complex of the Bangweulu Block is actually an extension of the Ubendian basement of northern Malawi and southwestern Tanzania. The oldest rocks in the Bangweulu basement occur as small schist belts (Fig.4.43) where quartzo-feldspathic rocks derived from semi-pelitic to psammitic and acid volcanics are believed to be the stratigraphic equivalents of the pre-

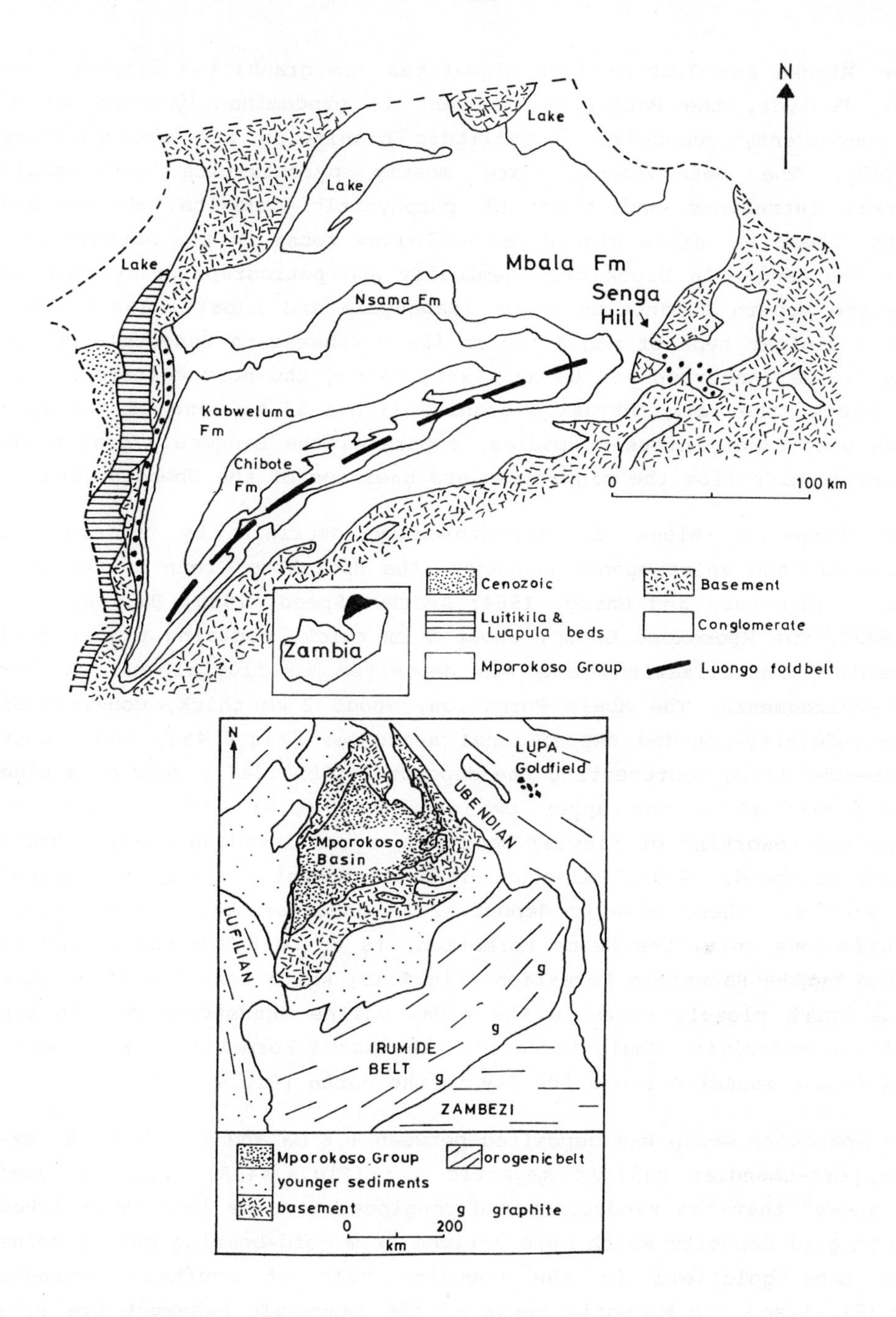

Figure 4.43: Geologic map and tectonic setting of the Mporokoso Basin. (Redrawn from Andrews-Speed, 1989.)

Ubendian Misuku and Jembia River migmatites and granulites already considered. However, the Bangweulu basement is predominantly composed of large concordant, composite, batholithic granitoids and metavolcanics (Fig.4.43). The metavolcanics are mostly pyroclastics with small hypabyssal intrusions and flows of porphyritic andesite, dacite and rhyolite. These extensive high-K calc-alkaline rocks in the northwestern part of the Bangweulu Block are chemically and petrographically akin to Andean-type modern subduction zones (Kabengele and Lubala, 1987). They suggest a similar type of magmatism in the northwestern Bangweulu region between 2.0 Ga (Anderson and Unrug, 1984). Along the northeastern margin of the block, the Kate intrusive granite (Fig.4.44,A), one of the late Eburnean discordant intrusive bodies, separates the Bangweulu granitoids and metavolcanics from the migmatites and gneisses of the Ubendian belt.

The Bangweulu Block is unconformably overlain by a cratonic siliciclastic and volcanogenic sequence, the Mporokoso Group of the Muva Supergroup (Anderson and Unrug, 1984; Andrews-Speed, 1989; Daly and Unrug, 1982). The Mporokoso Group, about 5 km thick, contains placer gold and uranium mineralization, and was deposited in fluvial and shallow marine environments. The Mbala Formation, about 2 km thick, consists of four unconformity-bounded depositional sequences (Fig.4.45), the lowest two sequences (A,B) representing the deposits of braided rivers on a wide fluvial plain, while the upper two sequences (C,D) each reflect the flooding and reworking of fluvial sediments by an invading shallow tidal sea (Andrews-Speed, 1989). Paleocurrent data suggest a southern sediment source area for these clastic deposits. An unconformable and extensive acid tuffaceous unit, the Nsama Formation, is up to 600 m thick, and is succeeded by the Kabweluma Formation (Fig.4.45) which consists of quartz-arenites which closely resemble the tidal marine sandstones at the top part of the underlying Mbala Formation. The Nsama Formation was probably derived from a magmatic arc which lay to the north (Fig.4.44,A).

The Mporokoso Group was deposited between 1.8 Ga and 1.1 Ga on an extensive post-Ubendian silicic magmatic arc (Fig.4.44,A). Andrews-Speed (1989) showed that the sandstones and conglomerates are locally enriched in placer gold deposits which were derived from gold-bearing quartz veins in the Lupa goldfield in the Ubendian belt of southern Tanzania (Fig.3.36). Also, the magmatic rocks of the Bangweulu basement are rich in uranium (10-60 ppm) which according to Andrews-Speed (1989) could have been redistributed into the Mporokoso Group during the evolution of the nearby Irumide belt (1.4-1.0 Ga).

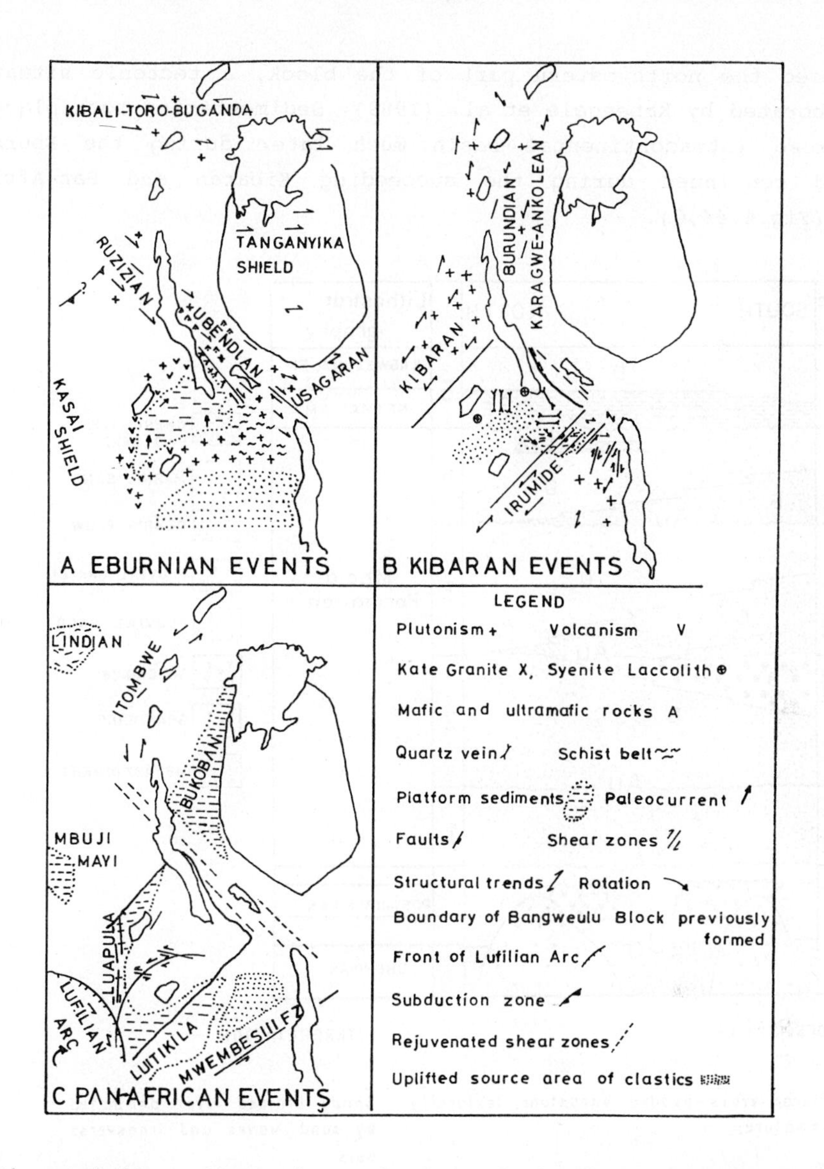

Figure 4.44: Tectonic evolution of the Bangweulu block during the Eburnean, Kibaran and Pan-African. (Redrawn from Anderson and Unrug, 1984.)

Figure 4.44 (Anderson and Unrug, 1984) shows a model for the geodynamic evolution of the Bangweulu terrane. The crystalline basement formed during the late Eburnean by the diapiric emplacement of granitoid batholiths in the southeastern part of the Bangweulu Block, whereas slightly later, a high-K calc-alkaline volcanic arc and associated intermediate and acid batholithic intrusives and basic plutons are believed to have occupied the northwestern part of the block, a tectonic situation also corroborated by Kabengele et al. (1989). Sedimentation took place in the Mporokoso intracontinental basin much later during the Eburnean event, and continued during the succeeding Kibaran and Pan-African orogenies (Fig.4.44,C).

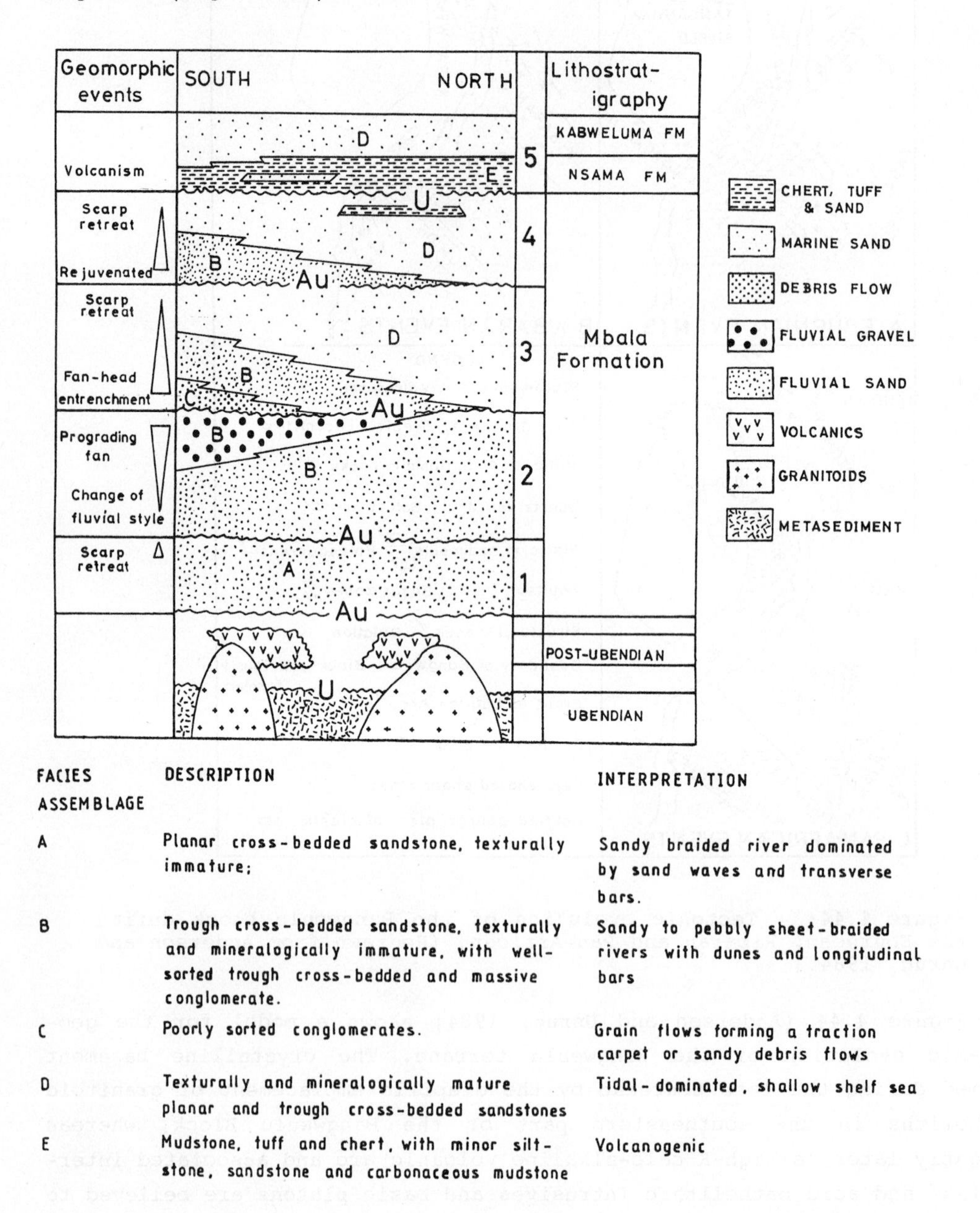

FACIES ASSEMBLAGE	DESCRIPTION	INTERPRETATION
A	Planar cross-bedded sandstone, texturally immature;	Sandy braided river dominated by sand waves and transverse bars.
B	Trough cross-bedded sandstone, texturally and mineralogically immature, with well-sorted trough cross-bedded and massive conglomerate.	Sandy to pebbly sheet-braided rivers with dunes and longitudinal bars
C	Poorly sorted conglomerates.	Grain flows forming a traction carpet or sandy debris flows
D	Texturally and mineralogically mature planar and trough cross-bedded sandstones	Tidal-dominated, shallow shelf sea
E	Mudstone, tuff and chert, with minor siltstone, sandstone and carbonaceous mudstone	Volcanogenic

Figure 4.45: Stratigraphy and gold and uranium mineralization in the Mporokoso Group. (Redrawn from Andrews-Speed, 1988.)

Deformation of the Mporokoso basin was largely confined to a narrow fold zone known as the Luongo fold belt (Fig.4.46). This folding was related to NW-directed thrust tectonics of the Kibaran age Irumide belt (Daly, 1986a).

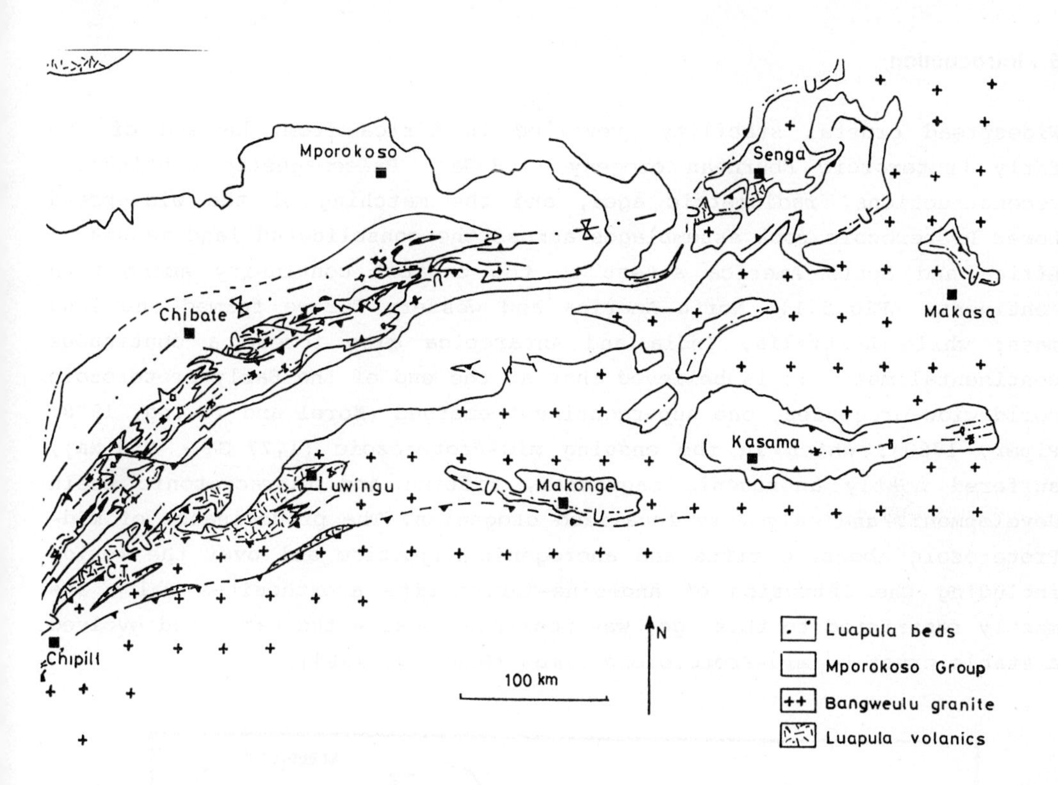

Figure 4.46: Structural sketch map of the Luongo zone. (Redrawn from figure supplied by M. C. Daly.)

Chapter 5 The Mid-Proterozoic Kibaran Belts

5.1 Introduction

Widespread crustal stability prevailed in Africa from the end of the Early Proterozoic Eburnean orogeny (1.8 Ga). Paleomagnetic continental reconstructions, radiometric ages, and the matching of the widespread Lower Proterozoic rock assemblages across the consolidated land masses of Africa and South America attest to the crustal continuity among both continents (Fig.5.1). North America and western Europe formed one land mass; while Australia, India and Antarctica also formed a continuous continental mass. It is believed that at the end of the Early Proterozoic world-wide orogenies, one supercontinent emerged (Morel and Irving, 1978; Piper, 1980), which in the ensuing mid-Proterozoic (1.77 Ga - 950 Ma), suffered mostly anorogenic magmatism, rifting and intracratonic basin development, and only very localized orogenies. The predominance of mid-Proterozoic abortive rifts and anorogenic magmatism all over the world, including the intrusion of andesine-labradorite anorthosites which are mostly restricted to this age, was possible because the Earth had evolved a stable crust by mid-Proterozoic times (Windley, 1984).

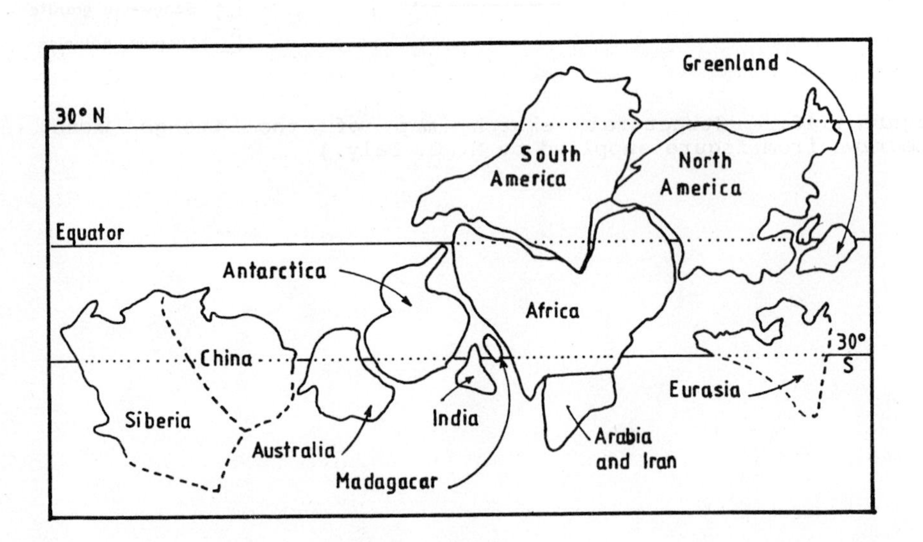

Figure 5.1: Paleomagnetic reconstruction of the supercontinent in the Late Proterozoic. (Redrawn from Condie, 1989.)

In Africa the mid-Proterozoic crustal quiescence was interrupted only in limited parts of the Eburnean terranes south of the Sahara, including the Namaqua and Natal mobile belts which, as already noted, stabilized as parts of the Kalahari craton after mid-Proterozoic orogenesis. The scene of mid-Proterozoic rifting, sedimentation, magmatism, metamorphism and deformation, was concentrated in the eastern part of central Africa, in the terranes between the Zaire, Tanzania and Kalahari cratons. In this region a mid-Proterozoic orogenic event, known as the Kibaran orogeny, took place in northeast-southwest-trending intracratonic mobile belts, the Kibaran, Irumide, and the Southern Mozambique belt (Fig.5.2). Kibaran rifting, sedimentation, magmatism, metamorphism and deformation transpired between about 1.45 Ga and 950 Ma, in central and southern Africa and did not extend northward into northeastern Africa (Vail, 1989a); it is also doubtful whether the Kibaran orogeny affected West Africa, (Black, 1984; Caby, 1989). However, outside the Kibaran mobile belts Kibaran-age sedimentary and magmatic rocks are known in Angola, the Sudan, the Tuareg Shield, the Benin-Nigerian Shield, and in Madagascar (Fig.5.2). Since these include volcano-sedimentary assemblages that actually belong to the initial phase of the Pan-African orogeny, some of these are considered in the next chapter.

From the mid-Proterozoic onward, the dominant type of ore mineralization in Africa changed from what it was in the cratonic terranes of the earlier Precambrian. Whereas the predominant mineralizations in the cratons are gold, iron, chromium, asbestos and diamond, the mid-Proterozoic and younger terranes in Africa constitute what Clifford (1966) considered the structural metallogenic domain that is characterized by major deposits of copper, lead, zinc, cobalt, beryllium, tin, tungsten and niobium-tantalum. In the Kibaran belt of Burundi, Rwanda, western Tanzania, and southwestern Uganda, there are economic deposits of tin, beryllium, tungsten, niobium-tantalum, gold and lithium.

5.2 Kibaran Mobile Belts

The Kibaran tectono-thermal event occurred in three parallel contemporaneous orogenic belts in central eastern Africa (Fig.5.3). The northerly Kibaran belt consists of two segments, the western type area and the eastern belt, and together with the Irumide belt to the south, these basins constitute intracratonic extensional basins which are transverse to the northwest-southeast-trending Ubendian belt. The Kibaran

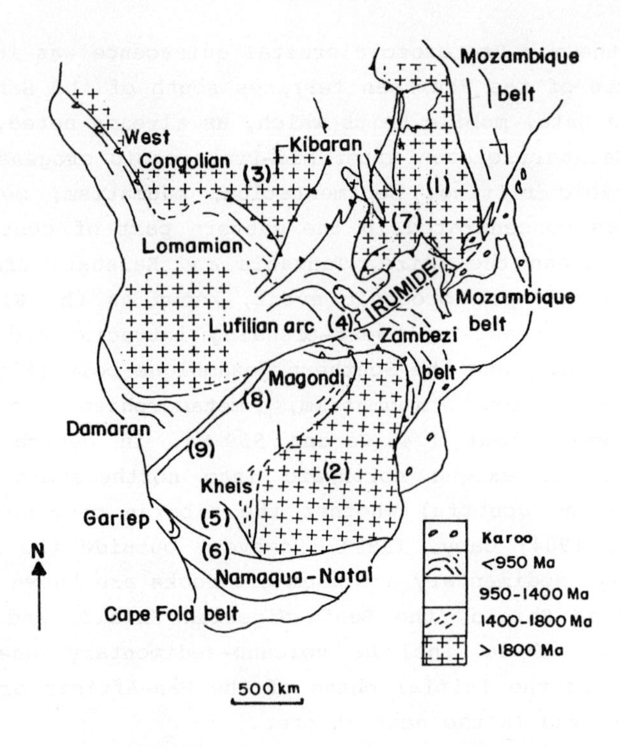

Figure 5.2: Mid- to Late-Proterozoic tectonic features of Southern Africa. (Redrawn from figure supplied by M. C. Daly.)

basins (Fig.5.3) are believed to have originated between 1.4 and 1.35 Ga during sinistral shearing along wrench faults in the Ubendian belt (Klerkx et al., 1987). Klerkx et al., (1987) likened the extensional origin of the Kibaran belts to that of the Basin-and-Range province of southern California, U.S.A., where strike-slip movements along wrench faults have produced oblique normal fault-bounded sedimentary basins.

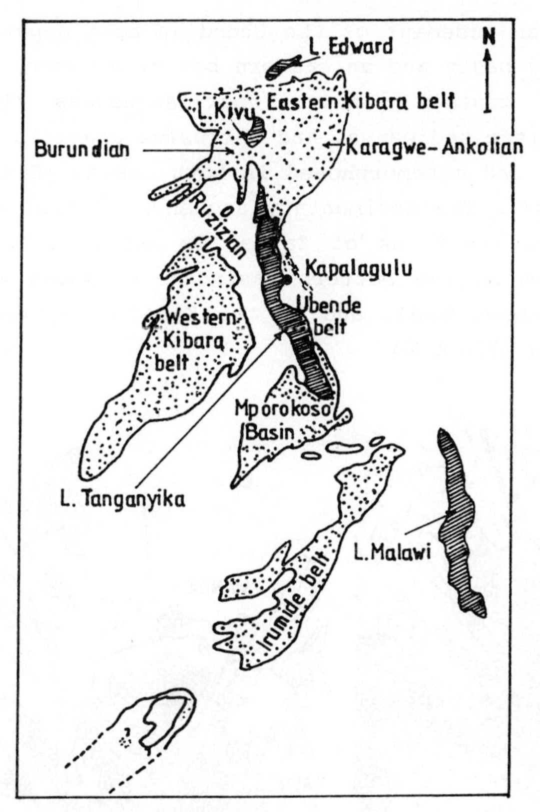

Figure 5.3: Kibaran basins of central Africa. (Redrawn from Klerkx, 1988.)

In contrast with the Kibaran intracratonic basins to the north, the Southern Mozambique belt of eastern Zambia, Malawi and northern Mozambique appear to have resulted from plate margin processes (Andreoli, 1984; Piper et al., 1989; Sacchi et al., 1984). The preponderance in this region of high-grade gneisses with slices of ophiolitic mafic and ultramafic rocks suggests that the Southern Mozambique belt, like the older Limpopo belt to the south, is a deeply exposed convergent margin and collision belt (Burke et al., 1977), where island arcs and oceanic crust accreted onto the African continent.

5.2.1 The Kibaran Belt

Lithostratigraphy

The Kibaran mobile belt extends for about 1,500 km from southeastern Zaire, through Burundi and Rwanda, to southwestern Uganda and western Tanzania. It is poorly developed in the middle part where the Lower

Proterozoic Rusizian basement of the Ubendian belt separates the Kibaran belt into a western basin and an eastern basin. However, both segments of the Kibaran belt contain similar rock sequences (Fig.5.3), mainly quartzites and pelitic sediments, and intrusive granitic rocks, which are similarly deformed and metamorphosed in both basins (Cahen et al., 1984; Klerkx et al., 1987). The sedimentary sequence in the western basin, in the Shaba and Kivu provinces of Zaire, is referred to as the Kibaran Supergroup, whereas in the eastern basin it is known as the Burundian Supergroup in Rwanda-Burundi, and as the Karagwe-Ankolean Supergroup in Uganda and Tanzania (Fig.5.4).

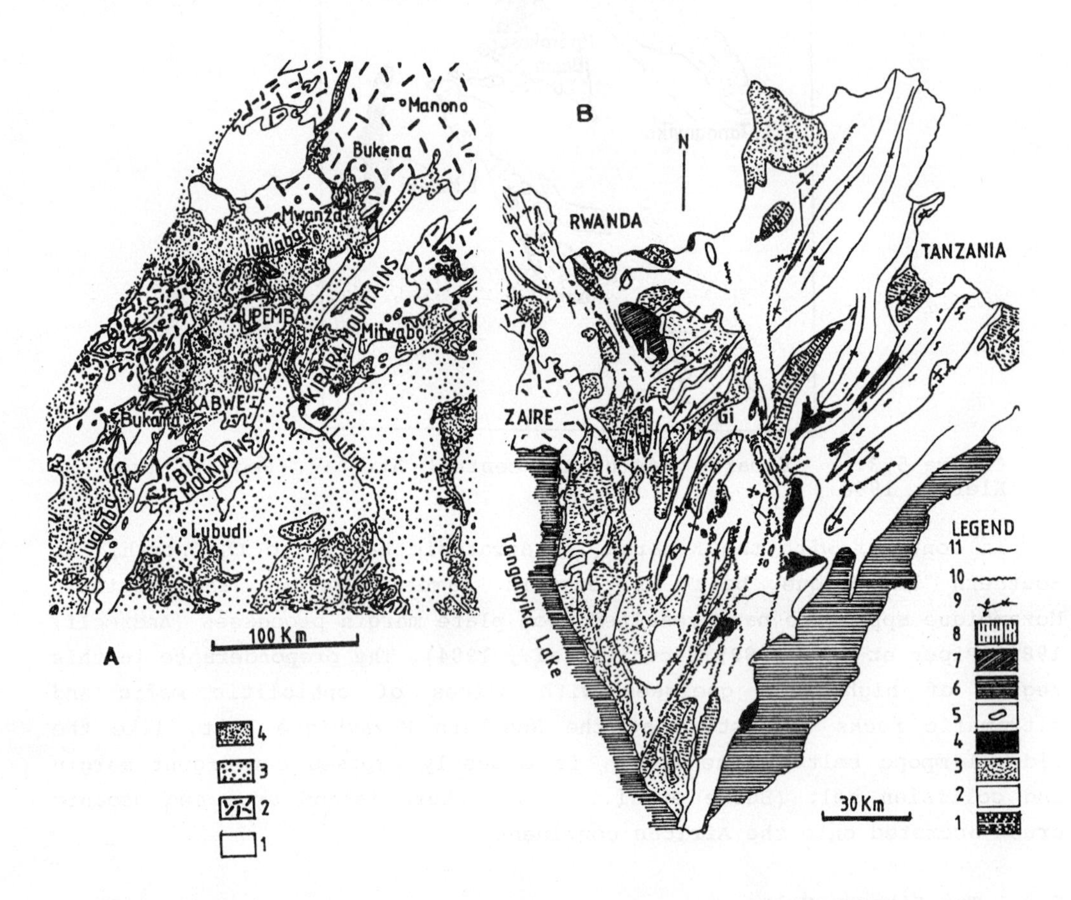

Figure 5.4: A, Kibaran belt in Central Shaba, Zaire; 1, Kibaran Supergroup; 2, Granites; 3, Katangan Supergroup; 4, Phanerozoic; B, the Kibaran belt in Burundi showing: 1, Archean; 2, Burundian metasediments; 3, Kibaran granitoids; 4, Mafic and ultramafic intrusions; 5, Late Kibaran alkaline intrusions; 6, Malagarasian (post-Kibaran) sediments; 7, post-Kibaran alkaline complex; 8, Cenozoic; 9, major axes of upright folding (D2); 10, major late-Kibaran shear zones ($D2^1$); 11, stratigraphic boundaries and structural trends. (Redrawn from Cahen et al., 1984; Klerkx et al., 1987.)

In the western Kibaran basin, the basal sequence, the Mt. Kiora Group which was deposited soon after the Ubendian orogeny, comprises dark monotonous phyllites, greenstones and carbonates. The Mt. Kiora Group is of a higher metamorphic grade than the unconformably overlying Lufira Group. The Lufira Group comprises basal conglomerates, quartzites, phyllites with occasional doleritic lavas at the top. This is overlain by the dark-coloured slates and conglomeratic quartzites of the Mt. Hakansson Group, which is disconformably succeeded by the Lubudi Group comprising basal conglomerates and arkosic graphitic shales and stromatolitic limestones and dolomite at the top.

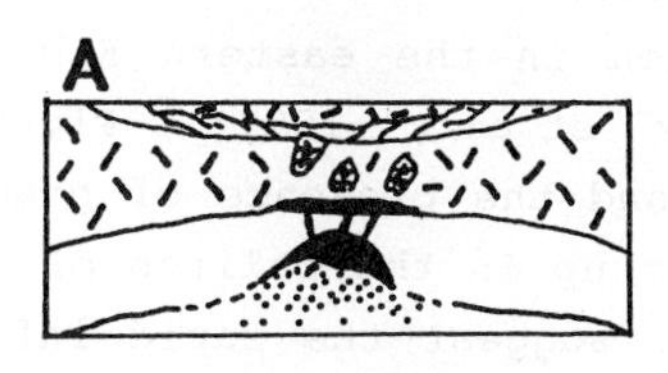

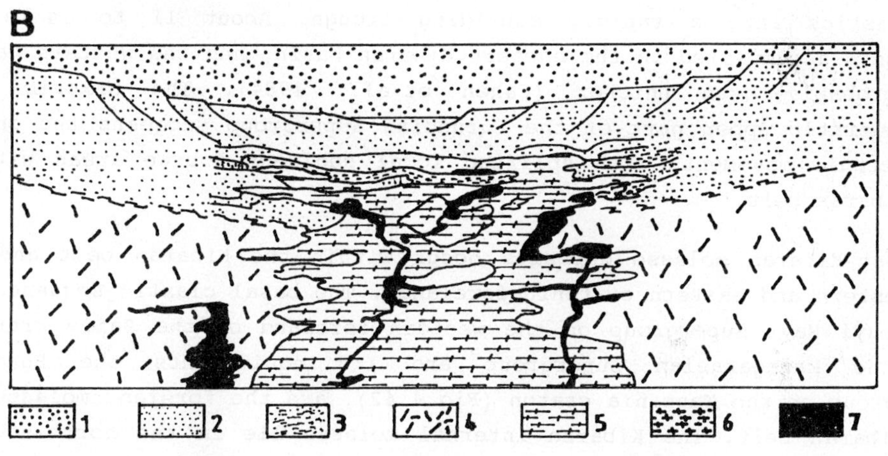

Figure 5.5: Cross-section through the Kibaran belt in Burundi depicting D1 phase deposition, deformation and magmatism. A, at lithospheric scale. B, across the crust of Burundi; 1, Upper Burundian sediments; 2, Lower and Middle Burundian sediments; 3, Quartzites in 2; 4, pre-Kibaran basement; 5, 6, granitoids intruded in D1 phase; 7, mafics intruded in D1. (Redrawn from Klerkx et al., 1987.)

In the eastern Kibaran basin the Burundian and the Karagwe-Ankolean Supergroups are also amenable to a three-fold stratigraphic subdivision (Fig.5.5). Here the lower groups are characterized by dark laminated turbiditic pelites with intercalations of mature quartzites, rare carbonates and volcanics; a more arenaceous, reddish middle group of phyllites, occasionally interbedded with conglomeratic quartzites and minor basaltic and dacitic volcanic rocks; and an upper disconformable group of immature arenaceous and conglomeratic deposits containing ferruginous quartzites. On the eastern margin of Lake Tanganyika, further south, the quartzites and phyllitic shales of the Itiaso series correlates with the Karagwe-Ankolean Supergroup. The Itiaso and the underlying Ubendian basement gneisses are intruded by the Kapalagulu layered complex (1.23 Ga) (Fig.5.3) comprising a basal zone of olivine cumulates with local sulphide and chrome-magnetite concentrations; a rhythmically layered intermediate zone; and a main zone with anorthosite (Wadsworth, 1963). The Kapalagulu complex is over 1,500 m thick.

Based on the age of an interbedded rhyodacitic volcanic rock at the base of the Burundian Supergroup, sedimentation in the eastern Kibaran basin began shortly after 1.40 Ga (Klerkx et al., 1987). Fluvial conglomerates at the base of the lower group and the presence of graded bedding, ripple marks and slump features higher up in the pelites of the lower group and rare stromatolitic carbonates, suggest the rapid influx of clastics into a rapidly subsiding trough. About 11 to 14 km of sediments accumulated in the eastern Kibaran basin; and the western basin was probably even thicker (Cahen et al., 1984). The arkosic and conglomeratic upper part of the Burundian Supergroup suggests the onset of tectonic movements with uplifted local sediment source areas within the Kibaran belt.

Post-Kibaran molasse deposits occur within the Kibaran belt and on its western and eastern cratonic forelands. The basal clastic sequence of the Mbuyi Mayi Supergroup on the western foreland on the Zaire craton, and the Malagarasian Supergroup and its equivalents the Bukoban Supergroup on the Tanzania craton (Fig.4.42), are the foreland molasse of the Kibaran belt. The Kibaran internal molasse lie in the north-south-trending Itombwe synclinorium.

Structure and Metamorphism

Detailed structural analysis of the eastern Kibaran belt by Klerkx et al., (1987) has revealed four main deformational episodes. The first phase of horizontal deformation (D1) occurs at deeper, more strongly metamorphosed structural levels near granite intrusions in anticlines or in the basement. D1 deformation produced bedding-parallel foliation, thin-skinned thrusting along the western part of the Kibaran belt in regions of intense granitic intrusions (Fig.5.5,B) and décollement of the sedimentary cover. In the eastern less metamorphosed parts of Burundi and Rwanda, D1 deformation caused basement mylonization. Major granite-gneiss domes were emplaced during the D1 event. The second phase of deformation (D2) affected all Kibaran rocks in both the eastern and the western Kibaran basins. It produced open, upright folds which are mainly oriented NE-SW, but swing NW-SE in the northern part of the eastern belt, where the regular structural trends are deflected around granite-gneiss domes. D2 deformation was compressive, but it was not sufficiently penetrative to obliterate D1 structures. A late phase of D2 is associated with shear folding and cataclasis which were superimposed on previous structures producing vertical NE-SW- and NW-SE-trending shear zones which also affected the alkaline intrusives emplaced during D2 deformation. Since it is located in a zone of persistent crustal instability, the Kibaran belt was also affected by later tectonic events ranging from the Pan-African orogeny to Cenozoic rifting.

Regional metamorphism in the Kibaran belt is generally of the greenschist facies, especially in the synclinoria, whereas amphibolite facies and migmatites occur near large batholithic granites and basement highs. There are two main phases of regional metamorphism which are associated with D1 and D2 deformation and syn-orogenic granite intrusives. Variable contact metamorphic aureoles were produced during these intrusions.

Intrusive Activity

The Kibaran belt from Kivu and Shaba in Zaire to Burundi and Rwanda is distinguished by ubiquitous granitoid and granite intrusions (Fig.5.4). Cahen et al., (1984) and Klerkx et al., (1987) discussed four types of granitic rocks in the Kibaran belt. The first two types, commonly referred to as the G1 and G2 granitoids, are syn-orogenic concordant intrusives found mostly within the lower Kibaran sequences. The last two sets of intrusives, G3 and G4 are post-orogenic alkaline granites.

The G1 batholithic granitoids are homogeneous, porphyritic gneissic biotite adamellites with low initial $^{87}Sr/^{86}Sr$ ratios which were emplaced around 1.35 Ga (Klerkx et al., 1987). The G2 granitoids are usually non-porphyritic peraluminous two-mica adamellitic orthogneisses with S-type chemistry. Such granites are typically syn-tectonic and were emplaced at depth where migmatites form as a result of anatexes (Condie, 1989). The G2 granitoids are more common in the Shaba province; they were emplaced during D1 deformation around 1.28 - 1.26 Ga (Klerkx et al., 1987). Where both G1 and G2 granitoids co-exist in the western Kibaran belt, the G1 cataclased porphyritic adamellites which are not gneissic, occupy the more central portions of the granitic massifs, and pass into completely gneissified G2 granitoids which are usually situated in the outward parts of the massifs. The G2 granitoids recrystallized under stress during D1 deformation. On the basis of their generally low initial $^{87}Sr/^{86}Sr$ ratios, in spite of strong crustal contamination, Klerkx et al., (1987) suggested that the primary magmas for these granites originated from the lower continental crust and became progressively contaminated with crustal material during their emplacement.

Although classified as post-orogenic G3 granites and dated between 1.19 Ga and 1.0 Ga by Cahen et al., (1984), Klerkx et al., (1987) subdivided this generation of granites in the eastern Kibaran belt into the late intrusives which are associated with D2 compressive deformation and those intrusives which are associated with shearing. The first group are typically intrusive, compositionally homogeneous and unfoliated, unlike G2 granites; they consist of two-mica granites with different compositions and also exhibit crustal strontium isotope signatures. The first group of G3 granites occupy the cores of the D2 anticlines with ages around 1.18 Ga, which also dates the formation of the anticlines. The second group of G3 granites comprises intrusions of alkaline granites which are spatially associated with NE-SW D2 shearing and are dated at about 1.10 Ga, the age of D2 shearing. The relatively low $^{87}Sr/^{86}Sr$ initial ratio suggests a deep crustal origin for these alkaline plutons, and thereby also imply a deep crustal origin for D2 shearing.

In the eastern Kibaran belt there is a string of mafic and ultramafic intrusions aligned northeast, parallel to both the major D2 folding and shearing direction (Fig.5.4,B). They extend from Burundi northeastwards to Lake Victoria in northwestern Tanzania. These are intrusive peridotites with or without associated gabbro, norite and leuconorite, in which the gabbro-noritic rocks exhibit some evidence of layering (Klerkx et al., 1987). Alternating layers of peridotite and norite in the ultramafics suggest that they were emplaced as a crystal mush of olivine crystals in a noritic liquid matrix, while the mafics were also derived form the ultramafic crystal-liquid mixtures. Klerkx et al., (1987) related these mafic and ultramafic bodies to mafic magmatism which was generated during an early phase of crustal extension.

The last major group of intrusives are the post-tectonic leucocratic equiangular alkaline "tin" granites or G4 granites (Cahen et al., 1984). These granites are of economic importance since they are invaded by mineralized pegmatites. The G4 granites are cataclastic, locally sheared and cut across country rocks of older granitoids and Kibaran sediments. They consist of quartz, microcline, albite, and muscovite (or biotite) with accessory apatite, zircon and tourmaline. Their roofs are invaded by mineralized pegmatites and quartz veins. The G4 granites were emplaced at about 976 Ma, probably from a deep-seated magma or from the fusion of G1 and G2 granitoids or pre-Burundian basement gneisses (Klerkx et al., 1987).

Tectonic Model

In their discussion of the tectonic evolution of the Kibaran belt Klerkx et al., (1987) considered this belt as a linear intracratonic trough, the origin and early development of which was controlled by crustal extension.

Like geologically younger regions with lithospheric stretching, such as the Late Paleozoic of Morocco and the Cenozoic of western United States, where bimodal mafic and acid magmatism are associated with rifting, the large number of granites and gabbros in the Kibaran belt which are associated with the first deformation (D1) may have been generated by crustal extension. The granitic magmas could have been produced by the partial fusion of the lower crust by the heat supplied by mafic magmas generated by the intrusion of hot asthenosphere (Fig.5.5,A). A slow and protracted extension of the Kibaran belt between 1.35 and 1.26 Ga probably triggered very little fracturing and thereby inhibited volcanism in the belt. Rather, heat from the mafic magmas accumulated and caused the partial fusion of crustal rocks thereby producing granitic magmas which were emplaced during the climax of deformation and crustal extension (D2). The granitic magmas could have acquired their strong peraluminous, S-type character by crustal contamination at the upper crustal levels while ascending.

Mineralization

Although production is not high by global standards, the Kibaran belts constitute a metallogenic province where "tin" granites have associated with them economic deposits of tin, tungsten, beryllium, colombo-tantalite and lithium-ores, in addition to minor but exploitable deposits of bastnaesite, uranium ores, mica and semi-precious stones. Small amounts of gold are won from Tertiary and Recent alluvial placers, which are derived from the basal conglomerates and quartzites of the lower Kibaran Supergroup (Radulescu, 1982; Tissot et al., 1982).

Mineralizations associated with the G4 post-tectonic Kibaran granites (Fig.5.6) have been interpreted by Bugrov et al., (1982), Radulescu

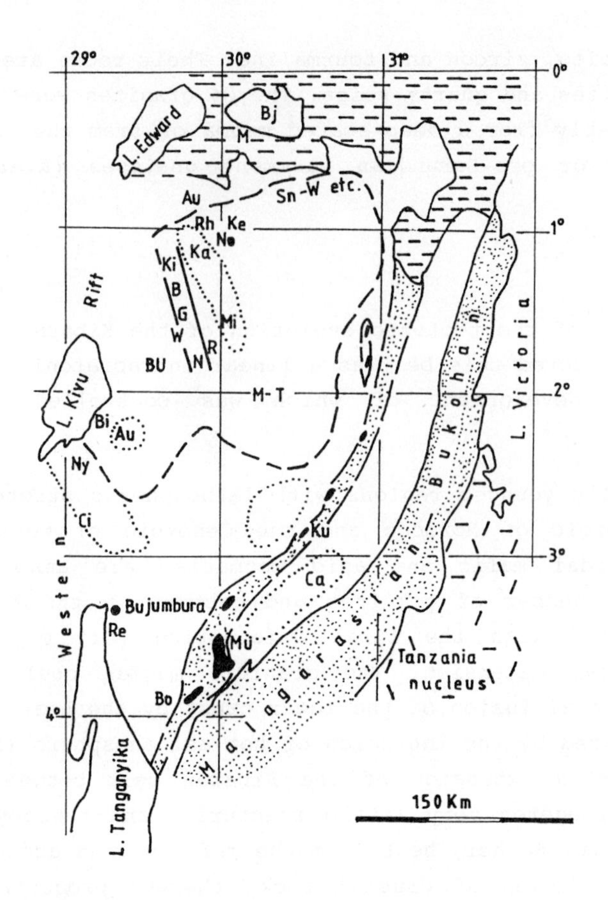

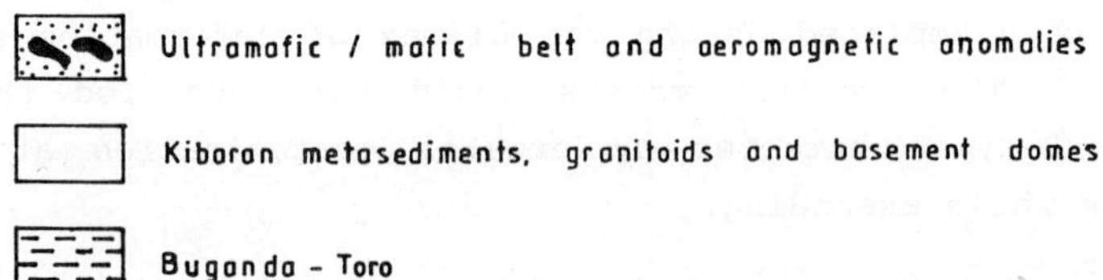

Figure 5.6: Geologic sketch map and mineral deposits in NE Kibaran belt. Mining districts in Uganda are: Bj-Buhewju plateau; Ka-Kamena Fe deposit; Ki-Kirwa W; Ke-Karenge G4 granite; M-Mashonga Au; No-Nyamulilo W; Rh-Ruhiza W; in Rwanda to the north are B-Bugarama W; Bu-Burange pegmatite; Mi-Miyoye Au; Bi-Bisesero Au; G-Gifurwe W; N-Nyamulilo W; Ny-Nyungwe Au; R-Rutongo Sn; M-T-Musha-Ntunga Sn; in Burundi they are Ci-Cibitoke Au; Ca-Cankuso Au; Bo-Buhoro Gabbro; Mu-Musongati Ni; Re-Kayonde Ree; in Tanzania they are Kb-Kabanga Ni, Cu, Co. Dotted lines enclose auriferous zones. (Redrawn from Pohl, 1987.)

(1982), Tissot et al., (1982) and Pohl (1987) as comprising pegmatites with mostly Sn and Nb/Ta (and subordinate Li, Be, W, Bi, U/Th, and non-metallics such as muscovite and kaolinite); quartz veins with mainly Sn and W (and pyrite, siderite, Bi, Au, U); limonitic silicification zones and quartz veins with gold. Pohl (1987) demonstrated that tin-bearing pegmatites and tungsten quartz veins are concentrated in the post-tectonic granites (Fig.5.7) and that these granites were the sources of the ores. In view of the apparent rarity of flourine during mineralization, the mineralizing fluids, some of which were sodium-rich, had mostly phosphorus, H_2O, and boron among the volatiles, and derived their metals from the granites rather than from the Kibaran metasediments.

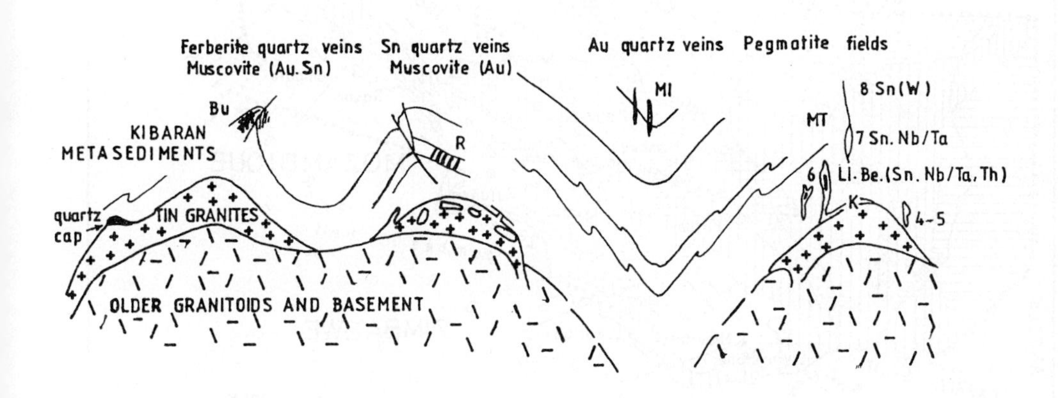

Figure 5.7: Kibaran mineralization related to G4 "tin" granites. Figures denote types of pegmatite. Bu-Bugarama; K-Karenge granite; Mi-Miyoye; M-T-Musha-Ntunga; R-Rutongo. (Redrawn from Pohl, 1987.)

In northern Burundi, northeast of Bujumbura, the Matongo carbonatite (1.30 Ga old) which is a part of an alkaline massif (Kampunzu et al., 1985), contains an igneous phosphate deposit with a reserve of about 40 million tons (Kurtanjek and Tandy, 1988). Apatite is the primary mineral.

5.2.3 The Irumide Belt

Stratigraphy

South of the Kibaran belt is a parallel coeval mid-Proterozoic intracratonic orogen, the Irumide fold belt. The Irumide belt is bounded to the south by the Late Proterozoic Lufilian arc, to the west by the Bangweulu block, and to the east the Irumide belt passes into the high-grade gneisses of the Southern Mozambique belt (Fig.5.8). The Irumide belt occupies most of eastern Zambia where it extends for about 700 km northeastward into northern Malawi (Fig.5.8).

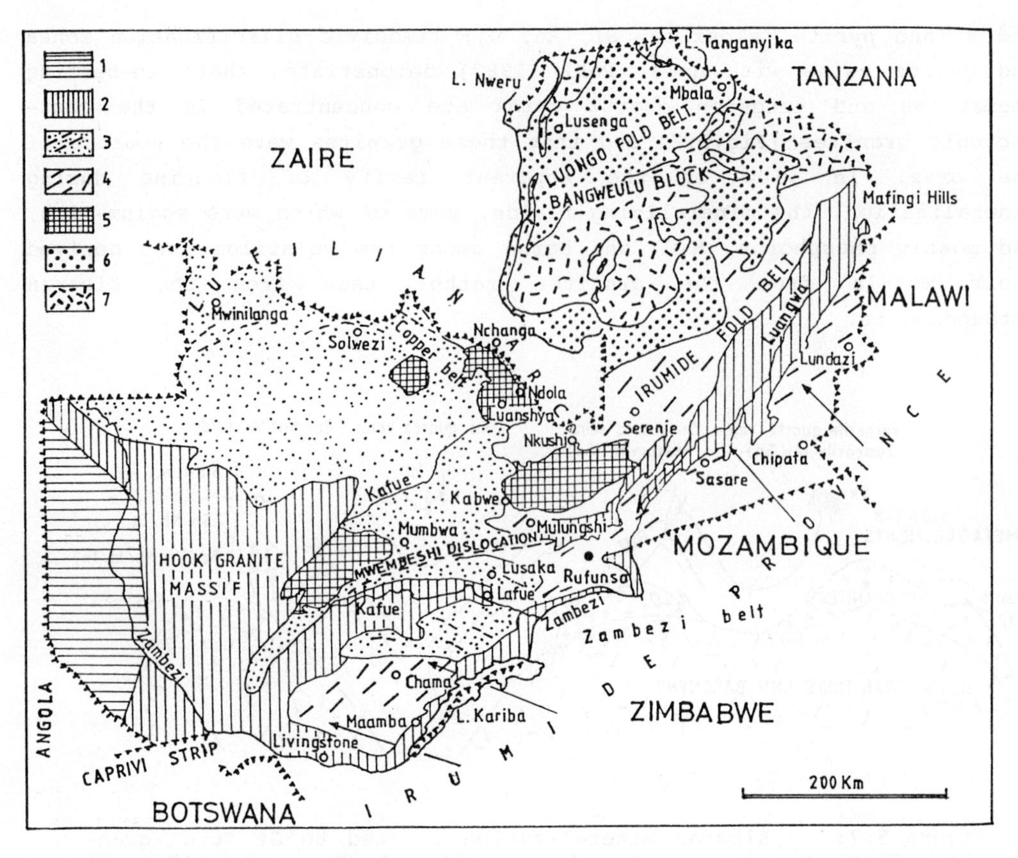

Figure 5.8: Outline geology of Zambia. 1, Cretaceous cover; 2, Karoo; 3, Katangan; 4, Irumide; 5, Katangan basement; 6, Mporokoso Group; 7, granites, gneisses, volcanics of the Bangweulu block. (Redrawn from Cahen et al., 1984.)

The basement in the Irumide belt consists of the circa 1.8 Ga Early Proterozoic Mkushi Gneiss complex. In the southern part of the Irumide belt, in the Rufunsa area (Fig.5.8), the basement is known as the Musensenshi Group. The Irumide basement and that of the adjoining Lufilian arc, the Lufubu schist (Table 5.1), and the basement schists and granitoids of the Bangweulu block, all formed during the Early Proterozoic Ubendian orogeny. The Mkushi Gneiss and the Musensenshi Group consist of amphibolite-facies gneisses and metasedimentary schists and migmatites with white, granular bands of quartzite.

Daly and Unrug (1983) attributed most Irumide metasedimentary sequences (e.g., the Mpanshya, Nwami, Sasare, Musofu, Fombwe, Mafingi Groups), which had been previously described in different parts of the Irumide belt (Table 5.1), to the Muva Supergroup (Fig.5.9). The Muva

Table 5.1: Suggested stratigraphic correlations in eastern Zambia. (Redrawn from Johns et al., 1989.)

RUFUNSA AREA	NYIMBA AREA	PETAUKE AREA	SASARE AREA	LUSANDWA R. AREA	CHIPATA AREA	MCHINJI AREA (MALAWI)	PROBABLE SUPERGROUP EQUIVALENTS
MPANSHYA GROUP: KANGALUWE FOMATION			UPPER SASARE GROUP	SASARE GROUP	MWAMI FORMATION	KACHEBERE FORMATION (UPPER PART)	KATANGAN
RUFUNSA METAVOLCANIC FORMATION	CHIMPI SEQUENCE		LOWER SASARE GROUP		- BASEMENT COMPLEX -	BASEMENT COMPLEX: MCHINJI GROUP: KACHEBERE FORMATION LOWER PART	MUVA
CHAKWENGA RIVER SCHIST FORMATION						MCHINJI RIDGE FORMATION	
KAULASHISHI QUARTZITE FORMATION		- SASARA GROUP -: CHITUNDULA SCHIST AND QUARTZITE FORMATION	FOMBWE GROUP: VARIOUS QUARTZITE FORMATIONS			PATE HILL FORMATION	
MULAMBA FORMATION						LIFUCHERE SCHIST FORMATION	
MUSENSENSHI FORMATION	HOFMEYER SEQUENCE (And Mkokomo sequence?)	CHIPIRINYUMA GNEISS FORMATION	SASARE GNEISSES GROUP	MVUVYE GROUP		PRE-MCHINJI GRANULITES AND GNEISSES	BASEMENT COMPLEX
		MVUVYE GNEISS FORMATION		LUSANDWA GROUP			
		MVUVYE MARBLE FORMATION		SINDA GROUP			
		MAMBO GNEISS FORMATION		NYANJI GNEISS FORMATION			

Supergroup is essentially a thick sequence of alternating quartzite and pelitic lithologies which have been strongly folded and thrust and subjected to varying degrees of metamorphism. The Muva Supergroup, in its expanded usage (Daly and Unrug, 1983), accumulated over a long span of geological time, from the end of Ubendian orogenesis (at the same time the Ndembara and Chocha Groups were deposited on the Tanzania craton) to the beginning of Irumide deformation, late in the Middle Proterozoic. Sedimentological analysis in the northwestern part of the Irumide fold belt, where the Muva Supergroup oversteps the Bangweulu block (Fig.5.10), shows a rapid thickening of the Muva Supergroup towards the Irumide trough (Anderson and Unrug, 1984; Daly and Unrug, 1983). In this region the thickness of the Muva ranges from 100-300 m on the Bangweulu block to about 8 km in the Irumide trough (Fig.5.10), suggesting the control of the Irumide basin margin by extensional normal faults (Anderson and Unrug, 1984; Daly and Unrug, 1983). In the northwestern part of the Irumide belt, where the Muva Supergroup has been subdivided into several lithostratigraphic units, the basal Mitoba River Group, which is the lateral equivalent of the Kasama Formation on the Bangweulu block (Fig.5.10), is overlain in the Irumide trough by the marine sediments of the Manganga River Formation and the Manshya River Group (Fig.5.10). The fluviatile deposits of the Kasama Formation and the Mitoba River Group were derived from the uplifted and eroded Mporokoso Group to the northwest on the Bangweulu block, whereas the overlying Manganga River

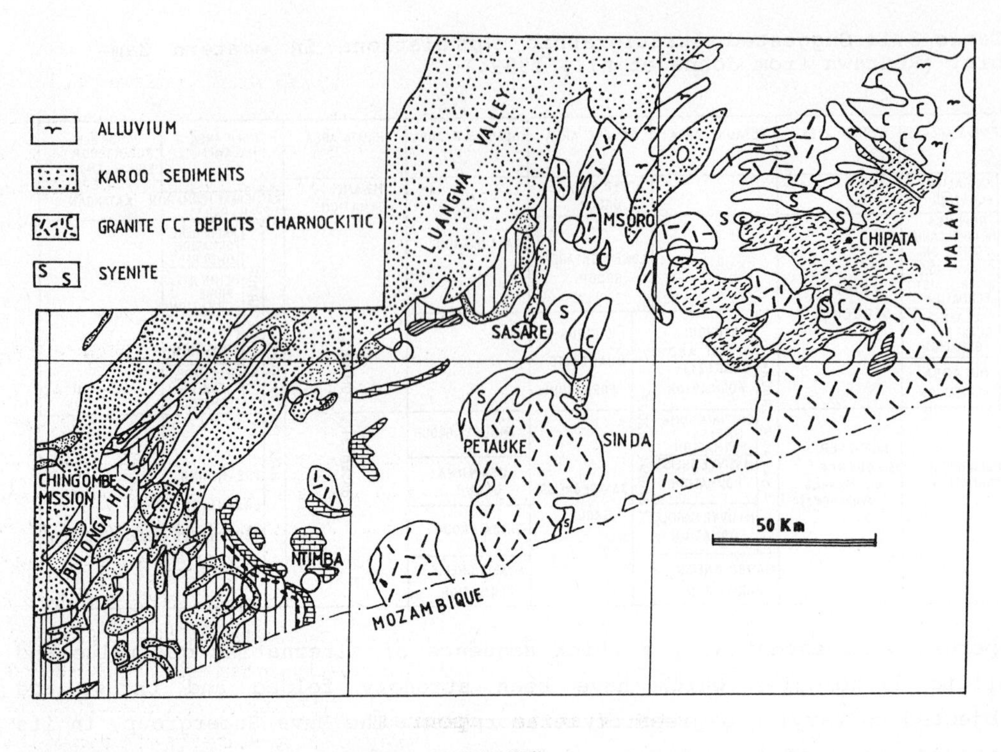

Figure 5.9: Outline geology of eastern Zambia. (Redrawn from Johns et al., 1989.)

Formation and the Manshya River Group were deposited by a marine transgression coming from the eastern part of the Irumide basin. As will be shown later, this paleogeographic setting has considerable regional tectonic implications. Further east beyond the Irumide trough, on the Malawi border other deposits of the Irumide transgression, the psammites and pelites of the Mafingi Group, rest directly on the crystalline basement (Fig.4.43).

Southward, in the region east of Lusaka, an atypical rock assemblage occurs in the Rufunsa area. These rocks which belong to the Mpanshya

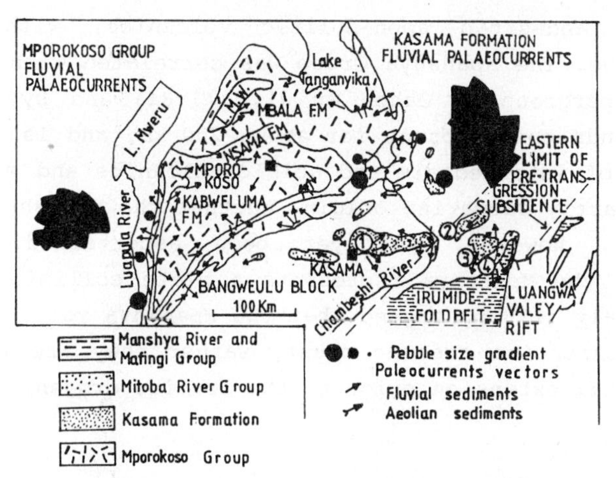

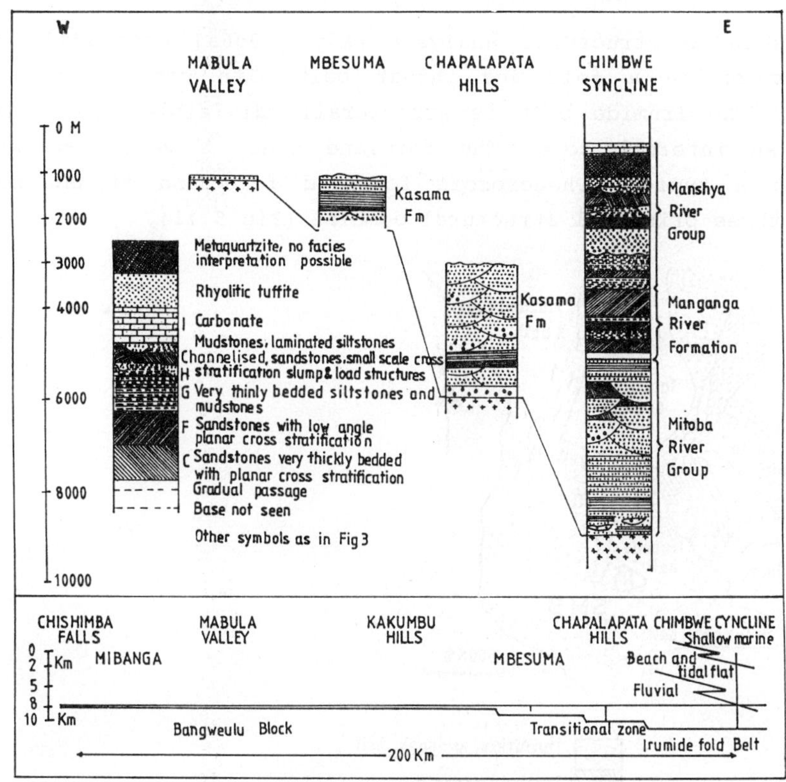

Figure 5.10: Geologic sketch map and stratigraphic columns for the Muva Supergroup. (Redrawn from Anderson and Unrug, 1984; Daly and Unrug, 1982.)

Group are neither similar to the Irumide basement lithologies nor to the Muva Supergroup as defined above. Rather, as shown by Barr (1976), the Mpanshya Group consists of a highly deformed and extensive sequence of tholeiitic and andesitic amphibolitic volcanics, with pillow lavas locally developed. The Mpanshya Group was correlated with the lower part of the Muva Supergroup by Cahen et al., (1984) and by Daly and Unrug (1983); it extends continuously for about 200 km, and is associated with a wide variety of intermediate and acidic volcanics and minor ultrabasic intrusives of partly submarine origin. Whether the Mpanshya volcanics are ophiolites has, however, not yet been ascertained. Although the likelihood that the Mpanshya represents a plate collision suture cannot be ruled out. Daly (1986b) argued that the Mpanshya volcanics are not the relicts of an extensive oceanic crust, rather they are the products of pronounced crustal extension prior to the Irumide orogeny.

Structure

After a detailed structural analysis Daly (1986a) concluded that the Irumide orogen is a fold and thrust belt with considerable crustal shortening. The Irumide belt is structurally divisible into a foreland zone and an internal zone. The foreland zone, some 350 km wide, is similar to a typical Phanerozoic foreland fold and thrust zone. It comprises three principal structural domains (Fig.5.11).

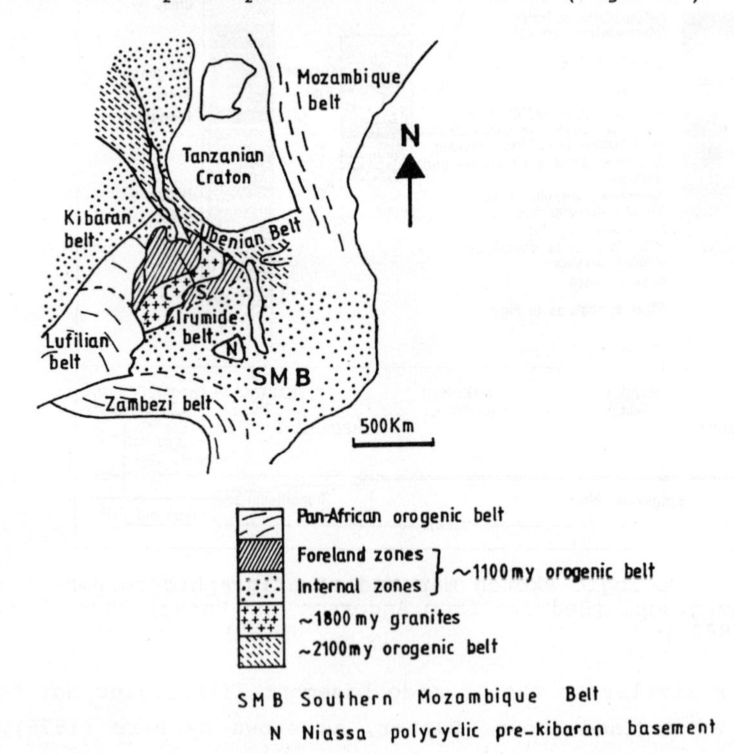

Figure 5.11: Distribution of Kibaran belts in central and East Africa. (Redrawn from Daly, 1986a.)

The westernmost structural domain is the Luongo fold and thrust zone (Fig.4.46). In this deformation belt, which is about 50 km wide, the folded and thrust Mporokoso Group of the Muva Supergroup overlies major arcuate thrust and shear zones in the basement of the Bangweulu block (Fig.5.12,A). The thrust and shear zones climb out of the basement and are located along the basement-Muva contact. This contact acted as a detachment surface along which the large asymmetric folds in the Muva Supergroup have been displaced. Throughout the Luongo fold and thrust zone fold vergence swings from WNW to NNW and the stretching lineation in the underlying basement shear zone always shows a southeasterly plunge suggesting NW-directed tectonic transport. About 10 km of crustal shortening was estimated across the Luongo zone (Daly, 1986a).

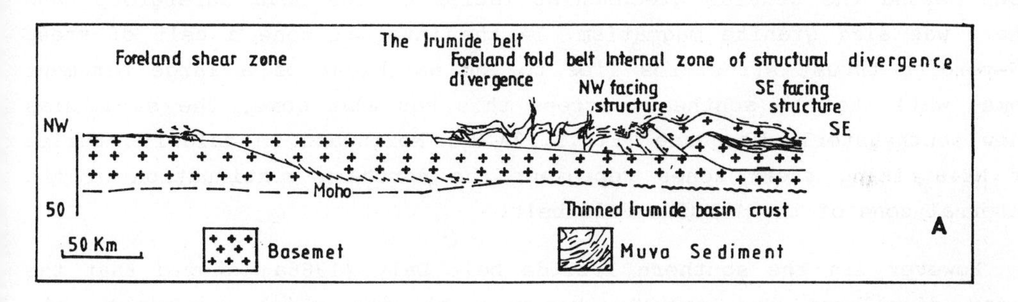

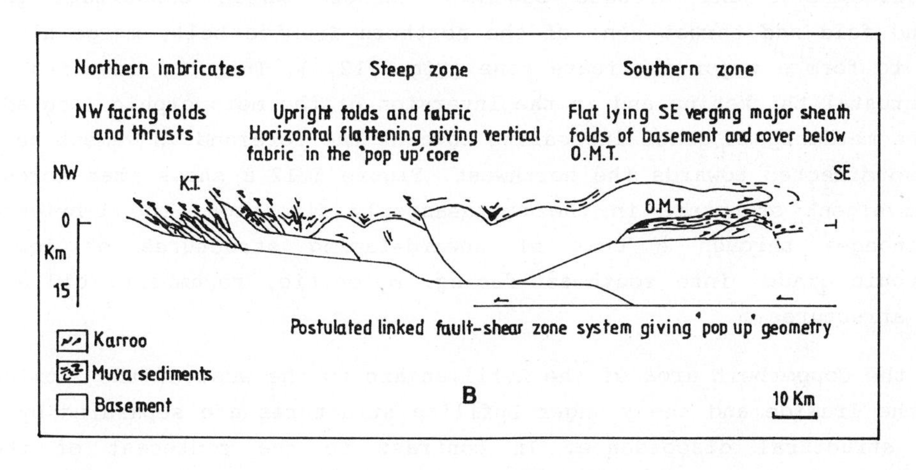

Figure 5.12: Structural sections through the Irumide belt; B crosses the Southern Irumide belt in the Mkushi area. (Redrawn from figures supplied by M. C. Daly.)

The Chambeshi fold and thrust zone, east of the Luongo zone, consists of open to tightly folded quartzitic sediments in which bedding plane thrusts have locally caused the repetition of beds. The folds and thrusts which are directed northwest are refolded by major SE-directed thrusts which developed in the basement (Fig.5.12,A). Eastwards, the Shiwa Ng'andu fold zone, about 100 km wide, comprises a sequence of major concentric upright and locally disharmonic folds which apparently developed above a basal décollment near the basement-sediment contact (Fig.5.12,A). These folds resemble those above the basal décollment in the Jura Mountains of Europe. About 40 km of crustal shortening has been estimated for the Shiwa Ng'andu fold zone where the décollment dips eastwards towards the internal zone of the Irumide orogen.

The internal zone of the Irumide orogen is a zone of structural divergence which can be traced throughout the Irumide belt in Zambia (Fig.5.13,A). Here the metamorphic grade attained the amphibolite facies, much beyond the general greenschist facies of the Muva Supergroup; and there was also granite magmatism. In the internal zone a belt of steep NW-verging thrust structures lies to the northwest of a large basement dome, while to the southeast across this basement dome, the structures show southeasterly vergence (Fig.5.12A,B). Northward in Malawi a series of NE-striking shear zones represent the northern continuation of the internal zone of the Irumide fold belt.

However, in the southern Irumide belt Daly (1986a) showed that the major décollment and arcuate basement thrusts which constitute the foreland fold and thrust zone of the northern Irumide belt, merge along strike to form a major imbricate zone (Fig.5.12,B). This has resulted in local crustal thickening and in the inversion of the metamorphic isograds with the metamorphic grade increasing towards the foreland. Movement here was also directed towards the northwest. Figure 5.12 B shows that across this divergent structure in the southeasterly direction, the imbricate zone changes through a belt of upward-facing structures of lower metamorphic grade, into southeast-facing, mylonitic, recumbent fold and thrust structures.

In the Copperbelt area of the Lufilian arc to the west of the Irumide belt, the Irumide and the younger Lufilian structures are separated by a marked structural discordance. In contrast to the southeast of the Copperbelt (Fig.5.8) where the Lufilian structures are superimposed on the Irumide structures, separation of these deformation has been possible using stretching lineations (Daly et. al., 1984). Here the Irumide fabrics contain a downdip extension lineation indicating southeasterly to northwesterly thrust direction, whereas the Lufilian structures which strike parallel to the Irumide structures reveal horizontal linear fabrics showing the characteristic Lufilian ENE to NE direction of tectonic movement.

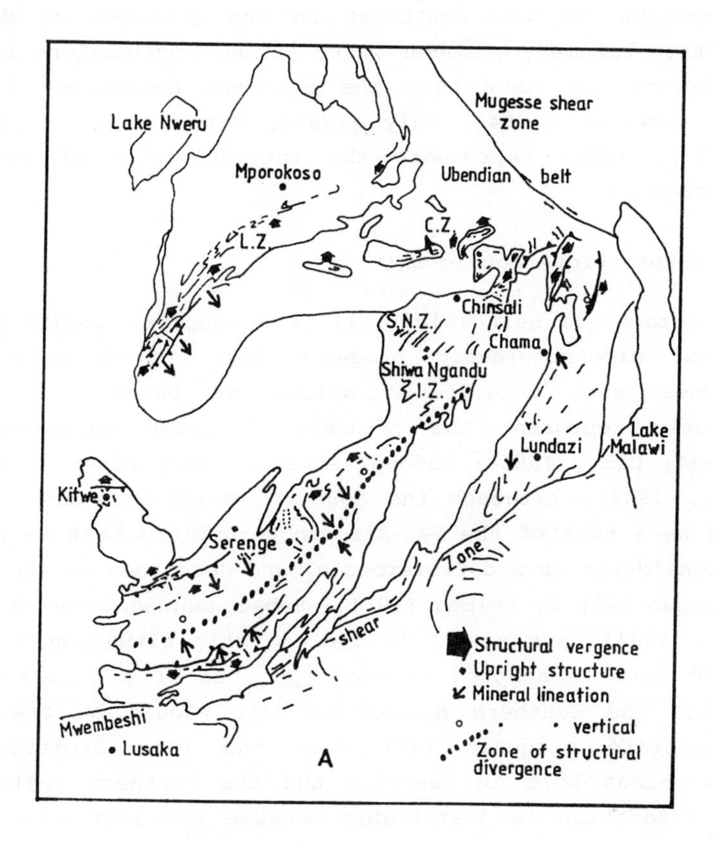

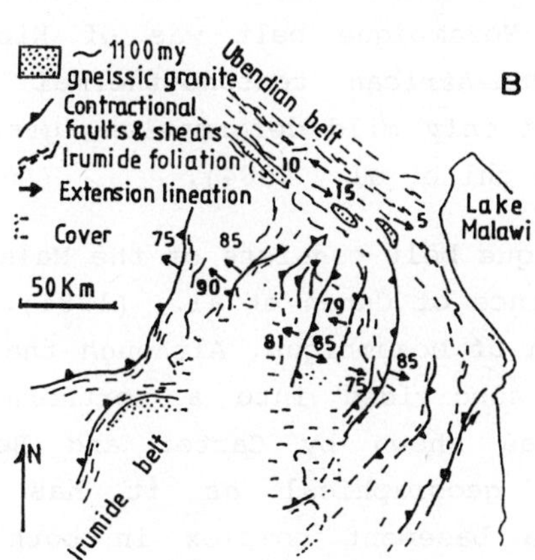

Figure 5.13: Sketch maps showing regional structures in the Irumide belt. (Redrawn from figures supplied by M. C. Daly.)

Irumide structures terminate against the Ubendian belt of Malawi and Tanzania (Fig.5.13). Syn-tectonic granites of Irumide age testify to Irumide displacement in this portion of the Irumide belt. Irumide structures continue to the southeast in the gneisses of Malawi and Mozambique. These basement gneisses which extend from eastern Zambia into Malawi and Mozambique constitute the Southern Mozambique belt which according to Andreoli (1984), Daly (1986), Piper et al., (1989), and Sacchi et al., (1984) represents the internal zone of the Kibaran collisional orogen.

5.2.4 Southern Mozambique Mobile Belt

The Southern Mozambique belt (Fig.5.11) is a complex, deeply eroded and deeply exposed mid-Proterozoic orogenic belt in which high-strain granulite zones with interleaved slices of basic and ultrabasic ophiolitic rocks represent the remnants of plate collision sutures (Andreoli, 1984; Daly, 1986a; Cadoppi et al., 1987; Piper et al., 1989; Sacchi et al., 1984). Although the Southern Mozambique belt has often been regarded as a part of the Pan-African Mozambique belt to the north, it had been considered as mid-Proterozoic, and described as the type area of the Mozambique belt by Holmes (1951). Subsequent workers (e.g., Daly, 1986a; Sacchi, 1984) have sought to redress this situation in which an older orogenic belt is adopted as the type area of a younger belt, by suggesting that the Southern Mozambique belt should be treated as a completely separate orogenic belt from the Late Proterozoic-Early Paleozoic Mozambique belt in Tanzania and the northern parts of East Africa. This separation is justifiable because the last major orogenic event in the Southern Mozambique belt was of Kibaran age, at about 1.10 Ga. During the Pan-African tectono-thermal event the Southern Mozambique belt suffered only mild deformation, granite emplacement and thermal rejuvenation (Sacchi et al., 1984).

The Southern Mozambique belt consists of the Malawi tectonic province and the Mozambique province of Cahen et al., (1984). The Malawi province includes the Tete region of Mozambique. Although the basement complex of Malawi was previously subdivided into a northern subprovince and a southern subprovince, as shown by Carter and Bennett (1973), this subdivision is merely geographical as it has no structural and petrological basis. The basement complex in both geographical areas comprises mostly amphibolite facies gneisses and granulites, with patches of Ubendian rocks occurring in the northern basement. However, based on the available detailed geochemical, structural and geodynamic interpretations (Andreoli, 1984; Piper et al., 1989), the basement complex in central and southern Malawi, and the Tete gabbro-anorthosite complex (Barr and Brown, 1987) are highlighted below.

Central Malawi Province

A recent geological synthesis for central Malawi by Piper et al., (1989) shows that the basement complex in this region comprises paragneisses and orthogneisses. The paragneisses (Fig.5.14) consist of biotite-hornblende gneisses interbanded with granulites (of probable volcanic or volcani-

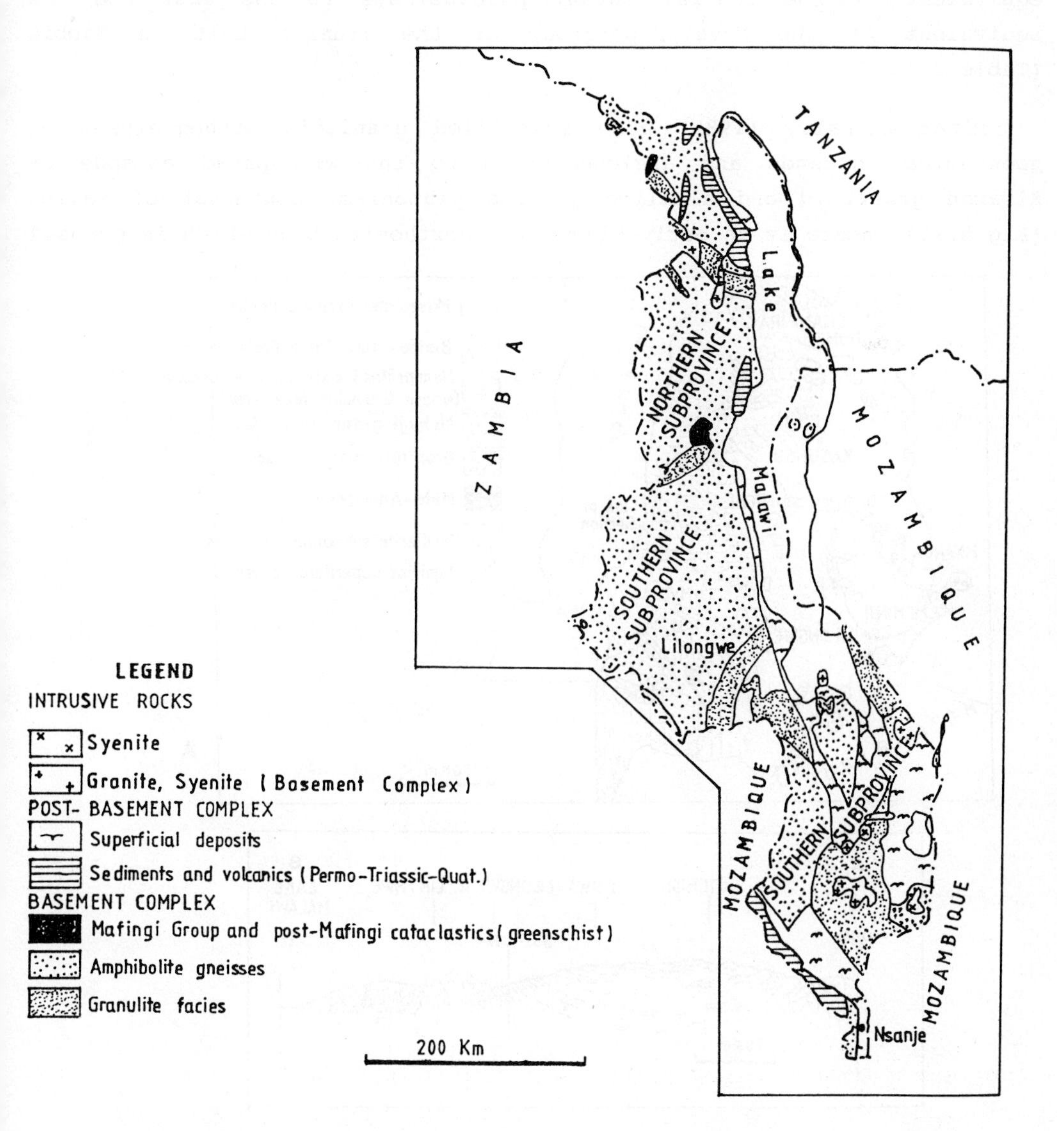

Figure 5.14: Outline geology of Malawi. (Redrawn from Carter and Bennett, 1973.)

clastic origin); biotite-hornblende gneisses which are not banded; and muscovite-graphite pelitic schists which are locally pyritiferous. These metasedimentary units are intruded by a suite of low TiO_2 olivine tholeiitic gabbros. Based on their geochemistry Piper et al., (1989) inferred that the protoliths of the central Malawi paragneisses were continental shelf-rise deposits, whereas a fourth metasedimentary sequence, the Mchinji Group which comprises metaconglomerates, psammites, semi-pelites and pelites, were interpreted as the proximal lithofacies equivalent of the central Malawi paragneisses to the east and the equivalent of the Muva Supergroup of the Irumide belt in Zambia (Table 5.1).

Piper et al., (1989) also identified granitoid orthogneisses on geochemical grounds and assigned them to the widespread episode of Kibaran granitoid and alkaline granite plutonism. Southwest of Salima (Fig.5.15) there is an early-kinematic anorthosite body which is exposed

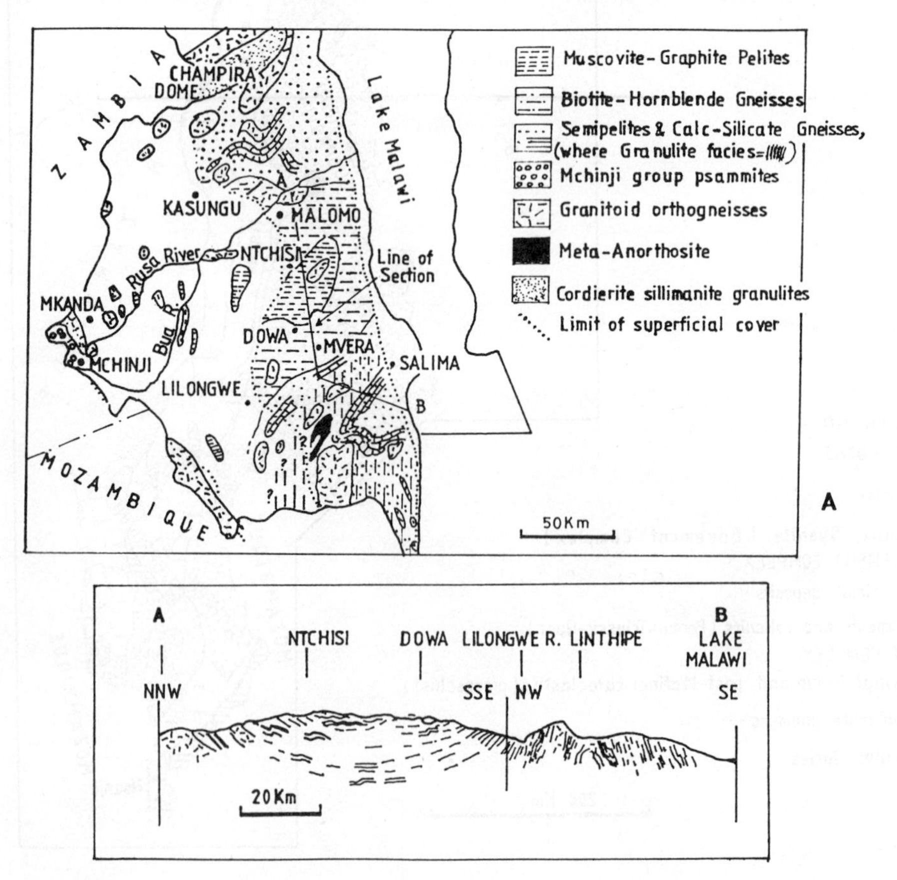

Figure 5.15: Generalized basement map of central Malawi (A). B, cross-section. (Redrawn from Piper et al., 1989.)

over an area of 250 km^2. Because this anorthosite body is spatially associated with perthite-rich adamellite-dominant granitoids, Piper et al., (1989) considered them genetically similar to the adamellite-charnockitic granite-anorthosite suites which are found in mid-Proterozoic belts in other parts of the world (Windley, 1984). The central Malawi anorthosite, unlike those of the Tete province and southern Malawi, probably intruded the metasedimentary pile during anorogenic plutonism along a continental margin in central Malawi.

Two phases of deformation have been recognized in central Malawi; an early phase with flat, isoclinal structures; and a second phase with asymmetric folding which shows vergence to the north, northwest or locally to the west, suggesting north and northwesterly tectonic transport directions compatible with the Irumide belt in the northwest. The paragneisses and gabbroic suites preferentially show ductile shearing. As depicted in Fig. 5.15,B, the asymmetric synclinal folds in the northern part change southwards into steep, upwardly- converging dips in the southeast with granulite-facies assemblages in the central part of the steeply dipping zone. Since these structures probably formed at deep crustal levels, ductile shear zones rather than thrust planes are evident. The structures and tectonic transport directions in central Malawi are shown in Fig. 5.13,B.

Southern Malawi Province

As in central and northern Malawi, amphibolite gneisses and granulites constitute the bulk of the basement rocks in southern Malawi. Andreoli (1984) demonstrated that the granulites which occur mostly east of Lilongwe and in the Blantyre-Zomba region (Fig.5.14) were derived from supracrustals, migmatitic and plutonic rocks, and also probably form alkali-olivine to high-alumina basalts. The supracrustal granulites of southern Malawi are generally interbanded with meta-pelites, calc-silicate rocks and marbles. Based on their major element data Andreoli (1984) concluded that a substantial part of the southern Malawi granulites were derived form island-arc volcanics and associated plutonic rocks and graywackes. The amphibolite-facies gneisses consist mostly of migmatites, paragneisses and occasional relicts of gabbro, dolerite and diabase.

On the Kirk Range relicts of serpentinized Alpine peridotites are preserved in the Lukudi River and Chimwadzulu areas. This ophiolitic assemblage comprises granulite- and amphibolite-facies basic rocks which are believed to have accreted onto an older granodioritic basement terrane (Andreoli, 1984).

Tete Province

In southern Malawi there are also gabbro-anorthosite plutons especially in the Linthipe area, but by far the largest concentration of anorthosite complexes is in the nearby Tete province of Mozambique, where the Tete gabbro-anorthosite complex, one of the largest of such bodies in the world, occurs (Figs.5.3;5.16). Barr and Brown (1987) presented a comprehensive account of the geology of the Tete complex.

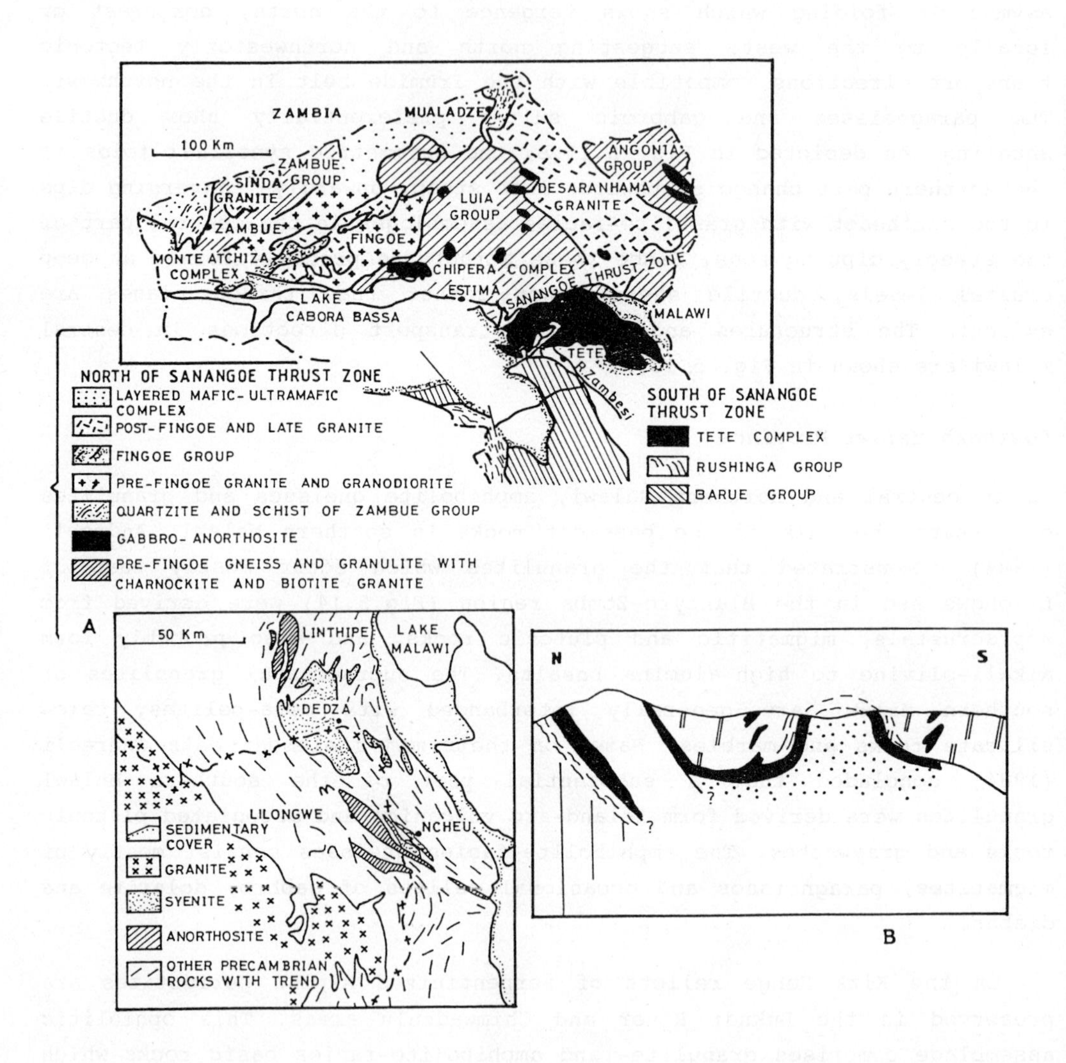

Figure 5.16: Geologic sketch maps and cross-section through the Tete Complex. (Redrawn from Barr and Brown, 1987.)

In the Tete province gabbro-anorthosite bodies are emplaced within country rocks comprising gneisses, granulites, and charnockites and biotite granites, all of which are surrounded by large bodies of granites and granodiorites (Fig.5.16,A). In the western part of the Tete province the gneisses and granulites are overlain unconformably by the metasedimentary rocks of the Fingoe Group and are intruded by younger suites of granite and ultramafic rocks. The Tete basic complex of anorthosites and norite-gabbros intruded the eastern gneisses and granulites and was in turn cut by north-south-trending basic dykes followed by mylonitization and uranium mineralization (Cahen et al., 1984). Other anorthosite bodies in this region including those of the Linthipe area and the layers of anorthosite gneisses which occur in the Kirk Range and in the Nsanje area of southern Malawi (Fig.5.14) are coeval with the Tete complex.

The Tete complex is a large sheet or lopolith (Fig.5.16,B), 10-20 km thick, extending over an area of about 6,000 km^2. It consists of mostly light-coloured gabbro and norite with subordinate anorthosite layers and numerous dolerite dykes. It has medium-to very coarse-grained or pegmatitic textures. The Tete complex extends from the big bend along the Zambezi River in Mozambique to the border with Malawi. Along its western margin where the Tete complex was intruded into flat-lying metasediments known as the Chidue Group, there are zones of country rock inclusions in the complex as well as skarns along the contact with the carbonate metasediments of the Chidue Group. Apparently the western part of the Tete complex was emplaced at shallower depth under amphibolite-facies metamorphism, whereas the eastern part of the complex was intruded into high-grade rocks and charnockites implying emplacement at a deep crustal level. Along its western margin the complex exhibits pneumatolytic alteration.

The Tete complex is similar to other mid-Proterozoic gabbro-anorthosite plutons such as the Adirondack and Morin anorthosites of the Greenville Province of North America. It is, however, very distinct from older large basic complexes such as the Great Dyke or the Bushveld complex. Unlike these African layered intrusives, the Tete complex is poorly banded and lacks rhythmic layering; it contains plagioclase rich in the albite end-members (andesine-sodic labradorite); it is poor in olivine and pyroxene which contains more of the iron end-members; and it contains abundant opaque minerals which are enriched in titanium and vanadium, but poor in chromium and cobalt (Barr and Brown, 1987). The geochemical contrast between the Tete complex, and the Great Dyke and the Bushveld complex includes the fact that the former is high in calcium and low in magnesium and also exhibits a different alkali enrichment and rare element pattern.

Throughout the world anorthosites represent a unique type of anorogenic magmatism which is peculiar to the Middle Proterozoic (Windley, 1984). These bodies were mostly emplaced between 1.70 and 1.0 Ga (Condie, 1989; Windley, 1984). The Tete complex and related anorthosite bodies in southern Malawi belong to the mid-Proterozoic anorogenic gabbro-anorthosite complexes. The coarse igneous texture and associated high-grade and charnockitic country rocks suggests intermediate crustal depths of origin for at least the eastern part of the Tete complex (Barr and Brown, 1987).

Mozambique Province

The Mozambique province contains two distinct collision suture zones, the Lurio belt and the Namama belt (Fig.5.17), which when considered in conjuction with the Chimwadzulu zone in the Malawi province, qualifies the Southern Mozambique belt as the collisional zone from which regional compressive displacements were generated (Fig.5.13) during the Kibaran orogeny (Daly, 1986b).

The Lurio belt is a collisional orogenic belt in the Mozambique province. It crosses Mozambique from southern Malawi from where it extends northeastward to the Indian ocean; there is a bifurcation near the eastern Malawi border, from which a northern branch originates (Fig.5.17,A). South of the Lurio belt, a major thrust belt, the Namama belt, displays easterly thrust vergence, unlike the Lurio belt which shows southeasterly vergence. The Lurio and the Namama belts are the products of the Kibaran orogeny which was locally referred to as the Lurian event and assigned an age of 1.10 Ga (Sacchi et al., 1984).

In the course of developing the mineral prospects (pegmatites with Li, Be, Nb-Ta), the Mozambique province has been mapped and investigated in detail by several mineral exploration teams (Cadoppi et al., 1987; Sacchi et al., 1984) who have furnished a clearer regional structural re-interpretation of this crucial part of the Southern Mozambique belt.

The pre-Kibaran basement rocks in the Mozambique province are the biotite- and biotite-hornblende-gneisses which exhibit diffuse pre-Lurian migmatitic textures. These are variously termed the Nampula and Namarroi Series or the Mocuba Complex (Table 5.2). Geochemical data suggests that the protoliths to these gneisses were volcanic rocks and subordinate diorite-granodiorite intrusives. Gneissic supracrustal cover rocks in the

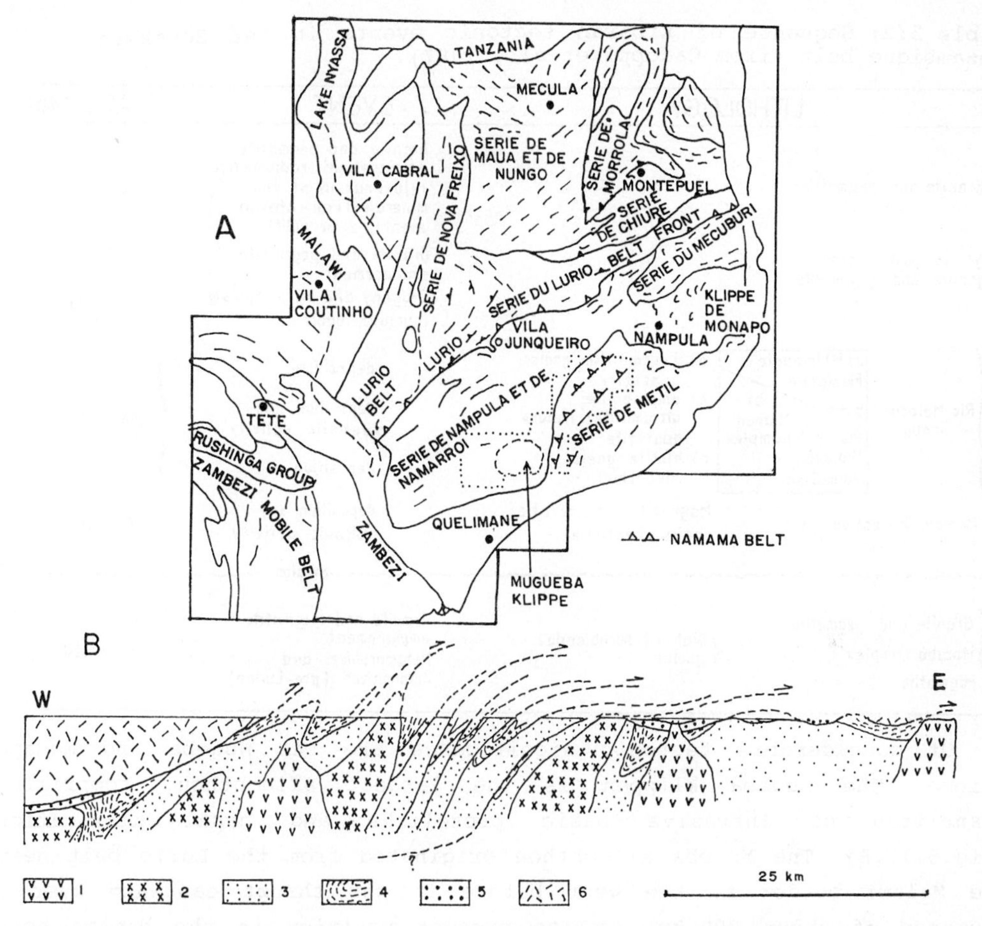

Figure 5.17: Tectonic map of the Southern Mozambique belt (A), and schematic cross-section across the Namama belt and Mugeba Klippe. 1, late- to post-kinematic granite; 2, pre- to syn-kinematic granite; 3, basement (Mocuba Formation); 4, various supracrustal covers - Mamala, Cavarro, Rio Molocue units; 5, ophiolitic subunit of the Rio Molocue Group (Morrua Formation); 6, Granulites of the Mugeba Klippe. (Redrawn from Sacchi et al., 1984.)

Mozambique province include the "Serie de Metil" comprising leucogneisses of mostly rhyolitic volcano-detrital origin (Mamala formation); and metasedimentary rocks such as mica schists, fine-grained gneisses, quartzites and ferruginous quartizites (Rio Molocue Complex). These supracrustals also contain local occurrences of metamorphosed ophiolitic ultramafic rocks. Syn-tectonic granitoids, dated at about 1.10 Ga, intrude both the basement and the supracrustal cover, while younger (about 500 Ma) Pan-African prophyritic granites intrude the earlier granitoids.

Table 5.2: Sequence of Kibaran tectonic events in the Southern Mozambique belt (from Cadoppi et al., 1987).

	LITHOLOGY		EVENT	AGE(Ma)
	Granite and pegmatite		Granite and pegmatite emplacement; radiometric rejuvenation of the minerals (Pan-African event); slow uplift	450÷500
	Syn-to-post-tectonic granite and pegmatite		Granite and pegmatite emplacement Orogeny and metamophism (Lurian); uplift	? 1000
covers	Rio Molocue Group: a) Mt.Inrepele Formation	a) Leucocratic gneiss; quartzite	deposition	1000-1100
covers	Rio Molocue Group: b) Morrua Complex	b) amphibolite; ultramafite; impure quartzite	deposition and magmatic activity	1000-1100
covers	Rio Molocue Group: c) Rio Nipiodi Formation	c) biotite gneiss; micaschist	deposition	1000-1100
covers	Mamala Formation	Magnetite-Leucocratic gneiss (Leptinite)	deposition and volcanic activity	1000-1100
			erosion	
basement	Granite and pegmatite Mocuba Complex Migmatite	Biotite (-hornblende) gneiss	Granite and pegmatite emplacement Metamorphism and deformation (pre-Lurian)	>1100

Of considerable tectonic significance are the Mugeba and Monapo klippe. The Mugeba klippen is the largest allochthon, comprising granulites of intrusive basic protoliths and basal ultramafics (Fig.5.17,B). The Mugeba allochthon originated from the Lurio belt near the Malawi border to the west before it was thrust eastward over a distance of about 200 km, to its present position in the Namama belt (Sacchi et al., 1984). The position of the root zone of the granulite Mugeba nappe in the central part of the Lurio belt near the Malawi border is structurally compatible with Andreoli's (1984) thrust belt in southern Malawi in which ultramafics suggest the existence of a collision suture in the region.

5.3 Regional Tectonic Model for the Kibaran Belts

A regional tectonic synthesis for the Kibaran belts of eastern central Africa must accommodate and account for the following salient geologic features in this region. There is crustal shortening across the Irumide fold belt. Irumide lithofacies and structures continue into the Southern Mozambique belt where higher grade granulite-facies paragneisses with slices of ophiolitic basic and ultrabasic rocks and island-arc volcaniclastic deposits are known in a number of places. In the Southern Mozambique belt a granulitic klippen which has been thrust southeastward from the Lurio belt overrides older pre-Kibaran basement gneisses and cover rocks in the Namama belt. Conditions appeared particularly favourable for the emplacement of plutons of gabbro-anorthosite in the Malawi province. The Kibaran orogeny in the Southern Mozambique belt which took place at about 1.10 Ga coincided with folding and thrusting in the Irumide belt, compressional deformation and alkaline magmatism in the Kibaran belt further north, and with northwest-trending sinistral strike-slip movement along the Ubendian belt.

Andreoli's (1984) plate tectonic model for the southern Malawi has offered a widely accepted framework for explaining the above tectonic features (Daly, 1986a,b; Klerkx et al., 1987; Piper et al., 1989; Sacchi et al., 1984). Andreoli postulated that the orogeny in southern Malawi was initiated by collision between the eastern passive margin of a Niassa craton (Fig.5.11) with an island arc which developed adjacent to the western margin of a Lurio craton (Fig.5.18). The strongly recrystallized

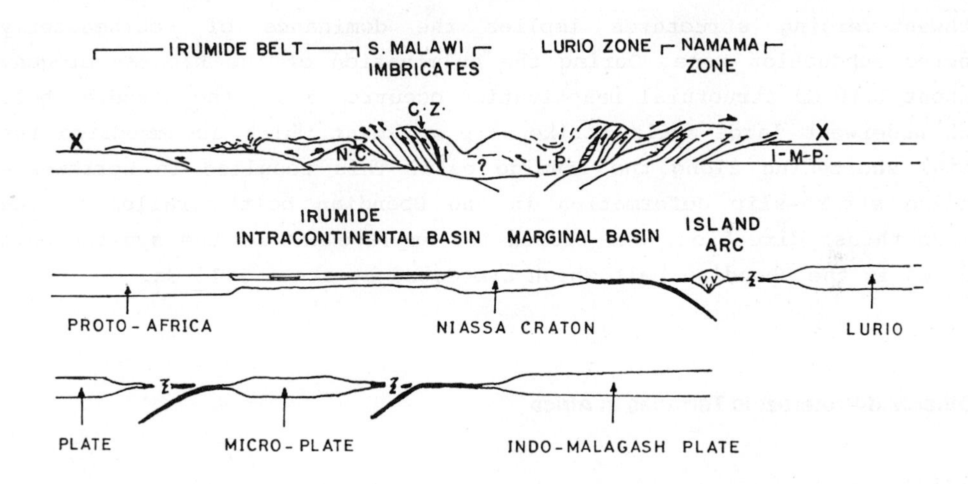

Figure 5.18: Schematic structural section across the Irumide-Southern Mozambique belt showing thrust zones, lithospheric plates and associated ophiolites. N.C., Niassa craton; L.P. Lurio plate; I-M-P, Indo-Malagash plate; C.Z. Chimwadzulu Zone. (Redrawn from figures supplied by M. C. Daly.)

polycyclic pre-Kibaran basement at Niassa and in the Namama belt (Fig.5.17) are regarded as the relict cratonic nuclei. The ophiolitic mafics and ultramafic slices in the Mpanshya, Chimwadzulu and Namama zones possibly represent relict ocean floor (Fig.5.18), while island-arc volcanics and associated sediments are represented in the amphibolites and gneisses of southern Malawi. The Muva Supergroup and its eastern high-grade equivalents, the paragneisses of central Malawi, are probably the continental shelf-slope assemblages which accumulated along the margin of the Niassa craton. Eastward subduction and underthrusting of the southern Malawi island arc probably generated the syn-tectonic granulites and amphibolites in the region while subsequent island-arc-continent and continent-continent collisions produced the fold and thrust belt tectonics in the Southern Mozambique belt (Fig.5.18). Anorthosite plutons which crystallized from mantle partial melts later recrystallized at the granulites facies and were isoclinally infolded within granulites and amphibolites.

As suggested by Daly (1986a,b) the complex structural and geologic relationships in the Southern Mozambique belt and the likelihood that several collision sutures existed in the region implies changes in subduction directions and the accretion of several crustal fragments in the region. The existence of northwest-southwest transport directions suggests pre-orogenic subduction directions while the predominance of northwest-verging structures implies the dominance of southeasterly directed subduction zone. During the culmination of the Kibaran orogeny at about 1.10 Ga structural reactivation occurred along the Ubendian belt which underwent large-scale strike-slip movement which accommodated the crustal shortening along the Irumide belt. This resulted in northwest-trending strike-slip deformation in the Ubendian belt parallel to the Irumide thrust direction, and caused the emplacement of the syn-tectonic granites in the Ubendian belt which are dated at about 1.13 Ga.

5.4 Other Mid-Proterozoic Terranes in Africa

Angola

In Angola sub-horizontal beds of Kibaran metasedimentary rocks occur in stable cratonic areas (Fig.3.32). Like the Kibaran belts of eastern central Africa the mid-Proterozoic cratonic platform cover sequences of Angola trend northeast-southwest. These sequences include the Chela Group and the Leba-Tchamalindi Formation of southern Angola, which extends into neighbouring Namibia; part of the Oendolongo Supergroup in west central Angola, and the Malombe and Luana Groups of northeastern Angola. The Chela Group consists of conglomerates, quartzites, sandstones, siltstones, shales, calc-alkaline volcanics and volcano-sedimentary units (Carvalho et al., 1987). Clasts of the Cuenene gabbro-anorthosite complex and the surrounding intrusive red grantites occur in the Chela Group conglomerates suggesting an age of 1.40 - 1.30 Ga for the Chela Group (Carvalho et al., 1987). The Leba-Tchamalindi Formation unconformably overlies the Chela Group and is composed of basal conglomerates and quartzites overlain by chemical sediments such as stromatolitic limestones and dolomites. The Leba-Tchamalindi apparently correlates with the stromatolitic dolomites of the Mbuyi Mayi Supergroup on the eastern part of the Zaire craton (Fig.6.59). The fact that the Leba-Tchamalindi is cut by noritic dolerite dykes, 1.10 Ga old, places it in the Kibaran.

Among the intrusives emplaced into the Eburnean basement complex of southern Angola is the vast Cunene gabbro-anorthosite complex in the extreme south of Angola. Emplacement was at about 1.5 Ga (Vermaak, 1981). This complex hosts an iron-titanium ore deposit. Although poorly exposed, it is believed to occupy an area of about 17,000 km^2 in southwestern Angola and northernmost Namibia. Its estimated thickness is up to 14 km (Simpson, 1970; Vermaak, 1981). The complex which comprises over 70 % anorthosite with granitic rocks and minor ultramafic border facies making up the remainder of its composition, contains titaniferous magnetite bodies (with an average of 49.5 % Fe, 18.7 % TiO_2) which are probably iron-titanium oxide segregations that are scattered through the north-central parts of the complex (Sawkins, 1990).

East Saharan Craton

Schandelmeier et al., (1990) and Vail (1988a) furnished a summary of the suspected Early-Middle Proterozoic assemblages within the basement of the Sudanese part (Fig.5.19) of the East Saharan craton. These include the Zalingei area in the northern part of South Darfur Province, the Bayuda Desert in southern Blue Nile Province, and the Equatoria Province, and the Red Sea Hills.

In the North and South Darfur Provinces probable mid-Late Proterozoic rocks, which are referred to as the Quartzite Group, consist of long infolded ridges of quartzite, flaggy biotite gneisses and sericite schists. Similar but highly folded amphibolite facies rocks of probable Middle Proterozoic age or older, occur near Zalingei town. These contain at the base (Kongyo Hills Sandstones) massive quartzites with psammitic intercalations, which pass upward through the Golba Siltstones and the Zalingei Semipelites, into the Tari Graphite-Quartz Schists. In southern Sudan and neighbouring Uganda and western Kenya a supracrustal sedimentary unit comprising massive quartzites and quartz schists rests unconformably upon basement gneisses, and is regarded as of probable Middle to Late Proterozoic age. The Madi Group and the Kinyeti Metasediments in the Sudan belong to this supracrustal sequence. The Pan-African reactivated basement in the Bayuda Desert is probably even of pre-Middle Proterozoic age. This comprises granitic gneisses (Abu Harik Series or Gray Gneiss Group) which are overlain by the Metasedimentary Group (Bayuda Formation), a geosynclinal succession with a basal sequence of quartzites, quartzo-feldspathic gneisses, mica schists and marble; a middle unit comprising acidic gneisses, biotite- and hornblende-gneisses

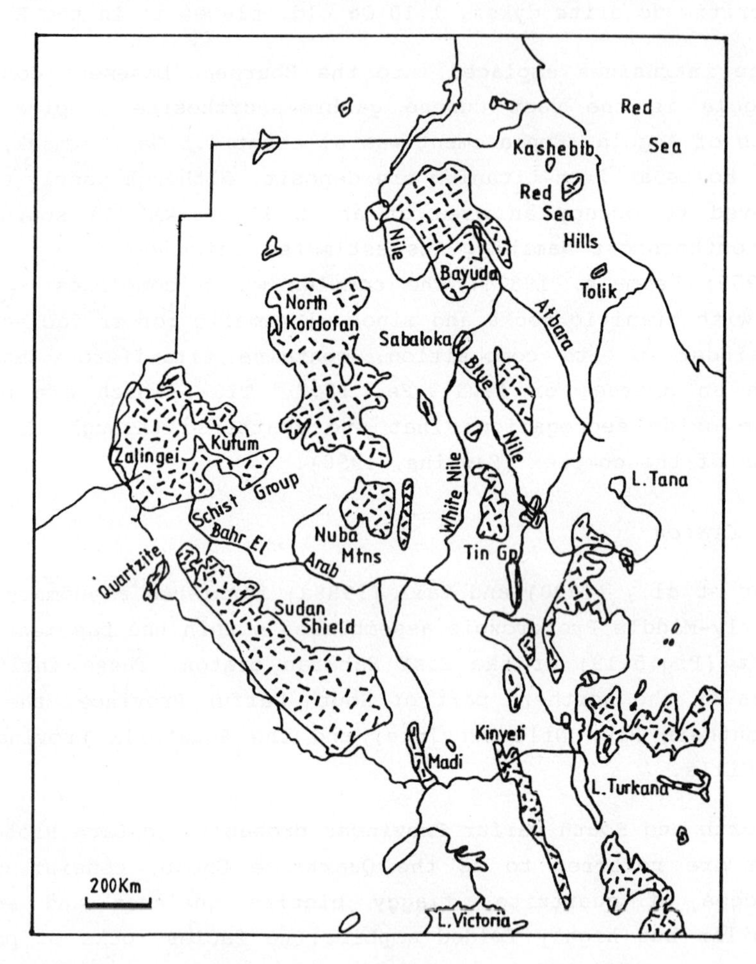

Figure 5.19: Lower-Middle Proterozoic basement rocks in the Sudan. (Redrawn from Vail, 1988a.)

and amphibolites; and an upper volcano-sedimentary suite. The latter consists of mica schists, ferruginous quartzites and marbles which are believed to represent an island-arc depositional setting with sediment-filled back-arc basins and shelf facies which existed prior to 1.0 Ga (Vail, 1988a).

In the southern Blue Nile Province Early to Middle Proterozoic basement rocks (Tin Group) comprise a lower unit (Selak Formation) of migmatitic gray gneisses, enclosing amphibolitic bands, and an upper supracrustal metasedimentary cover, the Gonak Formation which consists of paragneisses, pelites and calc-silicate rocks. In the Red Sea Hills Middle Proterozoic or older rocks are suspected to be the exotic basement terranes among Late Proterozoic rocks. The exotic basement includes acid gneisses, hornblende schists, chloritic slates and marbles, which are of the amphibolite grade.

Madagascar

Hottin (1972, 1976) and Vachette (1979) have demonstrated, based on geochronologic work, that Middle Proterozoic rocks occur in the northern and southern parts of Madagascar (Fig.3.45). Mid-Proterozoic rocks in Madagascar are predominantly supracrustals such as quartzites, mica schists, crystalline dolomites, volcanics, and metasediments of apparently deeper marine facies. These rocks rest unconformably upon the Archean basement of Madagascar. The 1.10 Ga orogeny which affected the Kibaran belt has also been recognized in Madagascar.

In his review of the occurrence of ophiolitic rocks in the Late Proterozoic of eastern Africa, Berhe (1990) referred to a probable ophiolitic suture zone in northeastern Madagascar. This zone of mafic-ultramafic rocks (1.4 Ga) occupies a narrow area 5-20 km wide over a distance of about 800 km, following a north-south trend. It consists of dunites, harzburgites associated with nickel and chromite deposits, gabbros and amphibolites. These rocks represent an ophiolite belt which probably originated after mid-Proterozoic rifting, during back-arc spreading, in an active subduction zone environment. The ophiolites were imbricated in a suture zone during the Kibaran continent-continent collision, thus corroborating the abundant evidence for Kibaran collision tectonics which we have seen in the Southern Mozambique belt.

Chapter 6 Late Proterozoic-Early Paleozoic Pan-African Mobile Belts

6.1 Introduction

Kennedy (1964) originally defined the Pan-African as a major and wide-spread tectono-thermal event that led to the structural differentiation of Africa into cratons and orogenic areas about 500 ± 100 Ma ago. Since Kennedy's time, refinements in geochronology, extensive field mapping, inter-continental correlations, and the concept of plate tectonics have led to the generally accepted view that Kennedy's definition of the Pan-African orogeny referred only to the final thermal episode of an orogenic cycle which spanned from at least 950 Ma to about 450 Ma (Kröner, 1984), almost the duration of the Phanerozoic! The Pan-African orogeny was not only of such magnitude involving several orogenic episodes in individual belts, but the regionally extensive Pan-African belts are also part of a world-wide system of mobile belts (Fig.6.1) which mark the limit between the Precambrian and the Phanerozoic (Black, 1984). It will be shown in this chapter that the Pan-African orogeny in the individual belts, starting from the initial rifting phase with related sedimentation and magmatism, through ocean opening and concomitant continental margin (geosynclinal) sedimentation, to subduction and plate collision, and post-collision magmatism, spanned the entire Late Proterozoic to Early Paleozoic.

The term Pan-African will be used here with a dual meaning. It will be used as a collective term for the orogenic cycles of Late Proterozoic-Early Paleozoic age, as well as for this age span, hence in a geochronological sense, equivalent to an era.

The Pan-African belts (Fig.6.1) display all the sedimentary, magmatic and structural facets of modern orogenic belts that are related to plate tectonics, and provide conclusive evidence for the operation of the Wilson Cycle in the Precambrian. It was after the identification of Andean-type and island-arc continental margin volcaniclastic sequences in the Hoggar (Caby, 1970), and dismembered ophiolites in Morocco (Leblanc, 1981) and in the Arabian-Nubian Shield (Garson and Shalaby, 1976) that it first became clear that modern plate tectonics had operated in the Precambrian. Also, the identification of Pan-African cryptic collision sutures in high-grade metamorphic terranes along the margins of the West-African craton and the recognition of their similarities with Himalayan-type collision belts

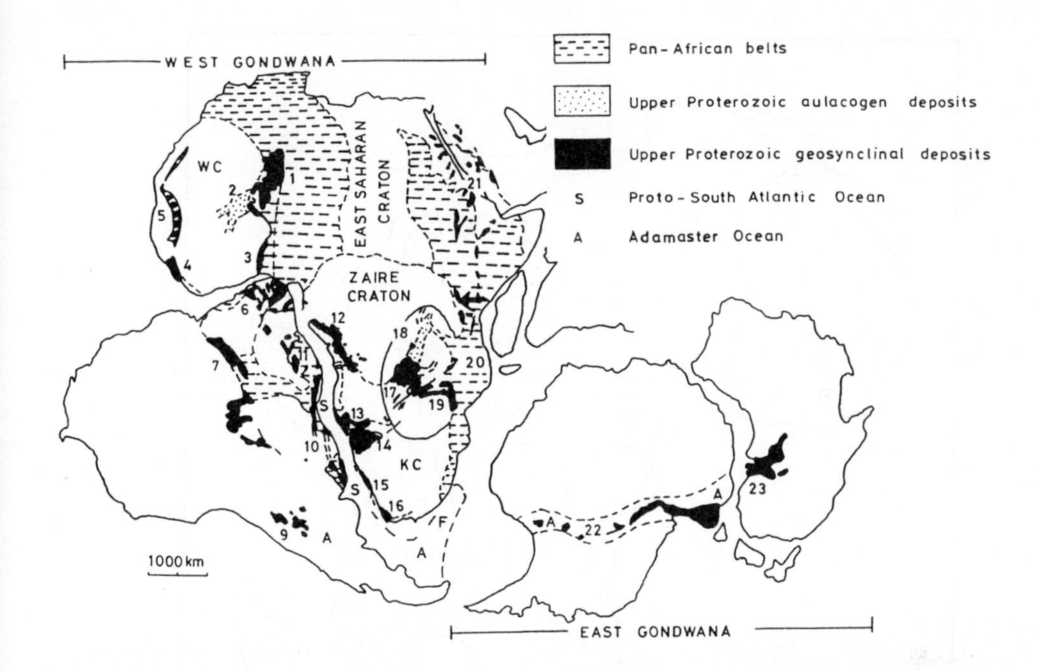

Figure 6.1: Pre-drift reconstruction of Gondwana showing Pan-African belts. 1, Pharusian belt; 2, Gourma aulacogen; 3, Dahomeyan belt; 4, Rokelide belt; 5, Maritanide belt; 6, Northeastern fold belt (Borborema province); 7, Araguaia belt; 8, Paraguay belt; 9, Sierras Pampeanas; 10, Ribeira belt; 11, Mantiqueira belt; 12, West Congolian belt; 13, Kaoko belt; 14, Damara belt; 15, Gariep belt; 16, Saldanhia belt; 17, Lufilian arc; 18, Shaba aulacogen; 19, Zambezi belt; 20, Mozambique belt; 21, Red Sea fold belt; 22, Transantarctic belt; 23, Adelaide belt. (Redrawn from Porada, 1989.)

(Burke and Dewey, 1972) ushered in the re-interpretation of similar high-grade terranes, such as the Mozambique belt of East Africa (Fig.6.1). Actually Clifford (1970) had earlier recognized two types of Pan-African orogenic belts which he characterized as zones of orogenically deformed upper Precambrian geosynclinal sediments (e.g. Lufilian arc, Damara, West Congolian, Pharusian belts), and zones of rejuvenated basement which he termed vestigeosynclinal belts (Mozambique belt, Zambezi belt, Nigeria-Cameroon province). What happened in the ensuing two decades was the fitting of both types of Pan-African belts into the plate tectonics paradigm.

Cahen et al. (1984) best summed up the significance of the Pan-African in the crustal evolution of Africa, when they stated that the crustal stability which had prevailed since the end of the Early Proterozoic orogeny (Eburnean) was only locally interrupted during the Middle Proterozoic (Kibaran), but was followed by widespread tectonism (Fig.6.2) after about 950 Ma to after about 600 Ma (Pan-African). The long phase of crustal

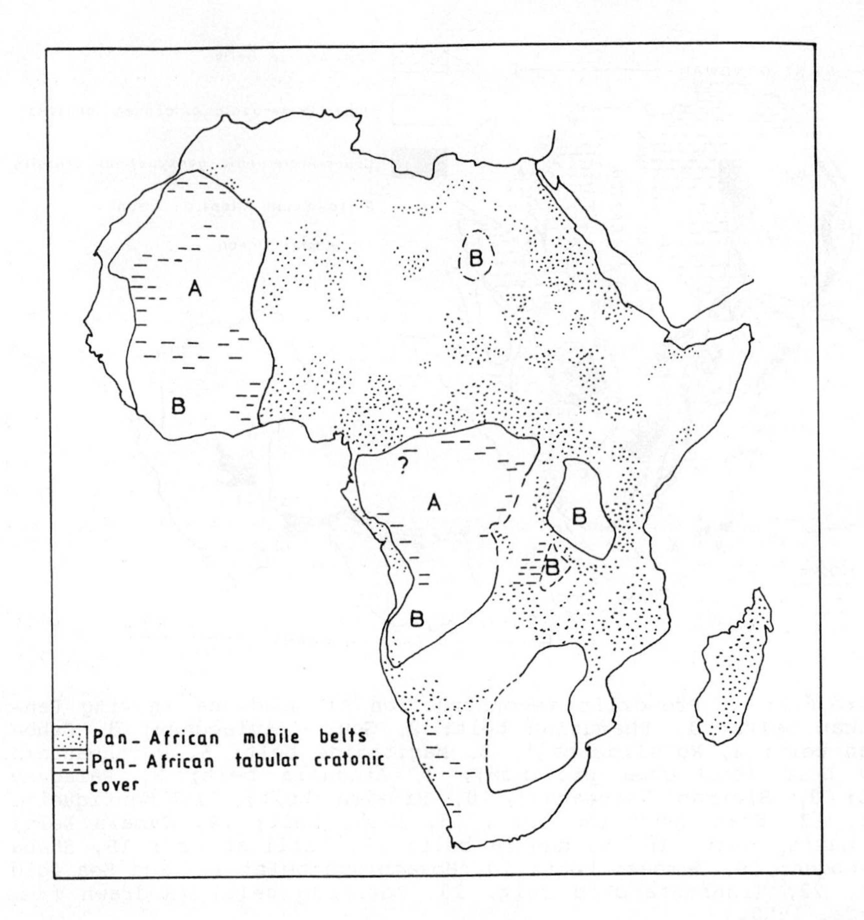

Figure 6.2: Africa showing Pan-African mobile belts and stable areas with cratonic cover. A, Cratonic areas where Pan-African supracrustals are covered by the Phanerozoic; B, cratonic areas stripped of Pan-African cover. (Redrawn from Cahen et al., 1984.)

quiescence after the Eburnean orogeny saw the amalgamation of continental blocks into the supercontinent Pangea I (Fig.5.1). Pangea I was fragmented during the Pan-African tectono-thermal event leading to the opening of a proto-North Atlantic Ocean known as the Iapetus, along the eastern margin of which the Mauritanides and the Rokelides orogens (Fig.6.1) evolved. Southward, the rifting and separation of Pangea I resulted in the formation of a proto-South Atlantic Ocean with a 3000-km chain of continental margin geosynclines which are now preserved in a series of re-entrants known as the West Congolian, Damara, Gariep, and Saldanhia orogenic belts (Fig.6.1). The opening and closing of Pan-African oceans are also now well documented in the vast Pan-African belt between the West African craton and the East Saharan craton, and in the Mozambique belt and the Arabian-Nubian Shield in eastern Africa (Fig.6.1). Plate collision at the very end of the Pan-African orogenic cycle resulted in the emergence of the Pangea II supercontinent, the Gondwana part of which is shown in Fig. 6.1.

The Pan-African era was of great significance in many respects. Apart from the structural differentiation of Africa into stable cratons and mobile belts, the subsequent rifting and break-up of Gondwana, and the initiation of a new Wilson Cycle in the Mesozoic were located along Pan-African mobile belts. The Pan-African marked the last period of widespread orogeny and the formation of extensive mountain chains in Africa. Subsequent orogenies only occurred in the northwest and southernmost parts of Africa, with the rest of Africa remaining cratonic. In terms of mineralization Clifford (1966) had distinguished two tectonic-metallogenic units. The younger orogens (Kibaran and Pan-African) are characterized by major deposits of Cu, Pb, Zn, Co, Sn, Be, Nb-Ta, whereas the older cratons contain important deposits of Au, Fe, Mn, Cr, asbestos and diamond.

The Pan-African was also an era of widespread glaciation in Africa with glaciogenic deposits appearing in many Pan-African belts almost at the same time. Organic evolution had gradually, throughout the course of the Precambrian, led to the widespread appearance of algal stromatolites, which by Pan-African times, had become the dominant agents of carbonate sedimentation in widespread early Pan-African epicontinental seas. By the close of the Precambrian soft-bodied metazoans had evolved. Their traces and body fossils are found in the Katanga orogen and in the Nama cratonic sediments which accumulated between the Damara and Gariep belts in southwest Africa.

6.2 The West African Polyorogenic Belt

6.2.1. Geological and Geophysical Framework

The West African polyorogenic mobile belt (Cahen et al., 1984) extends along the western margin of the West African craton (Fig.6.3). Known as the Rokelides in Liberia and Sierra Leone, the Bassarides in Guinea and Senegal, and as the Mauritanides in northern Senegal, Mauritania and the Western Sahara, this chain of mobile belts is covered by the Mesozoic-Cenozoic coastal basins on its western part, and is in turn thrust against the craton or tabular cratonic cover along its eastern flank (Fig.6.3). Repeated Late Proterozoic-Early Paleozoic orogenies have produced complexly deformed and metamorphosed rock assemblages in the Mauritanides and Bassarides belts which defy simple stratigraphic correlations. Thus, the lithostratigraphic units in the mobile zones of these orogens are treated as tectono-stratigraphic units in order to reflect the features imposed upon them by the tectonic processes they have undergone (Dallmeyer, 1989; Lécorche et al., 1989; Sougy, 1962).

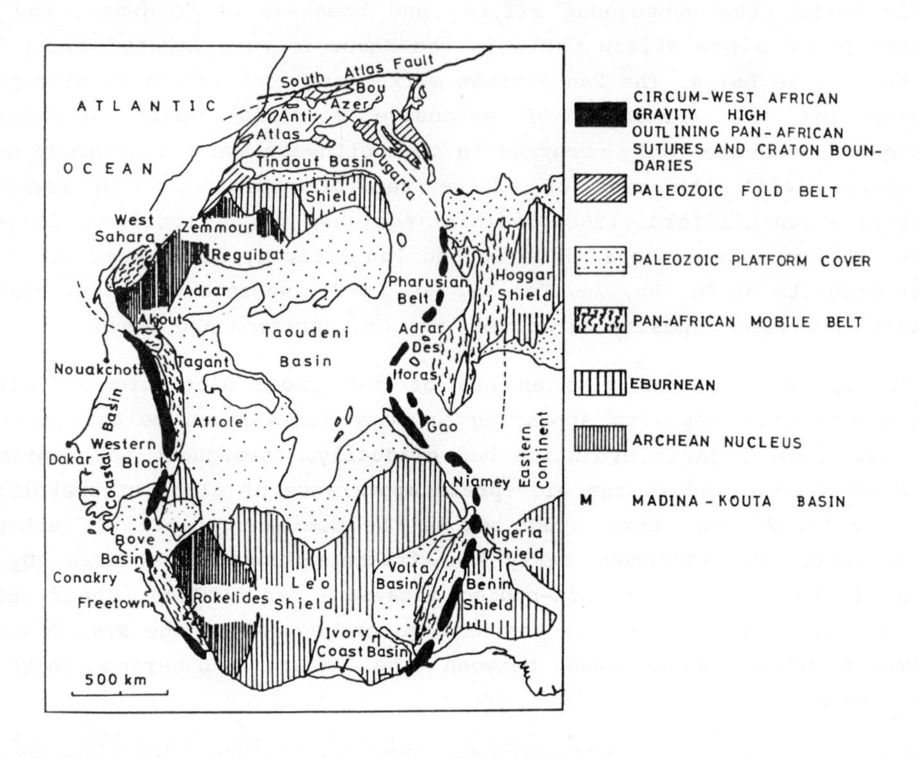

Figure 6.3: West African craton delimited by a belt of gravity highs (black), showing sutures and mobile belts. (Redrawn from Roussel and Lécorché, 1989.)

Regional crustal structure and terrane boundaries have been delineated around the entire West African craton using gravity and geoelectrical geophysical methods (Roussel and Lécorche, 1989; Ritz et al., 1989). A prominent regional belt of gravity highs (Fig.6.3) defines the axial portion and suture of the Pan-African orogenic belts surrounding the craton.

In the West African polyorogenic belt, segments of the circum-West African craton gravity high lie west of the Mauritanides and the Bassarides, and demarcate two gravimetrically contrasting crustal terranes (Fig.6.4,A). The eastern terrane corresponding to the craton, is defined by a broad regional negative anomaly (Fig.6.4,B) which is characterized by

NE-SW gravity trends. In contrast, the western basement terrane underneath the Mauritania-Senegal coastal basin is characterized by a generally positive Bouguer anomaly that is consistent with the existence of a western coastal basement block which is denser and thicker than the craton to the east (Fig.6.4,B). Separating the eastern and western crustal terranes is a prominent, nearly continuous, NNW-SSE-trending belt of positive anomalies. This density discontinuity is known as the Mauritanian anomaly. It runs parallel to the axis of the exposed parts of the Mauritanide-Bassaride orogen, but is displaced slightly westward. It dips westwards beneath the denser western block, and is believed to represent the remnant of a westward-dipping suture zone. In the Bassarides short wavelength gravity highs reflect unrooted dense bodies trapped along the Pan-African suture zone.

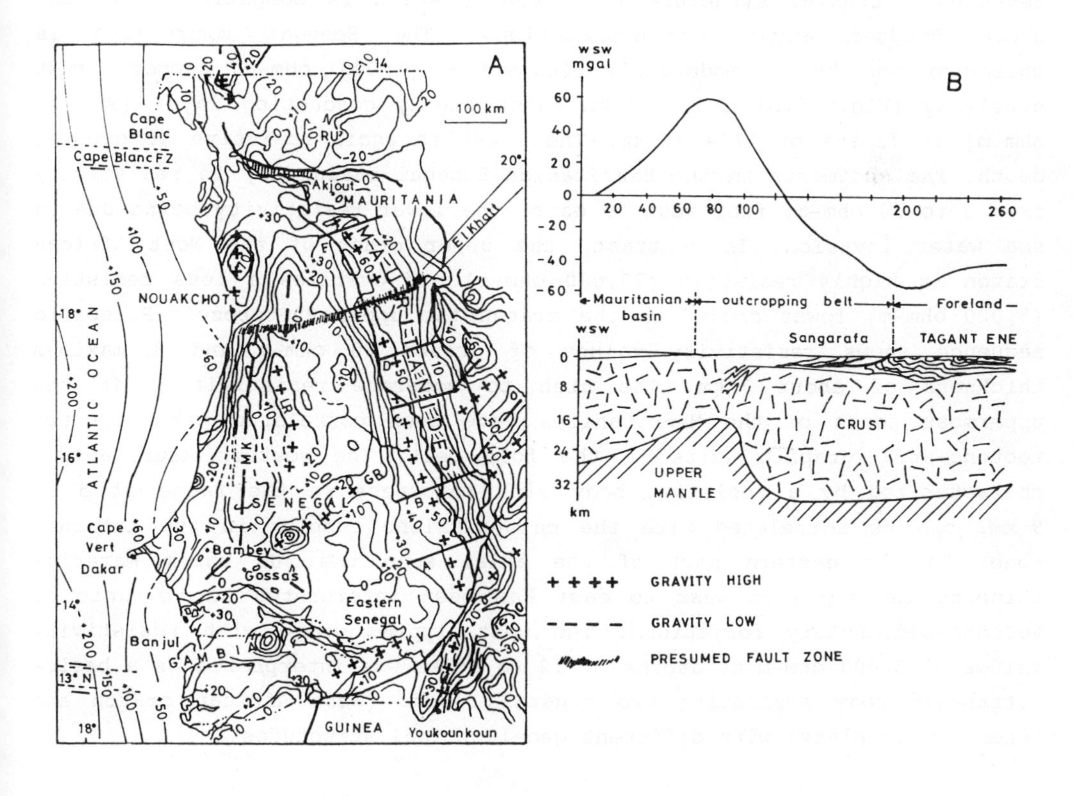

Figure 6.4: Bouguer anomaly map of the Mauritanides (A); B, gravity profile across the Mauritanides. (Redrawn from Lécorché et al., 1983.)

The long wavelength Mauritanian anomaly along the Mauritanide orogen has been interpreted as an asymmetric mantle-rooted mafic or ultramafic ridge with its crest at a depth of about 15 km and a significant westerly dip (Fig.6.4,B). Since the extension of the Mauritanian anomaly coincides with the segment of the Mauritanide-Bassaride orogen where the Late Paleozoic Hercynian orogeny is superimposed on earlier Pan-African deformations, Roussel and Lécorche (1989) inferred that the geometry of the Mauritanian anomaly is associated with compressional stress that was related to a considerable eastward translation of a western basement (Senegal microplate) during the collision of the West African craton against eastern North America (Venkatakrishna and Culver, 1988). As shown below this collisional model offers a mechanism that accounts for the evolution of the West African mobile belts.

A magnetotelluric survey of crustal resistivity across the various tectono-stratigraphic units in southern Mauritanides (Ritz et al., 1989) revealed a crustal structure (Fig.6.5B,C) which is compatible with the above Bouguer anomaly interpretations. The Senegal microplate is characterized by a moderately resistive (1,000 ohm-m) upper crust overlying (Fig.6.5,B) a 2 - 6 km thick highly conducting layer (2 - 15 ohm-m) at depths of 12 - 18 km. The crust is again resistive at greater depth. The sediments in the Mauritanian-Senegal basin vary in resistivity from 3 to 30 ohm-m, from west to east; the lower resistivity being due to sea water invasion. In contrast, the upper crust of the West African craton is highly resistive (30,000 ohm-m) and overlies a less resistive (5,000 ohm-m) lower crust. On the craton the upper Proterozoic-Paleozoic sequence shows resistivity values of around 80 ohm-m and a maximum thickness of about 7 km. The high to moderate resistivities in the uppermost zones of the Mauritanides have been correlated with the two tectono-stratigraphic units in the fold belt. The western Mauritanides characterized by a resistive body with thickness in the range of 5 to 9 km, can be correlated with the calc-alkaline complex in the internal zone. In the eastern part of the axial zone the 300 ohm-m material thinning rapidly from west to east has been interpreted as volcanic or volcano-sedimentary formations. The west-dipping wedge with resistivity values of 3,000 ohm-m at depths of 12 to 16 km was interpreted as a basic-ultrabasic body separating two crustal blocks (West African craton and Senegal microplate) with different geoelectrical structures.

6.2.2 Tectono-stratigraphic Units

Dallmeyer (1989) and Lécorche et al. (1989) defined several tectono-stratigraphic units across the West African polyorogenic belt. The West African craton is the basement or foreland zone which is unconformably overlain by horizontal mid-Late Proterozoic and Early Paleozoic cover

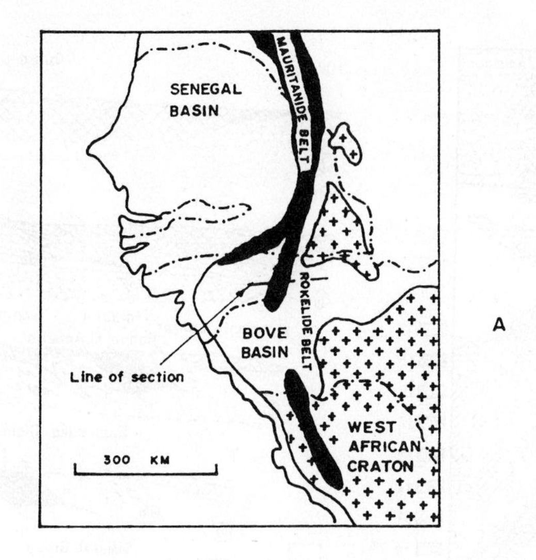

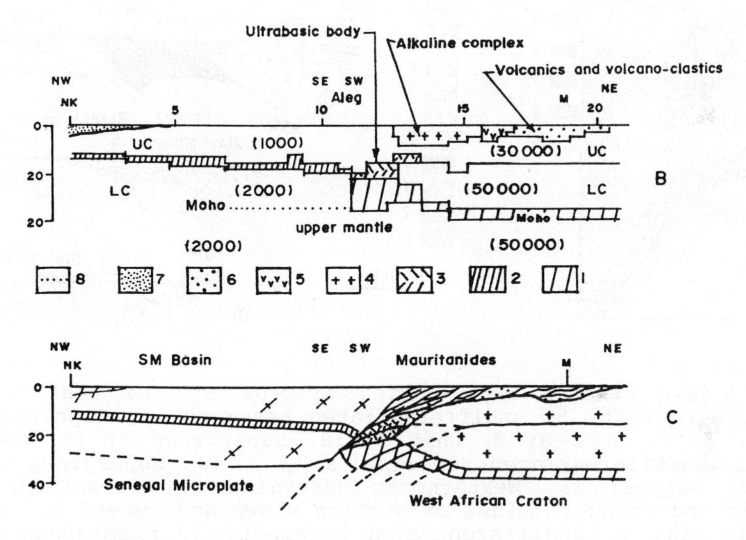

Figure 6.5: Crustal electrical resistivity interpretation across Southern Mauritanides based on the magnetotelluric method. (Redrawn partly from Ritz et al., 1989.)

sediments (Fig.6.6). The external zone to the immediate west contains geosynclinal clastic sequences; the axial zone consists of variably metamorphosed and deformed allochthonous rift volcanic-volcaniclastic rocks with dismembered ophiolites; whereas the internal zone furthest to the west comprises variably retrogressed gneissic terranes.

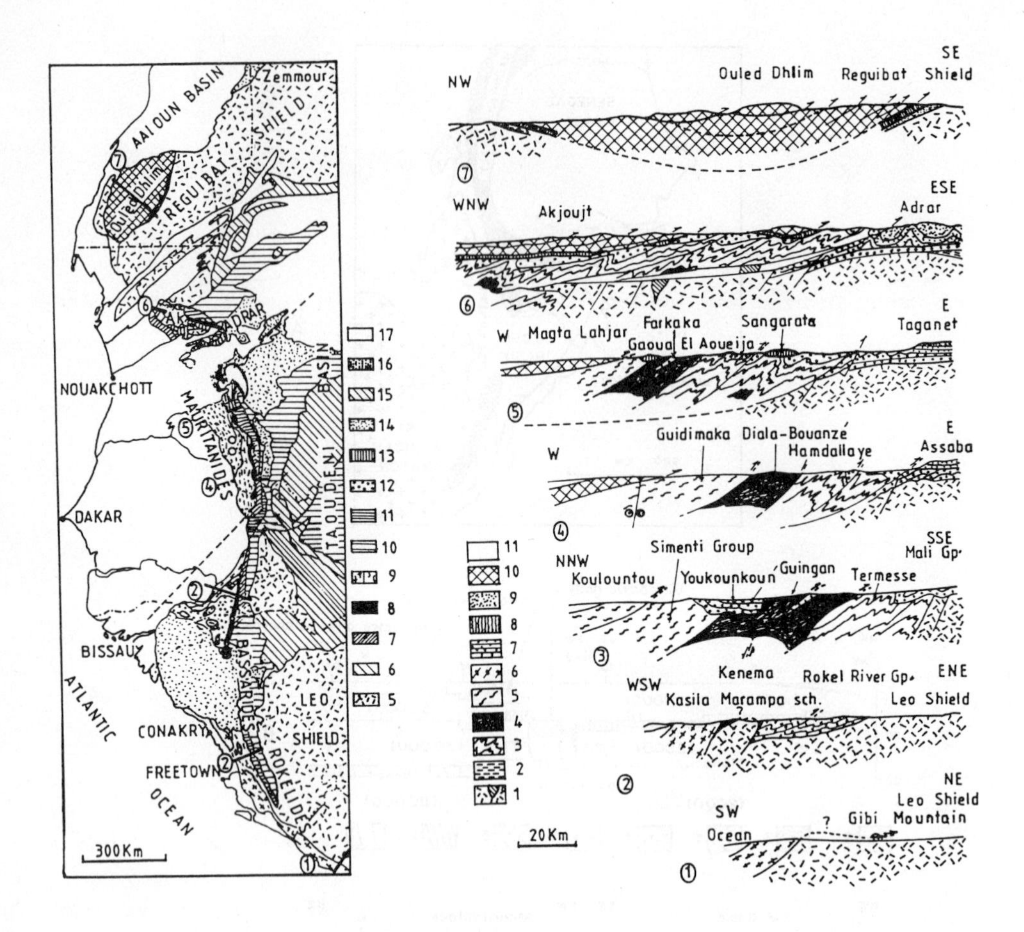

Figure 6.6: Sketch map showing orogens of the West African polyorogenic belt. 5, undifferentiated basement; 6, Supergroup 1; 7, Unit A; 8, Unit B; 9, Unit C; 10, Supergroup 2; 11, 2a; 12, S2b; 13, undifferentiated Supergroups 2, 3; 14, Supergroup 3; 15, internal nappes; 16, Westernmost Hercynian (?) formations; 17, Mesozoic and younger. Lines of section shown and legend for cross-sections are: 1, undifferentiated basement; 2, Supergroup 1; 3, Unit A; 4, Unit B; 5, Unit C; 6, exotic basement; 7, S2a; 8, undifferentiated S2b and S3; 9, Supergroup 3; 10, Late Paleozoic allochthons; 11, Mesozoic and younger cover. (Redrawn from Lécorché, 1989.)

Foreland Units

The foreland units are part of the Taoudeni cratonic basin. They are regionally extensive tabular epicontinental sequences. At the base an unconformable sequence termed Supergroup I by Trompette (1973), consists of quartzitic sandstones, shales and stromatolitic limestones; this sequence is 1,400 m thick. It is of mid-Late Proterozoic age. Supergroup I is unconformably overlain by Supergroup 2, a sequence of Late Proterozoic-

Cambrian epicontinental clastic deposits, about 1,200 m thick. Westward in the orogenic belt, Supergroup 2 overlies pre-deformed and metamorphosed tectono-stratigraphic units. Supergroup 2 has been subdivided into 2a, a sequence of upper Proterozoic tillite, baryte-limestone, chert and stromatolitic dolostone; and 2b, a sequence of cross-bedded, feldspathic red sandstones which succeed Supergroup 2a disconformably and is best developed in Mauritania. Inarticulate brachiopod faunas suggestive of the Cambrian-Ordovician boundary occur at the base of 2b, whereas at the top are conformable quartzitic sandstones and fine-bedded marine sandstones with the trace fossil *Scolithos*. Supergroup 2 is unconformably succeeded by Supergroup 3 throughout Mauritania. The latter begins with Late Ordovician tillites, followed by Silurian clastic beds, and by Devonian limestones, siltstones and shale (Fig.6.7,A). The Carboniferous has not been recognized in the West African polyorogenic belt.

The Bové basin in Guinea is a Middle Paleozoic basin containing relatively undeformed deposits which correlate with Supergroup 3. This basin which overlies the northern part of the Rokelide and the southern part of the Bassarides (Fig.6.3) is filled with Silurian shales which pass upward into Devonian sandstones and shale.

External Units

In the external parts of the orogens there are structural units containing deformed and weakly metamorphosed sequences which are the equivalents of Supergroup 2 on the craton. In the external zone these tectono-stratigraphic units are variously termed the Gibi Mountain Formation in the Rokelides in Liberia, the Rokel River Group in Sierra Leone, the Mali Group in the Bassarides, and have also been assigned various names such as the Bakel Group, in various parts of the Mauritanides (Fig.6.7). In the external zone of the fold belt these equivalents of Supergroup 2 are locally thrust eastward over deformed but unmetamorphosed foreland sequences. But towards the axial zone to the west, they become progressively imbricated with the tectonic units of that zone.

The Rokel River Group of the Rokelides consists of sediments and volcanics; it is about 5 km thick and occupies a broadly synclinorial structure within the Archean basement which it overlies with a marked unconformity. The sequence begins with conglomerates, feldspathic sandstones and clays which have been interpreted as tillites belonging to the regional tillite at the base of Supergroup 2 (Culver et al., 1978). A progressive deepening of the Rokelide basin is suggested by an upward change from littoral and sublittoral sandstones through neritic shales, subarkoses and orthoquartzites into turbiditic shales and sandstones. Overlying prodelta shales and mudstones indicate upward shoaling of the basin, which was followed by andesites, basalts and spilites with pillow structures. At the top are deltaic shales, siltstones, arkoses, subarkoses and orthoquartzites (Williams and Culver, 1982).

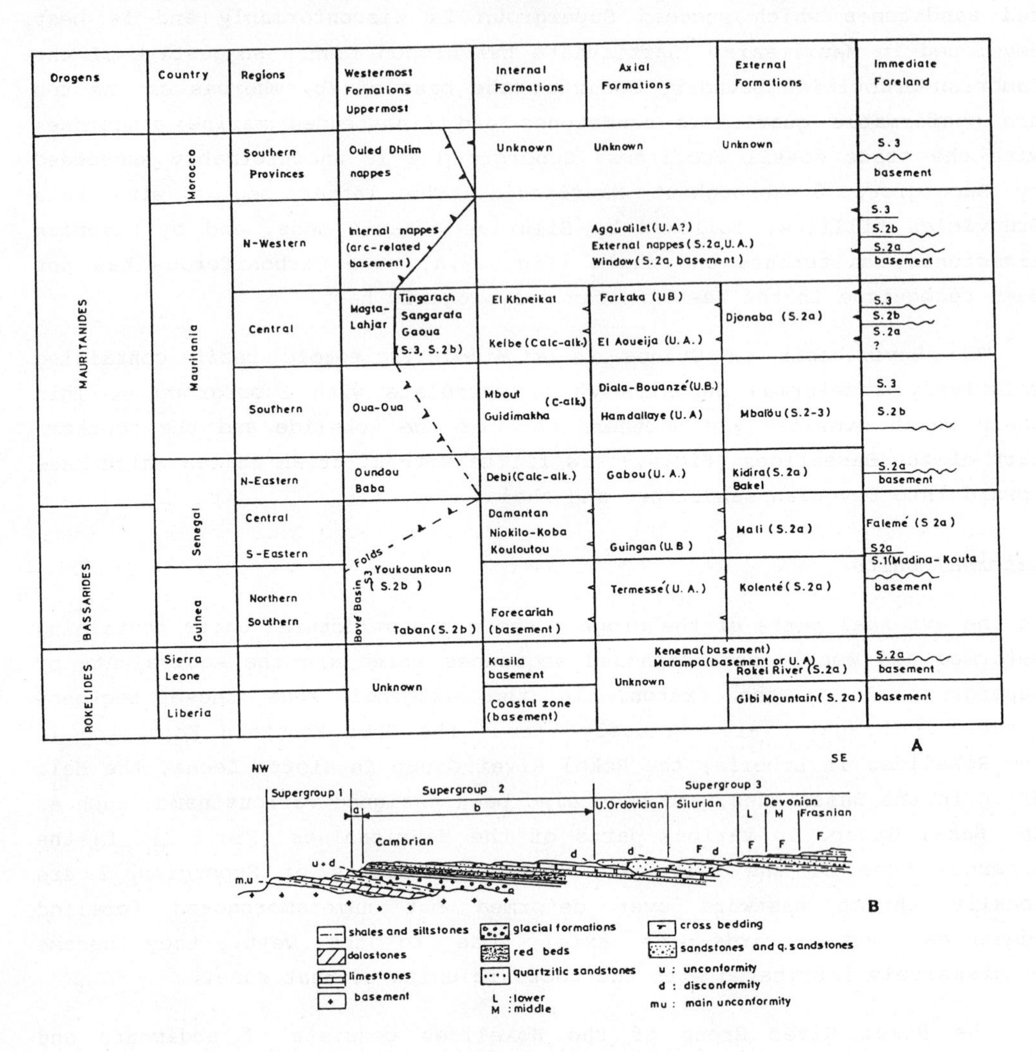

Figure 6.7: Tectonostratigraphic correlations in the Mauritanide, Bassaride, and Rokelide orogens (A); and (B) generalized tectonostratigraphic cross-section of the Mauritanian Adrar. (Redrawn from Lécorché et al., 1989.)

In the Bassarides of northwestern Guinea and southeastern Senegal, the Mali Group, an equivalent of the Rokel River Group, is also largely undeformed and unmetamorphosed and comprises basal conglomerates and sandstones of glacial origin. These basal tillites are succeeded by a sequence of shales, dolomites and quartzites which predominate over lavas and graywackes and indicate deposition on a passive continental margin where more open marine environments prevailed than in the Rokel River Group to the south. The Mali Group, unlike the Rokel River Group, is underlain by older eugeosynclinal graywackes, abundant volcanics and calc-alkaline intrusives which represent an earlier active continental margin that was mildly deformed and metamorphosed prior to the deposition of the Mali-Rokel River sequences. This older succession belongs to Supergroup I of mid-Late Proterozoic age.

Axial Units

Allochthonous units occur in parts of the Mauritanide-Bassaride orogen (Dallmeyer, 1989; Dillon and Sougy, 1974; Lécorche et al., 1989; Sougy, 1962). Lithologically these are variably deformed and metamorphosed volcanic and metasedimentary rocks which represent a Late Proterozoic rifted-margin prism that probably correlates with the upper parts of Supergroup I. Among the allochthonous units there are intercalated marine sequences in the form of graywacke, chert and jasper, serpentinite and tholeiitic and alkaline basalts. The various formations of the axial zone (Fig.6.7,B) have been grouped into the slightly metamorphosed tholeiitic to alkaline volcanic sequence (Unit A or U.A in Fig.6.7,B), and the higher grade (amphibolite) metavolcanic and metasedimentary western sequence (U.B). In the central Mauritanides, the Farkaka sequence (U.B) includes continental supracrustal components and intraplate rift-related meta-basalts. The paleogeographic setting for the U.B assemblages was probably a continental margin setting along the eastern side of a rifted continental block that probably existed west of the West African craton (Lécorche et al., 1989). Similarly Unit A represents coeval rifted-margin deposits that probably accumulated along the western margin of the West African craton during the same phase of ocean spreading.

There are uncertainties regarding the correlations of some of the metasedimentary units in the allochthons of the axial zone of the Mauritanide-Bassaride orogen, because their foreland equivalents are not known (Lécorche et al., 1989). For example, the sequence of very low-grade, deformed sandstones which contain Devonian faunas and structurally overlie pre-deformed high-grade axial units at Sangarata (Fig.6.6, 5), may be equivalent to Supergroup 2 or 3. Also uncertain are the correlatives of the extensive mylonitic quartzite nappes which are exposed in the Akjout region of northwestern Mauritania (Fig.6.6, 6). An extensive klippe of unknown origin rests on the western margin of the Reguibat Rise. A variety of variably metamorphosed and deformed rocks including anorthosite plutons are part of this internally imbricated nappe complex which is known as the Ouled Dhlim complex. The Ouled Dhlim complex may at least in part be structurally equivalent to the Akjout internal nappes. The latter nappes contain assorted rock assemblages including metasediments, meta-volcanics, and retrogressed gneissic rocks.

Internal Units

Variably retrogressed mylonitic gneissic rocks constitute the western internal structural units in the West African fold belt (Fig.6.6). In the Rokelides these are considered as exotic basement terranes that probably represent fragments of the Guyana craton of South America (Fig.6.8). These Pan-African reworked Archean granulite-greenstone basement rocks, the Kasila and Marampa Groups, constitute the internal zone of the Rokelide orogen. In the Mauritanide-Bassaride orogen the gneissic rocks contain calc-alkaline, variably deformed and metamorphosed igneous rocks which include felsic volcanic units and associated co-magmatic granitic plutons that are dated at about 685 - 675 Ma.

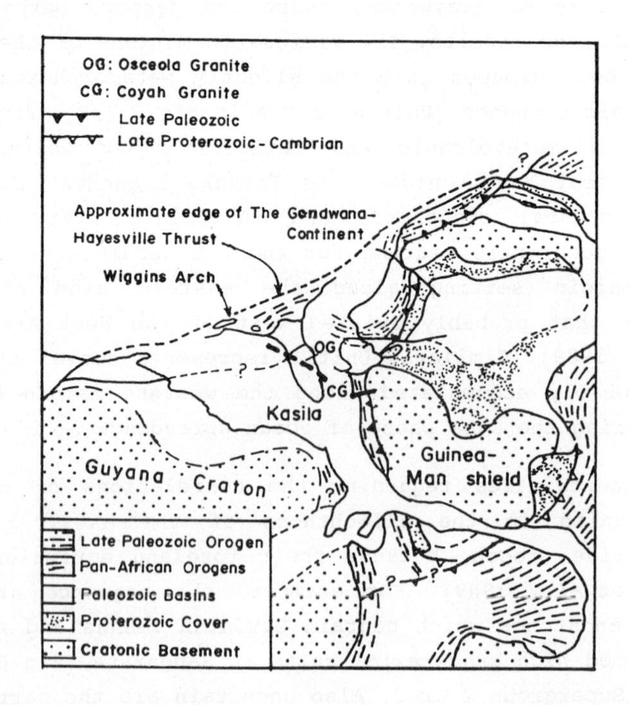

Figure 6.8: Pre-drift reconstruction of NW Gondwana showing the relationship between pre-Cretaceous lithotectonic units in the subsurface of SE U.S.A.; and correlative sequence in West Africa and NE South America. (Redrawn from Lécorché et al., 1989.)

6.2.3 Tectonic History

The polyorogenic West African fold belt was initiated during the mid-Late Proterozoic, between about 1.10 Ga and 770 Ma, by continental rifting of the Pangea I supercontinent (Fig.6.1) along the western part of Gondwana (Fig.6.1). Continental margin depositional wedges accumulated as sea-floor spreading opened a vast Late Proterozoic-Paleozoic ocean, known as the Iapetus (Fig.6.9,A) between Laurentia and western Gondwana. In this paleogeographic reconstruction (Condie, 1989) the Rokelides formed a southeast-trending orogen which probably linked up with the Araguaia orogenic belt in South America (Fig.6.1). Recurring orogenic activity in the West African polyorogenic mobile belt resulted from the convergence of the plates within and around the Iapetus (Fig.6.9A,B); and from several collisions of these plates at different times.

In the Mauritanide-Bassaride belt, Supergroup I and its equivalents (e.g. Termésé, Guigan, Diala-Bouanzé, Farkaka Groups) in the axial zone belong to the rift sedimentation phase (Fig.6.10,A). These rift-facies, with tholeiitic and alkaline basalts as well as continental supracrustal components accumulated to the west and interfingered with deep water sedimentary units. Limited sea-floor spreading produced an ocean of the Red Sea-type (Fig.6.10,A) which was narrow in the Bassarides but much wider northwards in the Mauritanides. Based on the age of the calc-alkaline complex in the internal zone, the ocean closed in the Late-Proterozoic (ca. 765 - 700 Ma) along a westward-directed subduction zone. This resulted in the formation of a western, ensialic volcanic arc over a western tectonic block of continental crust (Fig.6.10,B). In this tectonic model (Dallmeyer, 1989; Lécorche et al., 1989) continental collision during the Late Proterozoic (675 - 650 Ma) caused the first Pan-African episode (Pan-African I) which produced metamorphism and deformation in the Bassarides and Mauritanides (Fig.6.10,C).

By the latest Proterozoic, western Gondwana (Fig.6.9) had drifted southward towards the South Pole so that between about 650 Ma and 575 Ma the glacial and locally flyschoid sequences of Supergroup 2 (e.g. Rokel River Group, Mali Group) were deposited (Fig.6.10,D). A second orogenic episode took place in the Early Cambrian between about 575 Ma and 550 Ma, which had greater intensity in the south where there was continent-continent collision between the West-African craton and the Guyana craton. In the Rokelides this compression produced only slight deformation and very low-grade metamorphism in the western part of the Rokel River Group whereas in the eastern part this sequence is unmetamorphosed and only sporadically mildly deformed (Williams and Culver, 1988). The Kasila Group

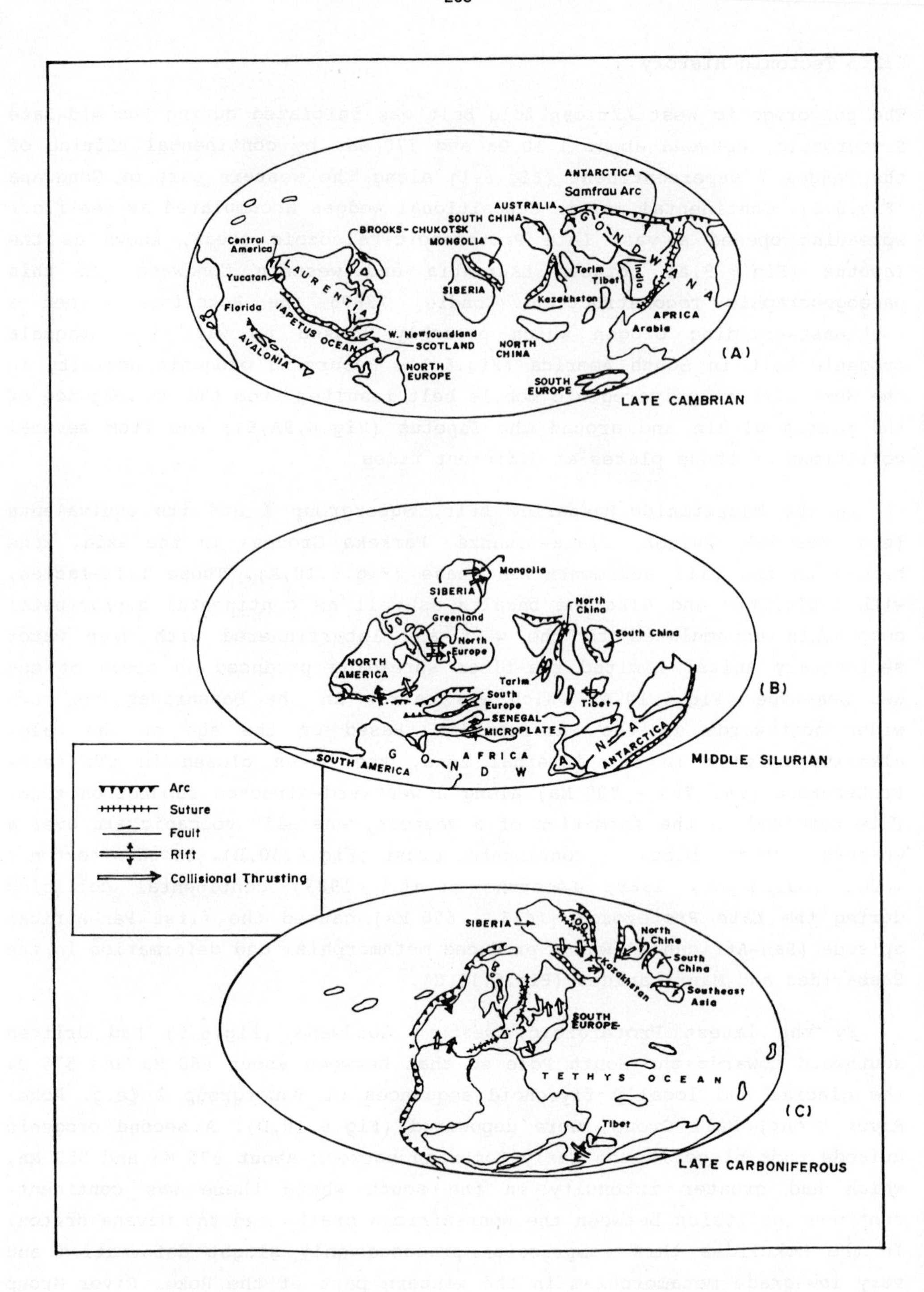

Figure 6.9: Phanerozoic continental reconstructions. (Redrawn from Frazier and Schwimmer, 1987.)

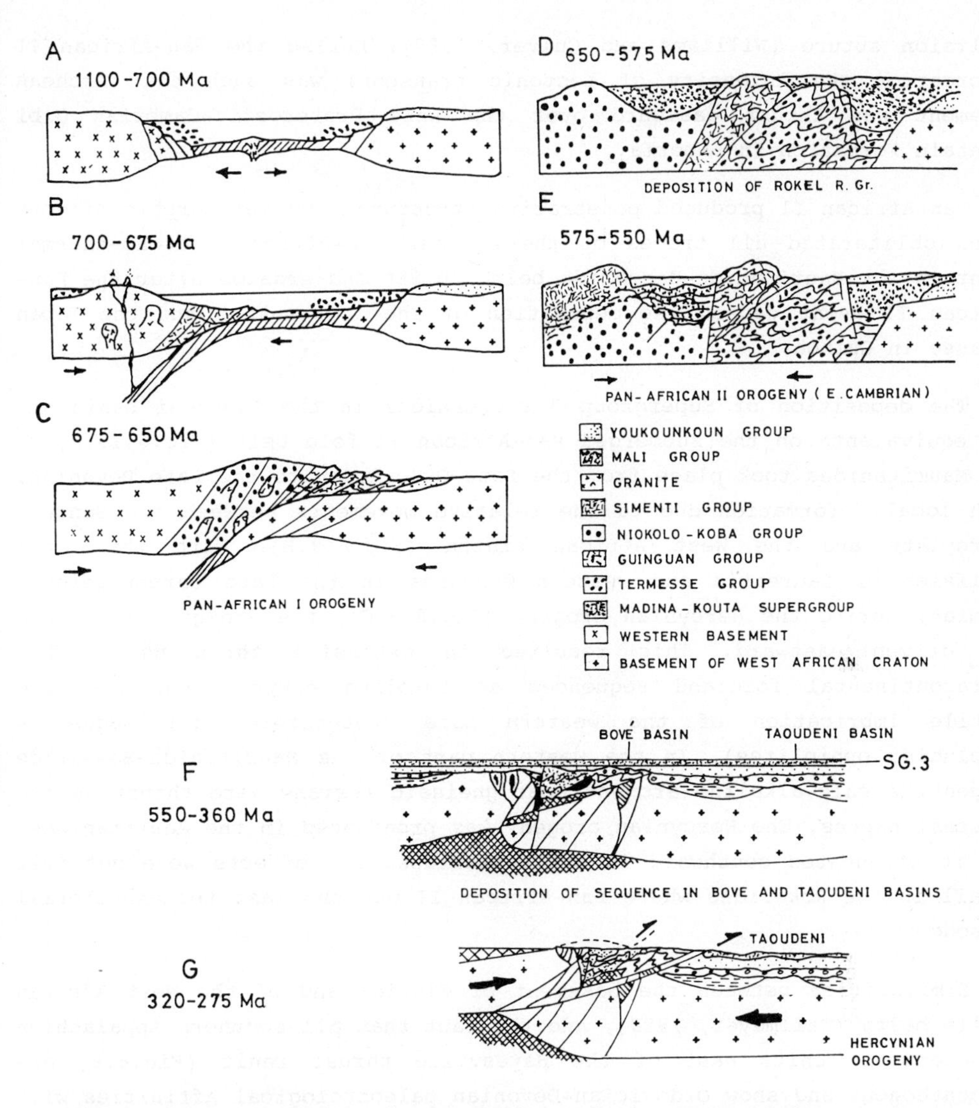

Figure 6.10: Geodynamic evolution of the West African polyorogenic belt. (Redrawn from Dallmeyer, 1989; Lécorché et al., 1989.)

to the west was reactivated during the Pan-African II event (Wright et al., 1985; Williams and Culver, 1988). The 5 km-thick zone of the sheared gneisses and mylonites which dip at moderate angles to the southwest with medium-angle thrusting onto the granitic basement (Figs.6.1,6.2), support Burke and Dewey's (1973) interpretation of the Kasila Group as a cryptic suture in a Pan-African reactivated basement which abuts against unreactivated basement. The mylonite zone, about 400 km long, continues southeastwards into Liberia as the Todi shear zone, although it is doubtful whether the mylonite zone in Liberia is part of a collision suture (Williams and Culver, 1988). During the Pan-African II deformation the intensity of tectonic transport was such that Archean basement was thrust eastward over the Late Proterozoic-Cambrian Gibi Mountain Formation in Liberia.

Pan-African II produced penetrative structural and metamorphic effects which obliterated all traces of the earlier Pan-African tectono-thermal event in the Mauritanide-Bassaride belt. Uplift and erosion after the Pan-African II event caused the deposition of the Youkounkoun and the Taban molasse in Guinea.

The deposition of Supergroup 3 successions in the Taoudeni basin and its equivalents on the submerged Pan-African II fold belt (Fig.6.10,F) of the Mauritanides took place from the Late Ordovician to the Late Devonian, with local deformation due to the relative movements between the Senegal microplate and the West African craton (Fig.6.9,B). With the final collision of Laurentia and western Gondwana in the Late Carboniferous-Permian, during the Hercynian orogeny (Fig.6.9,C), the Senegal microplate was driven eastward. This resulted in extensive thrusting of the intracontinental foreland sequences as foreland nappes, and the more ductile imbrication of the western Late Proterozoic rift sequences (including ophiolites). In the western part of the Mauritanide-Bassaride orogen the calc-alkaline arc and host gneissic terrane were thrust as the internal nappes. The Hercynian orogeny was pronounced in the Mauritanides, but it attenuated southward in the Bassarides. Its effects were not felt at all in the Rokelides where Pan-African II was the last tectono-thermal episode.

Similarities between the basement of Florida and of the West African mobile belts (Dallmeyer, 1989), and the fact that all southern Appalachian lithotectonic units east of the Hayesville thrust fault (Fig.6.8) are allochthonous and show Ordovician-Devonian paleontological affinities with West Africa rather than with North America, all point to their earlier location adjacent to West Africa, in the eastern part of the Iapetus Ocean (Fig.6.9A,B). Also, Dallmeyer (1989) stressed the fact that the various tectono-stratigraphic units of the West African fold belt and southern Appalachian, when traced stratigraphically upward, become increasingly unrelated to their underlying miogeosynclinal successions. Dallmeyer postulated that these exotic lithostratigraphic terranes represent sequences which could have accumulated along microplates (Fig.6.9,A,B) which were located in the Iapetus Ocean, away from both the Laurentian and African land masses. With the collision of Laurentia and Gondwana during the Hercynian orogeny, the Ordovician-Devonian lithologic units which were part of these exotic terranes, were thrust westward onto the North American margin, while the Senegal microplate was driven eastward against the West African craton. The latter movement generated the Ouled Dhlim klippe, and the Akjout and Magta Lahjar nappes in the northern part of the Mauritanides.

6.2.4 Trans-Atlantic Correlations with Southern Appalachian, U.S.A

The tectono-stratigraphic units outlined above have counterparts in the pre-Cretaceous basement in the subsurface of the southern part of the Atlantic Coastal Plain and the Gulf Coast of southeastern United States (Dallmeyer, 1989). Variably deformed and metamorphosed granite and host calc-alkaline, felsic volcanic-volcaniclastic rocks in the subsurface of Florida, dated at about 690 - 675 Ma, correlate with similar coeval calc-alkaline granites which are exposed in the internal parts of the Bassarides in West Africa.

The Osceola granite in the subsurface of the central peninsula of Florida is similar to the Coyah granite of the northern Rokelides in the Republic of Guinea (Fig.6.8). Dallmeyer (1989) suggested that the Osceola and Coyah granites were initially part of the same suite of post-kinematic shallow-level plutons which were emplaced along the northwestern margin of Gondwana. In the Suwanee basin in the subsurface of northern Florida a Lower Ordovician-Middle Devonian sandstone-shale sequence with Gondwana paleontological affinities is similar to the sequence in the Bove basin in Senegal and Guinea. Also, high-grade metamorphic rocks penetrated in a small area in the subsurface of east-central Florida, southeast of the Osceola granite, exhibit petrological similarities and a post-metamorphic cooling history similar to those in the Rokelide orogen in Sierra Leone and Liberia. Finally, an interlayered assemblage of gneisses and amphibolites in the subsurface Wiggins Arch in southwestern Mississippi and a low-grade metasedimentary sequence also penetrated in the Wiggins Arch (Fig.6.8), show reactivation dated 310 to 300 Ma, in contradistinction to the coeval basement rocks in the subsurface of the Atlantic and Gulf Coastal Plains, which show no apparent evidence of any tectono-thermal activity during that time span.

The above similarities with the pre-Cretaceous basement of southeastern United States suggest that the basement of the Atlantic and Gulf Coastal Plains were part of the Mauritanide, Bassaride, and Rokelide orogens of West Africa (Dallmeyer, 1989). They represent fragments of Gondwana, which after the amalgamation of Pangea II, were stranded during

Mesozoic rifting that was associated with the creation of the present North Atlantic Ocean.

6.3 The Moroccan Anti-Atlas

The West African polyorogenic mobile belt, though largely concealed beneath coastal basins, swings northeastwards around the Reguibat Shield and re-appears in the Moroccan Anti-Atlas as Precambrian inliers ("boutonniers") which lie in the cores of the Anti-Atlas Hercynian anticlinoria (Fig.6.11). Those inliers including those at Bas Dra, Zenaga, Ifni, Kerdous, Tagragras, Bou-Azzer, Siroua, Saghro and Ougnat (Fig.6.11), expose the Pan-African basement which underlies the Paleozoic sequences, which although mildly deformed in the Anti-Atlas, are severely tectonized north of the South Atlas Fault. Since it has been possible in the Moroccan Anti-Atlas to separate the Pan-African basement from younger Paleozoic stratigraphic sequences and orogenic events, it is appropriate here to consider only the Pan-African tectono-thermal events.

The Late Proterozoic orogenic cycle of the Anti-Atlas, according to Leblanc and Lancelot (1980), Leblanc (1981), and Cahen et al. (1984), began with platform sedimentation (Limestone and Quartzite Group), followed by an extensional phase during which the Bou Azzer ophiolites formed at about 787 Ma, followed by tectonism in which the Bou Azzer ophiolites were obducted at about 685 Ma. Subsequent subduction was followed by collision between the African plate and an unknown northern plate, between 615 Ma and 578 - 563 Ma. The latter dates which were obtained from an unconformable volcano-sedimentary sequence, the Ouzarzate Group, marks the initiation of the Caledono-Hercynian orogenic cycle (Cahen et al., 1984) which is not considered here.

6.3.1 Stratigraphy

The basal Limestone and Quartzite group rests unconformably upon Eburnean basement in the southwestern Anti-Atlas and are exposed in the Bas Dra, Kerdous and Zenaga Plain inliers. In these inliers this basal Pan-African sequence consists of several thousand m of quartzite with basal conglomerate and interstratified limestones which are occasionally stromatolitic (Cahen et al., 1984). Mafic sills have been injected within the succession. To the northeast in the Siroua massif and at Bou Azzer-El Graara, similar sediments belong to the Tachdamt-Bleïda Group which comprises black shales with interstratified siltstones and mafic pillow lavas accompanied by jaspers, graywackes and tuffs, quartzite and limestone horizons. The southwestern lithofacies has been interpreted as reflecting platform conditions whilst the northeastern facies, including the Bou Azzer ophiolite, is believed to represent formations deposited along a continental margin.

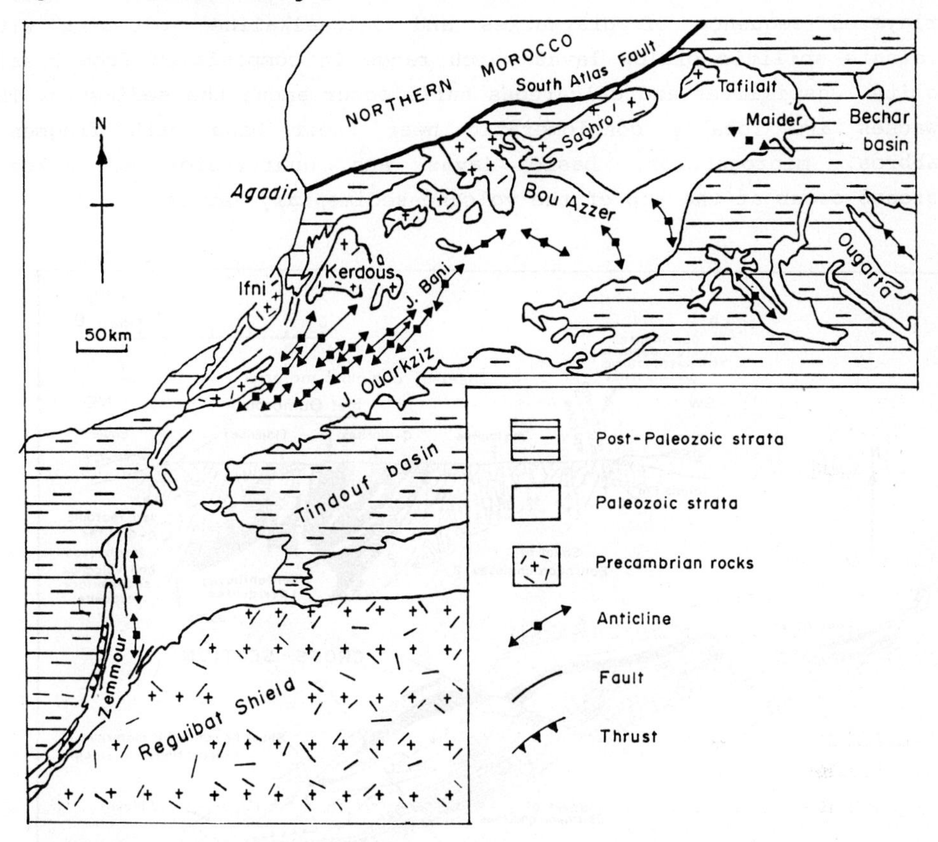

Figure 6.11: Tectonic sketch map of the Moroccan Anti-Atlas. (Redrawn from Piqué, 1989.)

6.3.2 The Bou Azzer Ophiolite

This has been considered to be among the oldest true ophiolites and among the earliest undoubted evidence for the operation of modern plate tectonics in the geological record (Leblanc, 1981; Bloomer et al., 1989). This ophiolitic complex, about 5 km thick, includes from north to south, a southward thickening sequence of calc-alkaline eruptive material, a dismembered basic alkaline volcanic-pluton complex and a mélange with sedimentary and mafic igneous blocks among a sheared argillite matrix (Bloomer et al., 1989). Leblanc (1981) reported (Fig.6.12) from bottom to top, about 2,000 m of serpentinized peridotites; about 500 m of discordant layered gabbros, large stocks of quartz-diorites; about 500 m of discordant layered gabbros; about 500 m of basic lavas with pillow lavas; and approximately 1,500 m of volcano-sedimentary rocks. The serpentinites constitute up to 40 % of the entire complex and were derived from dunites and harzburgites. In the volcano-sedimentary series there is an interlayered sequence of graywackes and calc-alkaline volcanics with subordinate spilitic pillow lavas which range in composition from basalt to dacite. Jaspillites and calcareous tuffs occur among the sediments. The graywackes are locally conglomeratic near their base with fragments of gabbros, microgabbros, basic lavas and quartz-diorite. Arkosic sandstones occur at the top of the volcano-sedimentary series.

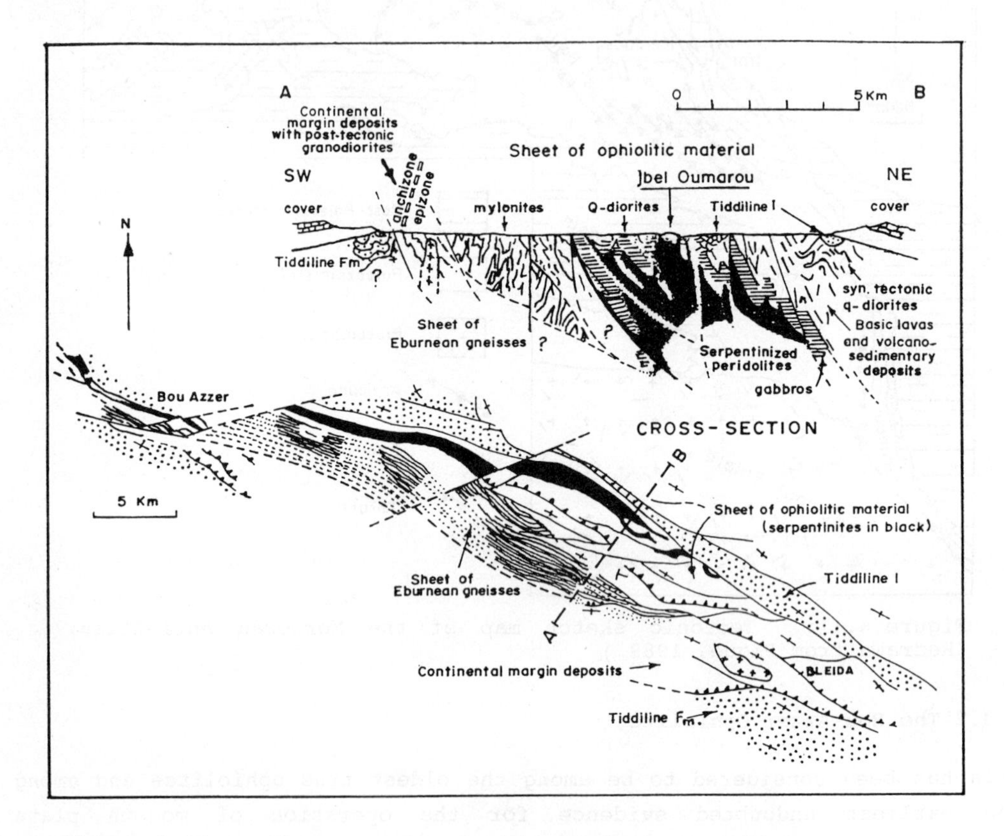

Figure 6.12: Sketch map and cross-section showing Bou Azzer ophiolite. (Redrawn from Leblanc, 1981.)

Whilst Leblanc (1981) considered the Bou Azzer ophiolite as representing oceanic crust, Bloomer et al. (1989), based on new structural and geochemical data, re-interpreted the ophiolite complex as originating not by Precambrian ocean-floor spreading, but rather, at a suture zone representing a mélange of a rifted continental margin, arc and fore-arc terranes which were stacked against the foreland of the West African craton during Pan-African plate collision.

During the Pan-African deformation at about 685 Ma, the ophiolite was obducted and dismembered into several sheets which were thrust against the continental margin (Fig.6.12). Deformation was accompanied by Barrovian-type greenschist metamorphism, and was followed by post-kinematic granodiorite intrusion at about 615 Ma. Similar deformation characterized by a northern vergence, north-south crustal shortening, ductile shearing and low-grade metamorphism, occurred in the western Anti-Atlas near Kerdous (Fig.6.11).

After this deformation, and the emplacement of the Bleïda granodiorite, the Tidilline Series was deposited, comprising basal conglomerates, graywackes (turbidites), laminated siltstones, mixtites with dropstones, feldspathic sandstones, and volcanic rocks. This sequence rests unconformably on the Tachdamt-Bleïda Group and on the post-tectonic Bleïda granodiorite, and was deformed without metamorphism by a late Pan-African event. In the Bou Azzer inlier this deformation produced folds with WNW-ESE trends and vertical axial planes, fracture cleavage and WSW-ENE faults.

6.3.3 Mineralization

In their analysis of the tectonic control of mineralization in Morocco, Bensaid et al. (1985) pointed out that although major metallic mineral deposits formed throughout the geological history of this region, some important mineralizations are associated with certain terranes. Thus, while the main copper and cobalt deposits are concentrated to the west of a deep crustal north-south lineament, all the major Pb-Zn deposits concentrate to the east (Fig.6.11). Based on copper mineralization in the volcano-sedimentary formation, Bleïda is the centre for copper production. The Bou Azzer ophiolite has deposits of cobalt, chromium and asbestos. The Middle Precambrian granodiorites in the Siroua, eastern Saghro and Ougnat inliers have Zn-, Ag-, Au-, Cu- and W-bearing veins and disseminations. Pan-African pegmatites bearing Be, Li and Nb traverse lower Precambrian alkaline granites in the Tazenakht, Iguerda Agadir, Tazeroualt and Kerdous areas. Also, the Precambrian mantle-derived Beni-Boucera peridotite in the Rif domain of northern Morocco has economic deposits of nickel, graphite and vermiculite. Among Morocco's metallic mineral exports are copper, silver, iron ore and antimony.

6.4 The Trans-Saharan Mobile Belt

6.4.1 Geodynamic Setting

As defined by Cahen et al. (1984) the Trans-Saharan mobile belt is the north-south Pan-African belt which borders the West African craton to the east (Fig.6.3). Its boundary with the craton is defined by the eastern segment of the circum-West African belt of gravity highs which according to Roussel and Lécorche (1989), marks an eastward-dipping suture comprising unrooted, dense mafic and ultramafic bodies. This suture zone can be followed over a distance of 2,000 km (Fig.6.3) along the entire length of the Trans-Saharan mobile belt. Prevailing geodynamic models (e.g. Black, 1984; Caby, 1987) relate this suture to Pan-African continental collision between the passive margin of the West African craton and an active codilleran-type margin of an eastern continent, the remnants of which are the Archean-Early Proterozoic terranes of the eastern Hoggar and the Benin-Nigeria Shield (Fig.6.13), which were reworked during Pan-African tectono-thermal events.

Although the Trans-Saharan mobile belt runs from southern Algeria across Mali, Niger and Chad in the Sahel, to the West African coast where it is exposed in eastern Ghana, Togo, Benin, Nigeria and Cameroon, it outcrops extensively only in the Hoggar Mountains, and in the Benin-Nigeria Shield. In the Sahel the mobile belt lies beneath a shallow Mesozoic-Cenozoic downwarp, the Iullemmeden basin. In Mali a major re-entrant into the middle of the West African craton is occupied by the Gourma aulacogen along 15°N latitude. The Trans-Saharan mobile belt extends for about 1,000 km across from 0° longitude to 15°E.

The Trans-Saharan mobile belt will be described in its three principal outcrop areas, the Tuareg Shield (Fig.6.14) in the north comprising the Pharusian, the polycyclic central Hoggar-Aïr and the eastern Hoggar-Tenéré domains; the Gourma aulacogen; and the Benin-Nigeria Shield. The mobile belt is characterized by roughly meridional shear zones which have sliced through all its tectonic domains. Stratigraphically (Fig.6.15), the widespread occurrence of pre-rift carbonate-quartzite platform sequences, rift-related magmatic suites and sediments, extensive continental margin calc-alkaline volcaniclastics, together with vast batholithic plutons showing subduction-related geochemical zonation, and the presence of paired metamorphic belts, strongly suggest the operation of a complete Wilson Cycle in the Trans-Saharan mobile belt (Black, 1984; Caby, 1987, 1989).

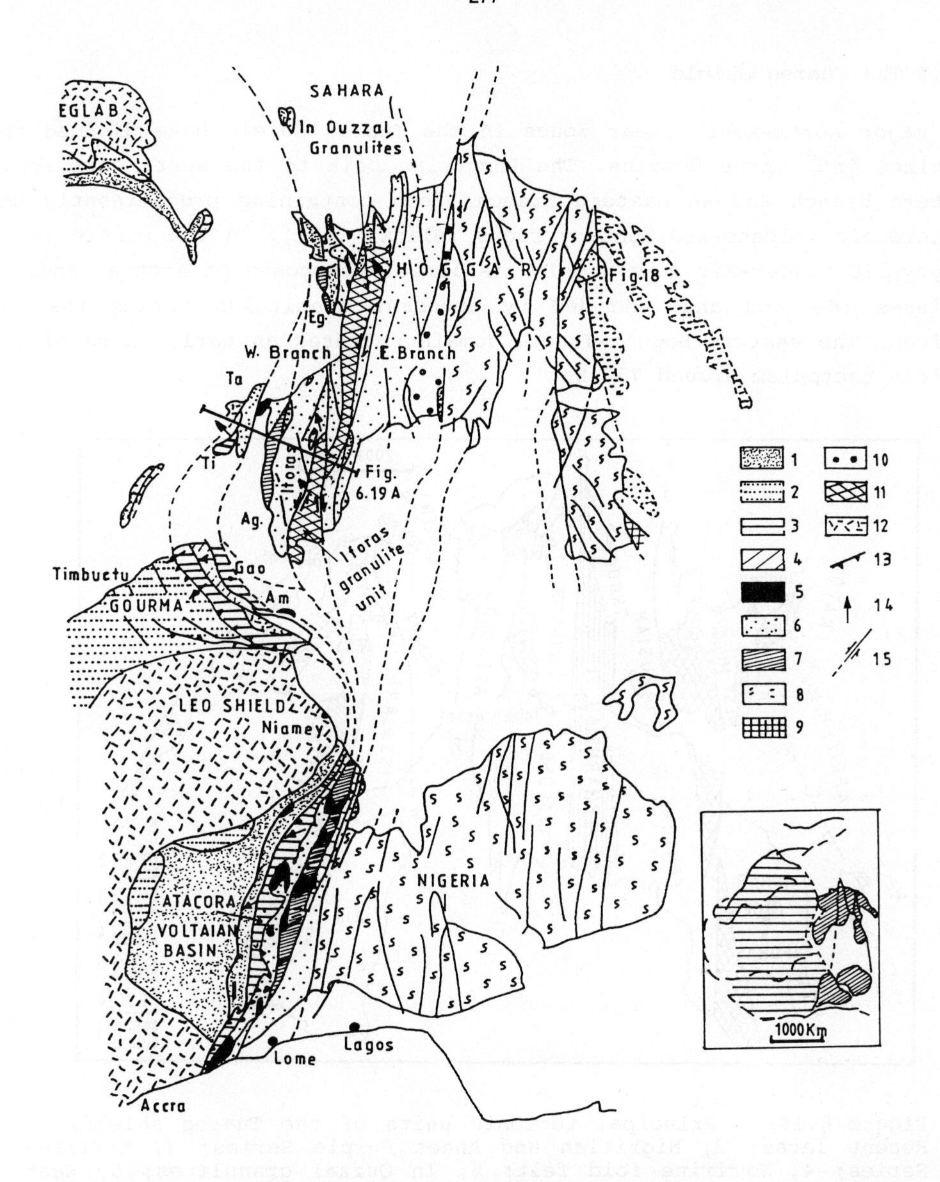

Figure 6.13: Geologic sketch map of the Trans-Saharan mobile belt. 1, molasse; 2, Upper Proterozoic passive margin sediments; 3, Atakora and Gourma metamorphic nappes (mainly quartzites and phyllites); 4, same as 3 with high pressure - low temperature metamorphic assemblages in Gourma; 4, major mafic-ultramafic massifs along the Pan-African suture; 6, metasediments and gneisses affected by the Pan-African; 7, high temperature - low pressure equivalent of 6; 8, undifferentiated gneisses; 9, rocks stabilized before 725 Ma in Djanet - Tafassasset domain; 10, undifferentiated rocks of central Hoggar, partly stabilized at about 840 Ma; 11, pre-Pan-African granulite basement 2.0 Ga old; 12, West African craton; 13, main frontal thrust of Gourma and Atakora nappes; 14, main direction of movement; 15, vertical shear zones and strike-slip faults. (Redrawn from Caby, 1987.)

6.4.2 The Tuareg Shield

The major north-south shear zones in the Tuareg Shield have divided this province into three domains. The Pharusian belt to the west, comprises a western branch and an eastern branch, both containing predominantly Late Proterozoic volcano-sedimentary rocks (Black, 1984). In the centre is the polycyclic Hoggar-Aïr domain which is largely composed of Archean-Eburnean gneisses reworked and intruded by abundant granitoids during the Pan-African. The eastern Hoggar-Tenéré domain suffered an early phase of Pan-African tectonism around 730 Ma.

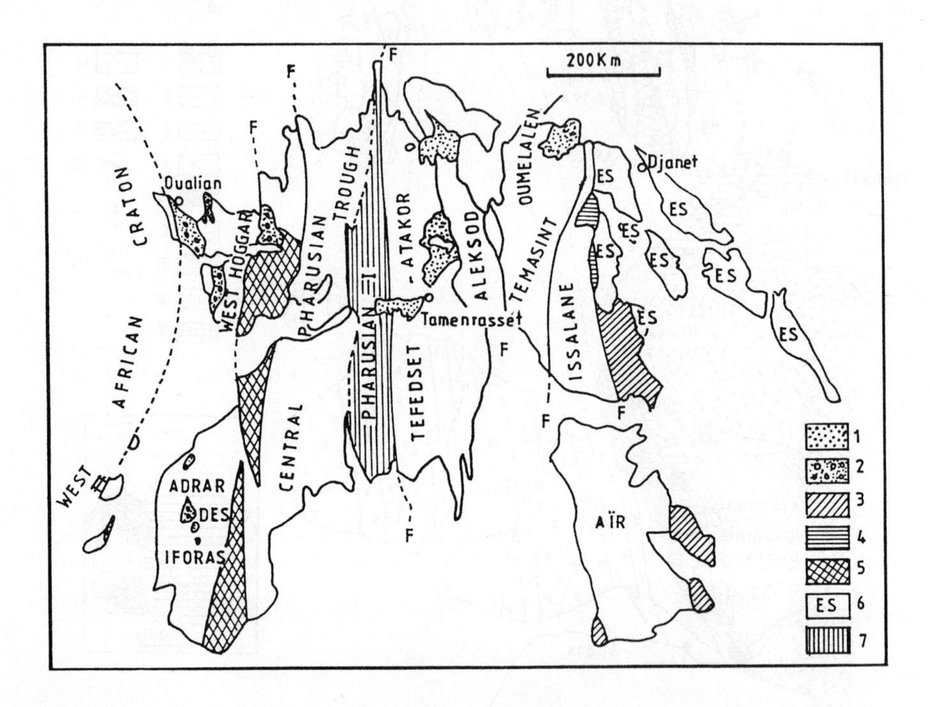

Figure 6.14: Principal tectonic units of the Tuareg shield. 1, Recent lavas; 2, Nigritian and Ahnet Purple Series; 3, Tiririne Series; 4, Tiririne fold felt; 5, In Ouzzal granulites; 6, East Saharan Craton (?); 7, Pharusian I. (Redrawn from Cahen et al., 1984.)

The polycyclic Archean-Early Proterozoic high-grade assemblages which are preserved as the In Ouzzal and Iforas granulites were last affected by the Eburnean orogeny (Fig.6.14), whereas other parts of the Eburnean terrane in the Tuareg Shield were profoundly reactivated during the Pan-African events. These granulites and the Pan-African reactivated rocks constituted the basement for the Pan-African orogenic cycle. The reactivated Eburnean basement rocks include the large orthogneissic masses of the Arechchoum

Series in east-central Hoggar; the metasediments of the Egéré and Aleksod Series, the polycyclic granulite and amphibiolite-facies gneisses of the Kidalian assemblage in Adrar des Iforas; and in Tamanrasset and Arefsa there are granulites and the Gour Oumelalen gneisses of central Hoggar.

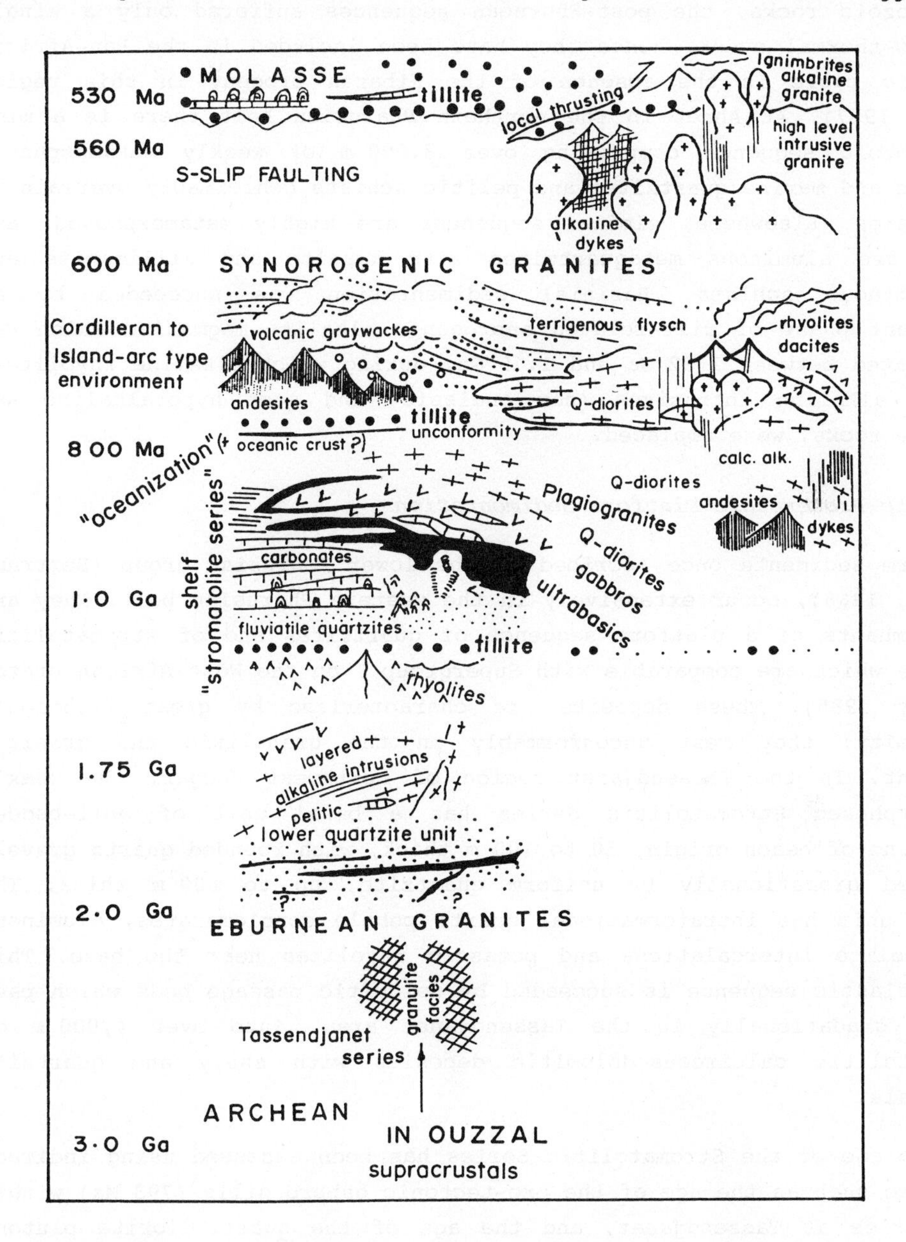

Figure 6.15: Summary of stratigraphic and tectonic events in the Tuareg shield. (Redrawn from Caby, 1987.)

Post-Eburnean Sedimentation and Anorogenic Magmatism

In the Pharusian belt there are piles of metasediments which were deposited after the Eburnean orogeny. Together with the overlying Upper Proterozoic rocks, the post-Eburnean sequences suffered only a single tectono-thermal event, hence they have been included in the Pan-African orogenic cycle in the absence of the Kibaran orogeny in this region (Caby, 1987). At Ahnet in the northern Pharusian belt there is a mid-Proterozoic sequence comprising over 3,000 m of weakly metamorphosed deltaic and marine quartzites and pelitic schists conformably overlain by carbonates. Elsewhere, similar sequences are highly metamorphosed, and there are aluminous metaquartzites with kyanite and sillimanite and peraluminous schists. Basinial sedimentation was succeeded by an important phase of rift-related anorogenic alkaline magmatism which has been dated between 1.82 Ga and 1.75 Ga, during which alkaline rhyolites, banded alkaline intrusives (now gneissic) and some hyperalkaline and syenite rocks, were emplaced.

Mid-Late Proterozoic Platform Sedimentation

Platform sediments once ascribed to the lower Pharusian group (Bertrand et al., 1966), occur extensively in the western Pharusian belt. They are the remnants of a platform sequence of quartzites and of stromatolitic marbles which are comparable with Supergroup I on the West African craton (Black, 1985). These deposits are characterized by great lithologic uniformity; they rest unconformably on the granulitic and granitic basement. In the Tassendjanet region of northwest Hoggar, the weakly metamorphosed Stromatolitic Series has a basal unit of well-bedded sandstone of beach origin, 50 to 100 m thick, with rounded quartz gravel, followed gradationally by uniform quartzite, up to 600 m thick. The latter unit has intraformational quartz pebble conglomerates, aluminous gray pelite intercalations and potassic rhyolites near the base. This siliciclastic sequence is succeeded by dolomitic passage beds which pass upward gradationally in the Tassendjanet area, into over 4,000 m of stromatolitic calcareous-dolomitic deposits with shaly and quartzite intervals.

The age of the Stromatolitic Series has been assessed using indirect evidence such as the age of the pre-tectonic gabbro sills (793 Ma) within the series at Tassendjanet, and the age of the quartz-diorite plutons (868 Ma) which cut the equivalents of the Stromatolitic Series in central Hoggar (Caby, 1987). Thus, the platform siliciclastic-stromatolitic carbonate sequence was deposited prior to these dates, probably in the

Middle to early Late Proterozoic. Paleontological correlation of these stromatolitic carbonates with Supergroup I of the Taoudeni basin is based on the presence of similar stromatolite biostromes with the form *Conophyton* occurring in both sequences. The Stromatolitic Series correlates with the Atar Group, (middle of Supergroup I) in the Mauritanian Adrar, which according to Bertrand-Sarfati and Moussine-Pouchkine (1988), was deposited between 1.0 Ga and 700 Ma, based on the stromatolite assemblages.

Mafic and Ultramafic Rocks Related to Crustal Thinning

Throughout the Trans-Saharan mobile belt there are small meta-mafic and meta-ultramafic masses which form concordant layers and lenses in high-grade metamorphic terranes. They occur within the gneisses and amphibolites or as isolated boudins of various sizes. Granulite meta-gabbros at Amalaoulaou in the Gourma aulacogen, and in the Kabré meta-gabbro massif in the Togo-Benin belt are believed to mark the suture of the Pan-African collision (Fig.6.13).

However, the other mafic and ultramafic masses in the Hoggar and Adrar des Iforas are not part of the above ophiolitic bodies. Rather, they were mantle-derived; and occur as diapiric layered or multiple intrusions in the form of asthenoliths, sills, lapoliths and laccoliths within the surrounding metasediments comprising quartzites, carbonates and pelitic schists. The mafic and ultramafic intrusives exhibit either contact metamorphism or tectonic contacts with the host metasediments. Examples of these intrusives include the serpentinite sills with magmatic layering, dunites, pyroxenites and wehrlite which occur in the Tassendjanet and Silet tectonic zones; and the cumulate peridotites, tholeiitic gabbros and calc-alkaline rocks in the Adrar Ougueda complex in the northwestern Pharusian belt. In these areas the mafic and ultramafic intrusions were emplaced into continental basement between about 1.0 Ga and 800 Ma (Fig.6.15).

Volcano-Sedimentary Sequences and Calc-alkaline Magmatism

These rock assemblages are thick and extensive in the Pharusian belt which they characterize. They were the island-arc and cordilleran volcaniclastics (Fig.6.16) representing the newly formed Pan-African continental crust which, during collision, were thrust onto the West African craton and onto the western margin of the East Saharan craton (Black, 1984; Caby, 1987; Liégeois et al., 1987).

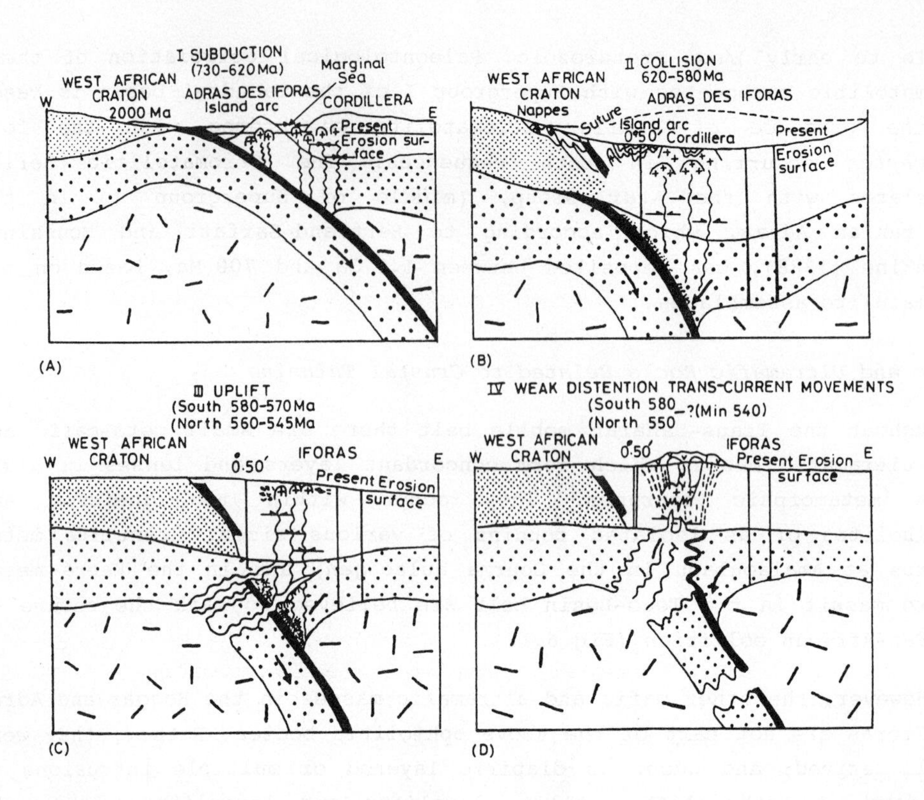

Figure 6.16: Tectonic model for the evolution of the Iforas batholith during the Pan-African orogeny. (Redrawn from Liégeois et al., 1987.)

In the Pharusian belt the platform sequence of quartzite and stromatolitic marbles are overlain by thick volcano-sedimentary sequences (Fig.6.15). The latter sequences display the typical features of an active continental margin, with characteristic island arc lithologies (Fig.6.16) which suggest the presence of an open ocean separating the West African craton from the East Saharan craton. The volcaniclastic sequence and associated calc-alkaline and magmatic rocks formed in the Late Proterozoic between about 850 and 700 Ma (Caby, 1987).

The volcaniclastic sequences are described below from east to west, that is, from the eastern to the western Pharusian belt, in a direction of increasing deepening, from an eastern marginal sea across an island arc, to the deep ocean which lay to the west (Fig.6.16,A).

Resting unconformably on deeply eroded and altered calc-alkaline and subalkaline batholiths dated between 880 Ma and 839 Ma, are several thousand m of Upper Proterozoic volcaniclastic rocks in the eastern Pharusian belt (Caby, 1987). The sequence begins with clastic units and lenticular dolomites of probable marine origin which interfinger with volcaniclastics. Upward, the succession includes rhyolites and andesite-dacites which are overlain by terrigenous pelites, that pass upward into predominantly rhyo-dacite volcanics. In the eastern and northern parts of the eastern Pharusian belt there are graywackes which are similar to the "Serie verte" of the northwestern Pharusian belt. These graywackes are mixed with terrigenous sediments and are intruded by many dolerite calc-alkaline batholiths comprising quartz-diorites, micro-diorites, granodiorites, and plagiogranites. A uniform terrigenous sequence covers the southwestern part of the eastern branch, and in eastern Adrar des Iforas a thick Upper Proterozoic pelitic detrital and volcaniclastic sequence with basal conglomerates rests upon mid-Proterozoic quartzites and alkaline orthogneisses.

In the northwestern part of the Pharusian belt the "Serie verte" is a thick monotonous flysch-like sequence of green volcanic immature graywackes which passes northwards into a subaerial calc-alkaline volcanic complex, about 6,000 m thick, probably representing part of the island arc itself. This complex consists of basic andesites, normal andesites, dacites with rhyolites near the top. Polygenetic conglomerates with basal tillite occurs at the base of the calc-alkaline complex. The geochemical characteristics of the "Serie verte" calc-alkaline complex suggests an active continental margin and the derivation of the andesites from a subduction zone, by the partial melting of the mantle, followed by low pressure fractionation (Black et al., 1979).

In the central part of Adrar des Iforas (Fig.6.14) the Upper Proterozoic volcaniclastic sequence belongs to the Tafeliant Group which rests unconformably on pre-Pan-African or older basement (Fabre, 1982) and comprises basal unstratified tillite with large local blocks, littoral calcareous sandstones and a unit of black pelites with clasts of marine tillites (Fig.6.17,A). The succession passes upward into volcanic graywackes with abundant andesitic material partially mixed with semi-pelitic siltstones, and with arkoses derived from the sialic basement; a bimodal suite of basalts and rhyodacites occurs at the topmost part. The Tafeliant Group was deposited in a shallow marginal basin bounded by north-south faults (Fabre, 1982). It may correlate westward with the Tessalit-Tilemsi volcanic graywackes which contain deep trough sedimentary features including a marine diamictite. The Oumassene Group, an andesite-basalt complex, located at about 100 km to the north of Tafeliant, may correlate with the Tafeliant Group.

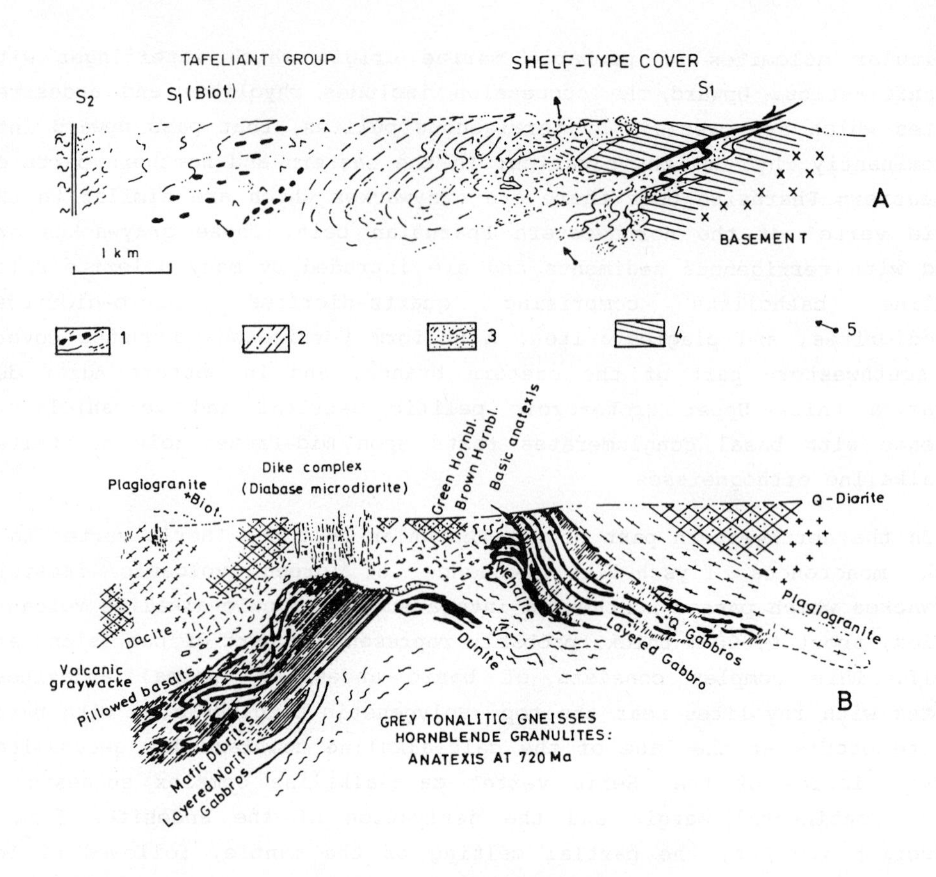

Figure 6.17: A, cross-section through the Tafeliant Group. 1, Tafeliant Group with basal glaciogenic sediments; 2, intrusive foliated metadiorite, 793 Ma old; 3, shelf-type cover of Upper Proterozoic age; 4, pre-Pan-African gneisses; 5, direction of younging strata. B, schematic section showing relationship of layered sequences and plutonic rocks in the Tessalit-Tilemsi island arc. (Redrawn from Caby, 1987.)

In the northwestern part of Adrar des Iforas the Tessalit-Tilemsi volcanic and volcaniclastic series (Fig.6.17,B) is of island-arc affinity. This sequence defines a broad belt of about 100 km wide that is characterized by metamorphic and structural features which are distinct from the rest of the Tuareg Shield. As depicted in Fig. 6.17,B the Tessalit-Tilemsi Series consists, in the lower part, of a bimodal volcanic sequence with meta-basaltic pillow lavas, dacites and younger andesites, followed by a thick sequence of volcanic graywackes with deep water and turbiditic sedimentary structures, and a large volume of several generations of plutonic rocks which make up about 70 % of the entire series. Among the plutonic rocks are meta-gabbros and ultramafic cumulates; microdiorite and dolerite bodies, stocks and dykes; and banded and differentiated gabbro-norite, tonalite, and granodiorite laccoliths and lopoliths. During the emplacement of these mantle-derived rocks there was early anatexis among the surrounding metasediments, from which were generated tonalitic gray gneisses and migmatites of low-pressure granulite facies, at about 720 Ma.

Since the geochemistry of the plutonic rocks attest to a sequence of magma evolution beginning with oceanic tholeiites to calc-alkaline volcanism, Caby (1987) concluded that the plutons represent deep-level magma chambers that are related to the calc-alkaline volcanism; and that they define the suture zone which has a positive gravity anomaly signature (Fig.6.3). The calc-alkaline batholiths were emplaced on the eastern side above the eastward-dipping subduction zone (Fig.6.16) at about 633 Ma, prior to collision. The distribution of the plutonic complexes, especially the pre-orogenic suite, and their progressive increase in K_2O and SiO_2 eastward is similar to the magmatic zonation across modern subduction zones. The Tessalit-Tilemsi complex was probably an island-arc environment that was active from about 800 Ma to 630 Ma. Since it had a magmatic root of gabbroic lopoliths and basic gray gneisses (Fig.6.17,B), it probably had no continental crust (Caby, 1987).

The Tessalit-Tilemsi accretion zone follows the sheared western margin of the Iforas (Fig.6.14). Notable also is the total absence, west of the Pan-African suture zone, of the Tessalit-Tilemsi graywacke formations and their kindred calc-alkaline magmatic rocks, which are extensively developed east of the suture. As postulated by Black et al. (1979) this spatial distribution of the rock assemblages can be accounted for by positioning an ocean basin between two continents in this region; the West African craton on the western side, and an eastern continent on the other side with an island-arc complex in between (Fig.6.16A,B).

Deformation and Metamorphism

Caby (1987) divided the post-Eburnian structural and metamorphic history of the Tuareg Shield into two phases--localized early Pan-African tectono-thermal events at 870 Ma, 730 Ma and 720 Ma; and the regional Pan-African event which affected most parts of the Trans-Saharan mobile belt at about 600 Ma.

Uncertainty surrounds the existence of a Kibaran or Aleksod event in the central Hoggar which was postulated by Bertrand and Lassere (1976) and by Cahen et al. (1984). Kibaran or Aleksod orogeny was believed to have produced kyanite-sillimanite and granulite facies metamorphism and subhorizontal foliation and recumbent folds in central Hoggar. But in the absence of reliable radiometric dates and in view of the lack of marked structural discontinuity between these probably older polycyclic basement rocks (at Aleksod, Arefsa and Tamanrasset) and the surrounding terranes in which a single Pan-African event is well established, Caby (1987) consigned the Aleksod and related metamorphic rocks to the Pan-African tectono-thermal event. However, U-Pb ages are required in order to conclusively unravel the thermal history of the polycyclic gneisses of east-central Hoggar which sometimes contain eclogites in garnet-amphibolite lenses that are enclosed in orthogneisses and metasediments, thus indicating the attainment of high-pressure conditions estimated at 15 kb at 700°C. Such extreme metamorphic conditions were reached prior to the low-pressure anatexis which affected the surrounding rocks. Also, in the central Hoggar, southwest of Tamanrasset, U-Pb ages suggest Pan-African retrogression and overthrusting of Eburnean granulite-facies rocks.

In the eastern part of the central Hoggar a tectono-thermal event at 870 Ma affected a platform quartz-carbonate sequence which correlates with the Stromatolite Series. This event which is supported by U-Pb ages of around 868 Ma from late-tectonic quartz-diorite and granodiorite batholiths, has not been recognized elsewhere in the Tuareg Shield (Caby, 1987). The Djanet-Tafassasset domain with deeper greenschist- and amphibolite-facies meta-pelites and graywackes occurring in vertical north-south-trending belts is cut by many syn-tectonic granitoids and by late-kinematic porphyritic granites dated at 730 Ma. This suggests that metamorphism, deformation, and stabilization of the Djanet-Tafassasset domain (or Tiririne fold belt (Fig.6.18,A)) occurred before 730 Ma, after which the Tiririne molasse was deposited. The high-grade gneisses of Issalane which overthrusts the Tiririne fold belt (Fig.6.18,C,D) represents early Pan-African metamorphic rocks which later overthrust the Tiririne belt in a Himalayan-style tectonics. High-level granitoids were emplaced by anatexis during this early Pan-African event.

The Tessalit-Tilemsi domain in the western Adrar des Iforas contains the vast island-arc calc-alkaline volcaniclastic complex which appears to have been subducted at about 720 Ma giving rise to the emplacement of basic plutonic rocks, metamorphism, and deformation which resulted in the intensely deformed plagioclase gray gneisses (Fig.6.17,B) metamorphosed at the hornblende-granulite facies. Anatexis also resulted from the subduction producing migmatites and foliated dioritic rocks with flow folding (Fig.6.17,B). The late Pan-African event at about 600 Ma caused the partial retrogression of these rocks to the greenschist facies and the formation of open folds with sub-vertical flow schistocity, especially in the low-grade metasediments.

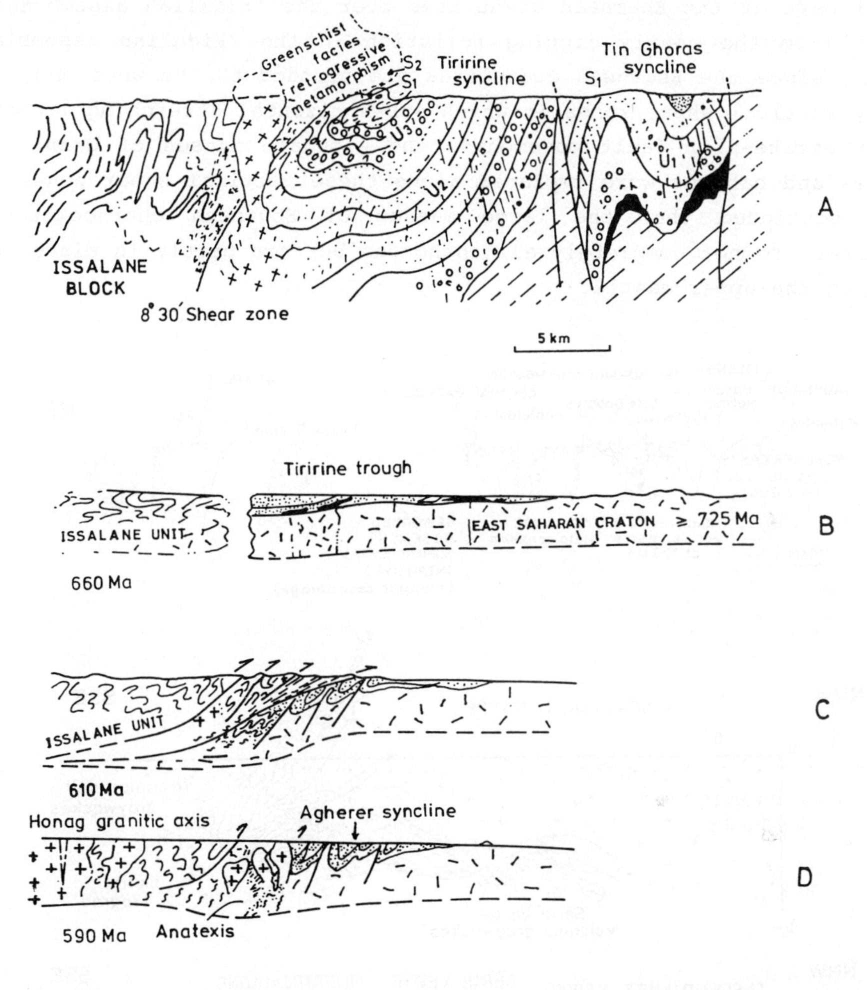

Figure 6.18: A, cross-section of the late Pan-African Tiririne fold belt (Eastern Hoggar); B, C, D and stages in the origin of this fold belt. (Redrawn from Caby, 1987.)

The late Pan-African event at about 600 Ma affected nearly all parts of the Tuareg Shield where early gently inclined schistocity in practically all domains resulted from large-amplitude horizontal movements. Metamorphism involved an initial intermediate-to-high-pressure phase followed by lower pressure and high temperature conditions.

In Adrar des Iforas (Fig.6.19,A) the middle-to-high-grade gneiss of central Iforas ("Kidalian assemblage") belongs to the middle and deep zones of the Pan-African and includes pre-Pan-African basement, cover and pre-metamorphic intrusives which were involved in Pan-African large-scale horizontal movements before 610 Ma. Limited northwestward thrusting of the frontal part of the Eburnean granulites over the "Kidalian assemblage" is evident from the gently dipping foliation in the "Kidalian assemblage". However, since the Eburnean granulites are bounded to the west and to the east by vertical shear zones in which sub-horizontal stretching lineations suggest strike-slip fault movements, and elongate masses of granitoides, diorites and gabbros were emplaced along these vertical shear zones, Caby (1987) concluded that the large masses of Eburnean charnockites and granulites are not completely allochthonous but are partly in place, being rooted in the upper mantle.

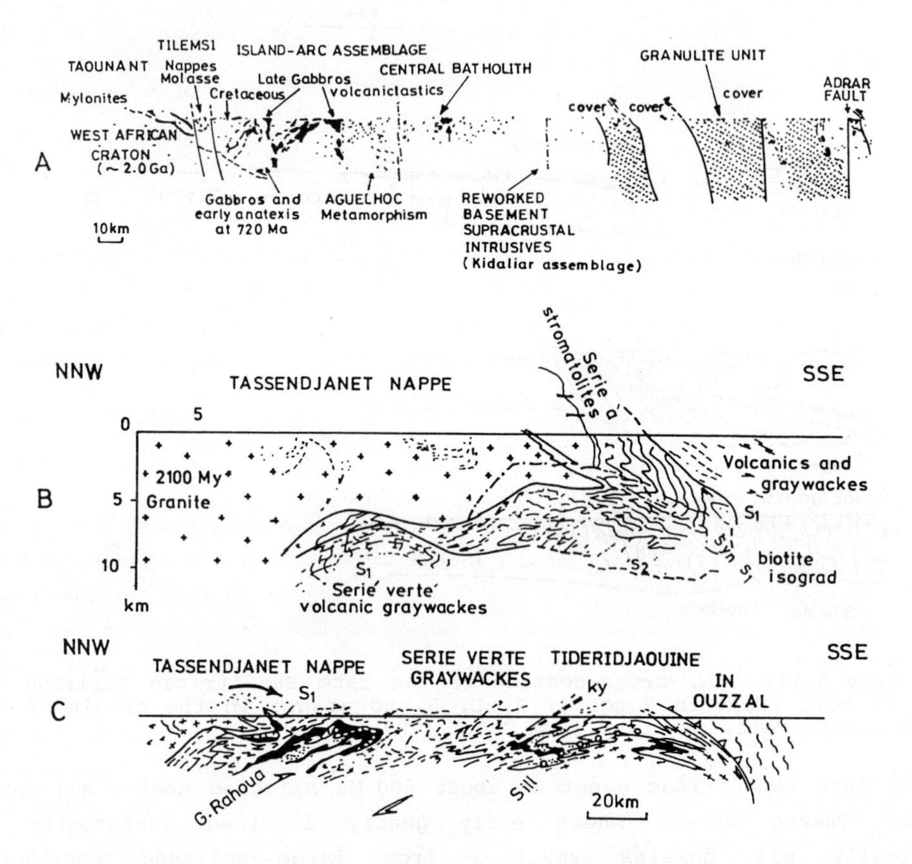

Figure 6.19: A, schematic East-West section through central and Western Adrar des Iforas showing nappes, central batholith bounded by vertical shears. B, C Tassendjanet nappe representing a Northern continent; black represents ultra-mafics. (Redrawn from Caby, 1987.)

Although polyphase deformation appears to be the case in several parts of northwestern Hoggar and in Adrar des Iforas, Caby (1987) was in favour of one continuous Pan-African deformation episode. The Tassendjanet nappe in northwestern Hoggar (Fig.6.19,B,D) shows large horizontal displacements at shallow depth which change to recumbent folds at deeper levels. The Tafeliant synclinorium (Fig.6.17,A), the youngest Pan-African structure, shows only a single fold generation with open folds trending north-northwest, and with up to 30% east-west crustal shortening. In the southeastern margin of this synclinorium, however, there are transitions from open folds with vertical axial planes to vertical recumbent folds at greater depth, which are similar to those in the older "Kidalian assemblage". Caby (1987) pointed out that these rocks were mostly affected by the regional Pan-African event between 620 Ma and 600 Ma during which gently inclined foliation and recumbent and sometimes re-folded folds and crustal thickening were produced under Barrovian metamorphic conditions.

East of the regional Pan-African suture zone at Egatalis and Aguelhoc (Fig.6.14), the syn-kinematic emplacement of gabbro-norite bodies have subsequently altered the Barrovian metamorphic grade and caused sillimanite to pseudomorph after kyanite.

Post-orogenic strike-slip movements occurred along north-south to N 20° shear zones and faults, between 566 Ma and 535 Ma (Lancelot et al., 1983); and a final post-collision deformation phase which occurred under brittle conditions, produced a conjugate set of strike-slip faults (Liégeois et al., 1987).

Syn-orogenic and Post-orogenic Magmatism

Pan-African granitoids make up to 50 % of the Pan-African outcrops in the Tuareg Shield, especially in the west-central part of the Pharusian belt. Syn-orogenic granitoids with migmatitic margins predominate in the amphibolite-facies terranes. These are often calc-alkaline in domains that have suffered only Pan-African tectonism or in terranes with abundant meta-graywackes and meta-volcanics. Some foliated masses range in composition from quartz-diorite to granodiorite, but the more potassic varieties are abundant in east-central Hoggar. However, the distinction between foliated truly syn-tectonic masses with diffuse margins and unfoliated masses with sharp contacts depends mostly on the level of intrusion.

A unique syn-orogenic granitoid complex is the composite central calc-alkaline batholith of Adrar des Iforas (Fig.6.20). This batholith extends for over 300 km and is over 50 km wide; it is bounded to the west by the Tessalit-Tilemsi island-arc accretionary terrane, and to the east by the Eburnean Iforas polycyclic granulitic basement nappes (Figs.6.20,6.21). According to Bertrand and Davison (1981) and Liegeois et

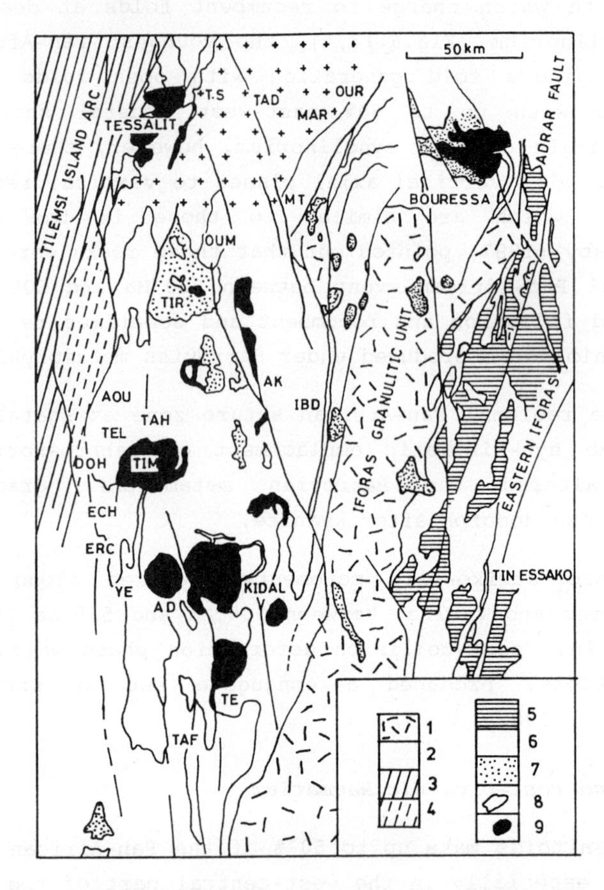

Figure 6.20: Geologic map of the region intruded by the Iforas batholith. 1, Eburnean granulites; 2, undifferentiated schists and gneisses; 3, volcanic and associated plutons of the Tilemsi island arc; 4, Aguelhoc gneisses; 5, Pan-African plutons; 7, rhyolite flows; 8, post-tectonic plutons of the Kidal assemblage; 9, alkaline ring complexes. (Redrawn from Liégeois et al., 1987.)

al. (1987) pendants of gneisses and belts of low-grade metasediments such as the Tafeliant and the Oumassene Groups occur in the central Iforas batholith (Fig.6.21). The batholith is capped by flat-lying rhyolitic lavas (Nigiritian) and cut by sub-alkaline and alkaline ring-complexes. As depicted in Fig.6.21 spectacular dyke swarms occur in the batholith. Two generations of dyke swarms exist. First, there are east-west-striking swarms comprising essentially subordinate basic dykes, quartz-microdiorites , micro-adamellites and felsites; and subsequent high-level adamellite and perthite granite. The second generation consists of north-south basic dykes and several sets of quartz micro-syenite, granophyres and rhyolites which can be traced within the Nigiritian rhyolite field for 250 km along the axis of the batholith and along the alignment of ring complexes.

Based on the petrology, geochronology and geochemistry of the batholith, Liegeois et al. (1987) concluded that the magmas from which this composite cordillera batholith was derived all share low initial $^{87}Sr/^{86}Sr$ ratios, suggesting derivation from the subduction of the Tessalit-Tilemsi island-arc calc-alkaline volcano-sedimentary sequence (Fig.6.16). Also, as shown in the model (Fig.6.16,D), some 50 million years after the beginning of collision, alkaline magmas were probably generated direct from the asthenosphere, thus producing anorogenic alkaline magmatism with alkaline ring complexes (Fig.6.20).

Late- and post-tectonic granitoids are common in west-central Hoggar. These include the Taourirt granites in Hoggar, dated at about 560 Ma, comprising a homogeneous suite of granites with restricted calc-alkaline chemistry and concentric structure. A suite of high-level granites with tungsten-tin mineralization cut the gneisses east of the 4°50 shear zone.

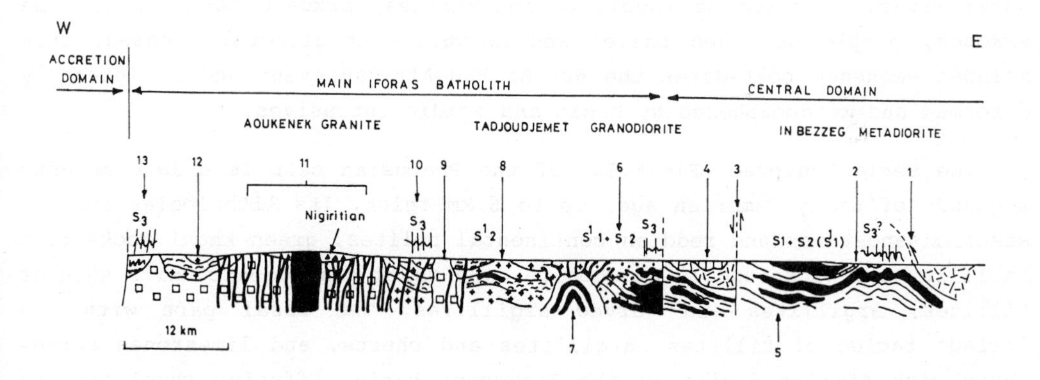

Figure 6.21: Cross-section through the Northern part of the Iforas batholith. 1, main thrust contact of the Iforas granulite unit; 2, Kidal assemblage; 3, mylonite belt; 4, low-grade upper Proterozoic supracrustals; 5, Kidal assemblage structurally below granulites; 6, deformation with folds and foliation; 7, Tedreq acid intrusives; 8, remnants of Kidal assemblage; 9, late pro-phyritic granite; 10, andesites; 11, central part of batholith with flat-lying rhyolites of the Nigiritian and alkaline ring complexes (solid black); 12, migmatites; 13, Tessalit Formation (clastic and volcanoclastic). (Redrawn from Betrand and Davison, 1981.)

Molasse Sequences

The oldest molasse sequence occurs in the eastern Hoggar, in the Tiririne fold belt (Fig.6.14), where the molassic Tiririne Series comprises up to 6 km of greenish arkose and graywacke. Paleocurrent indicators suggest derivation of the sediments from the Djanet-Tafassasset domain in the east, a terrane that had stabilized around 730 Ma and was being uplifted and eroded. Before its deformation the Tiririne Series was intruded at about 660 Ma by a variety of sills comprising rhyolites, dacites, andesites, gabbros, diorites and granodiorites.

The later Pan-African molasse are preserved either in north-south-trending grabens (Fig.6.14) bounded by major north-south vertical mylonite zones or in residual basins situated immediately east of the suture zone (Caby, 1987). In the northern Pharusian belt an early molasse sequence of arkoses and graywackes underlies the younger molasse, the Série Pourprée. An early molasse in the Pharusian belt is the Tessalit-Anefis molasse which occurs in a narrow graben which runs along the 1° shear zone for about 400 km. It consists of fine sandstones and khaki siltstones, polygenic conglomerates with abundant volcanic clasts which pass upward into a glacial facies and a volcanic sequence (basalt, andesites, rhyolites) at the top. The sequence is cut by the alkaline ring complexes of the Iforas batholith. A molasse sequence fills a narrow graben 300 km long, along the Adrar fault. Composed of rhyolitic ignimbrites, arkosic sandstones, fine arkoses, purple or green shales and polygenic conglomerate lenses, this molasse sequence post-dates the 600 Ma Pan-African event and is intensely deformed and metamorphosed by basic and acidic intrusives.

The Série Pourprée (Fig.6.14) of the Pharusian belt is a late molasse sequence of Early Cambrian age, up to 6 km thick. Its lithologies include arkosic sandstones and reddish continental pelites, green-khaki rocks rich in illite and occasionally glauconite, and thick glaciogenic units such as tillites, argillites and varved argillites. The basal part with the "triad" facies of tillites, argillites and cherts, and limestones correlates with similar facies in the Taoudenni basin. Effusive rhyolites and ignimbrites occur at the base of the Série Pourprée while alkaline basalts and syenites are present at the top.

6.4.3 The Gourma Aulacogen

Stratigraphy

Mid-way along the Pan-African suture zone of the Trans-Saharan mobile belt lies the Gourma belt of Mali (Fig.6.13). The Gouruma fold belt has the characteristics of a triple junction in that it marks the westward bifurcation of the roughly north-south-trending rift system which developed along the eastern margin of the West African craton around 850 - 800 Ma ago. The Gourma trough evolved as an aulacogen or failed-arm of the Pan-African rift system. Over 12 km of presumed Upper Proterozoic sediments accumulated in this rapidly subsiding basin. As depicted in Fig.6.22,B, the sedimentary succession (Black et al., 1979) begins with an

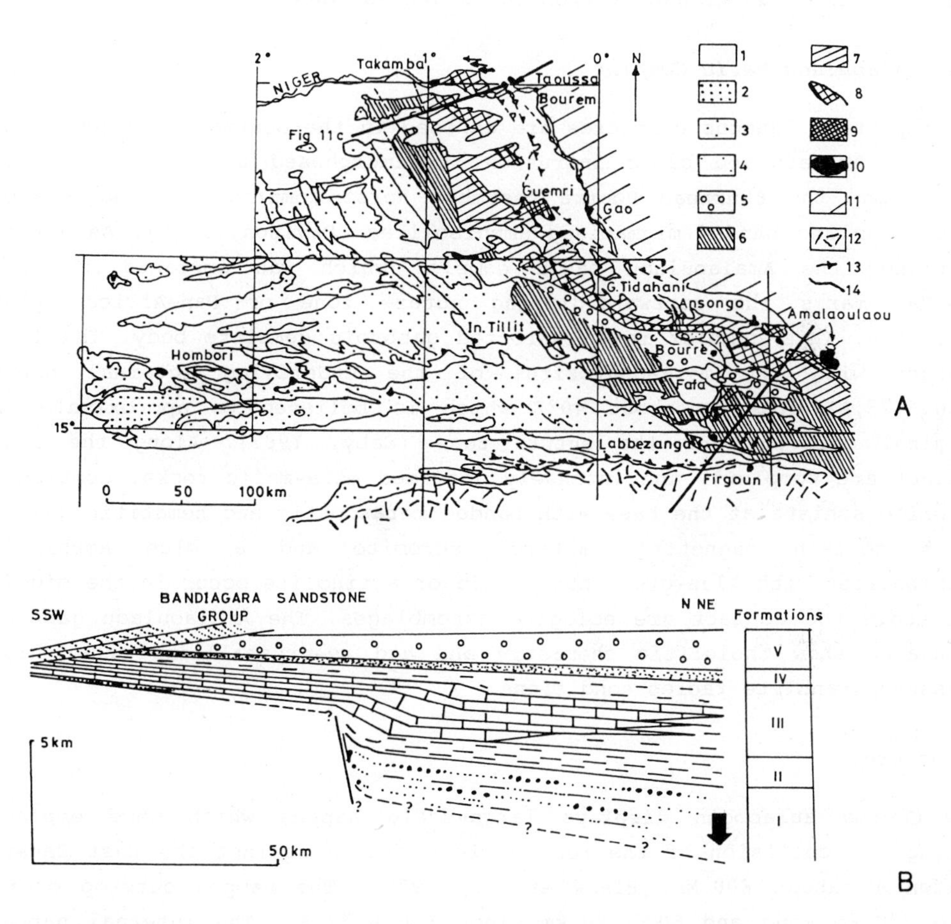

Figure 6.22: A, outline geologic map of the Gourma aulacogen. 1, Quaternary; 2, Bandiagara Sandstones (Cambrian ?); 3, undifferentiated possibly Upper Proterozoic formations; 4, same as 3 but parautochthonous; 5, Fafa parautochthonous unit; 6, exter-nal nappes with shallow greenschist metamorphism; 7, internal nappes; 8, internal nappes with high pressure - low temperature metamorphism including eclogite; 9, shales and quartzites of the Guemri window; 10, Amalaoulaou mafic-ultramafic massif marking suture zone; 11, Cenozoic of Gao trough; 12, undifferentiated basement, 2.0 Ga old of the West African craton and the Bourré inlier; 13, fold axis; 14, thrusts. B, section showing facies and thickness. (Redrawn from Caby, 1987; Black et al., 1979.)

early terrigenous phase (Units I and II), followed by a carbonate sequence which shows a lateral west-east facies progression from the cratonic platform through basin slope to trough (Unit III), ending with a typical passive continental margin non-marine clastic sequence (Units IV and V). Slope facies such as breccia and turbidites are well displayed in the carbonate sequence. Paleocurrent analysis in the overlying Bandiagara Sandstone Group reveals a converging drainage pattern centered on the Gourma trough with transport from the southwest and west.

The Amalaoulaou Mafic Complex

Two important igneous complexes are exposed in the Gourma aulacogen. These are the Amalaoulaou basic intrusion, metamorphosed under the granulite facies and the Eburnean Bourré basement horst comprising orthogneisses, pegmatites and banded migmatised amphibolites (Davison, 1980). As already mentioned the Amalaoulaou basic complex which was emplaced at about 800 Ma, marks the eastward-dipping suture zone of Pan-African plate collision. This complex is an unrooted gabbroic cumulate body. The lower contact with the meta-quartzites of the underlying internal nappes (Fig.6.23,A) dips at about 40°E which is believed to approximate the original direction of the suture zone (Caby, 1987). Along the lower contact are complexly and intensely deformed meta-mafic rocks, comprising chlorite schists at the base with banded metasomatic and hematitic jaspers which contain magnetite, altered chromite and a blue amphibole. Amphibolites with blue-green hornblende or actinolite occur in the middle; and above the contact are eclogite assemblages. The Amalaoulaou gabbroic cumulates show tholeiitic chemistry and had re-crystallized under high-pressure granulite facies conditions.

Structure

The Gourma aulacogen displays large-scale nappes which were emplaced during the collision of the West African craton against the East Saharan craton at about 600 Ma (Black et al., 1979). The nappes outcrop over a zone 300 km long and 50 - 80 km wide (Fig.6.22,A). The internal nappes, mostly mica schists and quartzites show flat-lying foliation and high-pressure/low-temperature (blueschist) metamorphic grade that increases to the southwest where it attains eclogite conditions in which jadeitic pyroxene is present. High-pressure/low-temperature initial metamorphic conditions (P = 18 Kb; T = 800°C) has been suggested (de la Boisse, 1981) for the internal nappes based on mineral associations (e. g. phengite + jadeitic pyroxene + pyrope-rich garnet + rutile) in the eclogitic aluminous mica schist at Takamba (Fig.6.23,B).

Earlier sharp folds in the Gourma trough are cut by foliation planes bearing phengite muscovite. Stretching lineations associated with NE-SW minor folds that are perpendicular to the general strike, occur within the internal nappes. In northern Gourma, along the banks of the River Niger (Fig.6.22,A) the internal nappes are in direct contact with the para-autochthonous folded greenschist facies formations by underthrusting towards the southwest (Fig.6.23,B); whereas to the south they are faulted against the Eburnean Bourré massif (Fig.6.23,A). The Bourré horst was up-thrown after the passage of the external nappes.

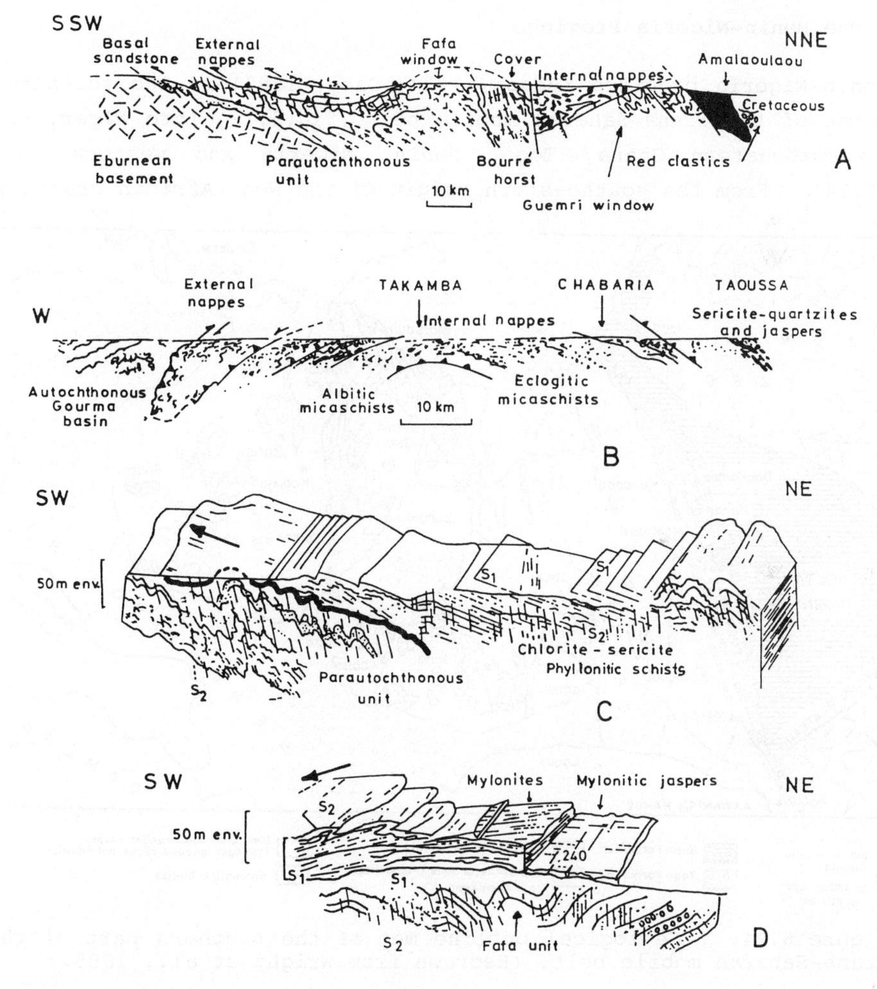

Figure 6.23: Sections through the Gourma region. (Redrawn from Caby, 1987.)

The external nappes are subhorizontal; they comprise mostly schistose greenschist basinal formations composed of quartzites, marble, siliceous dolomites and turbiditic sediments; and are devoid of high-pressure metamorphism. The Gourma formations (Fig.6.22,B) which lie in the aulacogen have been intensely deformed to form the para-authochthonous foreland (Fig.6.23,C,D). As evident from these diagrams (Caby, 1987), major folds in the external nappe are overturned to the southwest and are superimposed on early pre-nappe east-west folds. The non-metamorphic Bandiagara Sandstone (Fig.6.22,B) of probable Cambrian age, which unconformably overlies the Gourma sequence, was affected by a later phase of folding (Black et al., 1979).

6.4.4 The Benin-Nigeria Province

The Benin-Nigeria province (Black, 1984; Caby, 1987) is the southern continuation of the Trans-Saharan mobile belt (Fig.6.13) into Niger, Burkina Faso, southeastern Ghana, Togo, Benin, Nigeria and western Cameroon (Fig.6.24). From the southeastern margin of the West African craton east-

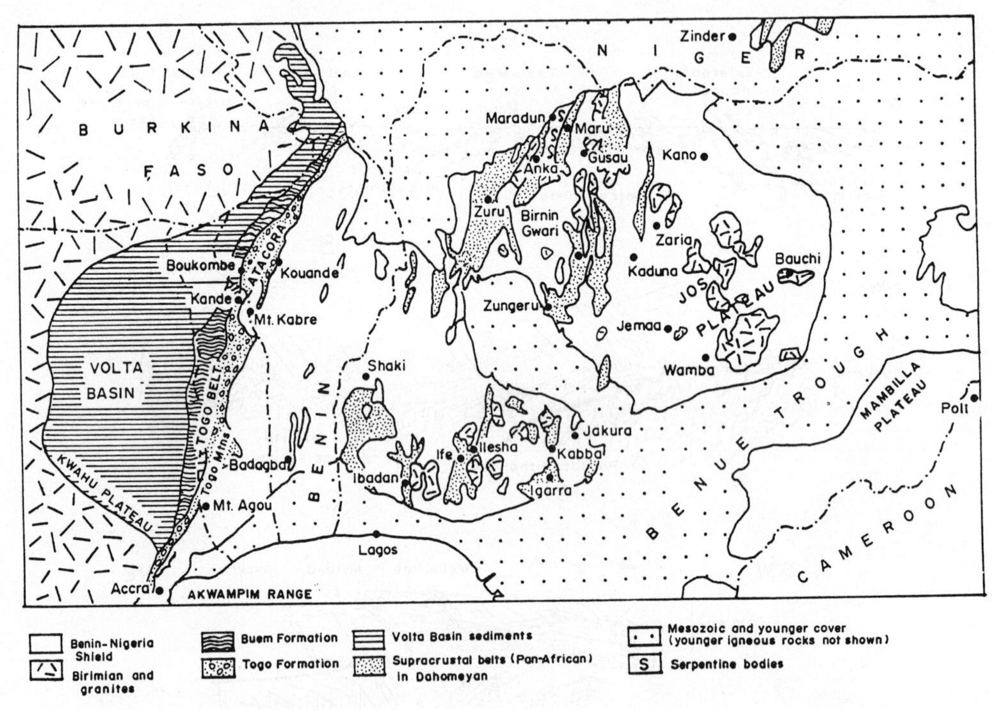

Figure 6.24: Geological outline map of the southern part of the Trans-Saharan mobile belt. (Redrawn from Wright et al., 1985.)

ward, three major tectonic domains occupy the Benin-Nigeria province. These are the Voltain foreland basin with flat-lying foreland nappes, the Beninian fold and thrust belt, and the Nigerian high-grade gneiss terrane

(Fig.6.25). In view of their structural and lithologic complexity, each of these domains is discussed separately below.

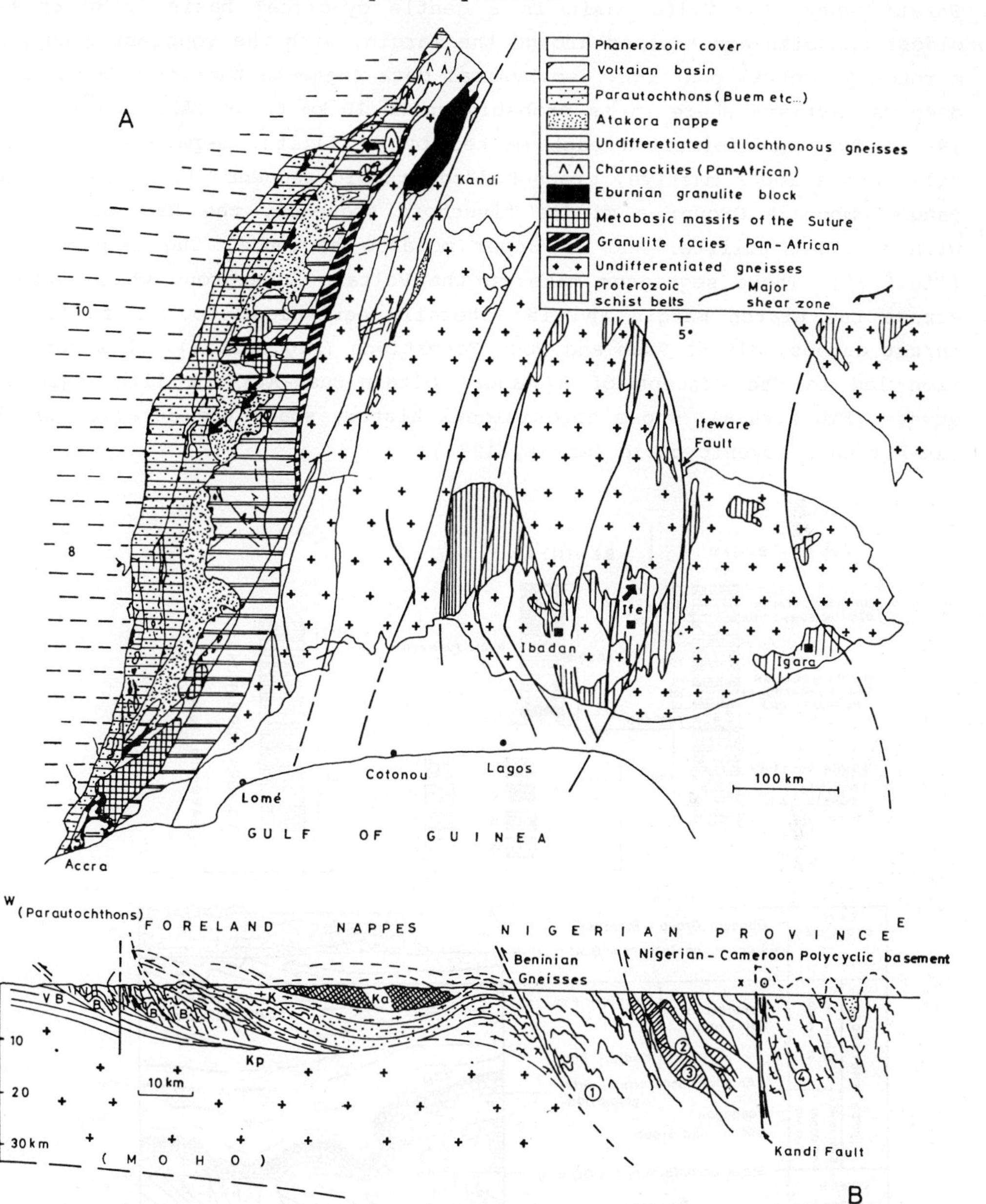

Figure 6.25: A, geological sketch map of the Beninian belt. B, cross-section; VB, Voltain basin; B, Buem parautochthon; Kp Kandé phyllites; A, Atakora metaquartzites; K, Lama-Kara leucocratic granite-gneisses; Ka, Kabié Klippe; 1, Beninian gneisses; 2, granulite belt; 3, garnet amphibolites; 4, polycyclic gray migmatitic gneisses. (Redrawn from Caby, 1989.)

The Volta Basin

Stratigraphy. The Volta basin is a gentle synclinal basin in which the oldest deposits are exposed around the margin, with the youngest occupying a roughly central position. Geophysical data suggests that the Volta basin deepens eastward where it is probably up to 10 km thick (Ako and Wellman, 1985). Two unconformities each marked by a tillite, separate the basin fill into a lower massive, cross-bedded arkosic sequence known as the Dapango-Bombouaka Group; a middle flyschoid sequence, the Pendjari Group, with a basal tillite; and an upper molasse sequence, the Obosum Group (Fig.6.26). These sequences make up the Voltain Supergroup which extends across the craton margin and is laterally equivalent to the folded and thrust eugeosynclinal Buem and Togo Formations (Fig.6.25,B). This correlation led to the concept of a major Voltain-Buem-Atakora lithologic sequence (Fig.6.26,A), whose depositional history spans the duration of the Pan-African orogenic cycle (Grant, 1973).

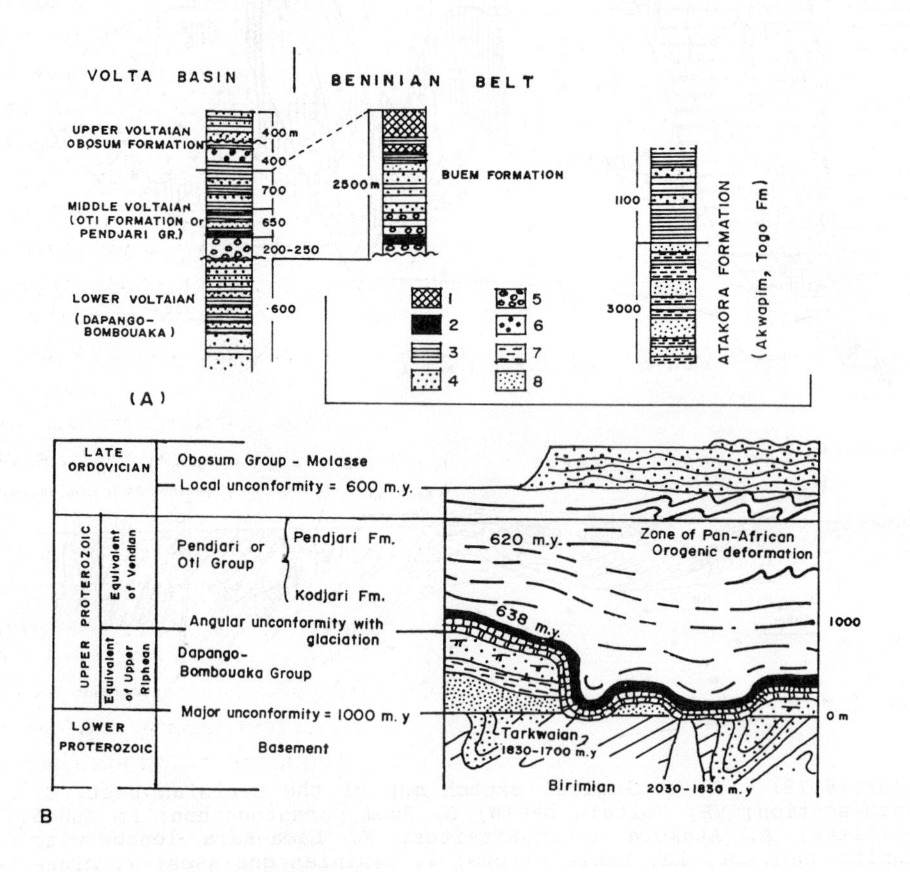

Figure 6.26: Stratigraphy of the Voltaian. (Redrawn from Affaton et al., 1980; Grant, 1973.)

The Dapango-Bombouaka or Lower Voltain Group rests unconformably on the Eburnean basement and has been correlated with Supergroup I of the Taoudeni basin (Wright et al., 1985). The Middle Voltain or the Pendjari Group (Oti Formation in Ghana) follows with a slight angular unconformity, showing basal erosional channels incised into the underlying Lower Voltain (Fig.6.26,B). The basal conglomeratic beds of the Pendjari have been interpreted as tillites which are succeeded by a sequence which includes brecciated or slumped carbonates (locally baryte-bearing and partly stromatolitic), chert, and silicified argillites bearing economic calcium phosphate deposits. The association of tillite, limestone, and baryte-silexite is known as the triad, which also occurs at the base of Supergroup 2 and the Rokel River Group in the West African polyorogenic belt. Among the principal lithologies of the Pendjari Group are shales, siltstones and sandstones which are glauconitic in places. The Pendjari Group, up to 2,000 m thick, accumulated on a rapidly subsiding marine shelf. As the Beninian fold and thrust belt is approached the Pendjari becomes increasingly deformed into gently NNE-SSW-trending asymmetric folds with southeasterly inclined axial planes. In the northern part of the basin in Burkina Faso the Pendjari Group is unconformably overlain by the Obosum molasse (Fig.6.26,B) which begins with a Late Ordovician tillite.

Phosphate Deposits. In the northern part of the Volta basin in Burkina Faso, Niger, and Benin there are economic phosphate deposits (Fig.6.27,A) which accumulated along the margin of the Pan-African epeiric sea which flooded the West African craton, where the sea descended into the deeper marine basin of the Beninian eugeosyncline (Lucas et al., 1986; Slansky, 1986; Trompette, 1988).

The richest phosphate interval is the Kodjari Formation in the lower part of the Pendjari Group (Fig.6.26,B) in Burkina Faso and at Tapoa in Niger Republic. At Kodjari the primary phosphate is fine-grained, massive or bedded and well sorted, with a thickness of 0 to 30 m and a P_2O_5 content of about 25 %; and a phosphate reserve of about 60 million t. At Tapoa (Fig.6.27,B), in a unit which is equivalent to the Kodjari Formation, fine to very fine, well bedded phosphorite occurs with an average P_2O_5 content of 18 to 35 %, and a reserve of about 100 million t. The phosphorites of the northern Volta basin were probably deposited between about 933 Ma and 660 Ma on a shallow marine platform with a low influx of clastic material under slightly reducing conditions as indicated by the presence of pyrite (Trompette, 1988). In the southern part of the Volta basin in Ghana, the stratigraphic equivalent of the phosphate-bearing Kodjari Formation is barren. But eastwards and basinwards there are phosphate shows in the epizonal units of the Buem Formation.

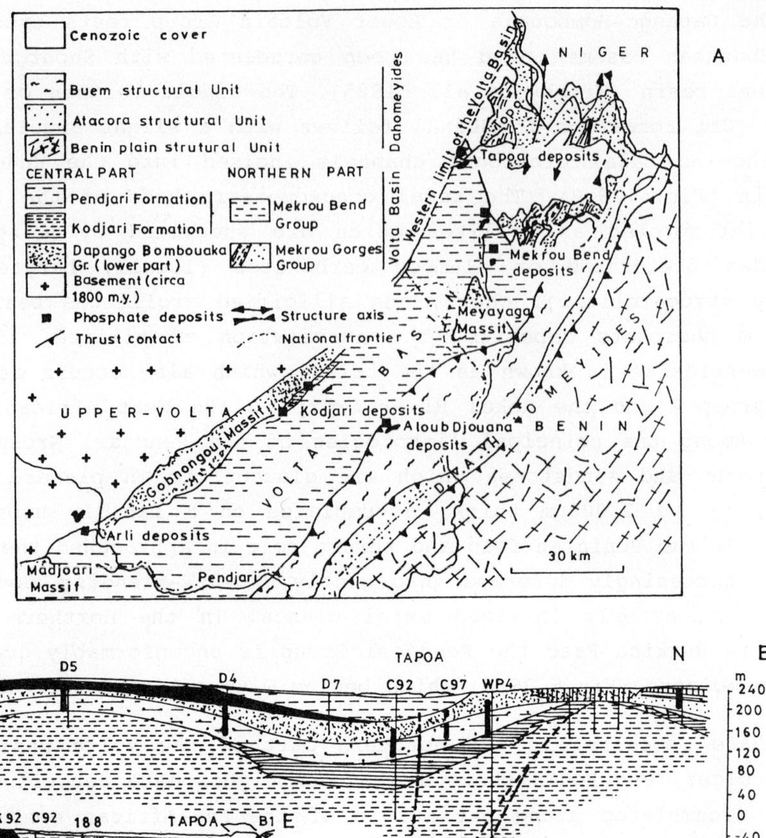

Figure 6.27: Phosphate deposits of the Volta basin. (Redrawn from Lucas et al., 1986; Trompette, 1988.)

The Beninian Fold Belt

This is the external zone of the mobile belt, sometimes referred to as the Dahomeyides, or the Togo belt (Wright et al., 1985). To the west the rocks of the fold belt are thrust against the Volta basin or onto the craton, whereas to the east the deformed metasediments of the fold belt are in thrust contact with the basement granulite gneisses of the Nigerian province (Fig.6.25,B).

Buem Formation. The Buem Formation outcrops in an area that is about 15 km wide and marks the eastern margin of the Volta basin. It is a southwestward-dipping largely unmetamorphosed sequence of mainly sandstones, siltstones and subordinate shales and mudstones. It is lithologically very similar to the Pendjari Formation, but differs in the presence of slices of dismembered ophiolitic rocks. The ophiolitic units of the Buem include alkaline and calc-alkaline basalts, occasionally pillowed, and with schistose and massive serpentinites, with chromite and cross-cutting dolerities. These lithologies are well developed in Ghana; but there, folding and the tectonic repetition of successions caused by the stacking of thrust sheets one upon the other (Fig.6.25,B), has masked the true thickness of the Buem Formation.

Atakora Formation. This sequence consists essentially of quartzites and mica schists which are equivalent to the Buem Formation and the Lower Voltain (Fig.6.26,A). It is synonymous with the terms Akwapim and Togo Formations. In the external part of the Beninian belt the Atacoran nappe consists of the Kandé phyllites and muscovites with basaltic volcanics (now greenschists) and marble; and the Atakora meta-quartzites (or Kirtachi Quartzite in southwestern Niger) which contains conglomeratic layers and some ferruginous meta-quartzites. The more internal units of the Atakora Formation attained high-pressure metamorphic grade in south Togo where lenses of kyanite-eclogite were crystallized at T = 700°C and P = 16 Kb, during the Pan-African orogeny at about 600 Ma (Caby, 1989). The foreland nappes of the Atakora includes leucocratic orthogneisses; the Lama-Kara granite-gneisses, which are considered to have been derived from Eburnean granites; and Atakoran meta-quartzites and kyanite-staurolite schists in northern Benin. These rocks which are now in the foreland were severely metamorphosed in the internal zone of the Beninian belt before being thrust westward towards the craton.

The Kabié meta-basic massif, a part of the Atakora in Togo, is believed to represent relict oceanic crust, now found caught up along the Pan-African suture of the Beninian fold belt (Fig.6.25,B). In the massif detached masses of granulitic meta-cumulates dip gently at 15° to 30° E.

The base of the massif is marked by garnet-amphibolites which separate the meta-mafic bodies from the orthogneisses below. The mafic bodies had probably intruded the base of the crust in an arc-type tectonic setting. During the orogeny meta-noritic gabbros of calc-alkaline character were subjected to deep amphibolite-to granulite-facies conditions before partial crystallization into kyanite eclogite. Blastomylonitic garnet-amphibolites with eclogites occur along the sole of the Kabié klippe. The klippe was probably thrust westward from the actual suture (Fig.6.25,B).

The Nigeria Province

Previously known as the Dahomeyan basement (e. g. Wright et al., 1985), the Nigerian province (Caby, 1989) includes the Beninian gneisses in the internal zone of the Pan-African Benin-Nigeria orogen, east of the suture (Fig.6.25,B), and the vast expanse of reactivated high-grade, probably Archean basement gneisses and Proterozoic supracrustals. Caby (1989) presented a regional synthesis and re-interpretation of this region. The Nigerian province is the southern continuation of the central Hoggar reactivated basement. Thrust and shear zones within the Nigerian province (Fig.6.25,A) allow the subdivision of this region into the Beninian gneisses and the Nigerian-Cameroon polycyclic basement.

Beninian Gneisses. These are high-grade amphibolite facies, predominantly quartzo-feldspathic and magmatitic rocks, often anatectic, and with migmatitic granitoids. The Beninian gneisses (Fig.6.25,B) flatly overthrust the Atacora unit, which is believed to be their equivalent (Black, 1984). These gneisses are in turn overthrust by an eastward elongated granulite belt composed mainly of kinzigites, two-pyroxene granulites and mafic amphibolites with relict kyanite-eclogite assemblages. These rocks could have attained high-grade conditions in the internal zone of the orogen simultaneously with the eclogites of the Kabié meta-basic massif. Whereas the granodioritic to tonalitic hornblende gray gneisses which outcrop in Benin Republic east of the Kandi fault (Fig.6.25,B) are lithologically similar to the adjoining reactivated Archean basement in Nigeria, the Badagba quartzites (orthoquartzite and marble sequence, and associated sub-alkaline to syenite orthogneisses overlain by pelitic gneisses) which are exposed along the Kandi fault probably represent a metamorphosed monocyclic Proterozoic cover.

The Migmatite-Gneiss Complex of Nigeria. The basement compplex of Nigeria consists of predominantly Archean polycyclic gray gneisses of granodioritic to tonalitic composition; remnants of unconformable Proterozoic cover now represented by variably migmatized metasediments which are preserved in synclinorial schist belts; and many syn-tectonic to late-tectonic intrusions (Ajibade et al., 1987; McCurry, 1976; Oyawoye, 1972). The Proterozoic metasediments have been classified into the Older Metasediments of Early Proterozoic age, and the Younger Metasediments of Pan-African age.

Reactivated Archean basement, often referred to as the migmatite-gneiss complex, occupies nearly a half of the surface area of Nigeria. It includes the migmatite-gneisses of the Zinder inlier in Niger Republic in the northeast (Fig.6.24); those of Obudu, and Oban (Ekwueme and Schlag, 1989) in southeastern Nigeria; and the migmatite-gneisses in neighbouring Cameroon Republic. The migmatite gneiss complex is dominated by quartzo-feldspathic biotite-hornblende-bearing gneisses; and schists and migmatites, in which minerals such as garnet, sillimanite, kyanite and staurolite suggest high-amphibolite facies metamorphism. Granulite-facies rocks are confined to charnockite bodies which are generally associated with granites of probably igneous origin.

In the Ibadan area of southwestern Nigeria (Figs.6.25,6.28) where the migmatite-gneiss complex has been studied in detail (Burke et al., 1976) the predominant rock types are banded gneisses, schists and quartzites (Fig.6.28) representing metamorphosed shales and graywackes with inter-bedded sandstones, whereas the interleaved amphibolite layers are probably metamorphosed tholeiitic basalts. Although the schists and quartzites intercalated within the Ibadan banded gneisses are part of relict supracrustals which are considered separately below, they are mentioned here in order to demonstrate the complex geological history of this part of the Nigerian basement complex. The Ibadan banded gneisses had a polycyclic history (Burke et al., 1976) which in spite of the paucity of reliable isotopic ages, is also true of most of the Nigerian basement.

The geological history of the Ibadan migmatite-gneiss complex began with the deposition of shale graywackes and sandstones with interbedded basalts. An early phase of folding and high-amphibolite facies metamorphism which formed the banded gneiss, quartzite, and amphibolite was succeeded by the emplacement of semi-concordant aplite schists in the banded gneiss at about 2.75 Ga, during the Liberian orogeny; and by the emplacement of microgranodiorite dykes. A second phase of intense folding followed, in which the Ibadan granite gneiss was emplaced at about 2.20 Ga, during the Eburnean orogeny. The Ibadan granite gneiss, a coarse granitic body of uniform composition probably originated through partial fusion of the banded gneiss country rock (Freeth, 1984). In Ife (Fig.6.28) Archean gray gneisses are also intruded by Eburnean orthogneisses similar

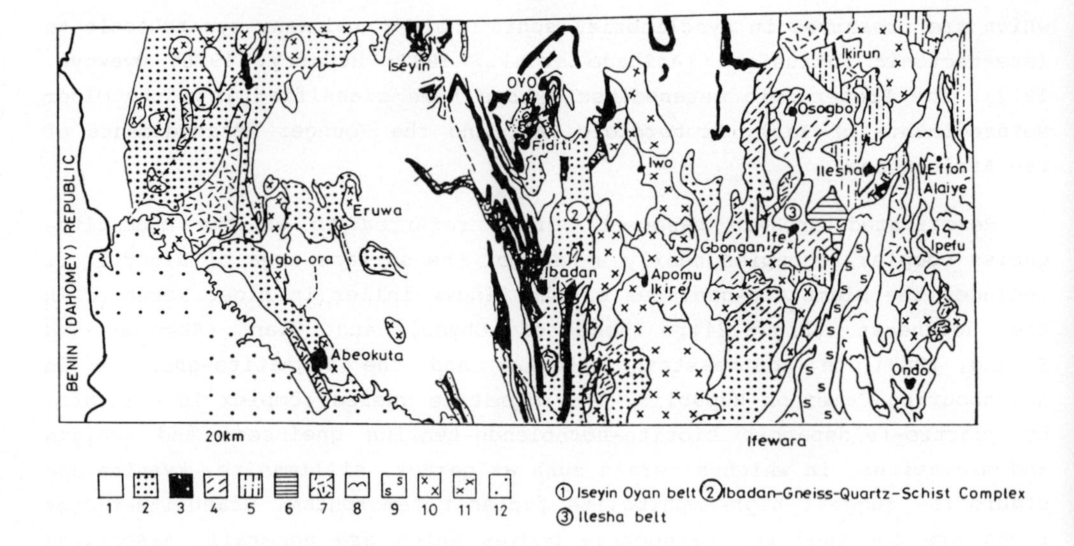

Figure 6.28: Sketch map of the basement of SW Nigeria. 1, migmatites and gneisses with amphibolite intercalations; 2, gneiss and schist complex; 3, quartzite and schist; 4, pegmatized schist; 5, schist and epidiorite complex; 6, epidiorite (Archean and Eburnean); 7, granite gneiss (Late Early Proterozoic); 8, quartzite; 9, feldspathic quartzite; 10, granites; 11, Late Proterozoic quartz syenite and charnockites; 12, Cretaceous-Recent cover. (Redrawn from Ajibade et al., 1987.)

to the Ibadan granite gneiss (Fig.6.29,B), which have yielded ages of about 1.82 Ga. The last tectonic event in the Ibadan-Ife area was the Pan-African event.

Older Metasediments. Supracrustal relicts in the Nigerian basement are either intercalated within reworked Archean gneisses as afore-mentioned or are unconformably disposed in synclinorial schist belts. The former category belongs to the so-called Older Metasediments which are again best known in the Ibadan-Ife region of southwestern Nigeria. In the Ibadan area these occur as schists and meta-quartzites within the banded gneisses and extend for about 75 km northward, into the Iseyin area and eastward into the Ife area (Fig.6.28). In the latter area the Older Metasediments appear to be bounded by the Ifewara fault system which seems to separate this older Iseyin-Ibadan-Ilesha schist belt from the presumably Late Proterozoic Igara-Kabba-Lokoja schist belt (Ajibade et al., 1987; Caby, 1989) to the east.

Caby (1989) observed that the Ibadan meta-quartzites overlain by pelitic schists which were intruded by Mg-rich mafic sills, may represent a monocyclic Early Proterozoic (Eburnean) sequence. In the Iseyin area pebbly meta-quartzites mark the basal unconformity with which the meta-quartzite-schist assemblage rests upon Archean gray gneisses. Around Ife aluminous mica schists with two micas and garnet and sillimanite, are exposed which often show well-preserved sedimentary bedding. Southeast of Ife these aluminous metasediments are overlain by layered gneisses of probable rhyodacitic origin. Massive amphibolite schist, talc chlorite schist, talc-tremolite rock and pelitic schists occur in the schist belt of the Ilesha area which also carries gold mineralization. The geochemistry of these amphibolites suggests derivation from subcrustal basalts and peridotites (Olade and Elueze, 1978). Since the Ife-Ilesha schist belt is cut by the 1.8 Ga granite gneisses (Fig.6.29,A), these supracrustals may therefore represent pre-Pan-African monocyclic sediments and magmatic rocks of probable Eburnean age (Ajibade et al., 1987; Caby, 1989; Hubbard, 1975).

Younger Metasediments. These are Late Proterozoic pelites (represented by phyllites, muscovite-schists and biotite-schists) with quartzites forming prominent strike ridges in several belts. Some belts contain ferruginous and banded quartzites, spessartite-bearing quartzite, conglomeratic horizons and marbles and calc-silicates. Igneous rocks, generally minor constituents in these belts, include amphibolites (originally lavas or minor intrusions), serpentinites and other ultramafics which were probably intruded along deep fractures during the deformation of the supracrustal (Bafor, 1982) belts. There are also small occurrences of acid meta-volcanics of dacite to rhyolitic composition (Wright et al., 1985).

The Younger Metasediments occur in southwestern Nigeria, but they are mostly found in the northwestern part of the country (Fig.6.24). They occur in synclinorial schist belts in which the low-grade rocks are characterized by tight to isoclinal folding and steeply dipping foliation with gradational, faulted or sheared boundaries with the surrounding migmatite-gneiss complexes. For example, a cataclastic belt, the Zungeru mylonites extends on both sides of the Birnin Gwari schist belt (Fig.6.30) and probably originated as a major thrust between the basement (Kusheriki Formation) and the supracrustals (Birnin Gwari, Kushaka, Ushama Schist Formations). The Birnin Gwari Formation includes phyllites, schists, and schistose mudstone conglomerates.

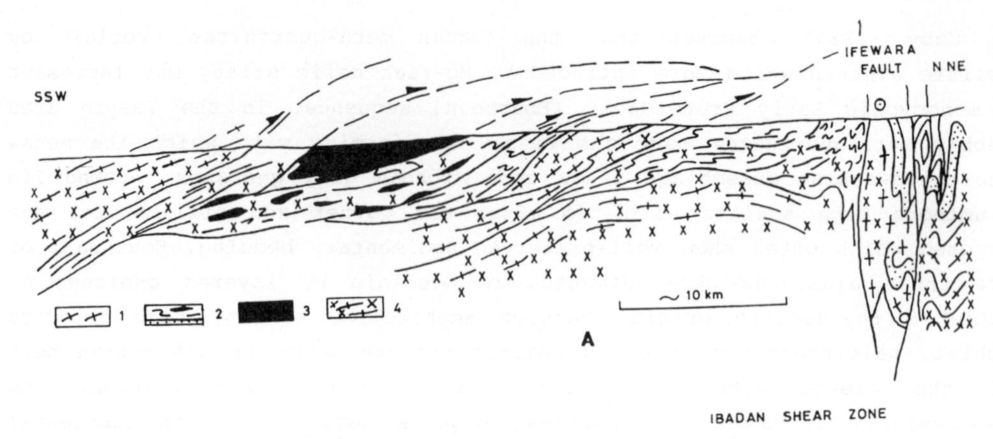

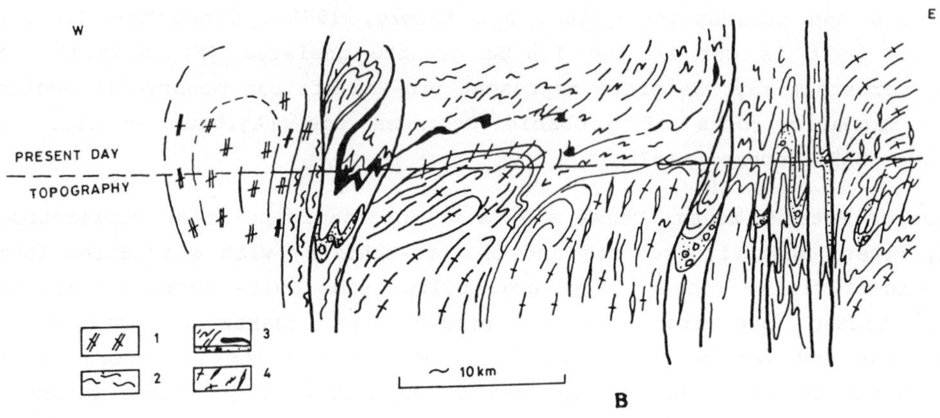

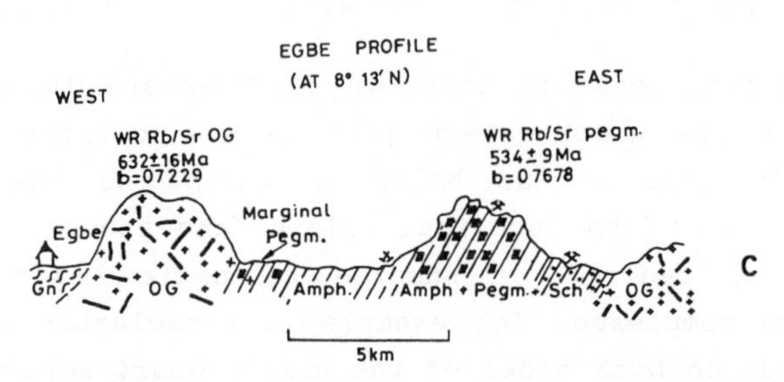

Figure 6.29: Geologic sections through SW Nigerian schist belts. A, Ife region: 1, 1.8 Ga orthogneisses; 2, Proterozoic sillimanite-schists and quartzite; 3, mafic amphibolites; 4, Archean gray gneisses. B, Ibadan region: 1, Pan-African syenite; 2, Proterozoic cover; 3, schists with lenses of mafic to ultramafics and basal, locally detached quartzites; 4, Archean gray gneisses. C, Egbe area. (Redrawn from Caby, 1989; Matheis, 1987.)

A major belt of Younger Metasediments extends from Igarra in southwestern Nigeria through Kabba and Okene (Annor,1983) to Lokoja (Fig.6.24).

The Igarra sequence includes basal polygenetic meta-mixtite layers with a semi-pelitic matrix and abundant crystalline angular cobbles, up to 1 m across (Caby, 1989; Odeyemi, 1976). The cobbles were derived from granodiorite, gneiss, pegmatite, amphibolite, and calc-silicates and marbles from the underlying units. The meta-mixtite is overlain by hornfelsic biotite schists containing calc-silicate concretions, and sedimentary relict features such as parallel-laminated meta-pellites, cross-bedding in arenites, and cross-cutting channels with pebbly infill, all attesting to a probable turbiditic origin. This turbiditic sequence grades laterally into migmatites which originated from the emplacement of voluminous sheets of late tectonic granitoids.

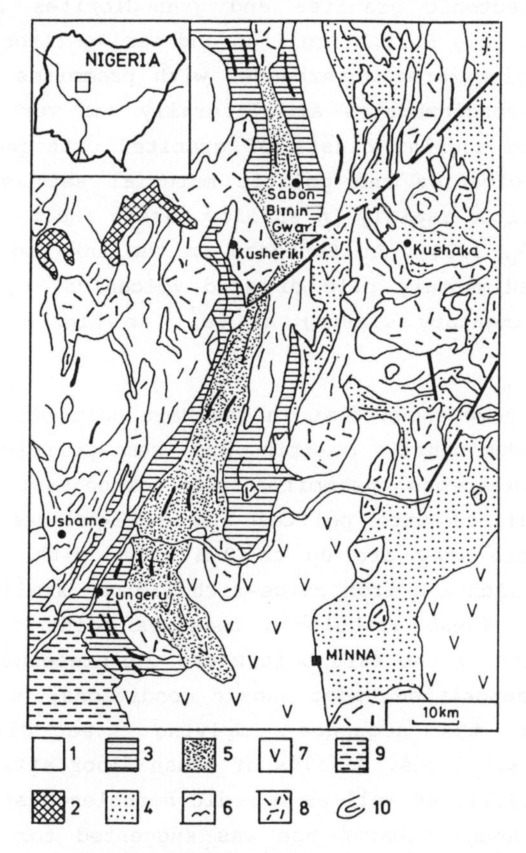

Figure 6.30: Geological map of Zungeru area. 1, migmatites and gneisses; 2, kyanite and sillimanite-bearing quartzites; 3, Zungeru mylonites; 4, Kushaka Formation; 4, Birnin Gwari Formation; 6, Ushama Schist Formation; 7, foliated tonalites and granodiorites; 8, Late Pan-African granites; 9, Cretaceous-Recent. (Redrawn from Ajibade et al., 1987.)

Southeast of Kabba in the northeastern part of the southwestern Nigerian basement, there are metamorphosed banded ironstones, rich in magnetite and hematite. Near Kabba and Jakura dolomitic marble occurs which is quarried. Numerous thin sheets and lenses of calc-silicates probably represent metamorphosed impure limestones.

Among the best preserved schist belts in Nigeria are those in the northwestern part of the country (Fig.6.24). Here about eleven meta-volcanic-sedimentary belts in greenschist to amphibolite facies are isoclinally folded into basement (Utke, 1987). The Anka and Maru belts are examples of these meta-volcanic-sedimentary belts. They comprise polycyclic migmatite-gneiss basement sequence with an infolded cover of low-grade metasediments; both basement and cover are intruded by a suite of syn-tectonic to late-tectonic granites and granodiorites (Holt et al., 1978). In the Anka belt (Fig.6.31,A) pelites and semi-pelites predominate, with polymict meta-conglomerates interbedded with psammites. Up to 150 - 250 m thick, the meta-conglomerates are laterally and vertically impersistent with rounded to angular clasts of granite, orthoquartzite, vein quartz, fine-grained volcanics and pelitic material set in a psammitic matrix. They were deposited rapidly in shallow basins (Holt et al., 1978), probably grabens (Black, 1984). At Bunkasau in the Anka belt there is a pre-tectonic volcano-sedimentary pile of acid volcanics, coarse clastics with volcanic and sedimentary clasts; as well as high-K calc-alkaline volcanics.

Holt et al. (1978) have shown that in the Maru belt (Fig.6.31,B), the metasediments are predominantly pelitic and semi-pelitic rocks with locally interbedded micaceous psammites and orthoquartzites. Banded ironstones are intercalated with pelites and semi-pelites (Fig.6.31,B) near Maru township where they are up to 2 m thick with characteristic alternations of quartz-rich and iron oxide-rich laminae similar to typical banded iron-formations. These lithologic features point to a basin of quiet water sedimentation in the Maru belt where the presence of pyrite in some pelites suggests deposition under anoxic conditions; but iron oxides predominate throughout the sequence implying free circulation of oxygenated waters in most places. Basalts of ocean floor affinity occur in the Maru belt (Ogeze, 1977), as well as a sodic hornblende syenite pluton, the Kanoma pluton. Although Kibaran age was suggested for the Maru and Anka belts (Holt et al., 1978), both belts have been assigned to Pan-African age, like similar widely developed supracrustals (Black, 1984) considered here.

Utke (1987) regarded the schist belts of NW Nigeria as typical volcano-sedimentary assemblages with tensional and compressional-related rocks. These rocks display the influence of paleo-rift systems.

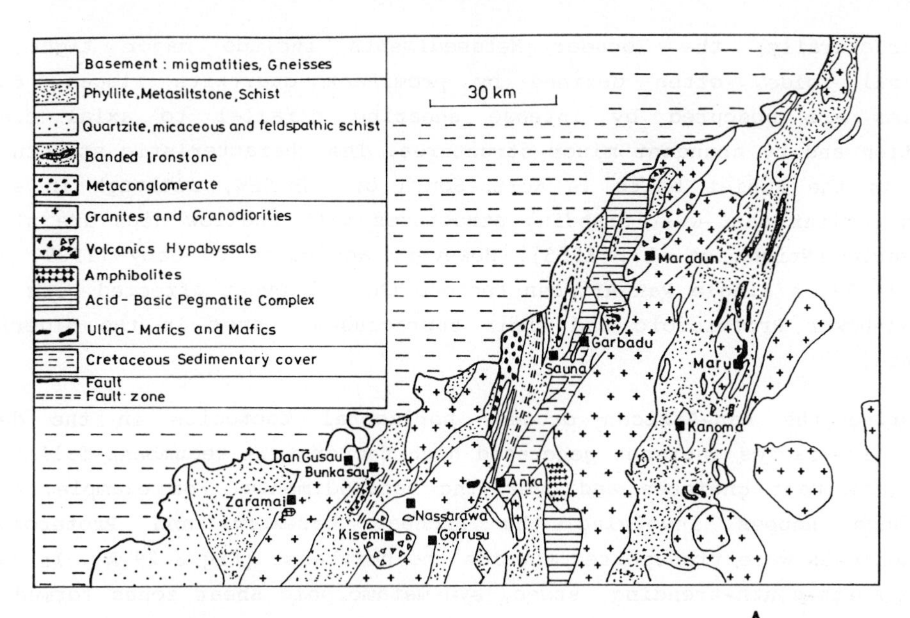

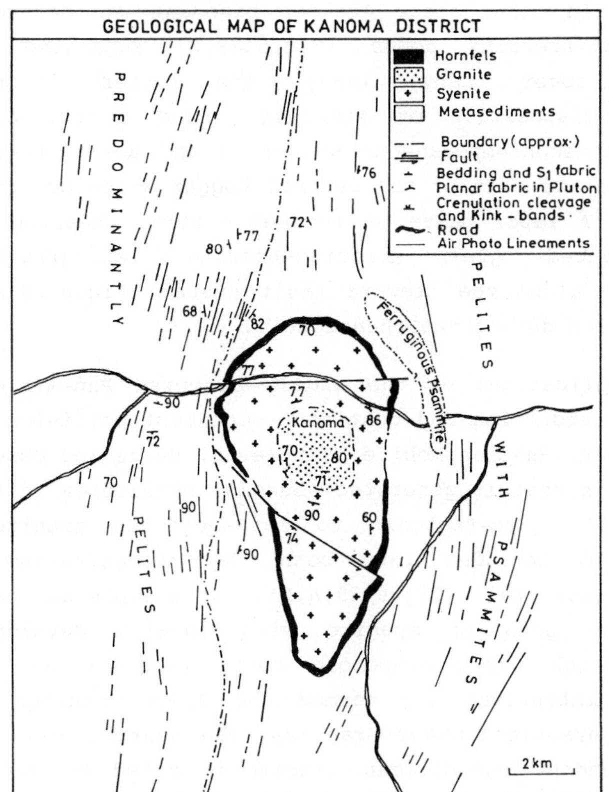

Figure 6.31: Sketch maps of Anka and Maru schist belts in NW Nigeria. (Redrawn from Holt et al., 1978.)

Structurally, the Younger Metasediments include major tight to isoclinal folds, often defined by prominent quartzite ridges. Fold closures are obscured by intense shearing parallel to axial plane foliation and by abundant minor structures. The characteristic structural trend in the schist belts is North-South or NNE-SSW, although in some places earlier east-west-trending structures with shallow dips are still noticeable (Wright et al., 1985). However, according to Caby (1989) and Grant (1973) a single Pan-African tectono-thermal event affected both the Archean-Lower Proterozoic and their supracrustal cover in the Nigerian province.

During the Pan-African orogeny horizontal tectonics in the deep amphibolite-facies gneisses generated high-temperature recumbent foliation in the basement gneisses and overlying metasediments. For example, near Ife huge nappes involving the Archean basement and Proterozoic metasediments were transported towards the northeast (Fig.6.29,A). In some places north-south-trending, steep, syn-metamorphic shear zones formed at high temperatures, whereas in other areas Archean basement was deformed into elongated north-south-trending domes (Fig.6.29,B) separated by synforms of Proterozoic cover (Caby, 1989). The latter is the characteristic structural disposition of Nigerian schist belts. Caby (1989) observed that the penecontemporaneous wrenching and northeasterly thrusting near Ife is similar to that of the central Hoggar which occurred between 629 Ma and 614 Ma. A later phase of renewed dextral shearing at lower temperatures produced green biotite-bearing retrogressive ultramylonites, for example along the Ifewara fault system (Fig.6.29,A). There was brittle reactivation during the Phanerozoic.

Granitoids. One of the implications of the widely accepted Pan-African tectonic model of eastward subduction and continent-continent collision in the southern part of the Trans-Saharan mobile belt (e. g. Burke and Dewey, 1973; Wright et al., 1985) is that it generated abundant granitoids in the Nigerian province (Fig.6.32). Syn-tectonic to late-tectonic granites, diorites and syenites were intruded into both the migmatite-gneiss complexes and overlying supracrustals (Fig.6.29,A,B). For example at Shaki in southwestern Nigeria a potassic syenite with locally developed granulite facies cuts through the Proterozoic cover (Fig.6.29,B). In Nigeria these Pan-African intrusives are termed the Older Granites to distinguish them from the Jurassic Younger Granites. The central Nigerian Shield is occupied by abundant Pan-African granitoids dated at 700 - 500 Ma. In northwestern Nigeria there are volcano-detrital materials including dacites and shonshonites of Cambrian age in the Anka and Maru belts which Black (1984) interpreted as probably representing Pan-African molassic grabens.

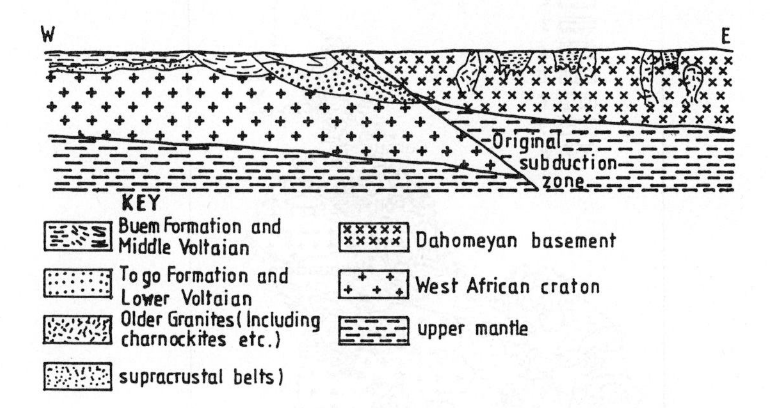

Figure 6.32: Plate tectonic explanation of the basement geology of the Beninian belt. (Redrawn from Wright et al., 1985.)

The Older Granites range in size from small subcircular cross-cutting stocks to large elongate concordant predominantly granodioritic batholiths. Some batholiths have adamellitic composition and coarsely porphyritic textures. Some of the post-tectonic granitic and dioritic intrusives, for example at Toro, exhibit ring-structures similar to the Jurassic Younger Granites. Charnockites also occur among the Older Granites in southwestern Nigeria, east of Ibadan, and in the northeastern part of the country where at Bauchi there is a distinctive fayalite-bearing variety known as bauchite (Oyawoye, 1972). Pan-African late to post-tectonic basalt and dolerite dykes also occur in the Nigerian basement.

The Cameroon Basement

The Pan-African belt in Cameroon , although belonging to the Trans-Saharan belt, marks the transition to the Zaire craton. The cratonic part is the Ntem complex of southern Cameroon (Fig.6.33) which continues southward into the Zaire craton. The rest of Cameroon Republic is considered as a Pan-African mobile belt (Cahen et al., 1984), comprising reactivated Archean, the Nyong gneisses of pyroxene gneisses and migmatites; the "intermediate" Precambrian groups including the Dja Group and mixtite complex; and syntectonic granites and migmatites. There are two generations of post-tectonic granites; and in northern Cameroon there are lower Paleozoic rocks as well.

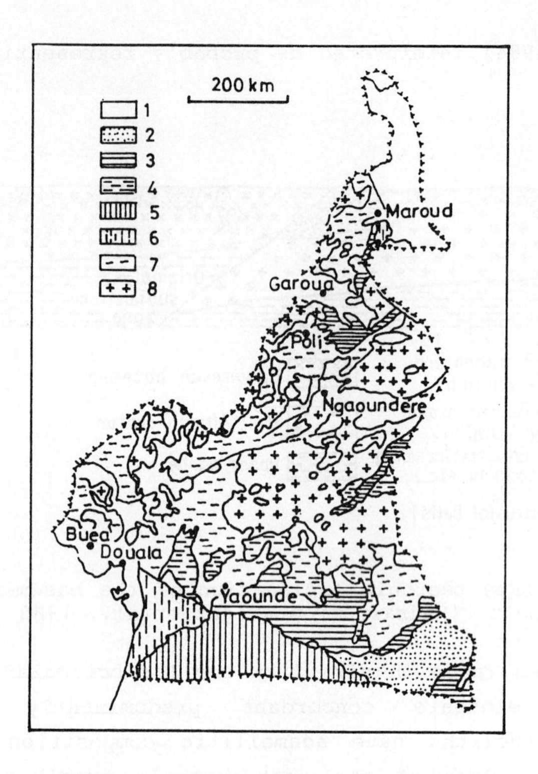

Figure 6.33: Geological map of Cameroon. 1, Phanerozoic cover and "granite ultimes"; 2, Lower Dja Group; 3, "Intermediate" groups; 4, formations of doubtful age; 5, Ntem Complex; 6, part of 5 reworked in the Pan-African; 7, migmatites; 8, granites. (Redrawn from Cahen et al., 1984.)

Although the so-called "intermediate groups" have not been dated, they have been placed in the Pan-African mobile belt. Among these are the weakly metamorphosed Poli-Matoua Group and the Lom Group which comprise rocks of sedimentary and volcanic origin, that were intruded by the first generation of post-tectonic granites at about 520 Ma. The Ayos-Mbalmayo-Bengbis Group and the Yokadouma Group are also sedimentary and weakly metamorphosed, the former being overlain by the lower Dja Group which is characterized by abundant sericite-chlorite schists. The lower Dja Group was less involved in the mobile belt and correlates with a similar unit, the Sembe-Ouesso Group in the Congo Republic to the south, and with the Nola Group in the Central African Republic to the northeast (Cahen et al., 1984). The lower Dja Group includes a mixtite complex which is equivalent to the lower mixtite of the West Congolian belt in the south.

In spite of age uncertainties and the lack of a deliberate attempt to correlate with the Nigerian province, there is no doubt that the Precambrian terrane in Cameroon extends across the Nigerian frontier

(Fig.6.34) into southeastern Nigeria where similar gneisses and migmatites occur. The Pan-African history of Cameroon involved the accumulation of the low-grade metasediments; migmatization during a tectono-thermal event at 600 - 565 Ma; the intrusion of very late Precambrian and early Paleozoic granites at 562 Ma and 516 Ma respectively; and the deposition of the Mangbai Group before 400 Ma (Cahen et al., 1984).

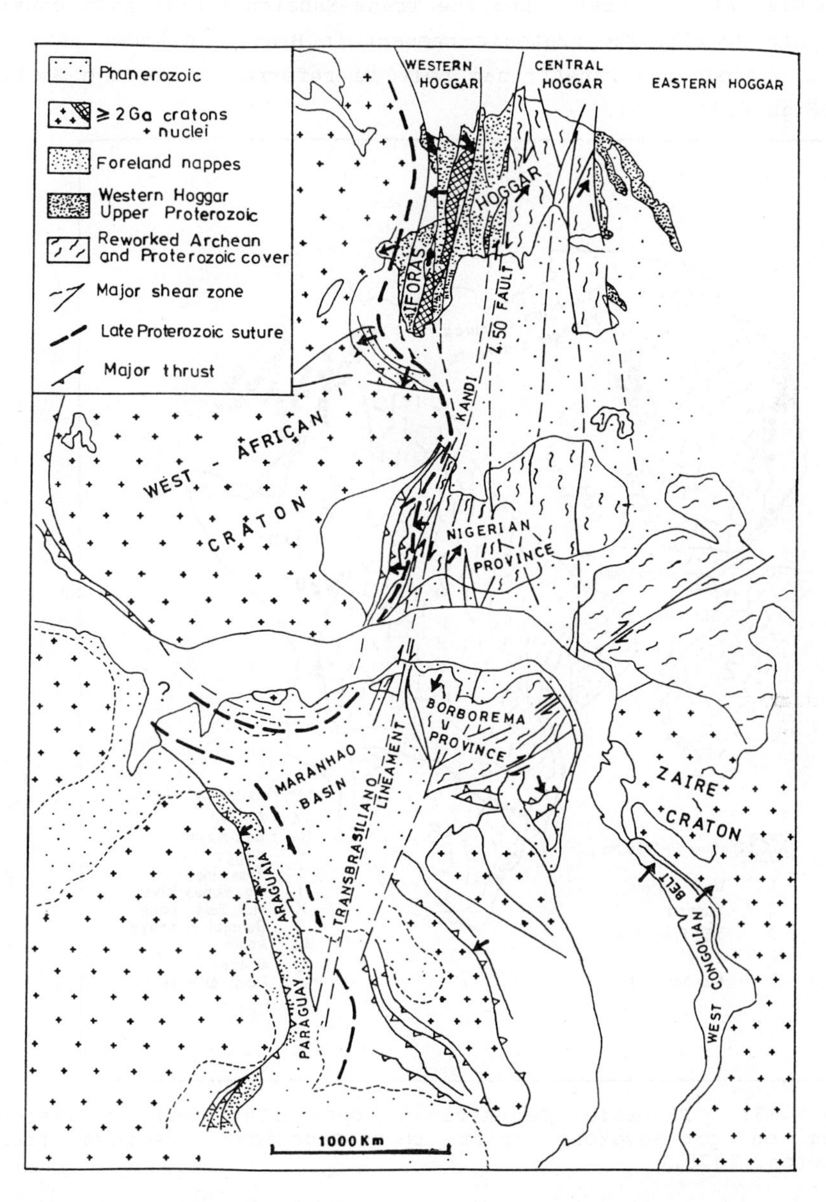

Figure 6.34: Continuation of the Trans-Saharan mobile belt into NE Brazil. (Redrawn from Caby, 1989.)

Trans-Atlantic Connections

Trans-Atlantic Precambrian geochronological correlations (Hurley and Rand, 1968), and geological, and structural (e. g. Caby, 1989; Shackleton, 1976) correlations between West Africa and Brazil have shown that prior to the break-up of Pangea II (Fig.6.35) and the opening of the South Atlantic Ocean, the West African craton and the Trans-Saharan mobile belt continued southward into Brazil. The cratonic remnant in Brazil is known as the São Luis craton whereas the Pan-African belt is referred to as the Borborema province (Figs.6.34,6.35).

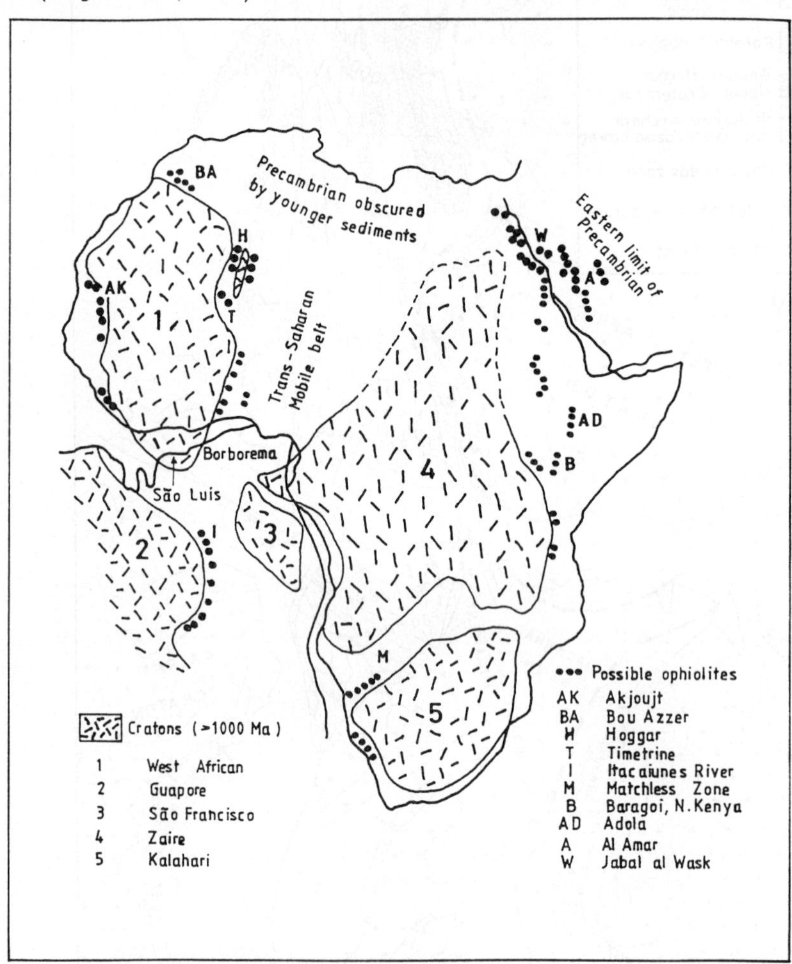

Figure 6.35: Probable Pan-African ophiolites and collision sutures on pre-Mesozoic drift reconstruction. (Redrawn from Shackleton, 1976.)

The Trans-Saharan mobile belt shows close geologic and structural continuity with the Brazilian Borborema province, suggesting that during the Late Proterozoic tectono-thermal events both regions behaved as one orogenic belt (Caby, 1989). The Brasiliano event is the equivalent of the Pan-African orogeny in Brazil. In his correlations of the high-grade Borborema and Nigerian provinces and comparison of their crustal evolution Caby (1989) stressed the following similarities . The Nigerian and the Borborema provinces are underlain by mostly gray gneisses and migmatites which include reworked Archean tonalites, trondjhemites and granodiorites, with minor supracrustals and greenstones. In both regions there are remnants of Early Proterozoic monocyclic metasediments consisting of aluminous meta-quartzites and pelitic schists, known as the Cera Group and the Jucurutu Formation in Brazil. As in Nigeria, these supracrustals are intruded by 2.0 - 1.80 Ga anorogenic granites. Younger synformal, disconformable, probably Late Proterozoic flysch-type supracrustals with turbiditic facies and Ca-Mg-Fe concretions, known as the Serido-Cachoeirinha sequence in Borborema, are very similar to the Igarra schist belt of southwestern Nigeria.

Like in the Nigerian province, Borborema has no convincing evidence for the existence of a Middle Proterozoic (Kibaran) orogeny. Rather, in both regions the effects of the Pan-African-Brasiliano orogeny are pervasive, and produced recumbent foliation, synchronous with major nappes of Archean rocks and Proterozoic supracrustals, under high temperature conditions. Both regions contain kyanite-bearing, muscovite-free assemblages in Proterozoic rocks as well as granulite and charnockite belts. In some linear schist belts metamorphism was at the lower greenschist facies in spite of lithological similarities with areas that attained higher metamorphic grade.

The sinuous and branched pattern of Pan-African shear zones also occurs in Borborema (Fig.6.34) where they are nearly all underlain by retrogressive mylonites and ultramylonites with horizontal lineations that suggest horizontal displacements. The Sobral fault or the Trans-Brasiliano lineament, continues northward into the Kandi fault and the 4°50′ fault of the Trans-Saharan mobile belt, thus representing a continuous transcontinental lineament of dextral lithospheric strike-slip movement. This fault system was generated in the final stages of the Pan-African-Brasiliano orogeny during continent-continent collision, and was reactivated later in the Phanerozoic (Caby, 1989).

However, whereas a Pan-African suture zone is evident in the western Pharusian belt and the Gourma aulacogen and the Beninian belt, no possible suture zones have been seen in Borborema (Caby, 1989). Also missing from Borborema are foreland basins which are equivalent to the Volta basin.

Mineral Deposits in the Trans-Saharan Belt

In contrast to the West African craton the Trans-Saharan mobile belt is not a rich mineral province, probably on account of its polycyclic nature which could have depleted it of economic mineralization, although crustal reactivation, could have produced local enrichment in certain elements (Wright et al., 1985). Also, unlike the craton, mafic and ultramafic rocks are volumetrically insignificant in the mobile belt. However, the variety and abundance of relatively minor metallic deposits and industrial minerals offer some scope for labour-intensive, small-scale mining as part of the rural development drive in debt-ridden West African nations.

Gold. Gold has been mined from the Nigerian schist belts since 1913 with about 12 t. produced (Woakes, 1989). With the recent renewed investment in mechanized gold-mining in Nigeria, production should rise and include the mining of primary gold occurring in quartz veins in the basement rocks. Woakes (1989) and Wright et al. (1985) observed that gold mining has been restricted to the Maru, Anka, Ilesha and Egbe schist belts (Fig.6.24) where gold is found in veins, stringers, lenses, reefs and similar bodies of quartz, quartz-feldspar and quartz-tourmaline rocks, in both the supracrustals and in the migmatite-gneisses. Primary gold veins are small ranging in thickness from several cm to a few m, with lengths of up to several hundred m. The veins sometimes have pinch-and-swell structures, steep dips and parallel or en echelon arrangements. They are often concordant and associated with fractures and shear zones (Elueze, 1986); some veins are cross-cutting. Placer gold was probably con-centrated mostly from disseminated pods and veinlets with low gold content.

In the Beninian belt gold is associated with Pan-African thrust zones. The largest deposit is at Pemba in northern Benin where mining has been carried out.

Chromite. This occurs in small serpentinite bodies in the Beninian fold belt. Other occurrences are in the basement to the east, where in northwestern Nigeria a linear belt with serpentinite bodies trends NE-SW along a major shear zone (Fig.6.24). These bodies are also potential sources of asbestos, talc, magnesite and nickel (Elueze, 1982).

Pegmatite Mineralization. Tin-tantalum-niobium-bearing pegmatites occur in a well defined ENE-WSW belt 400 km long, stretching from the Jos Plateau through the Wamba-Jemaa area (just south of Jos Plateau) through the Egbe schist belt, to the Ilesha area in southwest Nigeria (Fig.6.24). These pegmatites were emplaced between 580 Ma and 530 Ma, more or less conformably into their host rocks. Mineralization in these pegmatites was apparently influenced by the host rock lithology, in which case the pegmatites in southwest Nigeria emplaced into meta-sedimentary schist belts, are enriched in tantalum relative to niobium, whereas those emplaced into gneisses on the Jos Plateau are enriched in tin (Matheis, 1987). Pegmatites rich in Li, Sn, Nb, Ta, Cs also occur in the Oban massif, in association with tin which was mined in colonial times. Mineralized pegmatites provide nearly all the tantalum production in Nigeria, which averages about 5 t. per annum, but not exceeding 20 t. annual production. Small amounts of Nigerian tin and niobium production are from pegmatites, the bulk of these coming from the Younger Granites of the Jos Plateau.

Nigerian mineralized pegmatites are commonly massive bodies with pronounced pinch-and-swell structures, showing intense albitization and rich mineralization in the swelling. The non-mineralized or barren pegmatites are dominated by quartz and microline, often accompanied by varying amounts of oligoclase, biotite, muscovite and tourmaline. Mineralization is generally associated with late-stage Na-rich hydrothermal solutions that produced secondary albitization of the feldspars or quartz-mica greisens. Apart from cassiterite columbite and tantalite which are the main economic minerals, the mineralized pegmatites also carry a host of accessory minerals, including scheelite, beryl, apatite, monazite, the Li-rich mica lepidolite, black and pink-green tourmaline and gem-quality blue gahnite (Zn-rich spinel). Matheis (1987) pointed out that although Nigerian rare-metal-bearing pegmatites were emplaced simultaneously with the Older Granites during the last phase of the Pan-African event, there is no genetic link between the mineralized pegmatites and the Older Granites (Fig.6.29,C). Rather, repeated crustal reactivation probably caused greater re-cycling and concentration of elements in the basement and supracrustals from which the late-stage Na-rich hydrothermal fluids derived their enrichment.

Uranium. This occurs in the Pan-African granitoids of the Poli area in northwestern Cameroon (Fig.6.33). There are also uranium prospects in the Tessalit region of western Adrar des Iforas in Mali.

Iron Ore. Iron ore is of economic importance in some Pan-African schist belts in western Nigeria and in the Kribi region of southern Cameroon. In the latter region the grades are about 40 % Fe, with 70 % attained where there is secondary hematite enrichment. Banded ironstones also occur in the Buem and Atakora Formations in Togo; near Bandjeli 90 million t. of ore are known, whereas around Dako magnetite-hematite ores are associated with quartzite and mica schist with about 40 - 45 % Fe.

In the Maru schist belt in northwestern Nigeria quartz-hematite rocks with magnetite, associated with garnet-grunerite-schist, amphibolites, phyllites and quartzite, contain up to 40 % Fe. Richer and purer itabirite-type ores in the Older Metasediments interbedded among basement gneisses, occur as prominent ridges near Okene, southeast of Kabba and at Itakpe Hill, where iron ore mining has started. The Itakpe Hill has the form of an isoclinal fold oriented WNW-ESE and closing to the south-east, with an overall southerly dip (Olade, 1978). The ores are massive and banded, with magnetite and hematite as the main ore minerals. Proven reserves are at least 150 million t. at 35 - 50 % Fe. It will feed the Ajeokuta steel complex nearby.

Industrial Minerals. Among the commercial products from the Pan-African belts are the marble deposit at Jakura in southwestern Nigeria. There are also favourable prospects for kaolin, kyanite and sillimanite in this region.

6.5 South Atlantic Mobile Belts

The West Congolian, Damara, Gariep and Saldanhia orogens are all strung out along the South Atlantic margin of Africa (Fig.6.1), over a distance of 3,500 km. On a pre-Mesozoic drift reconstruction of western Gondwana, the West Congolian and Damara belts are continuous with the Brazilian Mantiqueira and Ribeira belts respectively (Fig.6.1). The continuity of the Pan-African-Brasiliano orogenic belts of southwestern Africa and eastern Brazil attests to their common geodynamic history which relates to the opening and closing of a Late Proterozoic proto-South Atlantic Ocean (Porada, 1989; Torquato and Cordani, 1981).

The fact that rift-related sedimentary and magmatic rocks in the lower part of the Pan-African-Brasiliano orogens are succeeded by fully marine strata suggests that these orogens originated from the rifting of a continental shield, followed by sea-floor spreading, and collision. As shown by Porada (1989) the original shield in this case was located in western Gondwana. The rifting of west Gondwana and the evolution of the proto-South Atlantic Ocean foreshadowed the subsequent development of the present South Atlantic which is aligned along the same persistent zone of Pan-African crustal weakness. As in the Mesozoic during which the South Atlantic opened progressively from south to north, the proto-South Atlantic may have followed a similar path, for it remained closed in the northern part where Africa and South America remained united along the São Francisco-Zaire cratonic bridge (Fig.6.1). The proto-South Atlantic evolved through a complete Wilson Cycle leaving behind a chain of orogens the West Congolian, Damara, Gariep, and Saldanhia belts. The latter belt, however, belonged more to a southern ocean, the Admaster Ocean which stretched eastward from Antarctica to Australia (Hartnady et al., 1985).

6.5.1 The West Congolian Orogen

Lithostratigraphy

The West Congolian mobile belt extends for over 1,300 km from Gabon southwards through Congo, Zaire, to northern Angola (Fig.6.36). It contains

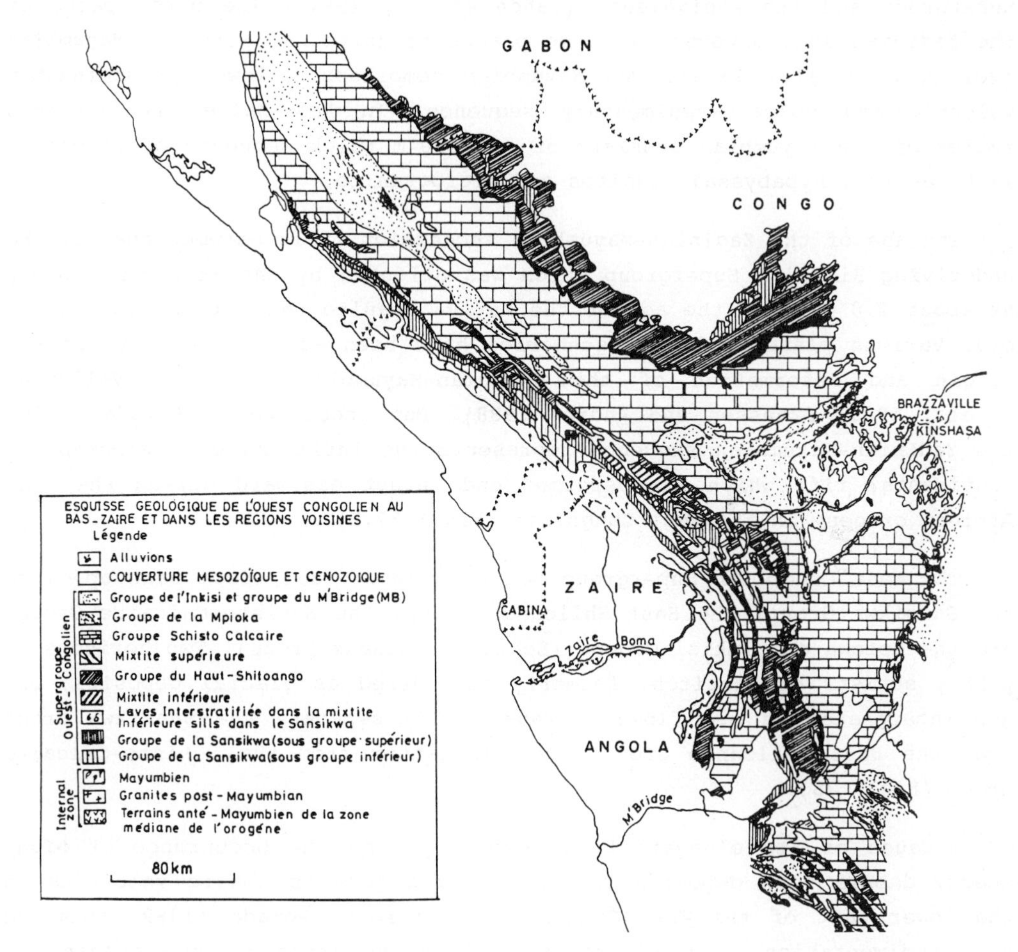

Figure 6.36: Geological map of the West Congolian orogen. (Redrawn from Cahen, 1978.)

three structural zones (Cahen et al., 1984; Porada, 1989), comprising from east to west, the external zone with subhorizontal strata; the median folded zone; and the internal zone to the west which, except for intrusive rocks, consists of pre-West Congolian basement rocks (Fig.6.37,A). The external and median zones contain the West Congolian Supergroup, a sequence of low-grade metasediments. Since older rocks considered are exposed in the internal zone, it is therefore appropriate to start with a description of that zone.

The older rocks in the internal zone are the Mayumbian and the Zadinian Supergroups (Fig.6.38). Throughout most of the West Congolian orogen the Mayumbian Supergroup unconformably overlies post-Eburnean-Middle Proterozoic metasediments and meta-volcanics, and the Zadinian Supergroup and its equivalents (Cahen et al., 1984). The upper parts of the Zadinian are, however, in some places transitional with the Mayumbian (Vellutini et al., 1983). The Mayumbian comprises a lower predominantly volcanic and volcano-sedimentary sequence. In Bas Zaire the volcanic facies of the Mayumbian consists of a 3,000-m thick sequence of rhyolites with dacites, hypabyssal granites and microgranite.

The age of the Zadinian-Mayumbian succession falls between that of the underlying Kimezian Supergroup which was affected by the Eburnean orogeny at about 2.0 Ga, and the age of the Mativa granite which is about 1.02 Ga old. Various plate collision scenarios have been advanced to explain the origin and deformation of the Zadinian-Mayumbian sequence (Vellutini et al., 1983; Franssen and André, 1988). But according to Porada (1989) the Zadinian-Mayumbian sequence represents the infilling of a Kibaran-age continental rift which was deformed and thrust eastward during the Pan-African orogeny in the West Congolian (Fig.6.37,B).

The West Congolian Supergroup has been subdivided, from base up, into the Sansikwa Group, the Haut Shiloango Group, the Schisto-Calcaire Group, and the Mpioka and Inkisi Groups (Schisto-Gresseux Group). Two horizons of pebbly schists or mixtites, formerly considered as glacial deposits, but now interpreted as mud flows (Porada, 1989) separate the Sansikwa Group from the Haut Shiloango Group, and the latter from the Schisto-Calcaire Group (Fig.6.38).

Because of its elongate basin geometry, and the occurrence of high-energy debris flow deposits (mixtites), red beds and basic volcanics in the lower part of the West Congolian Supergroup, Porada (1989) surmised that sedimentation in the external and median zones of the orogen was initiated in a fault-bounded continental rift, which he termed the West Congolian rift. The West Congolian rift was bounded to the west by an earlier rift structure of Kibaran age within which the volcanic and sedimentary rocks of the Zadinian and Mayumbian Supergroups accumulated before being intruded by anorogenic end-Mayumbian granites of Kibaran age. By the onset of West Congolian rifting and deposition, the Kibaran rift was already a positive feature supplying the basal coarse clastics of the Sansikwa Group.

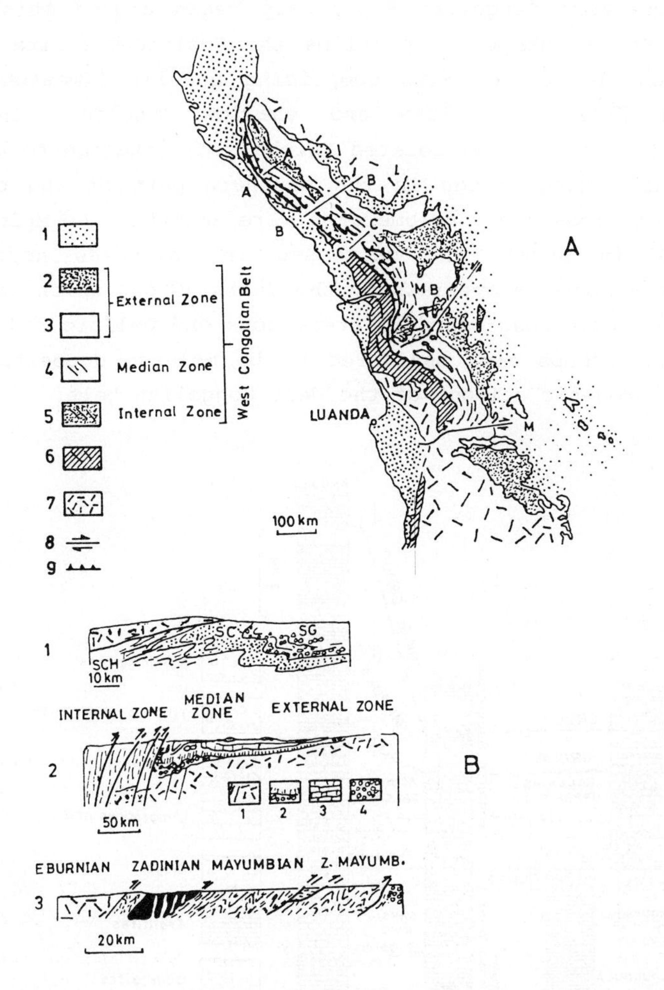

Figure 6.37: A, Tectonic map of the West Congolian orogen. 1, Paleozoic - Recent; 2, "Schisto-Greseux" Group; 3, "Schisto-Calcaire" Group; 4, "Schisto-Calcaire", Haut Shiloango, "Serie de la Louilla" and Sansikwa Groups of the median zone; 5, Mayumbian Supergroup; 6, Kimezian Supergroup; 7, undifferentiated basement; 8, transcurrent fault; 9, thrust zone. B, cross-sections; 1, Mayumbian and basement; 2, Sansikwa and Haut Shiloango Groups; 3, "Schisto-Calcaire"; SG, Schisto-Greuseux"; SCH, schists. (Redrawn from Porada, 1989.)

The Sansikwa Group is mostly a clastic sequence with subordinate carbonates; the Haut Shiloango contains mainly argillites; whereas the

Schisto-Calcaire Group marks the predominance of carbonate deposition which was followed by the in-pouring of the terrigenous sediments of the Schisto-Gresseux Group. Based on the age of the basic volcanic rocks in the Bouenza Series, the eastern equivalent of the Haut Shiloango Group (Cahen, 1978), which has an age of 1.11 Ga, it is believed that rifting and deposition of the West Congolian Supergroup began around this time during the Kibaran. The Mpioka Group overlies the Schisto-Calcaire Group unconformably with conglomeratic units comprising angular limestone and chert clasts which fill depressions and shallow synclines in the underlying carbonates. Cahen (1978) related Mpioka sedimentation to uplift of the M'bridge-Mbanza Ngungu ridge in the southern part of the orogen (Fig.6.36), from which limestone conglomerates were derived. The Mpioka is about 4,000 m thick in Angola and decreases in thickness northward (Fig.6.38). It is unconformably overlain by the Inkisi Group which is over 1,200 m thick and comprises conglomeratic arenaceous and pelitic red beds. The Mpioka and Inkisi Groups are considered to be molasse deposits laid down after the first orogenic episode of the West Congolian belt.

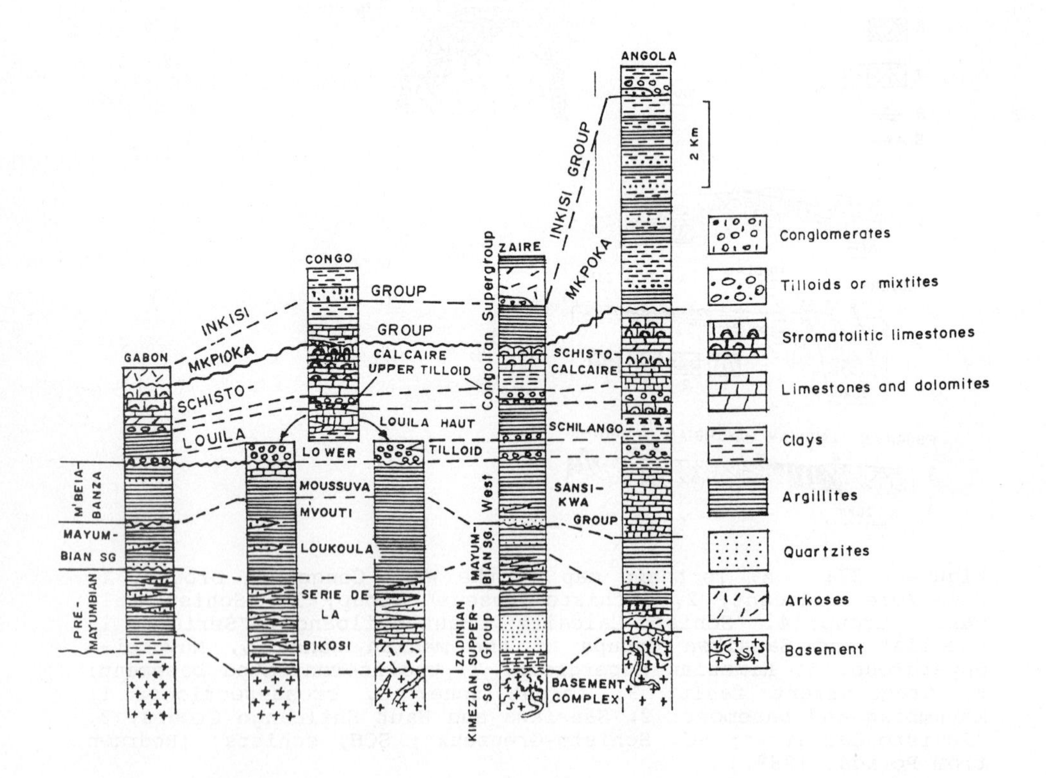

Figure 6.38: Stratigraphic correlations in the West Congolian orogen. (Redrawn from Vellutini et al., 1983.)

Tectonism

Deformation and metamorphism in the West Congolian increases from east to west and was in two stages, D1 and D2 (Porada, 1989). The D1 event produced the first cleavage and major fold structures in the median zone; thrusts, isoclinal folds and shear zones developed in the internal zone. The internal zone and the Eburnean basement granites and gneisses to the west were thrust northeasterly over the overturned sequences in the median zone (Fig.6.37,B). Dated at about 734 Ma (Cahen et al., 1984), D1 also caused the rehomogenization of older rocks. During D2, the D1 structures were coaxially refolded in the internal zone and locally pronounced crenulation cleavage and medium-to small-scale southwest verging folds were produced in the median zone as a result of back-folding (Coward, 1981). The second deformation phase has been dated from 625 Ma to 536 Ma, from the affected granites and migmatites (Cahen et al., 1984).

In the external zone metamorphism is at low-to very low-grade, but increases westward to the greenschist facies in the median zone and to the low-pressure amphibolite facies in the internal zone where pressure-temperature conditions have been estimated at 550° - 600°C at 2 - 3 Kbar (Franssen and André, 1988). In the median and external zones of the West Congolian magmatism was limited to sills and dykes which intruded the Sansikwa Group (Cahen, 1978). These intrusives are the feeders for the basaltic pillow lavas which accompany the lower mixtite. As already stated similar, but highly altered sills, as well as younger and fresher sills occur in the lower part of the Bouenza Series, near the du Chaillu foreland massif in Congo.

Based on the asymmetrical structure (Fig.6.37) and the metamorphism in the West Congolian belt, Porada (1989) suggested that plate convergence and collision probably took place west of the internal zone. However, the occurrence of ophiolites along the western margin of the internal zone is still a subject of debate (Vellutini et al., 1983; Franssen and André, 1988).

6.5.2 The Damara Orogen

Structural Framework

The Damara orogen is a major Pan-African belt located along the coast of Namibia. It comprises a northeast-trending intracontinental branch, 400 - 500 km wide, between the Zaire and Kalahari cratons; and a coastal or Kaoko branch which extends northwards into Angola and towards the West

Congolian belt (Fig.6.39). While the intracontinental branch is on trend with the Lufilian arc (Fig.6.1), the Kaoko branch along the coast and the Gariep belt align parallel to the Ribeira belt of Brazil and Uruguay. In view of the stratigraphic and structural continuity between the intracontinental branch and the Kaoko belt, both are considered as part of the same Damara orogen.

The Damara orogen originated through continental rifting between 1.0 Ga and 700 Ma (Porada, 1989). The existence of rifts at the beginning of the orogen (Fig.6.39,A) was postulated (e. g. Martin, 1983; Porada, 1983) based on lithofacies; the general sediment thickness distribution; the location of pre-Damara basement inliers; the location of hinge lines;

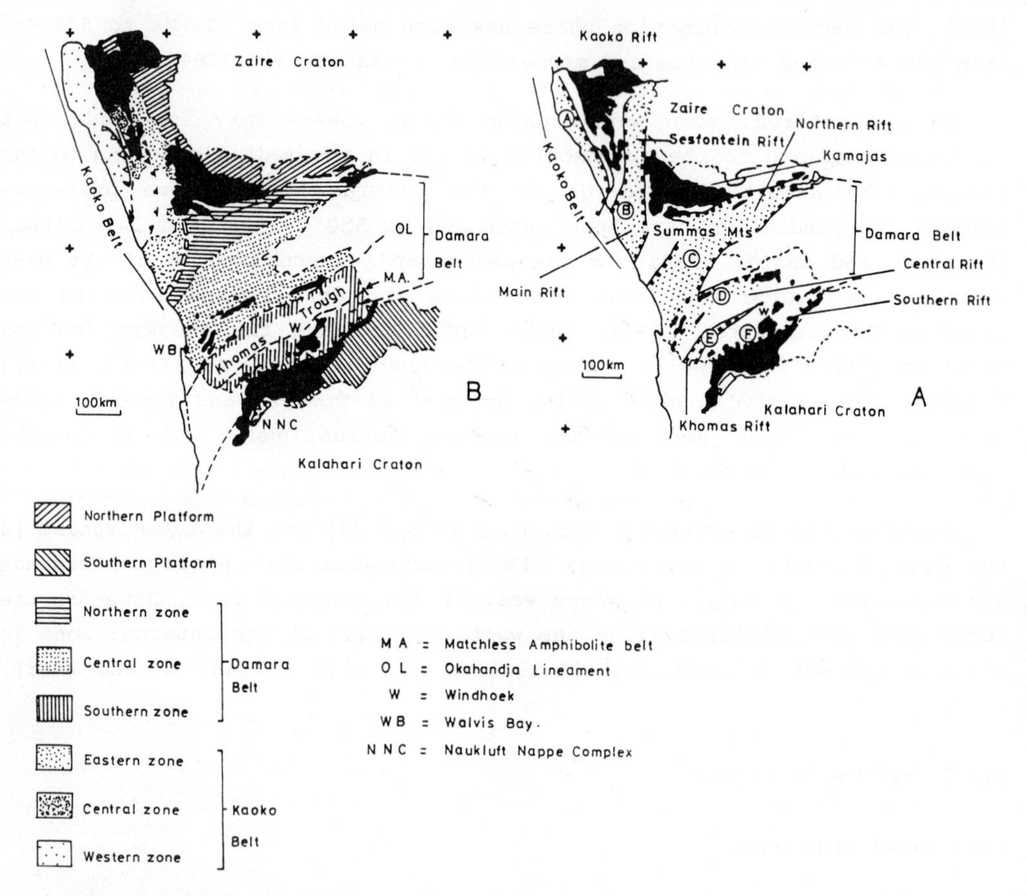

Figure 6.39: Tectonic subdivisions of the Damara orogen. (Redrawn from Porada, 1983.)

and the distribution of rift-related acid to alkaline volcanics. The intracontinental branch has been subdivided into three separate rift zones, each 50 - 70 km wide and over 200 km long. There is a Northern zone (graben), and Central and Southern zones which have been interpreted as half grabens. The latter zones are located on both sides of what looks like a fourth rift, the Khomas trough (Fig.6.39,B).

In the northern Kaoko belt, similarly, there is a north-south-trending Eastern zone, known as the Sesfontein graben, also about 50 km wide, bounded to the east by the Kamanjab basement inlier and to the west by a buried basement block. A Western rift zone, otherwise known as the Kaoko rift has been inferred based on pronounced westward sediment thickening. The above rift distribution has imposed a generally asymmetric morphology on the Damara orogen.

Rift Sedimentation and Volcanism

The Damara rifts were initially filled with terrestrial, mainly fluviatile and locally lacustrine or playa-like sediments, and acid to alkaline volcanics (Porada, 1983). These constitute the Nosib Group (Table 6.1) of the Damara Supergroup. Throughout much of the Damara orogen, Eburnean and Kibaran metasediments, gneisses and granitoids furnished the basement and elevated sediment source areas to the Nosib Group (Tankard et al., 1982)

In the Central zone thickness variations in the basal Etusis Formation and the distribution of conglomerate lenses, first-cycle sediments and fluvial erosion channels reflect the influence of fault-bounded blocks during the phase of rift sedimentation. Along the northern margin of the Northern graben a variety of mostly acid to alkaline ignimbrites and rhyodacites, consanguineous intrusives and agglomerates, belong to the Naauwport Formation which is over 6,000 m thick. The Naauwport Formation is associated with a large rift fault, the Summas fault, which marks the southern edge of the Kamanjab inlier (Fig.6.39,A). The Naauwport volcanics were extruded during rifting and are completely missing north of the Summas fault where pre-Damara basement is exposed.

In spite of intense deformation involving thrusting, the depositional environment in the early Damara rifts is still evident in the Southern zone. Here the Kamtsas Formation, a sequence of fluviatile fan and intermittent stream deposits such as quartzites and conglomerates, interfingers with playa sediments (pelites and siltstones) of the partly calcareous Duruchaus Formation. The playa origin of the Duruchaus Formation is based on sedimentary evidence such as cyclic sedimentation, fine bedding, desiccation cracks, and on the presence of large crystal pseudomorphs after evaporites (e.g. borax, trona, gypsum); and on geochemical evidence in the

Table 6.1: Stratigraphic correlations in the Damara orogen. (Redrawn from Porada, 1983.)

NORTH

GROUP	SUBGROUP	FORMATION	LITHOLOGY (MAX. THICKNESS)
MULDEN		OWAMBO 1000m	Shale, marl, siltstone, sandstone
		KOMBAT	Shale dolomite lenses
		TSCHUDI 3Km	Quartzite, conglomerate arkose, argillite
UNCONFORMITY IN NW			
OTAVI	TSUMEB	HÜTTENBERG 900m	Dolomite with chert, shale limestone, stromatolites, oolites
		ELANDSHOEK 1.1 Km	Dolomite with chert stromatolites
		MAIEBERG 950m	Dolomite, limestone, slump breccia
		CHUOS 700m	Mixtite, dolomite, limestone sandstone, itabirite, oolite
	LOCAL DISCORDANCE		
	ABENAB	AUROS 450 m	Dolomite, limestone, marl, shale, stromatolites
		GAUSS 750 m	Dolomite limestone oolitic chert sandstone
		BERGAUKAS 525 m	Dolomite, limestone stromatolites, arkose, greywacke
LOCAL DISCORDANCE			
NOSIB		VARIANTO	Mixtite, tuff, itabirite
		ASKEVOLD NAAUWPOORT 6Km	Rhyolite, tuff, agglomerate, andesite, epidosite bostonite
		NABIS	Quartzite, arkose, conglomerate

CENTRE

GROUP	SUBGROUP	FORMATION	LITHOLOGY (MAX. THICKNESS)
SWAKOP	KHOMAS	KUISEB 3 Km	Quartz-biotite schist, biotite garnet-cordierite schist, amphibole schist, quartzite, marble, calcsilicate rock
		KARIBIB 700 m	Marble, biotite schist, quartz schist, calcsilicate rock
		CHUOS 700m	Mixtite, marble, quartzite
	LOCAL DISCORDANCE		
	UGAB	ROSSING 700m	Marble, quartzite, conglomerate, biotite schist, biolite-hornblende schist, calcsilicate rock
LOCAL DISCORDANCE			
NOSIB		KHAN 1.1 Km	Calcsilicate rock, amphibole-pyroxene gneiss and quartzite
		ETUSIS 3.5Km	Quartzite, arkose, conglomerate, schist, rhyolite

SOUTH

GROUP	SUBGROUP	FORMATION	LITHOLOGY (MAX THICKNESS)
PARTLY EQUIVALENT TO NAMA GROUP			
SWAKOP	KHOMAS	KUISEB 10 Km	Biotite schist, biotite quartzite, graphitic schist, calcsilicate rock, amphibolite (matchless member)
		AUAS 1.8 Km	Quartzite, schist, marble amphibolite, itabirite
		CHUOS 1.65 Km	Pebbles schist, mixtite, quartzite, schist itabirite amphibolite, calcsilicate rock
	KUDIS	BLAUKRANS 1.7Km	Graphite schist, quartzite quartz-mica schist, conglomerate itabirite
		? HAKOS 2 Km	Quartzite schist
		CORONA 400m	Dolomite schist, coglomerate
LOCAL DISCORDANCE			
NOSIB		DURUCHAUS 5 Km	Phyllite, quartzite, conglomerate limestone
		KAMTSAS 6.7 Km	Quartzite, arkose, conglomerate

form of high K, Na, Li and B which suggests extreme chemical concentrations.

Mafic and felsic volcanics in the Kibaran Sinclair Group along the southern margin of the Damara belt (Fig.6.40,A) are also significant and relevant. The Sinclair Group consists of thick acid and andesitic lavas with interbedded conglomerates, quartzites, argillaceous sandstones, and rare limestones (Cahen et al., 1984). The Sinclair deposits are intruded by large amounts of high-level granites, rhyolites and basic-to-intermediate rocks. Apart from being emplaced on and in the basement along the southern margin of the Damara orogen, probably in a rift setting, the volcanics of the Sinclair Group which have been dated between 1.25 and 1.0 Ga are intercalated among rift sediments in the Southern zone of the Damara belt.

In the Kaoko belt, the Sesfontein graben shows the Nosib Group varying in thickness along strike from about 1,000 m on the western part of the graben to 5,000 m in the northern part. But on the Kamanjab basement inlier bordering the rift on the eastern side, the Nosib Group is either completely missing or greatly reduced in thickness. Westward, outside the Sesfontein rift, the Nosib is also missing or very thin; thus suggesting the presence of another basement high to the west. Along the western flank of the Sesfontein graben are bimodal acid and basaltic meta-volcanics and plutons which were associated with early rifting. Further north, the Chela Group of southern Angola with its abundant acid volcaniclastics and volcanics are at least partially equivalent to the Nosib Group (Kröner and Correia, 1980).

Regional Subsidence and Marine Transgressions

The rifts of the Damara orogen show two phases of subsidence which were accompanied by marine encroachments from the west (Fig.6.40,B), and clastic sediment transport from the east (Porada, 1983). The first marine transgression was more extensive in the northern rift zones where in the Sesfontein graben, and in the adjoining areas of the Northern platform (Fig.6.39,B), the dolomites and light-coloured shales of the Abenab Subgroup of the Otavi Group were deposited (Table 6.1). Because of differential subsidence the Abenab thickens from 0 - 30 m on the flanks of the graben to 500 - 700 m in the Sesfontein graben. The wide extent of the equivalent Ugab Subgroup in the Northern zone also suggests rapid subsidence in this region. However, in the Northern rift marine influence came earlier in the disconformably underlying Khan Formation at the top of the Nosib Group (Table 6.1;Fig.6.40,B). During this early phase of subsidence a carbonate shelf-to-basin setting (Fig.6.41) existed in the Northern

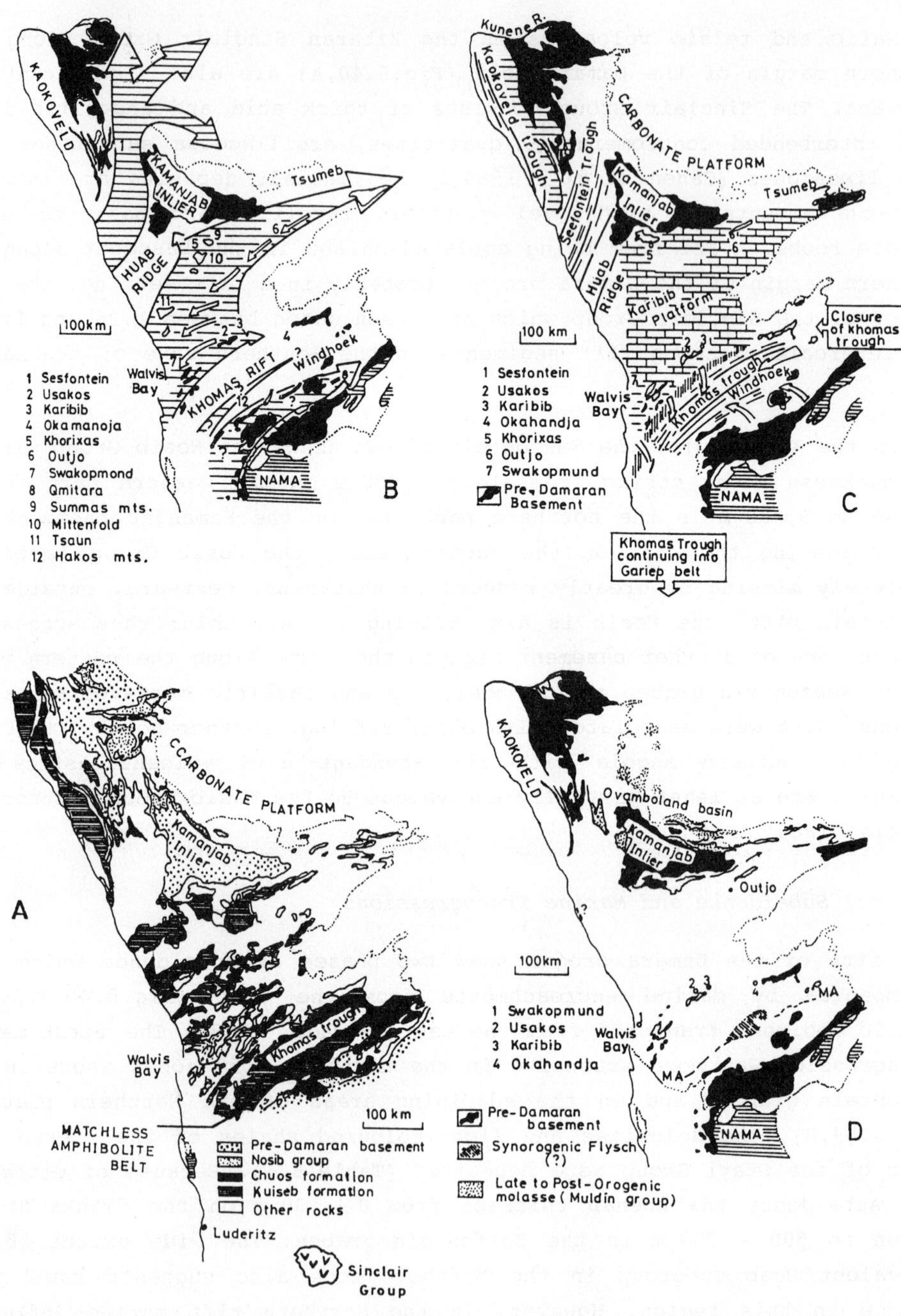

Figure 6.40: Sketch maps showing the depositional framework of the Damara belt. (Redrawn from Porada, 1983.)

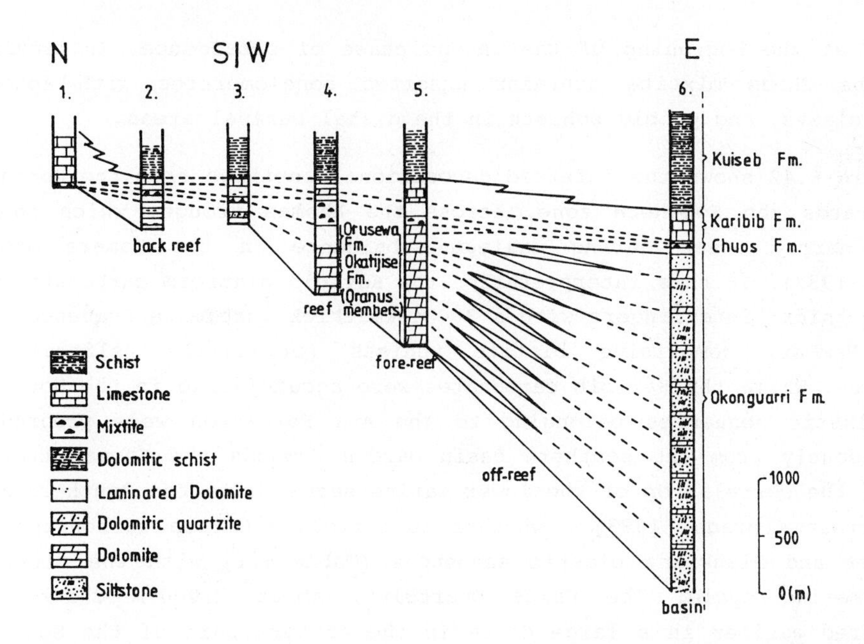

Figure 6.41: Facies relations in the Ugab Subgroup. (Redrawn from Porada, 1983.)

zone in which well developed back-reef, fore-reef, off-reef, and turbidite facies (Okonguarri Formation) accumulated. The Okunguarri turbidite formation contains siltstones and quartzite turbidites intercalated with calcarenites and rudites which originated from the reefs which were situated northward along the shelf edge of the Northern rift.

The next phase of subsidence and marine transgression was more profound than the earlier one. It altered the pre-existing paleogeography and basin configuration and created a wide basin between the Zaire and the Kalahari cratons. The shelf-to-turbidite basin in the Northern zone filled up and together with the Central zone became the Karibib carbonate platform (Fig.6.40,C) which continued southward towards a new and rapidly subsiding turbidite basin known as the Khomas trough. Northward the Karibib carbonate platform was bordered by another turbidite basin in the Sesfontein trough. Also, a northern carbonate platform was situated along the southern flank of the Zaire craton throughout the depositional phase of the Damara orogen. A western carbonate platform separated the Sesfontein trough from the more westerly Kaoko trough (Fig.6.40,C). A regional mixtite, the Chous Formation, underlies the Karibib and its equivalent, the Tsumeb Subgroup. Originally regarded as a tillite, the Chous Formation was reinterpreted (Porada, 1983) as slump-derived mass mud flow deposits which were triggered by rapid subsidence and syn-sedimentary faulting at the beginning of the second phase of subsidence. In proximal areas the Chous mixtite contains unsorted conglomerates with locally derived clasts, and pebbly schists in the distal basinal areas.

Figure 6.42 shows the inferred depositional settings from the Northern zone towards the Southern zone, across the Khomas trough, which opened into a narrow ocean during maximum subsidence in the Damara orogen (Miller, 1983). In this interpretation the Karibib platform carbonates, up to 700 m thick, interfingers with a 3 - 4-km thick turbidite sequence, the Tinkas Member, comprising biotite schists (originally pelites) and siltstones. While the Karibib carbonates were accumulating in the Northern zone, clastic sequences belonging to the Aus Formation were prograding simultaneously from the southern basin margin towards the Khomas trough. However, the correlation of the lower marine sequence in the Southern zone is uncertain (Porada, 1983) - whether to correlate Corona carbonates and the Hakos and Blaukrans clastic sequences (Table 6.1) with the first or second marine cycle. The Hakos Quartzite, about 2,000 m thick, had accumulated earlier in a large delta in the eastern part of the Southern zone. The Chausib turbidites in the Southern zone are the equivalents of the Tinkas turbidites in the Central zone.

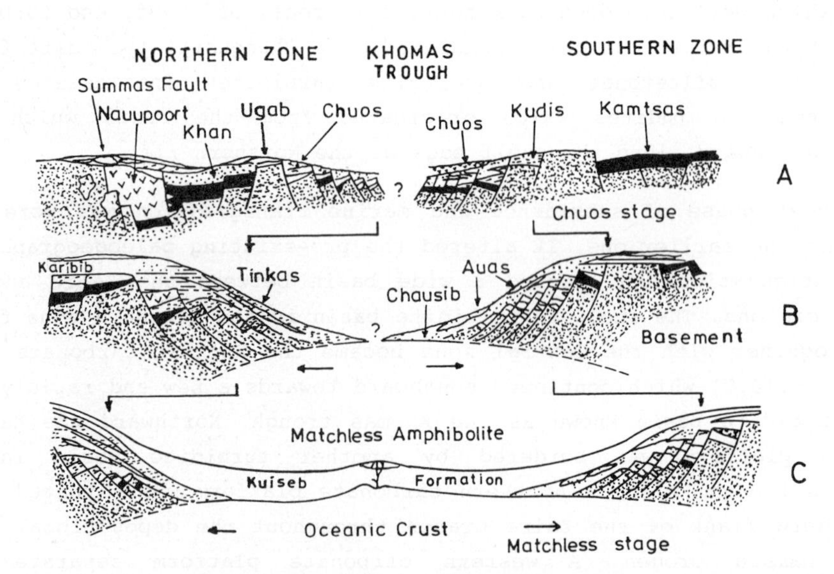

Figure 6.42: Cross-sections suggesting depositional settings for the Damara Supergroup. (Redrawn from Porada, 1983.)

With pronounced subsidence the Khomas trough extended into a deep basin in which about 10 km of pelites and quartzites (Kuiseb schist) accumulated (Fig.6.40,C). Near the top of the Kuiseb are tholeiitic basaltic rocks, the Matchless Amphibolites (Fig.6.42,C). The chemical characteristics of the Matchless Amphibolites suggest ocean-floor basalts; also, pillow structures and associated meta-gabbros and ultramafic chlorite and talc schist affirm oceanic origin; and support the view that an ocean arm opened along the axis of the Khomas trough (Fig.6.42,C) (Miller, 1983)), as was postulated earlier (Fig.6.35) by Shackleton (1976).

In the Kaoko belt the Tsumeb carbonates accumulated on a geanticline which separated the Sesfontein and the Kaoko troughs, while over 4 km of semi-pelitic schistose graywackes (equivalent to the Kuiseb), and rhyolites and feldspar porphyry were deposited in the Kaoko trough. Eventually the geanticline was overstepped eastward by the graywackes, whereas the Sesfontein trough which by then had filled up, became the site for the accumulation of the Tsumeb platform carbonates (Table 6.1). The widespread distribution of the Kuiseb Formation (Fig.6.40,A) implies regional submergence of carbonate platforms, the cessation of turbidite sedimentation, and the overstep of the Kuiseb onto erstwhile shelf and platform areas throughout the orogen. It also implies a tremendous supply of clastic sediments to fill the entire orogen. Syn-orogenic flysch 3 to 4 km thick occurs in the Khomas trough in the upper part of the Kuiseb Formation above the Matchless Amphibolites (Fig.6.40,D). The Kuiseb flysch is distinguishable from the underlying schists by the absence of beds and lenses of calc-silicate rocks which are typical of the Kuiseb schist; and by the absence in the flysch of kyanite, staurolite and andalusite which are found in the underlying Kuiseb.

Following Pan-African deformation the Northern platform became the site of a molasse basin, the Ovamboland basin (Fig.6.40,D) where the Mulden Group, about 4 - 5 km thick, was deposited. The Mulden comprises conglomerates and arkosic sandstones along the proximal part of the molasse basin, whereas the more distal and deeper parts of the basin were filled with siltstone and shale with a few stromatolitic carbonate layers. The Nama Group on the Southern platform is considered as another molasse sequence (Porada, 1983).

Tectonism

A recent account of the Pan-African orogeny in the Damara belt (Porada, 1989) shows that the orogeny was polyphase and lasted from about 650 Ma to about 480 Ma. In the Kaoko belt three main deformation events have been recognized, the first, (D1) which probably resulted from plate convergence (between Africa and South America) west of the present coast of Namibia, produced a bedding-parallel S1 foliation in the Kaoko rift. This was followed by the emplacement of 650 Ma old granitic rocks, uplift and erosion and by the deposition of the Mulden molasse. The D2 event was probably caused by the collision of the African and South American plates; it produced large-scale recumbent folds; and nappes formed by the Kuiseb Subgroup were thrust eastward over the Otavi Group carbonates and Mulden molasse (Fig.6.43,A). Granites dated at 590 Ma were emplaced following the D2 event, whereas a final D3 event caused west-vergent back-folding before the intrusion of other granites at about 570 Ma.

Deformation in the Central zone was also polyphase producing D1 and D2 structures which are characterized by bedding-parallel foliation, thrusting, and recumbent folding which appear to be related both to continental convergence and collision in the Kaoko belt, and to subduction in the Khomas trough (Porada, 1989). Post-D1 intrusive dioritic to granitic rocks in the Central zone have been dated at 650 Ma, whereas widespread D2 syn-tectonic granitic suites, the Salem granites of the Central zone are dated at 570 - 540 Ma.

In the Khomas trough continental collision occurred around 540 Ma which progressively deformed the sequence in this trough until about 480 Ma. This produced five successive deformation phases including thrusts onto the southern cratonic foreland (Fig.6.43,B). The collision caused D3 deformation in the Central zone, with characteristic upright folding, large-scale doming and concomitant syn-tectonic granitic intrusions and the partial melting of basement and sediment cover (Porada, 1989). The Southern zone preserves the deepest parts of the Khomas trough where the Kuiseb meta-pelites have been greatly thickened by recumbent folding and imbricate thrusting (Fig.6.43,B). The Okahandja lineament (Fig.6.43,C) marks the northern margin of the Southern zone along which the Donkerhoek granite was intruded at about 520 - 500 Ma (Tankard et al., 1982). A major feature in the evolution of the northeastern arm of the Damaran is the evidence for orogen parallel displacements as outlined by Coward (1983).

Metamorphism is at the greenschist facies in the northern part of the Damara orogen, but increases to high-temperature, medium pressure-temperature conditions in the Khomas trough. The metamorphic grade again decreases towards the Southern platform.

Mineralization

Base metals, mineralized pegmatites, and uranium are the principal types of deposits in the Damara belt of Namibia (Fig.6.44,A). Stratiform

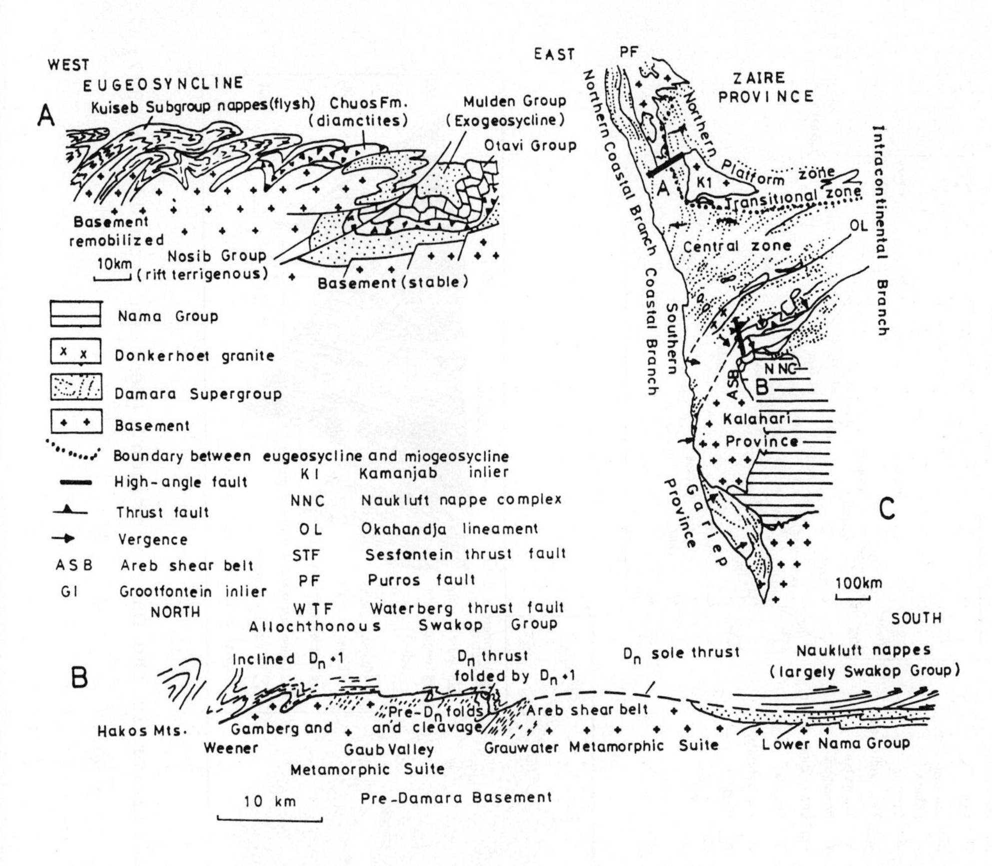

Figure 6.43: Structural styles in the Damara orogen. (Redrawn from Tankard et al., 1982.)

copper and iron sulphides and native silver are produced from the Oamites mine and elsewhere (Fig.6.44,A), where mineralization occurs in the upper part of the Nosib Group at the contact between the continental red beds and the marine beds (Tankard et al., 1982). Lead-zinc mineralization of the Mississippi Valley-type is hosted by the dolomitic limestones of the Otavi Group in the north, near Tsumeb and Aukas.

In Namibia pegmatite mineralization is extensively associated with the metasediments and Pan-African granitoid rocks (550 - 470 Ma old) of the Damara orogen (von Knorring and Condliffe, 1987). Figure 6.44,A shows that pegmatitic tin and tantalum deposits are restricted to three well defined, NE-SW-trending belts, each over 100 km long. At Uis a gigantic pegmatite, up to 1,000 m long and about 100 m wide is exploited, with columbite-tantalite as by-product. The pegmatites appear to be zoned, coarse to very coarse-grained bodies with microline, albite, muscovite and quartz. The accessory minerals in the Namibian tin belts include beryl and tourmaline.

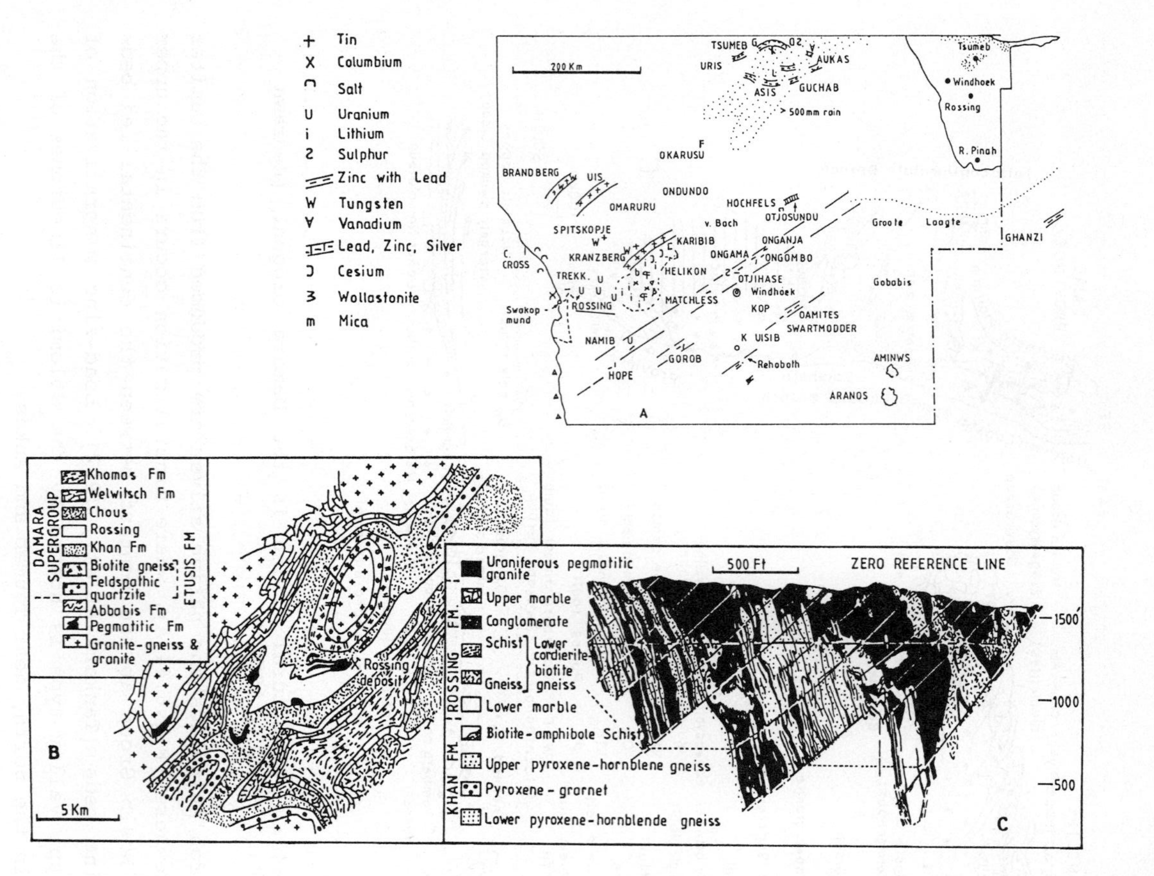

Figure 6.44: Mineralization in the Damara orogen. (Redrawn from de Kun 1987; Sawkins, 1990.)

The southern tin belt is characterized by wall-like pegmatite dykes which intrude the Damara metasediments from which they weather out in the desert plain as prominent wall-like features which can be followed over great distances.

Mineralized pegmatites in the Karibib area have for over 50 years yielded a host of rare-element minerals such as amblygonite-montebrasite, lepidolite, petalite, pollucite, beryllium-niobium-tantalum and bismuth minerals in addition to industrial mica, ceramic feldspar, quartz and a great variety of attractive gemstones. Here the larger and most productive lithium pegmatites are known as Rubicon and Helicon. Gem-quality tourmaline occurs in a zoned pegmatite in the Karibib-Usakos area.

Sawkins (1990) discussed the uranium deposit at Rossing (Fig. 6.44, A) as an example of vein-type uranium deposits that are associated with collisional (S-type) granites. The Rossing uranium deposit was believed to be the largest uranium producer outside the former communist block. As shown by Sawkins (1990) this deposit lies in the central, high-grade metamorphic zone of the Damara belt, where anatectic granites have intruded remobilized early Precambrian basement (Abbabis Formation) and granitized sedimentary rocks and possibly volcanics belonging to the Etusis and Khan Formations (6.44, B). The uranium mineralization occurs in a migmatite zone characterized by largely concordant relationships between uranium-bearing pegmatitic granites and the gneisses, schists, and marbles of the Khan and Rossing Formations (Fig. 6.44, C). Termed alaskites, the granitic rocks which contain the uranium display a broad range of textures, from aplitic to granitic and pegmatitic. The alaskite intrusions vary from large discordant intrusions to thin conformable dykes, typically disposed of in closely spaced arrays.

Most of the alaskite in the Rossing area are unmineralized or weakly mineralized, and uranium of economic grade is concentrated where the alaskite was emplaced into a garnet gneiss-amphibolite unit (northern ore zone) or into the amphibole-biotite schist, lower marble, and lower cordierite gneiss sequence in the central ore zone. The controls of this type of ore localization are not known. The primary ore mineral is uraninite which is confined to alaskite as very small grains of several microns to 0.3 mm, occurring either occluded within quartz, feldspar, and biotite, or within cracks or interstitially to these minerals. Uraninite also exhibits a preferential association with biotite and zircon. Small amounts of betafite are associated with the uraninite, while flourite, sulphides (pyrite, chalcopyrite, bornite, molybdenite, arsenopyrite), and oxides (magnetite, hematite, ilmenite) occur somewhat sporadically in the ore. Beta-uranophane and other secondary uranium mineral represent 40 % of the uranium in the orebody. Whilst a precise genetic model has not yet been advanced Sawkins (1990) suggested that the formation of the alaskite must have involved a concentration of uranium in those areas where the alaskite or the metasomatizing fluids were generated, especially as the uranium levels in the basement rocks are generally high.

6.5.3 The Gariep Belt

Stratigraphy

The Gariep belt lies along the coast of southwest Africa across the Namibian-South African boundary (Fig.6.45). Previously referred to as the Sperrgebiet (or forbidden territory because its beaches were strewn with diamond) the Namibian sector has consequently yielded less geological information than the South African or Richtersveld sector.

Like other South Atlantic Pan-African mobile belts the stratigraphic succession in the Gariep belt begins with a rift sedimentary-volcanic sequence, followed by a continental margin (miogeosynclinal-eugeosynclinal) sequence. A unique feature, however, is the presence of widespread indicators of a collision suture in the form of ophiolitic basic volcanic rocks (Fig.6.35), as well as some blueschist metamorphic rocks (Kröner, 1974).

The basin fill which is known as the Gariep Group rests nonconformably on the Namaqua Metamorphic Complex, in areas where the latter escaped reworking during the Pan-African orogeny (Tankard et al., 1982). Where there was reworking, the Namaqua-Gariep contact is a paraconformity. At such contacts the sheared and truncated basal greenschist facies rocks of the Gariep Group grade downward imperceptibly into Namaqua greenschists (schists and phyllites) representing retrograde amphibolite-facies rocks that had been altered during the Pan-African tectono-thermal event. However, whereas structural trends in the Gariep Group are oriented roughly north-northwest those of the Kibaran Namaqua Metamorphic Complex trend mostly westward.

The lithofacies of the Gariep Group is broadly divisible into an eastern rift and shelf (miogeosynclinal) sequence with continental beds, carbonates, shelf clastics, diamictites and subordinate volcanic rocks; and a western continental slope-ocean floor (eugeosynclinal) facies (Figs.6.45,6.46). As shown in Fig.6.46 both lithofacies either interfinger or are tectonically juxtaposed (Kröner, 1974; Porada, 1989; Tankard et al., 1982). Sedimentation in the Gariep orogen was controlled by an oscillating shoreline, hence the pronounced interfingering of the lithofacies.

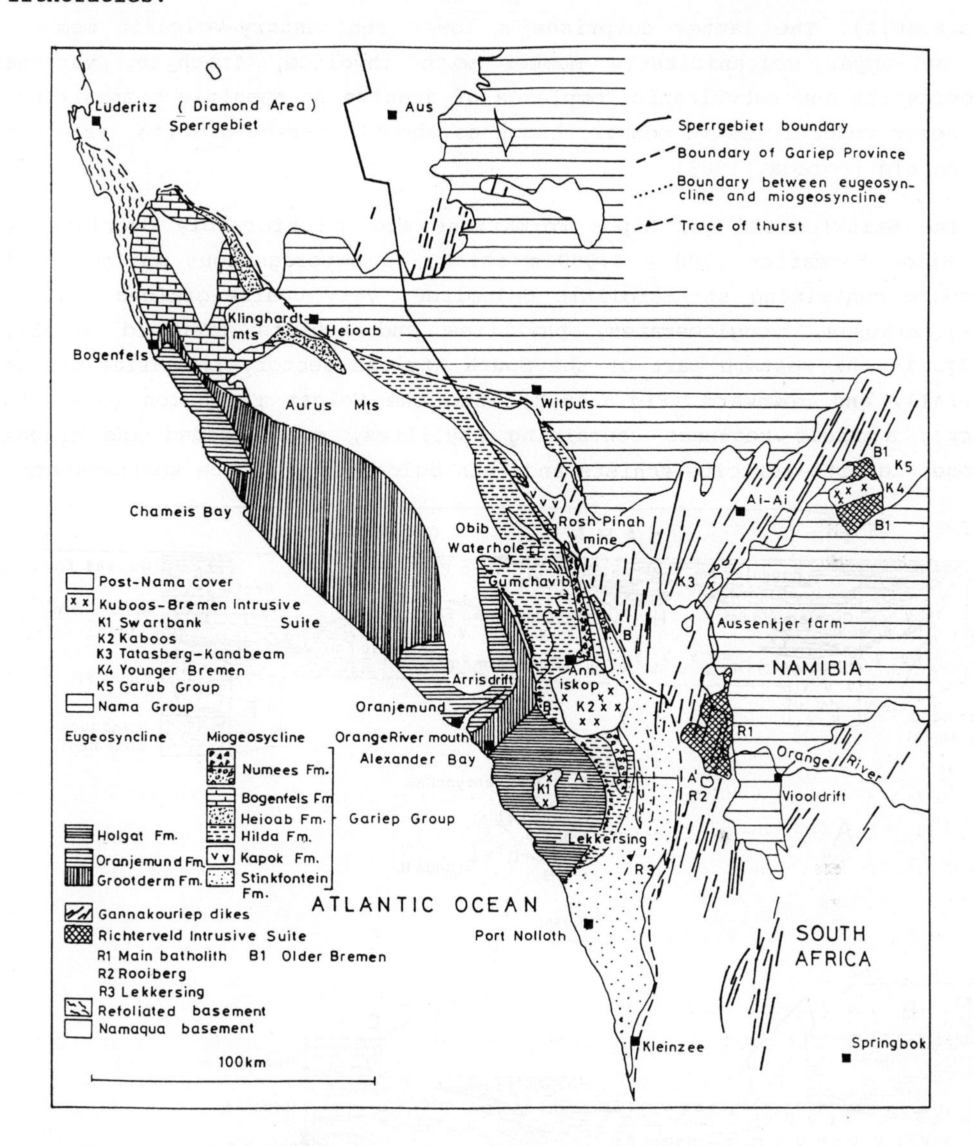

Figure 6.45: Geological map of the Gariep belt. (Redrawn from Tankard et al., 1982.)

The initial rift deposits are represented by the Stinkfontein Formation, a coarse arkosic and subarkosic fan sequence, 150 - 3,000 m thick, in which upward-fining fluvial sequences contain lenticular conglomerates and shelf dolomites with algal structures occurring at the top. In the South African sector the Stinkfontein rests non-conformably upon the Namaqua basement, but pinches out northward where it is replaced in Namibia by the Pb-Zn-bearing Kapok or Rosh Pinah Formation (Fig.6.46,A). The latter comprises a lower sedimentary-volcanic member, and an upper volcaniclastic member with rhyolite, trachyte, various pyroclastics and subvolcanic complexes of granite to syenitic composition. The upper volcaniclastic member, known as the Richtersveld Suite, is about 920 Ma old (Porada, 1989).

The Stinkfontein and Kapok Formations are unconformably overlain by the Hilda Formation (700 - 3,000 m thick), a heterogeneous marine shelf sequence containing stromatolitic dolomites, very coarse-grained quartzites, arkoses, conglomerates, phyllites and schists (Tankard et al., 1982). In the western part of the South African sector, the Hilda grades laterally and downward (Fig.6.46,B) into the Holgat Formation (5 - 6 km thick), a shelf sequence containing phyllites, cross-bedded and graded arkose, quartz-sericite schists and thin dolomites. In the southwestern,

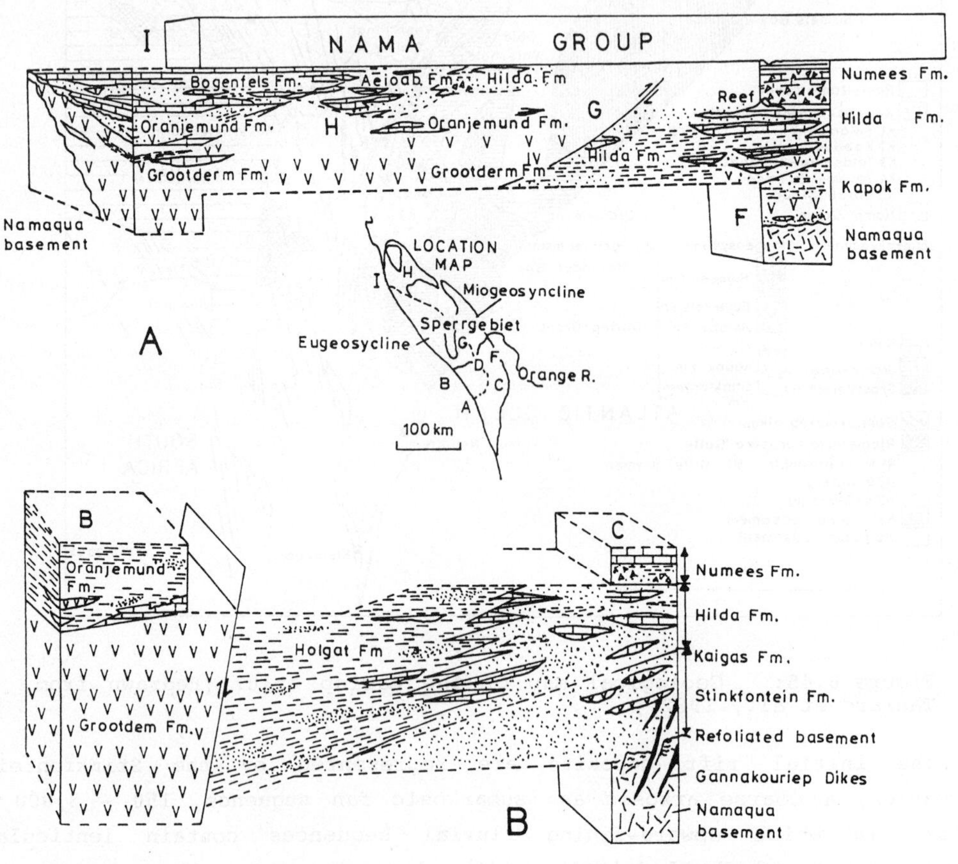

Figure 6.46: Geological sections through the Gariep belt. (Redrawn from Tankard et al., 1982.)

more basinal area, the Holgat rests unconformably on the Stinkfontein Formation. A diamictite, the Numees Formation (about 500 m thick) of probable glacio-marine origin, rests upon the Hilda and Stinkfontein Formations.

The eugeosynclinal equivalents of the Holgat Formation are the volcanic Grootderm Formation and the overlying volcano-sedimentary Oranjemund Formation. The Grootderm is a 4 - 5-km sequence of basaltic and andesitic lava, volcanic breccia, agglomerates and tuff; the Oranjemund (1,000 - 1,500 m thick) is predominantly graywacke with rare carbonates. The Grootderm Formation is believed to represent an obducted ophiolite mass. Regional metamorphism has greatly obliterated the sedimentary and volcanic features of both the Grootderm and the Oranjemund.

In his interesting reconstruction of the Oranjemund eugeosynclinal basin Kröner (1976) linked the close association of volcanic rocks and biogenic dolomites to possible carbonate reef growth on guyots or on quiescent submarine volcanoes. Deposition of the Oranjemund probably started with the building of organic reefs which were later displaced, crushed and mixed with the surrounding volcaniclastics to form olistostromes in which carbonate clasts are supported by a matrix of sheared volcaniclastic rocks or graywacke. In the Bogenfels area of Namibia (Fig.6.45) to the north, the Oranjemund Formation is thin and has a shelf character with dolomites and arenites (Martin, 1965) which pass upward into the onlapping Bogenfels Formation (Fig.6.46,A). The carbonate-rich Bogenfels Formation marked the end of basin filling and differentiation because it overlaps the eugeosynclinal Oranjemund and Grootderm Formations and the northward rising Namaqua basement. It grades laterally eastward into clastic lithofacies, the Heioab Formation which is probably the northern equivalent to the Hilda Formation, or it represents a non-volcanic facies of the Kapok Formation (Tankard et al., 1982).

Tectonism

Deformation in the Gariep belt is characterized by large-scale southeastward thrusting of the Gariep Group (Fig.6.46,A) especially the ocean-floor ophiolitic Grootderm Formation, and basement gneisses. These units are thrust as flat-lying sheets. A major plate collision event between 700 Ma and 530 Ma is believed to have caused the orogeny in the Gariep belt. The evidence for the existence of a collision suture includes the occurrence of basic volcanic rocks of ophiolitic affinities in the nappes, and the presence of glaucophane and other sodic amphiboles in the Grootderm meta-lavas suggesting high-pressure metamorphism (Kröner, 1974).

Elsewhere, metamorphism in the Gariep belt was commonly at the lower greenschist facies in the Stinkfontein Formation and at the upper greenschist facies in the Hilda Formation. The lower amphibolite grade was, however, attained in the Stinkfontein Formation in the Namibian sector.

The Richtersveld Intrusive Suite (granite and leucogranite batholiths), and the granitic to syenitic Old Bremen Intrusive Suite were emplaced before or after the onset of Gariep sedimentation around 920 - 911 Ma. A major dyke swarm with alkaline to tholeiitic composition, some of which cut the Richtersveld batholith, extends for some 100 km east of the Gariep belt (Fig.6.45). These dykes belong to the dyke injection phase which heralded continental break-up and initiation of the Gariep basin. Deformation and metamorphism were followed by the emplacement of a linear chain of plutons, the Kuboos-Bremen line (Tankard et al., 1982). The Kuboos-Bremen Suites, varying in composition from granitoid to syenitoid and carbonatite diatremes and sheets, were intruded between 550 Ma and 500 Ma.

Mineralization

In the Namibian sector the Rosh Pinah mine (Fig.6.45) exploits Pb-Zn strata-bound and stratiform deposits in the upper volcaniclastic member of the Kapok Formation. In addition to lead and zinc, silver is also recovered from this mine. The mineralization is hosted in carbonaceous chert and in dolomite with clastic intercalations. It is probably of sedimentary exhalative origin (Tankard et al., 1982).

6.5.4 The Saldanhia Belt

Located on the southwestern tip of the Republic of South Africa, the Saldanhia orogenic belt is exposed as a sequence of deformed and metamorphosed sedimentary rocks and granites (Cape granites) between Port Nolloth and Porth Elizabeth in southwestern Cape Province (Fig.6.47,A). The Saldanhia belt comprises a western meta-sedimentary sequence, the Malmesbury Group and eastern outliers, known as the Kango, Gamtoos and Kaaimans Groups (Fig.6.47,A). These low-grade metasediments accumulated between the end of the 1.0 Ga Kibaran event and about 500 Ma.

The Malmesbury Group consists of three distinct fault-bounded lithologic zones (Fig.6.47,B), namely; the southwestern zone with thick turbidite deposits (phyllites, graywackes) belonging to the Tygerberg Formation; the central zone with formations that are characterized by mica schists, fine-grained quartzites, quartz schists with limestones and dolomite lenses which suggest marine shelf environments; whereas the northeastern zone contains a complex sequence of coarse-grained quartzites, quartz schists and psammites with conglomerates and phyllite bands similar to those of modern near-shore depositional environments (Dunlevy, 1988).

Although poorly exposed, a somewhat similar lithofacies progression from south to north, occurs in the eastern outliers of the Saldanhia belt (Tankard et al., 1982). Here the Kango Group shows a succession of tidal flat-shallow shelf deposits overlain by turbidites which are in turn unconformably overlain by post-orogenic continental deposits. In the Gamtoos Group, a sequence of shelf clastics and carbonates are overlain by alluvial fan and fan delta deposits, whereas in the southernmost inlier, the Kaaiman Group, a flysch sequence is succeeded by shallow-water sandstones, shales with occasional thin carbonate beds.

The Malmesbury Group and the eastern outliers therefore display a classic continental margin sequence, which from ocean basin landwards consisted of pelagic sediments and turbidites; shelf-sea carbonates and clastics; and shallow-water and coastal arenaceous deposits and conglomerates. These depositional environments lay along the northern passive margin of the Adamaster Ocean (Hartnady et al., 1985). The location of the Saldanhia belt on a passive margin is suggested by the sparseness of basic and intermediate pyroclastic and extrusive rocks among the Malmesbury Group. Dunlevy (1988) suggested that the subduction of oceanic crust beneath South America on the opposite side of the Adamaster Ocean (Fig.6.1) eventually led to the collision of the Kalahari and South American cratons, and thus, the Saldanhia orogeny.

The Saldanhia orogeny was polyphase spanning from 610 Ma to 500 Ma. During the first phase, prior to the impact of the opposing cratons, the deep ocean and continental margin turbidite sequence (Tygerberg Formation) was deformed into a series of vertical isoclinal folds. Collision caused the tectonic front to migrate northward into the shelf-sea and eventually into the nearshore region, producing a deformational pattern of progressively decreasing intensity away from the suture (Dunlevy, 1988). The occurrence of ophiolitic greenstone and chert units in the Bridgetown Formation of the central zone is indicative of thrusting from a suture which was located somewhere to the southwest.

Post-collision isostatic uplift generated faulting and small post-tectonic basic and intermediate intrusives. The mountain ranges and fault-bounded intermontaine basins which formed during the Saldanhia orogeny, became the depocentres for the molasse-type Klipheuwel Formation of the

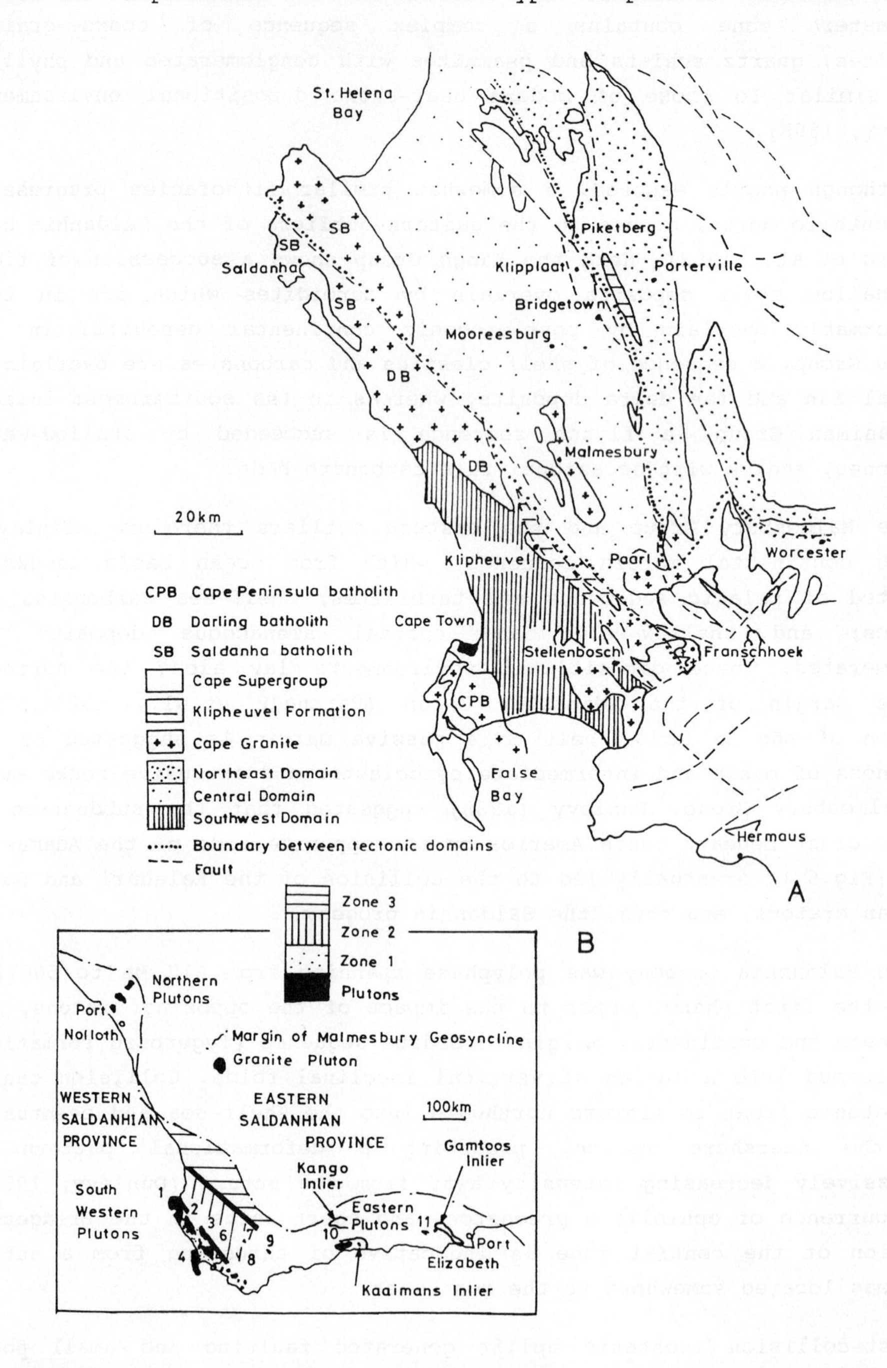

Figure 6.47: Geological sketch maps of the Saldanhia belt. (Redrawn from Tankard et al., 1982; Dunlevey, 1988.)

Malmesbury Group, which heralded the initiation of the Early Paleozoic Cape Supergroup deposition.

6.6.5 Platform Cover of the Kalahari Craton

The Nama Group

The Nama Group, a platform sequence on the Kalahari craton (Fig.6.48), accumulated between 650 Ma and 550 Ma based on the age of its Ediacara-type faunas (Tankard et al., 1982). It correlates roughly with the upper parts of the Damara Supergroup and the Gariep and Malmesbury Groups. The upper parts of the Nama Group represent Pan-African molasse. The northwestern and southwestern margins of the Nama basin were affected by Pan-African folding and thrusting in the Damara and Gariep belts (Fig.6.48). The Naukluff nappe complex on the northern margin of the Nama basin was thrust from the Damara orogen onto the Nama Group.

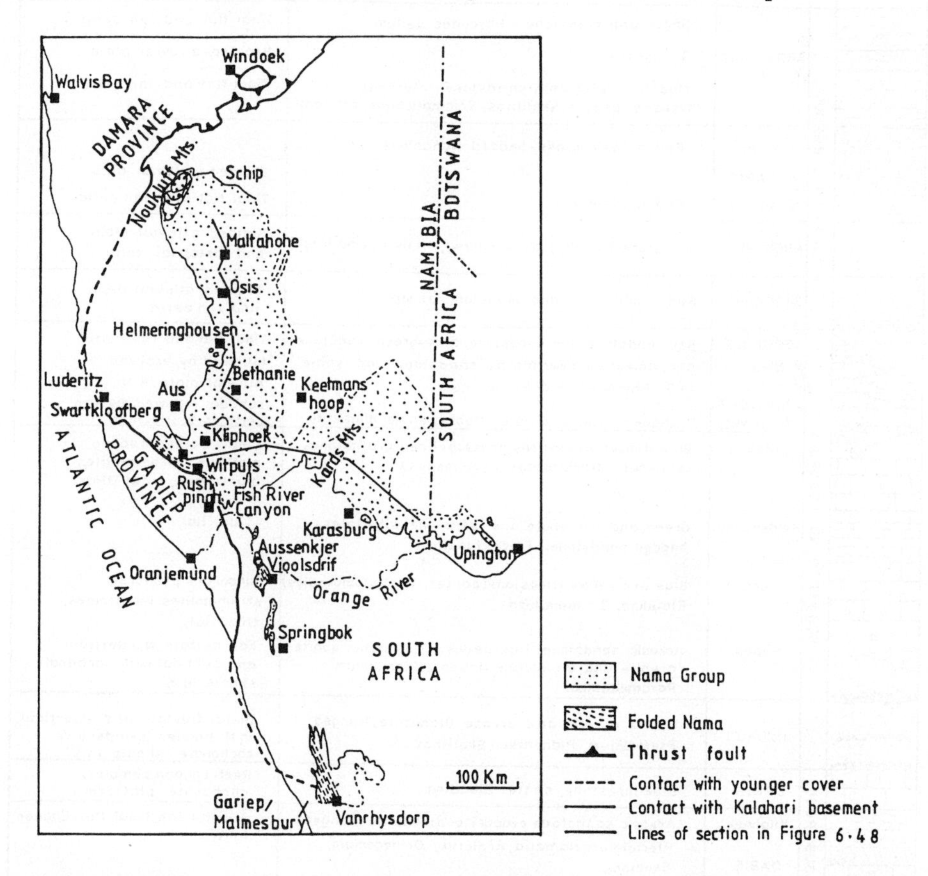

Figure 6.48: The Nama Group in Namibia. (Redrawn from Tankard et al., 1982.)

The Nama Group is of great paleontological significance because in it are well preserved the earliest metazoan organisms which attest to the appearance of soft-bodied organisms during the latest Precambrian. The Nama metazoan fauna which is discussed later in detail, belongs to the Ediacaran fauna, first discovered in the Nama Group, and later found in similar uppermost Precambrian strata in other parts of the world.

The Nama Group comprises the Kubis, Schwarzrand and Fish River Subgroups (Table 6.2). It thickens westward and displays an overall lithofacies pattern with shallow marine carbonate platforms in the west progressively intertonguing with intertidal argillaceous fan deltaic beds

Table 6.2: Stratigraphy of the Nama Group. (Redrawn from Tankard et al., 1982.)

Subgroup	Formation Member	Description	Interpretation
FISH RIVER SUBGROUP	GROSS AUB	Shale and mudstone. Phycodes pedum	Tidal flat and fan delta
		Sandstone	Braided alluvial plain
		Mudstone with thin sandstone interbeds. Phycodes pedum, Skolithos, Enigmatichnus africani	Tidal flat and fan delta
	NABABIS: Haribes	Red arkosic cross-bedded sandstone	Braided alluvial plain
	Zamnarib	Red mudstone	Tidal flat and fan delta
	BRECKHORN	Sandstone(trough cross-bedding, mudstone partings)	Braided alluvial plain and distal fan delta
	STOCKDALE	Red sandstone and mudstone at top	Braided alluvial plain and fan delta
SCHWARZRAND SUBGROUP	VERGESIG Nlep	Red mudstone and sandstone, blue-green mudstone and limestone overlain by sandstone and shale in S. Phycodes and Planolites	Tidal flat and fan delta followed by braided alluvial plain in N;
	NOMTSAS Kreyriver	U-shaped valleys, striated floors, shale, diamictite	subtidal followed by fan delta in S Glacial?
	Spitskop	Blue limestone(oolites, cross-bedded grainstones columnar stromatolites, Cyclomedusa	Clear-water shoaling platform, stromatolites biostromes, tidal flats
	URUSIS: Feldshuhhorn	Green and red shale and siltstone with interbedded sandstone. Trace fossils	Tidal flat
	Huns	Blue and yellow limestone(oolites, first stomatolites) Cloudina. Cyclomedusa	Carbonate platform, stromatolites biostromes, tidal flat,
	Nasep	Arkosic sandstone; interbedded limestone quartz arenite and diamictite. Nasepia, Pteridinium, Paramedusium	Fan delta in N, intertidal and subtidal with carbonate shoals in S
	NUDAUS	Quartz arenite and arkose Diamictite, Rangea, Pteridinium, Planolites Skolithos	Distal fluvial and intertidal in N; barrier islands and carbonate shoals in S
KUBIS SUBGROUP	ZARIS	Blue limestone, shale. Cloudina	Reef-Lagoon complex, carbonate platform
	DABIS: Kliphoek	Arkosic sandstone evaporite diamictite Rangea, Pteridinium, Namalia, Ernietta, Orthogonium, Skolithos	Alluvial fan, tidal flat, Lagoon in W
	Mara/Kanies	Conglomerate arkose, quartz arenite, organic-rich dolomite limestone, Cloudina	Braided fluvial plain, debris flows, bioherms and tidal flats in W

in the east (Fig.6.49). The fan deltas in turn interdigitate with coarser and less mature alluvial fan deposits further to the east. The paleogeography during the deposition of the Nama Group was dominated by shallow marine transgressions from the west which because of shoreline oscillations, created an alternation of clastic shoreline lithofacies and carbonate platform units. The siliciclastics show upward decrease in maturity.

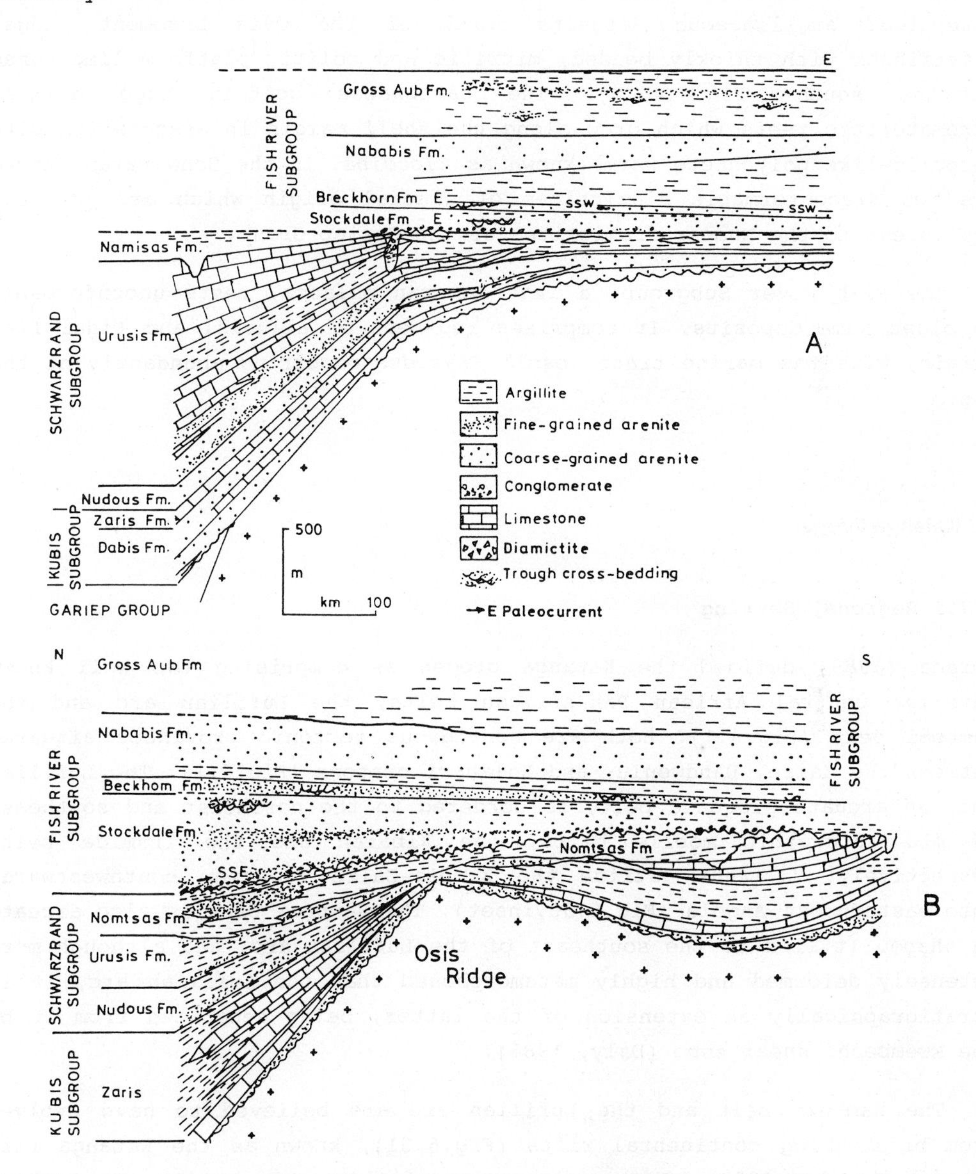

Figure 6.49: Stratigraphic sections through the Nama Group. (Redrawn from Tankard et al., 1982.)

A westward projecting basement high, the Osis Ridge initially separated the Nama basin into two parts. The Kubis Subgroup comprises a varied sequence of basal conglomerate, coarse-grained arkosic sandstone and quartz arenite which were deposited in an eastern braided fluvial environment; and more westerly shales and stromatolitic limestones deposited in intertidal and subtidal environments (Table 6.2, Fig.6.49). In the unconformably overlying Schwarzrand Subgroup, distal fluvial and intertidal argillaceous deposits north of the Osis basement ridge, interfinger with thickly bedded, micritic and oolitic platform limestones in the south (Fig.6.49,B). The carbonates contain huge domical stromatolitic reefs which grew along the shelf margin in association with serpulid-like polychaete worms known as *Cloudina*. In the Schwarzrand there are two disconformable diamictites of glacial origin which are probably equivalent to the Numees diamictites of the Gariep Group.

The Fish River Subgroup, a late Damaran molasse, rests unconformably on older Nama deposits. It comprises red-beds of alluvial and tidal flat origin, with the marine trace fossil *Phycodes* occurring abundantly at the top.

6.7 Katanga Orogen

6.7.1 Regional Setting

Porada (1989) defined the Katanga orogen as comprising two well known interior central African Pan-African belts, the Lufilian arc and the Zambezi belt (Fig.6.1). Both are contiguous tectonic provinces situated between the Zaire, Bangweulu, and Kalahari cratons (Fig.6.1). The Lufilian arc, an arcuate orogenic belt, is bordered to the northeast and southeast by mid-Proterozoic mobile belts, the Kibaran and the Irumide belts respectively. It runs astride the Zambia-Zaire frontier southwestwards into easternmost Angola (Fig.6.50,inset). The Zambezi belt is also arcuate in shape. It lies to the southeast of the Lufilian arc, and although more intensely deformed and highly metamorphosed than the Lufilian arc, it is stratigraphically an extension of the latter, being separated from it by the Mwembeshi shear zone (Daly, 1986).

The Zambezi belt and the Lufilian arc are believed to have evolved from bifurcating continental rifts (Fig.6.51), known as the Katanga rift system (Porada, 1989). The southern arm of this rift lay in the Zambezi belt; the central arms constituted the nuclei of the Lufilian arc; whereas the northernmost arm of the rift which extended northeastward at a high angle from the re-entrant of the Lufilian arc, developed into the Kundelungu aulacogen (Unrug, 1987).

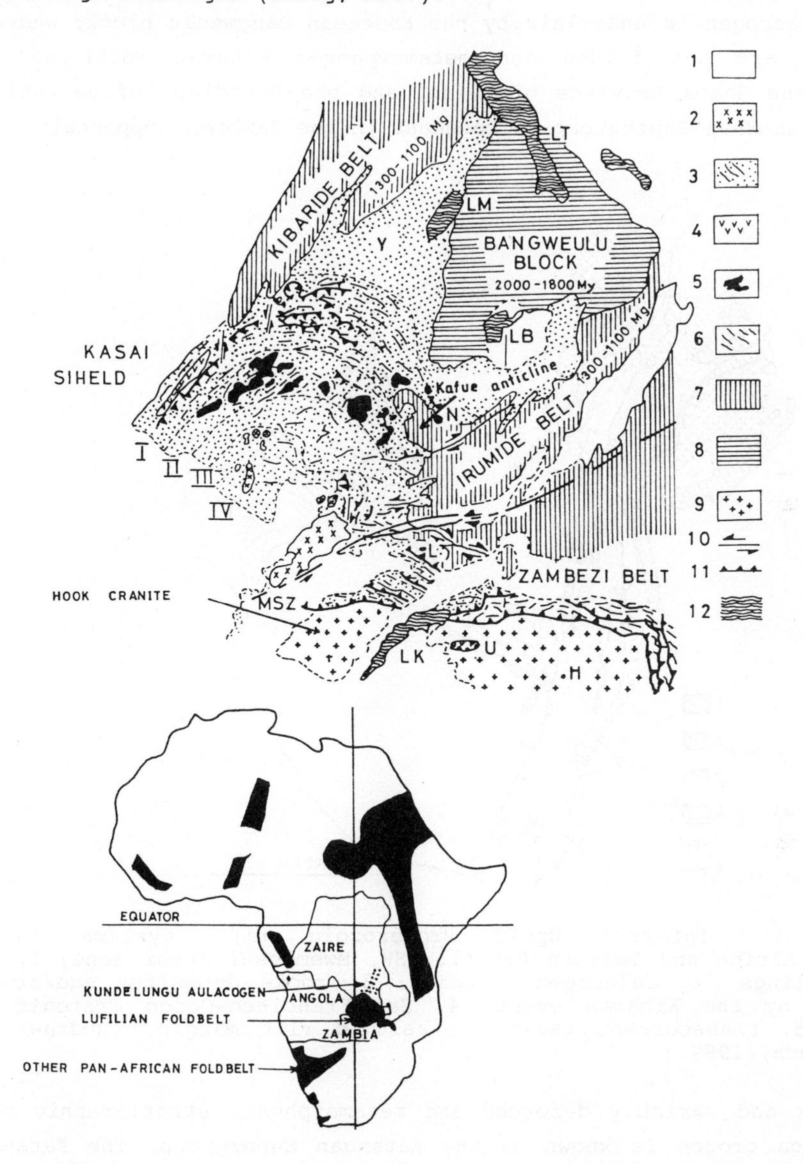

Figure 6.50: Geological map of the Lufilian arc. 1, Paleozoic-Recent; 2, granitoids; 3, Katanga Supergroup; 4, dolerite; 5, basement inlier; 6, metasediments and sheared basement in the Zambezi belt; 7, Kibaran belt; 8, Bangweulu block; 9, Archean-Lower Proterozoic basement; 10, thrust; 11, strike-slip faults; 12, lakes. H, Harare; L, Lusaka; N, Ndola; U, Urungwe klippe; MSZ, Mwembeshi Shear Zone; I, external fold thrust belt, II, Domes regions; III, synclinorial belt; IV, Katanga high; Y, Shaba aulacogen. (Redrawn from Porada, 1989.)

The Lufilian arc consists of four structural zones (de Swardt and Drysdall, 1964): the external arcuate fold and thrust belt; the domes region; the synclinorial belt; and the Katanga high (Fig.6.50). The Kundelungu aulacogen is underlain by the Eburnean Bangweulu block; whereas the Lufilian arc has folded and metamorphosed Kibaran rocks as its basement in the Shaba Province of Zaire, and the Ubendian Lufubu schists and the Kibaran Muva Supergroup as basement in the Zambian Copperbelt.

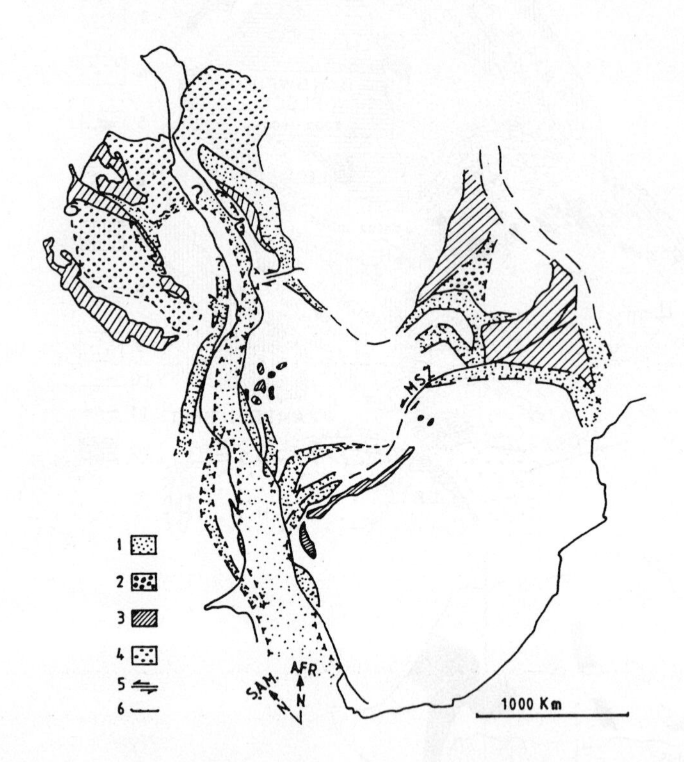

Figure 6.51: Inferred Upper Proterozoic rift systems in Southern Africa and Eastern Brazil. MSW, Mwembeshi Shear Zone; 1, rift fillings; 2, aulacogen deposits; 3, rocks deposited and/or affected by the Kibaran event; 4, Sao Francisco-Congo cratonic bridge; 5, transcurrent fault; 6, assumed rift margin. (Redrawn from Porada, 1989.)

The thick and variably deformed and metamorphosed stratigraphic pile in the Katanga orogen is known as the Katangan Supergroup. The Katangan Supergroup has attracted world-wide attention because it holds half of the world's cobalt reserves; it is among the world's largest copper mining areas; and with a host of other mineral deposits such as Zn-Pb sulphides, U oxides and noble metals, the entire Lufilian arc ranks as one of the greatest stratiform metallogenic provinces.

6.7.2 The Lufilian Arc

Stratigraphy

In the Lufilian arc the Katangan Supergroup comprises the lower heavily mineralized Roan Group, and the Lower and Upper Kundulungu Groups. Differences in the original rift morphology and the structural evolution of the Lufilian arc account for the stratigraphic variations between the different structural domains in the Lufilian arc. The stratigraphic sequence is thickest and most complete in the external fold and thrust belt. Fig.6.52 is a generalized stratigraphic section across the Lufilian arc; it shows an aggregate thickness of about 9 km for the Katangan Supergroup (Unrug, 1988).

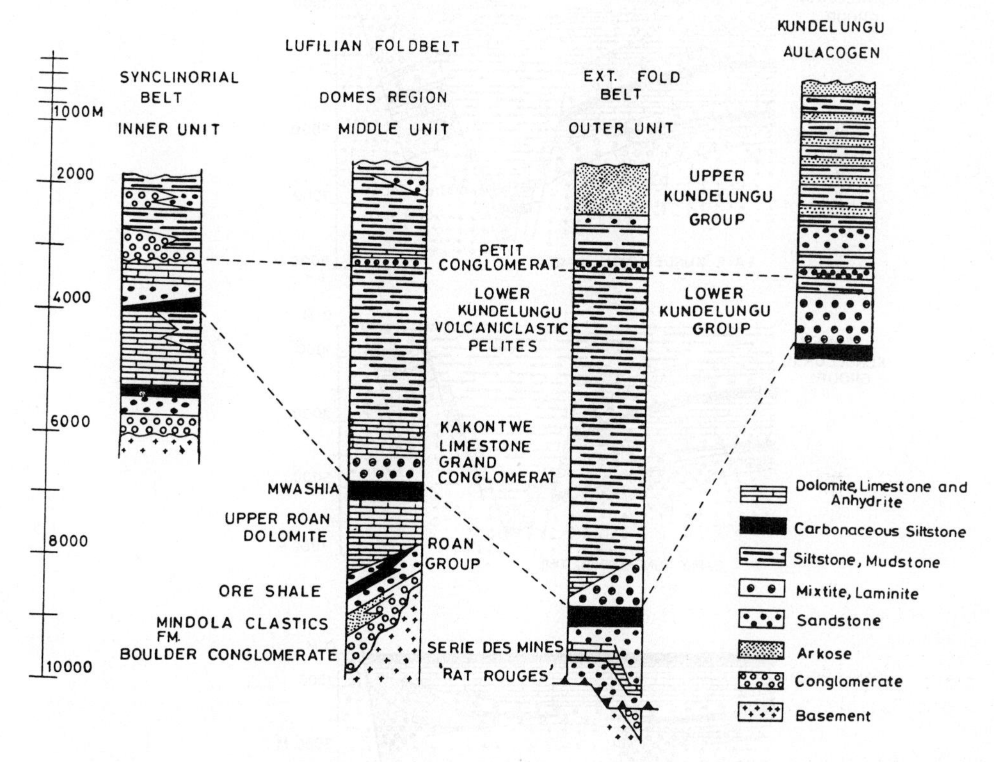

Figure 6.52: Stratigraphic columns for the Katanga Supergroup. (Redrawn from Unrug, 1988.)

In the external fold belt the Roan Group rests unconformably upon either Kibaran metamorphic rocks or upon Ubendian basement granites and gneisses. Since the Roan rests upon the Nchanga Red Granite which is about 1.2-1.1 Ga old, deposition presumably started after this time. The Roan Group represents the first phase of sedimentation (Fig.6.53,A), in which the initial uplift and rifting caused the deposition of a conglomeratic and arkosic unit, the Mine Series, followed by a marine transgression and the accumulation of shallow marine clastic rocks and the mixed carbonate platform-hypersaline lagoon evaporitic sequence, known as the Upper Roan (Fig.6.52). The Upper Roan is conformably overlain by the Mwashia Group, a sequence of carbonaceous shales and siltstones; and dolomitic shales with locally developed pyroclastic horizons. Many irregular sill-like bodies of gabbro occur in the Upper Roan and in the overlying Lower Kundelungu Group.

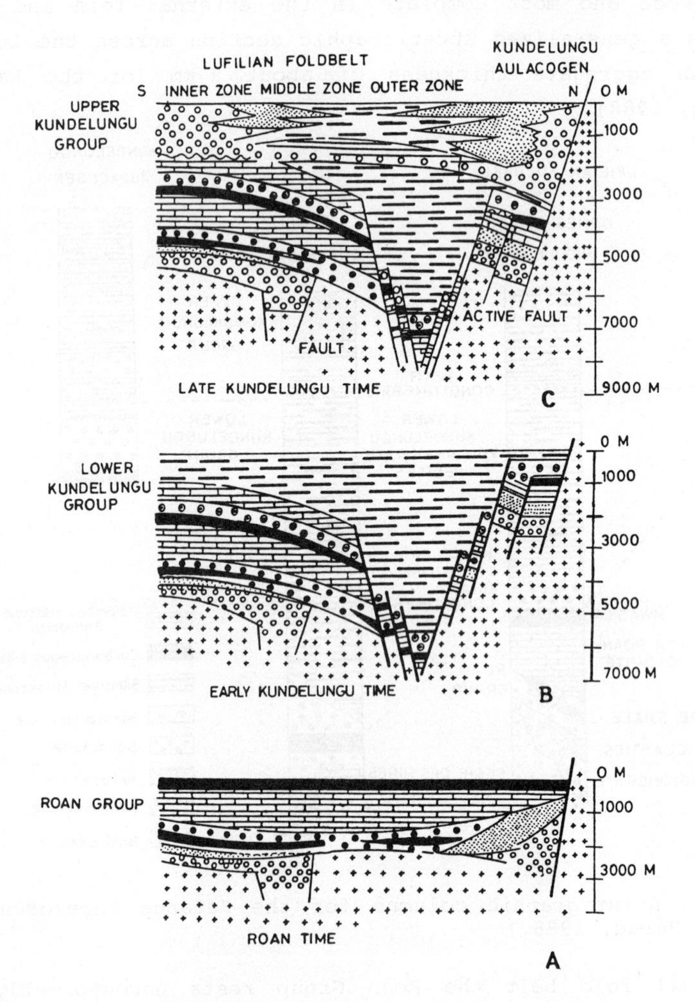

Figure 6.53: Basin evolution for the Katanga Supergroup. See Fig. 6.52 for explanation of symbols. (Redrawn from Unrug, 1988.)

According to Unrug (1988) the overlying Lower Kundelungu represents the second stage (Fig.6.53,B) in the evolution of the Lufilian basin, in which the northwestern and northeastern margins of the basin were uplifted and glacial conditions developed in the surrounding highlands (Cahen et al., 1984). Consequently, the "Grand Conglomerate" formed during this phase; and it consists of conglomerates, glacial diamicites, and glaciomarine laminites with dropstones. Since the conglomerates contain granite and pegmatite clasts which have been dated at 976 Ma, and the basalts at the base of the "Grand Conglomerat" have yielded an age of about 948 Ma, the latter date is probably close to the age of the glacial deposit. The overlying Kakontwe Limestone (Figs.6.52,6.53,B) represents the development of carbonate platforms on basement horsts in the southern part of the Katanga rifts. This limestone is up to 350-500 m thick in the Copperbelt, but thins out northward towards the external fold belt where there is a facies change into subgraywackes (Fig.6.52) which coarsen towards the northeast. Rifting and block faulting occurred in the external belt during the deposition of the Lower Kundelungu (Fig.6.53,B), leading to the accumulation of up to 5000 m of pelitic sediments; whereas north of this subsiding zone, in the Kundelungu aulacogen, the equivalent unit is only 400 m thick (Unrug, 1988).

The third phase of basin evolution is represented by the Upper Kundelungu Group. Because of uplift in the south, an unconformity developed on the Kakontwe Limestone, and the "Petit Conglomerat" --a conglomerate tongue-- spread northward into the basin. Pronounced uplift of the basin margins and subsidence of the basin floor in the Kundelungu aulacogen (Fig.6.53,C) resulted in the deposition of the thick clastic sequence of the Upper Kundelungu Group. Deposition of most of the Upper Kundelungu was syn-orogenic, but the sequence terminates with a distinctive purple arkose which may represent the molasse of the Lufilian orogeny. This molassic part has been traced far into Zaire and Angola.

Rift-related magmatism accompanied the evolution of the Lufilian basin especially during the deposition of the Roan and the Lower Kundelungu Groups. As outlined by Unrug (1988) volcanism was bimodal, acid, and basic and was localized along rift faults which delineated the western part of the Lufilian basin and in the north along the margins of the Kundelungu aulacogen. Trachyandesites are exposed in the Roan Group in a belt 180 km long in Alto Zambezi in Angola; rhyolite intrusions and tuffs occur in the Roan in the north-central part of the Lufilian arc; and clasts of both rhyolites and dolerites occur in the "Grand Conglomerat". Up to 50% comminuted volcanic rock fragment makes up the pelites of the Lower Kundelungu Group. These volcaniclastic material show evidence of regionally widespread hydrothermal alteration. Basalt lava flows occur at the surface along both margins of the Kundelungu aulacogen.

Tectonism

Based on structural interpretations by Cahen et al. (1984), Coward and Daly (1984), Daly (1986c) and Unrug (1988) two major (D1 and D2) phases of the Pan-African orogeny affected the Lufilian arc. These have been locally termed the Lusakan orogeny. The initial phase of the Lusakan folding, dated at about 850 Ma, caused the northeastward thrusting of Lufilian arc nappes towards the re-entrant of the Kundelungu aulacogen and onto cratonic areas. Thrusting also occurred in the synclinorial belt from where basement slices or nappes could have been thrust onto the domes region (Fig.6.54) which acted as a ramp during this orogeny. The external fold and thrust belt resulted from very intensive northeastward compression or crustal shortening in which the domes region was probably detached and forced northeastwards.

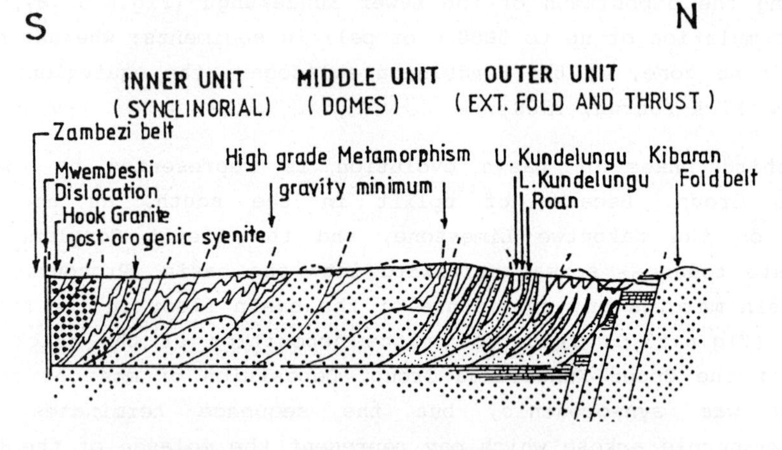

Figure 6.54: Schematic section across the Lufilian fold belt. (Redrawn from Porada, 1989.)

A later Lufilian orogeny (Cahen et al., 1984) at about 750-700 Ma (Porada, 1989) was probably responsible for the D2 deformation. It produced the present arcuate shape of the Lufilian belt. It also affected the Upper Kundelungu in which up to five deformational phases have been recognized (Cahen et al., 1984) at the re-entrant of the aulacogen.

External Fold and Thrust Belt. The external fold and thrust belt is characterized by an arcuate belt of thin-skinned outward-facing to recumbent D1 folds (Fig.6.54), and local nappes and overthrusts of the Roan over the Kundelungu. Over a distance of 800 km from Angola in the southwest to the Zambia Copperbelt, the southeast axial trend swings round NE-SW in Angola through east-west in southern Zaire to SE-NW in the Copperbelt. During D1 thrusting the Kafue anticline (Fig.6.50) controlled the external fold and thrust belt in the Zambia Copperbelt. It generated decoupling horizons parallel to bedding either at the cover-basement contact or within lower Roan sediments, in addition to producing basement-cover imbrication structures. Stretching lineations in the sheared basement and the prevalence of northeasterly-facing structures suggest movement towards the northeast.

During the D2 deformation D1 structures were re-folded in a northerly direction. The D1 structures which are best developed in the fold and thrust belt of the Copperbelt below the Upper Kundelungu, have been attributed to an earlier east-northeast-directed movement; whereas D2 structures which affected the younger strata in northwest Zambia and Zaire resulted from a northerly directed tectonic movement.

Throughout the external fold and thrust belt there is low-grade regional metamorphism at the greenschist facies. This increases slightly to the epidote-amphibolite facies in the south and west.

Domes Region. The domes region is characterized by thick-skinned tectonics with basement shearing and slices of basement rocks forming inliers in the thrust sheets. This region shows negative gravity anomaly due to crustal thickening by the superposition of basement slices (Fig.6.54). Tectonic transport was also towards the northeast during D1 event. This event produced quartz-muscovite and biotite schists which were subsequently folded to large recumbent east-west trending folds with southward dipping axial planes. The domes region has been interpreted as culminations above thrust ramps (Daly, 1986c).

In the domes region the Katangan sediments were metamorphosed to the amphibolite grade; and kyanite-rich mineral assemblages, for example in the vicinity of the shear zones of the Kabompo dome, indicate high-pressure conditions. But in the intervening Kundelungu Group, mainly exposed between the domes, metamorphism which hardly attained the almandine-amphibolite grade decreases westwards, southwards, and towards the external fold and thrust belt.

Synclinorial Region and Katanga High. The synclinorial belt was the marginal basin which lay to the south and west of the domes region. It was also completely folded during at least two deformation events. Although scanty structural information is available about the Katanga high, this region was also affected by the two deformation events. Almost all the granitic intrusions in the Lufilian arc are found in the Katanga high, with contact metamorphic aureoles developing in the otherwise very low-grade metasediments which surround the intrusives.

6.7.3 The Kundelungu Aulacogen

This is a wedge-shaped and northeastward tapering fault-bounded trough which branches from the northern re-entrant to the Lufilian arc. The Kundelungu aulacogen runs northeastward for 600-700 km between the Kibaran belt and the Bangweulu block. Nearly 7 km of unfolded Katangan Supergroup (Fig.6.53,C) lies in the Kundelungu aulacogen.

In the aulacogen the Roan Group consists of four alternating continental-marine cycles and is disconformably overlain by the "Grand Conglomerat" which along the southeastern margin of the Kibaran belt is up to 1,200 m thick probably reflecting active faulting, downwarping and the existence of active glacial fronts in that region. The Lower Kundelungu thickens from about 250 m along the edge of the aulacogen to about 400 m in the centre. There is a corresponding facies change from conglomerates and psammites along the margin to more pelitic deposits in the basin centre. The overlying "Petit Conglomerat" rests unconformably on the Lower Kundelungu and on the basement along the basin margin. The Upper Kundelungu Group which accumulated during a phase of compression and transcurrent faulting, consists of conglomerates and sandstones near the basin margin (Fig.6.53,C) and siltstones and shales towards the basin centre, with over 1,000 m of arkoses at the top.

Like the Gourma aulacogen in West Africa or the Middle Proterozoic Athapuscow aulacogen of northwestern Canada, the Kundelungu aulacogen evolved from a graben phase, through a downwarping stage, to a compressional phase. The clastics and volcanics of the Roan Group were deposited during the graben stage; followed by a phase of graben filling during which the argillaceous and calcareous upper part of the Mwashia Subgroup accumulated; whereas the "Grand Conglomerat" and the succeeding clastic sequence which overstep the trough margin represents the downwarping stage. Porada (1989) suggested that the downwarping stage was what Cahen et al. (1984) termed the Lomamian orogeny of about 976 Ma, which affected Katangan deposits along the western flanks of the Kibarides where it produced SW-NE-striking faults. The Upper Kundelungu Group accumulated after the beginning of the compressional stage which as already noted was caused by the first Lusakan deformation at about 850 Ma.

6.7.4 The Zambezi Belt

Regional Setting

The Zambezi belt occupies the northern part of Zimbabwe, southern Zambia, and a small portion of the western Mozambiquian pedicle (Fig.6.50, inset). The Zambezi belt is an arcuate thrust belt of high-grade and intensely deformed metasediments with intercalated basement gneisses which in the southern part truncate the Zimbabwe craton and the adjoining Magondi orogenic belt and the Umkondo basin (Fig.6.50). The Zambezi belt is separated from the Zimbabwe craton by an arcuate thrust zone. Along this thrust zone metasediments and basement gneisses have been thrust southward to the extent of detaching and transporting the Urungwe klippe (Vail and Snelling, 1971), from the Zambezi belt, and thrusting it over a distance of 40 km onto the craton.

To the east of the Zambezi belt lies the Southern Mozambique belt; and to the north the Mwembeshi shear zone separates the Zambezi belt from the Lufilian arc. The Mwembeshi shear zone marks a major sinistral transcurrent boundary which sharply separates the low-grade metamorphic rocks of the Lufilian arc with east-northeast tectonic transport, from the deep crustal medium-to-high grade metamorphic rocks of the Zambezi belt (de Swardt and Drysdall, 1965) which are characterized by west-southwesterly movement direction (Daly, 1988).

Stratigraphy

In spite of the difference in metamorphic grade Katanga metasediments have been traced across the Mwembeshi shear zone and are believed to have been metamorphosed to a higher grade in the Zambezi belt. The meta-sedimentary sequence of the Zambezi belt (Porada, 1989) has at its base the Chunga Formation comprising feldspathic quartzite, calcareous schists and dolomitic limestone. There are also epidosites, garnet-bearing amphibolites, and rhyolites which are probably the Kafue rhyolites. At the base of the Chunga sequence are schists probably representing altered basement gneisses. The overlying Cheta Formation consists of a thick basal limestone which is overlain by quartz-muscovite (chlorite) schist and quartzite. The next unit, the Lusaka dolomite, comprises a variety of carbonate rocks ranging from dolomite to limestone. The Kawena Formation at the top of the sequence is argillaceous at the base and psammitic in its upper part, and is unconformably overlain at a few localities by a coarse conglomerate which includes clasts from the underlying Katanga Supergroup (Porada, 1989).

The metamorphic grade in these rocks range from medium-to-high-grade with relict two-pyroxene granulite-facies assemblages in sheared basement rocks in southern Zambia, and sillimanite-bearing assemblages in Zimbabwe (Vail and Snelling, 1971).

Structure

The structure of the Zambezi belt consists of large-scale deep-level basement-cover imbrications with high-grade metamorphism and high-level thrust nappes which in Zambia consist of Katanga Supergroup schists and limestones at the epidote-amphibolite and greenschist metamorphic facies (Porada, 1989). Daly (1988) inferred a dominantly west-southwesterly tectonic transport direction along the entire Zambezi thrust belt and a general structural vergence towards the SW and WSW (Fig.6.55).

Two deformation phases similar to those of the Lufilian arc have been recognized in the Zambezi belt (Daly, 1986 c, 1988). The early D1 deformation caused the ENE-WSW-directed movement which was probably contemporaneous with the E-NE-directed D1 thrusting in the Lufilian arc between 950 Ma and 850 Ma. The D2 event was north-south directed (Fig.6.55) and caused emplacement of the Urungwe klippe of Zimbabwe around 900-800 Ma (Porada, 1989).

6.7.5 Mineralization in the Katangan Orogen

The Zambian-Zairean Copperbelt of the Lufilian arc holds more than a half of the world's reserve of land-based cobalt deposits and about 12% of the world's copper reserves. Of the 4.8 million tons of cobalt metal reserve in the Copperbelt, Zaire has the largest share, with about 3.1 million tons, while Zambia holds 1.7 million tons (Goossens, 1983). The Copperbelt is a polymetallic metallogenic province, the major types being stratiform, vein, and skarn, in which the dominant deposits are Cu-Co and Zn-Pb sulphide, Cl oxides and noble metals. Unrug's (1988) genetic model for stratiform and vein mineralization in the Copperbelt (Fig.6.56) offered the most comprehensive explanation for the sources of metals and their processes of concentration. The following outline is based mainly on Unrug's (1988) model.

Stratiform Mineralization

This occurs mainly in the external fold belt and in the domes region where the major deposits are copper, copper-cobalt and uranium. Other metals associated include nickel, gold, platinum group metals, selenium, cerium,

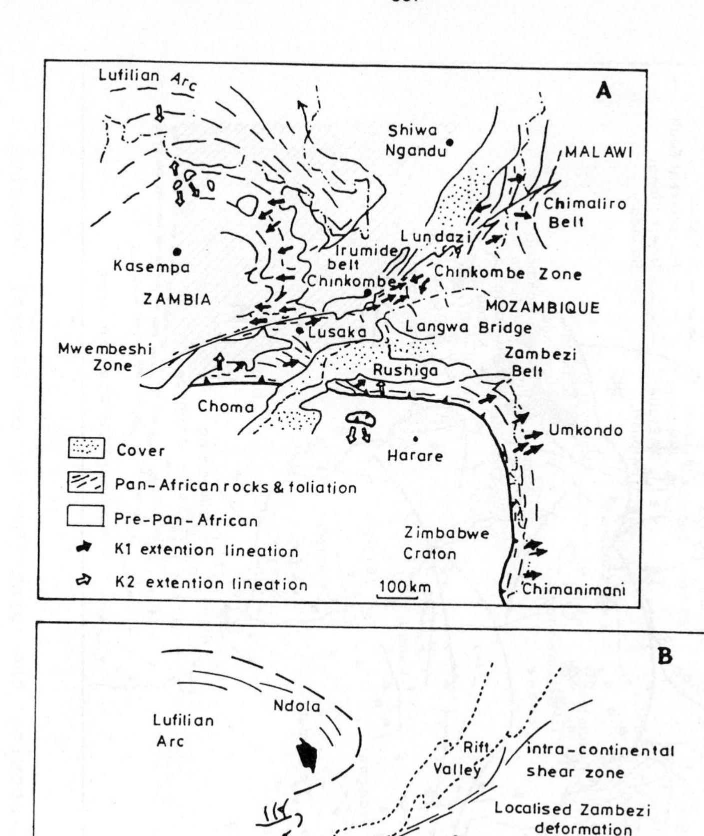

Figure 6.55: Mwembeshi Shear Zone and associated thrusts. (Redrawn from Daly, 1988.)

molybdenum, vanadium and tungsten. Uranium occurs as uraninite and in many secondary minerals. Copper sulphide such as chalcocite, bornite, chalcopyrite are present, and in the domes region there is a zonation of these minerals in the above sequence away from basement highs and from the base of the mineralized zone upwards where copper mineralization grades into pyrite. The copper sulphides in copper-cobalt deposits are accompanied by carrolite, cobalt pentlandite, and cobaltiferous pyrite. Stratiform zinc and lead sulphides occur along the fringes of some copper deposits, especially at Mufulira.

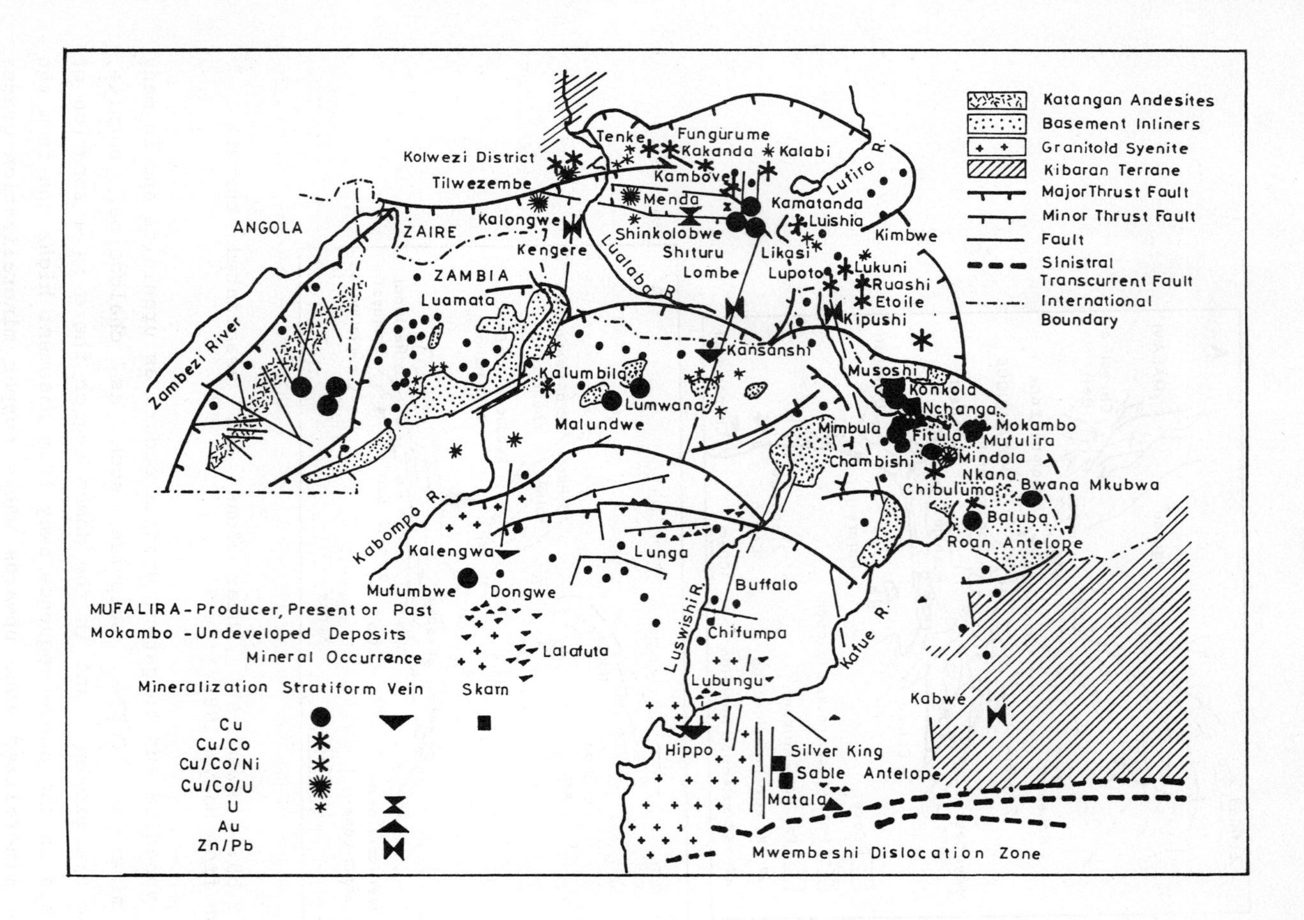

Figure 6.56: Lufilian arc showing the distribution of mineral deposits. (Redrawn from Unrug, 1988.)

Stratiform sulphide mineralization occurs in the entire Roan Group (Fig.6.52) in the permeable lithologies including conglomerate, arkose, pebbly arenite, feldspathic arenite, carbonaceous quartzite, dolomitic sandstone, siltstone, dolomitic siltstone, stromatolitic dolomite, and dolomitic silty argillite. Good stratigraphic control of the mineralization hanging wall occurs in most mines whereas the position of the footwall is unpredictable. Stratiform uranium mineralization is found around the basement inliers of the domes region where minor gold, copper sulphates and traces of nickel, cobalt, lead, chromium and molybdenum also occur. The upper boundary of stratiform mineralization in the Roan Group was provided by the impermeable layers of mixtites and laminites in the "Grand Conglomerat". Where mineralization extended upward into the karstified Kakontwe Limestone, the thick Lower Kundelungu pelites provided the impermeable layer.

Stratiform mineralization was also structurally controlled, by folds which determine the position of the Roan Group rocks that host the stratiform deposits at the surface at a depth that is accessible to mining. Also faults determine the position of the ore body.

The early hypothesis (e.g. Clemmey, 1978; Garlick and Fleischer, 1972) which were advanced to explain the origin of the Copperbelt stratiform mineralization emphasized the sedimentary origin of the ores. These models were based on the geochemical and diagenetic processes that prevailed during the transgressive - regressive cycles in the Roan Group (Fig.6.57). According to these models during the regressions saline lakes existed in the Roan basin where high concentrations of sulphides and borates took place and iron and cobalt sulphates formed in the lake mudflats or coastal sabkhas. During the transgressive phases when the lakes were re-established the sulphides in the sediments which had become sufficiently indurated were swept and re-deposited as detrital grains which were concentrated on foreset beds and in truncation planes along with other heavy minerals. Since terrestrial formation waters, with their low pH and high E_h values, can mobilize and transport large amounts of elements such as copper, lead, silver and zinc as they migrate through the hydrogen sulphide-charged algal mats of former lakes, such brines were believed to provide the mechanism for metal concentration in the stratiform deposits.

Although the above salient sedimentological features and perhaps some of the geochemical processes contributed to the Copperbelt stratiform mineralization, Unrug (1988) emphasized the importance of the convective re-circulation of basinal brines that were driven by the high thermal gradients in the Lufilian rifts. Unrug (1988) and Annels (1984) also stressed

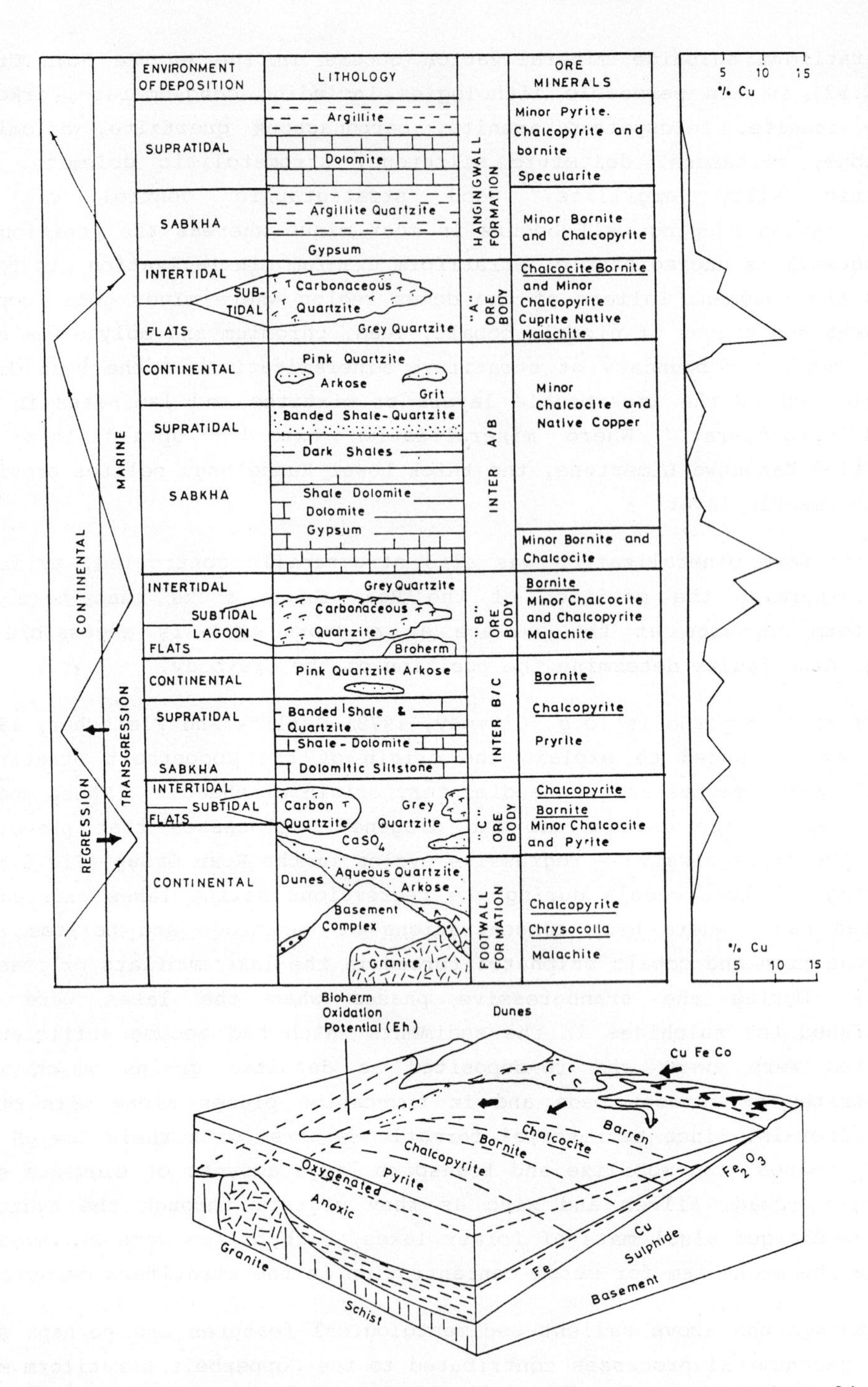

Figure 6.57: Depositional control model of copper mineralization in the Copperbelt. (Redrawn from Hutchinson, 1983.)

the role of igneous and volcaniclastic rocks, and hydrothermal contributions in sourcing the polymetallic mineral deposits. The following aspects of the Copperbelt mineralization were stressed by Unrug: the widespread mineralization in the entire Lufilian belt requires a basin-wide source of metals and emplacement mechanism; stratiform mineralization in the entire basin took place before folding; copper and cobalt were brought into the basin at a late stage of diagenesis during which it replaced pyrite; and fluid inclusions indicate the circulation of dense brines at temperatures of (200° to 250°C) during the mineralization.

In the brine circulation model primary pyrite constituted the metal trap and was replaced by copper and other metal sulphides during the interaction of metal-rich brines in early Kundelungu times. This explains why the earlier hypothesis of synsedimentary deposition of copper sulphides in the stratiform deposits of the Zambia Copperbelt proved successful in exploration. It also explains the origin of the famous mineralized cross-stratification foresets and truncation planes described from the Mufulira mine by Garlick and Fleischer (1972), meaning that with the association of primary pyrite with sedimentary structures, diagenetic replacement by metal sulphides accounted for the mineralization. The sources of the metals were the igneous and volcaniclastic rocks of the Lufilian arc, which with the presence of gabbro-dolerite sills, could account for the occurrence of cobalt, nickel, and platinum-group metals, which have mafic volcanic affinities. Since clasts of volcanic rocks show metasomatic alteration it is believed that hydrothermal processes operated, in which case the hydrothermal fluids provided the first pulse of metal input to the Roan Group aquifers during early Kundelungu times.

Hydrothermal activities in the Lufilian rift was probably analogous with those in the Recent East African Rift System where Pb, Cu and Zn sulphides precipitate from geothermal areas in Djibouti, Ethiopia, Kenya and Tanzania, and at the intersection of major faults under Lake Tanganyika. In the latter area Tiercelin et al. (1989) recovered sulphide deposits, hydrothermal fluids and hydrocarbons from shallow lake bottom sites (20 m water depth) on the northwestern part of Lake Tanganyika (Fig.6.58,A,B). Here high heat flow and deep and intense seismicity have induced hydrothermal circulation through a thick pile (up to 6 km) of sediments which have entrapped sulphides and other metallic and nonmetallic deposits.

During the second depositional phase in the Lufilian arc the Roan Group subsided by some 8 km and was overlain by a sequence of overpressured pelites (Lower Kundelungu Group) at least 5 km thick . With the dewatering of these pelites which are rich in volcaniclastic material another basin-wide pulse of metal-rich brines could have been generated which, under a high regional heat flow, could have circulated and deposited metal sulphides in the permeable Roan Group. (Fig.6.58,C).

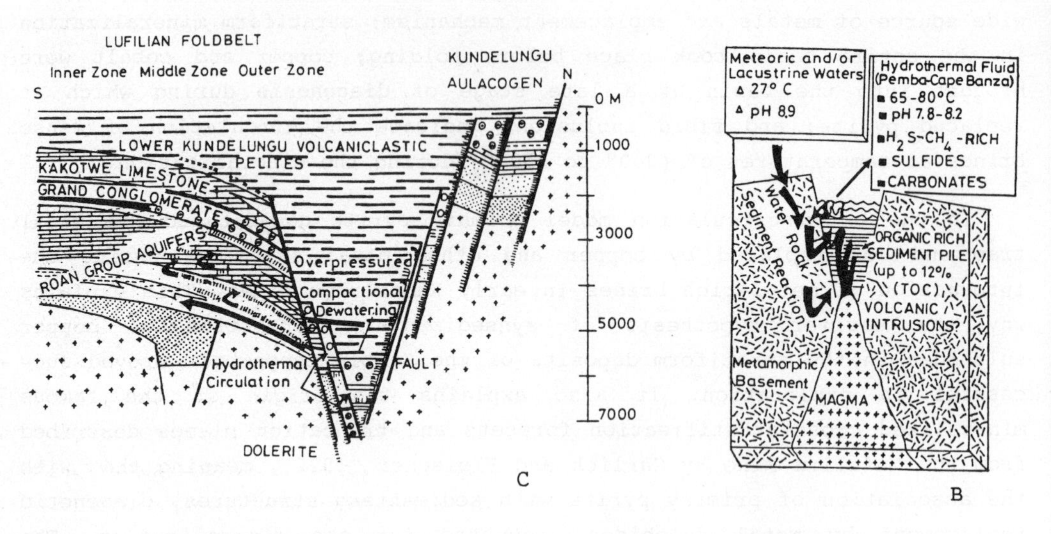

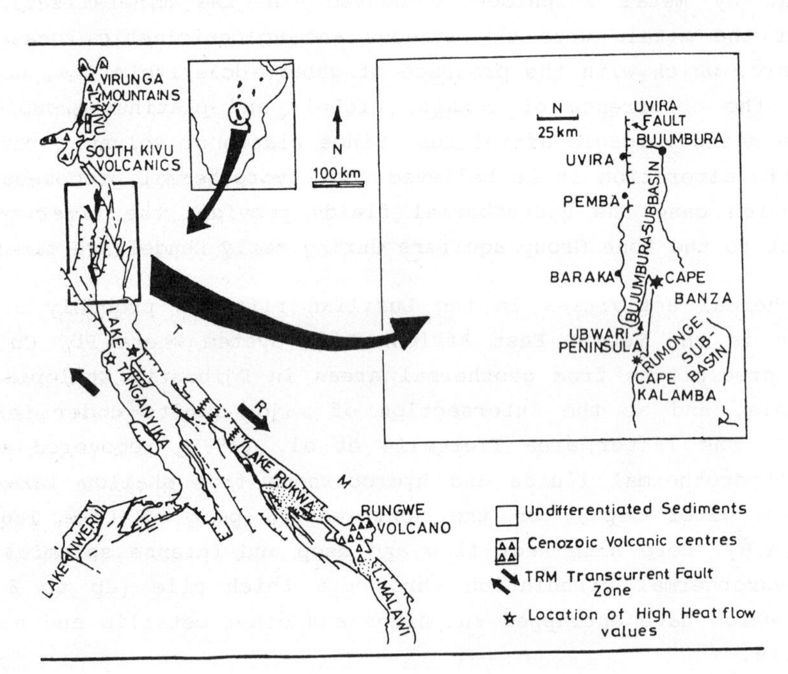

Figure 6.58: Tectonic models for the origin of mineralization in the Copperbelt. (Redrawn from Unrug, 1988; Tiercelin et al., 1989.)

Vein Mineralization

This occurs mainly in the external fold belt and in the synclinorial belt and in the Katanga high. The Kansanshi deposit is the only one located in the domes region. Major and minor deposits of zinc, lead, uranium, copper and gold are present in complex metal associations, and are exploited at Kipushi, Kabwe, Shinkolobwe and at Kansanshi (Fig.6.56). Gold-bearing quartz veins are associated with major shear zones and are located where the latter cut thermal aureoles around pre-Katangan granitoids which were rejuvenated during the Pan-African.

The association of vein mineralization with late faults which post-date folding and thrusting suggests the presence of considerable amounts of fluids in the basin after the major deformation phases. The location of large Zn/Pb and Cu vein deposits high in the stratigraphic section (Fig.6.52), above the impermeable mixitites and laminites of the "Grand Conglomerat", suggests that late faults created pathways that enabled the ore fluids to cross permeability barriers.

6.8 Western Rift Mobile Belt

6.8.1 Regional Setting

The Western Rift Rise, a highland and fold belt separating the Zaire and Tanzania cratons has already been mentioned (Chapter 4.10.2) as a poly-orogenic belt which was successively affected by the Ubendian and Kibaran orogenies (Cahen et al., 1984). During the Pan-African orogeny the Western Rift Rise constituted the foreland reactivation of pre-existing structures. The Western Rift belt starts from the northern extremity of the Lake Malawi where it branches from the Kibaran Malawi province of the Southern Mozambique belt (Fig.6.59). It extends northward along the Western Rift Valley as far as the northern end of the Lake Kivu where it merges with the vast Pan-African terrane which runs across the northern margins of the Zaire and Tanzania cratons (Fig.6.59). This latter belt shows mostly Pan-African basement reactivation spanning all the way from Cameroon Republic in the west, to eastern Uganda where it merges eastward with the Mozambique belt.

The Western Rift mobile belt is structurally complex and incorporates reactivated Ubendian and Kibaran basement which contain no Pan-African supracrustals. It also includes domains with deformed but well preserved supracrustal cover; as well as abundant post-tectonic intrusives. The geology of the Western Rift belt which was well summarized by Cahen et al. (1984) is presented below in two parts, the southern part covering northwestern Malawi and both sides of southern and central Lake Tanganyika; and a northern segment wherein lies the Itombwe synclinorium (Fig.4.42).

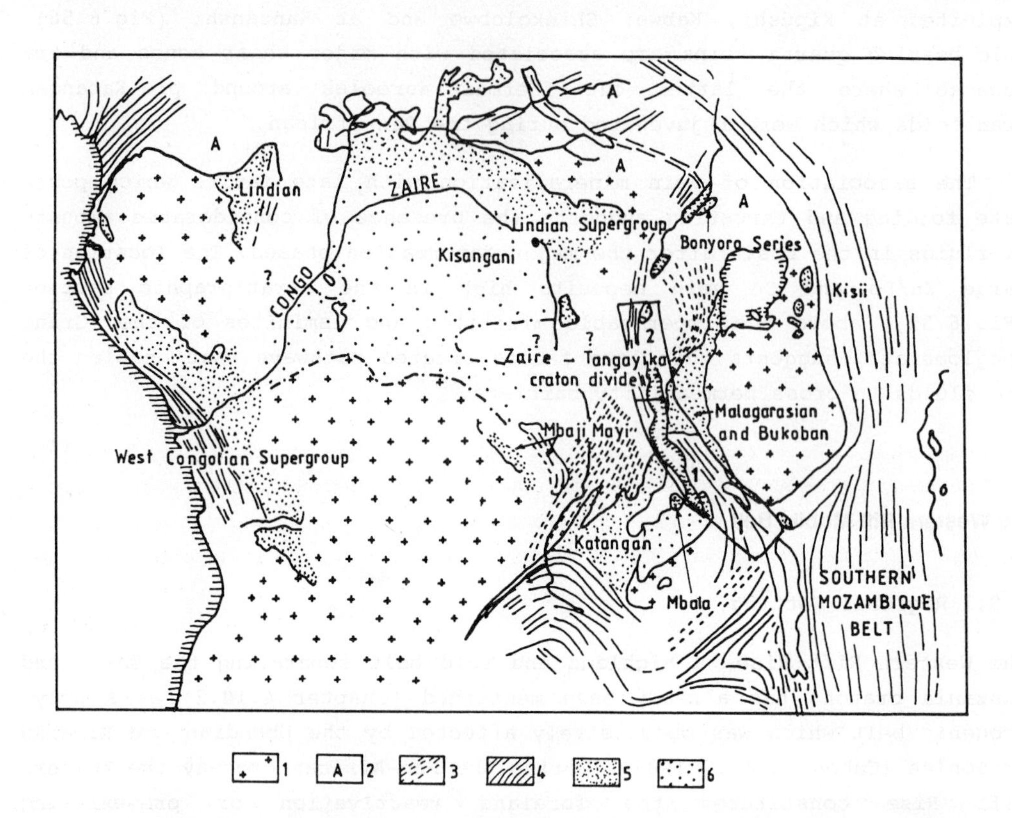

Figure 6.59: Platform sequences on the Zaire-Tanzania craton. 1, cratonic areas since Early Proterozoic; 2, Early Precambrian partly reactivated by the Pan-African; 3, Kibaran belt; 4, Pan-African mobile belt; 5, Pan-African stable zones; 6 as (5) but under Phanerozoic cover. (Redrawn from Cahen, 1982.)

6.8.2 The Southern Sector

The southern part of the Western Rift mobile belt starts from northwestern Malawi where Pan-African tectonism, regarded as part of the Mozambiquian of East Africa, was controlled by pre-existing structures. Pan-African tectonism caused the tightening of pre-existing Ubendian folds; re-alignment of most Irumide structures; and important shear movements such as along the Mugesse shear zone. Diastrophism took place under greenschist-to-amphibolite facies. No metasediments associated with the various deformation episodes have been reported. However, these deformations were accompanied by the emplacement of syenites, nepheline seynites and pegmatites. These intrusions probably took place at the same time with those in the neighbouring Katanga orogen.

Along both sides of Lake Tanganyika (western Tanzania and eastern Zaire) metasediments are preserved in symmetrical external zones in Tanzania and northeast Shaba, on both sides of an internal zone where closely folded metasediments are found. The metasediments in the eastern external zone in Tanzania belong to the Buanji Group which extends northwestwards towards the tabular cratonic cover, the Bukoban or Malagarasian Supergroup. However, the Buanji and the Bukoban-Malagarasian are not continuous.

The Buanji Group, about 1,700 m thick, rests unconformably upon the Kibaran Ukinga Group. It consists of a lower alternating series of shales, mudstones, siltstones, and arkosic sandstones; a middle sequence of shales, siltstones, quartzitic and conglomeratic sandstones and graywackes, with palynomorphs of Vendian age; and an overlapping sequence containing amygdaloidal lavas, pelites, sandstones, conglomerates, and dolomitic limestones. The Buanji is strongly folded with a northeastern vergence with folds dying out towards the foreland. The sequence on the western foreland belongs to the Marungu Supergroup. The lower part of this Supergroup is a sequence of coarse-grained detrital rocks which are folded along NNW axes, with folds which disappear towards the southwest. The rest of the Marungu Supergroup are probably equivalent to the Katangan Supergroup. The internal zone is poorly known, but is believed to contain folded metasediments which have been structurally followed up to Kalemie on the western shore of Lake Tanganyika, where it appears to connect with the Itombwe synclinorium of northeastern Zaire (Fig.4.42).

6.8.3 Itombwe Synclinorium

This represents the northern part of the Western Rift mobile belt. The Itombwe synclinorium extends approximately north-south from Kalemie to Lake Kivu and is occupied by a deformed meta-sedimentary sequence belonging to the Itombwe Supergroup. The Itombwe Supergroup comprises the Nya Kasiba Group (1,000-1,500 m thick) and the disconformably overlying Tshibangu Group (2000 m thick). Resting unconformably on the Burundian Supergroup, the Nya Kasiba Group comprises basal conglomerates and quartzites which pass upward into graphitic phyllites, quartzites and conglomerates, and black or gray slates and phyllites. The Tshibangu Group is underlain by a mixtite, and consists predominantly of pelites.

The Bukoban or Malagarasian Supergroup on the western margin of the Tanzania craton is the foreland equivalent of the Itombwe Supergroup. The correlatives of the Itombwe Supergroup on the Zaire craton are the patchy cratonic tabular formations which are generally classified as the southern outcrops of the Lindian Supergroup on the northern parts of the Zaire craton (Fig.6.59).

The Itombwe Supergroup was deformed three times along north-south axes. The first deformation took place at the end of the deposition of the Nya Kasiba Group; the second deformation was after the deposition of the Tshibangu Group during which symmetrical open folds and axial plane flow schistocity formed. The last deformation produced shear folds along north-south axes under east to west regional compressive stress. The Itombwe Supergroup was intruded at about 660 Ma by alkaline massifs comprising syenites, gabbros and granites, on both sides of the Western Rift, before the shear folding which cataclased and sometimes recrystallized the intrusions.

6.9 Platform Cover of Zaire and Tanzania Cratons

6.9.1 Regional Distribution

Tabular cratonic sedimentary formations rest on the Zaire craton along its margins (Fig.6.59), representing the remnants of a once extensive flat-lying Pan-African cover sequences which are equivalent to the deformed successions in the craton-encircling Pan-African mobile belts. Cahen (1982) observed that the deposition of these tabular sequences on the stable craton generally began after the Kibaran orogeny and was preceded in some areas by felsic volcanism. The basal parts of the Pan-African cratonic cover therefore contain Kibaran molasse as well, while the upper part includes Pan-African molasse. The parts of the cratonic cover which were located directly on the foreland of the Pan-African orogens were mildly deformed during Pan-African tectonism. Since only few radiometric ages are available from the cratonic sequences, stratigraphic correlations have depended on the stromatolites and acritarchs in these sediments.

The Pan-African platform cover of the Zaire craton crop out around the Phanerozoic intracratonic sequences in the Zaire basin which they also underlie. On the southeastern part of the Zaire craton is the Mbuyi Mayi Supergroup; the northern tabular sequence is known as the Lindian Supergroup (Table 6.3). Between the Zaire and Tanzania cratons southerly

Table 6.3: Correlation of platform cover sequences on the Zaire and Tanzania cratons. (Redrawn from Cahen et al., 1984.)

West Congo (Zaire, Angola Congo, West Congolian Supergroup)	Kasai and Western Shaba (Zaire) Mbuji Mayi Supergroup	Shaba and Zambian Copper Belt Katangan		N.E. Zaire Lindian Supergroup	
≥625 Ma		602		>590	Aruwimi Group
Inkisi Group			III		Banalia arkoses
734					Lowest part of "Banalia arkoses"
Mpioka Group		Upper KundeLungu	II		Alolo shales
					Galamboge quartz
Schisto calcaire Group: CIV, CIII 84, CII, CI			I	788	
		"Petit conglomerate" mixtite Katangan			Lokoma Group
Upper mixtite Haut Shiloango Group		Lower Kundelungu			
ΔΔΔΔΔΔ Lavas and lower mixtite	ΔΔΔΔΔ 948 Lavas	"Grand conglomerate" and Lavas ΔΔΔΔΔ		950	Akwokwo mixtite
		976			
Sansikwa Group	BII Group: BII e, BII d, BII c, BII b, BII a	Roan	Mwashya, Upper Roan, Lower Roan 83		Ituri Group
ca 1050	BI Group				
Mayumbian volcano sediments	BO Group	ca 1100-1200 Nchanga Red Granite			
	1310	1310			

Central African Republic	Eastern Kivu (Zaire) Itombwe Supergroup	SE Burundi and W. Tanzania Malagarasian (Bu) and Bukoban (T) Supergroup		
	660			
		Kibago Group (Bu)	Manyovu Red beds (T)	Uha Group (T)
700 Dialinga Formation, Bakouma Formation, Bondo 'fluvio-glacial conglomerate', Nakondo quartzite	774	Mosso Group (Bu)	Bugongo(Bu)=Ilagala(T) carbonate rocks; Kabuye(Bu)=Gagwe(T) amygdaloidal lavas	
		ΔΔΔΔΔ 813		
Bougboulou Group	Tahibangu Group with basal mixtite	Nkoma Group (Bu) ≥900 ± 24		
		ΔMusindozi Group (Bu) = Kigonero Flags (T) Δincluding carbonate Δrocks		
Mixtite				
	978			
		Kavumbwe	Masontwa	
			Itiaso (Kibaran)	

patchy outcrops of the Lindian correspond to the folded Itombwe Supergroup as already noted. Eastward on the western margin of the Tanzania craton, the Bukoban or Malagarasian Supergroup is the eastern tabular equivalent of the Itombwe Supergroup. The Mpioka and the Inkisi Groups which are the foreland tabular sequences of the West Congolian orogen, have already been described.

6.9.2 Sequences on the Zaire Craton

Mbuyi Mayi Group

Formerly referred to as the Bushimay Supergroup, the Mbuyi Mayi is exposed in a narrow belt which extends from eastern Kasai Province in Zaire southeastwards into western Shaba Province (Fig.3.31). It is nonconformable upon the Eburnean basement of Kasai, and rests unconformably upon Kibaran rocks in Shaba where it is equivalent to the Roan Group. As shown in Table 6.3 the base of Mbuyi Mayi Supergroup in Shaba comprises a basal group termed B0 (Cahen, 1982) which has mostly conglomerates and quartzites that are considered as the Kibaran molasse. The middle group (BI) comprising siltstones, shales and dolomitic shales with some dolomitic beds at the top, constitute the base of the Mbuyi Mayi in Kasai where it is about 550 m thick, whereas the combined thickness of B0 and BI in the Shaba Province to the east is nearly 3,000 m. The upper group BII is essentially composed of dolomitic and stromatolitic carbonates with species of the stromatolite *Conophyton*, *Baicalia*, *Tungussia*, and *Gymnoselen*. These forms have been reported from the lower part of Supergroup I in the western Taoudeni basin by Bertrand-Sarfati (1972) and correlated with the Upper Riphean (1.0-700 Ma old). Based on the acritarchs recovered from boreholes which penetrated BI and BII, Baudet (1987) concluded that BII is diachronous, being older in the west and getting younger towards the east (Fig.6.60,C). This was because the stromatolitic sequence (BII) was deposited in a transgressive Upper Riphean epicontinental sea which progressively encroached from the west towards the east.

The paleogeographic setting during Mbuyi Mayi deposition (Baudet, 1987) was one in which uplifted Kibaran mountains were eroded and deposited as molasse (B0) in bordering troughs (Fig.6.60,A); this was followed by the filling of the trough with finer clastics (BI). With the cessation of terrigenous sedimentation an epicontinental sea suitable for colonization by stromatolites invaded the region as from about 1.0 Ga. Deposition of the Mbuyi Mayi Supergroup terminated in Kasai Province with volcanism at about 948 Ma, when a widespread amygdaloidal basalt unit, a prominent regional marker horizon (Fig.6.60,C), accumulated in the basin. The Mbuyi Mayi Supergroup was deformed during the Lusaka folding episode in the nearby Lufilian arc at about 850 Ma.

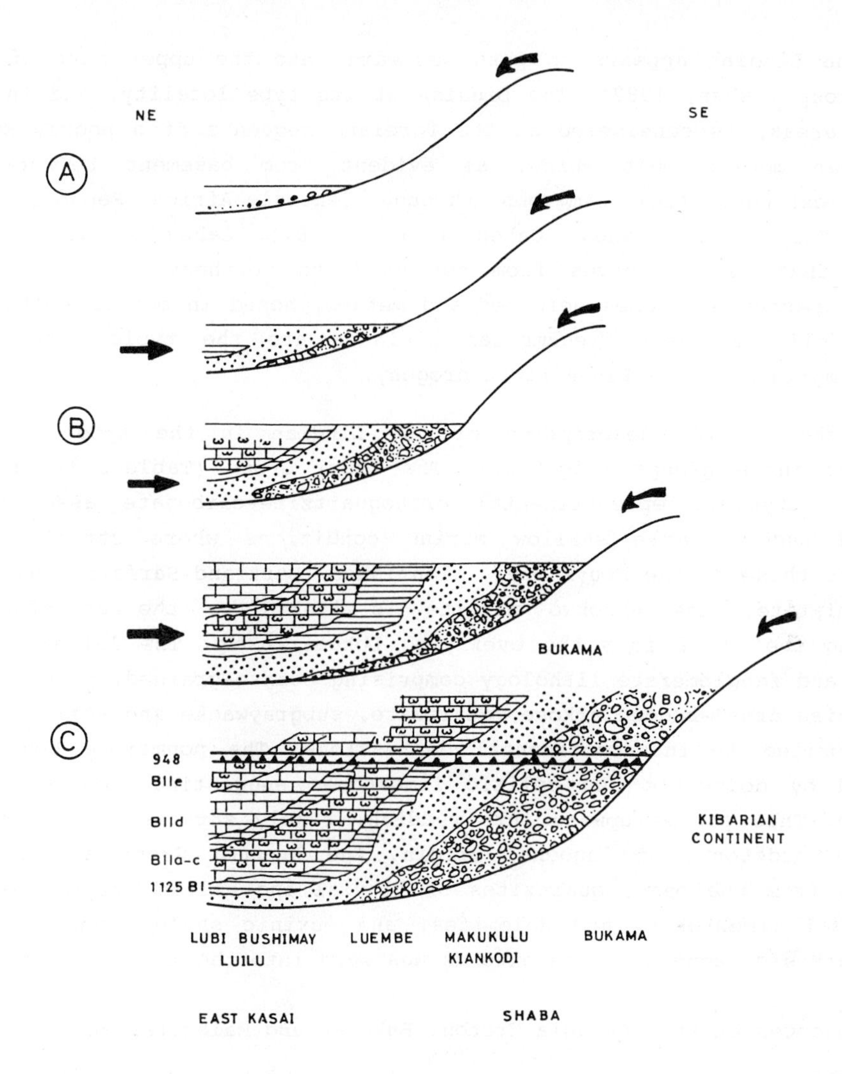

Figure 6.60: Tectonic model for the deposition of the Mbuyi Mayi Group. Arrows to the left depict direction of marine transgression; right arrows depict dominant sediment source area. (Redrawn from Baudet, 1987.)

Lindian Supergroup

This is a sub-horizontal non-metamorphosed sequence, up to 2,500 m thick, the type area of which lies north and northeast of Kisangani in northern Zaire (Fig.6.59). Outliers occur northward in Central African Republic, and westward in Cameroon and Congo Republics. In the subsurface of the western part of the Zaire intracratonic basin, beneath the Phanerozoic cover, the Lindian appears to pass westward into the upper part of the Inkisi Group (Cahen, 1982). The Lindian at its type locality, and in its northern areas, is considered as the foreland sequence of a poorly known Pan-African mobile belt which, as evident from basement re-working, probably extended from Cameroon through Central Africa Republic and northern Zaire into Uganda (Cahen et al., 1984). Cahen et al. (1984) observed that as one moves from northwest to northeast or east, the Lindian Supergroup becomes deformed and metamorphosed in mobile belts, of which refoliated lower Precambrian gneisses are the typical tectono-thermal imprints of the Pan-African orogeny.

Verbeek's (1970) description of the Lindian in the type area identified three groups (Fig.6.61). The Ituri Group (Table 6.3) at the base is a typical epicontinental orthoquartzite-carbonate association deposited under stable shallow marine conditions where stromatolites similar to those of the Mbuyi Mayi, flourished (Bertrand-Sarfati, 1972). A middle mixtite, the Akwokwo mixtite is linked to the unconformity separating the Ituri form the overlying Lokoma Group. The latter has a piedmont and fanglomerate lithology comprising coarse-grained, pink, lilac and greenish cross-bedded arkose, quartzite, subgraywacke and shale, which were deposited in intracontinental depressions. The nonmarine beds are succeeded by dolomitic and oolitic limestones suggesting marked marine influence. This passes upward into reddish, finely stratified, micaceous and silty mudstones of lagoonal origin. The Aruwumi Group at the top comprises from the base, quartzites of fluvial and eolian origin; marine or lagoonal limestones, and dolomites, and euxinic shales; and a thick deltaic arkosic sequence which extends westward into the Inkisi Group.

6.9.3 Sequences on the Tanzania Craton: Bukoban and Malagarasian Supergroups

On the western part of the Tanzania craton the Bukoban Supergroup of Tanzania and its equivalent the Malagarasian Supergroup of Burundi occur in an arcuate exposure belt where they are slightly folded along the Western Rift mobile belt. The folds attenuate towards the Tanzania craton. The Bukoban is over 3,000 m thick. In Burundi the oldest unit, the Kavumwe Group sits unconformably on the Kibaran whereas in Tanzania, the oldest unit the Masontwa Group, unconformably overlies older pre-Bukoban clastic beds belonging to the Kibaran Itiaso Group (Table 6.3). The age of the Bukoban Sandstone (=Kavumwe of Burundi) is slightly greater than that of its gabbro intrusive which is dated at about 1.02 Ga. The unconformably overlying Musindozi dolomitic limestone and quartzite group (= Kigonero

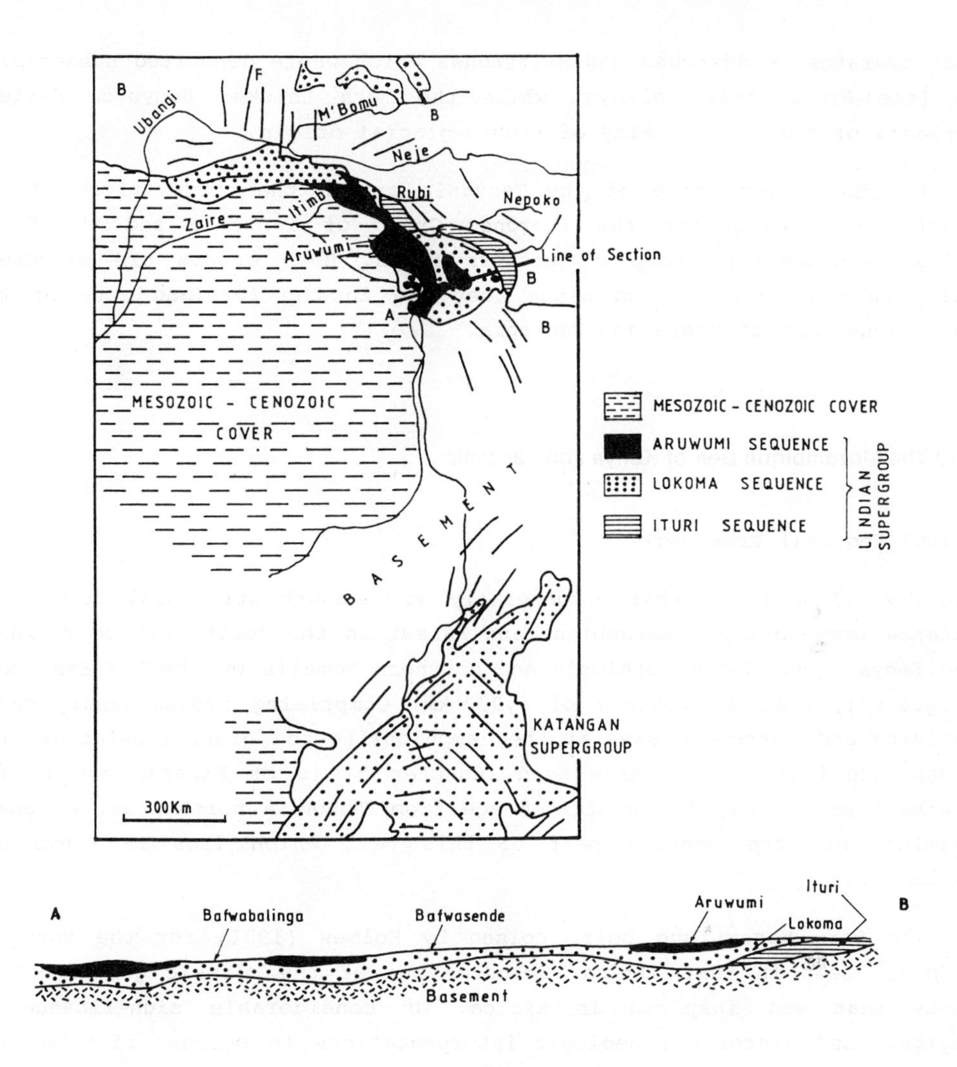

Figure 6.61: Sketch map and geologic section for the Lindian Supergroup. (Redrawn from Verbeek, 1970.)

Flags of Tanzania) has columnar stromatolites which correlate with BII of the Mbuyi Mayi Supergroup, thus suggesting that the Musindozi is older than 948 Ma. The upper Bukoban comprises conglomeratic sandstones disconformably overlain by dolomitic and silicified limestones, with a disconformable clastic sequence at the top.

In Uganda the correlatives of the Bukoban include the Mityana and the Bunyoro "Series" (Jackson, 1980). The former is very thin and consists of conglomerates, sandstones and siltstones which where deposited subaerially in Late Proterozoic valleys, while the much thicker Bunyoro "Series" consists of clastic deposits of fluvio-glacial origin.

On the eastern side of the Tanzania craton the correlatives of the Bukoban Supergroup are the Ikorongo Group of Tanzania and the Kisii volcano-sedimentary group of western Kenya. Both groups are discussed fully in the next section since they lie in the foreland zone of the Mozambique belt of Kenya and Tanzania.

6.10 The Mozambique Belt of Kenya and Tanzania

6.10.1 Regional Framework

The Mozambique belt exhibits generally north-south structural trends and extends from northern Mozambique and Malawi in the south through Tanzania and Kenya, into Sudan, Ethiopia and northern Somalia on the Red Sea coast (Fig.6.62), over a distance of 6,000 km. Comprising predominantly paragneisses and orthogneisses at the amphibolite-to granulite-facies, the Mozambique belt, as we have seen (Chapter 5) is of Kibaran age in the southern part, and of Pan-African age from Tanzania northwards. We shall examine here the central part of this belt which lies in Kenya and Tanzania.

The term Mozambique belt, coined by Holmes (1951) for the youngest orogenic belt in East Africa, embodies one of the classic geologic concepts that was inspired in Africa. Of considerable significance to regional and historical geologic interpretations in Holmes' time was his observation that the Mozambique belt demonstrated the concept that cross-cutting structural relationships can be used to determine the relative ages of Precambrian terranes (Cahen et al., 1984). In what was the earliest application of radiometric dating to the African continent, Holmes (1951) furnished actual ages which proved that the high-grade Mozambique belt, dated at about 1.3 Ga, and characterized by a north-south structural trends (Fig.6.62), was younger than the adjoining low-grade greenstones of the Tanzanian Shield with east-west structures, and the intervening Ubendian belt which has NW-SE structural trends.

The Mozambique belt was originally perceived as a meridionally trending orogenic belt along the eastern margin of southern and equatorial Africa. However, Cahen et al. (1984) subdivided the Mozambique belt into a

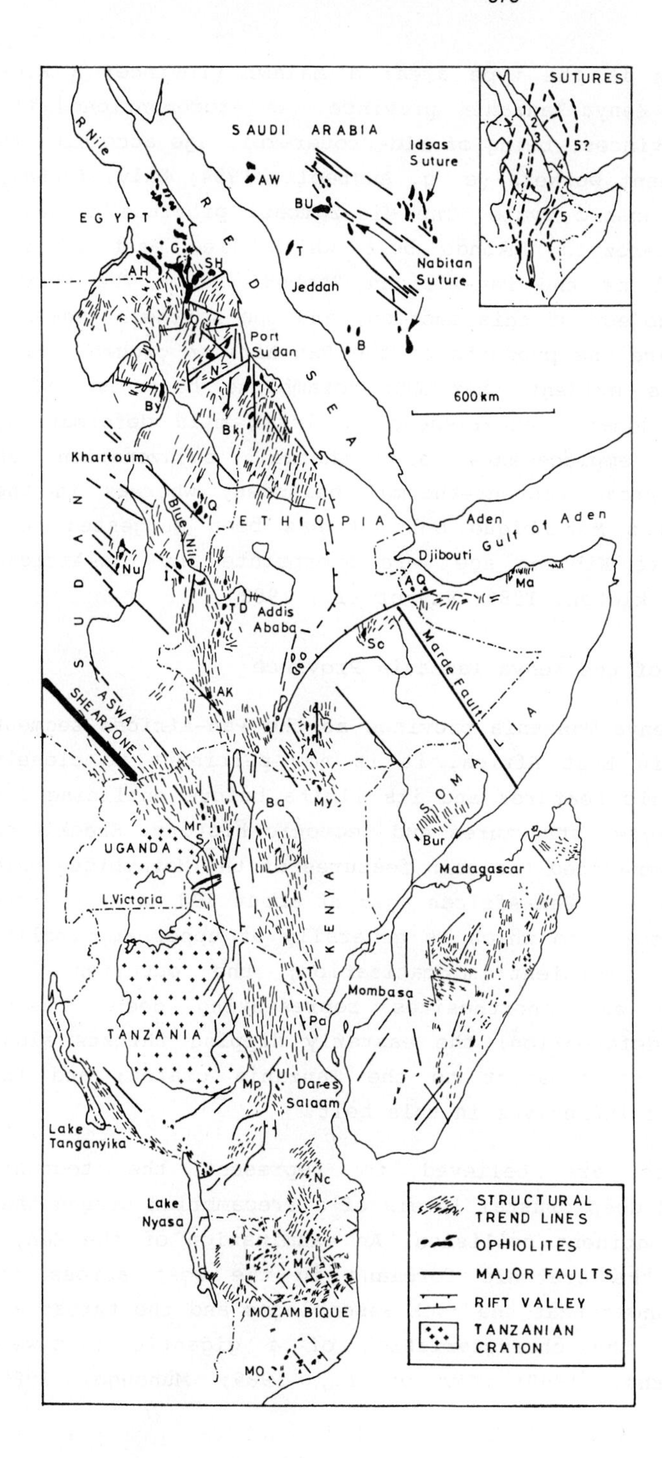

Figure 6.62: Structural trends in the Mozambique belt. (Redrawn from Behre, 1990.)

Mozambique province which is the type area; a Malawi province; a Mid-Zambezi province; and a Kenya-Tanzania province. As aforementioned the Malawi and Mozambique provinces, being of mid-Proterozoic age according to Holmes (1951) and subsequent workers (e. g. Andreoli, 1984; Daly, 1986a), belong to the Kibaran orogenic cycle. The Mid-Zambezi province actually includes the Early Proterozoic Umkondo belt which lies east of the Zimbabwe craton, as well as the Pan-African Zambezi belt. The Kenya-Tanzania province, the subject of this section, and the northern segments of the Mozambique belt, are the products of the Pan-African orogeny. From the aforementioned it is evident that the Mozambique belt is indeed polyorogenic. The older domains experienced at least mild deformation, granite and pegmatite emplacements and thermal rejuvenation as manifestations of Pan-African tectono-thermal episodes, whereas in the Pan-African domains of the Mozambique belt it has been suggested that older rocks of Archean to Kibaran age were overprinted by Pan-African events (Almond, 1984; Shackleton, 1986; Key et al., 1989).

6.10.2 Tectonic Features of the Kenya-Tanzania Province

Having singled out the Kenya-Tanzania province as the Pan-African segment of the Mozambique belt in East Africa, it is now pertinent to closely define its salient geologic features and its limits before outlining its distinctive rock assemblages, structures and tectonic history. Shackleton (1979) and Pohl (1988) summarized the main features of the Mozambique belt as: the ubiquitous presence of Pan-African ages of about 600 Ma; the high-grade and polyphase metamorphism which is generally of upper amphibolite facies with ubiquitous incipient migmatization; the occurrence of granulites, charnockites and anorthosites representing deep crustal environments; polyphase deformation; the easterly dipping thrusts along the western orogenic front adjacent to the Tanzania craton; and the general rarity of granitic intrusives in this belt.

These characteristics are believed to represent the tectonic signatures in the exposed deep crustal levels of a Precambrian orogen that had suffered continent-continent collision. An examination of the Kenya-Tanzania province from the cratonic foreland in the west across the internal or mobile zone shows that the rock assemblages and the thrust and fold structures display the characteristics of a gigantic east-west collisional suture (Berhe, 1990; Key et al., 1989; Muhongo, 1989; Shackleton, 1986).

Except along the foreland, on the eastern margin of the Tanzania craton where small tabular cratonic supracrustals have been assigned to the Pan-African of the Mozambique belt, Pan-African stratigraphic sequences in the Mozambique belt have either been metamorphosed to very high-grade, or as implied in the deeply eroded collision belt model, they have been eroded away. Almond (1984), however, raised the interesting question of what happened to the eroded sediments because, as he rightly pointed out, the stratigraphic records of such sediments are not known. Rather, the narrow belt of reworked Archean and Lower Proterozoic rocks along the Tanzania craton, for example, has been included in the foreland and external zones of the Mozambique belt (Cahen et al., 1984). The stratigraphic record of the Mozambique belt is indeed scanty.

Several orogenic fronts occur along the southeastern margin of the Tanzania craton near Iringa (Fig.6.63). This illustrates the polyorogenic nature of the Mozambique belt and how the western fronts or external zones of successive orogenic belts (Shackleton, 1979) have been telescoped along the margin of the Tanzania craton. Shackleton identified a most westerly front or thrust zone separating Archean rocks to the west from the Konse geosynclinal paragneisses (actually the lower sequence of the Usagaran Supergroup) which were deformed and metamorphosed during the Ubendian orogeny at about 1.9 Ga. The Konse gneisses are overlain unconformably by the Ndembera acid volcanics which although undeformed on the western part are strongly tectonized east of the post-Ndembera front (Fig.6.63). The first two fronts which are of Ubendian age were succeeded by a third or Pan-African orogenic front which is located along the western limit of deformed post-Ubendian granites. The Pan-African front lies in the zone where K/Ar mineral ages, which generally decline eastwards from about 2.0 Ga near the cratonic margin, attain values of about 500 Ma which are the Pan-African ages. Here then lies the cardinal principle and pitfall in the stratigraphy of the Mozambique belt.

The stratigraphy of the Mozambique belt is not based on well dated stratigraphic sequences that are exclusively of Pan-African age. Rather, the available radiometric ages are those of Pan-African metamorphism, deformation, magmatism and uplift. The original ages of the rocks affected remain conjectural; they could have been Archean (e. g. Shackleton, 1986) or Kibaran (e. g. Key et al., 1989). As stated by Shackleton (1986) "it thus appears that several different but lithologically similar transgressive sequences may be represented in the metasediments in the Mozambique Belt". However, in spite of the uncertainties surrounding the original ages of the metasediments in the Mozambique belt, Shackleton (1986) summarized the lithofacies in the belt (Fig.6.63) and stressed their gross paleogeographic and tectonic implications. Fig.6.63 shows that quartzite and marble shelf lithofacies are exposed mainly in the western part of the Mozambique belt, while marbles occur in the central and eastern supposedly more basinal parts of the belt.

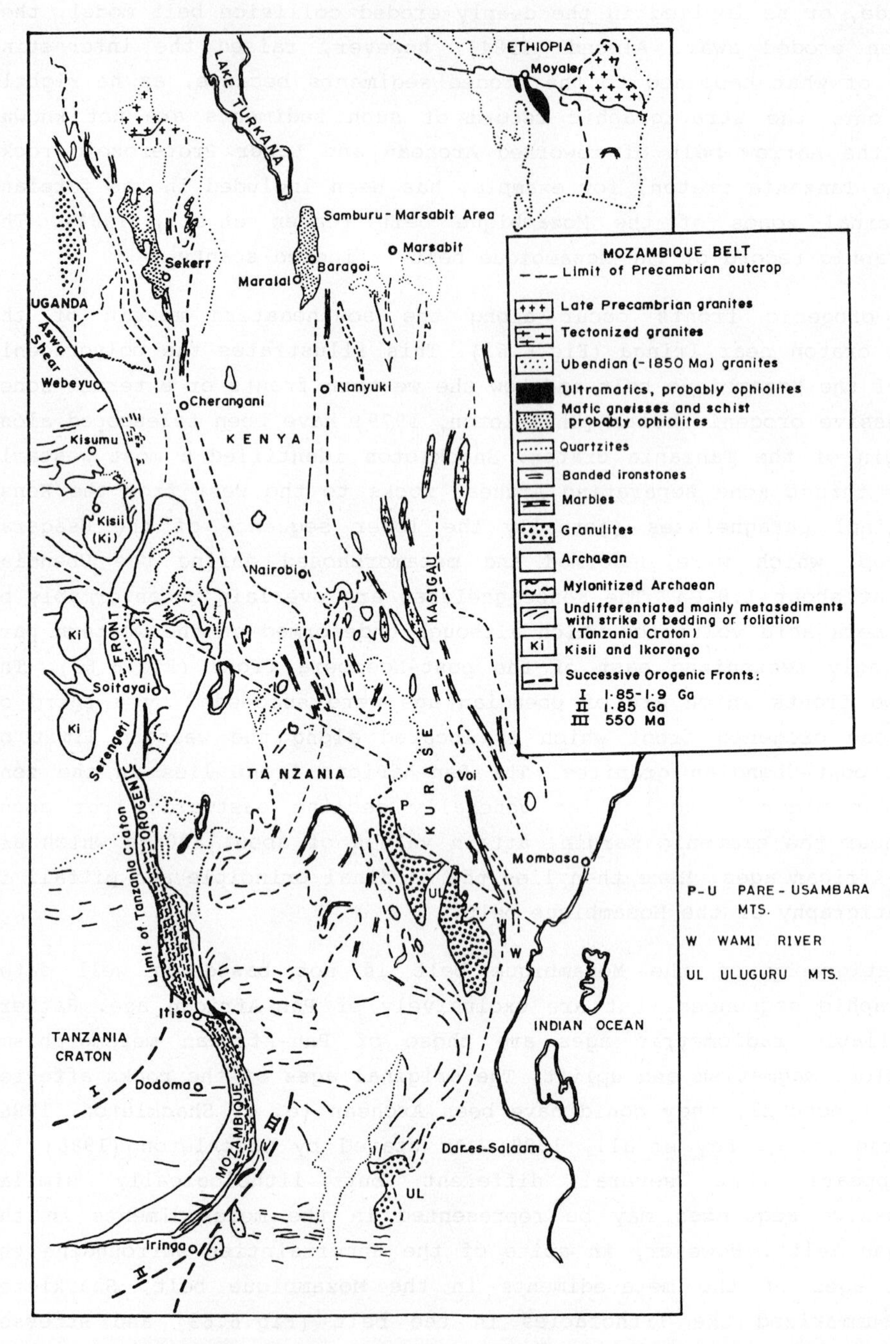

Figure 6.63: Structural and stratigraphic sketch map of the Mozambique belt. (Redrawn from Shackleton, 1986.)

6.10.3 Foreland and External Zones

Resting unconformably along the eastern margin of the Tanzania craton between about 3°S and the equator are low-grade and tabular metasediments which have traditionally been correlated with the Bukoban Supergroup of northwestern Tanzania (Cahen et al., 1984). These are the Ikorongo and the Soit Ayai Groups of Tanzania, and the Kisii Group of the adjoining parts of Kenya, which are the putative foreland representatives of the formations that were deposited in the Mozambique orogen to the east. In central Serengeti in northwestern Tanzania (Fig.6.63) the Ikorongo Group is only slightly metamorphosed, and comprises about 1,500 m of quartzitic metasediments. The Soit Ayai Group, the more highly metamorphosed equivalent of the Ikorongo Group, lies some 60 km eastward.

The lower formation of the Ikorongo Group is a purple quartzite with ripple marks and cross-bedding and locally developed coarse conglomerate containing boulders mostly derived from the underlying Archean rocks. A thick sequence of siltstones, slates, and sandstones constitute the upper formation which correlates with meta-gneisses in the Soit Ayai Group. There is a progressive increase in deformation and metamorphism from the Ikorongo Group to the Soit Ayai Group, with the Ikorongo attaining the greenschist facies while the Soit Ayai Group having attained a higher metamorphic grade, contains kyanite, garnet and staurolite. Based on its paleomagnetic pole which falls between that of the Bukoban Sandstone (c. 1.0 Ga) and that of the Kigonero Flags (c. 900 Ma) the deposition of the Ikorongo Group is believed to have occurred within this interval (McElhinny and McWilliams, 1973; Piper, 1975).

The Kisii Group is exposed near Kisii township in southwestern Kenya (Fig.6.63) as a succession of relatively flat-lying rocks, about 1.3 - 1.0 Ga old (Shackleton, 1986), resting non-conformably upon the Archean Nyanzian and Kavirondian metasediments. The Kisii Group forms shallow basins with north-south axes which are slightly tilted towards the west. The group comprises three subdivisions, the oldest being the Kisii soapstone, some 15 - 20 m thick; non-porphyritic basalts (170 - 500 m thick); and porphyritic basalts (0 - 200 m thick). The middle subdivision (15 - 150 m) sometimes resting with a blanket-type conglomerate on the eroded basement in places where the lower group is missing, is a sequence of ferruginous siltstone and chert; and quartzites with thin pebble beds. An upper subdivision of considerable thickness consists of rhyolites and rhyolitic tuffs; intercalated feldspathic sandstones and conglomerates; andesites and dacites; and porphyritic and non-porphyritic felsites.

6.10.4 The Internal Zone

Granulite Complexes

Granulites and associated amphibolite-facies rocks occur in the vast region east of the Tanzania craton. Hepworth (1972) delineated several domains in the granulite terrane of central and eastern Tanzania using aerial photos, and also adopted for this region the blanket term Usagaran, a stratigraphic term which in its type area denotes rocks of Early Proterozoic age. Similarly, Sanders (1965) applied the common term Turokan system to the Mozambiquian rocks of western Kenya, which rest with a marked unconformity on the cratonic foreland. However, Cahen et al. (1984) rightly insisted that the term "Usagaran Supergroup" should be restricted to its type area near Iringa where, as aforementioned, the Early Proterozoic Usagaran Supergroup rests on the craton and forms a distinct orogenic belt which, however, served as the tectonic front for subsequent Precambrian orogenies.

The internal or mobile zone of the Mozambique belt in the Kenya-Tanzania province consists of predominantly high-grade metamorphic assemblages (granulite- and amphibolite-facies) and minor but tectonically significant ophiolitic rocks. Both the granulite and ophiolitic complexes are discussed below in areas where they are best known. In Tanzania the granulite complexes form north-south-trending discontinuous domains. Recent radiometric ages from the granulites of Tanzania (Muhongo, 1989) and Kenya (Key et al., 1989) range from 2.75 Ga to 538 Ma, suggesting the reworking of Archean to Kibaran rocks during the Pan-African tectono-thermal events.

Hepworth (1972) had grouped the granulites of Tanzania (Fig.6.63) into western or foreland granulite complexes which lie on the craton and do not exhibit Pan-African ages (Muhongo, 1989); central granulite complexes which are characterized by the abundance of quartzitic lithofacies; and eastern granulite complexes which lack or seldom contain quartzites, but contain mostly marbles.

Central Granulite Complexes of Tanzania

According to Malisa and Muhongo (1990), the central granulite complexes of Tanzania, represented by the Loliondo, Longido, Ifakara, Fura and Songea complexes, have fault-bounded structures, and consist lithologically of pelitic and psammitic metasediments and their migmatized equivalents. They are predominantly biotite-hornblende and garnet-pyroxene granulites;

feldspathic micaceous quartzites; and chlorite schists. Metamorphosed equivalents of igneous rocks include meta-gabbros, meta-pyroxenites, meta-dolerites, and amphibolite lenses. Pegmatites are abundant. Commonly occurring quartzites suggest sedimentation of the granulite protoliths under the influence of granitic and gneissic source areas which lay on the Tanzania craton.

The central granulites have been metamorphosed at the almandine-amphibolite grade, with granulite-facies reached in certain places. The complexes occur as synformal and antiformal structures. Their general east-west-trending axes swing towards the southeast; and mineral stretching lineations dominantly plunge to the southeast and northeast, suggesting the refolding of earlier folds in the Mozambique belt. Faulting was predominantly in the NE-SW direction. The original ages of the rocks of the central granulite complexes are not known, but they have yielded only Pan-African ages (Malisa and Muhongo, 1990).

Uluguru Mountains Granulite Complex

The Uluguru Mountains of Tanzania (Fig.6.63) belong to the eastern granulite complexes which are large fault-bounded structures partly concealed beneath Cenozoic volcanics and sediments. The eastern granulite complexes include also the Pare Mountains, Usambara Mountains, Wami River, and Nachingwea complexes. These polymetamorphic complexes have yielded Archean to Pan-African ages (Muhongo, 1989).

The striking peculiarities of the eastern granulite complexes include: the preponderance of marbles; the probable occurrence of charnockites; and the occurrence of eclogites and metamorphosed ophiolite rocks including labradorite anorthosites, pyroxenites, dunites, serpentinites, amphibolite dykes and magnesitiferous peridotites and dolerites in shear zones, suggesting the preservation of cryptic sutures in this region.

The Uluguru Mountains have a long and complex Precambrian history, and comprise three groups, a granulite group, an acid group, and a crystalline limestone group (Cahen et al., 1984). Since they probably have a more complex history, the granulite and acid groups are believed to be older than the limestone group. A meta-anorthosite body occurs mostly in the granulites showing conformable contacts with the granulite foliation, but which on a wider scale, appears to be intrusive. The rocks of the granulite group including the meta-anorthosite were initially at the granulite facies metamorphism, but they later underwent retrogressive metamorphism in the amphibolite facies resulting in units of the acid gneiss group which contain relicts of granulites.

The complex tectonic history of the Uluguru Mountains (Cahen et al., 1984) involved: isoclinal folding on north or northeast axes resulting in the regional strike of the foliation, with smaller-scale structures on the same pattern; smaller-scale folding on major NNE-SSW axes and minor ESE-WNW axes involving all three groups of rocks in the complex; subsequent gentle folding giving rise to broad anticlines and synclines with mainly north-south axes; and lastly the intrusion of mica pegmatites.

Pare-Usambara Mountain Granulite Complex

These form horst-like mountain ranges in northeastern Tanzania, and consist of high-grade rocks known as the Pare-Usambara Mountain Group, an assemblage with enderbites, granulites and anorthosites (Cahen et al., 1984). The Pare-Usambara horst is flanked to the northwest by the Masai Steppe Group and to the east by the Umba Steppe Group. Both groups are amphibolite-facies gneisses and migmatites, and of younger age than the Pare-Usambara Mountain Group. Structurally the latter group consists of well-defined flat layers dipping gently to the northeast. Pre-existing structures are rare. Large-scale folds seem to swing from northeast in the south, to NNE in the north, and the youngest structural trends are NNW. The Masai Steppe and Umba Steppe Groups contain relicts of pyroxene granulites; the granulite structures are not flat as in the Pare Mountains but are irregularly folded about northeast axes. These younger groups contain relicts of pre-granulite structures.

Kurase and Kasigau Groups of Kenya

Key et al. (1989) outlined the main features of the internal zone in Kenya. Throughout this region a migmatite basement to the metasediments has been recognized. Part of this basement (Turbo migmatites of Sanders (1965)) in western Kenya represents the reworked Archean of the Tanzania craton, which has been traced for about 100 km eastwards into the Mozambique belt (Vearncombe, 1983). Elsewhere in Kenya mid-Proterozoic ages have been obtained from the migmatite basement. In southeastern Kenya the overlying Kurase-Kasigau group of metasediments shows facies change from shallow water shelf lithofacies eastward into deeper water sediments. The uppermost shelf metasediments to the east of Nairobi contain an evaporite component in the form of scapolite-bearing gneisses.

The Kurase-Kasigau Group of metasediments occurs in the Voi-Tsavo area of southeastern Kenya (Fig.6.63) as a continuation of the Pare-Usambara Mountain terrane. The Kurase Group comprises miogeosynclinal lithologies such as marble and quartzite; and graphite, sillimanite and kyanite-gneiss and schists; biotite-hornblende gneiss; and amphibolites (Gabert, 1984). The overlying Kasigau Group represents the eugeosynclinal facies, and consists mostly of graywackes which have been metamorphosed to quartz-feldspar-biotite-hornblende gneiss, with intercalations of ortho-amphibolites. Facies transitions occur between the Kurase and Kasigau Groups; and basic and ultrabasic rocks were emplaced along regional faults or thrusts zones.

Three phases of deformation have been recognized in the Kurase and Kasigau Groups, the last of which controlled the emplacement of late pegmatites and the joint pattern in this region (Gabert, 1984). Isoclinal and open folds with NNE and NNW-striking axes are prevalent in the Karuse and Kasigau Groups. In the area southwest of Voi stretching lineations show predominantly northerly dips which are parallel to the axes of the minor folds; there are also NNE-dipping foliations in the area in which the stretching lineations are mostly parallel to the roughly north-south trend of the Mozambique belt.

Although the succession of geologic events was not clearly defined, Gabert (1984) outlined the late phases of the geological history of the Kurase-Kasigau area as comprising: Barrovian medium-to high-grade metamorphism; hydrothermal activity leading to syngenetic base metal mineralization; pegmatite emplacement; and widespread re-setting of radiometric ages during Pan-African tectonism at about 550 Ma.

North-Central Kenya Granulite Complex

This is a vast exposure of the Mozambique belt south of Lake Turkana in the Samburu-Marsabit area of north-central Kenya (Fig.6.63). Key et al. (1989) unravelled the complex geology of this region in such detail (Fig.6.64) that it warrants a close examination of its tectonic features and scenario in order to gain more insight into the polyphase evolution of the Mozambique belt as a whole.

The lithostratigraphy of the Samburu-Marsabit area (Fig.6.65) consists of the basal Mukogodo Migmatites which are unconformably overlain by metasediments such as banded gneisses, into which the migmatites have been thrust as subconcordant sheets. Continental clastic units (e. g. Ndura, Loroki, Kotim gneisses) comprising meta-arkoses, meta-quartzites and manganiferous sandstones, with locally preserved sedimentary structures, are among the banded gneisses. These continental units show facies change into more pelitic metasediments, the Don Dol Gneisses, in the centre of the Samburu-Marsabit area. The basal metasediments are overlain by a

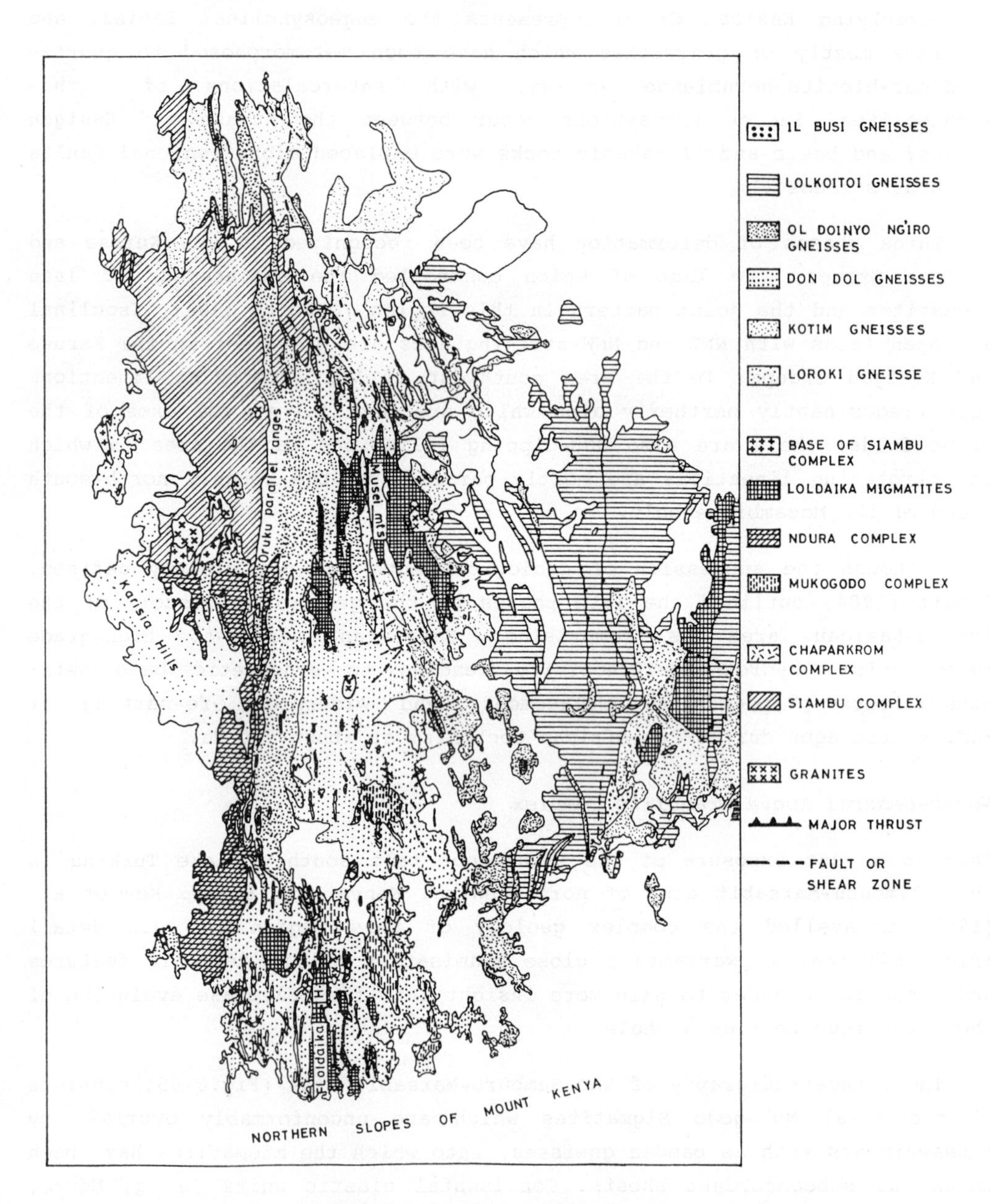

Figure 6.64: Geological map of Northern Kenya (Samburu-Marsabit area). (Redrawn from Key et al., 1989.)

The lithostratigraphic groups

Lithostratigraphic group	Lithologies	Maximum thickness (m)	Remarks
Chaparkrom Complex and Korr Complex	Melanocratic hornblende + quartz gneisses fissile leucocratic gneisses, disrupted marbles, metaquartzites, amphibolites ultramafics	Unknown	Slice of ophiolitescomplex (?) tectonically emplaced over Kotim and Lolkoitoi Gneisses
Siambu Complex	Assorted melanocratic hornblende-rich gneisses, and fissile biotite gneisses marbles, metaquartzites and sheeted dykes: migmatites and allochthonous plutons derived from the host complex during subsequent metamorphisms	+4000	Defines the Morilem Synform: thrust over the Ol Doinyo Ngiro and Loroki Gneisses: local migmatisation along contact.
Il Busi Gneisses	Fine-grained, dark-grey graphitic gneisses marbles, metaquartzites	~1000	Forms the highest synform in a listric fault complex in the east.
Matoni Gneisses	Fissile garnet + sillimanite gneisses and schist and minor biotite gneisses muscovite gneisses, marbles and calc-silicate	~1000	Local unit above the Kotim Gneisses
Lolkoitoi Gneisses	Banded fine-grained and biotite gneisses	~1000	Overlies the Kotim Gneisses
Ol Doinyo Ngiro Gneisses	Grey biotite gneisses with major marbles metacherts, metaquartzites graphite gneisses, corundum + garnet gneisses, quartzo-feldspathic gneisses and rare meta-ironstones	~4000	Overlies Ndura Complex and Don Dol Gneisses, tectonically interleaved with Ndura and Siambu Complexes
Don Dol Gneisses	Grey biotite gneisses and darker hornblende gneisses, quartzo-feldspathic gneisses, metaquartzites, graphite gneisses		Overlies Mukogodo migmatite in centre of map area
Kotim Gneisses	Massive quartzo-feldspathic gneisses, metaquartzites, charnockitic gneisses (NW) manganiferous metapsammites, biotite gneisses	~4000	Major thrusts cut the area
Loroki Gneisses	Quartzo-feldspathic gneisses.	~2000	Lithologically similar to the Kotim Gneisses
Ndura Complex (Cover)	Leucocratic, banded grey gneisses with from 15 to 50% concordant quartzo-feldspathic sheets (stromatic migmatites) with mafic orthogneisses and garnetiferous migmatitic gneisses	+1500	Underlies the Loroki and Ol Doinyo Ngiro Gneisses
Mukogodo Migmatites (Basement)	Agmatites and nebulitic migmatites with quartzo-feldspathic palaeosomes as well as K-feldspar and garnet porphyroblasts, mafic dyke swarms and numerous felsic vein phases	Unknown	Basal unit

Figure 6.65: Suggested stratigraphy of the Samburu-Marsabit area. (Redrawn from Key et al., 1989.)

sequence of marbles, meta-pelites and vanadiferous graphitic gneisses (Ol Doinyo, Ng'iro, Lolkoitoi, Makoni, Il Busi Gneisses). Meta-volcanic slices, some of which are ophiolitic (Chaparkrom, Korr, Siambu Complexes), were emplaced along subhorizontal thrusts into high tectonic levels in the meta-sedimentary pile.

The granulites of the Samburu-Marsabit area extend westward underneath the Cenozoic strata of the East African Rift Valley where the root zone of the Mozambique belt was probably located. This zone corresponds to the axis of bilateral symmetry which Sanders (1965) had postulated as the one where the nappes of western Kenya face northwest and those in the eastern part of the Samburu-Marsabit area face to the southeast. According to Key et al. (1989) this was the root zone of the Mozambique orogen marking the separation of the southeastwards backthrust units in the east from the overthrust western units located on the external and foreland zones near the craton.

Several syn-tectonic and post-tectonic granitoid intrusives occur in the Samburu-Marsabit area. These consist of small trondhjemite sheets which are either concordant or discordant with the gneisses they intrude; low level concordant plutons; and high level discordant plutons. There are also vertical and narrow micro-granite plutons which are concordant with the foliation of the adjacent gneisses and with narrow partial melt migmatite zones rimming some of the larger stocks; and common minor felsic dykes and veins of various ages.

The following sequence of tectonic events was recognized by Key et al. (1989). First there was plate collision which caused a tectono-thermal event in the amphibolite-granulite facies at about 820 Ma, and produced major recumbent folds with ductile thrusting which interleaved the basement, meta-sedimentary cover, and slices of ophiolitic meta-volcanic complexes. The first phase of this deformation involved the emplacement of crustal melt granites and metabasic dykes. Between 620 Ma and 570 Ma there was post-collisional greenschist-amphibolite facies deformation which produced regional uptight folds and vertical ductile strike-slip shear zones which strike subparallel to the orogenic strike. This culminated in the intrusion of syntectonic granites. High level open folding and brittle shears mark the terminal orogenic events. The final uplift and cooling has been dated at about 500 - 480 Ma.

Karasuk-Cherangani Group

The basement of northern Uganda comprises mostly granulite grade migmatitic gneisses and hornblende schist of Archean or Early Proterozoic age which have been strongly overprinted with Pan-African ages (Almond, 1969; Cahen et al., 1984). An extensive meta-sedimentary unit known as the Karasuk Group in eastern Uganda, and the Cherangani meta-sedimentary group in western Kenya, overlies this basement sometimes with thrust contacts. Although known by different names in each country, the Karasuk-Cherangani metasediments extend across the Uganda-Kenya boundary and even northwards into southern Sudan (Fig.6.66). The metasediments are made up of quartzites, quartzo-feldspathic psammites, mica and graphitic schists and calc-silicate gneisses and minor crystalline limestones. They have been interpreted as miogeosynclinal deposit along a stable continental margin to the west. An active outer eugeosynclinal volcanic-arc belt lies to the east and is represented by the Sekerr ophiolite-island-arc volcano-sedimentary assemblage (Fig.6.66).

The Karasuk-Cherangi area has yielded significant structural information that relate to the tectonics of the western Mozambique belt (Shackleton, 1986). The western half of the area consists of a northeast-dipping mylonite zone which lies along the margin of the craton; and an imbricate stack of thrust slices of Archean basement followed eastward by another imbricate zone of Proterozoic metasediments (Fig.6.67). The Sekerr ophiolites are thrust westward at a higher structural level. The complicated imbricate zone at Sekerr dips gently to the east with structures facing alternately upwards and downwards, and includes an extensive zone with repetitions of dismembered ophiolites now occurring as lenses. These are the features of a major crustal shear zone which has been interpreted as a suture on account of the well-defined zone of ophiolites (Shackleton, 1986; Vail, 1988b).

6.10.5 Ophiolitic Rocks

The internal zone of the Mozambique belt contains imbricate slices of mafic and ultramafic rocks whose ophiolitic nature has survived the high-grades metamorphism and intense deformation in the mobile zone. Although these ophiolitic rocks have been widely reported by several previous authors (e. g. Kazmin et al., 1978; Prochaska and Pohl, 1984; Shackleton, 1986; Vearncombe, 1983), Behre (1990) has added more geochemical and structural data which have further elucidated their geodynamic significance. Behre also traced northward these ophiolitic rocks which define cryptic sutures in the Mozambique belt, and linked them with the

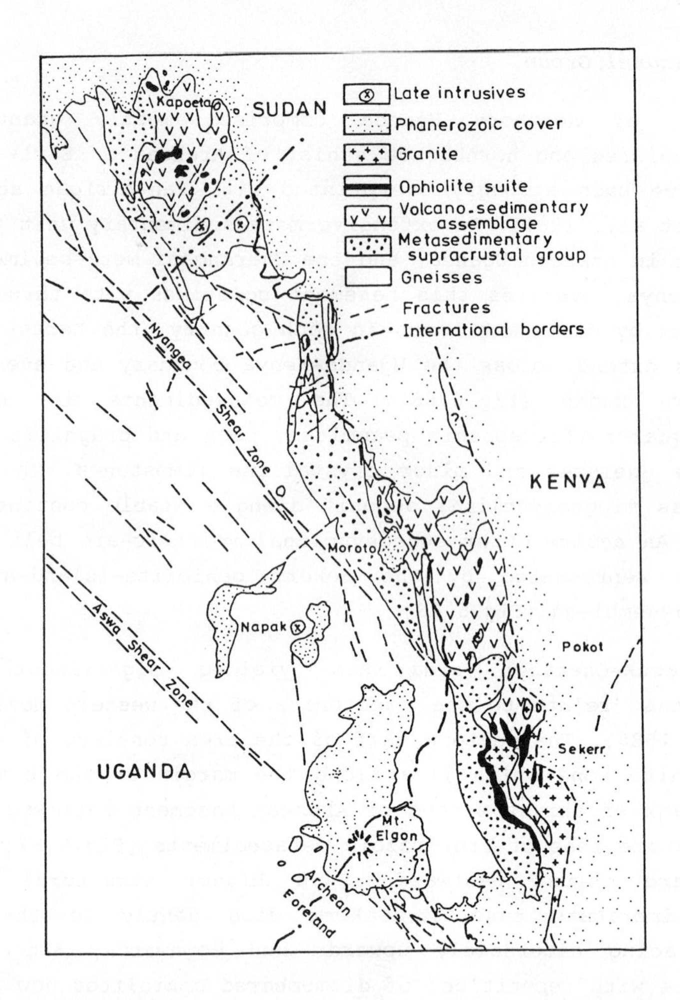

Figure 6.66: Geological sketch map of the Kapoeta-Karamoja-Sekerr area. (Redrawn from Vail, 1988b.)

extensive ophiolites of the Arabian-Nubian Shield (Fig.6.62), thus demonstrating that the Mozambique belt and northeast Africa were affected by the same regional collision events. The Sekerr, Baragoi and Moyale ophiolites of northern Kenya (Fig.6.63) provide the links between the Mozambique belt and the Arabian-Nubian Shield ophiolites.

Sekerr and Itiso

As aforementioned there is a complicated imbricate zone of ophiolites which rests upon the metasediments of the Sekerr area (Vearncombe, 1983) which has been followed northward into the Karamoja district of southern

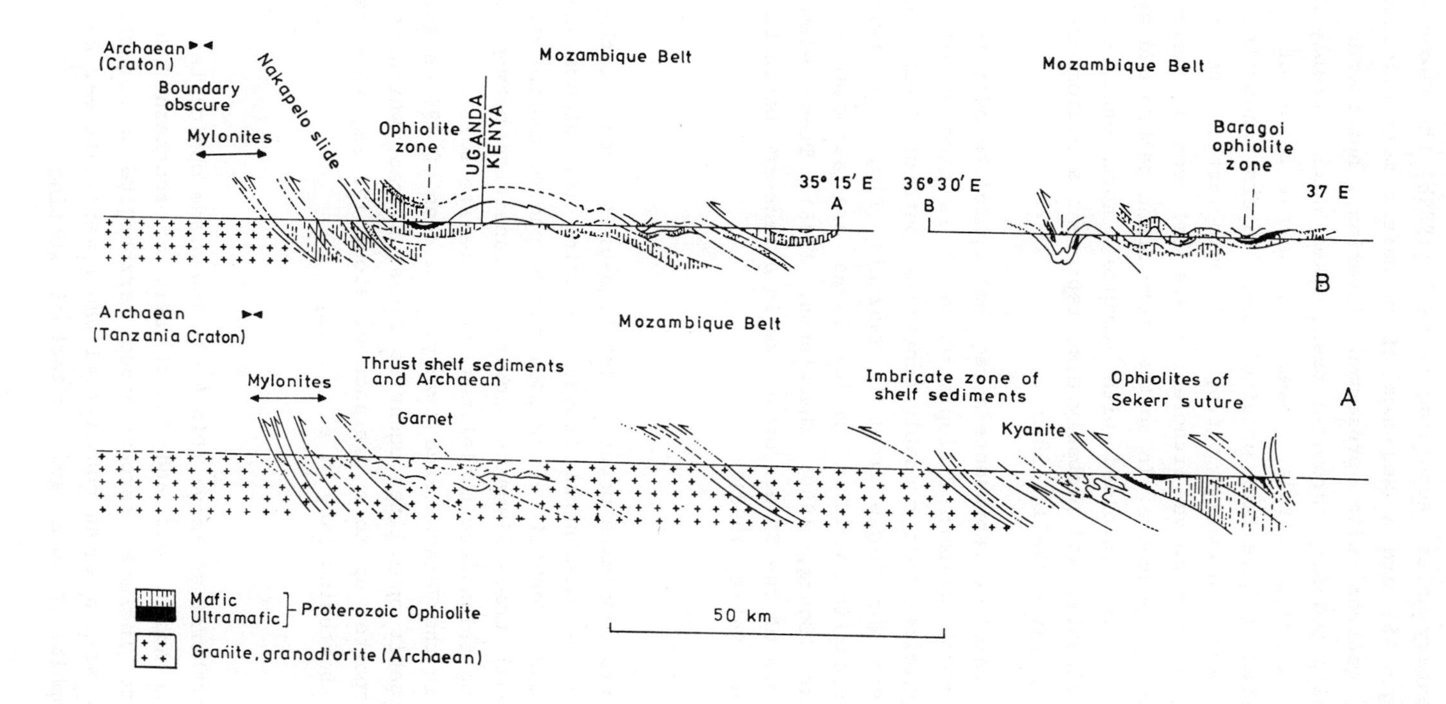

Figure 6.67: Cross-sections through the Northern Mozambique belt. (Redrawn from Shackleton, 1986.)

Sudan (Vail, 1988b) where there is strong thrusting of the ophiolites onto the meta-sedimentary units. According to Vail (1988b) the Sekerr-Karamoja ophiolites (Fig.6.66) are a sequence of andesitic meta-volcanic rocks; pillow lavas; gabbros with preserved layering; hornblende schists; serpentinites with podiform chromite; basic dykes which probably represent a sheeted dyke complex; marble lenses; and narrow bands of quartzites which are believed to have been original chert layers; psammites and mica schists which were probably tuffs; and pyroclastic and turbiditic sediments. All of the above lithofacies are believed to belong to an island-arc ophiolite sequence. In southeastern Sudan gabbro and pyroxenite fragments are tightly infolded with marbles, hornblende schists and gneissic metasediments, and probably also represent a dismembered island-arc and ophiolite suite (Vail, 1988b).

The Sekerr ophiolite represents a regional ophiolite belt that extends all the way through western Ethiopia where it is exposed in the Akabo region. This ophiolite belt probably continues south of Sekerr into Itiso along the eastern Mozambique front in Tanzania (Fig.6.62) where in the Itiso mafic-ultramafic complex, pillow lavas are associated with the ultramafic rocks (Behre, 1990; Shackleton, 1986). Trace element data suggest the origin of the Sekerr ophiolite in a back-arc basin between 1.0 Ga and 663 Ma ago (Behre, 1990).

Baragoi

Several ophiolitic complexes occur in the high-grade rocks of the Samburu-Marsabit area in north-central Kenya, including the Siambu complex at Baragoi. The Baragoi ophiolite includes metamorphosed mantle dunites and sheeted dykes with trace elements showing a transition between mid-ocean ridge basalts and island-arc tholeiites. Two separate suites of ophiolitic gabbroic rocks in the Baragoi area have yielded ages of 796 Ma and 609 Ma. The Samburu-Marsabit ophiolites occur as thrust slices, which at Marsabit have been transported up to a distance of about 100 km, thus reflecting severe crustal shortening (Key et al., 1989)

Moyale

At Moyale in northernmost Kenya (Fig.6.63) low-grade ophiolite occur which extend into the Adola fold and thrust belt of southern Ethiopia. The Moyale ophiolite includes serpentinized harzburgite and gabbros which occur as thrust slices among continental shelf meta-pelites. Behre (1990) assigned this ophiolite to a back-arc tectonic setting.

Pare Mountains

A highly dismembered and scattered ophiolite occurs here comprising serpentinites, meta-pyroxenites, meta-gabbros, and amphibolites. Ophiolites also occur in the neighbouring Taita Hills of southeast Kenya in what probably represents the continuation of a suture zone which Behre extrapolated northward into the Adola-Moyale belt (Fig.6.62).

6.10.6 Molasse

Low-grade metasediments which may represent Pan-African molasse rest probably unconformably upon high-grade rocks in the Mozambique belt of Kenya (Pohl, 1988). These are known as the Ablun Series in northeastern Kenya and as the Embu Series in the central part of the country (Pulfrey and Walsh, 1969). The Ablun Series consists of tightly folded, slightly metamorphosed rocks, including conglomerates, sandstones, phyllites, and limestones. The Embu Series consists of mostly pelitic rocks, with some limestones, sandstones and conglomerates. Tourmaline is predominant among the accessory minerals in both groups of metasediments.

6.10.7 Madagascar

Pan-African ages are widespread in the island of Madagascar and suggest the reworking of the extensive Archean to Kibaran terranes, during which the characteristic north-south foliation trends of the Mozambique belt were imposed on the island. Cahen et al. (1984) therefore considered Madagascar as an integral part of the Mozambique belt. This link is reinforced by plate reconstructions of the paleo-position of Madagascar alongside the coast of East Africa (Fig.6.62) based on the evidence provided by magnetic anomalies and other geophysical and regional geologic correlations (e. g. Rabinowitz et al., 1983; Reeves et al., 1987).

Most of what is known about the Pan-African of Madagascar is based on its abundant radiometric dating, rather than on detailed structural and regional field geologic investigations. The scheme for Pan-African events (Cahen et al., 1984) involved metamorphism at about 845 Ma which attained the granulite facies in some places but was only at the greenschist facies in the Early Proterozoic Vohibory metasediments in the southernmost part of Madagascar (Fig.3.45). Intrusions of granites and another metamorphic processes took place at about 740 Ma. There were granite and pegmatite intrusions between 750 Ma and 600 Ma during an episode of mild folding. Granites and pegmatites were again emplaced as from 600 Ma to 480 Ma.

6.11.8 Geodynamic Model

The fact that the Mozambique belt represents the lower crustal level of a huge continent-continent collision orogen is now widely accepted (Behre, 1990; Burke and Senghor, 1986; Key et al., 1989; Muhongo, 1989; Shackleton, 1986). The multiplicity of structural and petrological evidence that have been adduced in support of this model is summarized below. Whether or not there is enough evidence for the operation of a complete Wilson Cycle, and what plates collided, are pertinent questions raised by this model. The latter issues are also addressed in what follows below.

Shackleton (1986) argued convincingly that the fold and thrust structures so glaring in the Kurasuk and Cherangani Groups (Fig.6.67) are the products of plate collision, a view shared by Key et al. (1989) who placed the granulite internal nappes which were generated at the root zone of the collision somewhere beneath the Gregory Rift of Kenya. The eastern granulite complexes of Tanzania have been interpreted as tectonic thrust slices (Maboko et al., 1985) on the evidence (Muhongo, in press) of the hornblende-pyroxene granulites and amphibolite gneisses which are commonly thrust over graphitic marbles. There are also major thrust-zones at lithologic boundaries in which strong schistocity, mylonites and pseudotachylites have developed between granulites and meta-anorthosites, for example in the Uluguru and Pare Mountains. The occurrence of many local thrusts or shear-zones within schists, gneisses, granulites and meta-anorthosites are further evidence of intensive deformation among the eastern granulite complexes.

Although the actual number of sutures and the age of suturing have not been precisely determined, (e. g. Behre (1990) suggested two suture zones), the position of some of these sutures is roughly known based on the dismembered ophiolites which decorate them.

The stretching lineations in the western part of the Mozambique belt commonly plunge to the southeast and trend NW-SE, like the brittle shear zones and the Aswa wrench fault (Fig.6.62), thus implying approximately NW-SE plate motion. However, the stretching lineations in the central zones show a rather persistent roughly north-south trend which probably resulted from a late orogenic relative motion of plates parallel to the plate boundaries (Shackleton, 1986).

A possible Wilson Cycle scenario for the Mozambique belt was proposed by Behre (1990) who suggested an early stage of rifting, followed by a phase of subduction and island-arc accretion, ending with continent-continent collision . Although the stratigraphic record of early rifting in the Mozambique belt is quite tenuous and circumstantial compared with coeval Pan-African mobile belts already reviewed, the rifting of a Kibaran continent along East Africa would account for the presence of Archean to Kibaran rocks in central Kenya and in Madagascar, east of the suture zones; and also explain the existence of passive continental margin sediments (Kisii, Ikorongo, Soit Ayai Groups) along the foreland of the Mozambique belt. Subduction and island-arc accretion are evident from the ophiolites of the Mozambique belt, while continent-continent collision, as already mentioned, generated nappe-type folds and thrusts on a regional scale.

Paleomagnetic evidence (McWilliams, 1981) in the form of a large misfit between East and West Gondwana suggest that a large ocean separated both regions and that Madagascar and parts of eastern Tanzania, Kenya, Ethiopia and Somalia belonged to East Gondwana, whereas the Tanzania craton lay in West Gondwana (Fig.6.1). The tectonic evolution of the northern segments of this eastern Pan-African ocean will be examined next in the Arabian-Nubian Shield.

6.10.9 Mineralization

Metamorphic processes strongly controlled mineralization in the Mozambique belt. In general economically important mineralization is rare in this region. As pointed out by Pohl (1988), because of the heterogeneous nature of this belt, the mineral deposits vary widely (Fig.6.68). Metamorphic minerals which are economically important include flake graphite and kyanite in the metasediments; amphibole asbestos in ultra-mafics, green gem-quality vanadium grossularite in calc-silicate graphite schist; and blue zoisite in impure marbles.

The eastern granulite complexes of Tanzania have a high potential for gemstones such as ruby, varieties of garnet, green tourmaline, sapphire, scapolite, hornblende, tremolite, emerald, alexandrite, varieties of quartz, zircon, apatite and enstatite (Malisa and Muhongo, 1990). These gemstones are commercially exploited by small-scale mining. They are found in pegmatites either as conformable folded bands and layers of metamorphic mobilisates or as undeformed cross-cutting veins (Pohl, 1988).

Of lesser economic importance so far are the small syngenetic stratiform ores such as the Fe- and Mn-quartzites in Kenya; and the magnetite associated with the basic meta-volcanics of the Kasigau Group.

The ophiolites contain small deposits of chromium, nickel, platinum and copper. There are also ore deposits which are related to intermediate magmatic rocks. These include copper in the diorite-rhyolitic gneisses of the Voi area in Kenya; and magnetite and ilmenite in charnockitic gneisses, anorthosites, and associated pegmatites in the Pare Mountains of Tanzania.

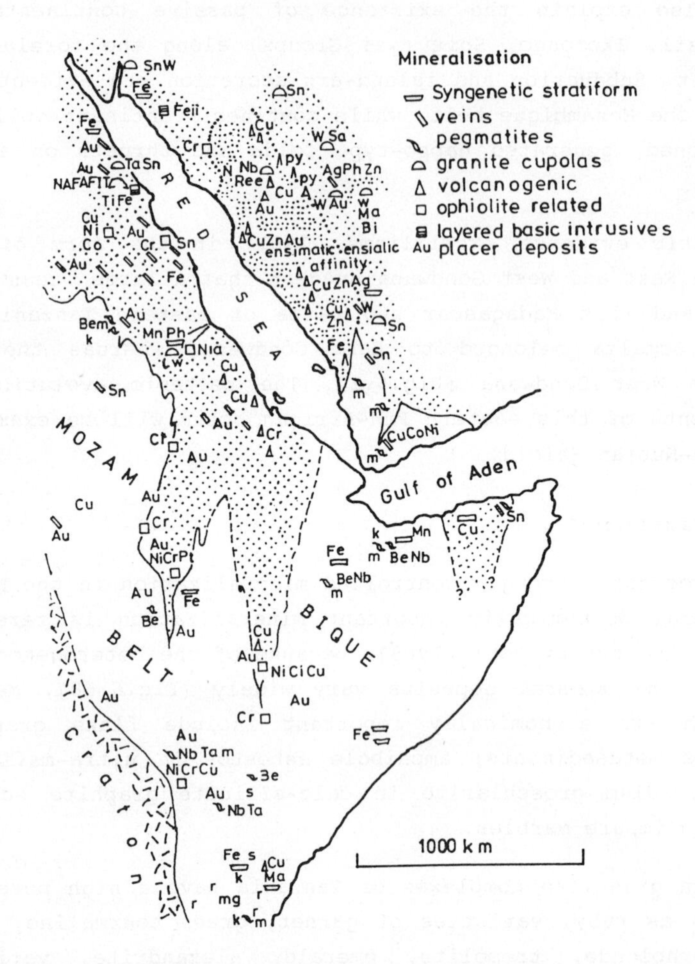

Figure 6.68: Metallogenic map of the northern Mozambique belt and the Arabian-Nubian Shield. (Redrawn from Pohl, 1988)

6.11 The Arabian-Nubian Shield

6.11.1 Tectonic Framework

Beyond the Kenya-Tanzania province the Mozambique belt splits into two segments (Fig.6.68). One segment continues northeast through southeastern

Ethiopia and Somalia into Yemen across the Gulf of Aden, while the other segment runs northwestwards into the Sudan. Both segments, characterized by high-grade gneisses and metasediments, are separated by large wedges of low-grade volcano-sedimentary rocks containing ophiolites (Fig.6.68). In this northern region the Mozambique belt becomes part of another tectonic province, the Arabian-Nubian Shield (ANS). The Nubian or African part of the ANS includes Egypt, Sudan, Ethiopia and Somalia, whilst Saudi Arabia and Yemen constitute the Arabian part of the shield.

A Late Proterozoic paleo-tectonic sketch map of NE Africa (Fig.6.69) in which the Red Sea, a much later geological feature is removed, displays the tectonic continuities across the ANS. Vail (1987) defined the ANS as extending from about 28° E to 50° E and from the equator to 30° N (Fig.6.70), thus encompassing those pre-Phanerozoic rocks previously referred to as the Precambrian Basement Complex, which are exposed in Sinai, the Egyptian Eastern Desert, Sudan, northern Uganda, northern Kenya, Somalia, and in Saudi Arabia and Yemen. Thus defined, the ANS includes the low-grade ensimatic (volcano-sedimentary and ophiolite) assemblages that constitute the Red Sea fold belt (Kazmin et al., 1978) which extends into Saudi Arabia where it is even more extensively developed. The ANS also includes the surrounding Pan-African or Mozambiquian continental shelf metasediments overlying Pan-African reworked Archean gneissic terranes. As stated by Schandelmeier et al. (1990), in Sudan, Ethiopia and Saudi Arabia, the continental polymetamorphic rocks of the Mozambique belt are in sharp thrust contact with Pan-African juvenile oceanic volcano-sedimentary and ophiolite assemblages (Fig.6.69).

The ANS is characterized by two principal lithological-tectonic units reflecting two contrasting tectonic environments (Vail, 1988b). First, are the gneisses and meta-sedimentary rocks including both reworked pre-Pan-African rocks, as well as Pan-African continental margin sequences. These occur as scattered basement inliers east of the East Saharan craton and as exotic basement terranes further eastwards within the volcano-sedimentary and ophiolite assemblages. The latter constitute the Red Sea fold and thrust belt, and are by far the most extensive lithological-tectonic unit in the ANS. They represent oceanic margin and island-arc settings and their associated syn- to post-tectonic intrusions. The volcano-sedimentary and ophiolite units are low-grade rocks at the greenschist facies, in contrast to the amphibolite-facies quartzo-feldspathic and foliated rocks in the gneisses and meta-sedimentary units.

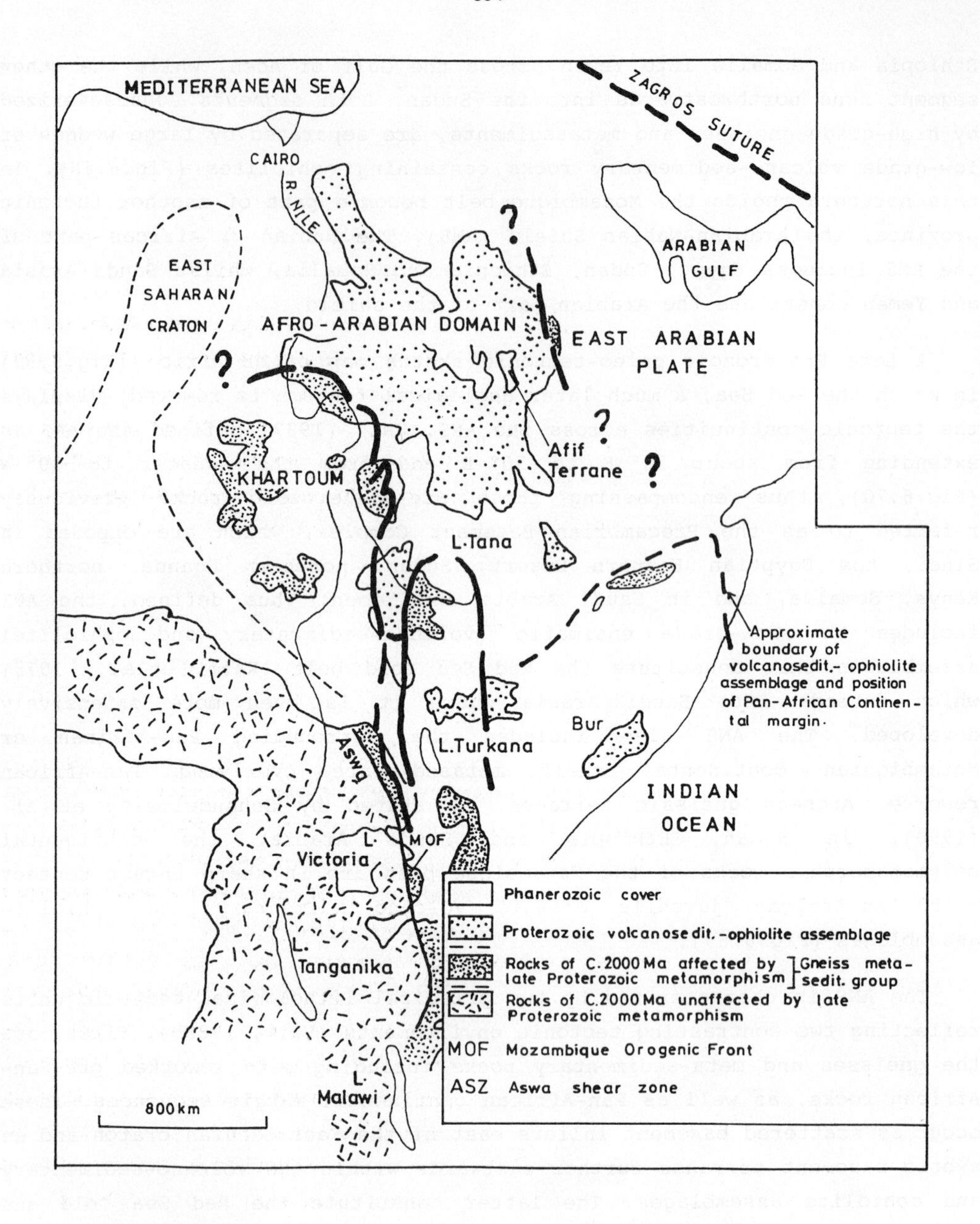

Figure 6.69: Tectonic setting of the Arabian-Nubian Shield on the pre-drift reconstruction. (Redrawn from figure supplied by N. J. Jackson)

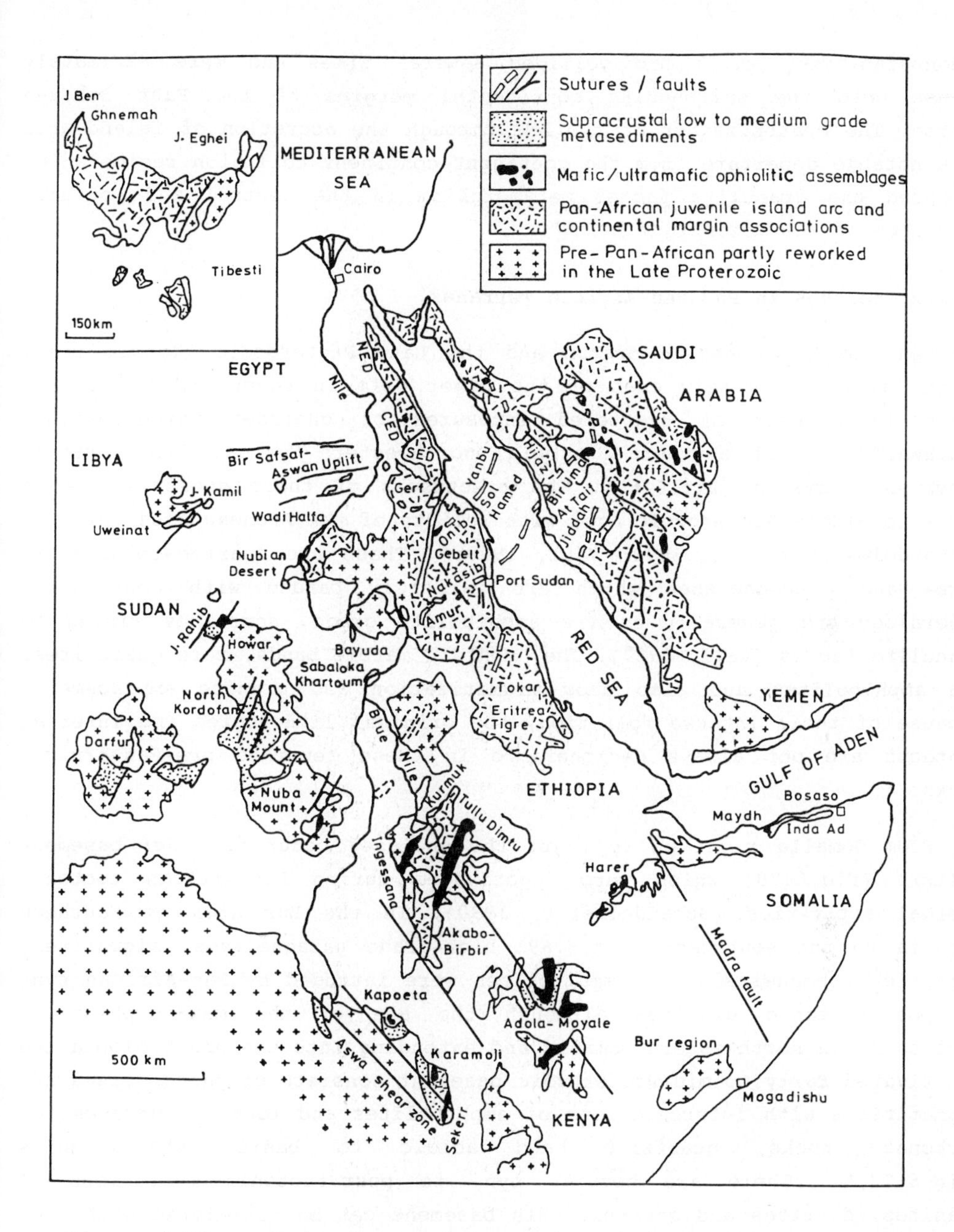

Figure 6.70: Major basement tectonic units of the ANS. (Redrawn from Schandelmeier et al., 1990)

The ANS evolved through the Wilson Cycle and displays the stages of rifting, oceanization with multiple island-arc complexes (Fig.6.71) of the

Indonesian-type, which had collided several times and were ultimately thrust onto the surrounding continental margins of the East Saharan craton. The cratonization of the ANS through the accretion of island-arcs is a notable departure from the continent-continent collision regime which produced the granulite facies metamorphism in the central and southern provinces of the Mozambique belt.

6.11.2 Gneisses in Pre-Pan-African Terranes

Between the East Saharan craton and the Late Proterozoic oceanic realm which lay approximately east of the River Nile in Egypt and the Sudan, there is a belt of scattered exposures of gneisses which extends southwards through southern Ethiopia, northwestern Uganda, into northern Kenya and Somalia (Fig.6.69). The precursors of these gneisses are of Early to Middle Proterozoic age, with relicts of the Archean also involved (Schandelmeier et al., 1990; Vail, 1988b). These are quartzo-feldspathic para- and orthogneisses which are variously banded with contrasting mineralogy and generally at the amphibolite grade, sometimes rising to granulite facies (Vail, 1987). They contain marble bands, rare quartzites, and amphibolites and also show migmatization and granite emplacement. Because of their intense foliation, the original lithologies and internal contacts are not always decipherable in these reworked pre-Pan-African rocks.

From Somalia right up to Egypt the gneisses occur as older basement inliers (Fig.6.70) which were reactivated during Pan-African tectono-thermal activities (Schandelmeier, 1990). In the Bur area of Southern Somalia to the southeast (Fig.6.69) high-grade paragneisses, migmatites, granitic orthogneisses and amphibolites were intruded by Pan-African syn- to post-tectonic granites although the age of the metamorphism is uncertain. In northwestern Somalia and extending into eastern Ethiopia the reactivated Early to mid-Proterozoic basement consists of paragneisses and migmatitites with intercalations of amphibolites and local occurrences of carbonate rocks, quartzite, and acidic to basic meta-volcanics (Fig.6.72,A). There are several syn- to post-tectonic intrusions of granites, diorites and gabbros. This basement can be correlated with that of Yemen on the opposite side of the Gulf of Aden (Fig.6.70).

Pan-African reworked basement in the Sudan includes those of the Nuba Mountains, the Darfur block, and the Bayuda and Nubian deserts. In the latter area high-grade granitoid gneisses with minor inliers of high-grade metasediments were migmatized and intruded by voluminous granitoids late in the Pan-African. Small patches of reworked migmatitic gneisses occur on

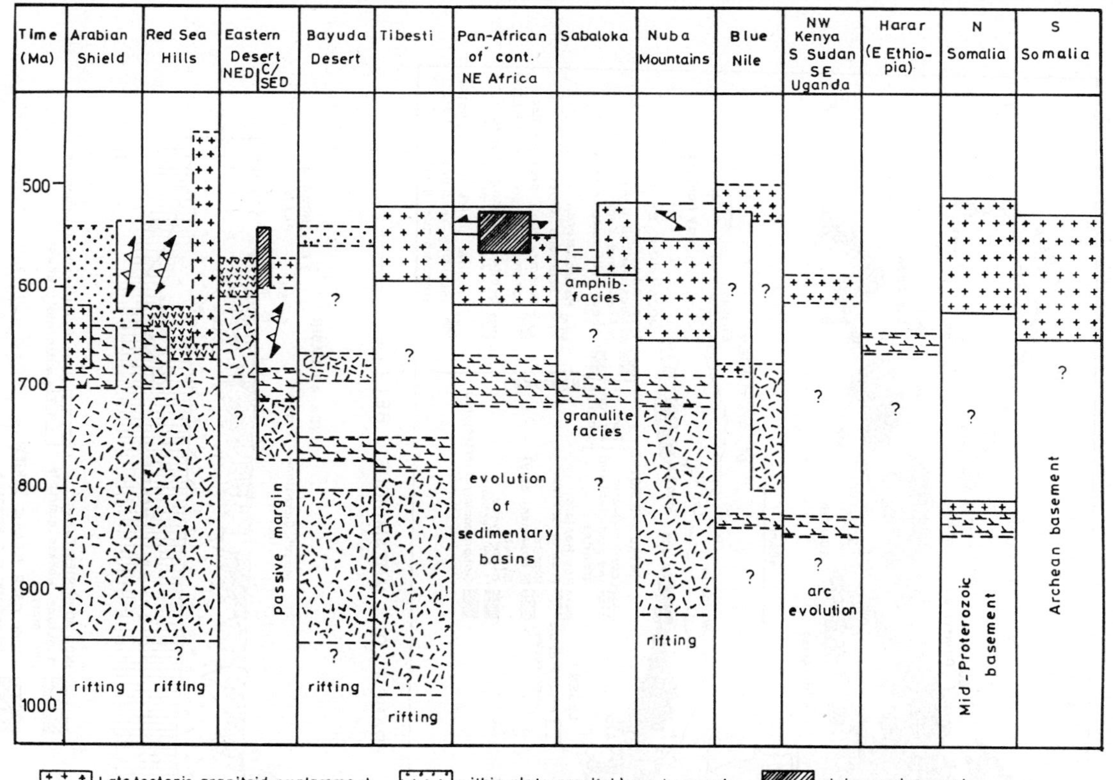

Figure 6.71: Correlation of principal tectonic episodes in the basement of the ANS. (Redrawn from Schandelmeier et al., 1990)

the extreme eastern part of Jebel Kamil in the Egyptian part of the Uweinat complex and as the high-grade granitoid basement with minor intercalations of metasediments in the Bir Safsaf-Aswan uplift (Fig.6.70).

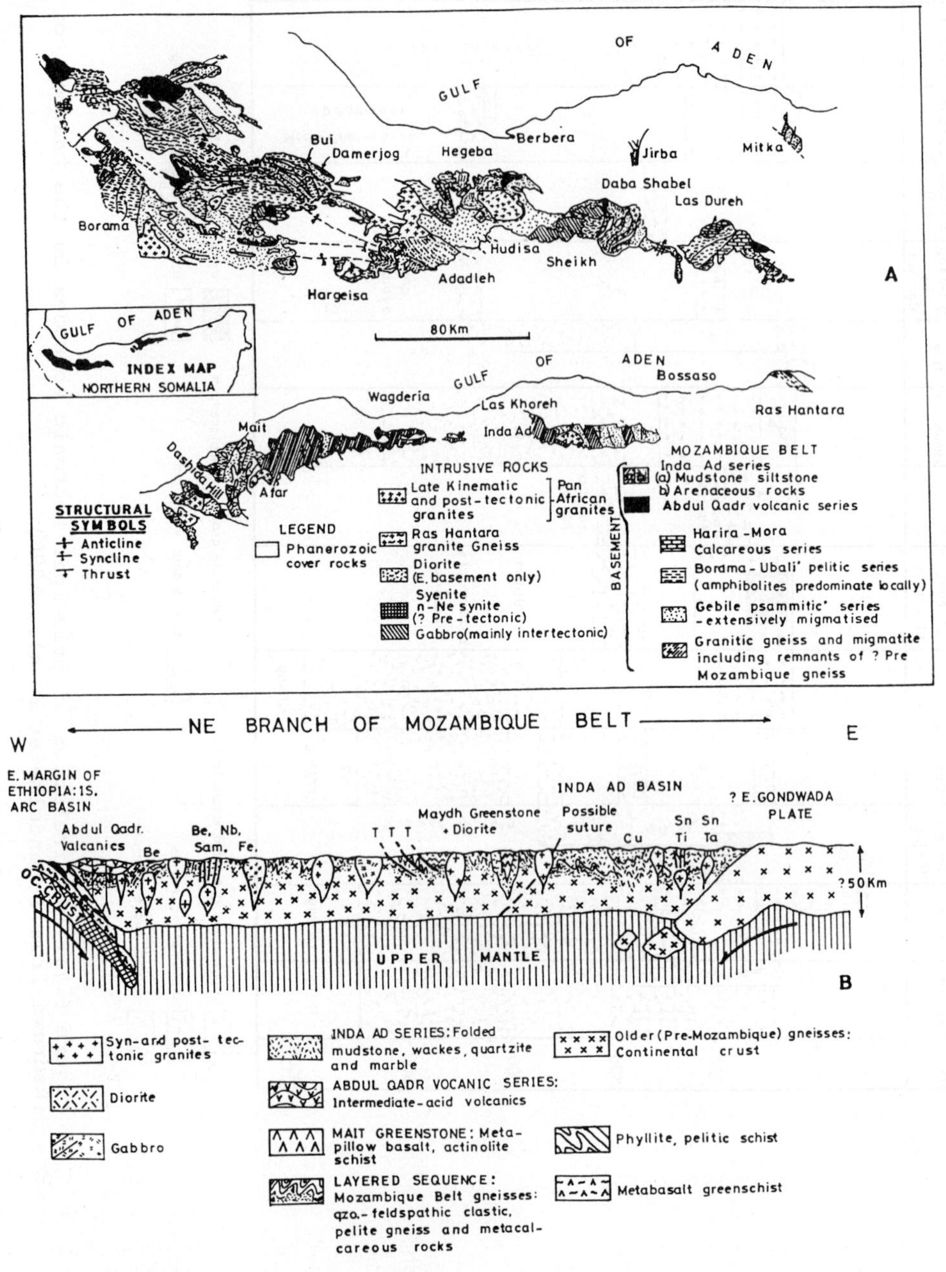

Figure 6.72: A, geological sketch map of N. Somalia, B, Pan-African plate tectonics for this region. ((Redrawn from Warden and Horkel, 1984.)

Also, small inliers of amphibolite-facies gneisses and metasediments occur as exotic terranes within the lower-grade volcanogenic ophiolite assemblages in the Southern Eastern Desert of Egypt, in the adjacent Red Sea Hills of the Sudan, and in Ethiopia and Saudi Arabia.

6.11.3 Meta-Sedimentary Belts Around the Red Sea Fold Belt

Schandelmeier et al. (1990) presented an interpretation of the rock assemblages and tectonic evolution of the meta-sedimentary belts scattered between the East Saharan craton and the Red Sea fold and thrust belt (Fig.6.70). They emphasized the fact that these belts represent a zone of early rifting that developed along the eastern margin of the East Saharan craton during the initiation of a Late Proterozoic Pan-African ocean in the ANS. The outline presented below together with the paleotectonic setting for the region are largely culled from Schandelmeier et al. (1990).

Southern Uweinat Belt

South of the Uweinat block in the northernmost extremity of the Sudan is exposed a belt of psammitic and pelitic rocks with minor marble, amphibolite and banded ironstones, all of which have been metamorphosed under low- to medium-grade conditions. Bimodal volcanics are intercalated within the metasediments in a manner that suggests early synsedimentary basin extension. Although no age data are available even from the syn- to post-tectonic granitoids which sharply truncate older structures in this belt, this basin is believed to be of Pan-African age on the basis of the lithologic, metamorphic and structural similarities with the Jebel Rahib belt nearby (Fig.6.70). Folds in the southern Uweinat belt strike along NE-SW axes and are open to isoclinal. The entire belt owes its characteristic sygmoidal outline to dextral wrench faulting late in the Pan-African.

Jebel Rahib Belt

This belt contains complexly deformed ultrabasic and basic igneous rocks and a thick sequence of arenaceous and subordinate carbonaceous metasediments which have been interpreted as deposits of a Red Sea-type Pan-African rift basin. Since no arc-type magmatic rocks and related volcaniclastic derivatives have been found, subduction of oceanic lithosphere was probably not involved during the closing of the Jebel Rahib basin. An age of 570 Ma from post-orogenic granitoids which were not affected by the penetrative NNE-SSW strike-slip shearing in this belt, sets the minimum age for its deformation and low-grade metamorphism.

An ophiolite assemblage with ultramafic rocks, pyroxenite, podiform chromites, massive and layered gabbros, dykes, pillow-lavas of clear mid-oceanic ridge affinity, and chert deposits, furnish the evidence supporting the appearance of oceanic crust in the Jebel Rahib rift. These ophiolitic rocks impose some constraint on the geodynamic evolution of this area, and imply that juvenile Pan-African rocks were generated in the Nubian Shield outside the Red Sea fold and thrust belt.

North Kordofan Belt

In its depositional setting and structural style the North Kordofan belt is similar to the Jebel Rahib belt, except that ophiolites have not been found. Although the ages of the deposition, metamorphism and deformation of the meta-sedimentary pile in the North Kordofan belt have not yet been ascertained, among the intrusive granitoids a tourmaline-bearing granite has been dated at about 590 Ma. Also late Pan-African shear zones which are sealed by mica-bearing pegmatites have yielded an age around 560 Ma.

Darfur Belt

The low-grade meta-sedimentary unit structurally overlying basement gneisses in the southeastern Darfur block may also be equivalent to the North Kordofan and Jebel Rahib metasediments. Intrusive granitoids have yielded ages of about 590 Ma and 570 Ma in the Dafur belt.

Eastern Nuba Mountains Belt

In the eastern Nuba Mountains a NE- to NNE-striking belt (Fig.6.70) of low-grade volcano-sedimentary rocks is exposed which contains fragments of highly dismembered ophiolites and basic to acidic plutons. These arc ophiolite assemblages were metamorphosed around 700 Ma, with post-tectonic magmatism ceasing around 550 Ma. Since the eastern Nuba Mountains do not represent the boundary with the volcano-sedimentary and ophiolite belt of the Red Sea fold belt, the Pan-African juvenile terrane of the eastern Nuba Mountains represents either a klippe thrust over a considerable distance from the east, or more likely it represents a minor ocean basin behind a large probably rifted-off continental fragment.

Bayuda Desert

Here Pan-African rocks occur as two different tectono-stratigraphic units. First, on the eastern part along the Nile, is a narrow strip of low-grade metasediments, meta-volcanics and granitoids which range compositionally from early tonalites through granodiorites to large peralkaline granites. The tectonic evolution of the area involved a main metamorphic event which followed plate collision at about 761 Ma. Before then, granitoids were emplaced above a subduction zone at about 898 Ma, followed by the emplacement of other subduction-related granitoids at about 678 Ma; and by anorogenic within-plate magmatism at about 549 Ma.

An extensive meta-sedimentary sequence of marbles and intercalated meta-quartzites is exposed between the Nile and the Red Sea Hills west of Gabgada (Fig.6.70), which according to Kröner et al. (1987) may represent the only autochthonous continental margin deposit that reflects the approximate position of the former continental margin (Fig.6.69). This would imply that the arc assemblage along the Nile south of Abu Hamed (Fig.6.70) belongs to an independent basin, separated from the major oceanic basin which was located in and beyond the Red Sea Hills (Schandelmeier et al., 1990).

Exotic Meta-Sedimentary Terranes

According to Kröner et al. (1987) rifted fragments of the East Saharan craton occur as high-grade meta-sedimentary exotic terrane among the volcanogenic-ophiolite-granitoid assemblages in the Egyptian Central Eastern Desert, the Southern Eastern Desert, and the Sudanese Red Sea Hills, notably at Meatiq, Hafafit and in the Sasa Plain of Gebeit, and near Haya, southwest of Port Sudan (Fig.6.73). Also, the Afif terrane (Fig.6.70) in the eastern Arabian Shield was identified as an exotic block of ancient gneisses (Stoesser et al., 1984) which though remobilized by the Pan-African tectono-thermal events, bears resemblance to African cratonic gneisses.

The exotic meta-sedimentary terranes, termed the "older shelf sequences" and regarded as the oldest rocks found in the Eastern Desert, were fully discussed by Kröner et al. (1987) within a Pan-African paleo-tectonic setting. As exposed in the composite dome structure of Hafafit (Fig.6.73), these metasediments consist of meta-quartzites and quartzitic schists which are mostly feldspathic and locally cross-bedded where the sedimentary structures had survived the intense metamorphism in these rocks. Some of those metasediments were aluminous as attested by the presence locally of sillimanite. The clastic metasediments of the central and Southern Eastern Desert have yielded U-Pb zircon ages as old as 2.06 Ga old, which suggests that their provenance lay in an ancient continental crust exposed probably along the margin of the East Saharan craton.

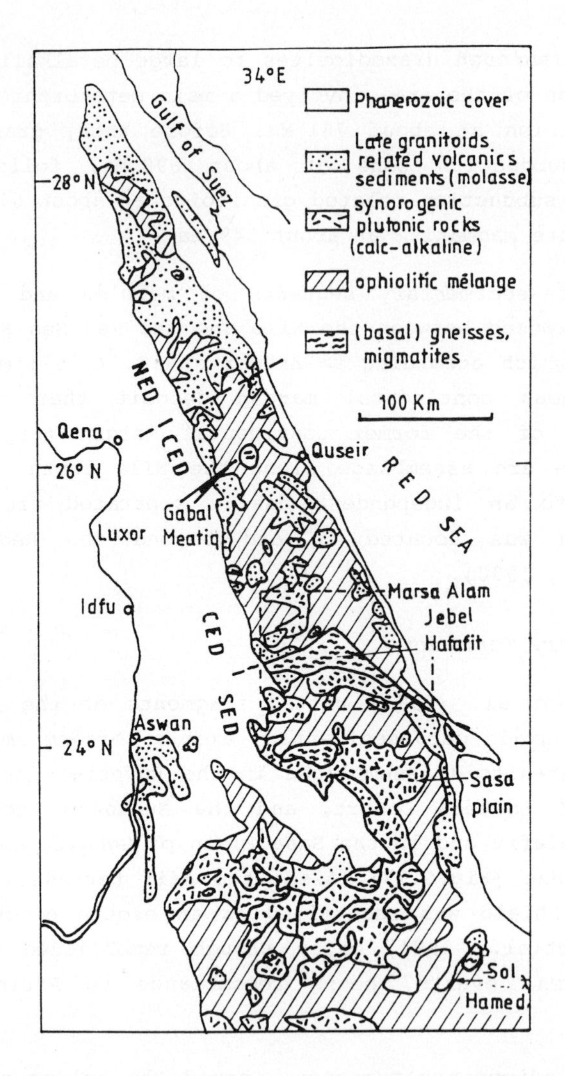

Figure 6.73: Precambrian rocks in the Egyptian Eastern Desert. NED, North Eastern Desert; CED, Central Eastern Desert; SED, South Eastern Desert. (Redrawn from Greiling et al., 1988.)

Local volcanic components among the Hafafit metasediments were probably derived from a primitive magma source, perhaps during the initial rifting and formation of a passive continental margin at about 900 Ma to 800 Ma ago. Continental margin deposits also occupy the tectonically lowest positions in the Red Sea Hills of the Sudan. These include the quartzites with associated marbles found in small outcrops in the Sasa Plain south of Gebeit and in extensive areas south of Wadi Amur; as well as the high-grade and partly aluminous metasediments near Haya, southwest of Port Sudan.

Inda Ad Group (Northern Somalia)

In northern Somalia the Inda Ad Group of metasediments (Fig.6.72,A) forms the eastern border of the local volcano-sedimentary and ophiolite sequence of the "Maydh Greenstone Belt" (Fig.6.72,B). The pelitic and psammitic rocks of the Inda Ad Group which are intercalated with marbles, are folded along N-S-trending regional fold axes and metamorphosed in the greenschist facies. The Inda Ad Group extends northward into southern Yemen where it is equivalent to the Ghabar Group. A granodiorite and post-tectonic granitoids have yielded late Pan-African ages. The tectonic setting of the Inda Ad Group - "Maydh Greenstone Belt" resembles that of the Jebel Rahib belt in the Sudan.

Tibesti Mountains (Chad-Libya)

Although located far out on the western part of the East Sahara craton, on the Chad-Libya frontier (Fig.6.70), the Tibesti Mountains contain magmatic and metamorphic rocks which represent the relics of an ocean basin (Ghuma and Rodgers, 1978) which began in the mid-Proterozoic and developed fully during the Pan-African, in a style and at a period similar to the Jebel Rahib rift to the east. Lithologically, the Precambrian of the Tibesti has been divided (Klitzsch, 1966) into a Lower Tibestian (medium-grade metasediments intercalated with basic volcanic rocks such as mica schists, micaceous quartzites, hornblende schists, amphibolites and pyroxenites), and an Upper Tibestian (low-grade quartzites and arkoses alternating with slates and rhyolitic lavas). Although these metasediments are also like those of the Rahib basin, volcaniclastics are more abundant in the Tibesti area and calcareous deposits are less prominent.

Subduction and metamorphism occurred on the eastern side of the Tibesti basin between 1.0 Ga and 750 Ma, but only the late Pan-African granitoids, ranging in age from 590 Ma to 520 Ma are well dated. These granitoids and late rhyolitic to basaltic dykes are petrologically and geochemically akin to coeval intrusives from southern Egypt and northern Sudan. The late Pan-African granitoids from Jebel Ben Ghemah on the western Tibesti basin are believed to have been induced by subduction.

Paleo-Tectonic Setting for the Meta-Sedimentary Belts

Vail (1988b) summarized the paleogeographic implications of the meta-sedimentary belts of the ANS. He regarded them as representing either a narrow continental margin with a shallow-water miogeosynclinal wedge (Jackson, 1980), which rested upon a gneissic cratonic foreland, or as supracrustal infillings of early Pan-African continental margin rifts, a view shared by Schandelmeier et al. (1990). Some of these rifts developed into small ocean basins that later closed, for example the Jebel Rahib, Nuba Mountains, and Inda Ad basins (Fig.6.72,B), or as in the Southern Uweinat, Darfur and North Kordofan basins, the mini-ocean stage was not attained. Thus, prior to or contemporaneously with extensive oceanization in the Red Sea fold belt and in Saudi Arabian parts of the ANS, processes of crustal extension, lithospheric thinning and the development of abortive rifts transpired extensively in other parts of the ANS (Jackson, 1987).

6.11.4 Volcano-sedimentary and Ophiolite Assemblages

Volcano-sedimentary Assemblages

These are heterogeneous piles of oceanic island-arc, and plate margin Andean-type volcanic and associated pyroclastic volcanogenic and shallow-water shales, siltstones and limestones (Vail, 1988b). As shown in Fig.6.70 they occupy the Sinai peninsular, most of the Central and Southern Eastern Deserts of Egypt, the Red Sea Hills, most of the basement of Ethiopia, and the western Arabian Shield and Yemen basement, thus forming the core area of the ANS. The volcanic rocks are predominantly calc-alkaline, ranging compositionally from basaltic and andesitic to rhyolitic types. Because of their characteristic greenschist to lower amphibolite facies, Vail (1976, 1979) grouped those in the Sudan into what he termed the Greenschist Assemblage (Table 6.4).

Jackson (1980) summarized the stratigraphic terms that have been assigned to the volcano-sedimentary units which he collectively termed the "younger meta-volcano-sedimentary units" (Table 6.4). In Egypt, those are found in the upper formations of the Abu Ziran Group; they are referred to as the Jiddah, Samran, Halaban and Hulayfah Groups in Saudi Arabia; the Thalab and older volcanic rocks in Yemen; the Tambian and Tsaliet Groups in northeast Ethiopia; and are included in parts of the "Older Series" of northeast Somalia.

Ophiolites

Closely associated with the volcano-sedimentary assemblages are linear masses of tectonized mafic-ultramafic complexes, comprising a succession which from base upward typically contain (Kröner et al., 1987; Vail, 1988b) serpentinized pyroxenites and peridotites, layered gabbros, sheeted dyke complexes, pillow lavas, and rare siliceous bands and plagiogranite, all pointing to an ophiolite suite (Fig.6.74). In Egypt the dismembered

Table 6.4: Correlation of the Late Proterozoic of the ANS. (Redrawn from N. J. Jackson, 1980.)

	? UNNAMED	METAMORPHIC	UNITS	Saramuj 'Series'	JORDAN
HALI GROUP		Ajal Group Bahah Group Baish Group Hali Group Jiddah Group	Samran Group Halaban Group Hulayfah Group Urd Group Ablah Group	Fatima Group Shammar Group Murdama Group Jibalah Group	SAUDI ARABIA
MITIQ GNEISSES	Abu Ziran Group Geosynclinal Metasediments	Ceosynclinal Metavolcanics Shadli and other volcanics	Rubshi Group	Dokhan Hammamat Group Awat, Asteriba, Homagar	EGYPT
	Metasedimentry Group (Kashebib Group)		Greenschist Assemblage or Nafirdeib Gr. etc.		SUDAN
	Mormora Group Adola Group		Tsaliet Group Tambian Group	Didykama Formation Shiraro Formation Matheos Formation	ETHIOPIA
	Aden Metamorphic Group		Older volcanics Thalab Group	Ghaber Group	PEOPLES DEMOCRATIC REPUBLIC OF YEMEN
	Thaniya Group unnamed units		unamed units	? ? ?	YEMEN ARAB REPUBLIC
	Older 'Series'		Older 'Series'	Inda Ad 'Series'	SOMALIA
	Kisii 'Series'	Mozambiquian 'System'		Ablun 'Series' Embu 'Series'	KENYA
			Bukoban 'System'	Mityana 'Series' Bunyoro 'Series'	UGANDA
		Bukoban System Busondo Group Uhu Group Ikorongo Group Kuvimba Group			TANZANIA

1000 900 800 700 600

CRUDE RADIOMETRIC SCALE (Ma)

ophiolites belong to the Rubshi Group; they belong to the Urd Group in Saudi Arabia. The ophiolite belts of the Nubian Shield include those of the Sudan, namely: the Sol Hamed-Wadi Onib Complex (Fig.6.70), the Khor Nakasib-Oshib Complex, and the Wad Wadela-Ingessana Complex. Others are the Tullu Dimtu-Akabo-Birbir belt of western Ethiopia, the Adola belt in eastern Ethiopia and the "Maydh greenstones" in northeastern Somalia.

As already pointed out the volcano-sedimentary and ophiolite zones of the ANS extend southward in two main prongs (Fig.6.68) into the Mozambi-

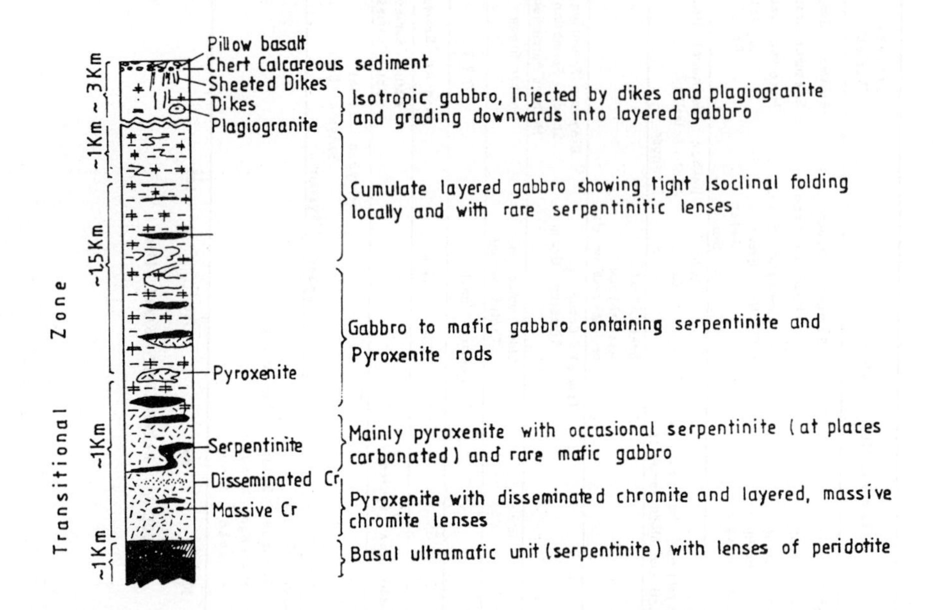

Figure 6.74: Schematic section through Wadis Onib and Sudi ophiolites of the northern Red Sea Hills of the Sudan. (Redrawn from Kröner et al., 1987.)

que belt of East Africa where they define the Pan-African collision sutures in the latter region (Behre, 1990; Vail, 1988b). One prong lies in the Blue Nile region of Sudan and Ethiopia (Fig.6.70) as an approximately north-south-trending belt of volcano-sedimentary and ophiolite and granitoid assemblage bordered to the west by the Ingessana-Kurmuk ophiolite zone of eastern Sudan and to the east by the Tullu Dimtu-Akabo-Birbir zone of western Ethiopia. Both ophiolite zones have been correlated with the Sekerr dismembered ophiolites of Uganda-Kenya. However, the ophiolites of the Blue Nile region have so far not been dated; but they are probably older than 850 Ma, the age of some of the syn-tectonic magmatic rocks and metamorphism in the surrounding high-grade migmatitic orthogneisses (Selak Formation) and paragneisses (Tin Group). Late- to post-orogenic granitoids in the area have been dated between 520 Ma and 500 Ma. The volcano-sedimentary rocks and ophiolites of the Ingessana-Kurmuk area are in thrust contact with the Selak and Tin basement rocks.

Another southward-extending ophiolite belt is in the Adola area of eastern Ethiopia (Fig.6.75,A) which continues into the Moyale belt of northern Kenya (Fig.6.62). The Adola ophiolite belt contains imbricated mafic-ultramafic rocks which have been intensely thrust over each other and onto basement gneisses and migmatites (Fig.6.75,B) with tectonic transport towards the east involving a minimum of 30 to 40 km, and considerable crustal shortening (Baraki et al., 1989).

Ophiolitic Mélange and Olistostromes

Two types of subduction-related lithologies occur among the ophiolites of the Eastern Desert and the Red Sea Hills. Both represent a chaotic mixture of heterogeneous rock material in a pelitic matrix. A mélange is a mappable body of deformed heterogeneous rock material consisting of pervasively sheared, fine-grained commonly pelitic matrix thoroughly mixed with diverse and angular, poorly sorted inclusions. An olistostrome is also a mappable lens-like chaotic unit of intimately mixed heterogeneous material that lacks true bedding but is intercalated among normally bedded sequences (AGI, 1972).

The ophiolites of the Eastern Desert form part of an extensive tectonic mélange which resulted from the complete dismemberment and total disruption of their original stratigraphic character and distribution. As stated by Hassan and Hashad (1990) the mélange of the Eastern Desert are characterized by the presence of a significant proportion of serpentinites either as matrix or as variably sized blocks, in addition to other ophiolitic fragments, deep-sea sediments and calc-alkaline volcanics. Other components such as granitic rocks, carbonate rocks, quartzites and mudstones attest to the characteristic heterogeneity of the mélange which nevertheless, still constitute mappable lithostratigraphic entities. The mélange are commonly thrust sheets or slices which were incorporated within allochthonous belts of metasediments.

At Wadi Ghadir in the Central Eastern Desert near Jebel Hafafit (Figs.6.73;6.76,A) there is a large ophiolitic mélange with proximal and distal facies (Hassan and Hashad, 1990). The proximal facies consists of rolled and fragmented rock-debris of highly variable sizes in a sheared matrix of scaly and schistose mudstones; abundant serpentinized peridotite blocks, some of which are surrounded by sheaths of schistose talc-carbon-

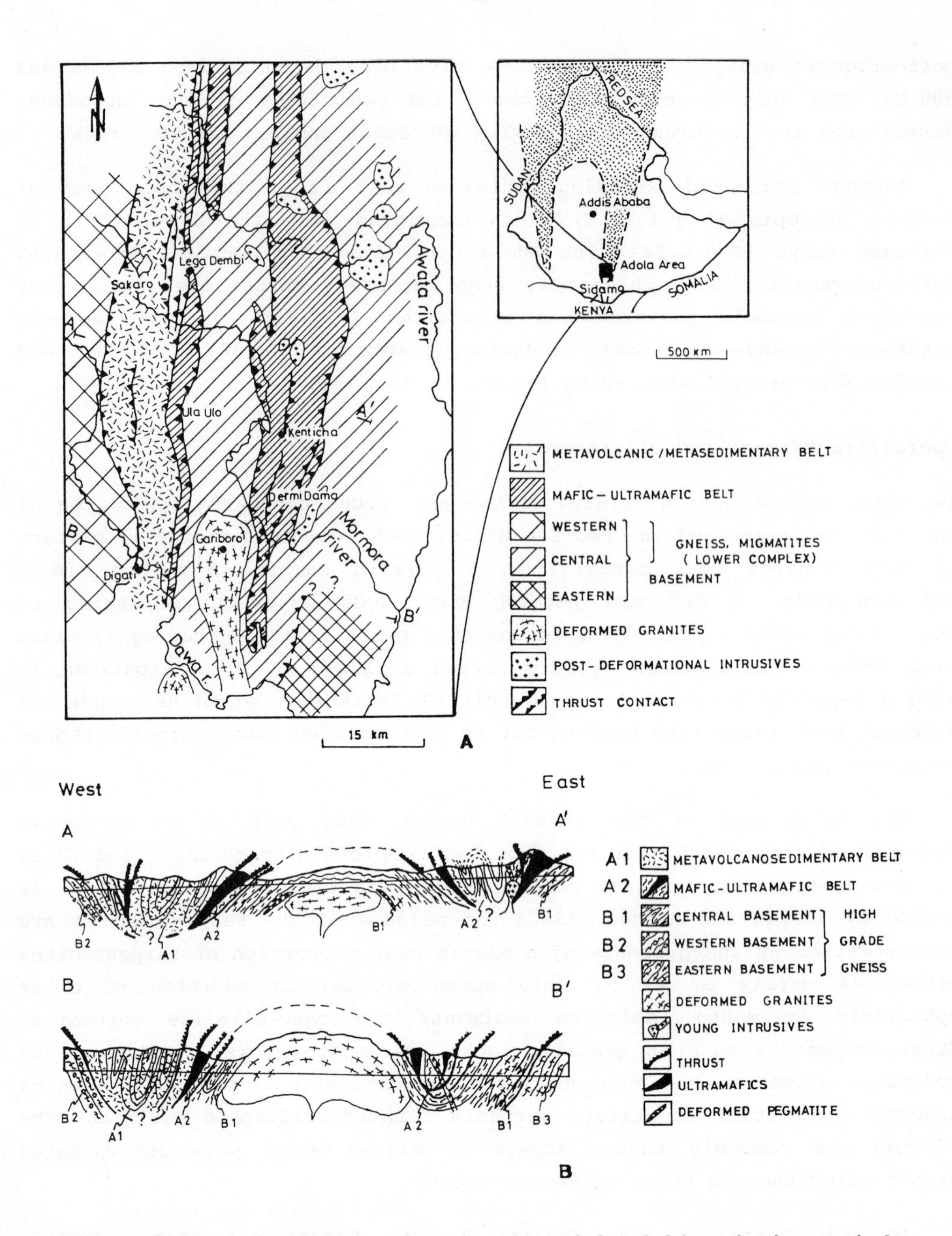

Figure 6.75: Tectonic units in the Adola fold and thrust belt of southern Ethiopia (A); B, schematic sections showing structural relationships. (Redrawn from Baraki et al., 1989.)

ate rock produced by squeezing and rolling of the blocks; and other rock debris including volcanic material, graywackes, quartzites, chert, granite, and amphibolites. The distal facies is a low-grade pelitic schist with pockets and lenses of highly schistose talc carbonate rock. A genetic model (El Bayoumi, 1984) for the Wadi Ghadir ophiolitic mélange involved westward subduction of oceanic crust resulting in gravity sliding of disrupted ophiolites and continental margin sediments into the trench where they mixed and formed a chaotic mass (Fig.6.77). The area was later intruded extensively by dykes and calc-alkaline granites and leucogabbros.

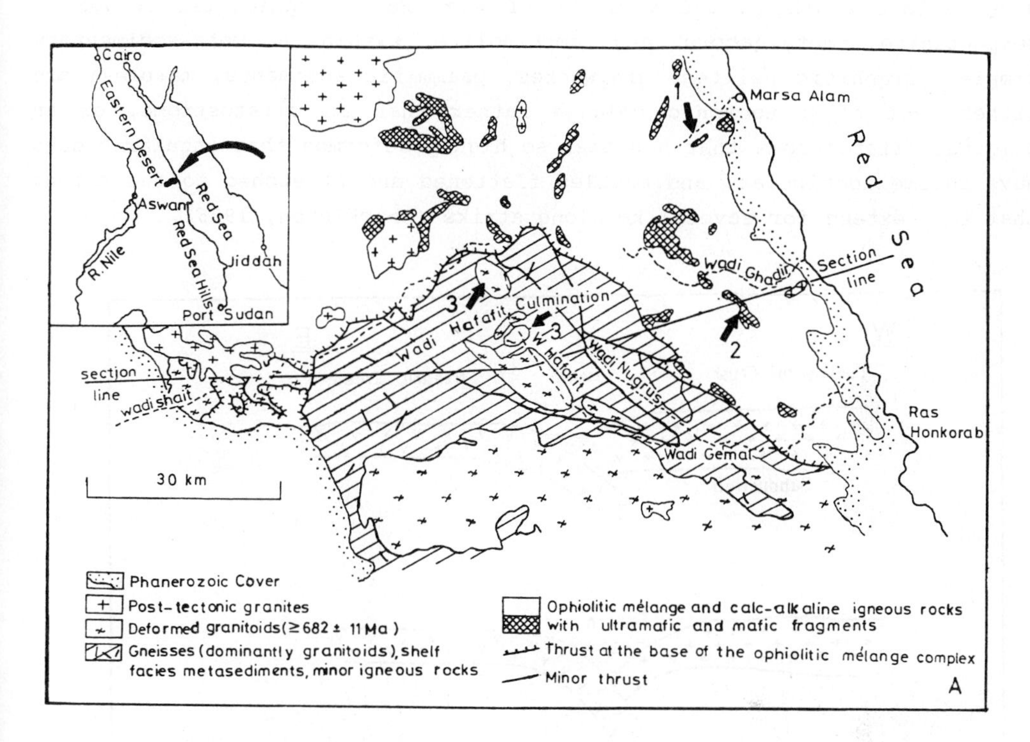

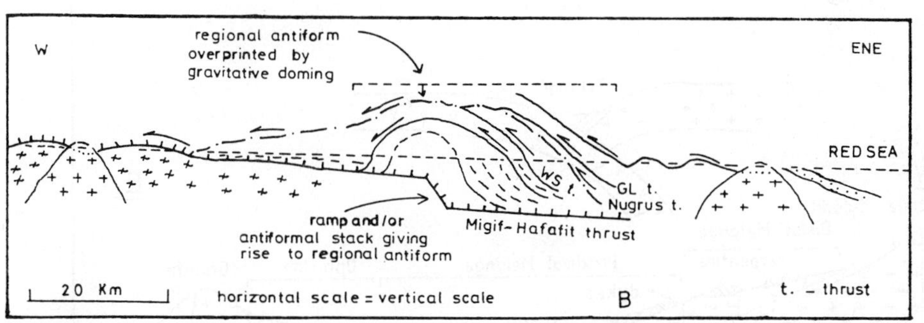

Figure 6.76: Schematic map (A) and section (B) through Wadi Hafafit Culmination. 1, volcanic rocks near Marsa Alam; 2, Wadi Ghadir ophiolite; 3, Hafafit igneous suite. (Redrawn from Greiling et al., 1988.)

In the Wadi Mubarak area the mélange developed initially as an olistostrome (Shackleton, 1986). Attesting to this origin are the unstratified, mainly pelitic matrix with little sign of deformation other than late cleavage; enclosed angular blocks and large masses of ophiolites and sediments; the sharp contrast between the mélange and normal sediments; and an extensive mass of ophiolitic mélange which in one locality rests with normal sedimentary contact on turbidites and pelites. Much further south in the vicinity of Wadi Haimur ophiolitic lenses of serpentinite, meta-gabbro and amphibolite within a meta-sedimentary complex (graphitic pelites, graywackes, psammitic sediments, marbles) are either part of a tectonic mélange rather than an olistostrome, or an original olistostrome that has been so highly deformed that angular blocks have become lenticular, and marbles flattened and stretched to the extent that they extend for several km along strike (Shackleton, 1986).

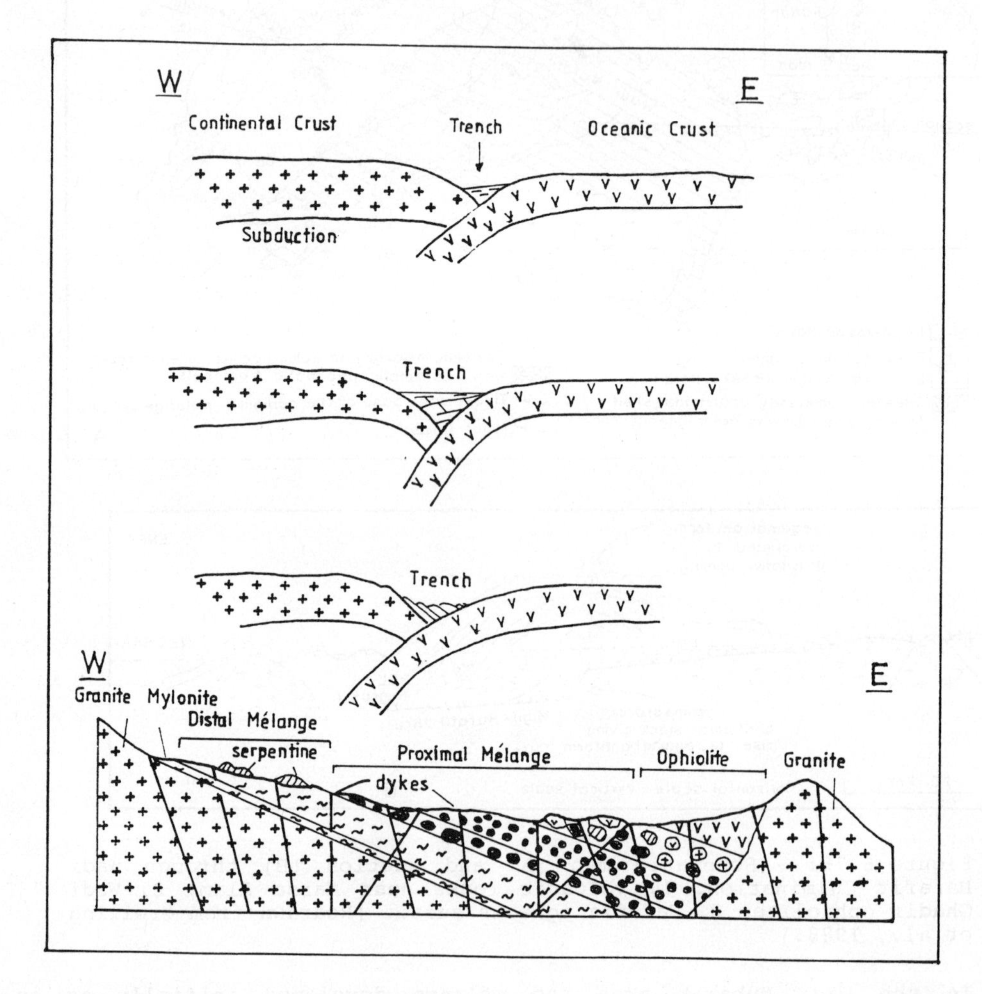

Figure 6.77: Model for the origin of Ghadir mélange. (Redrawn from Hassan and Hashad, 1990.)

6.11.5 Syn- and Post-orogenic and Anorogenic Magmatism

Intense plutonic activity suggestive of plate margin environments is associated with the volcano-sedimentary and ophiolite assemblages of the ANS. Igneous rocks occur as large heterogeneous batholiths and plutonic complexes of diorite-gabbros, granodiorites, tonalites and adamellites-granites, previously termed the "older granitoids" in Egypt and the "Batholithic Granites" in the Sudan (Vail, 1987; 1988b). They are most extensively developed in the Northern Eastern Desert of Egypt (Fig.6.73). High-level late- and post-tectonic plutonic bodies with ring complex structures are also developed which are characteristically bimodal calc-alkaline gabbro-granite complexes, as well as alkaline granites and syenites. An important later magmatic development in the Northern Eastern Desert and the Red Sea Hills of the Sudan is the suite of anorogenic granite, syenite and rare foid syenite ring complexes and plutons, in what is a major alkaline province which extends from northern Uganda to southern Egypt (Vail, 1989a).

6.11.6 Molasse

Jackson (1980) referred to the uppermost, slightly metamorphosed depositional sequences in the ANS as the "Infracambrian volcano-sedimentary and sedimentary units", which rest unconformably on older successions, and are generally of subaerial or very shallow-marine origin. As shown below, the sedimentary units of this assemblage exhibit the characteristic features of molasse. In Egypt older metamorphosed units are unconformably overlain by the Dokhan volcanics (Table 6.4), which are sequences of purple-coloured, porphyritic acid and intermediate volcanics and equivalent pyroclastics, with minor components of volcaniclastic sediments. The youngest Pan-African sequence exposed mainly in the Central and Northern Eastern Desert is the Hammamat Group, about 4,000 m thick at the type locality, comprising thick sequences of conglomerate, arkose, graywacke, limestone, slate and minor volcanics. The Hammamat Group is a typical molasse sequence, deposited in alluvial fan-braided stream complexes and playa lakes in disconnected intermontane basins as a result of rapid uplift and erosion (Hassan and Hashad, 1990).

In the Sudan the equivalent to the Hammamat Group are termed the Abu Habil Series and the Amaki Series (Vail, 1988a); and in Ethiopia the Didykama, Shiraro and Matheos sedimentary formations (Table 6.4) are composed of conglomerates, sandstones, slates and stromatolitic limestones. Lithologic units equivalent to the Hammamat Group are well developed in Saudi Arabia and have been variously designated (Table 6.4).

6.11.7 Tectonism

Tectonic Model

Before examining a few examples of the deformational styles found in the Red Sea fold and thrust belt, it is illuminating first to consider the tectonic setting that has been postulated for this structural province. This approach of going from the tectonic model to the resulting structure is preferred here partly because there is considerable unanimity regarding the plate tectonics regime that operated in the Pan-African of the ANS (e.g. Burke and Sengör, 1986; El-Gaby and Greiling, 1988; Jackson, 1987; Kröner et al., 1987; Schandelmeier et al., 1988; Shackleton, 1986; Stoesser and Camp, 1985); and also because the deformation mechanisms are more readily understandable within the plate tectonics framework.

The analogy between the island-arc setting of the volcano-sedimentary and ophiolite assemblages of the ANS and the Recent southwest Pacific plate tectonic setting is now widely accepted. A microplate arc-back-arc ocean basin existed between 900 Ma and 600 Ma in the Red Sea fold belt and in Saudi Arabia similar to the situation in Indonesia today (Kröner et al., 1987). The modern island-arc setting is characterized by volcanic arcs and associated volcanic flows, pyroclastic deposits, tuffs; volcanic fronts which occur some 80-150 km inland from the trench where tholeiitic and calc-alkaline magmas are found with mostly andesites and basaltic andesites; active back-arc basins over subduction zones where sediments range from volcaniclastics to pelagic, with hemipelagic sediments and turbidites in the distal parts of the basin; and ophiolites representing fragments of back-arc oceanic crust (Condie, 1989). Although disparate subduction directions have been proposed for the ANS, some favouring eastward subduction (e.g. Riess et al., 1983; Shackleton, 1986), while most workers favour westward subduction (e.g. El-Gaby et al., 1988; Greiling et al., 1988), there is a consensus that the ophiolites of the Red Sea fold and thrust belt represent sutures which resulted from arc-arc and arc-continent collisions at various times. The structures which resulted from these collisions are considered below before examining the timing of the collision events.

Red Sea Hills

Unlike the Egyptian Eastern Desert which is characterized by extensive low-angle thrust regimes, the major tectonic boundaries in the Red Sea Hills of the Sudan are often steep with large shear zones which contain highly sheared lensoid mafic-ultramafic **bodies** which represent dismembered ophiolites (Kröner et al., 1987). The ophiolites define the sutures separating successively accreted island arcs. Kröner et al. (1987) correlated the major northeast-trending ophiolite belts in the Red Sea Hills with those in Saudi Arabia (Fig.6.70) and used them as sutures to define common tectonic terranes in the region. The Onib-Sol Hamed suture zone which separates the Midyan and Hijaz terranes shows steep to vertical dips and faces the southeast (Fig.6.78,A) suggesting obduction from the southeast to the southwest (Kröner et al., 1987). The structural grain in the northwestern part of the Red Sea Hills trends dominantly SW-NE with NW- and SE-verging folds and minor thrusts. Prominent late north-south regional shear zones with large sinistral and locally dextral displacements also occur in the Red Sea Hills. These shear zones are characterized by mylonites (Almond, 1987).

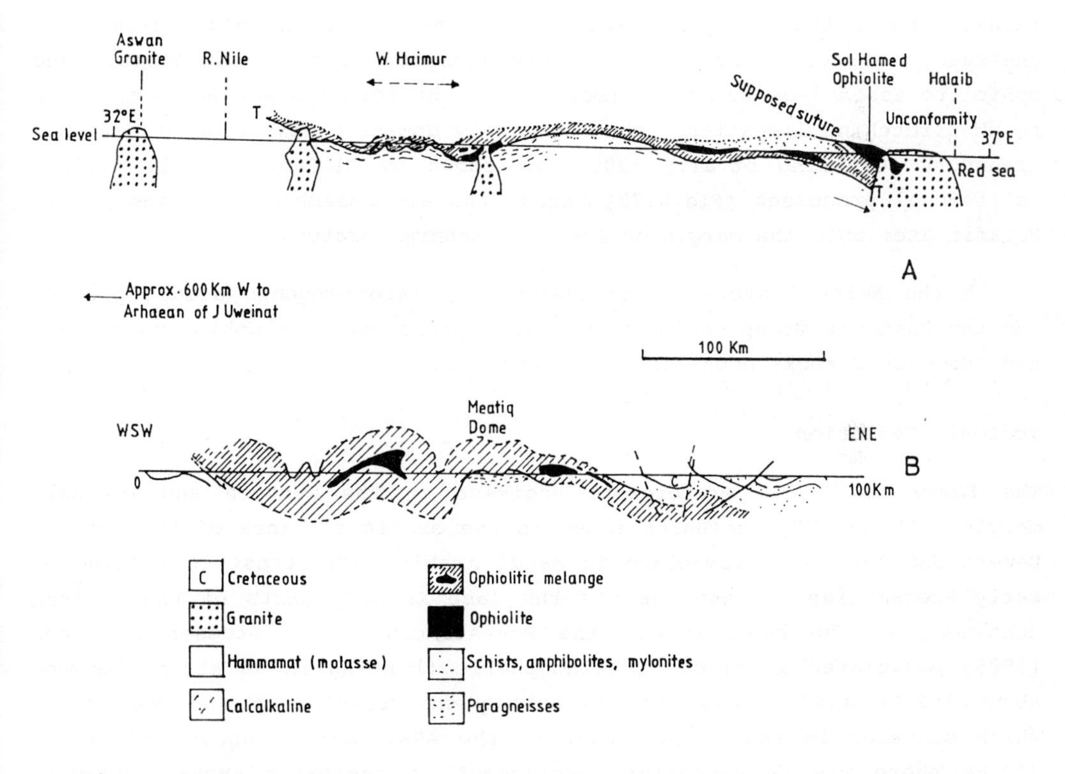

Figure 6.78: Sections across the Eastern Desert of Egypt. (Redrawn from Shackleton, 1986.)

Central and Southern Eastern Desert

This is the fold and thrust belt *sensu stricto*, which appears to be one composite allochthonous thrust sheet (Fig.6.78,A), the leading edge of which is located along the Nile (Shackleton, 1986). The thrust sheet contains huge fragments of oceanic crust and upper mantle, enclosed in a regional ophiolite mélange underlain by a ductile thrust zone. This ductile zone shows recumbent mylonites which are underlain by the older pre-Pan-African high-grade gneissic basement. Calc-alkaline volcanics with transitional island-arc chemistry are thrust over the ophiolite mélange; and the entire complex is unconformably overlain by the molasse-facies Hammamat Group (Fig.6.78,B). The underlying older shelf-facies and high-grade metasedimentary group and granitoid rocks are themselves allochthonous and are exposed in tectonic windows (exotic terranes) at the Meatiq Dome and at the Migif-Hafafit Dome. Detailed structural investigations (Greiling et al., 1988) of the Migif-Hafafit Dome shows that this feature, which is known as the Migif-Hafafit Culmination, is an antiformal composite stack of westward-directed thrusts (Fig.6.76) above a major basal Migif-Hafafit thrust. The footwall of the basal thrust consists of the older high-grade gneisses, whereas the younger greenschist volcano-sedimentary and ophiolite assemblage forms the roof of the antiformal stack which shows SE to NW stretching lineations, the regionally dominant direction of tectonic transport (Greiling et al., 1988). As shown by El-Ramly et al. (1984), collisional processes (Fig.6.79) thrust the arc assemblages of the Migif-Hafafit area onto the margin of the East Saharan craton.

In the North Eastern Desert and Sinai, late orogenic acidic plutons and the Hammamat Group sediments are widespread, whereas ophiolite mélange and associated rocks occur as minor remnants.

Tectonic Evolution

The Early to Middle Proterozoic gneisses on the western and southern margins of the ANS, including those in the exotic terranes of the Eastern Desert and the Afif microplate in Saudi Arabia, all attest to rifting of early Precambrian terrane, as was the case in many parts of the African continent at the beginning of the Pan-African cycle. Stoeser and Camp (1985) postulated a period of lithospheric thinning in NE Africa between about 1.2 Ga and 950 Ma, and the subsequent creation of an ocean basin which centered in the eastern part of the ANS. Between about 950 Ma and 700 Ma there was an extensive development of ensimatic arcs which were situated between the Afif terrane in the east and the African continental margin (Fig.6.80,A). But ensimatic conditions persisted east of the Afif terrane until about 640 Ma ago (Fig.6.80,B). The main arc microplate accretion and collision events took place between about 700 Ma and 640 Ma (Figs.6.79;6.80,B) and post-collisional granitoid plutonism occurred between about 680 Ma and 620 Ma. Cratonization of the ANS was then completed. From about 630 Ma to about 540 Ma intracratonic granitoids intruded the cratonized shield.

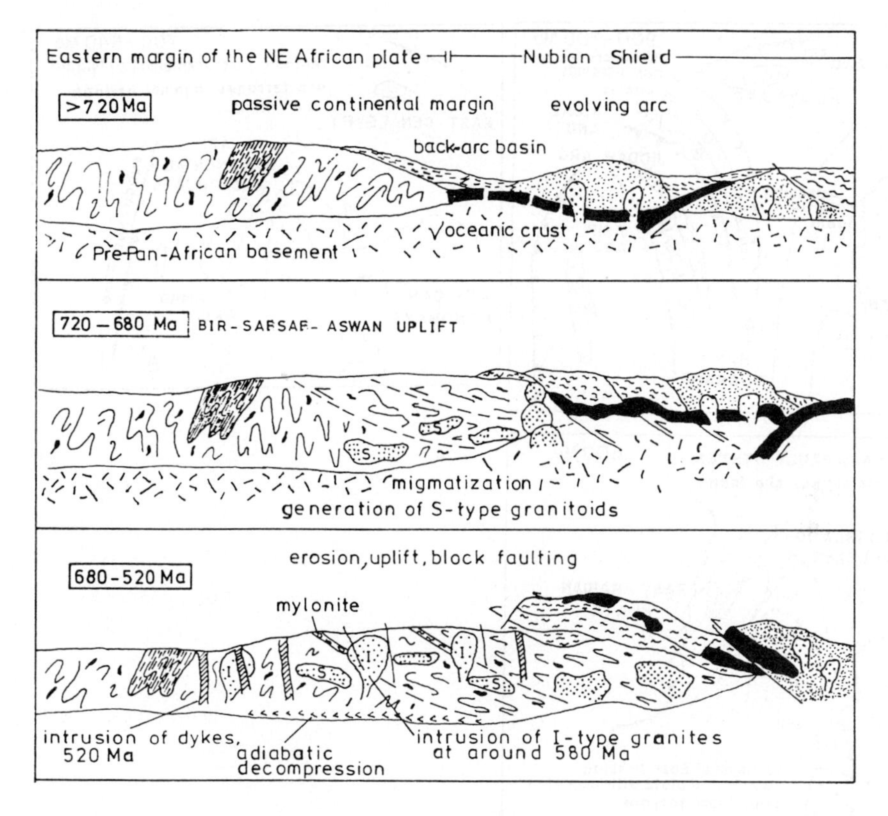

Figure 6.79: Plate Tectonic model for the development of the ANS. (Redrawn from Schandelmeier et al; 1988.)

However, available radiometric ages reveal that the island-arc terranes or microplates did not all develop simultaneously (Fig.6.80). Rather, terrane correlations from the Arabian Shield to the Nubian Shield show that different island-arcs developed and accreted at different times. Thus, in the Red Sea Hills the Haya terrane which is believed to be continuous with the At Taif-Jiddah terrane in Saudi Arabia (Kröner et al., 1987) evolved between 900 Ma and 800 Ma (Fig.6.80,A), while the Gebeit terrane to the north which correlates with the Hijaz terrane evolved between 800 Ma and 700 Ma. The Bir Umq-Nakasib-Amur suture between both microplates resulted from collision at between about 700 Ma and 680 Ma. Following the welding together of both microplates (Haya and Gebeit terranes), the Red Sea Hills underwent an extensional tectonic regime between 670 Ma and 620 Ma as indicated by the occurrence of volcanics which do not show the penetrative NE-SW structural grain produced during the microplate collision along the Bir Umq-Nakasib-Amur belt.

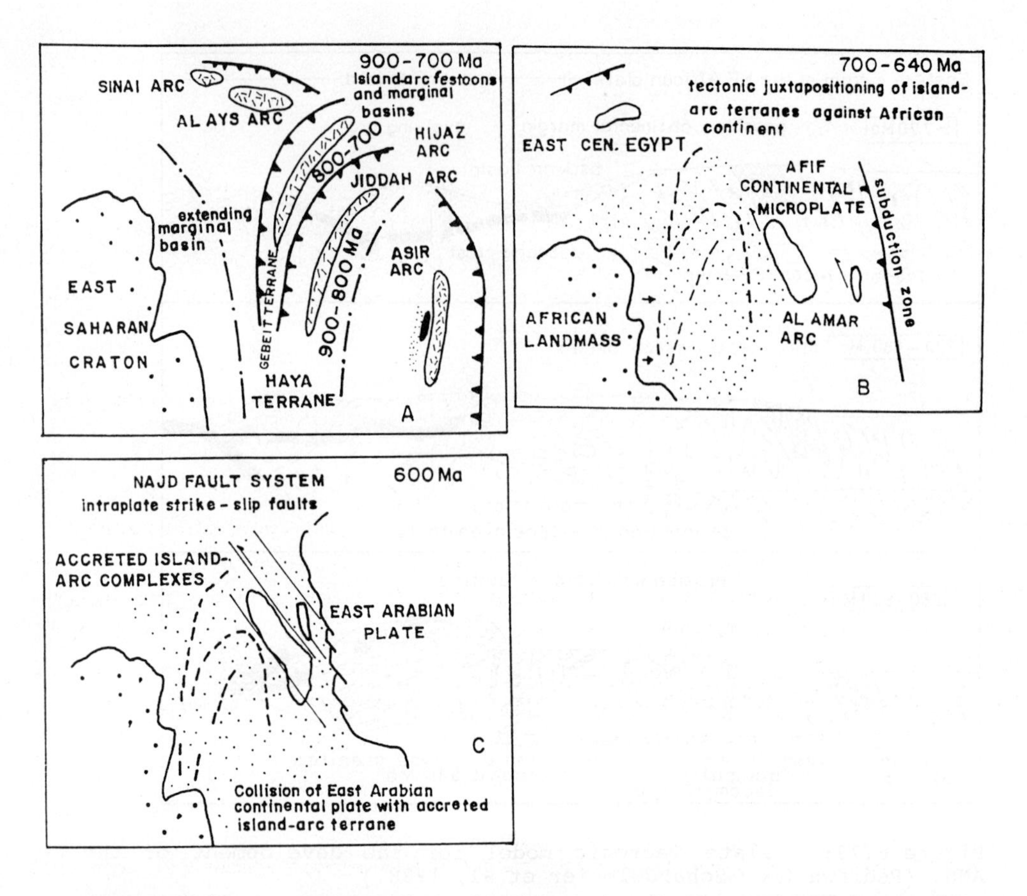

Figure 6.80: Progressive development of the ANS (Redrawn from figure supplied by N. J. Jackson.)

In the Tokar terrane further south in the Tigre and Eritrea provinces of northern Ethiopia low-grade meta-volcanics of island-arc character predominates, which consists mainly of andesites and associated volcaniclastics that accumulated in a shallow-water setting (Kazmin et al., 1978). These island-arc rock suites contain strongly deformed bodies of syn-tectonic diorites and granodiorites, and intrusions of late-to-post-tectonic granites and granodiorites with cooling ages ranging from 700 Ma to 450 Ma.

In the Eastern Desert of Egypt a passive margin seemed to have been installed in the central and southern parts until about 800 Ma. Here the emplacement of initial arc volcanics and granitoids between about 770 Ma and 710 Ma suggests a change to subduction, hence conversion to an ensimatic tectonic regime with ophiolite subduction leading to the development of arc systems which lasted until about 680 Ma ago when the arcs collided and were accreted onto the margin of the Nile craton (Fig.6.79). Lastly, following molasse-type deposition, there was low-angle thrusting, strike-slip faulting and the emplacement of late-tectonic plutons at about 600 Ma to 570 Ma (Stern, 1985).

In contrast, the Northern Eastern Desert evolved mainly in the strong compressional regime which followed arc accretion. Excluding Sinai, the oldest rocks in the Northern Eastern Desert are granodiorites, 680 Ma to 610 Ma old; the bimodal Dokhan volcanics and their intimately associated clastic Hammamat molasse formed between 600 Ma and 575 Ma, during an extensive phase of rifting. Late-tectonic granitoids were emplaced between about 600 Ma and 570 Ma; and bimodal dyke swarms intruded from about 590 Ma to 540 Ma.

The microplate collision and accretion events of the Red Sea fold and thrust belt were felt in the Mozambique belt as well. The island-arc systems in the southern terminations of the ANS, though poorly dated, also were folded and thrust onto the surrounding basement areas, for example along the Adola-Moyale belt, and along the Ingessana-Kurmuk and Tullu Dimtu and Sekerr sutures. Further south, the Mozambique belt of Kenya and Tanzania was also involved in severe continent-continent collision and suturing at about the same time, between 900 Ma and 600 Ma ago.

6.11.8 Mineralization

Syntheses on the mineralization in the ANS have been presented by Pohl (1984, 1988) and by Vail (1979, 1985, 1987), in addition to numerous and widely dispersed publications, and inaccessible technical reports in the various countries that make up the ANS. For example, Hussein (1990) has furnished an elaborate account of the mineral deposits in Egypt. The following synopsis is largely based on the genetic descriptions of Pohl (1988) as shown on Fig.6.68.

The ANS is generally not considered to be a very productive metallogenic province, although gold mining dates from antiquity, especially the Pharaonic times; and a wide range of metallic and industrial minerals have been exported in small quantities from the Sudan and Ethiopia, for example platinum, chromite and mica. However, the development of a large number of mineral prospects in the ANS will depend on the recovery of the international metal commodities market from its present depression (Pohl, 1988). A genetic classification of the mineral deposits of the ANS shows that syngenetic stratiform ores, ophiolite-related deposits, volcanogenic base-metal sulphides, and magmatic deposits including extensive pegmatite mineralization, are quite promising.

Syngenetic Stratiform Ores

In Egypt and Saudi Arabia magnetite and hematite with banded iron-formation characteristics occur which are probably of volcanogenic-hydrothermal origin. Some of these ferruginous-banded cherts associated with these deposits host gold as well. There are magnesite deposits in sedimentary carbonates in Saudi Arabia where there are Mn-Zn-Cu-barite lenses as well in the terrigenous metasediments.

Ophiolite-related Deposits

In Egypt, Sudan and Ethiopia there are chromium and platinum ores in ultramafics ; magnesite veinlets and stockwork bodies occur in dunite and serpentinites. In Egypt high-grade talc deposits are found in some shear zones, in addition to the occurrence of low-grade talc-carbonate rocks.

Volcanogenic Base Metal Sulphides

The most prominent metallogenic environment in the ANS are the ensimatic and Andean-type arcs. Among the deposits associated with them are replacement magnetite ores at the contacts with gabbros, diorites and granodiorites. Stockwork and proximal massive sulphide deposits containing Cu, Zn, Pb, Au and Ag are associated with acidic subvolcanic domes and breccias. Also present are the more distal lenticular Zn-Pb-Cu sulphide beds with a more hydrothermal-sedimentary character, and with bands of exhalites, calc-dolomite and graphitic tuffs, for example at Nuqrah in Saudi Arabia. There are quartz veins and stockwork bodies with Au-Ag mineralization in or around acidic subvolcanic intrusions. The numerous small gold fields of Egypt are of this type.

Magmatic Deposits

As already shown, the tectonic evolution of the ANS was so complex that while post-orogenic magmatism was obtaining in one part, orogenic deformation and syn-tectonic plutonism were still active elsewhere. However, by about 640 Ma island-arc accretion and suturing had largely ceased allowing magmatic rocks to intrude at shallow levels until about 540 Ma. Among these late intrusives are granodiorites, monzogranites, alkali-feldspar granites including alkali granites and syenites, often with equivalent volcanic rocks and layered gabbroic rocks, especially in the southern part of the ANS. The Najd fault system in Saudi Arabia remained active intermittently during this period resulting in greenschist metamorphism and coarse molasse deposits.

Among the important types of mineralization are Ta-Nb, Sn, Be and W associated with small copolas of highly evolved granites which had been involved in albitization and greisenisation. These deposits, for example those at Abu Dabbab in Egypt, include disseminations within the copolas, marginal pegmatitic phases and external quartz veins and stockworks. Alkali granites and their pegmatitic-hydrothermal suite are mineralized with Nb, Zr, Y, R_{EE}, U and Th; and ilmenite and magnetite occur in layered mafic complexes. In the Baish Group of Saudi Arabia (Table 6.4) scheelite with quartz and calc-silicates in hornblendite and amphibolite occur in the immediate vicinity of a post-tectonic muscovite-biotite granite thus indicating a genetic link with acidic magmatism.

The coarse-grained muscovite pegmatites of the Bayuda desert in northern Sudan, formerly mined for mica, carry two different generations of muscovite: Rb-Cs-Sn-Nb-rich varieties which were emplaced at about 698-644 Ma; and Mg-Ti-Li-rich phengite muscovites, reflecting a lower-temperature event dated at 552-526 Ma (Küster et al., 1990). These pegmatites are believed to be the anatectic products of regional metamorphism which formed contemporaneously with syntectonic granitoids during the main tectonic phase of the Pan-African. Other important pegmatite fields lie in northern Somalia (Küster et al., 1990), in the Berbera region in the northwest (former mining district for beryl, columbite, mica), and in the Bosaso area to the northeast (former mining district for tin and tantalum). According to Küster et al. (1990) the rare metal-bearing vein-type pegmatites of northern Somalia were emplaced into amphibolite-grade metamorphic basement units and into the greenschist-grade meta-sedimentary Inda Ad Group between 497 Ma and 392 Ma, after Pan-African granites had triggered the circulation of fluid phases in a tectonically reactivated terrane.

Gold-quartz and gold-carbonate veins, with pyrite in which there may be Ag, Cu, As, Pb and Zn, are widespread in the ANS. These are hosted by intrusive volcanic and ophiolitic rocks, including post-tectonic granites, and quite often, a direct relationship with cooling intrusives may not be evident. The gold veins are usually thin, less than one mm and several hundreds of m long. These veins often form systems several km long, showing strong tectonic control by ductile or brittle shearing which sometimes caused boudinage of the veins. Almond et al. (1984) explained these veins as originating from large hydrothermal systems which were either induced by metamorphism or by the cooling of unexposed intrusives.

Chapter 7 Precambrian Glaciation and Fossil Record

7.1 Precambrian Glaciation

A major aspect of the Precambrian stratigraphy of Africa, mentioned earlier in passing, is the widespread occurrence of glacial deposits especially in the Late Proterozoic. Evidence for continental glaciation abounds in the Precambrian of other continents, except Antarctica (Fig.7.1). From a compilation of the Earth's pre-Pleistocene glacial deposits Hambrey (1983) and Harland (1983) determined that the intervals of world-wide expansions of continental ice sheets can roughly be grouped into glacial eras, periods and epochs as shown below:

(I) Late Proterozoic Glacial Era:	
(i) Late Sinian Glacial Epoch:	610-580 Ma
(ii) Varangian Glacial Period (with 2 main epochs):	650-610 Ma 720-660 Ma
(iii) Sturtian Glacial Period (with 2 main epochs):	790 Ma 800 Ma
(iv) Lower Congo Glacial Period (with 2 main epochs):	820 Ma 950 or 865 Ma
(II) Middle Proterozoic Glacial Era:	2.0 - 1.0 Ga
(III) Late Archean-Early Proterozoic Glacial Era: Huronian Glacial Period (with 3 or more epochs):	2.3 Ga
Witwatersrand Glacial Period (with 4 or more epochs):	2.65 Ga

Though sometimes ambiguous, the case for ancient glaciation involves direct evidence such as glacial deposits (tillite, tilloid, diamictite, mixtite), striated and polished basement surface which commonly shows friction cracks, roches moutonnés and other geomorphic forms; as well as indirect evidence such as post-glacial rapid and pronounced sea-level rise (Crowell, 1983). Till and tillite (consolidated till) consist of unsorted rock debris, ranging in grain-size form clay to boulders, with some of the larger stones having been transported by ice over great distances in which case they are sometimes faceted and striated and can be traced to their source areas. Tilloid refers to tillite-like rocks of doubtful origin. Diamictite is a general term for an unsorted deposit with boulder beds, clays and sand, pebbly sandstones, and mudstones.

Tillites and tilloids are sometimes termed mixtite. The origin of some African mixtites is controversial.

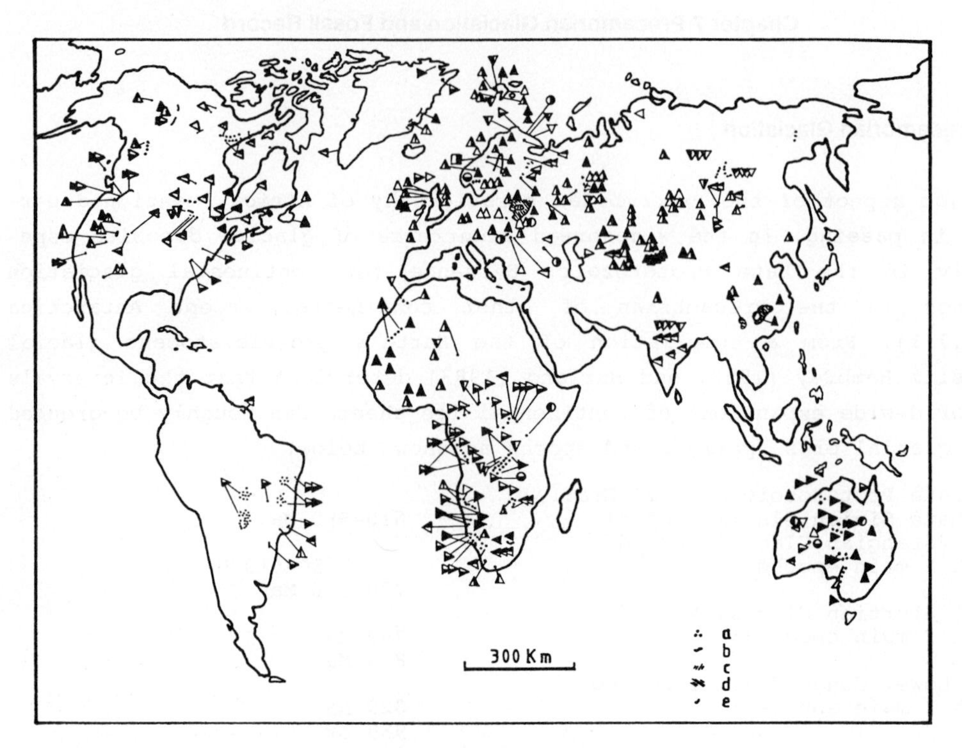

Age: Geological	V + R 4	R 3	R 2 + R 1	PR 1 + A	?
Age: Ma	680 - 560	1000 - 680	1650 - 1000	>1650	?
Tillite	▲	▶	—	◀	▲△
Other gl. rocks	◑	⊖	◐	⊖	⦶
Mixtites	◭	▷	▽	◀	△
Non gl. mixtites	△	▷	▽	◁	△△

Figure 7.1: Global distribution of Precambrian tillites. (Redrawn from Windley, 1984.)

For example, whereas the mixtites of the West Congolian orogen have been attributed to debris flows triggered by strong subsidence at the beginning of new sedimentary cycles (Cahen and Lepersonne, 1976; Porada, 1989; Stanton et al., 1963), a mode of origin also invoked for the Chous Formation in the Damara Supergroup (Porada, 1983), these deposits are believed to be of glacial origin by some workers (e. g. Harland, 1983; Salop, 1983; Tankard et al., 1982). There has been a general tendency, however, to correlate African Precambrian mixtites regionally, and with glacial deposits in other parts of the world, especially when their inferred ages so permit (Chumakov, 1981; Deynoux, 1983; Deynoux et al., 1978). Since their precise ages are often difficult to ascertain directly, glacial deposits are usually assigned approximate ages based on their stratigraphic position above and below radiometrically dated intervals.

7.1.1 Late Archean-Early Proterozoic Glacial Era

The Witwatersrand Supergroup contains the earliest known glaciation (Harland, 1983; Tankard et al., 1982), estimated to be older than the overlying Ventersdorp lavas dated at about 2.64 Ga and younger than an underlying granite which is about 2.66 Ga old. These glacial deposits belong to the Witwatersrand Glacial Period. They consist of diamictites with striated pebbles associated with alluvial fan deltaic and distal shelf deposits at two or three stratigraphic levels in the West Rand Group (Fig.4.5A). Tankard et al. (1982) postulated that the most likely agent of deposition for the West Rand Group diamictites was submarine debris flow triggered from accumulations of ice-rafted moraines.

Named after the Huronian tillites of Ontario, Canada, the Early Proterozoic Huronian Glacial Period is represented in South Africa by diamictites which occur sporadically beneath the regional unconformity within the Postmasburg and Pretoria Groups of the Transvaal Supergroup (Fig.7.2). These glacial diamictites contain remnants of a striated pavement and associated conglomerates, cross-bedded sandstones, siltstones, mudstones and varved shales which have been interpreted as of glacio-fluvial and glacio-marine origin (Tankard et al., 1982). Salop (1983) considered the glaciogenic deposits in Kimberley (N.W. Australia), Brazil, and Wyoming (U.S.A.) to be roughly equivalent to those of South Africa (Fig.7.2).

7.1.2 Mid-Late Proterozoic Glacial Eras

Mid-Proterozoic tillites are known below the Stromatolitic Series in the Tuareg Shield (Fig.6.15). But by far the most extensive glacial period on Earth, compared even with later glaciations (during the Ordovician-Silurian, mid-Late Devonian, Late Paleozoic, and the Pleistocene), was the Late Proterozoic Glacial Era. The glacial deposits of this era account for the bulk of the Precambrian glacial deposits plotted in Fig.7.1.

African Late Precambrian platforms and mobile belts were awash with tillites.

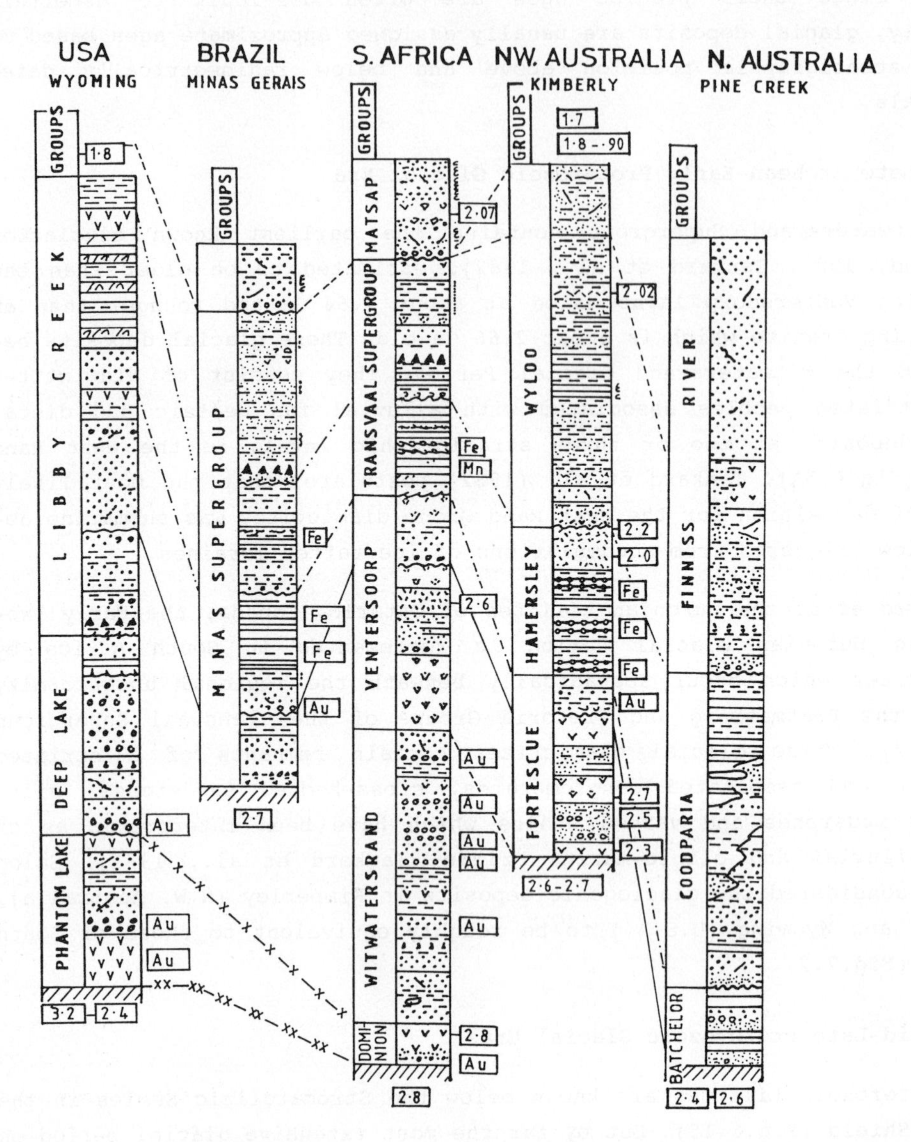

Figure 7.2: Geologic columns showing correlations of Precambrian diamictite-bearing supracrustals. (Redrawn from Salop, 1983.)

In the West Congolian mobile belt, the type area for the Lower Congo Glacial Period, the "Tillite inférieure du Bas Congo" and the "Tillite supérieure du Bas Congo" underlie the Louila and the Schisto Calcaire Groups respectively. The age of the lower tillite is believed to be about

950 Ma, while that of the upper tillite is probably 820 Ma (Harland, 1983). The equivalents of both tillites are the Grand Conglomérat and the Petit Conglomérat of the Katangan Supergroup (Table 6.3). The Akwokwo mixtite of the Lindian Supergroup in the NE Zaire Precambrian platform basin correlates with the Grand Conglomérat and the "Tillite inférieure du Bas Congo".

In the Damara Supergroup of Namibia the diamictites of the earlier Sturtian epoch occur in the Nosib Group at the base, whereas those of the later epoch include the widespread Chous mixtite, and its equivalent the Numees mixtite in the Gariep Group (Figs.6.42, 6.46). Sturtian tillites in the Tuareg Shield (Fig.6.15) include the tillites of the "Siere Verte" and of the Tafeliant Group (Fig.6.17A).

Deynoux (1983) presented a synthesis on the Late Precambrian glacial deposits in West Africa. These are exposed as a thin ribbon along the northern and western parts of the Taoudeni basin (Fig.7.3), and belong to the Varangian Glacial Period. Varangian glacial deposits in West Africa include the tillites of the Tabe Formation at the base of the Rokel River Group (Culver et al., 1978); the Kodjari tillites in the Volta basin; and their equivalents dated at about 675 Ma in the Buem Formation of the nearby Beninian mobile belt. In the Taoudeni basin the Varangian tillites fall between the age of the upper middle part of Supergroup I (Atar Group) in the Adrar region, dated at about 775 Ma, and the age of the green shales (595 Ma) in the overlying Supergroup II (Fig.7.4A).

The "Jbeliat Group" is the collective lithostratigraphic term proposed for the Adrar tillites and other Varangian tillites in the Taoudeni basin (Deynoux and Trompette, 1981). A detailed description of the Jbeliat Group was presented by Deynoux (1983). This deposit, up to 50 m thick in its type area in the Adrar, consists of two unconformable phases of terrestrial tillite accumulation, each overlying an erosional surface. The erosional surface represents the pre-glacial substrate for the lower tillite, and an irregular surface with "roches moutonnées" for the second tillite (Fig.7.4B). The tillites are succeeded by fluvial sandstones and lacustrine or marine shales with dropstones which were deposited during glacial retreats. The interglacial deposits (fine-medium sandstones, rarely conglomeratic, argillaceous siltstones) between the two tillites exhibit slump structures related to friction or the ploughing of ice-blocks on muddy tidal flats. The Jbeliat glacial deposits are capped by a thin and extensive disconformable sandstone horizon containing polygonal structures within sandstone wedges. This is overlain by post-glacial marine transgressive deposits belonging to the Teniagouri Group (Fig.7.4A).

Two regionally persistent and characteristic post-glacial lithologies occur immediately above the polygonal sandstone horizon. A thin baryte-rich calcareous dolomite horizon, 3 - 5 m thick, is overlain by marine bedded chert (Fig.7.4B). The mixtite, dolomite with baryte, and chert constitute the triad, a regional marker for the Varangian tillite in West Africa.

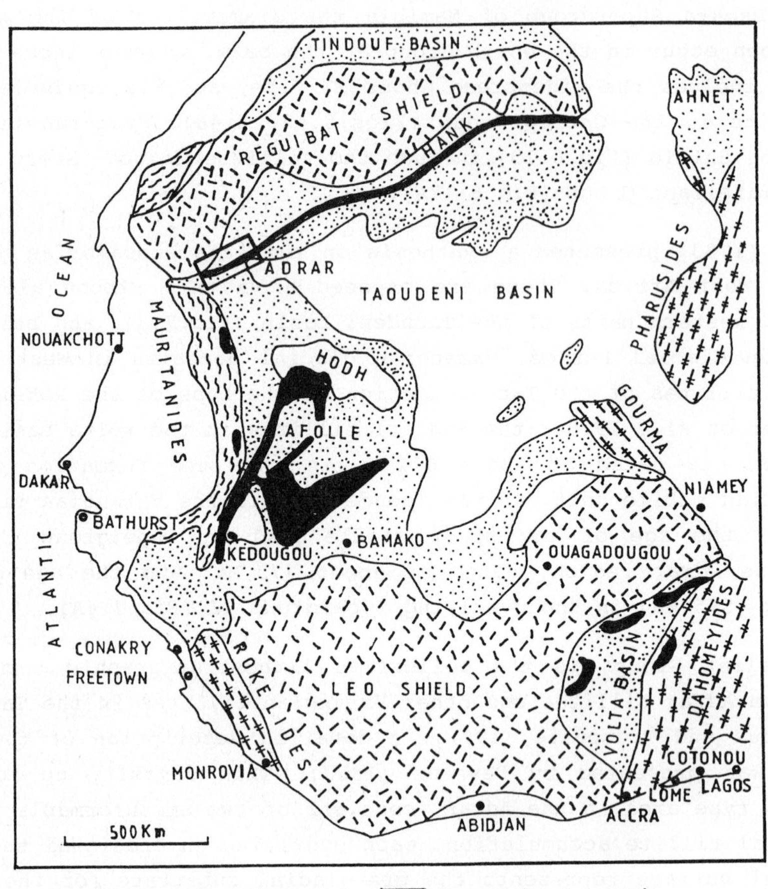

Figure 7.3: Distribution of late Precambrian tillites in West Africa. (Redrawn from Deynoux, 1983.)

The glacial deposits of the Late Sinian Epoch are believed to occur in South Africa and Namibia in the lower parts of the Nama Group (Table 6.2), according to Harland (1983) and Tankard et al. (1982).

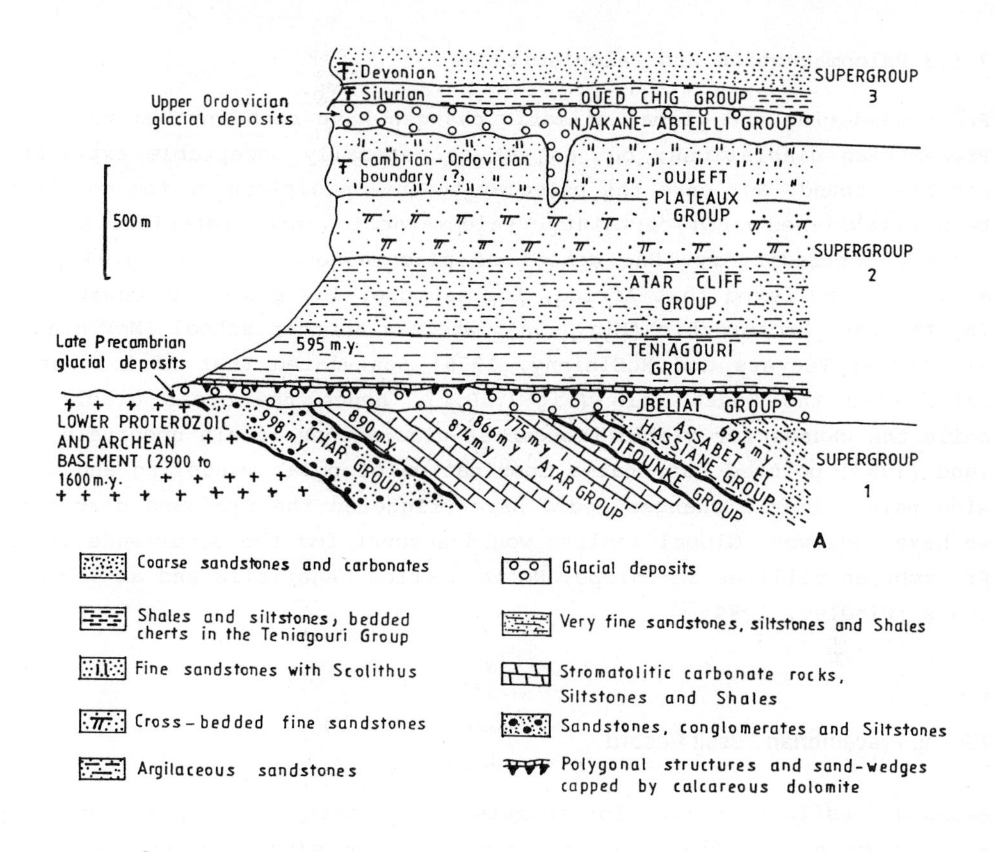

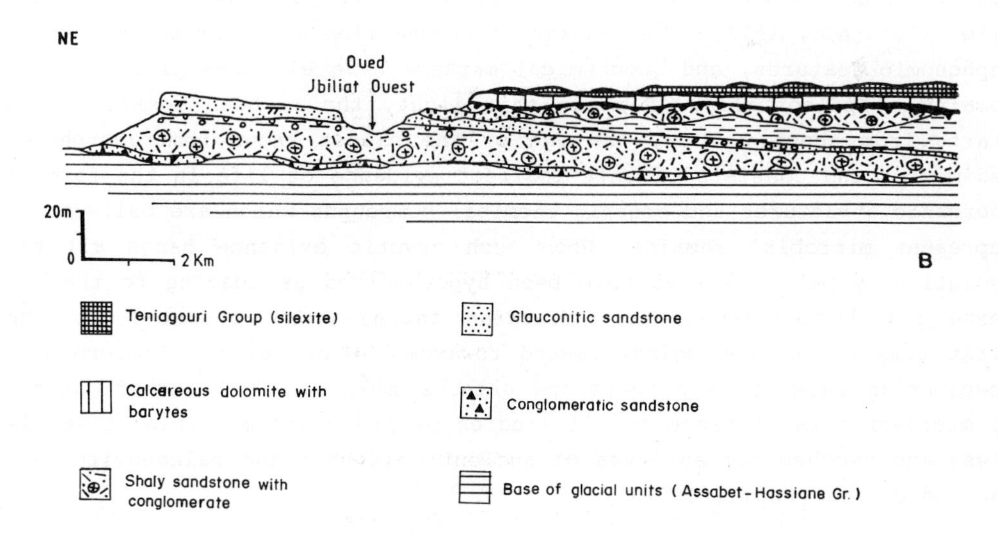

Figure 7.4: Late Precambrian and Early Paleozoic sequence of the Taoudeni basin. (Redrawn from Deynoux, 1983.)

7.1.3 Paleomagnetism and Paleolatitudes

Polar wandering and global climatic changes have been invoked to explain Precambrian glaciations; but as yet no generally acceptable explanation has been found. Nor have the paleolatitudinal positions of the continents been established with certainty. Paleomagnetic interpretations are completely contradictory. One school of plaeomagneticists (e. g. Piper et al., 1973) believed that Africa, for example, lay along the equator during the Late Proterozoic (Fig.7.5A), whereas another school (McElhinny et al., 1974; Veevers and McElhinny, 1976) postulated that Africa was located near the South Pole (Fig.7.5B,C), hence the great glaciations. While the causes for global Precambrian glaciations remain uncertain, Harland (1983) pointed out that a combination of polar wandering and worldwide paleoclimatic changes could have triggered the profound glaciations we have reviewed. Global cooling would account for the occurrence of Late Precambrian tillites in Europe, North America, Australia and at low latitudes (Windley, 1984).

7.2 The Precambrian Fossil Record

Because fossils by definition embrace "any remains, trace, or imprint of a plant or animal that has been preserved by natural processes in the Earth's crust since some past geological time; any evidence of past life ...." (AGI, 1972), the remains of Precambrian micro-organisms, their taphonomic features, and geochemical markers have all been placed in the domains of paleontology. Seen in this light, the Archean fossil record starts from about 3.5 Ga, the age of the oldest known unmetamorphosed sediments; and consists of only indirect evidence of life in the form of inorganic structures and organic chemical compounds which are believed to represent microbial remains. Upon such cryptic evidence hangs all the evolutionary pathways that have been hypothesized as leading to the diverse soft-bodied metazoans (Ediacaran fauna) which appeared for the first time in the geological record towards the end of the Proterozoic. Precambrian paleontology therefore entails morphological investigations of micro-organisms; trace fossil studies of preserved microbial communities; and geochemical analyses of ancient metabolic and paleoenvironmental indicators (Knoll, 1990).

Frazier and Schwimmer (1987) grouped most Precambrian fossils older than 1.0 Ga into: spheres and bacilliform structures (possible bacteria, blue-green algae, bacterial or fungal spores, fungi); filaments possibly

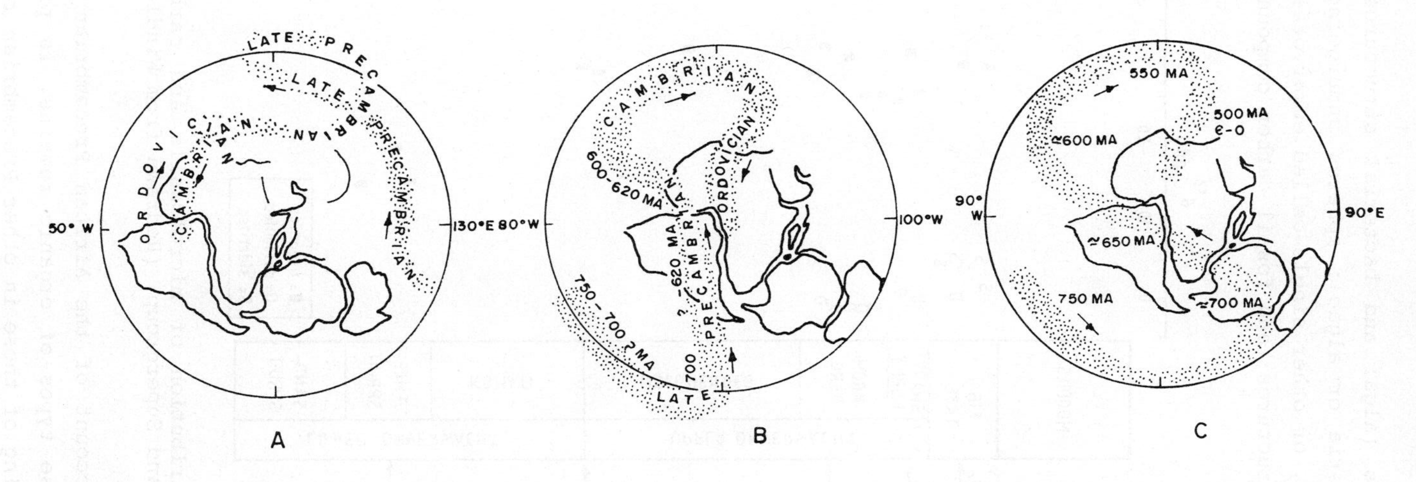

Figure 7.5: Late Precambrian - Early Paleozoic apparent wandering of the South Pole. (Redrawn from Deynoux, 1983.)

of algae; stromatolites (algal and bacterial structures); clusters of spheres (colonial bacteria or algae); spheres undergoing cell division (bacteria, algae, fungi, or other single-celled eukaryotes); irregularly-shaped single-celled structures; and fossil carbon compounds.

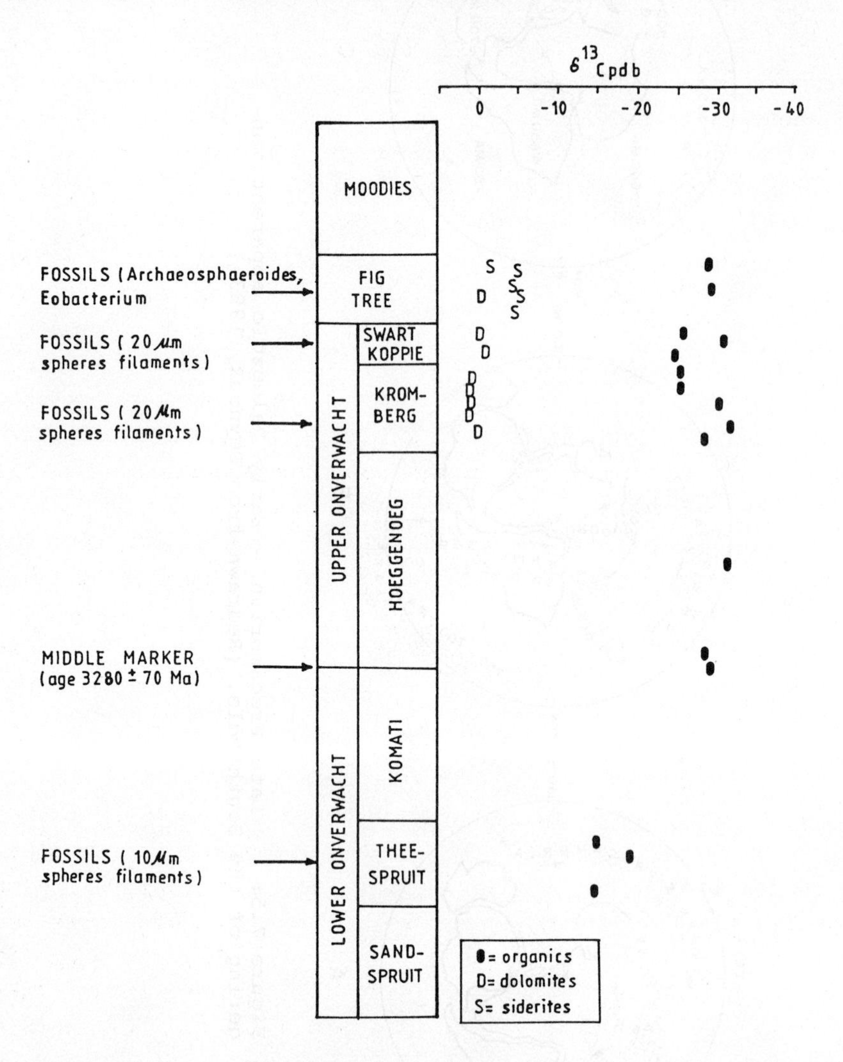

Figure 7.6: Distribution of microfossils and carbon isotope data for the Swaziland Supergroup. (Redrawn from Windley, 1984.)

A chronological account of the African Precambrian fossil record, comprising some of these types of organic remains, is presented below. Mention is made in passing of those in other Precambrian regions in order to fill the missing gaps in the African record.

7.2.1 The Archean Fossil Record

Windley (1984) presented a comprehensive survey of the known Archean-Proterozoic microfossils of Africa. Three stratigraphic levels in the Swaziland Supergroup greenstones (Fig.7.6) in the Kaapvaal province have yielded microfossils at least 3.5 - 3.4 Ga old. These probable microfossils are carbonaceous cell-like spheroids, rod-shaped bacterium-like bodies, and filamentous thread-like structures. They are found in cherts in the Lower Onverwacht Group (Theespruit Formation), in chert and argillite in the Upper Onverwacht Group, and in the organic-rich black cherts and shales of the Fig Tree Group (Fig.7.6). Ranging in size from 1 micron to 55 microns, these microfossils increase in size upward in the stratigraphic section, so that those in the Lower Onverwacht are half the size of the forms higher up in the section. In the Upper Onverwacht some of the spheroids contain evidence of binary cell division, and some of the carbonaceous spheroids in the Fig Tree Group resemble algae and cysts of flagellates. Also, the Pieterburg greenstone belt of South Africa, dated at about 2.6 Ga, contains spherical carbonaceous aggregates which probably represent vegetative diaspores of primitive columnar plants. Some of the materials ascribed to microfossils have, however, been questioned by Schopf and Walter (1983). But among the arguments in favour of the presence of organic matter in the Swaziland Supergroup are the occurrence of possible algal bodies with coatings of sulphur and compounds of Cu, Fe, Ni, and Ca which were probably precipitated by metabolic processes; and isotopic evidence suggesting carbon fractionation through photosynthesis.

The Swaziland microfossils are believed to have carried out photosynthesis, a vital process which could even have started earlier and liberated oxygen into the anoxic primordial evironment. The above geological evidence and findings in other Archean greenstone belts such as the Warrawoona Group in western Australia suggest that the earliest Archean life consisted of single-celled procaryotes (cells without nuclei). Three groups of procaryotic organisms probably existed, comprising eubacteria which include cyanobacteria (blue-green algae) and most of the commoner species of bacteria; the archaebacteria (a distinct primary kingdom which contains bacteria that can thrive in hot, acid or salty environments); and a third group of organisms which were probably the ancestors to the modern eucaryotes (micro-organisms with nuclei). It is believed that cyanobacteria were the builders of the earliest stromatolites which by the Late Archean had appeared in great abundance. The cyanobacteria could initially have utilized H_2S for photosynthesis without generating oxygen, but later they were able to exploit both sunlight and water as alternative sources of energy for food manufacture, thereby liberating oxygen. The early biochemical pathways in these primitive systems have been discussed in detail by Nisbet (1987). Our main concern here is the fossil record, of which cyanobacteria made their impressive contribution in the form of stromatolites.

Since they are among the best-preserved stromatolites, those in the Cheshire Formation in the Upper Bulawayan greenstones of Zimbabwe are discussed here in detail. While the occurrence of true stromatolites have been doubted in older Archean strata such as the Fig Tree Group and in the Middle Archean Nsuze Group of the Pongola basin in South Africa, morphological studies of the Late Archean Cheshire stromatoliles (Martin et al., 1980) and geochemical studies suggest that these were similar to modern stromatolites, built by blue-green algae or cyanobacteria. Modern stromatolites are carbonate or chert structures which are produced by mats of these micro-organisms; the mats trap and bind detrital materials with organic filaments during growth. The wavy laminations and large domes (with radii of up to 400 mm) in the Cheshire stromatolites (Fig.7.7) and their well-preserved microstructures attest to their truly algal origin. Like modern stromatolites which are commonly of lagoonal and intertidal origin, the Cheshire forms are enclosed in shales and siltstones of intertidal origin.

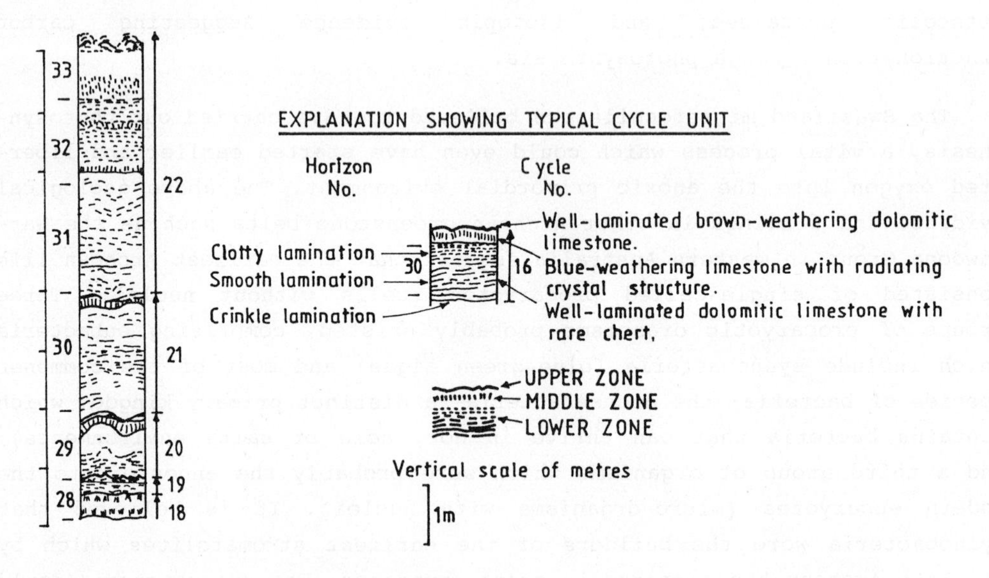

Figure 7.7: Stratigraphic columns for the main stromatolite outcrop in the Cheshire Formation, Belingwe greenstone belt, Zimbabwe. (Redrawn from Nisbet, 1987.)

The Cheshire stromatolites show cyclical laminations reflecting local fluctuations of energy conditions, and temperature and salinity, as supported by carbon and oxygen isotopic variations (Abell and McClory, 1987).Various forms of stromatolites similar to modern forms such as *Baicala*, *Conophyton*, *Irregularia*, and *Stratifera* have been identified from the Cheshire Formation.

7.2.2 The Early-Mid Proterozoic Fossil Record

By far the most extensive profliferation of stromatolites was in the Early Proterozoic, when,as in the Transvaal epeiric sea of the Kaapvaal province, thick accumulations of low relief and domical stromatolites built the intertidal to shallow subtidal carbonate platforms in the Chuniespoort and Ghaap Groups of the Transvaal-Griqualand West Supergroup. These stromatolites are characterized by large elongate stromatolitic mounds which ranged in width from 5 to 200 m and were up to 40 m long, with laminae relief of over 3 m (Tankard et al., 1982).

Also, as previously noted in connection with the Witwatersrand gold reefs (Chapter 4.2.2), kerogen seams with remarkable remains of micro-organisms have been discovered, which include bacteria, algae, fungi, and lichen-like plants. These forms are believed to have existed as carpet-like encrustations which extracted gold and uranium from the surrounding water, in a manner similar to modern fungi and lichen.

From the coeval Gunflint Iron Formation in Ontario, Canada (dated at about 2.0 Ga), and the younger Beck Springs Dolomite in eastern California (1.4 - 1.2 Ga old) evidence has been uncovered for the earliest eucaryotic cells. The Gunflint biota includes forms which resemble modern soil micro-organisms, and also contain some of the strongest geochemical indications for the presence of chlorophyll at an interval in geological history when the widespread accumulation of banded iron-formations suggests enhanced oxygen levels through photosynthesis. The Mid-Proterozoic Bitter Springs chert in central Australia (1.0 Ga - 900 Ma old) also contains diverse algae including green algae which were found at the stage of mitotic cell division. This finding further attests to the appearance of sexual reproductive processes, an improtant evolutionary step which established the mechanism for the transfer of genetic material, thus enhancing evolutionary diversification in the Late Proterozoic.

7.2.3 The Late Proterozoic Fossil Record

By the Late Proterozoic, stromatolites, which had witnessed unrivalled proliferation earlier in the Proterozoic, attained the acme of their dominance and evolutionary radiation throughout the world. This situation rendered it feasible to subdivide and correlate Late Precambrian sedimentary supracrustals using stromatolite assemblages - an undertaking for which Russian geologists blazed the trail. Russian biostratigraphers subdivided the Riphean period (1.65 Ga - 680 Ma) into four major units (Fig.7.8) based on distinctive biostratigraphic assemblages of stromatolites which are now used routinely all over the world.

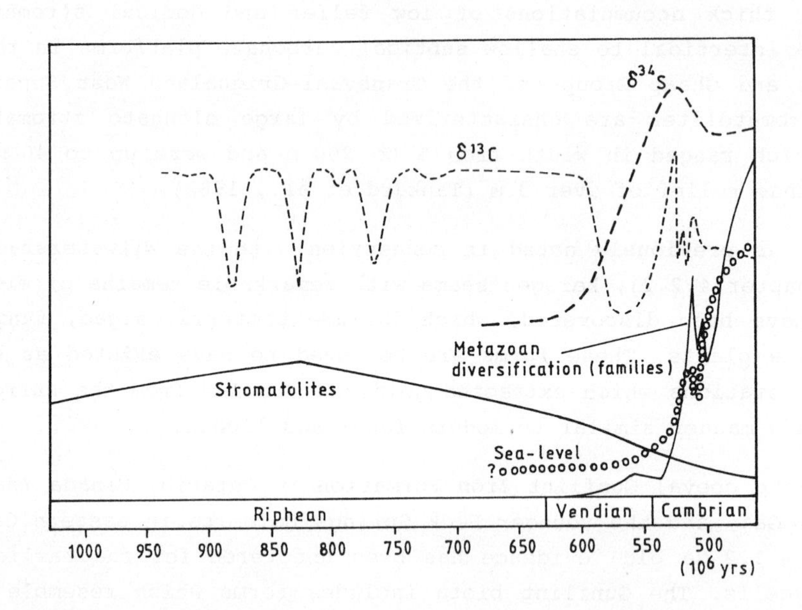

Figure 7.8: Diversity of stromatolites and metazoans during the late Precambrian and global changes in ocean chemistry and sea-level. (Redrawn from Morris, 1990; Windley, 1984.)

For example, the Late Riphean (Fig.7.8) has been recognized throughout the African Late Proterozoic stromatolitic sequences (Fig.7.9) and utilized to correlate from the Taoudeni basin on the West African craton, to the lower part of the Mbuyi Mayi Supergroup on the Zaire craton, and to the Bukoban Supergroup on the western part of the Tanzania craton (Bertrand-Sarfati, 1972). Bertrand-Sarfati and Walters (1981) pointed out that in spite of taxonomic uncertainties Early Proterozoic stromatolites are morphologically different from the Late Proterozoic forms. Late Proterozoic stromatolites are also of great paleoenvironmental value since biological (Trompette, 1982) and paleoecological parameters such as sunlight and tidal range influenced their maximum growth (Cloud, 1968). Bertrand-Sarfati and Moussine-Pouchkine (1988) used the extensive columnar stromatolite biostromes and bioherms in the northwestern part of the Taoudeni basin to interprete subtidal epeiric sedimentary environments.

Other microbiotas underwent evolutionary changes during the Late Proterozoic. A compilation of available evidence by Trompette (1982) showed that up to about 1.45 Ga ago procaryotic cells, sheaths (cyanophytes and bacteria) and filaments (bacteria and oscillatoraceans) were small; after that time there was a marked increase in size, accompanied by the diversification of taxa. Acritarchs appeared in the Late Proterozoic. These are organic-walled microfossils that occur in fine clastic sediments. They range in size from a few microns to several tens of microns in diameter. Generally spheroidal in shape and often unicellular, their affinities with modern taxonomic groups (e.g. algal cells, spores, cysts, protistans, annelid eggs) is uncertain. But being abundant in Late Proterozoic sediments all over the world, their biostratigraphy, also initiated by the Russians, has enabled more precise correlation of Late Riphean, Vendian, and Early Cambrian strata than has been feasible using stromatolites. For example, acritarchs have been utilized in equatorial Africa to establish the age of the Mbuyi Mayi Supergroup (Baudet, 1987), and the age of the Roan Group in the Katangan orogen (Binda, 1972).

7.2.4 The Ediacaran Fauna

The Ediacaran fauna, comprising the imprints and moulds of soft-bodied organisms which are mainly preserved in shallow marine terrigenous strata, were the first multicellular (metazoan) organisms to appear in the geological record, during the Vendian (Fig.7.8). As stated in Chapter 2.2 the age of the Ediacaran assemblage, in its broadest definition, ranges from 650 to 570 Ma, although Morris (1990) considers a span of about 620 - 570 Ma preferable. However, the first appearance of Ediacaran organisms in the fossil record is not precisely known. The presence of animal burrows at an interval dated at about 1.05 Ga in the basal beds of the Roan Group in the Zaire-Zambian Copperbelt, suggests that metazoans probably appeared this early (Clemmey, 1976).

First discovered in Namibia, prior to the Second World War, at a time Namibia was a German colony known as Deutsch Sud-West Africa, the Ediacaran fauna was announced and christened in South Australia where faunas similar to those of Namibia were later uncovered. The Ediacaran fauna was truly cosmopolitan; for it appeared at the same geological age in America, Asia, and Europe, without many provincial characteristics in any of these regions.

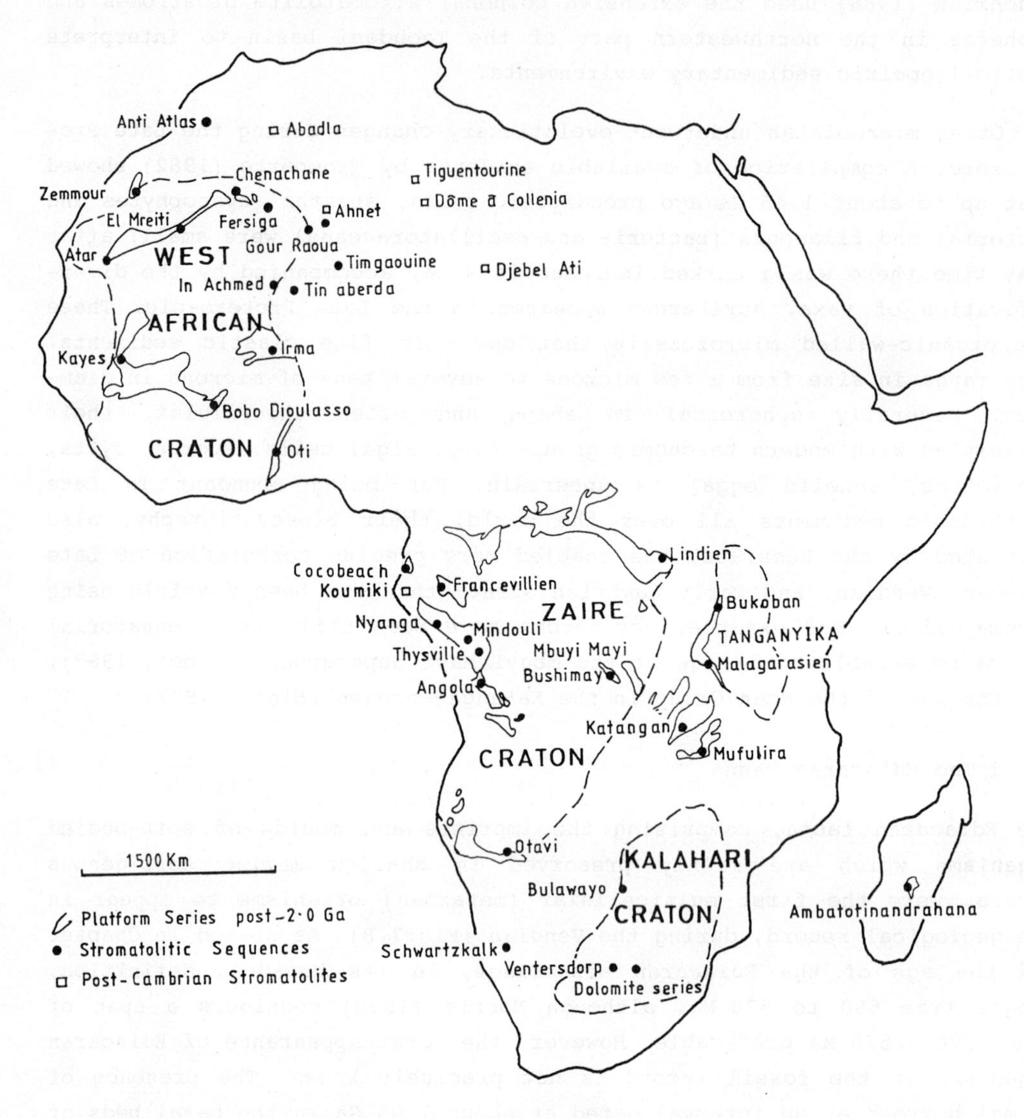

Figure 7.9: Known distribution of Precambrian stromatolites in Africa. (Redrawn from Bertrand-Sarfati, 1972.)

Generally, the unknown organisms whose imprints and moulds constitute the Ediacaran fauna were morphologically closest to coelentrates (sometimes shaped like feathers, combs, fans, and brushes); others were bilaterally symmetrical but non-segmented organisms like platyhelminth worms; segmented forms resembled annelids; and still, forms were there that bore no resemblance to living creatures (Seilacher, 1984). Two broad

Ediacaran faunal assemblages are known, namely: shallow-water dwellers such as those in Namibia and the Ediacaran Hills in Australia; and a probable deep-water assemblage, also found in Australia, and in Charnwood Forest, U.K.

The Nama platform group in Namibia (Table 6.2) contains a diverse Ediacaran assemblage which inhabited tidal flats, shallow seas and also lived among stromatolitic biostromes (Tankard et al., 1982). Coelentrates (*Rangea*, *Pteridinium*, *Namalia*, *Ernietta*) and worm-like forms flourished on tidal flats in front of braided alluvial plains, while the polychaete *Cloudina* scavenged the stromatolites for food. In addition to these body fossils, traces known as *Phycodes* and *Skolithos* represent other soft-bodied organisms that burrowed through the sediments.

The Ediacaran faunas are of great scientific interest because they mark a major step in the evolution of life -the appearance of the earliest metazoans. This cursory review of Precambrian life shows that Archean organisms consisted of procaryotic micro-organisms with stromatolites dominating the scene throughout the Proterozoic. Stromatolites flourished so extensively, perhaps on account of contributions from eucaryotic algae, which later joined the stromatolite mat-building community. The arrival of acritarchs suggests the existence of other organic communities. It has been speculated that the ancestors of the Ediacaran faunas was found among acritarchs (Trompette, 1982). However, ironically, the Ediacaran metazoans are not believed to have been the ancestors of the important invertebrate groups (Archaeocyatha, Mollusca, Brachiopoda, Echinodermata) which appeared in the Early Cambrian, already so advanced as to suggest the existence of well established phyletic lines in the Precambrian (Fedonkin, 1990). Their advanced evolutionary stage indicates that these invertebrates had existed in the latest Proterozoic, but as generally believed, they had no mineralized or preservable skeletons.

Fedonkin (1990) advanced an interesting hypothesis for biotic evolution near the dawn of the Phanerozoic. Low species diversity among Ediacaran assemblages suggests that this group lived for a geologically brief duration. In many parts of the world Ediacaran faunas are found after the last Precambrian glaciation. The world-wide post-glacial marine transgression probably caused the Ediacaran faunas to rapidly diversify and attain their peak of evolutionary radiation, after which they declined. Their decline is evident in their general lack of endemism, meaning that Ediacarans did not evolve into new niches. Another indication of their decline is their increase in individual body size compared to the first very small shelly fossils which appeared at the base of the Cambrian. Mass extinctions among Ediacaran communities could have set in as from the Middle Vendian, probably because of competition from the many small and more efficient feeders which were the ancestors to the Cambrian invertebrates.

Chapter 8 Paleozoic Sedimentary Basins in Africa

8.1 Structural Classification of African Sedimentary Basins

Africa lay at the centre of Gondwana at the close of the Precambrian. The Pan-African orogeny had joined other continents to its eastern and western margins. Throughout most of the Paleozoic North Africa occupied the southern seaboard of the Iapetus Ocean (Fig.8.1), and South Africa, too, was bordered by a shelf sea. After the Iapetus closed in the mid-Devonian and the Hercynian orogeny had in the Late Carboniferous brought the remaining northern continental blocks into Pangea, Africa assumed an even more interior location in this new supercontinent. Africa remained in this position until Mesozoic-Cenozoic times when Pangea fragmented and each continent went its separate way, leaving Africa surrounded once again by the ocean, as was the case before Gondwana began.

Seen against this global Phanerozoic paleogeographic backdrop, the classification of African sedimentary basins into four main types can be readily appreciated. Clifford (1986) and Picha (1989) grouped African basins into: divergent passive marginal basins; intracratonic sag basins; intracratonic fracture basins; and cratonic foreland basins. Figure 8.2 shows the distribution of these primary types of sedimentary basins, and the secondary or modified basins, all of which cover about a half of the continent. Depending on their structural location and on the dominant depositional process, the passive or marginal sag basins have been further classified as wrench, deltaic sag, and fold belts. Although this chapter deals principally with the Paleozoic it is considered necessary, however, to place all basins including those of Mesozoic-Cenozoic age in their structural settings, especially considering the fact that some basins had polycyclic structural settings and histories, having suffered modifications since their inception. Some Paleozoic basins also assumed new structural settings in the Mesozoic-Cenozoic.

The above classification of African Phanerozoic basins has been determined by their stage in the plate tectonic (Wilson) Cycle: rifting, drifting and sagging, and subduction and continental collision (Picha, 1989). Only two parts of Africa were marginally involved in Phanerozoic collision tectonics. Northwest Africa comprising the Moroccan Hercynide foreland and thrust belt and the Mesozoic-Cenozoic Atlas and Rif Alpine orogenic systems with thin-skinned thrust belts, and the Early Paleozoic

Cape fold belt in South Africa, were affected by Phanerozoic orogenies. The basins in the rest of the continent are at the first two stages in the plate tectonic process. Paleozoic intracratonic sag basins constitute the Saharan platform which extends from Mauritania and Morocco in the west to Egypt and the Sudan in the east.

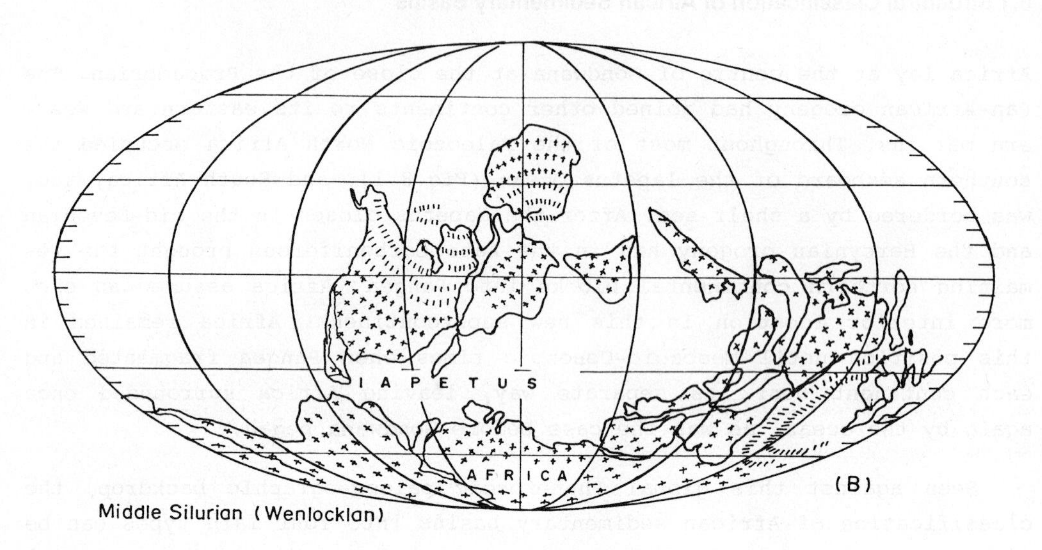

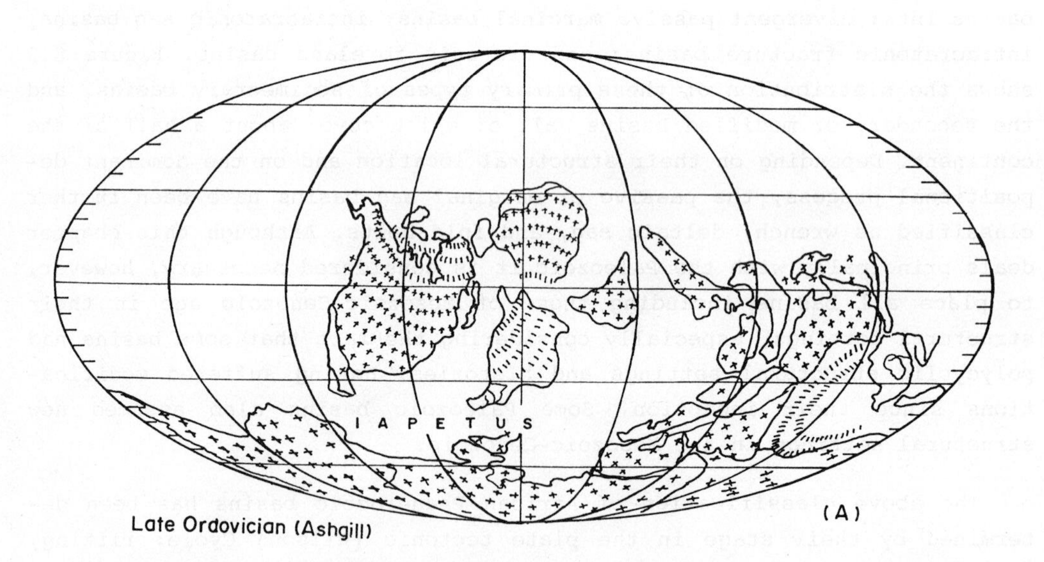

Figure 8.1: Global paleogeography of the Ordovician-Silurian. (Redrawn from Scotese, 1986.)

In sub-Saharan Africa the Zaire, Okawango and Etosha intracratonic sag basins, although containing Precambrian to Phanerozoic sediments,

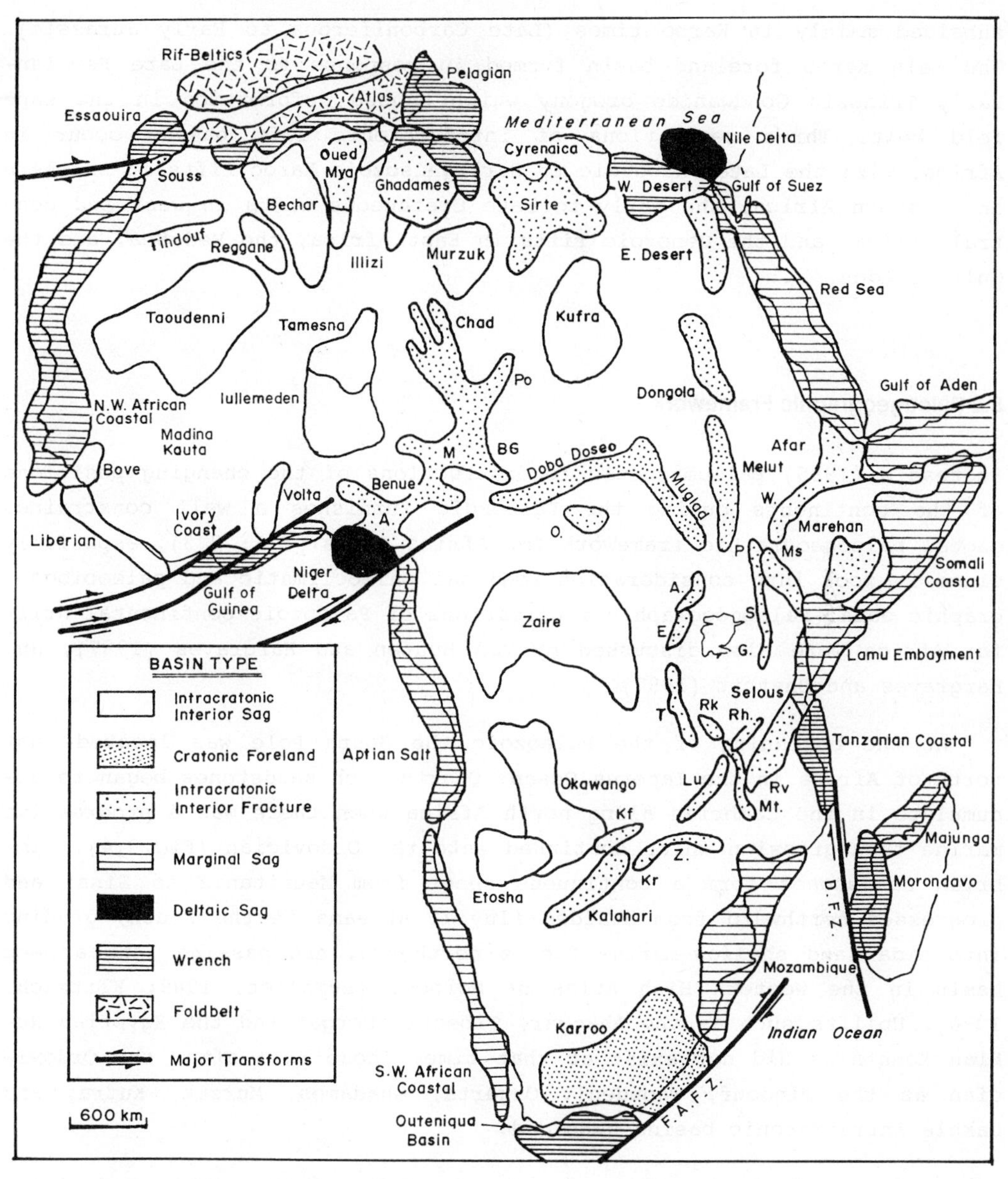

D.F.Z. DAVIES FRACTURE ZONE
F.A.F.Z. FAULKLAND-AGULHAS FRACTURE ZONE

Figure 8.2: Types of African sedimentary basins. (Redrawn from Clifford, 1986.)

subsided mainly in Karoo times (Late Carboniferous to Early Jurassic). The main Karoo foreland basin formed in response to the Late Permian-Early Triassic Gondwanide orogeny which caused deformation in the Cape fold belt. Three generations of intracratonic rift basins occur in Africa, viz: the Late Paleozoic to Early Mesozoic Karoo rifts in southern and eastern Africa; the Early to Late Cretaceous rifts in west and central Africa; and the Cenozoic rifts in East Africa, the Red Sea, and the Gulf of Aden.

8.2 Paleogeographic Framework

Scotese's (1986) paleomagnetic reconstructions of the changing positions of the continents during the Paleozoic furnishes a well constrained global paleogeographic framework for Africa (Figs. 8.1; 8.3), especially since it took into consideration regional paleoclimatic and paleobiogeographic data. Paleogeographic implications of Paleozoic continental drift for Africa were also discussed by Van Houten and Hargraves (1987), and Hargraves and Onstott (1987).

At the beginning of the Paleozoic the South Pole was located just north of Africa in the Iapetus Ocean. Quartz-rich sandstones began to accumulate in the Cambrian along north Africa when there was a progressive marine transgression which continued into the Ordovician (Fig.8.4). Cambrian sandstones form a continuous sheet from Mauritania to Sinai and progressed northward from braided fluvial streams in the south, grading into tidal and shallow-marine facies northward, and passing into a deep basin in the Western High Atlas of Morocco (Burollet, 1989; Klitzsch, 1990). Uplifts such as the Ahaggar, Tibesti, Gargaf and the Egyptian Nubian Mountains did not exist at that time. These rose after the Ordovician as the Tindouf, Taoudeni, Ougarta, Ghadames, Murzuk, Kufra, and Dakhla intracratonic basins subsided.

Gondwana drifted over the South Pole during the Ordovician (Neugebauer, 1989) and by the Late Ordovician the South Pole was located far inland in northwestern Africa (Fig.8.1A), leading to widespread continental glaciation in Africa (Fig.8.5A). Rapid northward drift in the Early Silurian (Fig.8.1B) brought the northern margin of western Gondwana back into a warm climatic realm, causing the melting of the polar ice caps, and a profound marine transgression (Fig.8.4) which inundated the entire Saharan platform with marine embayments reaching as far south as the Cape region of South Africa (Fig.8.5B).

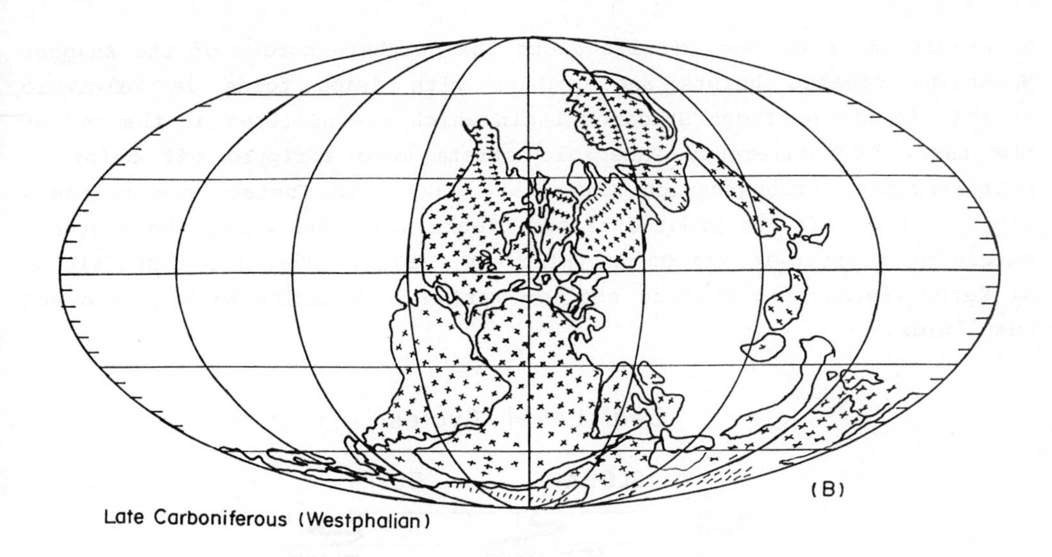

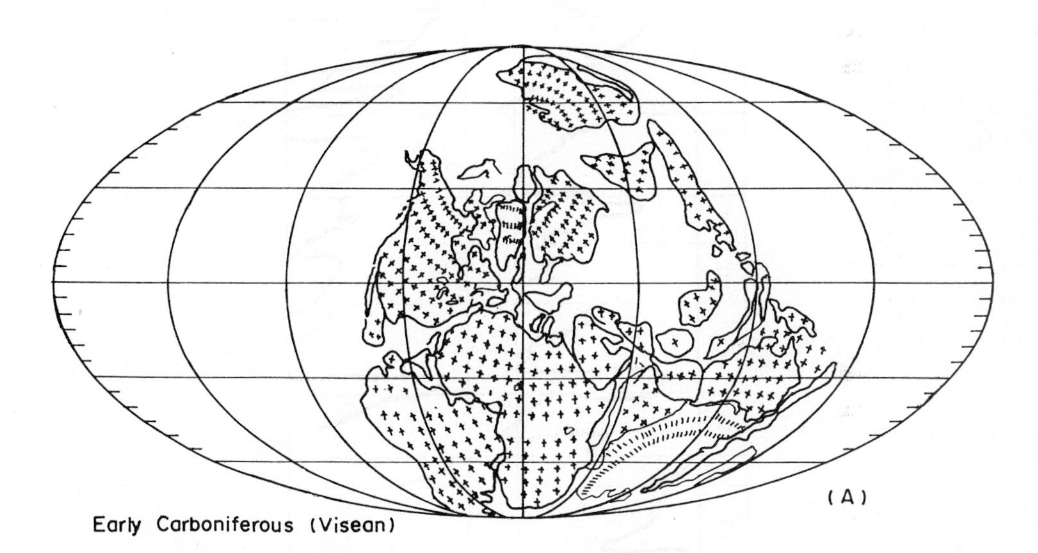

Figure 8.3: Global paleogeography of the Carboniferous. (Redrawn from Scotese, 1986.)

Major marine transgressions also took place in the Mid-Devonian (Fig.8.5C) and Early Carboniferous. Basin subsidence accentuated the Silurian and Devonian transgressions, and caused the northward progradation of sand bodies over marine shales. Hercynian deformation affected Morocco in the Middle to Late Carboniferous. It also affected the Saharan platform including the Ougarta ranges which were faulted and uplifted. An east-west uplift developed across the eastern margin of the Kufra basin.

Rejuvenation of basement faults along the northern border of the Ahaggar Mountains created horsts and grabens with drape folds in Paleozoic strata. In the northern Ghadames basin which was uplifted at the end of the Early Carboniferous, Paleozoic strata were stripped off before a post-Hercynian transgression (Burollet, 1989). In Tunisia and northern Libya a set of faults created tilted blocks and steps along the southern margin of a proto-Tethys Ocean (Klitzsch, 1971), causing the deposition of Carboniferous and Permian shallow marine lithofacies with corals and fusulinids.

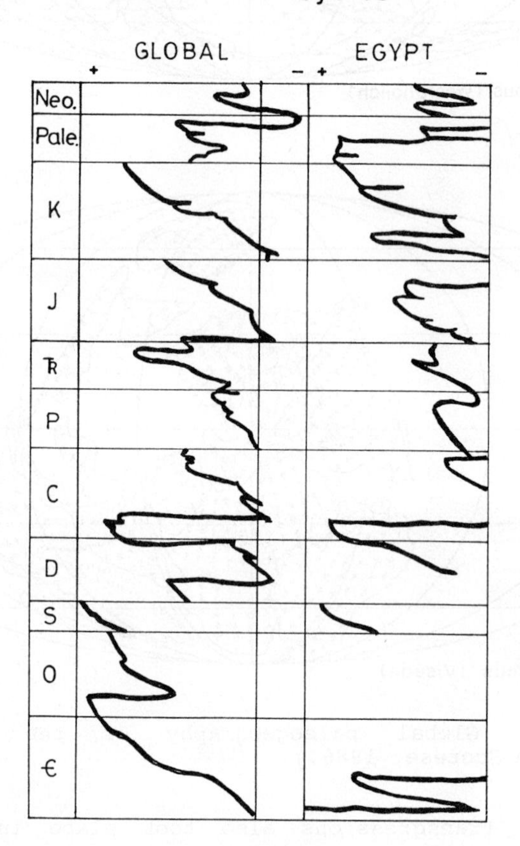

Figure 8.4: Global and Egyptian sea-level curves. (Redrawn from Morgan, 1990.)

Southern Africa moved over the South Pole during the Late-Carboniferous (Fig.8.3B), thus witnessing the most pronounced glaciation in geological history, the Permo-Carboniferous glaciation. Tillites which accumulated during this glaciation covered all the southern continents from South America, southern and eastern Africa through the Arabian peninsular to Antarctica. The Permo-Carboniferous glaciation initiated the predominantly nonmarine Karoo depositional cycle (Klitzsch, 1986) in Africa, with widespread rifting, basin subsidence and deposition of thick fluviatile and lacustrine clastics. Karoo-type deposits with their characteristic Gondawanan *Glossopteris* flora (Fig.8.6) accumulated throughout southern Gondwana during Late Carboniferous to Early Jurassic times. Since the Karoo depositional cycle began in the Late Paleozoic it is included in this chapter.

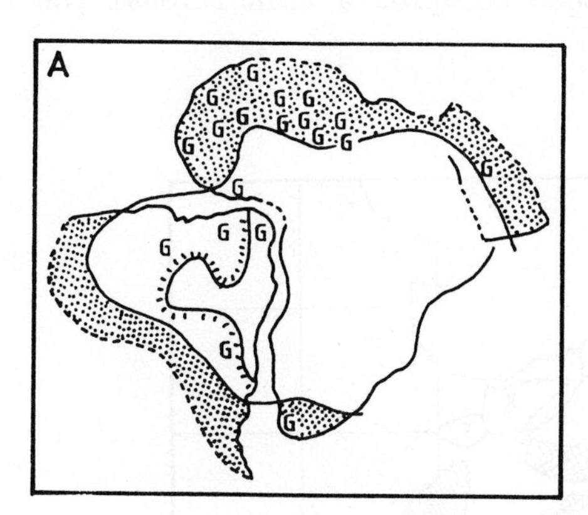

Mid - Ordovician shoreline
G = Late Ordovician - earliest Silurian glacial deposits

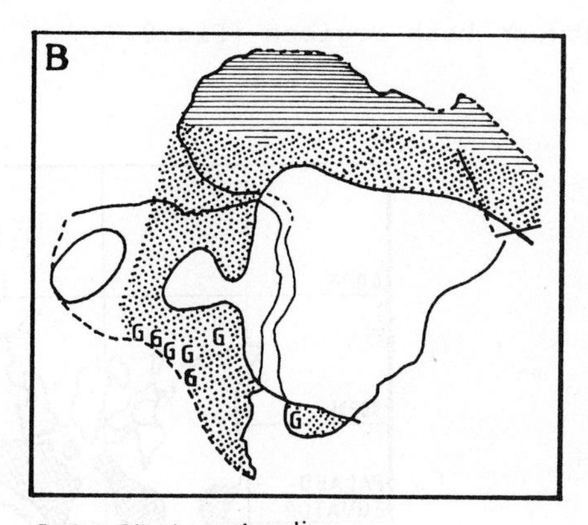

Early Silurian shoreline
G = Early Silurian glacial deposits

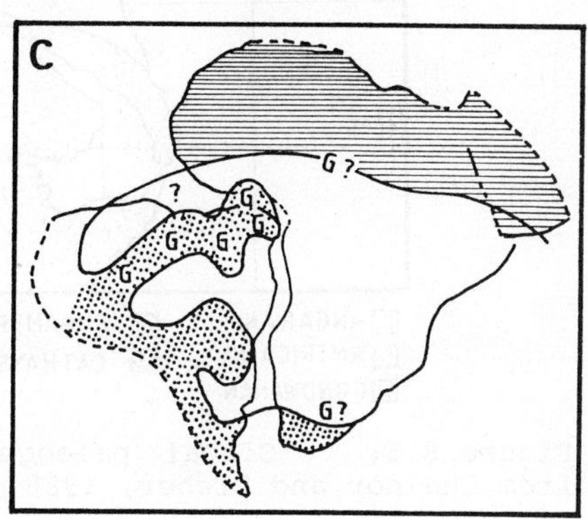

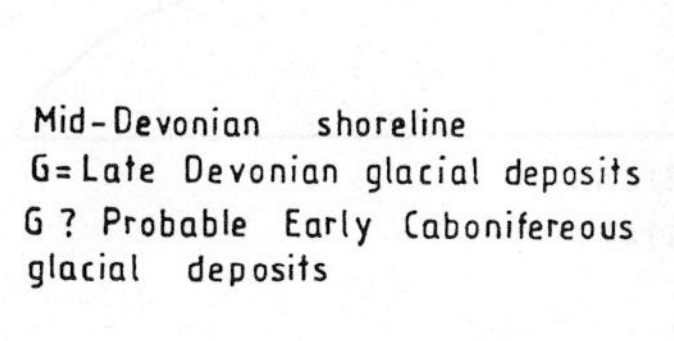

Cold-water (Malvinokafric) realm

Warm-water realm

Figure 8.5: Ordovician - Devonian paleogeography of western Gondwana showing glacial deposits. ((Redrawn from Hargraves and Van Houten, 1987.)

8.3 The Moroccan Hercynides

8.3.1 Structural Domains

Because of its structural and stratigraphic complexity it is necessary to consider first the structural framework of Morocco before examining its tectonic and stratigraphic evolution during the Paleozoic. Various structural subdivisions and terminologies have been applied to Morocco in an attempt to relate Morocco to the tectonic domains of the African continent and southern Europe, since Morocco occupies a transitional position between both regions (Fig.8.7).

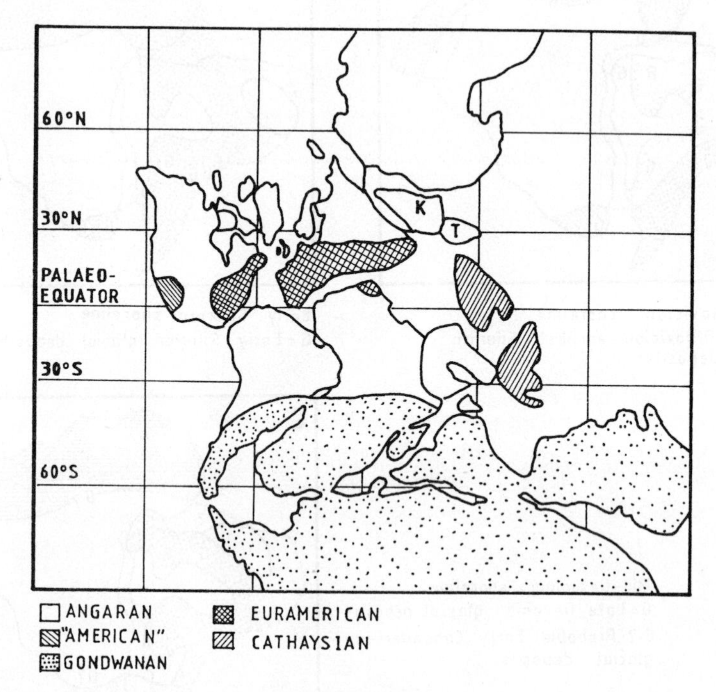

Figure 8.6: Global paleogeography of the Permian. (Redrawn from Chalnor and Creber, 1988.)

According to Bensaid et al. (1985) Morocco consists of two main tectonic domains, Mediterranean and African Morocco. Mediterranean Morocco consists of the Rif Mountains (Fig.8.7) which are part of the North African Maghrebide Mountain chain to which the Tell Mountains of Algeria also belongs. The Maghrebide, a belt of alpine nappe tectonics differs fundamentally from middle and southern Morocco. South of the Rif, the rest of Morocco (African Morocco) belongs to the relatively stable West

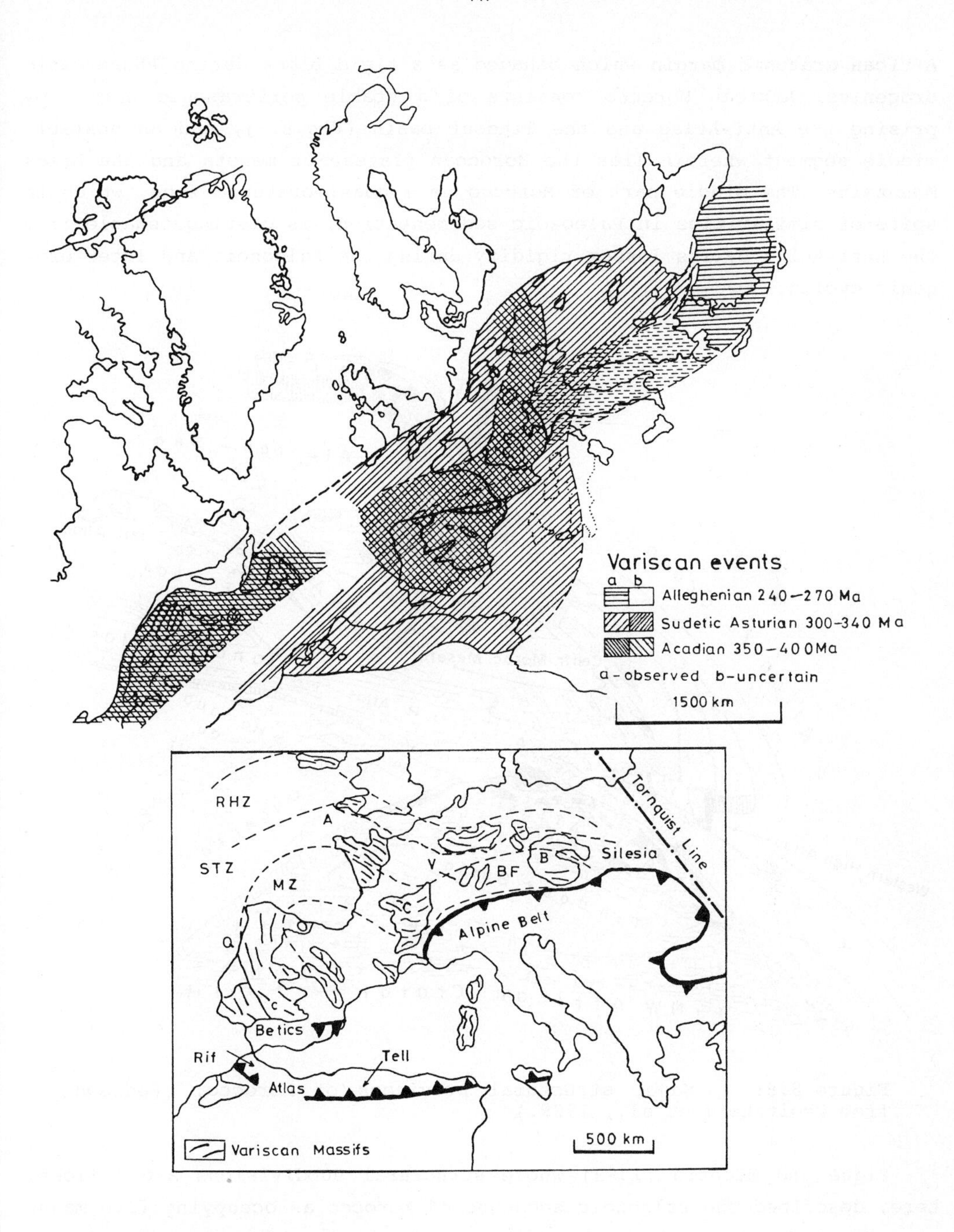

Figure 8.7: Hercynian orogenic belts of Europe and North Africa. (Redrawn from Windley, 1984.)

African cratonic margin which behaved as a rigid block during Phanerozoic orogenies. African Morocco consists of a stable pericratonic part comprising the Anti-Atlas and the Tindouf basin (Fig.8.8), and an unstable middle segment wherein lies the Moroccan plateau or meseta and the Atlas Mountains. The middle part of Morocco is a quasi-cratonic zone, which in spite of similarities in Paleozoic sedimentation, is distinguishable from the Anti-Atlas by its lesser rigidity during the Paleozoic and later orogenic cycles.

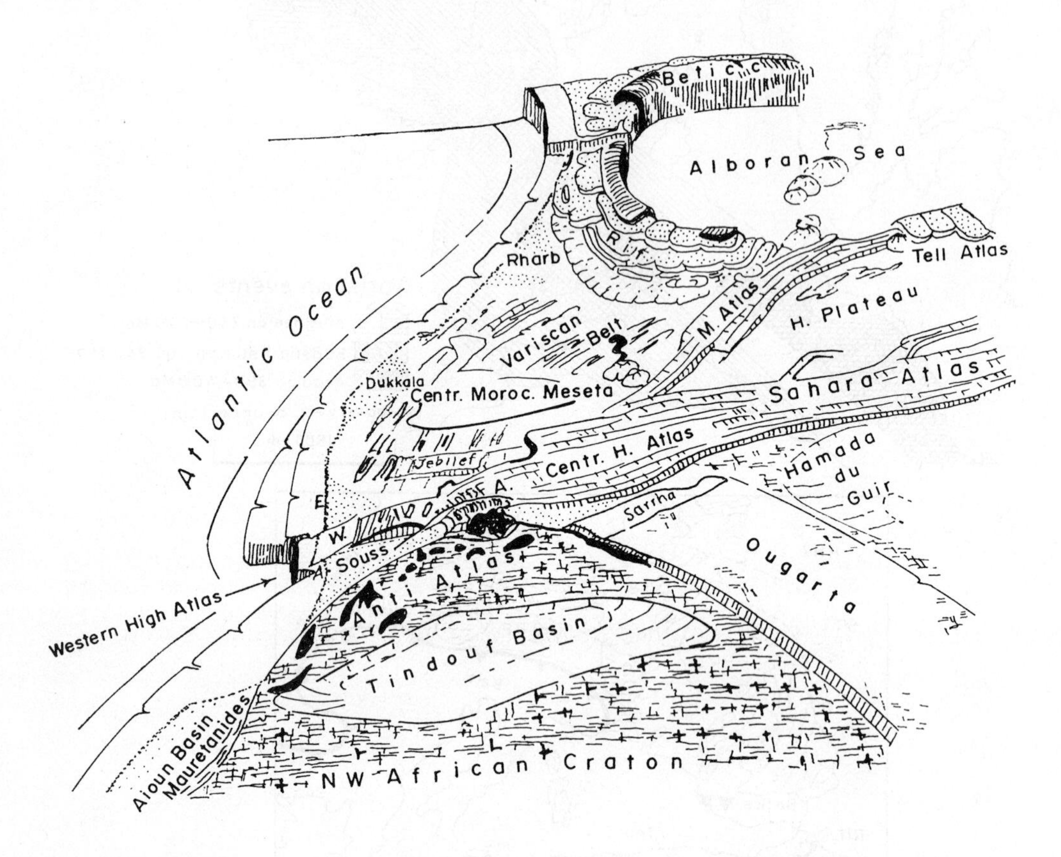

Figure 8.8: Major structural provinces of Morocco. (Redrawn from Froitzheim et al., 1988.)

Piqué and Michard (1989) whose structural subdivisions are followed here, described the Paleozoic sequence of Morocco as occupying five major structural domains, viz., the Saharan, the Anti-Atlas, the Meseta, the Atlas, and the Rif domains. Whilst the Tindouf basin, the main structural feature of the Saharan domain (Figs. 8.9, 8.10) is a large syneclise filled with Paleozoic epicontinental sequences, the Anti-Atlas occupying the northern cratonic border, is an anticlise or anticlinorium. The Anti-

Atlas is characterized by thinner Late Proterozoic to Paleozoic strata and by domal uplifts where the underlying Pan-African basement is exposed (Fig.8.10). Another major pericratonic feature is the Ougarta block to the southeast (Fig.8.8), with Paleozoic sediments and extensive Late Precambrian lava flows and volcaniclastic sequences.

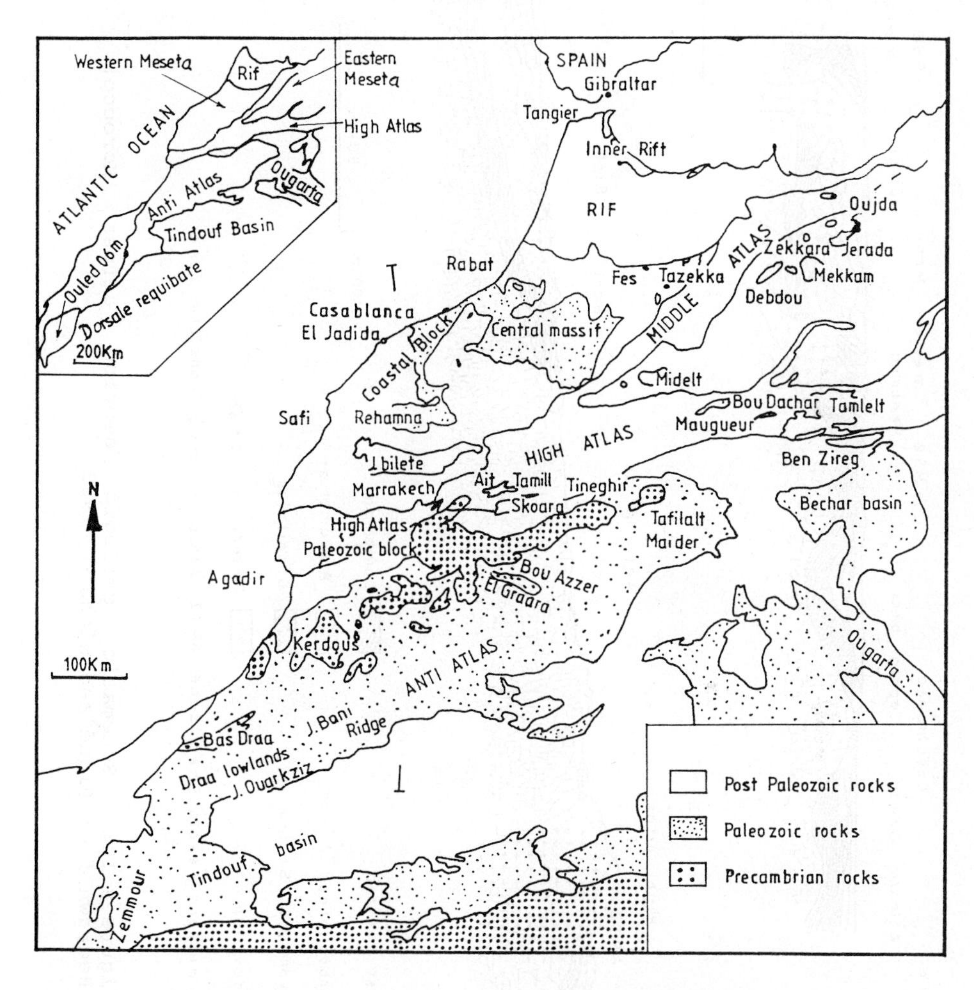

Figure 8.9: Paleozoic massifs and structural domains of Morocco. (Redrawn from Piqué and Michard, 1989.)

The Meseta domain which is separated into the Moroccan or Western meseta and the Oran or Eastern meseta by the Middle Atlas fold belt (Fig.8.9), is a denuded segment of a Hercynian (Variscan) belt which was accentuated by late-orogenic granite intrusions that are concentrated in the central anticlinoria. Undeformed Mesozoic-Cenozoic strata unconformably overlie Paleozoic rocks in the Meseta, and where eroded, Paleo-

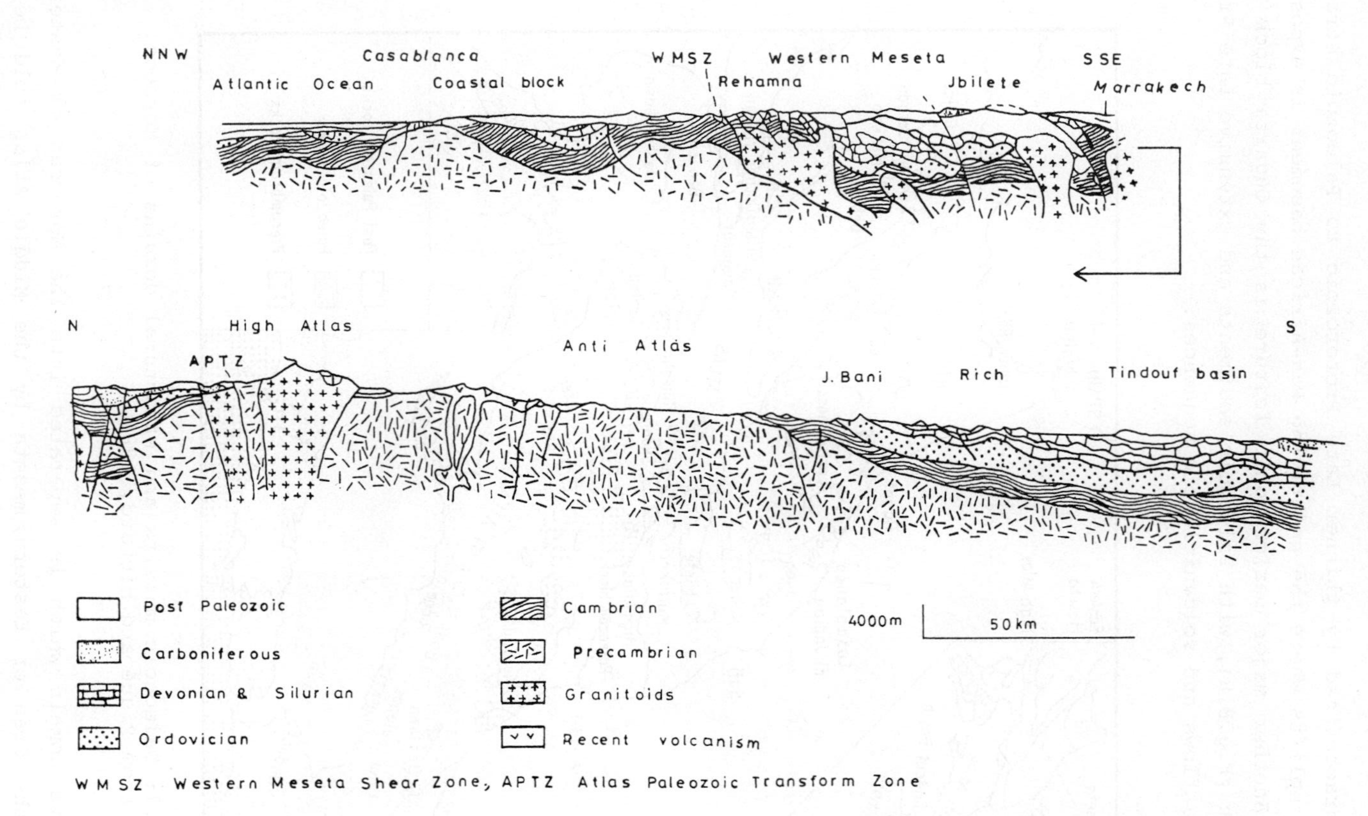

Figure 8.10: Schematic geologic section across Morocco. (Redrawn from Piqué and Michard, 1989.)

zoic massifs--Central massif, Rehamna, Jbilete--are exposed in the Moroccan meseta. In the Oran meseta the younger strata are less eroded and hence Paleozoic massifs are more restricted (eg. Jerada, Debdou, Mekkam, Midbelt, etc.).

Running from the Atlantic coast of Morocco in the west to Tunisia in the east, over a distance of 2,000 km, are the Atlas Mountains which appear to be an autochthonous terrane where essentially Mesozoic strata are folded over Paleozoic and Precambrian basement (Schaer, 1987). The Atlas domain is known for its post-Triassic rift tectonism which controlled the opening of the central Atlantic Ocean. This domain consists of the Middle Atlas, and the High Atlas which separates the Mesetas from the Anti-Atlas. The Atlasic (Alpine) orogeny with its Late Jurassic and Tertiary deformation phases, folded the Atlas domain. Large Paleozoic inliers are exposed at Aït-Tamlil, Mougueur, Tamlet, and in the Middle Atlas (Fig.8.9). Paleozoic terranes also occur in the inner part of the Rif domain where they constitute the Nappes paléozoïque or Ghomarides.

Since the above tectonic domains were all affected by the Hercynian orogeny (Fig.8.7) during the Middle to Late Carboniferous they were collectively termed the Moroccan Hercynides by Piqué and Michard (1989). However, pre-Hercynian deformation of Late Ordovician age (Caledonian orogeny) is known in the Sehoul block of northwestern Morocco.

8.3.2. Stratigraphy and Tectonic Evolution

Piqué and Michard (1989) published a synthesis on the stratigraphy and tectonic evolution of the Moroccan Hercynides, an outline of which is presented below. From the Late Proterozoic to the Middle Devonian the Anti-Atlas and Western Morocco (except the Sehoul zone) generally belonged to the same depositional setting. This phase was characterized initially by post-Pan-African molasse deposition of redbeds and by post-collisional volcanism. Later, southern and middle Morocco constituted one vast epicontinental shelf during the Early Paleozoic with terrigenous sediments coming from the Sahara. A carbonate shelf was installed in this region during Early to Middle Devonian, while a deep turbidite basin occupied the northwestern part of Morocco, south of the Sehoul block. The main Hercynian deformation took place in this northern deep basin during the Late Devonian, at a time when the western part of Morocco and the Anti-Atlas platform disintegrated into fault-bounded basins. These later basins progressively deformed and closed during the Middle to Late Carboniferous, the main phase of Hercynian movements in north Africa.

The Precambrian-Cambrian Transition (Infracambrian)

In the Anti-Atlas a major angular unconformity separates Precambrian supracrustals (shallow shelf quartz arenites and stromatolitic limestones) affected by Pan-African II orogeny, from a thick Late Proterozoic-Cambrian sequence (Fig.8.11). The term Infracambrian (Pruvost, 1951) has been used in the Anti-Atlas for the post-Pan-African conformable nonmarine sequence below the lowest fossiliferous level which bears Cambrian trilobites. Also termed Precambrian III (Choubert, 1963), the lithostratigraphic successions belonging to this age interval include the Anezi-Siroua-Saghro Group and the Anela-Ouarzazate Group, both of which generally contain thick sequences of predominently redbeds with major intercalations of volcanic materials. Upward-finning clastic sediments and carbonates mark the initiation of epicontinental sedimentation which is represented by the conformably overlying Adoudounian Group (Fig.8.11). As shown in Fig.8.11 this group comprises Lower Limestones, a middle sandstone and slate sequence ("Série Lie de vin"), and Upper Limestones which carry the earliest Cambrian trilobites in Morocco.

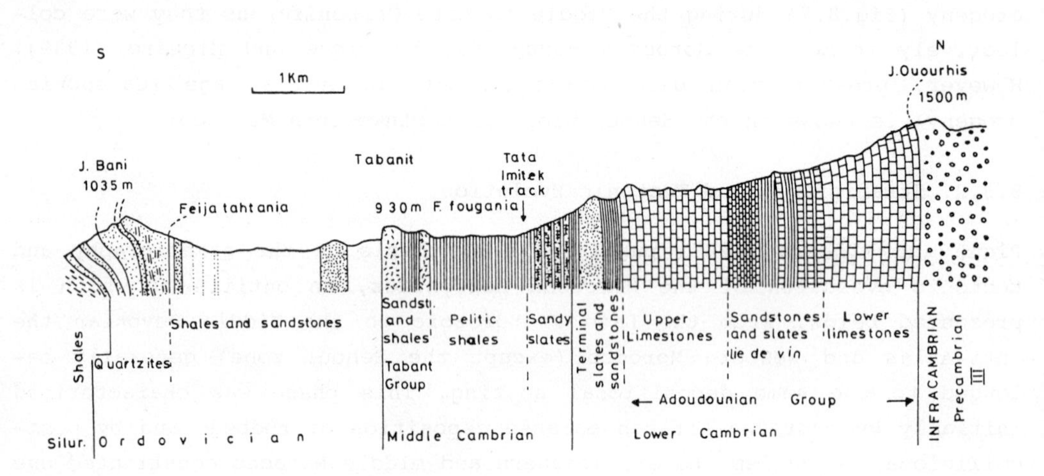

Figure 8.11: The lower Paleozoic section near Tafa. (Redrawn form Legrand, 1985).

The Adoudounian Group is about 1,400 to 2,100 m thick in the Western High Atlas where the entire Infracambrian to Carboniferous sequence is up to 10 km thick (Froitzheim et al., 1988). The Adoudounian in the Western High Atlas is well exposed in a succession that reflects Late Precambrian rifting, differential subsidence and marine transgression. According to Killick (1988) the sequence records a marine transgression (Ighir Formation); coastal stillstand (Tighricht Formation); rift-related progradational regression with basin-margin alluvial fans and fan deltas which interfinger with basin-centre fluvial and peritidal deposits (Al Makhzen Formation); another stillstand (Taghbar Formation); and renewed marine transgression (Talat-N-Ou-Lawn Formation).

Elsewhere, unfossiliferous strata similar to the Adoudounian Group are exposed in the cores of some anticlines in parts of western and eastern Morocco, thus suggesting the northward extension of Infracambrian sequences beneath the Lower Paleozoic.

The position of the Infracambrian-Cambrian boundary is however, controversial. Sdzuy (1978) and Schmitt (1979), for example, placed this boundary at the top of the "Série lie de vin". But Buggish and Flügel (1988) reported the Early Cambrian calcareous algae *Kundatia composita* from the "Série des calcaire inférieurs" (Lower Limestone) below (Fig.8.11), and assigned the lower Adoudounian Group to the Early Cambrian. Buggish and Flügel postulated that the sea encroached from the region of Agadir in the west, into an Anti-Atlas gulf that occupied what was probably a graben or an aulacogen that was linked to the Iapetus Ocean further to the west (Fig.6.9, A), a reconstruction that generally agrees with the Infracambrian-Cambrian stratigraphy of the Western High Atlas (Killick, 1988).

Cambrian subsidence and Volcanism

The classical Moroccan Cambrian stratigraphic sections are found in the Western Anti-Atlas (Fig.8.11). Most of the shale-graywacke sequences in Fig.8.11 are of Middle Cambrian age (Destombes et al., 1985) while the overlying Tabanit sandstone and shale group is of Middle to Late Cambrian age. Generally Cambrian strata disapear or thin eastwards in the Anti-Atlas, so that in the eastern Anti-Atlas the Lower Cambrian occurs as thin conglomeratic detrital layers or is absent. Local outcrops of Lower and Middle Cambrian strata occur in the Atlas and Meseta domains, with facies similar to those of the Western Anti-Atlas. Traces of Middle Cambrian volcanism occur everywhere in Morocco and are represented by thick complexes of andesitic and basaltic flows, volcanic breccias and tuffites.

Ordovician Platform and the Sehoul Terrane

Ordovician strata are exposed at the southern flank of the Anti-Atlas from Jbel Bani to Ougarta (Fig.8.9, 8.10), and in northern Morocco where, like Cambrian strata, they occur in Hercynian anticlinoria. They consist essentially of fossiliferous argillites and sandstones (Fig.8.11) deposited in a marine clastic epicontinental shelf which extended from Morocco to the Tindouf basin in Algeria. On this shelf 2,000 to 2,500 m of clastics accumulated in the Anti-Atlas and neritic faunas (graptolites, trilobites, brachiopods, echinoderms, pelecypods, orthoconic nautiloids, ostracodes, gastropods), bioturbated bottom conditions, and storm deposition occured with minor sea level oscillations causing numerous local unconformities. A notable feature of Moroccan Ordovician sedimentation was its overwhelming clastic nature and the rarity of carbonates. Turbidites were however, deposited in eastern Jbilete and Tazekka (Fig.8.9).

Late Ordovician glacial deposits are mostly sandstones and micro-conglomerates. Unweathered detrital biotites which are common in Moroccan diamictites attest to slow weathering rates in ice-covered polar or sub-polar areas of the Late Ordovician Sahara.

Volcanic flows are intercalated among Ordovician strata, but in the northern part of Morocco the sequences with volcanic flows in the Meseta are separated from the Sehoul zone by a tectonic contact. In the Sehoul zone the Cambrian is overlain by fine sandstones and siltstones containing Lower Ordovician graptolites. But more significant is the fact that the Ordovician in the Sehoul zone had been folded and subjected to low-grade metamorphism before the intrusion of the Rabat-Tiflet granite at about 430 Ma.

Silurian Post-glacial Transgression

Following the melting of the Late Ordovician Saharan ice caps, the entire Moroccan shelf and the Saharan platform were inundated. A widespread uniform graptolitic shale and siltstone facies attests to this unique drowning event. In the Anti-Atlas the peak of this transgression was attained in the Middle to Late Llandoverian. The Draa lowlands in the western Anti-Atlas shows a typical succession of Lower Silurian platy sandstones, followed by uniform facies of dark graptolitic shales and siltstones, which are overlain by an upper shale sequence. The upper shale sequence bears carbonate concretions and the nautiloid *Orthoceras* (Fig.8.12), with *Scyphocrinites* limestones representing the uppermost transgressive unit. Silurian black shales were probably deposited under euxinic conditions in relatively shallow water, like coeval black shales in other parts of the world (Berry and Wilde, 1978).

In western Morocco (Ouled Abbou region) thin basaltic flows are intercalated within Lower Silurian strata. Silurian strata in Morocco exhibit striking thickness variations, being thickest (1,300 m) in the northern half of the Tindouf basin, but usually not exceeding 200 m elsewhere. Thickness variation has been attributed to a north-south structural differentiation between the southern Anti-Atlas domain and the northern Hercynian domain which began since the Middle Cambrian.

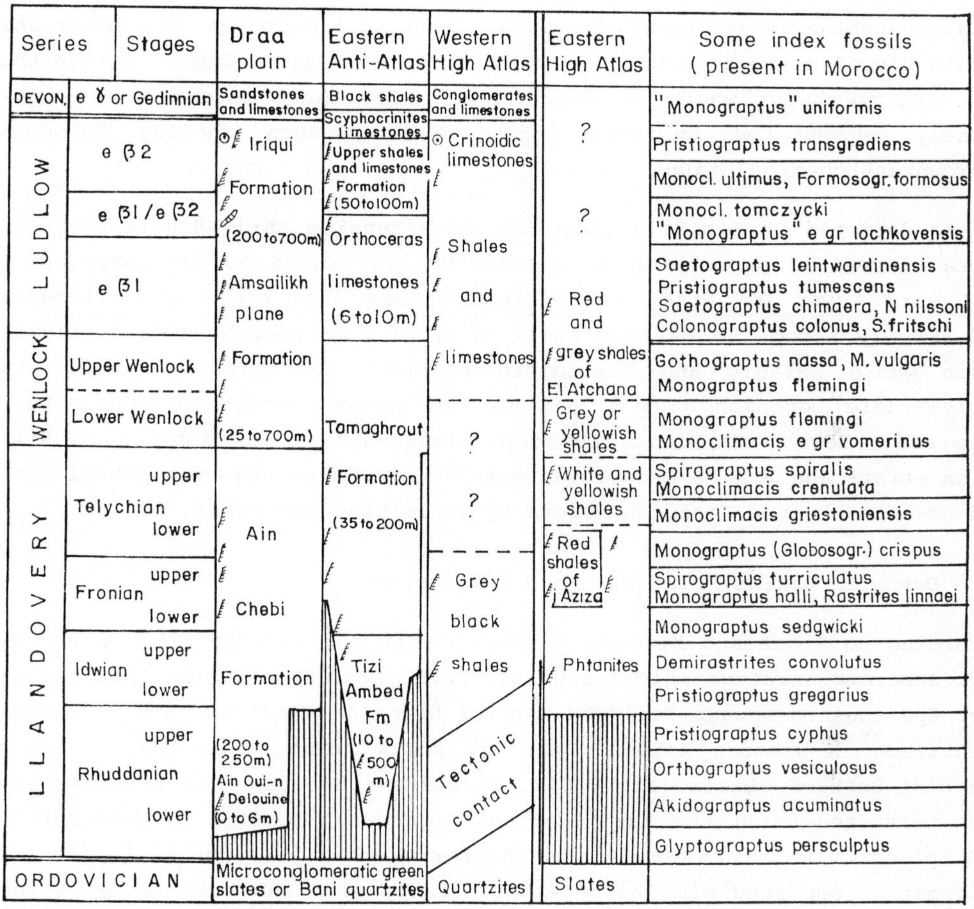

Series	Stages	Draa plain	Eastern Anti-Atlas	Western High Atlas	Eastern High Atlas	Some index fossils (present in Morocco)
DEVON.	e γ or Gedinnian	Sandstones and limestones	Black shales	Conglomerates and limestones		"Monograptus" uniformis
LUDLOW	e β2	Iriqui	Scyphocrinites limestones; Upper shales and limestones	Crinoidic limestones	?	Pristiograptus transgrediens
		Formation	Formation (50 to 100m)			Monocl. ultimus, Formosogr. formosus
	e β1 / e β2	(200 to 700m)	Orthoceras		?	Monocl. tomczycki; "Monograptus" e gr lochkovensis
	e β1	Amsailikh plane	limestones (6 to 10m)	Shales and	Red and	Saetograptus leintwardinensis; Pristiograptus tumescens; Saetograptus chimaera, N nilssoni; Colonograptus colonus, S.fritschi
WENLOCK	Upper Wenlock	Formation		limestones	grey shales of El Atchana	Gothograptus nassa, M. vulgaris; Monograptus flemingi
	Lower Wenlock	(25 to 700m)	Tamaghrout	?	Grey or yellowish shales	Monograptus flemingi; Monoclimacis e gr vomerinus
LLANDOVERY	Telychian upper		Formation		White and yellowish shales	Spiragraptus spiralis; Monoclimacis crenulata
	Telychian lower	Ain	(35 to 200m)	?		Monoclimacis griestoniensis
	Fronian upper				Red shales of Aziza	Monograptus (Globosogr.) crispus
		Chebi		Grey		Spirograptus turriculatus; Monograptus halli, Rastrites linnaei
	Fronian lower			black		Monograptus sedgwicki
	Idwian upper	Formation	Tizi Ambed Fm (10 to 500 m)	shales	Phtanites	Demirastrites convolutus
	Idwian lower					Pristiograptus gregarius
	Rhuddanian upper	(200 to 250m)		Tectonic contact		Pristiograptus cyphus
	Rhuddanian	Ain Oui-n Delouine (0 to 6 m)				Orthograptus vesiculosus
	Rhuddanian lower					Akidograptus acuminatus
						Glyptograptus persculptus
ORDOVICIAN		Microconglomeratic green slates or Bani quartzites		Quartzites	Slates	

Figure 8.12: Graptolite biozonation and Silurian correlations in Morocco. (Redrawn from Legrand, 1989.)

Early Middle Devonian Platforms and Trough

Lower and Middle Devonian strata are well exposed in southern Morocco on both sides of the Tindouf basin and in the Tafilalt-Maider area in the eastern Anti-Atlas. Here Wendt (1985, 1988) identified Middle to Late Devonian carbonate platforms and basins. Following the retreat of the Silurian sea, fluvial and regressive conditions prevailed on the Saharan platform, south of the Tindouf basin. In the Draa lowlands on the northern margin of the Tindouf basin, the Lower Devonian is represented by about 1,300 m of mostly detrital cyclical deposits known as the "Rich" sequence. Each "Rich" cycle is composed of platy limestones followed by alternating argillite beds and fine-grained sandstones with coarse and more massive and lenticular sandstones occurring at the top. Here the Middle Devonian is represented by black pyritic limestones. The Lower and Middle Devonian are progressively more calcareous and pelagic towards the east and northwest where in the Tafilalt area these intervals are completely represented by thin goniatite limestones showing European (Hercynian-Bohemian) faunal affinity.

As depicted in Fig.8.13 there was an extensive Early-Middle Devonian reefal limestone deposition in western Morocco which occurs above about 1000 m of Lower Devonian dark pyritic shales. The reefs which trended NNE-SSW lay along the western shelf of a deep Marrakech-Oujda turbidite basin where pelagic and turbiditic sediments accumulated during the Early-Middle Devonian (Fig.8.13). To the extreme northwest the Sehoul zone shows Lower and Middle Devonian limestones above Cambrian to Silurian strata and a Late Ordovician granite, implying that the Sehoul zone had accreted to the Moroccan Meseta by Early Devonian times.

Late Devonian Basins, Platforms and Deformation

According to Piqué and Michard (1989) eastern and western Morocco underwent separate stratigraphical and structural evolution in the Late Devonian. Epeirogenic upwarp followed earlier Devonian platform carbonate sedimentation in western Morocco, part of which strongly subsided to form the Sidi-Bettache basin (Fig.8.14). In eastern Morocco marine sedimentation persisted until the Early Carboniferous when Eovariscan deformation took place in what previously was the Marrakech-Oujda trough. Eovariscan deformation occurred almost contemporaneously with the subsidence of the Side-Bettache basin. The main phase of Eovariscan deformation was in the Late Devonian.

In the western Anti-Atlas (Draa lowlands) the Upper Devonian consists of about 1,300 m of an essentially pelitic sequence with silty layers and calcareous concretions. On the contrary, a carbonate platform and clastic basin setting developed in the eastern Anti-Atlas. Here, as demonstrated by Wendt (1985, 1988), about 800 m of clays with calcareous and sandy interbeds and basin margin slump deposits accumulated in the Maider basin, whilst about 100 m of siliciclastic limestones and condensed cephalopod limestones covered the Tafilalt platform (Fig.8.14). Deltaic sandstones were uniformly deposited in the Tafilalt-Maider region at the Devonian-

Carboniferous transition, with some olistostromes generated in the north-western margin of the Tafilalt basin, east of the Tafilalt carbonate platform.

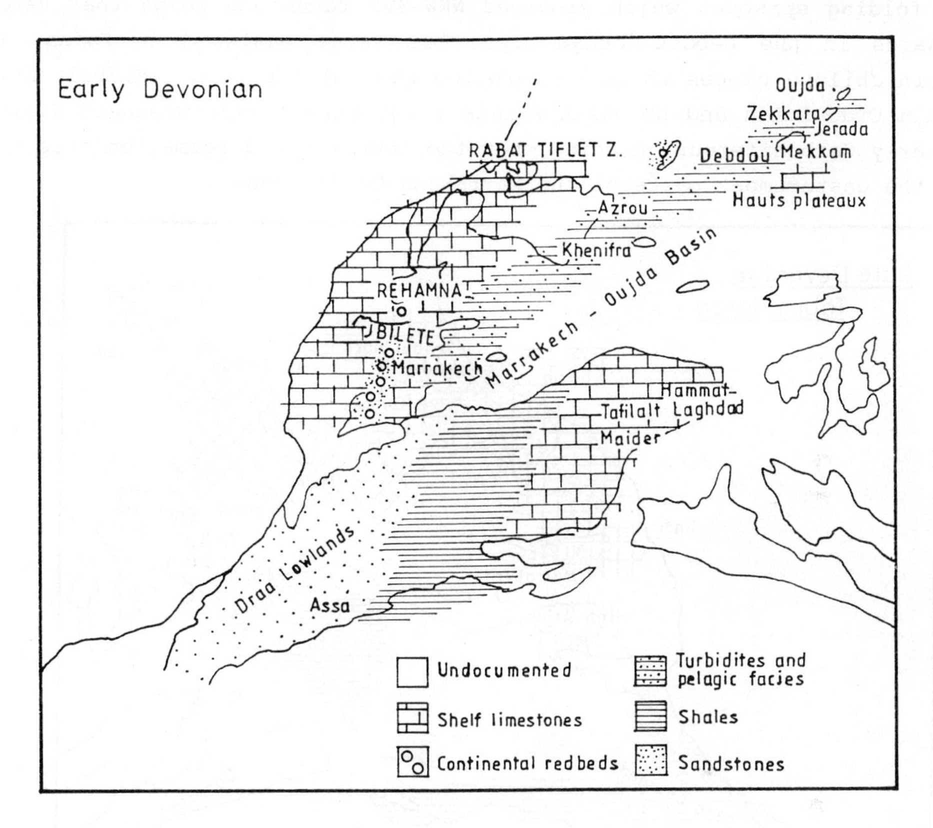

Figure 8.13: Early Devonian lithofacies of northern Morocco. (Redrawn from Piqué and Michard, 1989.)

North of the Anti-Atlas, basins opened in western Morocco by extensional processes which involved normal faulting. Syn-sedimentary faulting and graben subsidence took place generating gravity sliding of shelf deposits down into the grabens. Thus, while the Moroccan Coastal block, the Zaër rise and the Sehoul zone went up yielding proximal chaotic lithofacies (olistostromes, mudflows) and Middle Devonian limestone breccias down into the Sidi-Bettache trough, the latter subsided and received fine-grained graywackes and pelites into its distal facies (Fig.8.14).

Repeated regional deformation in the Marrakech-Oujda basin between the Late Devonian and the Late Carboniferous removed Upper Devonian strata from most of this region except in places such as the Debdou inlier (Fig.8.9). Here Eovariscan deformation caused regional metamorphism dated at about 366 Ma in the Midelt and Debdou areas (Fig.8.14), an event which according to Piqué and Michard (1989), is also known in western France (Fig.8.7). In Morocco metamorphism was accompanied by two successive folding episodes which produced NNW-SSE recumbent folds that verge westwards in the Debdou-Mekkam area, but verge eastward in Zaïan. In eastern Jbilete nappes which were rooted west of the Debdou-Midbelt axis contain Ordovician and Devonian strata which were thrust westward during the Early Carboniferous, meaning that the Eovariscan deformation affected only the easternmost parts of the Marrakech-Oujda trough.

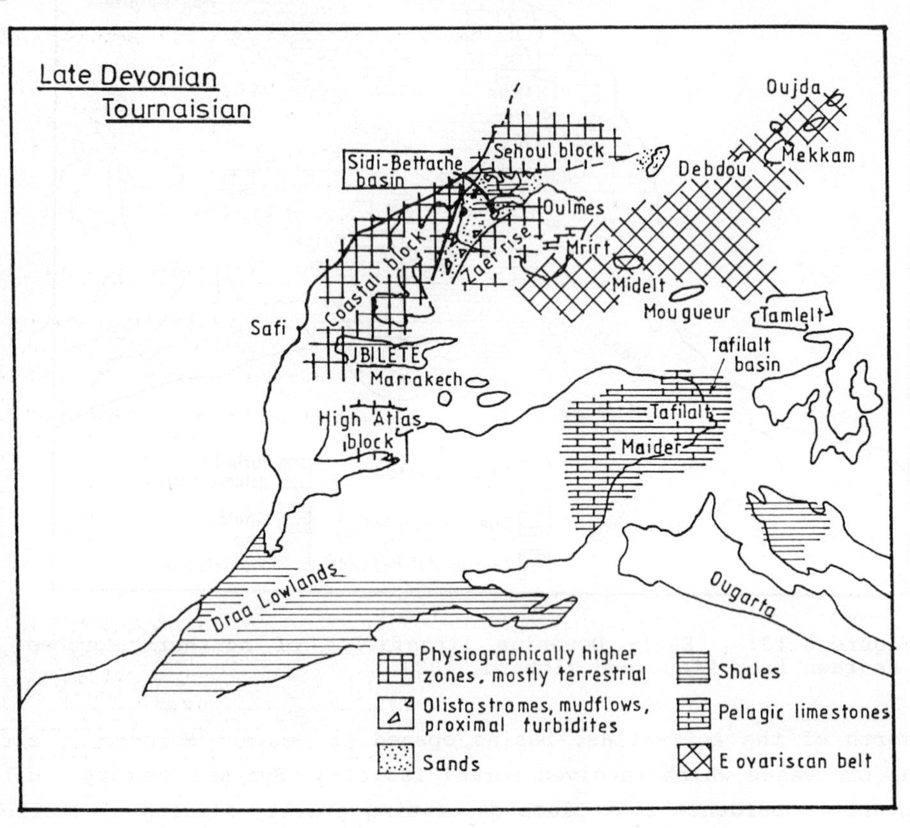

Figure 8.14: Late Devonian lithofacies of northern Morocco. (Redrawn from Piqué and Michard, 1989.)

Carboniferous Basins and Hercynian Deformation

Earliest Carboniferous time was marked by a shallow marine transgression which covered most of the Anti-Atlas during which pelagic facies (Gattendorfia beds) were deposited, followed by regressive sandstones in the Draa and Ougarta areas. The western and southern parts of the Tindouf basin and probably the Tafilalt also emerged and shed clastics into northern depocentres. Late Devonian paleogeography, however, persisted in western Morocco with uninterrupted Upper Devonian-Early Carboniferous clastics (Korifla-Bou-Rzim sequence) accumulating in the subsiding Sidi-Bettache basin (Table 8.1). Along the margins of this basin olistostromes, mudflows, and proximal turbidites suggest re-deposition down active fault scarps. Contemporaneous tholeiitic to alkaline magmatic rocks intercalated within the Korifla-Bou-Rzim sequence indicate crustal extension which created rifts or back-arc basins.

Table 8.1: Upper Paleozoic correlations and stratigraphic terminology for Morocco. (Redrawn from Piqué and Michard, 1989.)

Southwestern Anti-Atlas	Sidi - Bettache basin			Central Jbilete		Northeastern Morocco
	western limit	western margin	center	western part	eastern part	
Betana						Jerada beds
Ouarkziz	Bou-Rzim			Sarhlef	Kharrouba	
Betaina			Korifla			
Tazoul						
Daraa lowland sequences	Ben-Slimane	Foulzir				Debdou slates

Renewed marine transgression occurred in the Draa lowlands during the late Early Carboniferous (Visean) producing goniatite-bearing pelites and silts (Betaina beds) and reefs in the Tafilalt area. But continental conditions prevailed southwestwards with the strandline of this transgression located near Jbel Ouarkziz (Fig.8.15). The Sarhlef series (Table 8.1) comprising fine clastics in which pyrrhotite and chalcopyrite suggest euxinic paleoenvironments, accumulated in the western part of central Jbilete massif, in a basin that extended eastward over the Eovariscan fold and thrust belt of the Central massif. Here conglomerates, limestones and argillites were deposited, including calc-alkaline andesitic flows and breccias in the Azrou-Khenifra region. A turbiditic sequence (Kharrouba series) accumulated in the eastern deeper part of central Jbilete. Middle Carboniferous (Hercynian) deformation triggered olistostromes followed by allochthonous Ordovician and Devonian pelagic limestones and turbidites which were emplaced from the east as gravity nappes. At Khenifra the Late Visean is also overlain by nappes that were emplaced by gravity sliding at the end of the Visean.

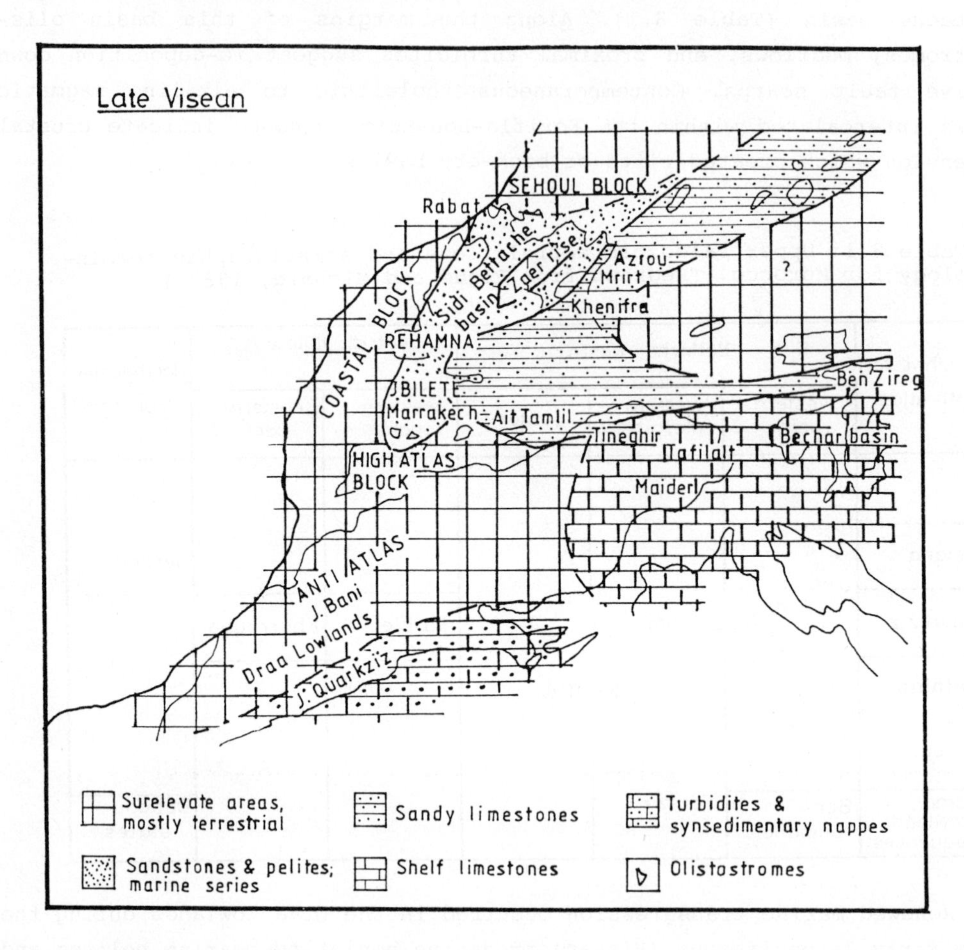

Figure 8.15: Late Visean lithofacies in northern Morocco. (Redrawn from Piqué and Michard, 1989.)

In the Anti-Atlas the Hercynian deformation attenuated and disappeared completely in the Tindouf basin (Fig.8.10). It generated open folds in the Anti-Atlas with NE-SW sinistral strike-slip faults, and associated en-échelon folds in the western Anti-Atlas. NW-SE dextral strike-slip faults occured in the Ougarta belt in Algeria. The Anti-Atlas acted as the southern foreland to the Hercynides during the Middle to Late Carboniferous deformation, while the Coastal block and the Sehoul zone served as the northwestern foreland. Major shear zones such as the Western meseta shear zone, and the Tizi n'Test fault zone and Tineghir-

Béchar thrust which constitute the so-called South Atlas fault zone, separate the central part of the Hercynides from the forelands (Fig.8.16 A). These shear zones are characterized by intense deformation with reclined or recumbent folds (Fig.8.16B), and by granite intrusions and abrupt stratigraphic changes. To the northeast the Tazekka-Bsabis-Khenifra thrust zone corresponds to the limit between the Eastern meseta (deformed during Eovariscan and Middle Carboniferous) and the Western meseta (deformed in the Late Carboniferous).

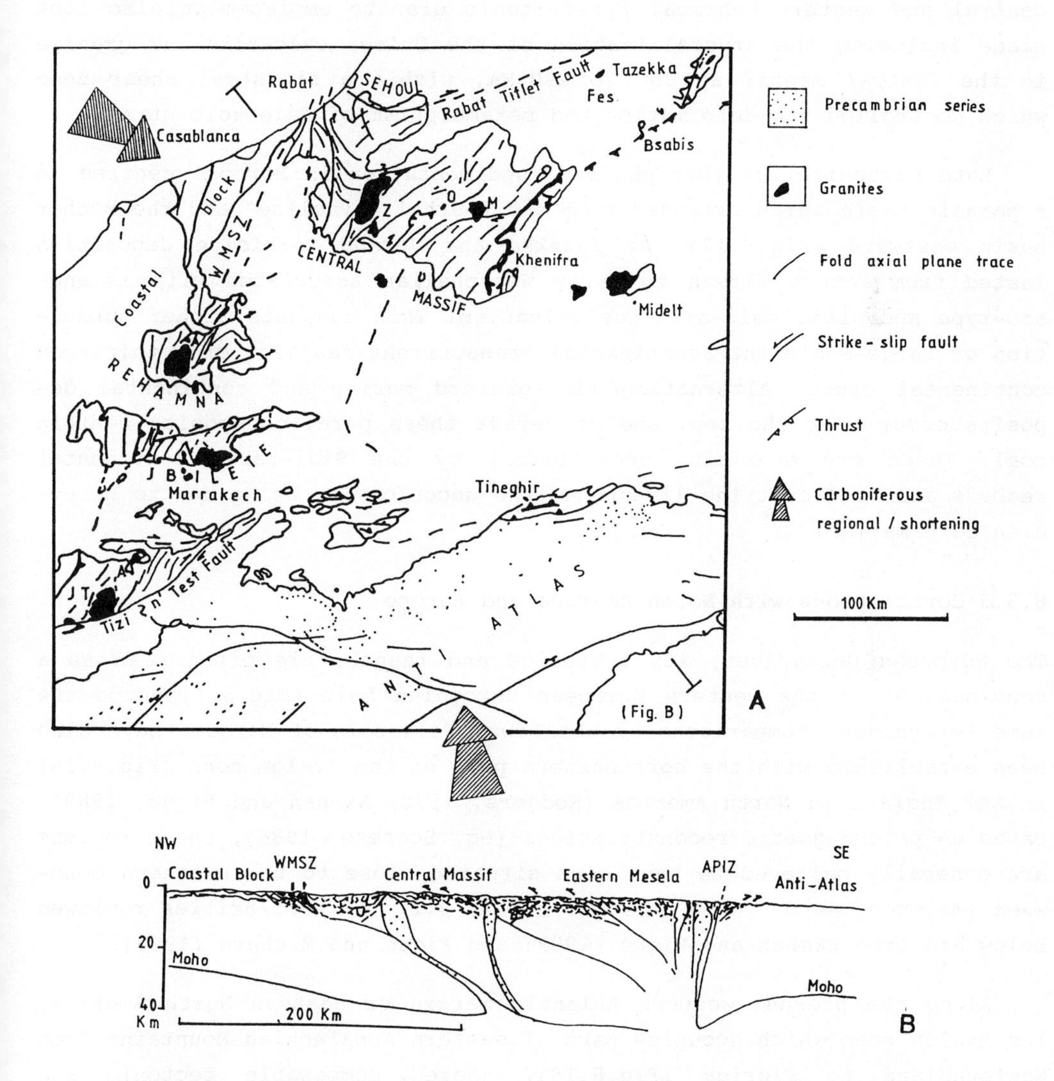

Figure 8.16: Structural map (A) and section (B) of the Moroccan Hercynides. For (A); Z, Zaier; O, Oulmes; M, Ment; JT, Jebel Tichka; A, Azigour. ((Redrawn from Piqué and Michard, 1989.)

Whereas the vergence of Eovariscan folds (Late Devonian) is to the east and west, those of the Carboniferous Hercynian event verge to the northwest and to the southeast. The latter deformation involved NW-SE crustal shortening (Fig.8.16 B) which increased from 15% in the western Coastal block to 50% in the Central massif, and also entailed the displacement of the Coastal block for about 100 km towards the Anti-Atlas. Since most of the Carboniferous deformation was by ductile processes regional metamorphism was also involved, reaching the amphibolite facies in central and western Rehamna. Syn-tectonic granite emplacement also took place including the crystallization of the Oulmès calc-alkaline granite in the Central massif at 300 to 290 Ma, within a sinistral shear zone which controlled the deformation and metamorphism of Paleozoic strata.

Late Carboniferous (Westphalian) sedimentation in Morocco centred in a paralic basin which extended from the Fourhal "syncline" and the Béchar basin eastward (Fig.8.17). At Tazekka and Jerada turbidite deposition lasted from latest Visean to Early Westphalian associated with island-arc-type andesitic calc-alkaline volcanism. This suggests either subduction or large-scale intracontinental transcurrent faulting of a thickened continental crust. Alternating fine-grained marine and continental deposits occur near the top, and at Jerada these paralic deposits contain coal. These are overlain unconformably by the Sidi-Kacem continental redbeds and by lacustrine limestone, the unconformity being due to Hercynian deformation.

8.3.3 Correlations with North America and Europe

The Moroccan Hercynides, its tectonics and faunas, are often seen as a continuation of the western European Hercynian belt (Fig.8.7) hence the name Hercynides. Comparisons of the Lower Paleozoic of Morocco have also been established with the northeastern part of the Avalon zone (Fig.8.18) in New England in North America (Rodgers, 1970; Skehan and Piqué, 1989). Based on paleomagnetic reconstructions (eg. Scotese, 1986), these regions are generally believed to have been situated close to northwestern Gondwana (Morocco) as depicted in Figs. 6.9 A, 8.2. The similarities reviewed below are from Skehan and Piqué (1989) and Piqué and Michard (1989).

Along the present western Atlantic margin or eastern North America, the Avalon zone which occupies part of eastern Appalachian Mountains from Newfoundland to Florida (Fig.8.18), shared comparable tectonic and stratigraphic evolution with Morocco from the Late Proterozoic to the Early Paleozoic, suggesting African connections. Parts of the Avalon zone are believed to have arrived in North America at different times during the Paleozoic.

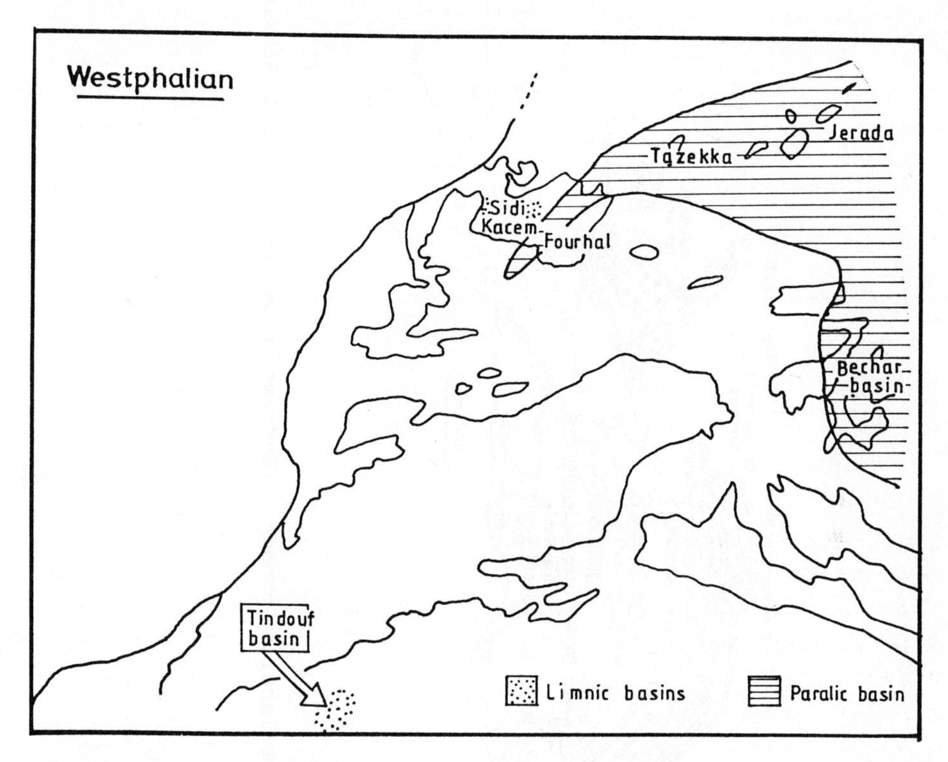

Figure 8.17: Westphalian lithofacies of northern Morocco. (Redrawn from Piqué and Michard, 1989.)

Similarities shown in Fig.8.19, between the Avalon zone of New England and Morocco, affect the Middle Proterozoic stromatolitic carbonates and quartzites which overlie the Eburnean basement in Morocco. The equivalents in New England belong to the Blackstone Group. Other similarities are the Bou Azzer ophiolites of Morocco (Middlesex volcanics in Avalon); the Pan-African I orogeny (Avalonian orogeny); the Ouarzazate-Tanalt Group and the "Lie-de-vin" unit (Mattapan volcanics of the Boston Bay Group in Avalon); and the calcareous Adoudounian which is probably equivalent to the upper olistostrome unit in the Newport Group of the Avalon zone. The Upper Adoudounian Group and the overlying fine clastic sediments have been equated with a sequence in New England that consists of basal conglomeratic quartzite (Hoppin Quartzite) overlain by green and maroon slates and thin trilobite-bearing limestones (Weymouth Formation). The Middle Cambrian graywacke and siltstone sequence in the Anti-Atlas containing the trilobite *Paradoxides* is equivalent to a similar sequence in Boston and Rhode Island where the *Pardoxides* fauna has close affinity

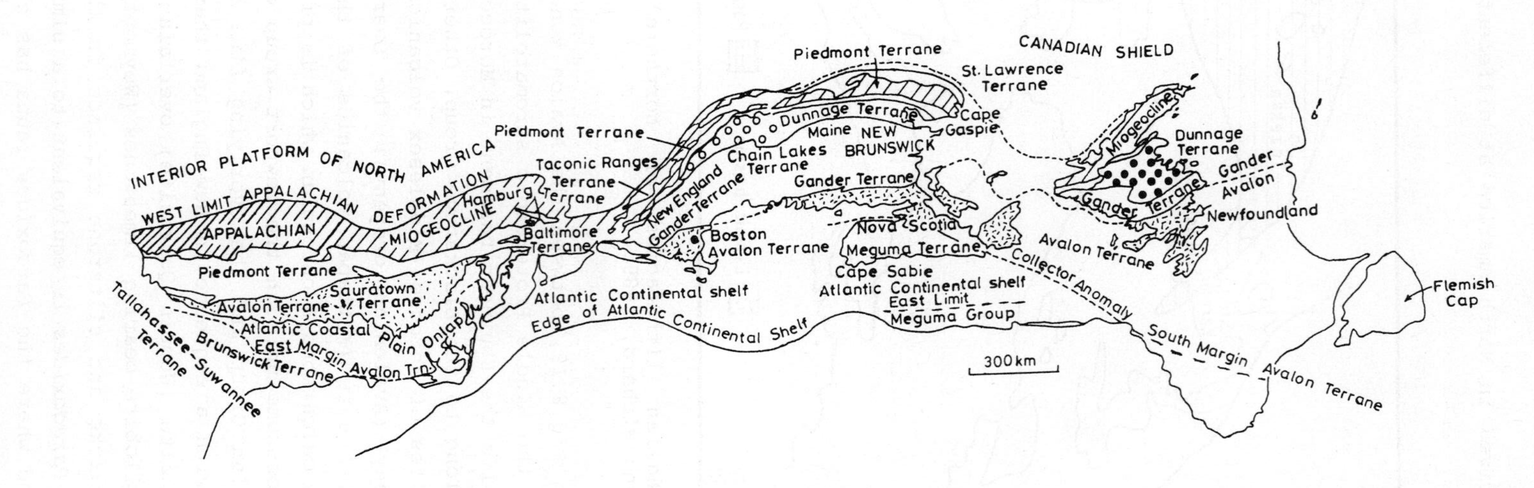

Figure 8.18: Some terranes in the Appalachian orogen. (Redrawn from Frazier and Schwimmer, 1987.)

with Morocco and Spain. However, the sequence is thinner in New England unlike Morocco where it is up to 8 km and contains volcanics. Also, whereas the Ordovician in New England has been recognized in a sequence of distal turbidites known as the Dutch Island Harbour Formation in Rhode Island, equivalent units in Morocco are shallow water upward coarsening clastic sequences which contain diamictites.

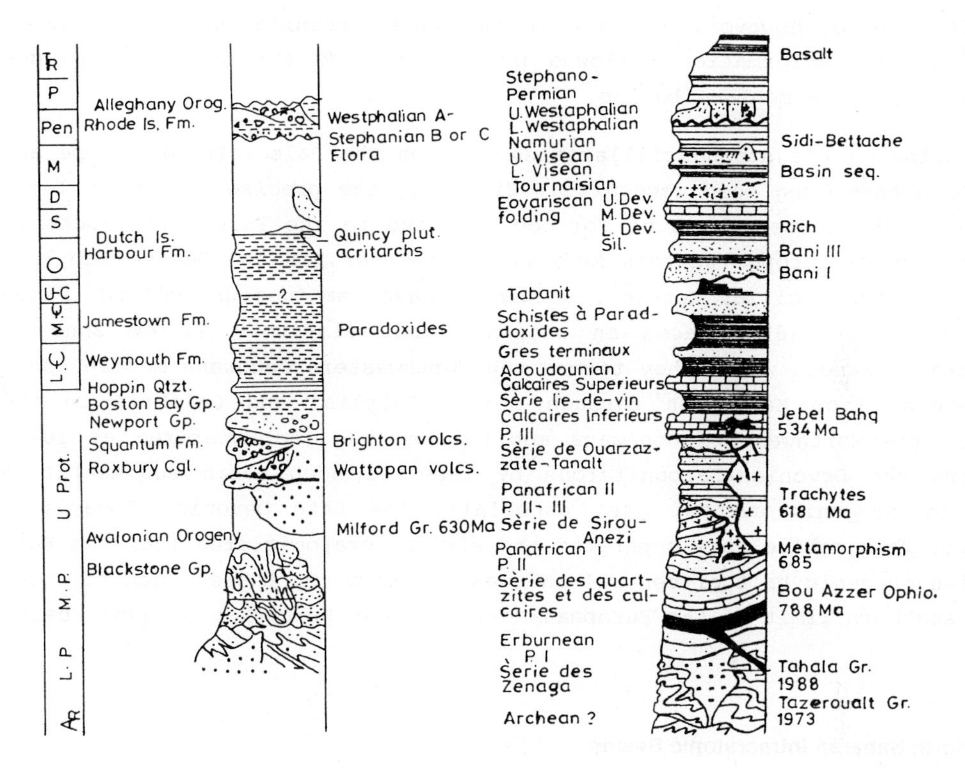

Figure 8.19: Comparisons of the Late Proterozoic-Paleozoic of the Avalon Zone of New England with that of Morocco. (Redrawn from Skehan and Piqué, 1989.)

The Meguma terrane on the southeastern part of Avalonia in Nova Scotia (Fig.8.18) has been correlated with the Sehoul zone in Morocco and with western Mediterranean Paleozoic blocks. In Nova Scotia the Cambro-Ordovician Meguma Group comprises a very thick clastic progradational sequence that was deposited in the deep-sea environment with the provenance of the clastics lying somewhere on the West African craton. Similar Cambro-Ordovician deep-sea fans occur in the Sehoul zone. Piqué and Michard (1989) suggested that somewhere in the Cambro-Ordovician, after the Meguma terrane and southern Portugal had been part of the northwestern African continental margin, the Meguma terrane began to separate from

Africa (Fig.6.9A). From then on, the similarity between the Avalon zone and northwest Africa largely ceased.

The Silurian through Devonian stratigraphy and the tectonic evolution of Morocco and New England are disparate with rifting and the production of alkaline plutonic rocks occurring earlier in New England (Ordovician-Silurian, Devonian) and during the Early Carboniferous in Morocco. Similarities were, however, restored in the Late Carboniferous with sedimentation and deformation followed by coal-bearing strata that were deposited in intermontane basins.

Although the strong similarities between the Paleozoic of Europe and Morocco have long been recognized (Fig.8.7), the precise Paleozoic location of the eastern part of Morocco relative to the Paleozoic blocks of southwestern Europe (western Mediterranean-Inner Rif and Betic, Kabylia, Calabria and Sicilia, Corsica, Sardinia) have remained unresolved. Piqué and Michard (1989) traced an elongate and relatively narrow zone of crustal weakness which they termed the Northwestern Gondwana mobile zone, extending from Marrakech through Oujda, Kabylia, and Calabria in the east. The Northwestern Gondwana mobile zone was characterized at least during the Devonian-Carboniferous by repeated deep basin sedimentation and by orogenic events, most especially the Late Devonian Eovariscan phase. This mobile zone separated the African cratonic area from the Hercynian orogenic belt of central and western Morocco, thereby constituting the southern limit of the European-North African Hercynian orogenic belt.

8.4 North Saharan Intracratonic Basins

Giraud et al. (1987) pointed out that "from southern Morocco to the Egyptian-Libyan border one can observe a sequence of basins, troughs, horsts, blocks or uplifted areas of various orientations". These basins were termed the North Sahara basins by Giraud et al., among which they included the "Tindouf, Timimoun, Oued Mya, Ghadames, Mursuk, Hun, Dor el Goussa, Syrte and Kufra". This suite of basins constitutes the North Saharan platform basins which contain extensive Paleozoic sequences.

8.4.1 Tectonic Control of Basin Development

The post-Pan-African structural picture of the North Saharan platform (Fig.8.20) including the structural effects of the Caledonian and Hercynian orogenies can be gained from the Phanerozoic structural evolution of northeast Africa presented by Klitzsch (1971, 1981, 1986), Klitzsch and Wycisk (1987), and Schandelmeier et al. (1987).

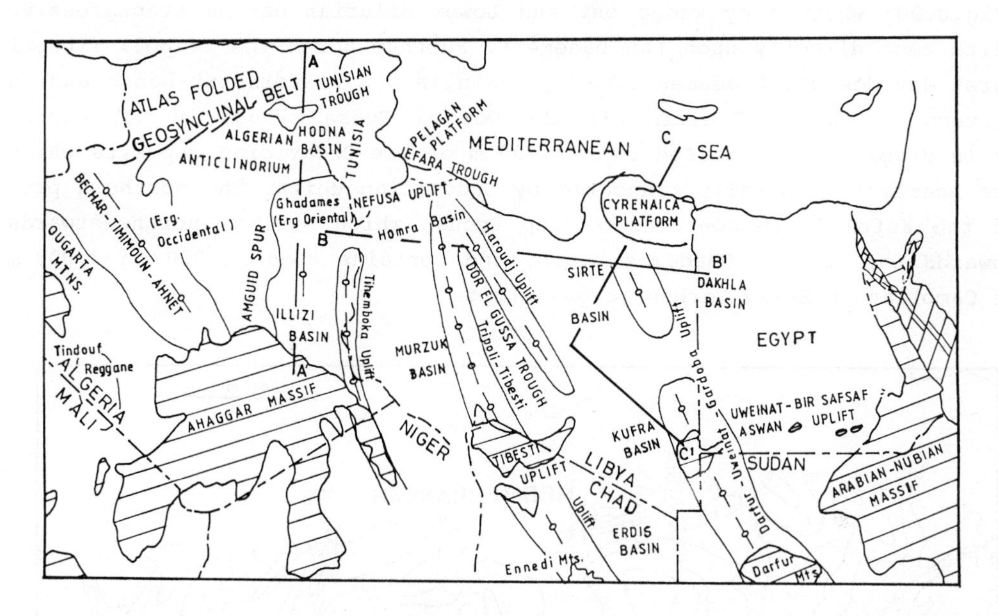

Figure 8.20: Structural sketch map of northern Africa. (Compiled from figure supplied by J.A. Peterson; Klitzsch, 1981; Schadelmeier et al., 1987.)

Regional peneplanation prevailed in North Africa after the final phases of the Pan-African orogeny. But unlike northwest Africa where, as we have seen, continuous continental-shallow marine deposition took place across the latest Proterozoic-Cambrian, a marked hiatus occurs at the base of the Cambrian all the way from the Tindouf basin in the west to Egypt and Sudan in the east, implying regional arching. In Egypt the distribution of Lower Cambrian strata suggests considerable structural relief. The stratigraphic thicknesses in different parts of the North Saharan platform reveal broad graben-type structures and smaller horst blocks which trend NNW-SSE (Figs. 8.20, 8.21). These were the products of ENE-WSW-oriented post-Pan-African intraplate tensional forces which broke up the Saharan shield. By Middle to Late Cambrian times sedimentation in Libya, northern Chad and northern Niger was already taking place in well defined NNW-trending grabens or troughs which were bounded by horsts (Fig.8.20). These structures were accentuated by the Caledonian orogeny (Ordovician-Silurian) after which they continued to determine sediment depocentres, and the centres of magmatism especially along the borders of the Murzuk basin, the southwestern margin of the Kufra basin, and in the eastern part of the Ennedi Mountains.

In the eastern Murzuk basin Cambrian and Ordovician strata are up to 1,000 m thick but very thin on the neighbouring Tripoli-Tibesti uplift (Fig.8.20) where they wedge out and Lower Silurian marine transgressive units rest directly upon the basement. Further north the Tripoli-Tibesti horst divides the Ghadames (Homra) basin into a western sub-basin and an eastern sub-basin. Faulting in the Dor el Gussa trough in the Murzuk basin displaced and tilted over 1,700 m of pre-Ordovician deposits which are nearly horizontally truncated by Ordovician units. The northern part of the Kufra basin contains a deep trough which strikes northwestwards towards the eastern Ghadames basin, and contains about 2,000 to 3,000 m of Cambrian to Early Carboniferous strata.

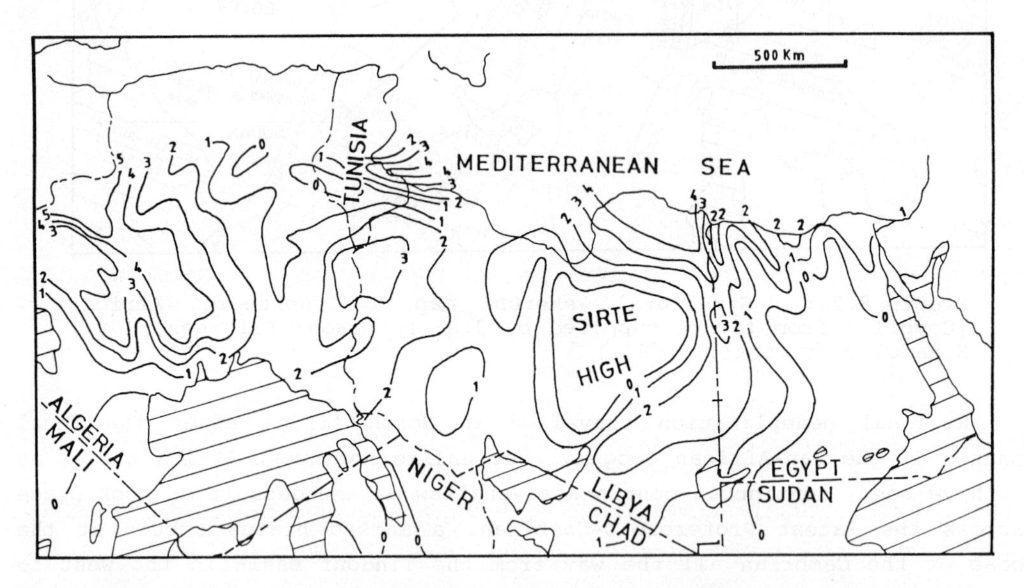

Figure 8.21: Approximate thickness of Paleozoic sequences in northern Africa. (Redrawn from figure supplied by J.A. Peterson.)

The southwestern part of the Kufra basin is known as the Erdis basin (Fig.8.20). The Erdis-Kufra trough runs northwards towards the Sirte basin, where after Early Carboniferous times, most of the Paleozoic sequences were eroded away. Along the eastern border of the Sirte basin lies northern prolongation of the Darfur-Uweinat uplift (Gardeba uplift). East of the Gardeba high lies the Dakhla basin, the northern part of which experienced the greatest subsidence during the Paleozoic. On the Darfur-Uweinat-Gardeba uplift Cambrian and Ordovician strata, which are generally thin on this horst, progressively disappear southward where they are overstepped by the more extensive Silurian marine units. Except in northern Egypt the Uweinat uplift effectively prevented Ordovician, Silurian, Devonian, and Early Carboniferous marine transgressions from reaching eastwards into north-central Sudan. Nevertheless, the NNW-trending Saharan platform troughs allowed Paleozoic marine transgressions to encroach southwards upon the northerly tilted African plate, leading to the deposition of thick sequences in the troughs and a thin sheet of Paleozoic strata (Ordovician to Carboniferous) on the structural highs.

During the post-Visean Hercynian tectonism when NE-Africa collided with Europe, ENE-trending structures resulted on the North Saharan platform. This was on account of the reactivation of deep-seated Precambrian dextral wrench faults which became conjugate sinistral shear faults. North of the Egyptian-Sudanese border this reactivation produced the prominent Uweinat-Bir-Safsaf-Aswan uplift (Fig.8.20). Similar structures include the Nefusa uplift in northwestern Libya, southern Tunisia and eastern Algeria, and the Algerian anticlinorium, whereas sagging developed in the Gulf of Suez graben. Because of widespread post-Hercynian erosion on the North Saharan platform Permian rocks are generally missing or very thin on the platform especially on the uplifts, but markedly thicker north of the Nefusa uplift in the Jefara trough (Fig.8.20) segment of the Pelagian platform (Peterson, 1985). A major Permo-Triassic continental basin developed across southeastern Libya, northern Sudan and southern Egypt in response to the Uweinat-Bir-Safsaf-Aswan uplift.

Rocks of all the Paleozoic systems occur in the Saharan platform. Maximum thickness of Paleozoic strata is in the Tindouf basin in Algeria where the Silurian-Devonian succession is about 14 km thick (Guerrak, 1989). Paleozoic rocks are about 5 km thick in the northern part of the Dakhla basin (Peterson, 1985) and over 3 km thick in northeastern and northwestern Libya, eastern Algeria and in the Jefara trough (Fig.8.21). Among the important natural resources in the Paleozoic of North Africa are vast reserves of oolitic ironstones, and oil and gas especially in Algeria and Libya. The widespread Silurian graptolitic shales and Carboniferous strata provided very rich petroleum source rocks. Sandstone hydrocarbon reservoirs of variable quality abound throughout the Paleozoic of North Africa.

8.4.2 Tindouf and Reggane Basins

The Tindouf basin is a large post-Silurian ENE-WSW-trending asymmetrical synclinorium, 800 km long and 200 to 400 km wide (Deynoux et al., 1985), extending from southeastern Morocco into southwestern Algeria, where most of the basin lies. Eastward, the Tindouf basin merges with the Reggane basin (Fig.8.22). Cambrian to Carboniferous strata fill the Tindouf and

Reggane basins and the sequence is much thicker and more complete in the northern part of the Tindouf basin than along its southern flank where the Cambrian is absent and the Lower Ordovician is also often missing. Glacial Late Ordovician deposits and sometimes Silurian shales rest direct on the basement complex along the southern flank of the Tindouf basin. Elsewhere, the bulk of the sequence comprises marine Silurian shales, and alternating marine carbonates and clastics of Devonian to Carboniferous age (Fig.8.23A). Being further removed from marine influence the Carboniferous in the Reggane basin was more continental (Fig.8.24) than in the Tindouf basin.

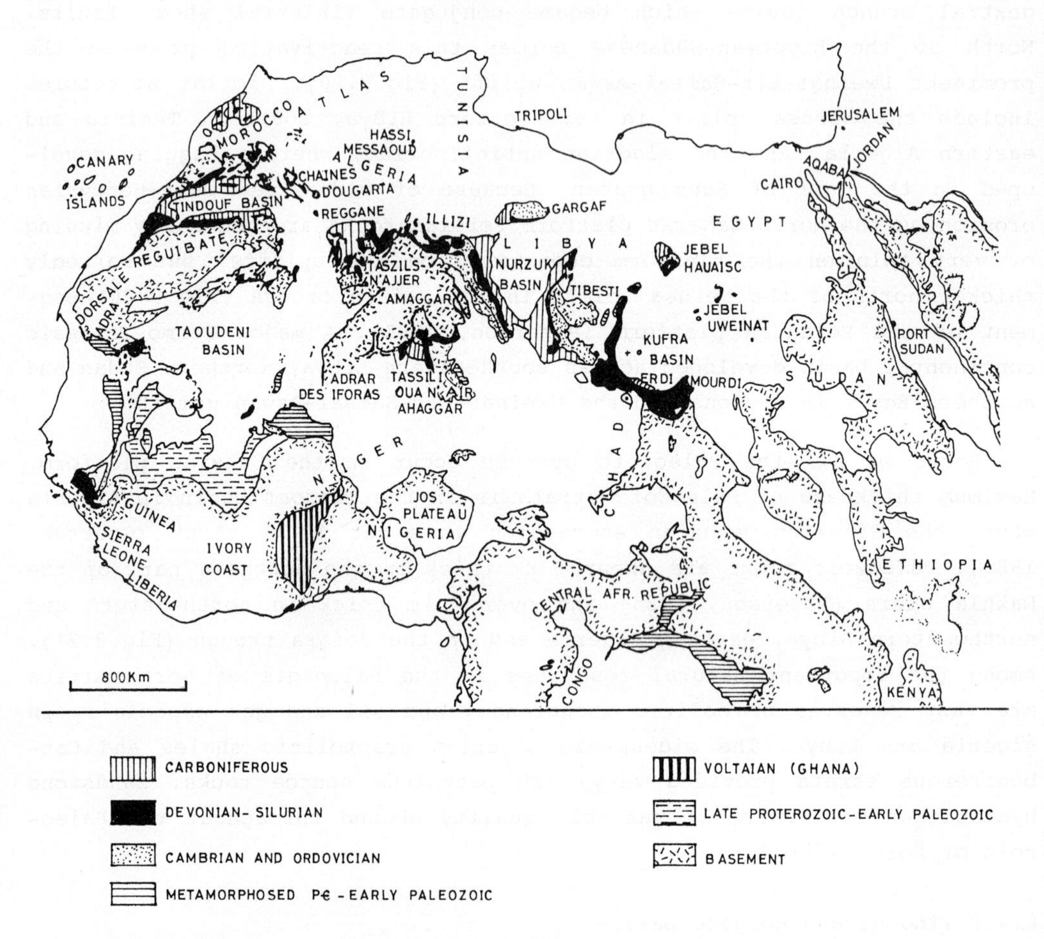

Figure 8.22: Distribution of some major Paleozoic exposures in northern Africa. (Redrawn from Whiteman, 1972; Bellini and Massa, 1980.)

In the Reggane basin littoral and shallow marine clastics occur in the Lower Carboniferous, followed by continental beds which marked tempo-

		DJEBILET SUB-BASIN		GENERALIZED LITHOLOGY	IGUIDI SUB-BASIN	
		FORMATIONS	Thick.(m)		FORMATIONS	Thick.(m)
CARBONIFEROUS	NAMURO-WESTPHALIAN	HASSI AOULEOUEL	350	Sandy shales, Fine sandstones; Shales	CONCEALED AREA	?
	VISEAN	AIN EL BARKA	600	Shales, Limestones and Dolomites; Anhydrite; Shales and Limestones		
		KERB ES SEFIAT	310	Shales and Limestones		
	TOURNAISIAN	KERB ES SLOUGUIA	80-160	Shales and Limestones		
DEVONIAN	FAMENNIAN	KERB EN NAGA	100-140	Silstones and Shales	MECHERI	200-250
		OUED GHAZAL	100-150	Argillaceous Silstones		
	FRASNIAN	OUED TSABIA	80-160	Silstones and Limestones	BOU BERNOUS	70
	L.DEV.	OUED TALHA	40-100	Limestones and Shales	Upper FEDJ MLEHAS	70
		DJEBILET	50-100	Limestones and Sandstones		
	SILURIAN	SEBKHA MABBES	80-200	Shales	Lower FEDJ MLEHAS	90
	C-O	GHEZZIANE	0-70	Sandstones	GARA SAYADA	70-1000
	Precambrian	YETTI		Granites	EGLAB	

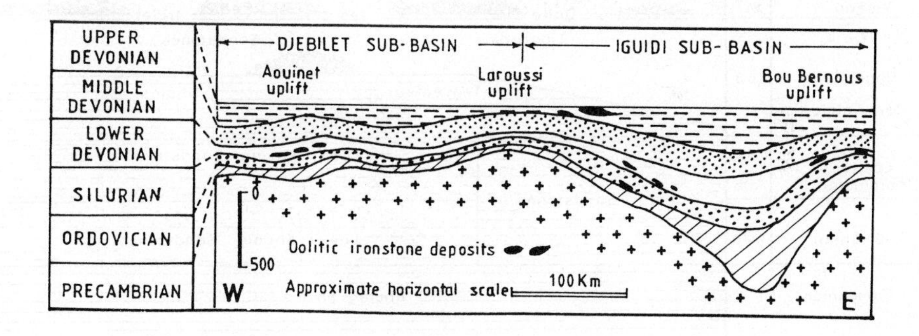

Figure 8.23: Stratigraphic sections of the Tindouf basin. (Redrawn from Guerrak, 1989.)

rary emergence (Conrad, 1985). A Late Visean and Namurian transgression is represented by terrigenous deposits containing foraminifera and conodonts and by extensive carbonates with an upper gypsiferous regressive facies. Upper Carboniferous regressive strata include red shales, sandstones, micro-conglomerates, and carbonate intercalations containing marine and sometimes lacustrine algae and ostracodes. A thin continental Cretaceous to Recent (Hamada Formation) unconformably overlies the Paleozoic sequence in the Tindouf and Reggane basins.

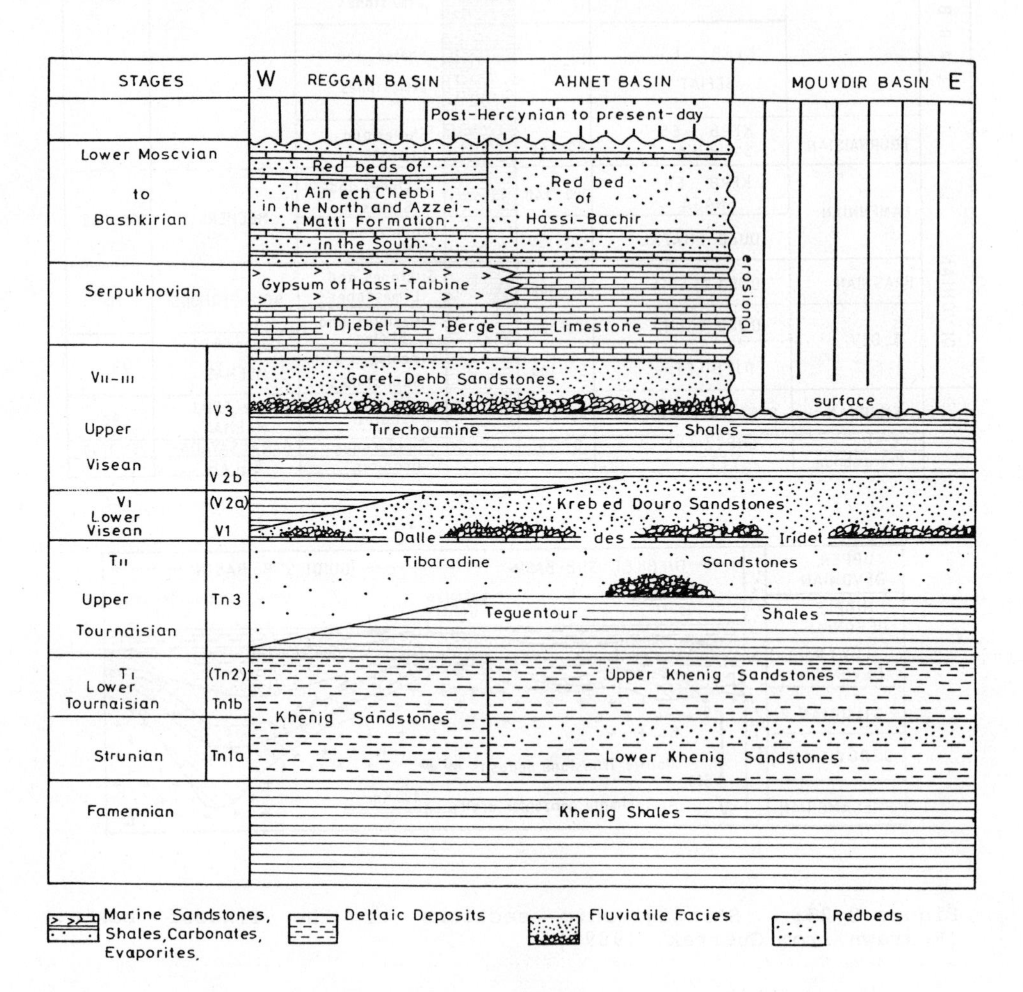

Figure 8.24: Stratigraphic succession of the Carboniferous NW of the Ahaggar. (Redrawn from Conrad, 1985.)

Along the southern part of the Tindouf basin, in the Djebilet and Iguidi sub-basins, numerous oolitic ironstones lenses are interbedded within argillaceous and arenaceous deposits (Fig.8.23B) of mostly

Devonian age. These ironstones are part of a major North African Paleozoic oolitic ironstone belt that extends for over 3,000 km from Morocco to Libya. An estimated total reserve of about 10 billion tons of ironstone lies in these deposits with about 1.5 billion tons in the Silurian and 9.2 billion tons in the Devonian (Guerrak, 1989). The ironstones are of the Local Ironstone Deposition (LOID) type having formed in shallow Paleozoic epicontinental seas in settings such as coastal barriers and deltas (Guerrak, 1989). LOID in the Tindouf basin is characterized by: (i) paleogeographic control of sediments which accumulated on the flanks of uplifts; (ii) the occurrence of ironstones at the top of coarsening-upward sedimentary cycles, mostly located towards the end of major regressive cycles; (iii) ooid growth in iron-rich mud in quiet environments such as lagoons or embayments; (iv) southern relatively near iron source areas; and (v) paleolatitudinal location in cold and temperate climatic regimes after North Africa had recovered from the Late Ordovician continental glaciation.

8.4.3 Central and Southern Algerian Basins

Paleozoic rocks underlie most of Algeria. They outcrop along a NNW-SSE uplifted zone that runs along the Ougarta ranges through the Touat and Bled-el-Mass anticlinorium (Figs. 8.22; 8.25A); in the adjoining Bechar basin to the northeast; along the northern margins of Yetti Eglab (Reguibat Shield) which also constitutes the southern flank of the Tindouf and Reggane basins; and also form a girdle around the Tuareg Shield with the Ahnet and Tassili N'Ajjer among the largest Paleozoic exposures (Deynoux, 1983). Between the uplifted zones of central Algeria extensive Paleozoic sequences fill great depressions, the largest of which are the Bechar-Timimoun (Erg Occidental) basin; the Ghadames (Rhadames or Erg Oriental) basin; the Illizi (Polignac) basin; and the Ahnet basin. Detailed stratigraphic descriptions of these basins were presented by Legrand (1985) and Whiteman (1972).

Bechar-Timimoun Basin

According to Conrad (1985) the Bechar basin contains one of the most complete marine sequences in Algeria, therefore constituting a stratigraphic reference point which has been correlated with the Ougarta chain nearby (Fig.8.26). The Cambrian consists of up to 1,500 m of arkosic sandstones (the Ougarta Sandstone Group) which is glauconitic in the upper part because of the onset of a marine transgression. A graptolitic argillaceous Ordovician sequence succeeds unconformably (Fig.8.26) and passes upward into a sandstone complex with numerous mid-Ordovician fossiliferous shale

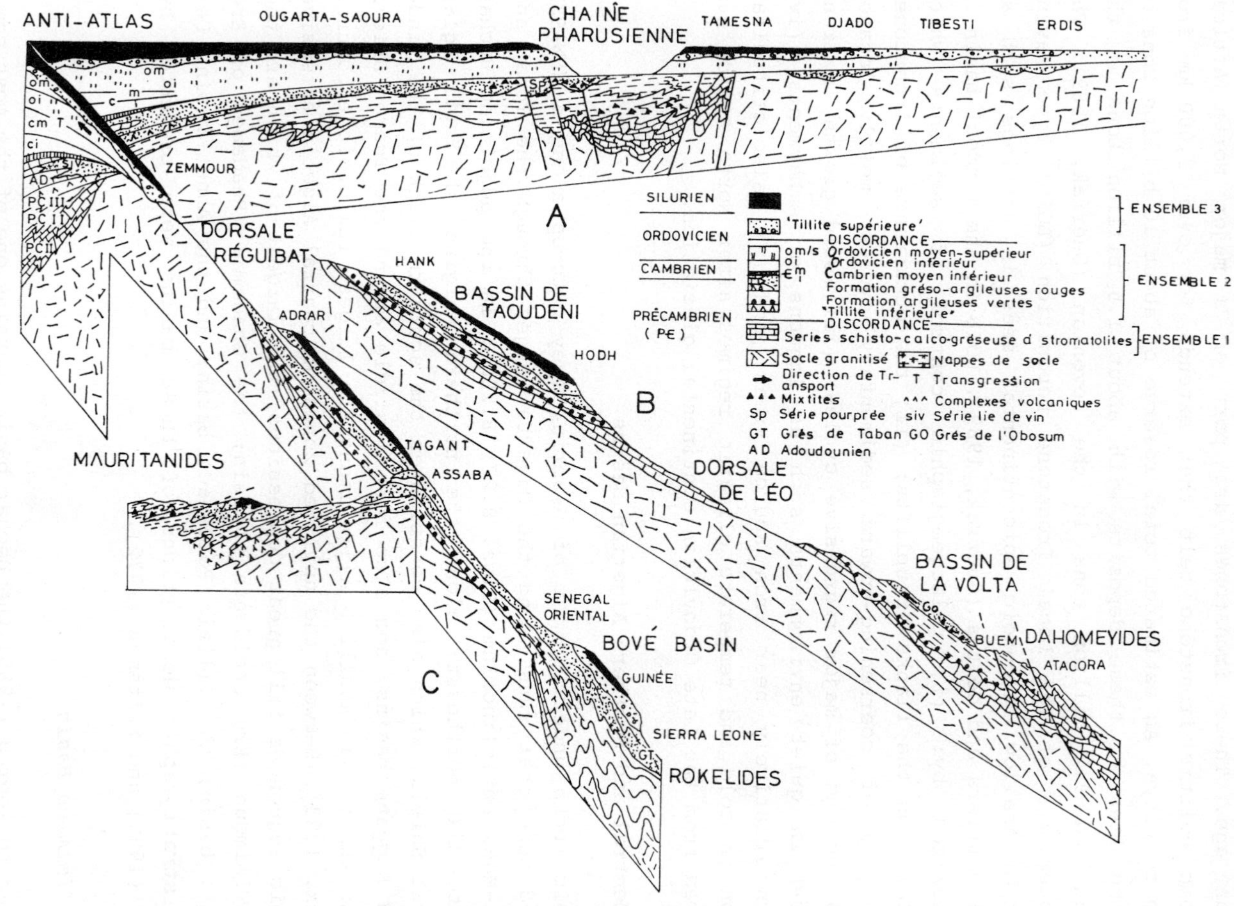

Figure 8.25: Late Proterozoic - Early Paleozoic geologic sections across western Africa. (Redrawn from Deynoux, 1983.)

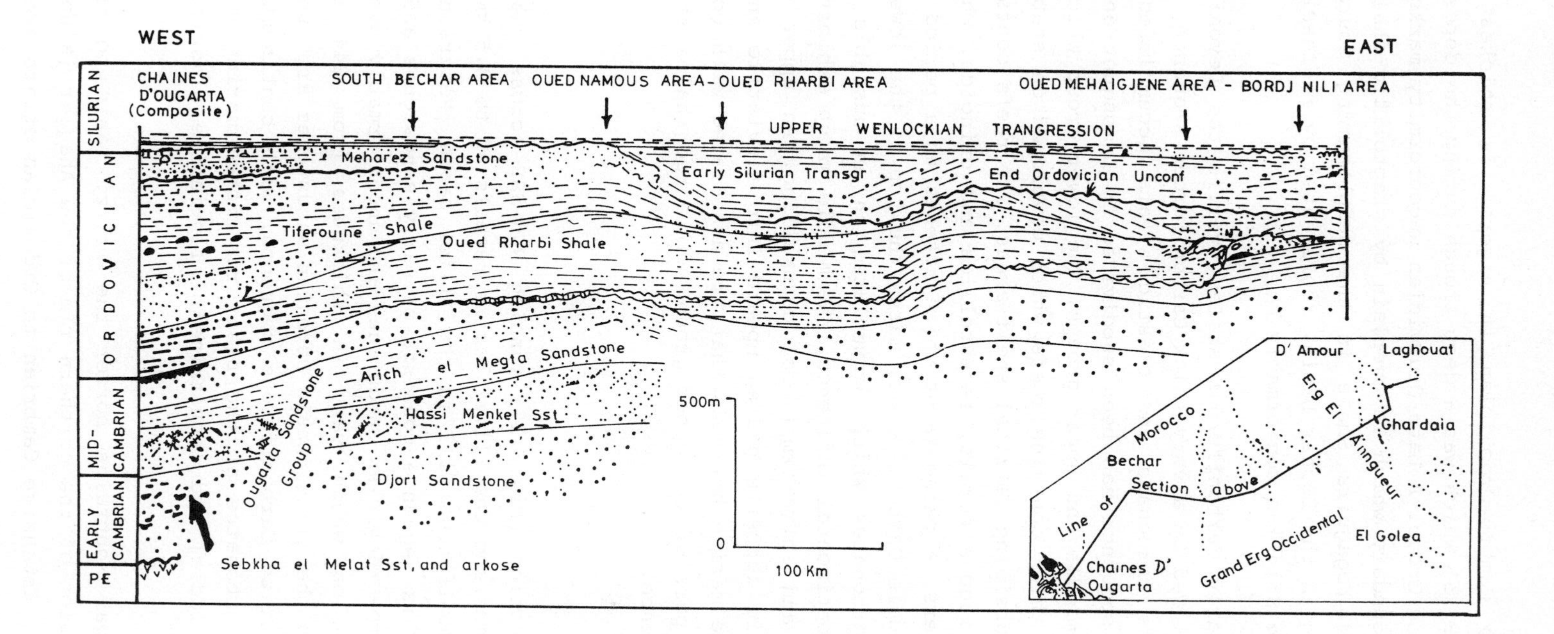

Figure 8.26: Stratigraphic section of the Ordovician of the Algerian Sahara. (Redrawn from Legrand, 1985.)

intercalations. The Cambro-Ordovician in Algeria is thickest in the Ougarta chain (Fig.8.25) which was a deep trough during the deposition of the Ougarta Sandstone Group. A Late Ordovician unconformity marks the Ordovician-Silurian boundary which is overlain by graptolitic shales (Oued Ali Clay Formation; Fegaguira Shale in the neighbouring Gourara subbasin). The Devonian in the Bechar basin, represented by regressive sandstones, succeeds the Silurian conformably.

Transgressive Early Carboniferous shales overlie the Devonian Sandstones, and are followed by a very thick Carboniferous sequence. Lower to Middle Carboniferous carbonate and clastic strata accumulated in the Bechar basin. This sequence has been subdivided by Lemosquet and Pareyn (1985) into lower, middle and upper groups. The lower group which is locally up to 4.4 km thick contains reefs which developed on stable platforms bordered by subsiding detrital shelf areas. The reefs contain stromatactis, sponges, rugose corals, fenestellids and crinoids, which were the main reef builders. A paleo-karst which attests to a period of emergence during the Middle Carboniferous, marks the top of the lower group. Characterized by carbonates which accumulated on an unstable platform that was prone to continental influence, the middle group exhibits detrital intercalations and intraformational channelling. An upper group of predominantly deltaic deposits grades upward into fluviatile and lacustrine deposits. Some marine horizons that are associated with coal seams occur in the lower part of the upper group, while red shales at the top mark regional emergence.

Illizi Basin

A regional cross section (Fig.8.27) running NNW-SSE across the Bechar-Timimoun, the Ghadames and the Illizi basins reveals up to 3.5 km of Cambrian to Upper Carboniferous fill in the Illizi basin (Peterson, 1985). The lower part of this sequence consists of marly sandstones and shales, in which thick Cambro-Ordovician sandstones are unconformably overlain by Silurian marine shales which vary in thickness from 244 to 305 m (Clifford, 1986). In the Illizi basin Devonian sandstones are overlain by a very thick and varied Carboniferous sequence, with continental sandstones and marine carbonates at the base; restricted marine evaporitic carbonates in the middle; and lacustrine of fluviatile clastics at the top.

Since the Ahaggar Mountains border the Illizi basin to the south (Fig.8.27) a discussion of the northern rim of the Ahaggar is not out of place at this point. Extensive Cambrian to Ordovician strata are exposed

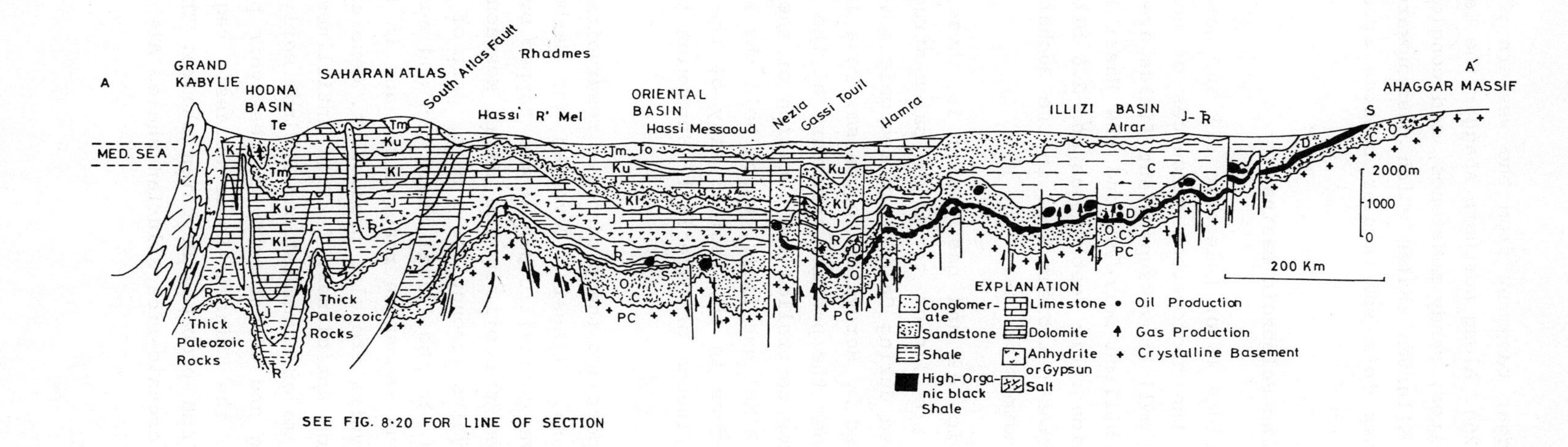

Figure 8.27: Stratigraphic cross-section of Algeria. (Redrawn from figure supplied by J.A. Peterson.)

around the northern Ahaggar basement from the western part to Djado in the east (Figs. 8.22; 8.25). Along northern Ahaggar the Cambro-Ordovician comprising mostly quartzites, sandy mudstones, and conglomerates are unconformably overlain by Silurian shales which pass upward into Devonian sandstones. The sandstones form spectacular cuestas around the Ahaggar Mountains (Fig.8.25).

8.4.4 Petroleum in Algerian Paleozoic Basins

Algeria is among the world's major Paleozoic oil and gas provinces. Peterson (1985) appraised the hydrocarbon resources of north-central and northeastern Africa. Two well known supergiant fields are the Hassi Messaoud oil field with 9 billion barrels, and the Hassi R'Mel gas field (Fig.8.27) with 70 trillion cubic feet of gas and 2.6 billion barrels of crude. The bulk of Algerian petroleum lies in the Bechar-Timimoun basin and in the Illizi and Ghadames basins.

Most of the Paleozoic oil and gas fields in these basins are in structural traps which are linked to major paleo-structural features (Fig.8.27) that originated during Caledonian tectonic movements and were later strongly accentuated by Hercynian tectonism. To a large extent the latter movements determined the present structure of the western Sahara. Two large domal structures controlled the location of the Hassi Messaoud oil field and the Hassi R'Mel gas field, which hold the substantial part of the oil and gas reserves in the region. Many of the smaller fields show evidence of the influence of especially Hercynian paleo-structural growth.

In their order of importance the productive reservoirs in Algeria are in Cambro-Ordovician, basal Triassic, Devonian and Carboniferous sandstones. Fractured Cambrian quartzitic sandstones capped by impervious Ordovician clastics trapped the oil in the Hassi Messaoud field. Basal Triassic sandstone reservoirs account for a major part of the gas reserve in the Hassi R'Mel field. In the Illizi basin gas and some oil occur in Devonian and Carboniferous sandstone reservoirs, and in the Ahnet basin Devonian sandstone reservoirs are also productive. Generally the oil and gas fields are located on anticlines, faulted anticlines, or on domes. Important aspects of the major accumulations include early stratigraphic-paleo-structural trapping and unconformity or reservoir pinch-out traps. The sealing beds include the Triassic-Lower Jurassic regional evaporite sequence, Carboniferous and probably Devonian shales. The major source rock was the widespread organic-rich Silurian shale. Also, marine shales of Devonian and Carboniferous age sourced hydrocarbon accumulations in the Illizi and Ahnet basins.

8.4.5 Ghadames Basin

This basin occupies northwestern Libya and extends into Tunisia and Algeria (Figs. 8.28A; 8.29) where it is also known as the Erg Oriental basin. Paleozoic strata in the Ghadames basin extend into the Murzuk basin in the south (Table 8.2). Cambrian to Carboniferous sequence fills the Ghadames basin with major unconformities caused by Caledonian and Hercynian movements. Lower Paleozoic strata are well exposed near Jebel Gargaf in the southern part of the Ghadames basin (Fig.8.28 A), otherwise they are mostly covered by Mesozoic-Cenozoic strata (Fig.8.29). The description of the Ghadames basin presented below applies mainly to the Libyan part and is culled from Bellini and Massa (1980), Goudarzi (1970), and Klitzsch (1981).

Cambro-Ordovician strata, known as the Gargaf Group, rest nonconformably upon basement with basal conglomerates and sandstones (Hassaouna Formation) containing *Tigillites* at the top. This is unconformably overlain by Middle Ordovician sandstones and Late Ordovician shales (Table 8.2). The Late Ordovician Melez Chograne Shale is fossiliferous and interbedded with siltstones and fine-grained sandstones with tillite. Set in a matrix of green shale, the tillite contains clasts such as granite gneiss and re-deposited boulders and pebbles of shale and sandstone. The Silurian Tanezzuft Shale succeeds and passes upward into the Acacus Sandstone. The upper part of the latter formation, though fossiliferous like the Silurian shale below, shows increasing continental influence as a result of Caledonian movements, and is unconformably overlain by continental sandstones (Tadrart Sandstone) of Early Devonian age (Fig.8.30). Devonian facies in the Ghadames basin belongs mostly to the Aouinet Group which consists of ferruginous sandstones, oolitic ironstones, and fossiliferous sandstones and carbonates with brachiopods, trilobites, bryozoans, and corals. This sequence grades upward into Carboniferous sandy and shaly facies in which fine to very fine sandstones with plant remains, feldspars and frequent ferruginous and hematitic levels suggest deltaic environments. The Carboniferous in the Ghadames basin also contains limestones with fusulinids and brachiopods, and an upper homogeneous red-brown dolomitic shale of Late Carboniferous age with shaly dolomites and anhydrite in the lower part. Petroleum occurs in the Devonian of the Ghadames basin (Fig.9.29).

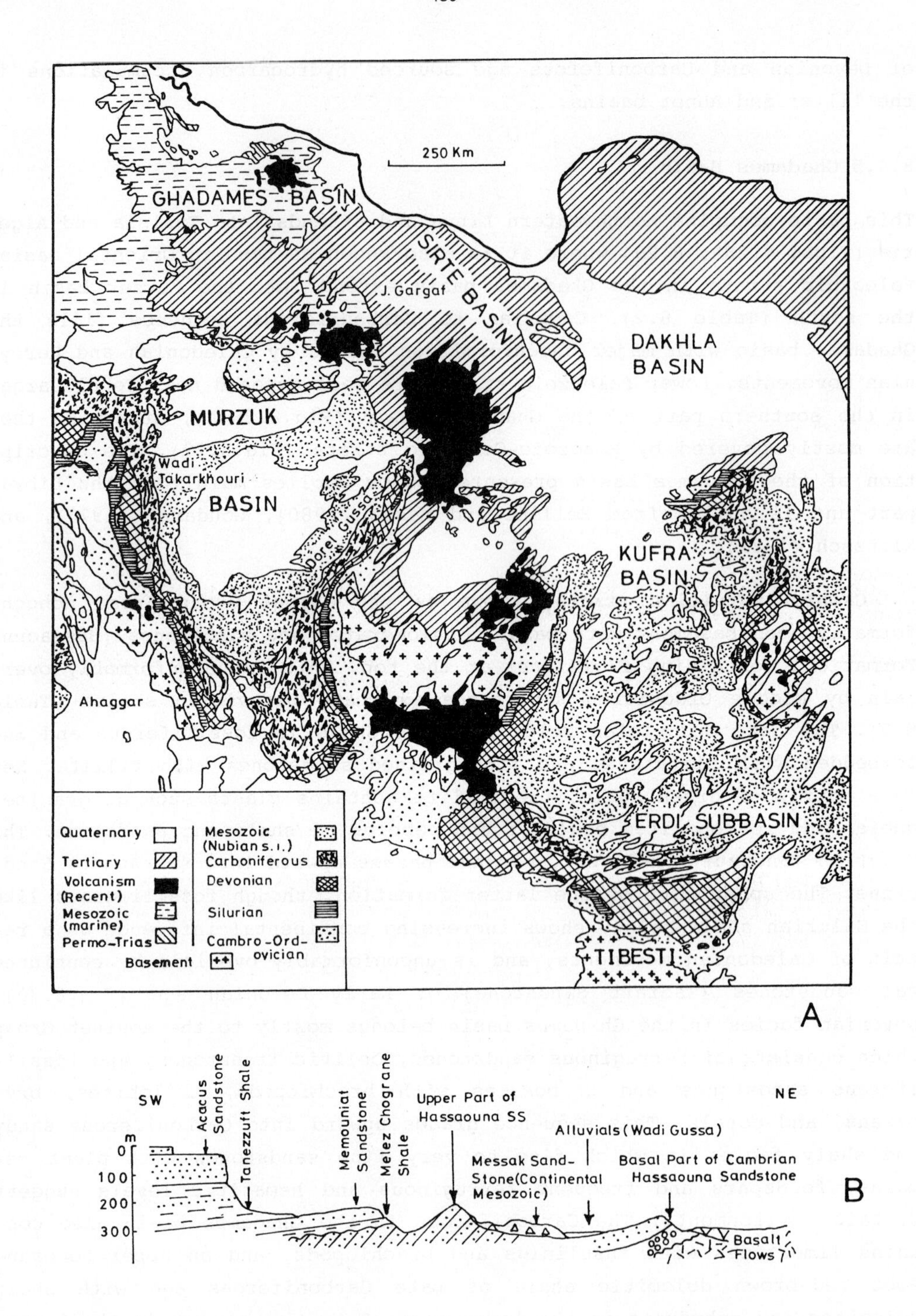

Figure 8.28: Geologic sketch map of Libya (A) and schematic section across the northern edge of Dor el Gussa, eastern Murzuk basin. (Redrawn from Bellini and Massa, 1980; Klitzsch, 1981.)

Table 8.2: Correlation of the Lower Paleozoic of the Homra, Murzuk and Kufra basins. (Redrawn from Klitzsch, 1981.)

	SOUTHERN HOMRA BASIN (JEBEL GARGAF)	WESTERN MURZUK BASIN (JEBEL ACACUS)	EASTERN MURZUK BASIN (DOR EL GUSSA)	WESTERN KUFRA BASIN (EGHEI/DAHONE)	NORTHERN KUFRA BASIN (JEBEL GARDEBA)	EASTERN KUFRA BASIN Arkenu Border-Lybia-Egypt-Sudan
LOWER DEVONIAN	TADRART SANDSTONE					
SILURIAN	ACACUS SANDSTONE TANEZZUFT SHALE					
ORDOVICIAN	MEMOUNIAT SANDSTONE		Locally Tillite	MUNCHAR FORMATION	HAUAISC SANDSTONE	MEMOUNIAT SANDSTONE
	MELEZ CHOGRANE SHALE	←Locally Present→	MELEZ CHOGRANE SHALE			
	HAOUAZ SANDSTONE	←Locally Present→	HAOUAZ SANDSTONE			
CAMBRIAN	HASSAOUNA SANDSTONE		INFRACAMBRIAN SANDSTONE South of 24° North	TEDA SANDSTONE ZOUMA SANDSTONE	?	HASSAOUNA SANDSTONE
PRECAMBRIAN	PRECAMBRIAN METAMORPHICS AND INTRUSIONS					

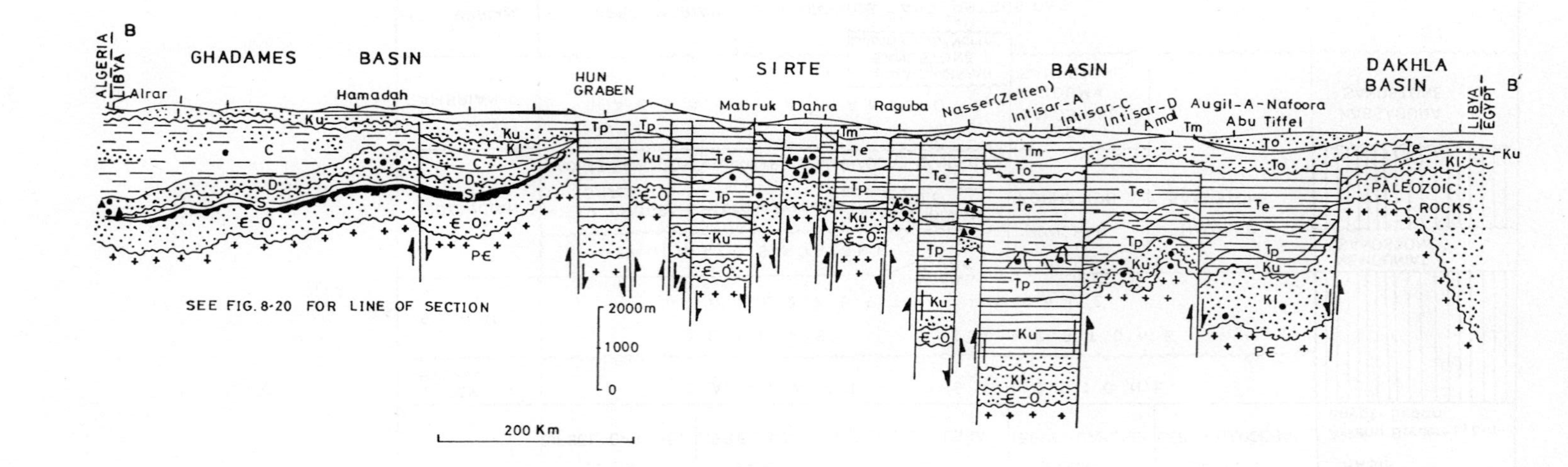

Figure 8.29: Structural-stratigraphic section across West-central Libya and NW Egypt. (Redrawn from figure supplied by J.A. Peterson.)

8.4.6 Murzuk Basin

The Murzuk basin is a vast intracratonic basin located between the Ahaggar and the Tibesti Mountains (Fig.8.20). Its lower fill comprises Cambrian to Carboniferous strata. A system of cuestas rim the western, eastern, and southern margins exposing some of the Libyan Lower Paleozoic stratotypes at Dor el Gussa in the eastern sub-basin (Fig.8.28B), and at Wadi Takarkhouri in the western sub-basin. Split into two sub-basins by the Tripoli-Tibesti uplift (Fig.8.20), the Murzuk basin consists of an eastern predominantly clastic sub-basin (exposed at Dor el Gussa) and a more marine western sub-basin which is best exposed at Wadi Takarkhouri. Restricted marine environments were established in the eastern sub-basin in which there was heavy oxidation and continuous detrital sedimentation which originated from the surrounding landmasses from where vegetal debris with *Lycophytes* were derived.

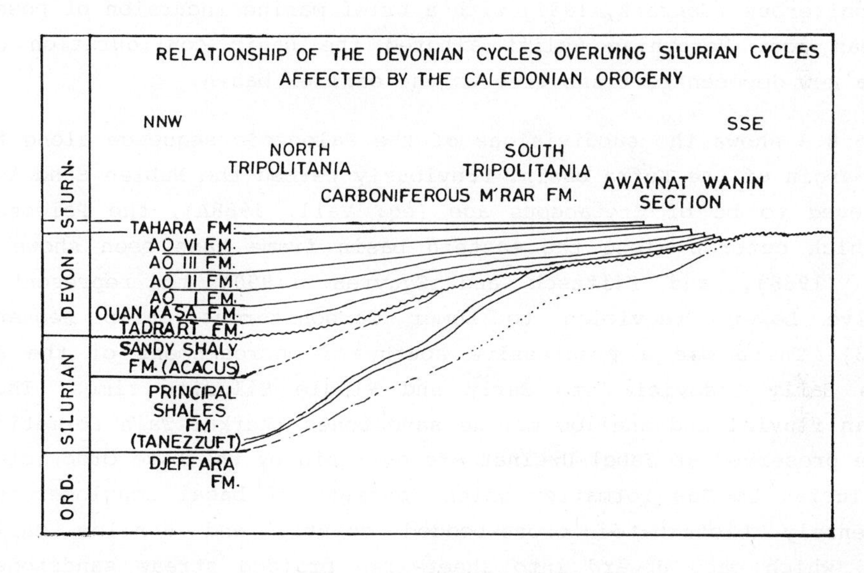

Figure 8.30: Schematic section depicting the Silurian-Devonian of northern Libya. (Redrawn from Bellini and Massa, 1980.)

As aforementioned the Lower Paleozoic succession in the Murzuk basin is essentially the same as that of the Ghadames basin to the north (Table 8.2). The Devonian is also continental at the base passing upward into marine units with an Upper Devonian regressive paralic cycle which contains extensive oolitic ironstones (Van Houten and Karasek, 1981). The Early Carboniferous is transgressive with mixed marine and lagoonal facies, while in the Late Carboniferous red continental sediments accumulated.

8.4.7 Kufra Basin

This is a large Paleozoic basin which extends from southeastern Libya into Chad where it is known as the Erdis basin, and marginally into southwestern Egypt and northwestern Sudan (Fig.8.28A). It is bordered to the northeast by the Dakhla basin of Egypt and northeastern Libya. Except along its eastern margin where the Silurian oversteps onto basement (Fig.8.31), the Kufra basin has a thick sequence of Cambrian to Ordovician strata (Table 8.2). The Silurian marine transgression which was widespread in this basin, extended as far as eastern Egypt and northwestern Sudan (Fig.8.32) and terminated with the deposition of the Acacus Sandstone under shoal water conditions. Following Caledonian movements, Devonian sedimentation began under predominantly continental conditions and passed upward into open but shallow marine environments with locally developed lagoonal conditions. Mostly nonmarine beds were deposited in the Carboniferous (Schrank, 1987) with a brief marine incursion of possibly Visean age. Hercynian uplift altered the basin configuration and created a new depocentre transverse to the earlier basin.

Table 8.3 shows the subdivisions of the Paleozoic sequence along the eastern margin of the Kufra basin. Previously termed the Nubian Sandstone and believed to be of Cretaceous age (eg. Vail, 1988A), the Paleozoic strata which outcrop along the eastern basin frame have been shown by Klitzsch (1986), and Klitzsch and Squyres (1990) to represent a progressive Lower Ordovician to Lower Carboniferous onlap sequence (Fig.8.33). There was a progressive southward encroachment of the sea from the Early Ordovician to Early and Middle Silurian times. Thus, Ordovician fluvial and shallow marine sandstones (Karkur Talh Formation) which are preserved at Jebel Uweinat are overlain by the Late Ordovician-Early Silurian Um Ras Formation which consists of basal conglomerates, and intensely bioturbated cross-bedded coastal and shallow marine deposits, which pass upward into sheet-like braided stream sandstones. Unconformable Devonian sandstones (Tadrat Formation) are succeeded also unconformably by Lower Carboniferous fluvial to coastal plain and tidal-subtidal clastics. These are overlain by Late Carboniferous glacial lake deposits, which represent one of the rare northernmost occurrences of Permo-Carboniferous glacial deposits which are more widespread in the Karoo Supergroup of southern Africa.

Post-Hercynian sedimentation in northwestern Sudan and neighbouring parts of Egypt and Libya took place in a new depocentre south of the Uweinat-Bir Safsaf-Aswan uplift (Fig.8.34) under Karoo-type continental conditions. This sequence begins with Late Carboniferous tillites

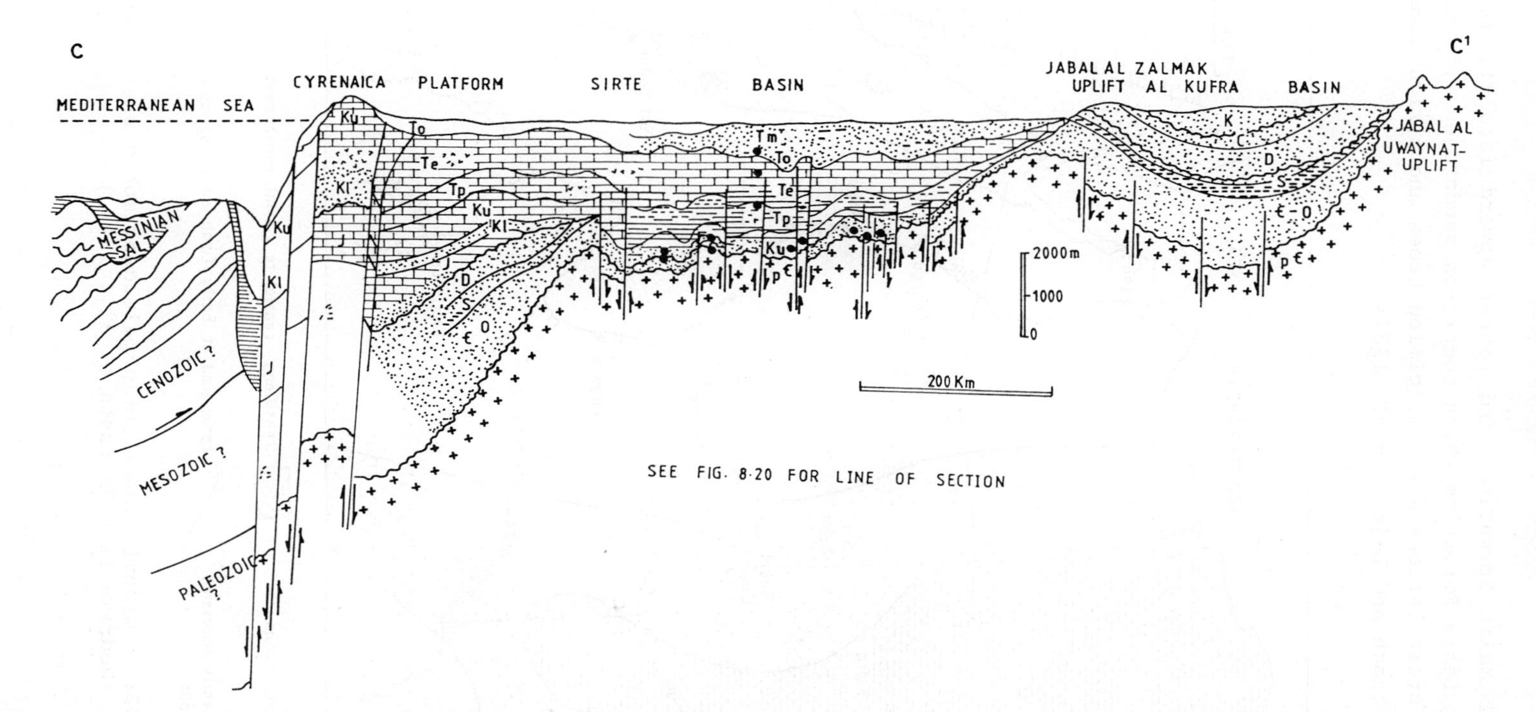

Figure 8.31: N-S section across eastern Libya. (Redrawn from figure supplied by J.A. Peterson.)

(Northern Wadi Malik Formation) and passes upward into the Permian to Early Jurassic Lakia Formation which comprises sandstones with paleosols, and mudstone intercalations with silicified trees and a rich Lower Jurassic flora (Klitzsch and Lejal-Nichol, 1984).

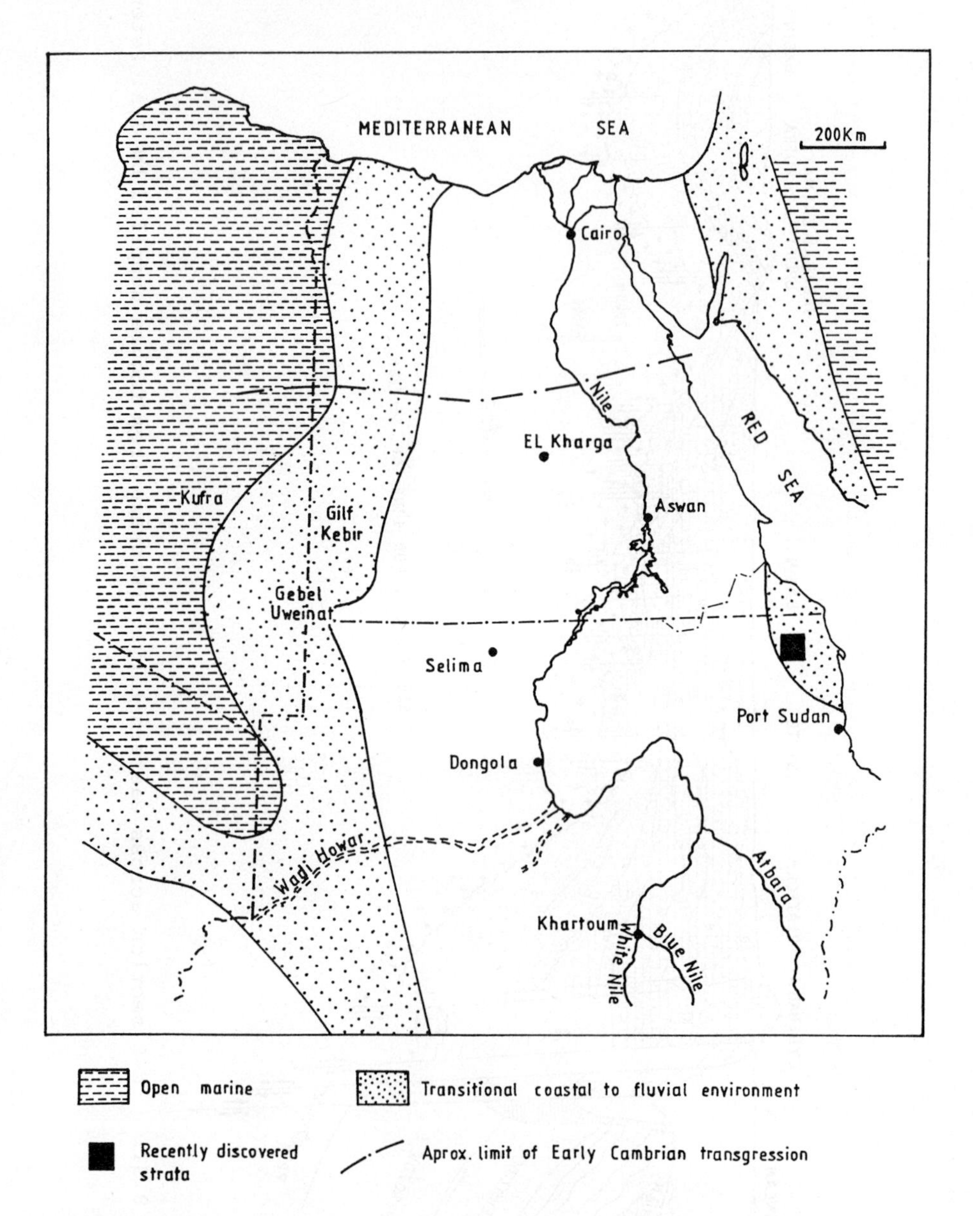

Figure 8.32: Ordovician-Silurian paleogeographic sketch map of NE Africa. (Redrawn from Klitzsch et al., 1990.)

Table 8.3: Subdivisions and correlation of the former Nubian Sandstone, and stratigraphic chart for Egypt. (Redrawn from Klitzsch, 1986.)

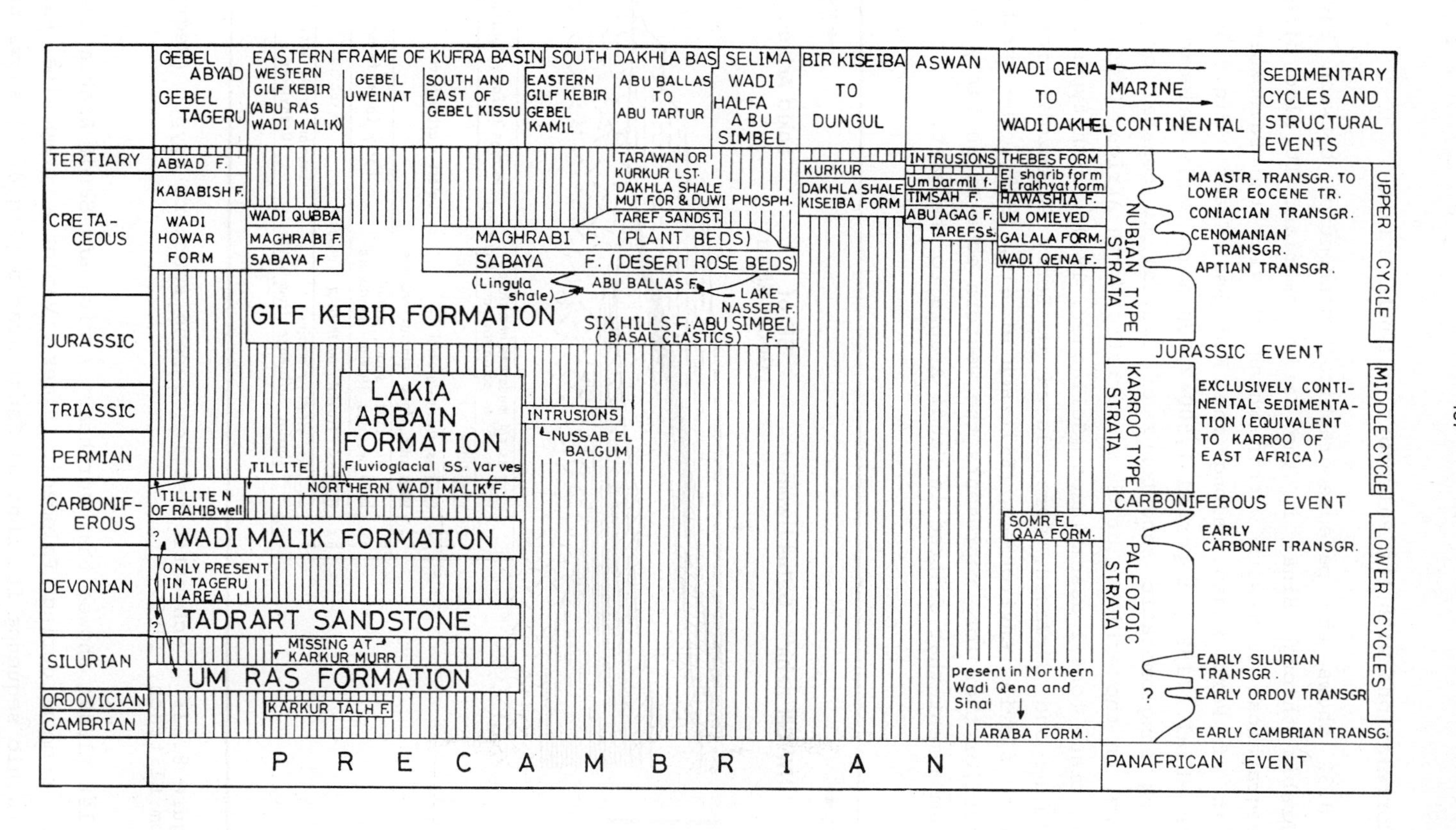

8.4.8 Correlations with the Paleozoic of Saudi Arabia

Figure 8.32 shows the paleogeographical connection between Egypt and Saudi Arabia through Sinai in the Early Cambrian. It also offers the paleogeographical framework within which to view the Paleozoic outlier northwest of Port Sudan, along the Red Sea Coast (Fig.8.32). Here Klitzsch et al. (1990) reported 1,500-2,000 m of Early Paleozoic strata comprising conglomeratic sandstones of fluvial to coastal origin. The sandstone, on the evidence of characteristic North African Ordovician and Silurian trace fossil assemblages (Seilacher, 1990), with types belonging to *Cruziana* and *Harlania*, were placed in Early Paleozoic transitional coastal to fluvial environment (Klitzsch et al., 1990). The northeastern Sudan outlier therefore represents the western edge of an Early Paleozoic marine embayment that centred in Saudi Arabia to the east.

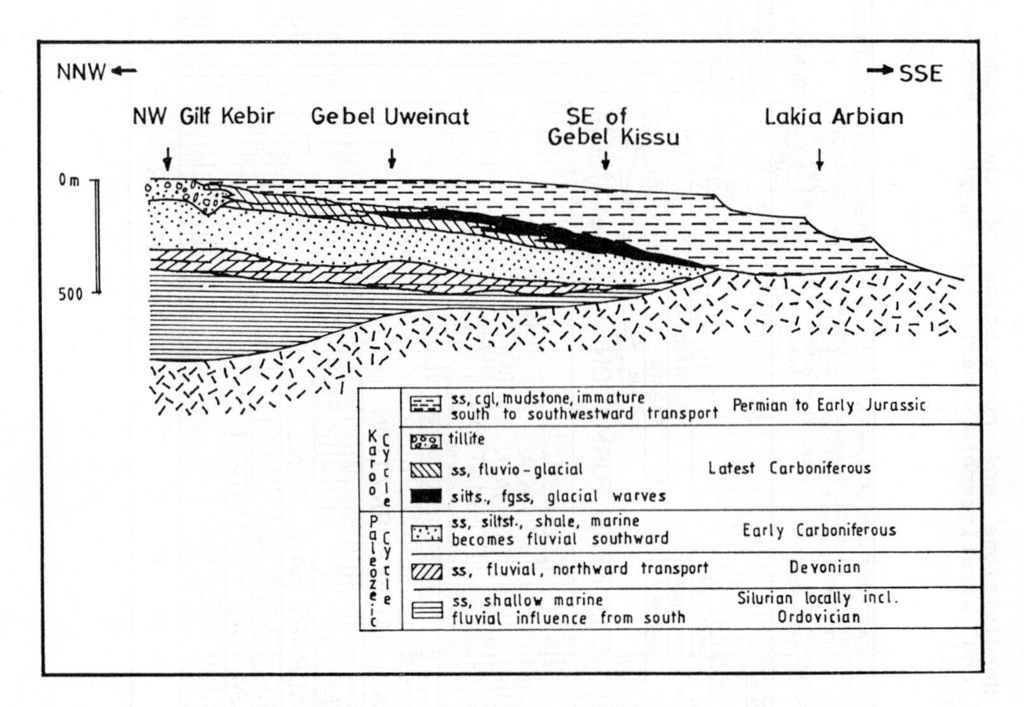

Figure 8.33: Schematic section across southern Egypt. (Redrawn from Klitzsch, 1986.)

Vaslet (1989) showed that northeast Africa and Saudi Arabia had a lot more in common during Paleozoic times. Vaslet's composite Paleozoic stratigraphic sequence for central Saudi Arabia (Fig.8.35) shows the following similarities with Egypt and the Sudan. Early Paleozoic strata in central Saudi Arabia were deposited on a major unconformity over the

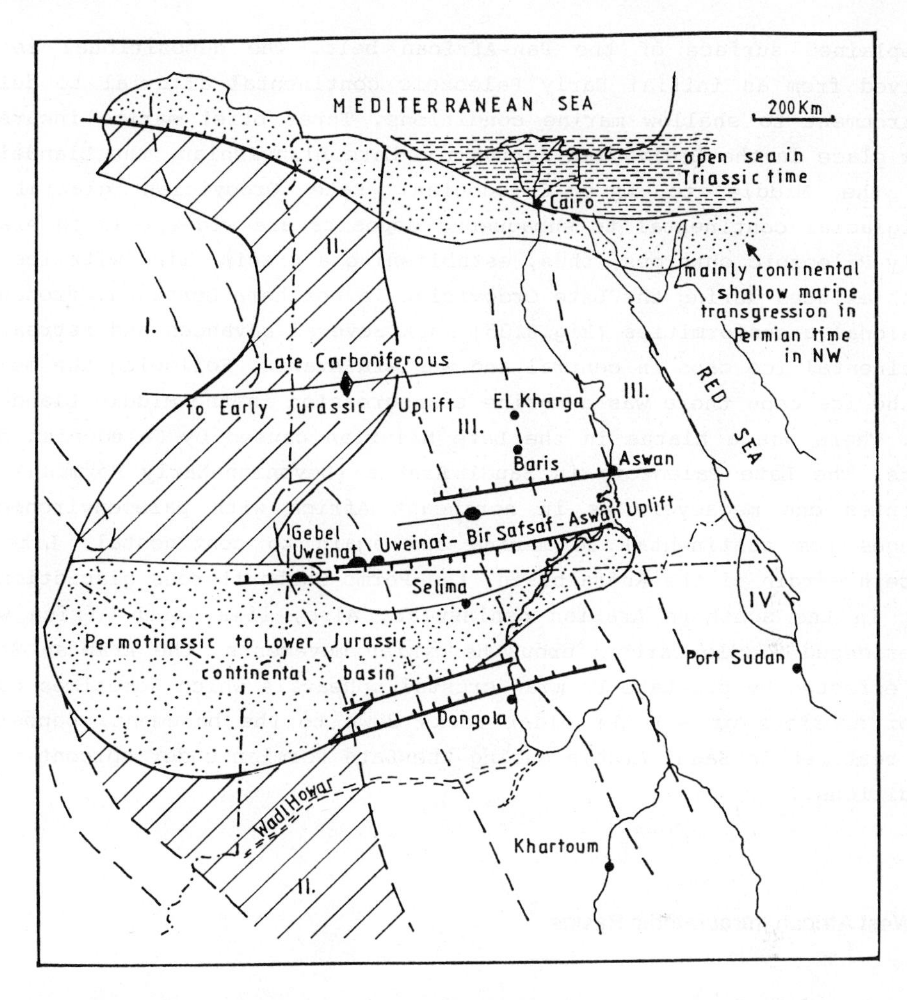

Figure 8.34: Post-Hercynian structural re-organization of NE Africa. (Redrawn from Schandelmeier et al., 1987.)

peneplained surface of the Pan-African belt. The depositional setting evolved from an initial Early Paleozoic continental alluvial to deltaic environment to shallow marine conditions. Three brief marine incursions took place in the Ordivician during the Late Llanvirnian, the Llandeilian and the Middle to Late Caradocian. Late Ordovician glacial and periglacial continental to subaqueous deposits are widespread in Arabian Early Paleozoic outcrops, thus, establishing a precise link with ice caps which existed during the Late Ordovician in northern Gondwana. Pronounced erosional unconformities (Fig.8.35) mark several advances and retreats of continental ice caps in central and western Arabia. Following the melting of the ice caps there was a marine transgression in the Middle Llandoverien. There was a hiatus in the Late Silurian caused by Caledonian movements. The Late Paleozoic in Saudi Arabia (Devonian-Early Permian) constitutes one megacycle as in northeast Africa with paleoenvironmental changes from continental to marine and again to continental. Like the eastern margin of the Kufra basin, the Permo-Carboniferous glaciation was felt in the southern Arabian peninsular. Also, like North Africa which experienced Middle Carboniferous Hercynian movements, the Arabian Shield was effected by pre-Late Permian crustal upheavals which sometimes caused major erosions of all the older rocks down to the basement. Deposition was restored in Saudi Arabia during the Late Permian under epicontinental conditions.

8.5 West African Intracratonic Basins

8.5.1 Taoudeni Basin

Centred on the West African craton is the vast Taoudeni basin (Figs. 8.36; 8.25B), which having served as the Late Proterozoic-Early Paleozoic foreland to the encircling Pan-African mobile belts to the west and to the east (Fig.6.3), contains tabular cratonic sequences of Late Proterozoic to Carboniferous age (Fig.8.37). Thus, during this extremely long period of geologic time shallow seas periodically flooded the Taoudeni basin and persisted long after the Pan-African events had ended.

The Taoudeni basin contains the most extensive outcrops of Lower Paleozoic strata in West Africa (Deynoux, 1983; Deynoux et al., 1985) and underlies most of Mali, and extends into Algeria, Burkina Faso, and continues into the two Guineas and Senegal in the southwest, where it is known as the Bove Basin (Fig.8.25C). Structurally the Taoudeni basin is a simple shallow interior sag basin with very low dips of one degree or less (Fig.8.37). It lies unconformably upon the Eburnean basement which is exposed to the northwest and southeast. It is flanked and overthrust to the west by the Mauritanides along a narrow (10-50 km) tectonized margin with folds and fractures which resulted from Hercynian deformation, whereas to the southwest its extension, the Bové basin, sits discordantly upon the Rokelides and the Bassarides (Fig.8.25C). In the region of Adrar des Iforas to the northeast the Taoudeni basin is bounded by the Tuareg Shield. Cyclical epeirogenic movements in the Taoudeni basin (Petters, 1979) resulted in unconformity-bounded stratigraphic sequences (Fig.8.37). There are synsedimentary faults with low throws as well as dolerite injections (Fig.8.37).

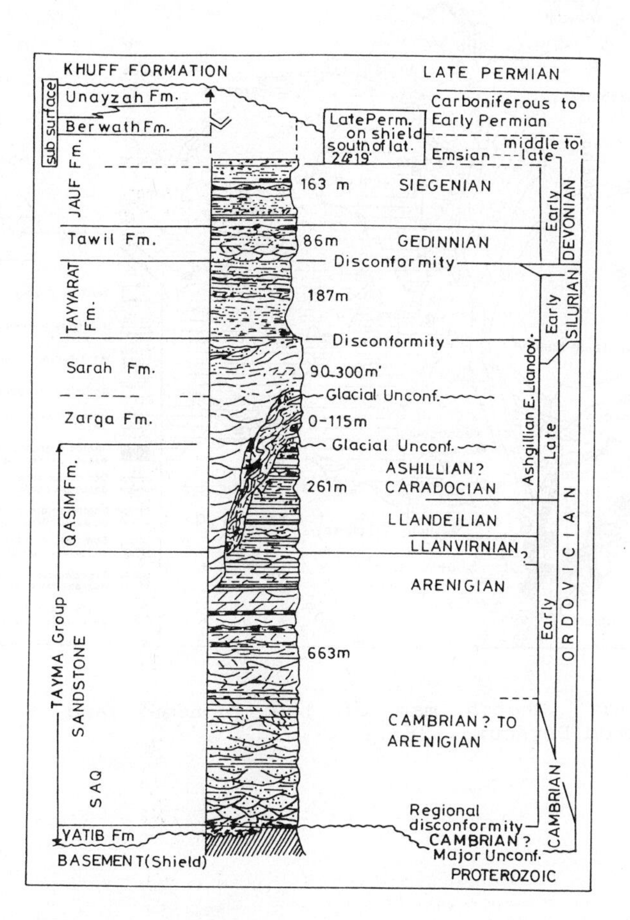

Figure 8.35: Schematic Paleozoic succession for central Saudi Arabia. (Redrawn from Vaslet, 1989.)

Stratigraphically, the Taoudeni basin consists of fine-grained clastics and carbonates, 2,000-3,000 m thick, the type sections of which are

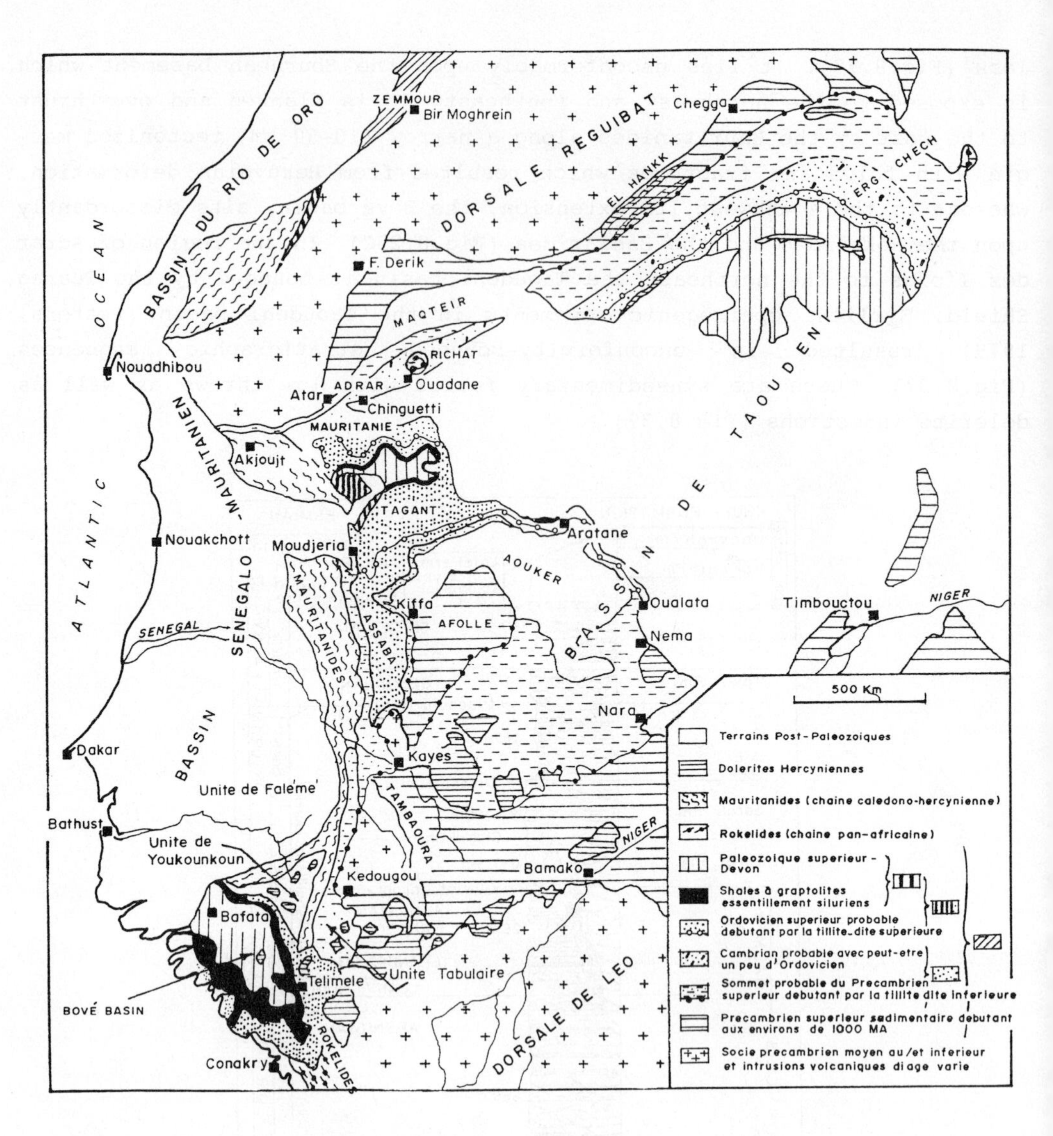

Figure 8.36: Sketch map of the Taoudeni and Bové basins. (Redrawn from Deynoux, 1983.)

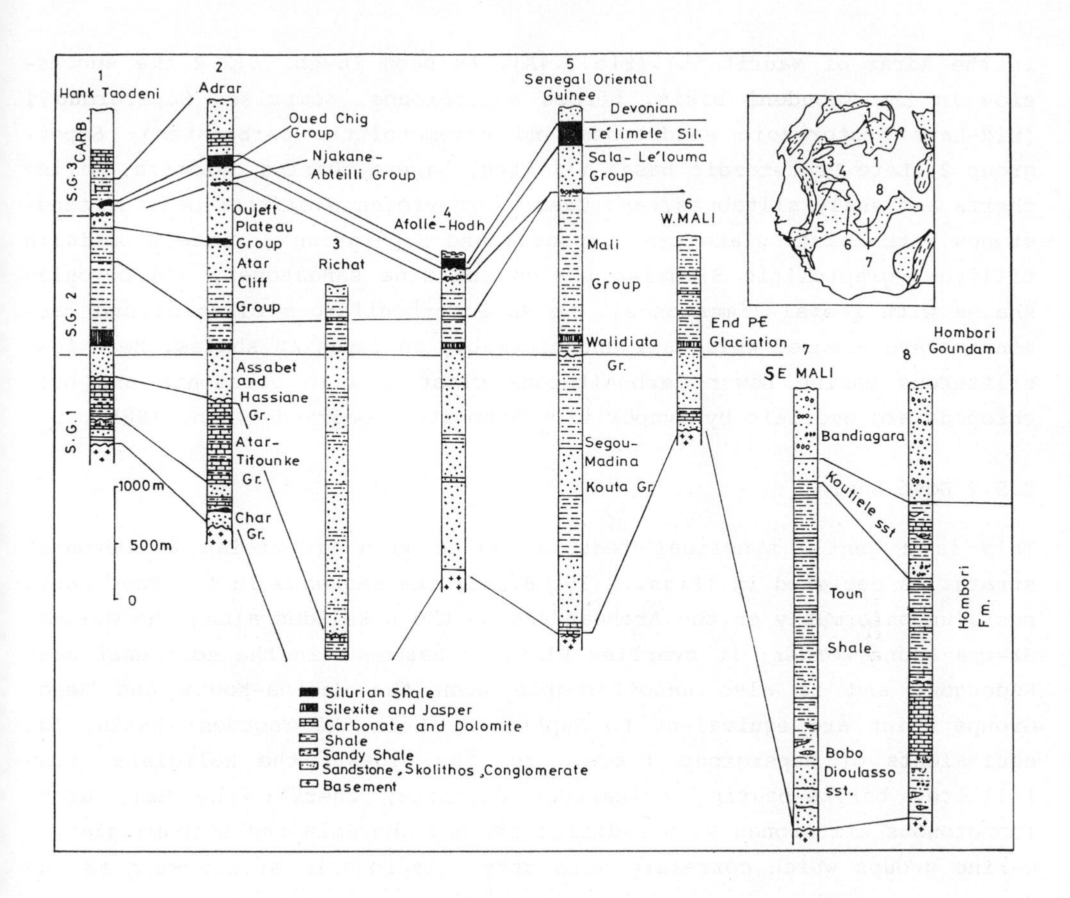

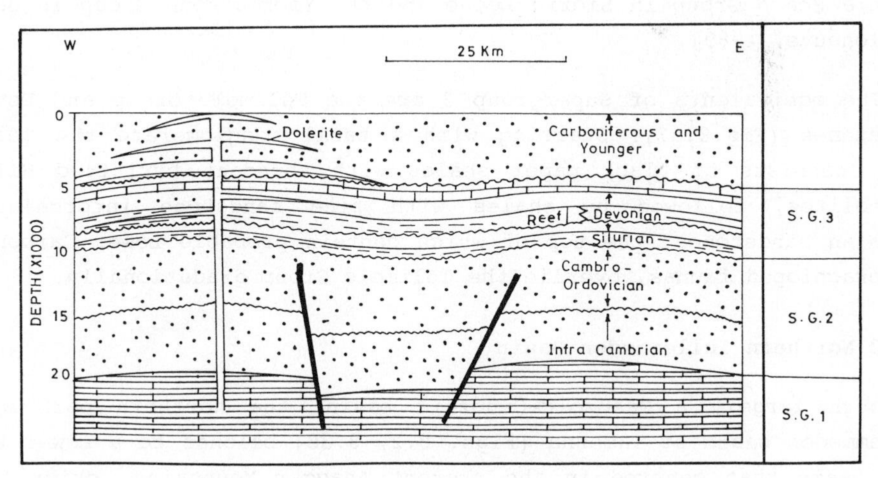

Figure 8.37: Geologic sections across the Taoudeni basin. (Redrawn from Deynoux, 1983; Clifford, 1986.)

in the Adrar of Mauritania (Fig.7.4A). As seen in Ch. 6.2.2 the succession in the Taoudeni basin, termed supergroups, comprises Supergroup 1 (Mid-Late Proterozoic sandstones and stromatolitic carbonates); Supergroup 2 (Late Proterozoic basal tillites, baryte-bearing dolomite, marine cherts and shaly siltsones, and Cambro-Ordovician *Skolithus*-bearing sandstones with inarticulate brachiopods); and Supergroup 3 (Late Ordovician tillites, graptolitic Silurian shales and fine sandstones, and Devonian shales with reefal limestones). In Mali Carboniferous clastics and carbonates are exposed which sit unconformably on Devonian shales. Here fossiliferous marine Lower Carboniferous clastics with conodonts and brachiopods are overlain by evaporitic carbonates (Legrand-Blain, 1985).

8.5.2 Bové Basin

This is a gentle synclinal feature filled with Ordovician to Devonian strata. As depicted in (Figs. 8.36; 8.25C) the sequence in the Bové basin rests unconformably on the Archean and on the Rokelides along the Guinea-Sierra Leone border; it overlies Birimian basement in the northeast near Kédougou; and is also unconformable upon the Madina-Kouta and Ségou Groups which are equivalent to Supergroup I in the Taoudeni basin. The equivalents of Supergroup 2 are, from the oldest, the Walidiata Group (tillite, baryte-bearing calcareous dolomite, chert); the Mali Group (monotonous siltstones with radiolaria); and the Sala and Lélouma glaciomarine groups which correlate with other glaciogenic strata such as the Saionia Scarp Group in Sierra Leone and the Youkounkoun Group in Senegal (Villeneuve, 1989).

The equivalents of Supergroup 3 are the Télimele Group and Devonian sandstones (Fig.8.37). Starting with a basal conglomerate the Télimele Group consists of black paper shales with abundant pyritized Silurian graptolites, followed by shales with fine sandstone intercalations. Devonian sandstones, the top of which contain probable Lower Carboniferous brachiopod faunas, overlie the Télimele Group gradationally.

8.5.3 Northern Iullemmeden Basin

Although largely a Mesozoic-Cenozoic basin, the northern part of the Iullemmeden basin at Tamesna (Figs. 8.2, 8.38) belongs to a Lower Paleozoic basin that centred in the present Ahaggar Mountains, prior to the post-Hercynian uplift of the Ahaggar. Generally Paleozoic strata in the Tamesna sub-basin thicken towards the Ahaggar (Fig.8.38) and thin southwards where they disappear. The Ahaggar Mountains did not exist in the Early Paleozoic. Cambro-Ordovician strata are up to 500 m thick in eastern Tamesna where they comprise basal conglomerates, sandstones with *Cruiziana* and *Skolithus,* and are unconformably overlain by glacial deposits. These are followed by Early Silurian graptolitic shales, up to 100 m thick, and discordantly overlain by Devonian sandstones and shales, and Late Devonian-Early Carboniferous deltaic sandstones. A transgressive Late Tournasian coal-bearing paralic sequence is disconformably overlain by Visean shales and limestones (Fig.8.38) containing corals and conodonts.

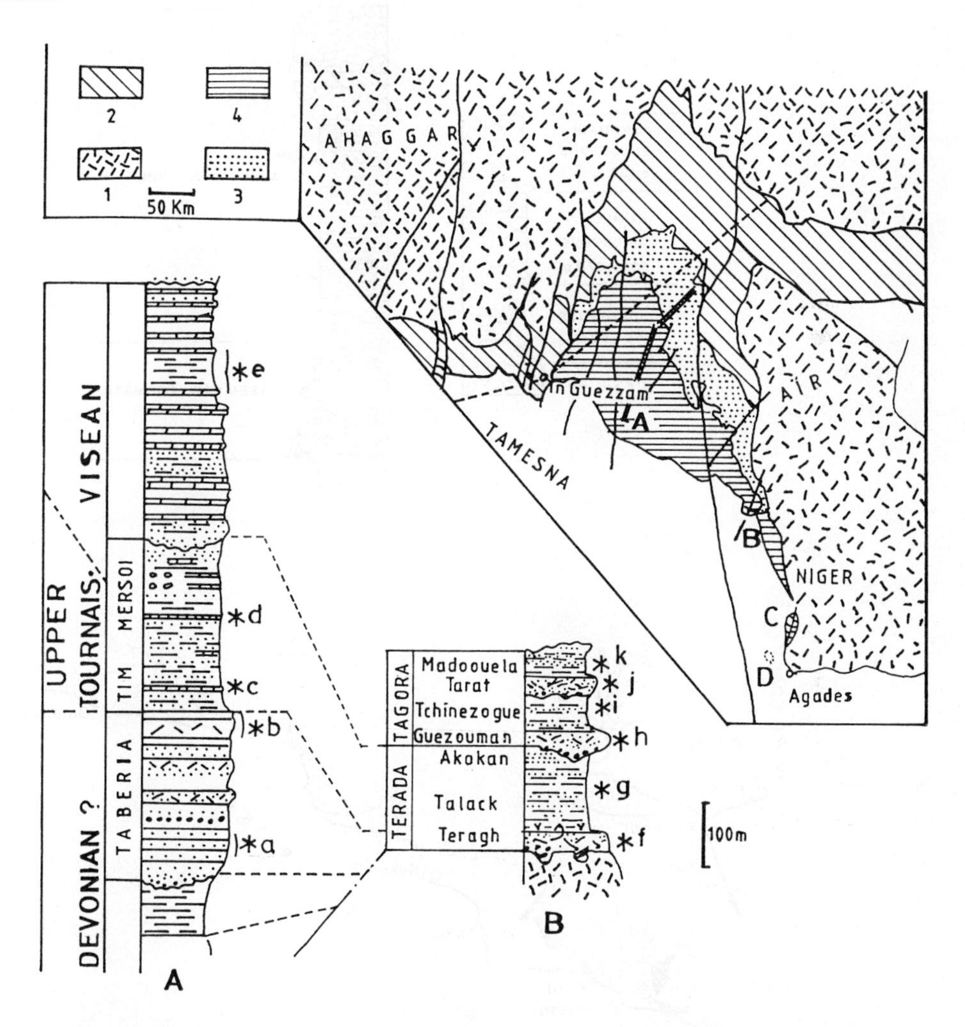

Figure 8.38: The Paleozoic of northern Iullemmeden basin. 1, Precambrian; 2, Lower Paleozoic-Middle Devonian; 3, Upper Devonian-Lower Carboniferous; 4, Carboniferous. a-k are fossiliferous horizons. (Redrawn from Legrand-Blain, 1985.)

8.5.4 Paleozoic Exposures Along the West African Coast

Few exposures of Paleozoic strata occur near Monrovia, Liberia, and near Takoradi and Accra in Ghana (Fig.8.39). These are believed to represent more extensive Paleozoic sequences which lie in the coastal basins. Near Monrovia the Paynesville Sandstone, up to 1,000 m thick, represents probable Devonian-Carboniferous continental deposits.

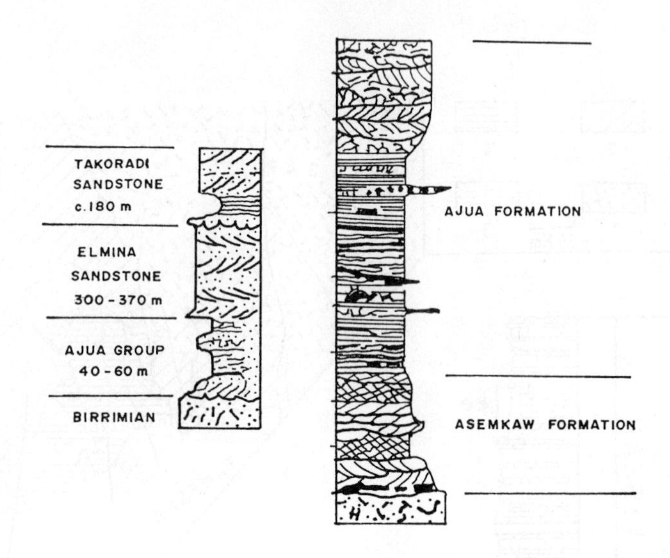

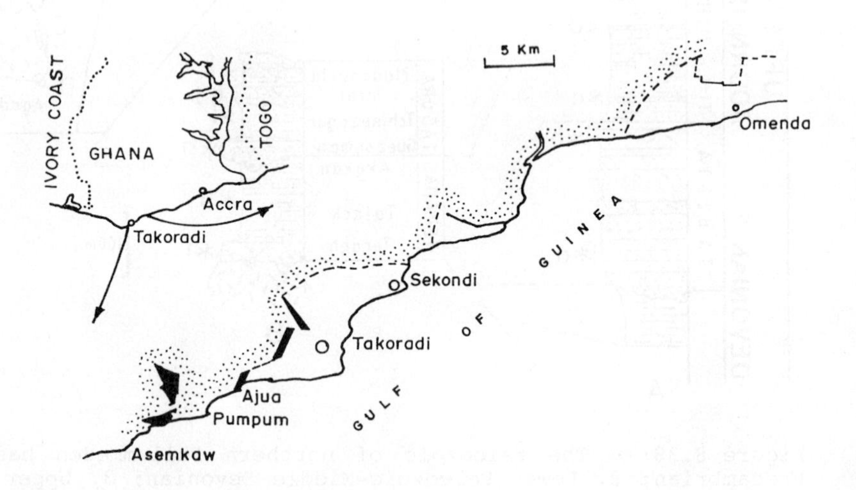

Figure 8.39: Paleozoic exposures and sequences along the West African coast. (Redrawn from Talbot, 1981.)

Preserved along the coastal strip in Ghana, to the east and west of Takoradi, are small discontinuous Paleozoic sections (Fig.8.39), referred to as the Sekondi Series. The Sekondi Series, 1,245-1,325 m thick, occurs in faulted blocks and is predominantly a sequence of sandstones and shales resting unconformably upon the Birimian (Fig.8.39). Most notable in these sections is the basal Ajua Group, an intertidal to shallow marine or lacustrine deposit that accumulated under locally freezing or glacial conditions probably during the Late Ordovician glaciation (Talbot, 1981). The oldest biostratigraphically dated horizon is, however, at the base of the Takoradi Sandstone (Fig.8.39) where basal shales have yielded Late Devonian microflora from a horizon 300-400 m above the Ajua glaciogenic group. Poorly preserved brachiopods, pelecypods and fish remains also occur at this level.

Further east near Accra, the Accraian Group is exposed in a small area of about 11.7 km^2. This is the best dated Paleozoic section on the West African coast. Believed to be of Early to Middle Devonian age on faunal and palynological evidence, the Accraian Group comprises from its base: coarse pebbly cross-bedded sandstones; alternating fine sandstones and shales; thicker fossiliferous shales with trilobites and brachiopods; massive cross-bedded sandstones and alternating shales and thin-bedded micaceous sandstones (Kesse, 1985). Based on its paleobiogeographic affinity with the Appalachian brachiopod fauna, the Accraian brachiopod assemblage was assigned to the Appalachian paleobiogeographic province (Johnson and Boucot, 1973), thus placing the West African coastal Paleozoic strata in the northern part of a Devonian seaway that came from North and South America (Fig.8.40).

8.6 The Cape Fold Belt

8.6.1 Aborted Rifts and Glaciations

Two dominant factors determined basin development in South Africa during the Paleozoic. First, the Lower Paleozoic Cape Supergroup, a phenomenally thick sequence of nearshore and shallow shelf sandstones, accumulated in the initial rifts along which southern Gondwana attempted to break up in the Early Paleozoic. Figure 8.41 (inset) shows the incipient plate boundaries which formed a triple junction between the African, South American, and Antarctic plates. However, the development of a subduction zone some 1,000 km further south (Fig.8.41) aborted further crustal extension in the Cape region of South Africa. The Cape region instead remained as the passive continental margin of what Du Toit (1937) termed the Samfrau geosyncline. Northward-directed flat-plate subduction (Lock, 1980) subduction generated compressional forces that deformed the Cape Supergroup clastic wedge which then became the Cape fold belt. The Cape fold belt belongs to the Gondwana orogenic belt. Other segments of this orogenic belt are now widely dispersed in remote regions such as Bolivia, Peru and Argentina in South America, and in Antarctica, and eastern Australia (Tankard et al., 1982).

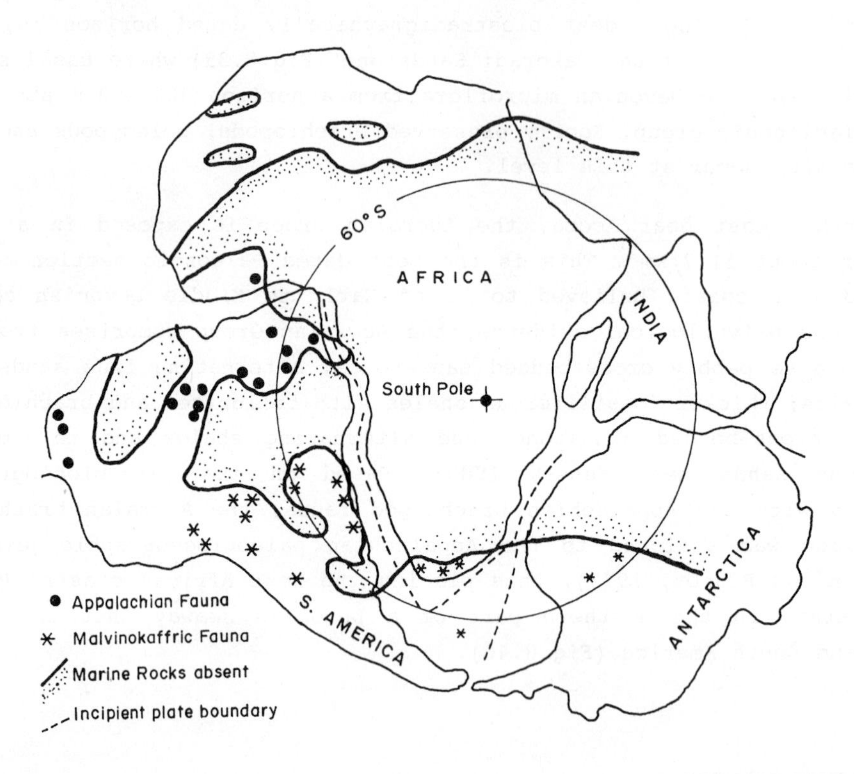

Figure 8.40: Early Devonian paleogeography of Gondwana. (Redrawn from Tankard et al., 1982.)

As already mentioned in the introduction to this chaper, South Africa witnessed spectacular environmental changes during the Paleozoic. It experienced the Late Ordovician glaciation, and later lay at the centre of the great Permo-Carboniferous glaciation of southern Gondwana.

8.6.2 The Cape Supergroup

This is an 8-km thick Early Ordovician to Early Carboniferous clastic sequence which forms folded mountain ranges along the coast of South Africa (Fig.8.42A). Its equivalent, the Natal Group, is exposed along the east-

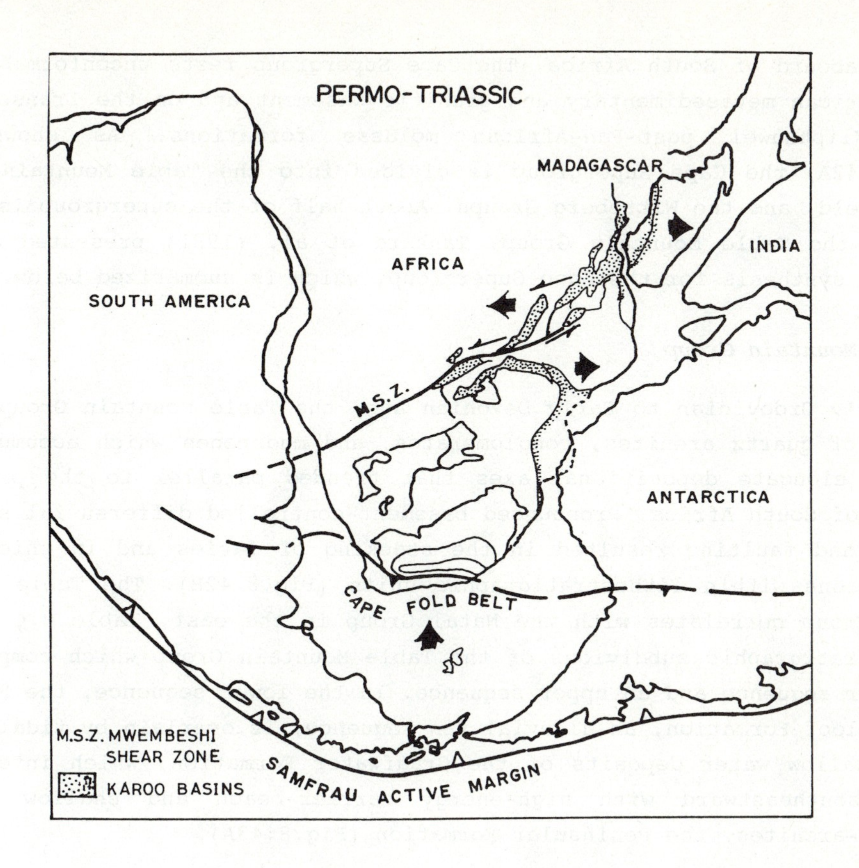

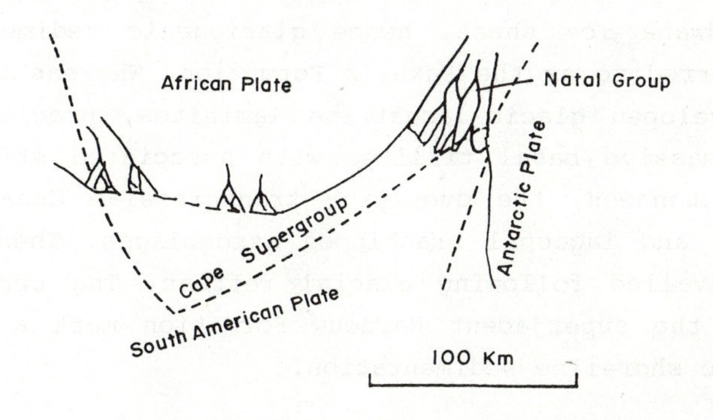

Figure 8.41: Tectonic model for the Karoo basins and the Cape fold belt; paleogeographic setting for the Cape Supergroup on a pre-drift reconstruction of Gondwana. (Redrawn from Daly et al., 1989; Tankard et al., 1982.)

ern seaboard of South Africa. The Cape Supergroup rests unconformably on Pan-African metasedimentary and granitic basement and on the Franschhoek and Klipheuwel post-Pan-African molasse formations. As shown in Fig.8.42A, the Cape Supergroup is divided into the Table Mountain, the Bokkeveld, and the Witteberg Groups. About half of the supergroup is made up of the Table Mountain Group. Tankard et al. (1982) presented a detailed synthesis for the Cape Supergroup, which is summarized below.

Table Mountain Group

Of Early Ordovician to Early Devonian age, the Table Mountain Group consists of quartz arenites, conglomerates, and mudstones which accumulated in an elongate depositional axes that trended parallel to the present coast of South Africa. Pronounced basement-controlled differential subsidence and faulting resulted in the stacking of facies and in thickness variations within lithostratigraphic units (Fig.8.42B). The Table Mountain Group correlates with the Natal Group in the east. Table 8.4 shows the stratigraphic subdivions of the Table Mountain Group which comprises a lower sequence and an upper sequence. In the lower sequence, the Piekenierskloof Formation, an alluvial fan sequence, is overlain by tidal flat and shallow water deposits of the Graafwater Formation, which interfingers southeastward with high-energy barrier-beach and shallow shelf quartz-arenites, the Peninsular Formation (Fig.8.43A).

During the Late Ordovician the Cape basin lay along the margin of an extensive Gondwana ice sheet, hence glaciogenic sediments accumulated which are referred to as the Pakhuis Formation. Whereas the Pakhuis contains well-developed glacio-lacustrine laminites, proglacially reworked tillites and massive basal tillites with associated striated pavements and roches moutonnées, the overlying transgressive Cedarberg Formatiom contains tidal and lagoonal brachiopod assemblages. These environmental conditions prevailed following glacial retreat. The upper part of the Cedarberg and the superjacent Nardouw Formation mark a return to preglacial clastic shoreline sedimentation.

Natal Group

In the Natal embayment the Natal Group, about 1000 m thick, was deposited under greater wave and tidal current intensity than the Table Mountain Group. Since the Natal embayment was funnel-shaped (Fig.8.41, inset) stronger tidal currents were generated and directed perpendicular to the shoreline resulting in lenticular tidal sand bars (Fig.8.43B) which occur in the stratigraphic sequence as truncated, stacked and en échelon sandstone units. Otherwise, as shown in Fig.8.43, the Natal and Table Mountain Groups have similar stratigraphic characteristics.

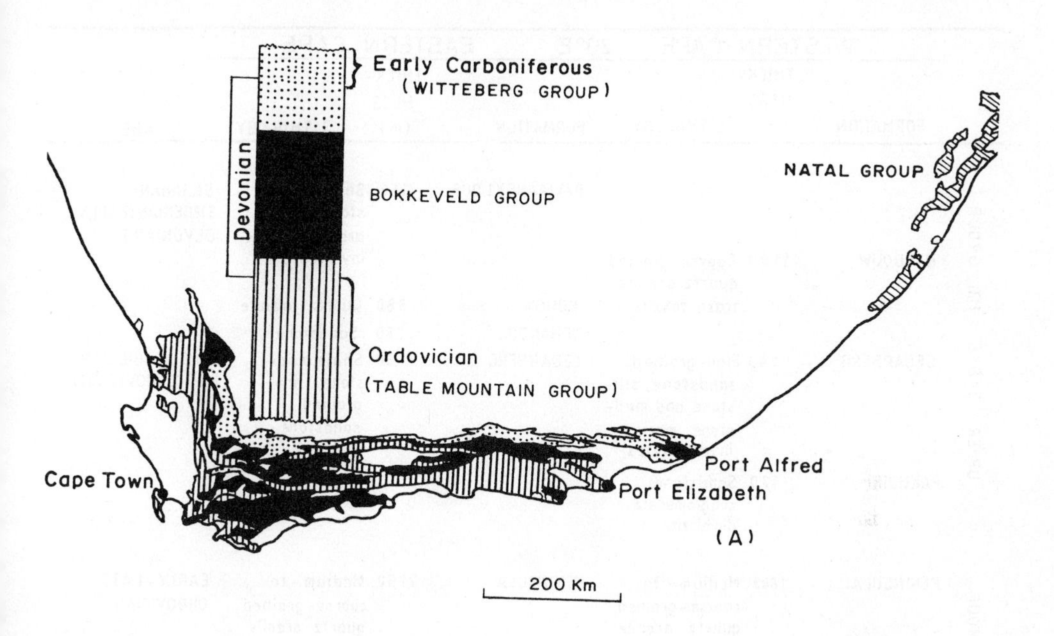

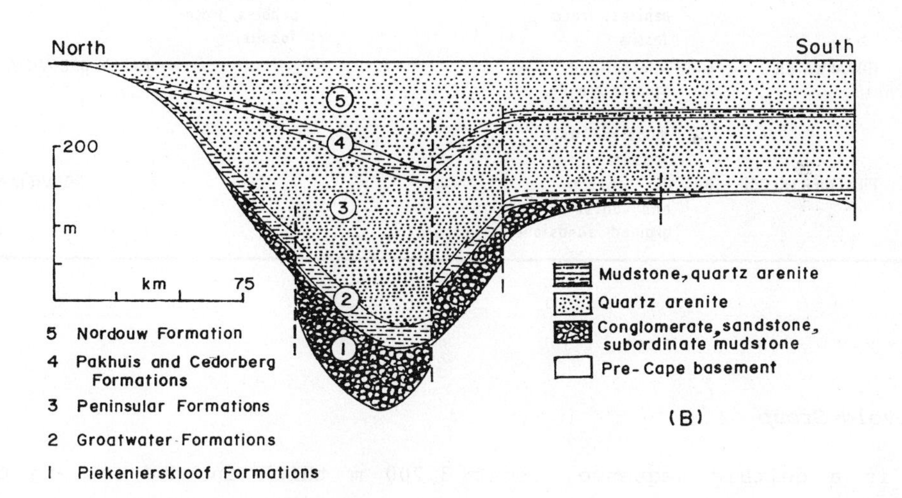

Figure 8.42: Occurrence of the Devonian in South Africa (A); and N-S section of the Table Mountain Group. (Redrawn from Hiller and Theron 1988; Tankard et al., 1982.)

Table 8.4: Lithostratigraphy of the Table Mountain Group. (Redrawn from Tankard et al., 1982.)

	WESTERN CAPE		20°E		EASTERN CAPE		
	FORMATION	THICKNESS (m)	LITHOLOGY	FORMATION	THICKNESS (m)	LITHOLOGY	AGE
UPPER TABLE MT. GROUP				BAVIAANSKLOOF	150	Shale, mudstone, quartz arenite, marine invertebrates	SILURIAN-SIEGENIAN (EARLY DEVONIAN)
	NARDOUW	1100	Coarse-grained quartz arenite, trace fossils	KOUGA	380	Quartz arenite	
				TCHANDO	280	Sandstone	
	CEDARBERG	140	Fine-grained sandstone, siltstone, and mudstone, marine invertebrates	CEDARBERG	50	Shale, mudstone, fine-grained sandstone	LATE ASHGILLIAN (END ORDOVICIAN)
	PAKHUIS	120	Sandstone, conglomerate, diamictite				
LOWER TABLE MT. GROUP	PENINSULA	1800	Medium- to coarse-grained quartz arenite with quartz pebbles, trace fossils	PENINSULA	2150	Medium- to coarse-grained quartz arenite with quartz pebbles, trace fossils	EARLY-LATE ORDOVICIAN
	GRAAFWATER	440	Interbedded quartz arenite, siltstone, and mudstone, trace fossils				EARLY ORDOVICIAN
	PIEKENIERSKLOOF	800	Conglomerate and coarse-grained sandstone				EARLY ORDOVICIAN

Bokkeveld Group

This is a deltaic sequence, about 3,200 m thick in the eastern Cape (where subsidence was greatest) and at least 1,500 m thick in the western Cape. According to Hiller and Theron (1988) the Bokkeveld consists essentially of argillaceous horizons which alternate with arenaceous units, each of which is a formation (Fig.8.44A). These alternations represent the vertical stacking of five or six upward-coarsening deltaic cycles (Fig.44B) caused by tectonically-controlled marine transgressions and regressions. Hiller and Theron (1988) adopted the sedimentological

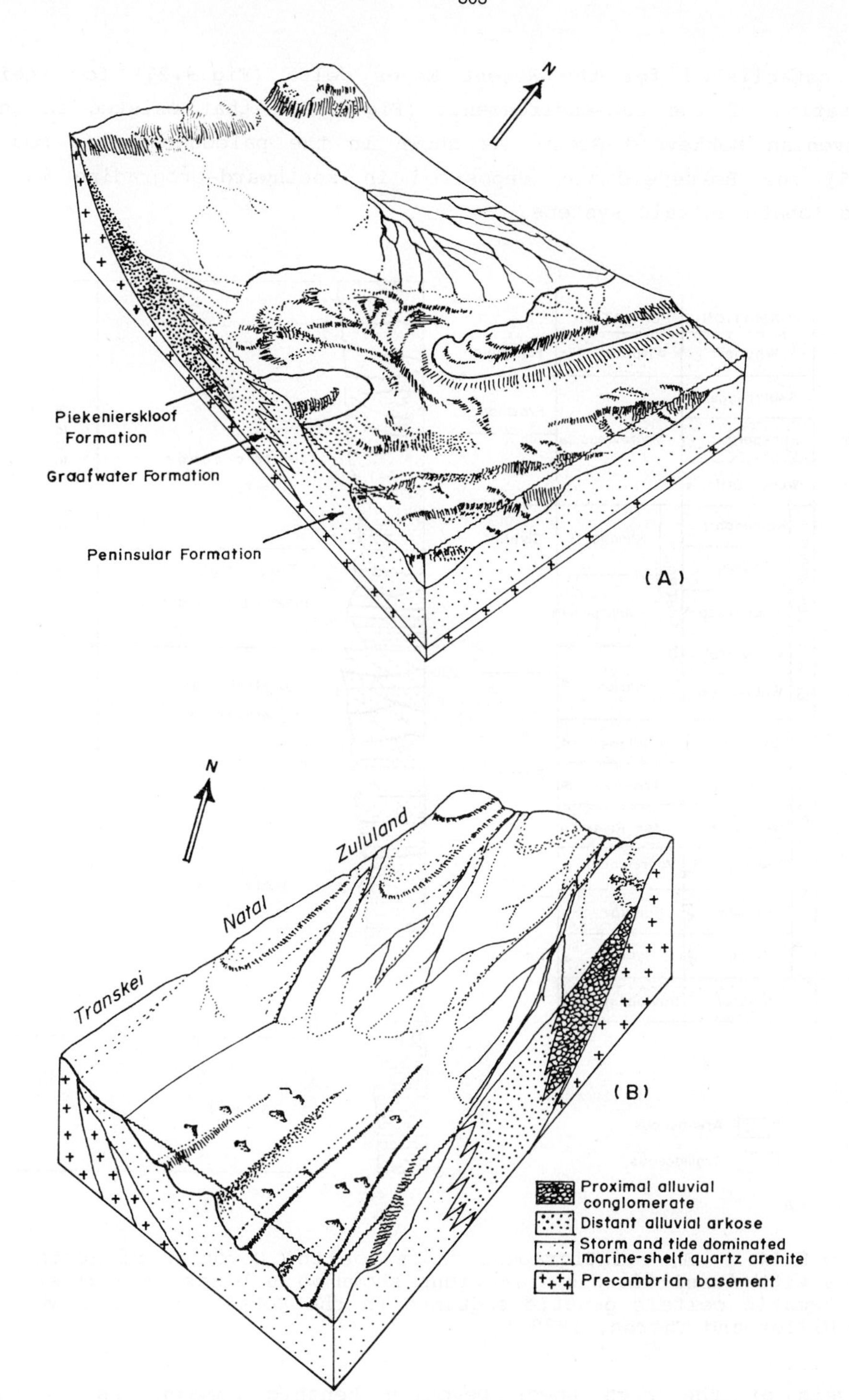

Figure 8.43: Depositional environments of the Table mountain Group (A); and the Natal Group (B). (Redrawn from Tankard et al., 1982.)

criteria established for the Recent Niger delta (Fig.9.27) for their interpretation of the sub-environments (Fig.8.44B) that existed in the Early Devonian Bokkeveld Group. As shown in the paleogeographic model (Fig.8.45) the Bokkeveld was deposited in southward-prograding wave-dominated lobate deltaic systems.

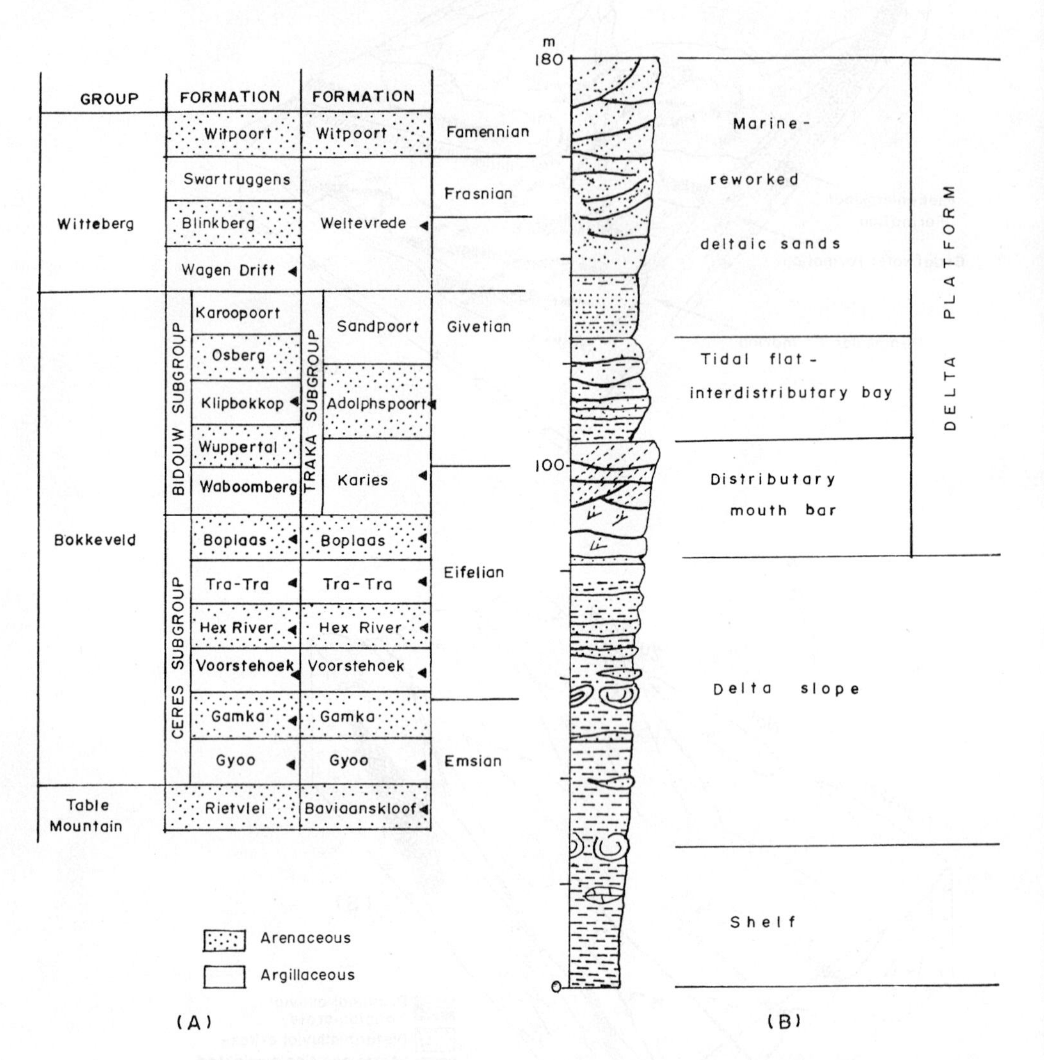

Figure 8.44: A, stratigraphic table for the Devonian of South Africa with fossiliferous formations shown with black triangles; B, schematic celtaic genetic sequence in the Bokkeveld. (Redrawn from Hiller and Theron, 1988.)

Analysis of the rich Lower Devonian benthic communities in the Bokkeveld Group considerably refined paleoenvironmental deductions based on sedimentological criteria (Hiller and Theron, 1988). The Bokkeveld benthic communities belonged to the Malvinokaffric faunal province, unlike coeval communities in southern Ghana which were part of the Appalachian fauna (Fig.8.40). In the Bokkeveld the pro-delta benthic community (Fig.8.46A) was characterized by the most diverse fossil assemblage dominated by thin-shelled, free-lying brachiopods, infaunal pelecypods, gastropods, trilobites, crinoids, and hyoliths. Higher in the stratigraphic sequence, the delta slope paleoenvironment with coarse-grained sandstones and interbedded siltstones contain mostly thicker-shelled brachiopods, and fewer trilobites (Fig.8.46B), whereas the upper shallow deltaic environments such as distributary mouth bars and tidal flats had lower diversity communities (Fig.8.46C). Large thick-shelled brachiopods dominated the distributary mouth bars where they were fixed to the substrate by functional pedicles; while infaunal bivalves and inarticulate brachiopods dominated the tidal flats.

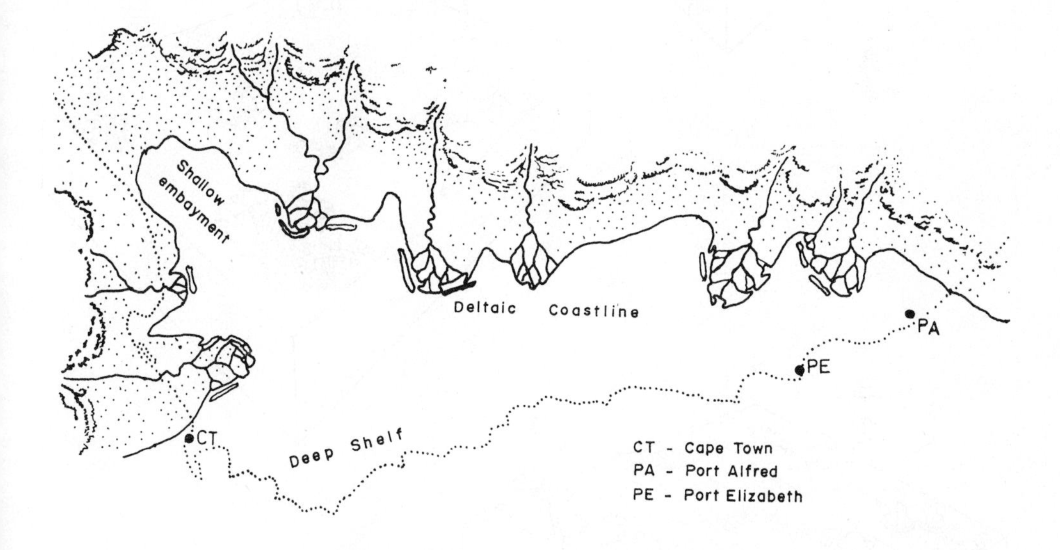

Figure 8.45: Paleogeography of the Bokkeveld Group. (Redrawn from Hiller and Theron, 1988.)

Witteberg Group

Named after prominent mountain ranges in the Cape region where quartz arenites are well exposed, the Witteberg Group is a clastic sequence, over 2,000 m thick. It occupies a transitional stratigraphic position between the Bokkeveld Group below, and the basal Dwyka Formation of the Karoo Supergroup above. Alternating transgressions and regressions, established since Bokkeveld times, also controlled Witteberg deltaic progradation. Thick shelf, delta slope, delta platform with barrier beach,

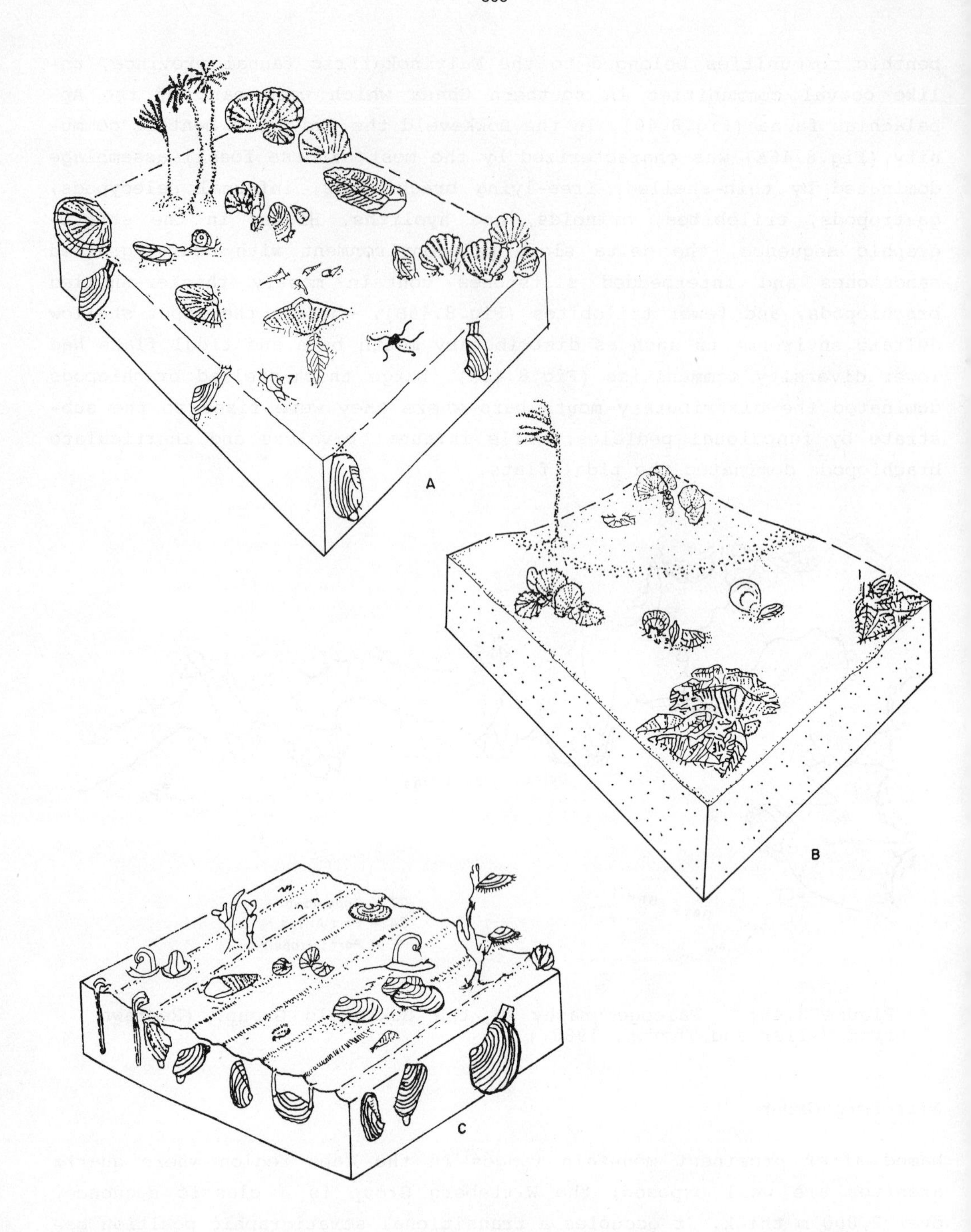

Figure 8.46: Devonian faunal communities of South Africa. (Redrawn from Hiller and Theron, 1988.)

lagoonal, and tidal flat deposits, are found in the Witteberg succession (Fig.8.47) which represents the vertical stacking of these facies.

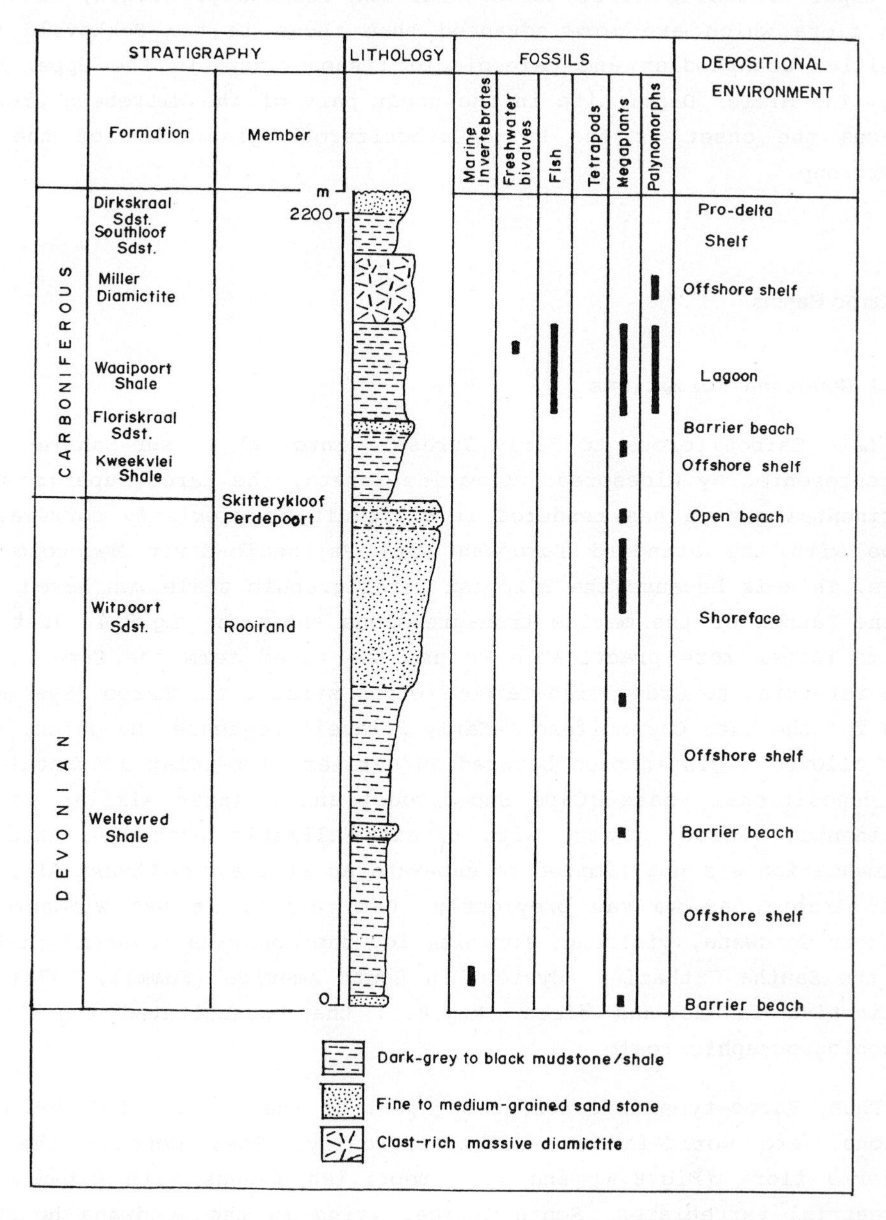

Figure 8.47: Stratigraphy of the Witteberg Group. (Redrawn from Loock and Visser, 1985.)

Because of the paucity of fossils in the Witteberg Group, its age has been somewhat uncertain. But Loock and Visser (1985) assigned the Witteberg to the Late Devonian-Late Carboniferous, on account of its depauperate terminal Malvinokaffric assemblages which occur in the basal part. The upper Witteberg is of terrestrial and freshwater origin, with abundant flora which are more advanced than those in the Bokkeveld below. Primitive acanthodian and paleoniscid fishes occur in the upper Witteberg. The Miller Diamictite in the upper part of the Witteberg Group announces the onset of the Permo-Carboniferous glaciation of the Karoo Supergroup.

8.7 Karoo Basins

8.7.1 Gondwana Formations

The Late Carboniferous to Early Jurassic interval in sub-Saharan Africa is represented by widespread nonmarine strata, the Karoo Supergroup. Its continental nature has rendered it difficult to precisely correlate the Karoo with the standard European Late Paleozoic-Early Mesozoic timescale. This is because the European stratigraphic scale was based on the marine faunas of the marine transgressions shown in Fig.8.4. Just as it became rather more practicable to use the broad term the Cape "System" when referring to Ordovician-Carboniferous strata, the Karoo "System" was used for the Late Carboniferous-Early Jurassic sequence (Haughton, 1969). This allowed a distinction between an earlier rift-related coastal clastic depositional phase (Cape Supergroup) and a later similar but more continental (Karoo) phase, with greater climatic extremes. Karoo-type sedimentation was not limited to sub-Saharan Africa, northeast Africa,and Saudi Arabia, as we saw previously. But rather, it was widespread in southern Gondwana, yielding, for example, the Gondwana "System" in India, and the Santha Catharina "System" in South America (Kummel, 1970), with distinctive faunas and flora (Fig.8.6) that constituted the Gondwana paleobiogeographic realm.

Thus, Karoo-type deposits, being part of the so-called Gondwana formations, are world-famous paleontologically. They contain the *Glossopteris* flora (Fig.8.6) and rich reptilian faunas with pre-mammalian terrestrial vertebrates. South Africa, lying in the Gondwana heartland, had the greatest vertebrate diversity with regional reptilian biozones established for the correlation of Karoo strata. Before examining the habitats of these unique vertebrate assemblages, and other Karoo depositional settings which favoured coal formation, let us first consider the tectonic framework of these basins.

8.7.2 Regional Tectonic Settings

The term "Karoo" is entrenched in African geologic literature, being used in a generic sense for Late Carboniferous to Early Jurassic tectonically and climatically controlled continental sequences. A typical Karoo succession comprises, from the base: tillites; coal-measures; fan-deltaic clastic wedges which interfinger with lacustrine deposits; fluvial and eolian beds; and extensive basalt flows (Haughton, 1969; Kreuser and Markwort, 1989; Nichols and Daly, 1989; Tankard et al., 1982; Wescott, 1988). Karoo basins are of three types. The main Karoo basin which extends in an east-west direction across South Africa (Fig.8.48) subsided as a foreland basin because of prolonged regional compression and uplift in the Cape fold belt (Daly et al., 1989). Outside the Karoo foreland basin are shallow, broad intracratonic sag basins to the west (Rust, 1975). These sag basins are known in South Africa, Botswana, Namibia, Angola, Zaire and Gabon. The third and eastern group of Karoo basins according to Rust (1975) are the narrow grabens and half-grabens or troughs which occur in eastern Africa, for example in Tanzania, Kenya, Uganda, Zambia, Zimbabwe and Madagascar (Fig.8.48). These Karoo troughs resulted from a long period of regional crustal extension which preceeded the fragmentation of Gondwana in the Late Jurassic-Early Cretaceous.

The most extensive Karoo rifts in eastern Africa are the Mid-Zambezi and the Lower Zambezi rifts which trend roughly east-west; the NNE-SSW-trending Luano, Lukusashi and Luangwa rifts; and the Maniamba rift and the Ruhuhu trough of Tanzania, with the latter connecting with coastal Karoo basins (Kent et al., 1971).

Daly et al. (1989) attributed the Karoo rifts in eastern Africa to crustal extension caused by sinistral strike-slip motion along the Mwembeshi shear zone (Fig.8.41) during the Permo-Triassic Gondwana orogeny. Strike-slip motion caused the complex fault pattern found in Karoo rifts. For example, the ENE-trending fault-bounded Kafue, Luano, Ruhuhu and Maniamba rift basins are asymmetric half-grabens with major border faults. They also contain smaller internal faults and half-grabens which are pull-apart structures caused by ENE-WSW sinistral strike-slip motion along the Nwembeshi shear zone. Also compatible with this transcurrent movement are the en échelon half-grabens found in the Lukusashi and Luangwa rifts. Geophysical estimates of stratigraphic thickness in Karoo rifts suggest a minimum of 5 km for the Kafue, Luano, Ruhuhu, Lukusashi, Luangwa, and Lower Zambezi basins (Daly et al., 1989; Orpen et al., 1989).

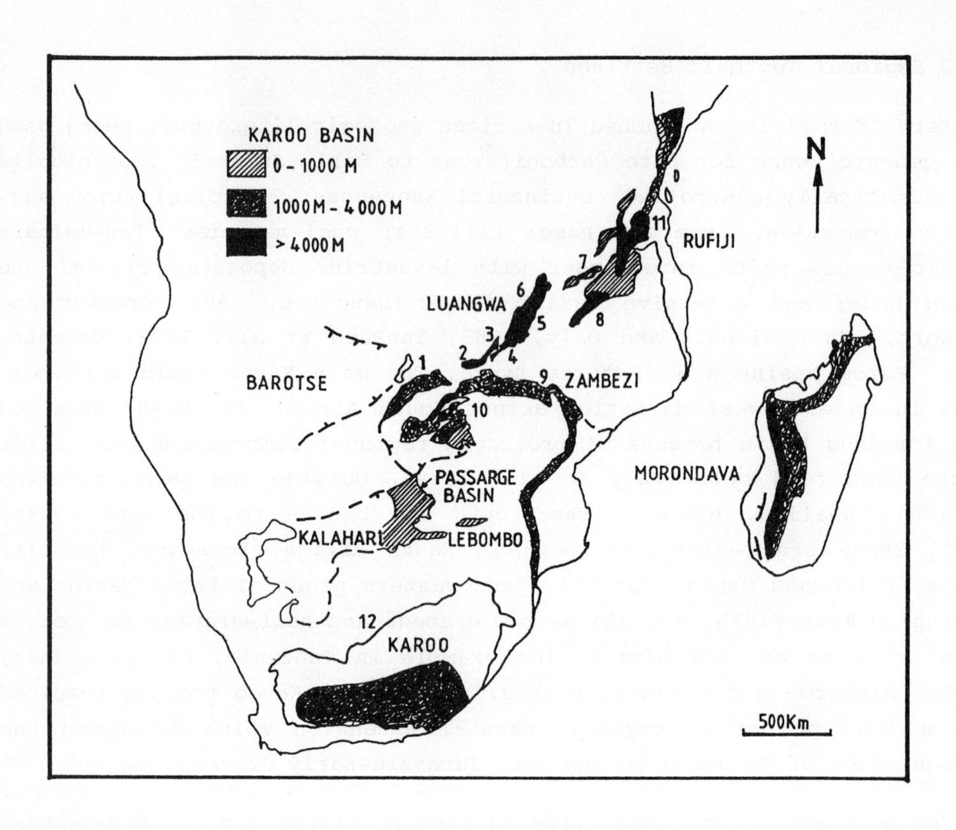

1. KAFUE
2. LUANO
3. LUKUSASHI
4. SOUTH LUANGWA
5. MID- LUANGWA
6. NORTH LUANGWA
7. RUHUHU
8. MANABA
9. LOWER ZAMBEZ
10. MID - ZAMBEZI
11. RUFIJI
12. KAROO FORELAND

Figure 8.48: Distribution of major Karoo basins of southern Africa. (Redrawn from Daly et al., 1989.)

8.7.3 The Karoo Foreland Basin of South Africa

The type area for the Karoo Supergroup is in the Cape Province of South Africa (Fig.8.49A) where nearly horizontal continental sandstones and shales are exposed intersected by dolerite sheets and dykes in a region referred to as the "Karoo" by early explorers. Being the type area, this region which lies in the Karoo foreland basin, will be described first.

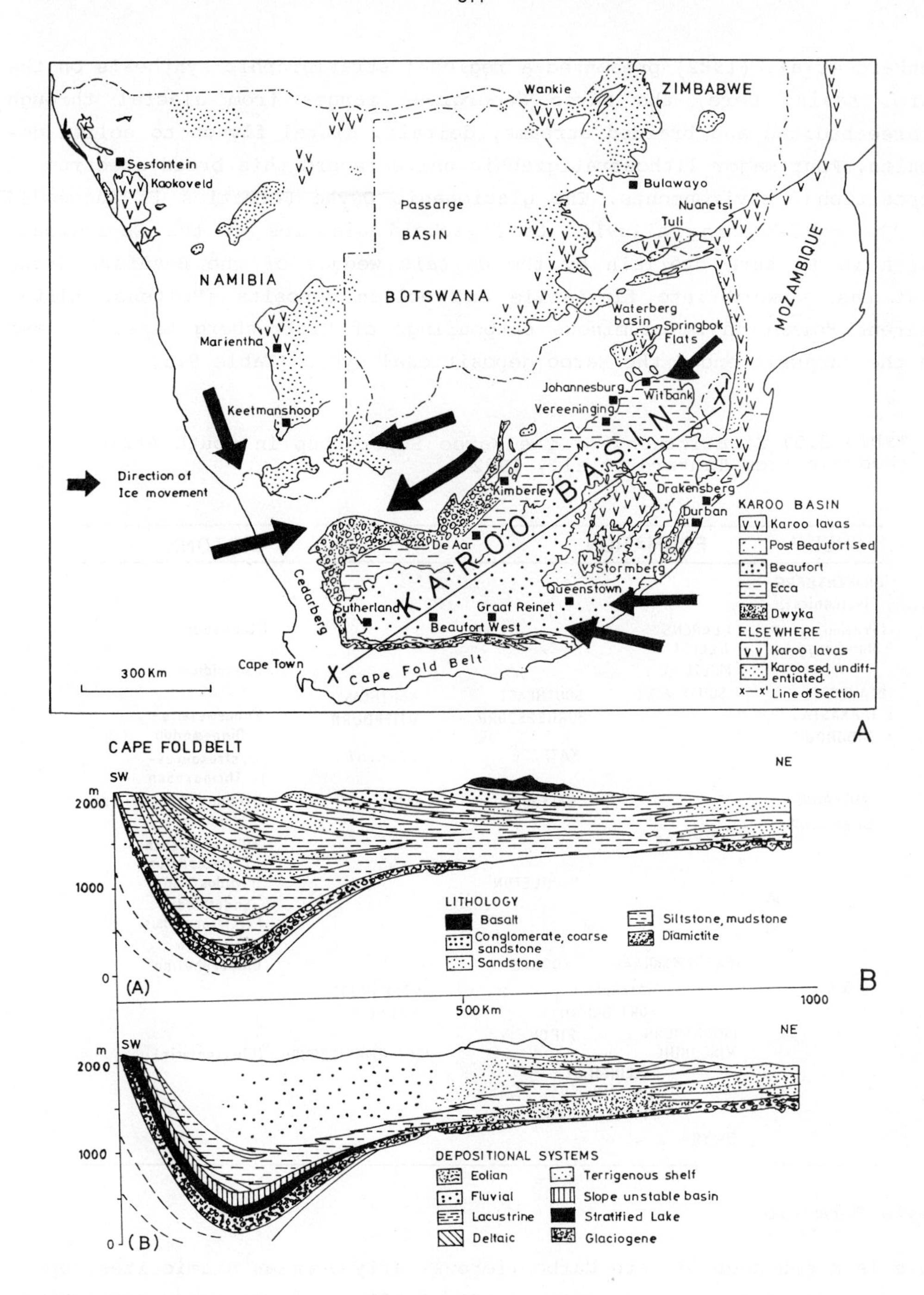

Figure 8.49: Karoo basins of South Africa (A) and cross-sections showing major geologic units (B). (Redrawn from Tankard et al., 1982.)

Tankard et al. (1982) presented a regional stratigraphic synthesis on the Karoo basin. Here, the Karoo Supergroup ranges from glacial through coarse bedload and braided streams, deltaic, distal flysch to eolian deposits. Four major lithostratigraphic units record this broad spectrum of depositional environments. The glaciogenic Dwyka Formation is succeeded by the coal-bearing alluvial to flyschoid clastics of the Ecca-Group, which is in turn overlain by the deltaic wedges of the Beaufort Group that pass upward into fluviatile and eolian deposits (Molteno, Elliot, Clarens Formation). Voluminous outpourings of Drakensberg basaltic lava in the Jurassic ended the Karoo depositional cycle (Table 8.5)

Table 8.5: Stratigraphy of the Karoo Supergroup in South Africa. (Redrawn from Tankard et al., 1982.)

GROUP	FORMATION			BIOZONE
DRAKENSBERG (volcanics)				
(Previously Stormberg)	CLARENS			(Dinosaurs)
	ELLIOT			
	MOLTENO			(Dicroidium)
BEAUFORT	SOUTHWEST	SOUTHEAST	NORTHEAST	
TARKASTAD SUBGROUP		BURGERSDORP	OTTERBURN	Kannemeyeria-Diademodon
		KATBERG	BELMONT	Lystrosaurus-Thrinaxodon
ADELAIDE SUBGROUP		BALFOUR	ESTCOURT	Dicynodon lacerticeps-Whaitsia
	TEEKLOOF			Aulacephalodon-Cistecephalus
		MIDDLETON		Tropidostoma-Endothiodon
				Pristerognathus-Diictodon
	ABRAHAMSKRAAL	KOONAP		Dinocephalian
ECCA	WATERFORD		VOLKSRUST	
	FORT BROWN		VRYHEID	
	LAINGSBURG	RIPON		
	VISCHKUIL		PIETERMARITZBURG	(Glossopteris)
	COLLINGHAM			
	WHITEHILL			
	PRINCE ALBERT			
	DWYKA			

Dwyka Formation

This is a sequence of Late Carboniferous-Early-Permian diamictites, up to 700 m thick in the Cape depocentre (Fig.8.49B), but wedging out northward over Precambrian basement. The diamictite, sometimes up to six units, include pebble- and boulder-size clasts of local and distant origin, supported by silty and clay matrix; fine-grained cross-laminated glacial outwash and glacio-lacustrine siltstones, and varved carbonaceous shales. These accumulated mostly as ground moraines and as terrestrial and subaqueous terminal moraines, recording several cycles (up to nine) of glacial advance and retreat (Fig.8.50,B). Each cycle comprises an upward-fining diamictite overlain by glacio-lacustrine shales. Sporadic occurrences of plant and palynomorph-bearing shales within the diamictites especially in the northern part of the Karoo basin suggests that the Karoo basin and the surrounding areas locally supported Tundra conditions with plant cover.

Based on ice movement directions inferred from glaciated pavements, the orientation of tillite-filled glacial valleys, exhumed roches moutonnées with smoothly polished and striated upstream surfaces and jagged glacially plucked downstream sides, the centres of maximum glaciation from which ice moved are believed to have been located in Namibia, Transvaal, Natal, and possibly in the Atlantic area also (Fig.8.50,B).

Ecca Group

Major changes in the paleoclimatic, paleo-tectonic and paleogeographic setting of the Karoo basin took place in the Early to Middle Permian. This was during the deposition of the Ecca Group. This group accumulated in a wide range of paleoenvironments: from alluvial plain through fluvio-deltaic to a deep trough (Fig.8.50,C,D). First, as South Africa drifted from the South Pole and ice sheets melted, a broad epicontinental sea encroached into the Karoo basin from the west. This transgression is represented by an extensive marker unit of fossiliferous and phosphatic shales, the Whitehill Formation (White Band) which rests conformably on the Dwyka Formation, and contains paleoniscid fish, gastropods, the pelecypod *Eurydesma*, *Orthoceras*, and *Eosianites*, brachiopods, echinod spines, crinoids, radiolaria, and foraminifera.

While flyschoid deposits (Collingham, Ripon, Laingsburg, Pietermaritzburg Formations) with turbidites accumulated in the Karoo trough (at about 500 m maximum bathymetric depth), stable shelf prograding deltaic complexes of the Vryheid Formation were deposited along the margins of the Karoo trough (Fig.8.50C). Further landward, along the northern basin margin, lay extensive alluvial coal swamps.

Of the nineteen Karoo coalfields that are mined in South Africa, with 81 billion tons of recoverable bituminous coal, eighteen of these lie in the Ecca Group; one lies in the Molteno Formation above (Tankard et al., 1982). The Ecca coal basins in South Africa include the Limpopo, Sout-

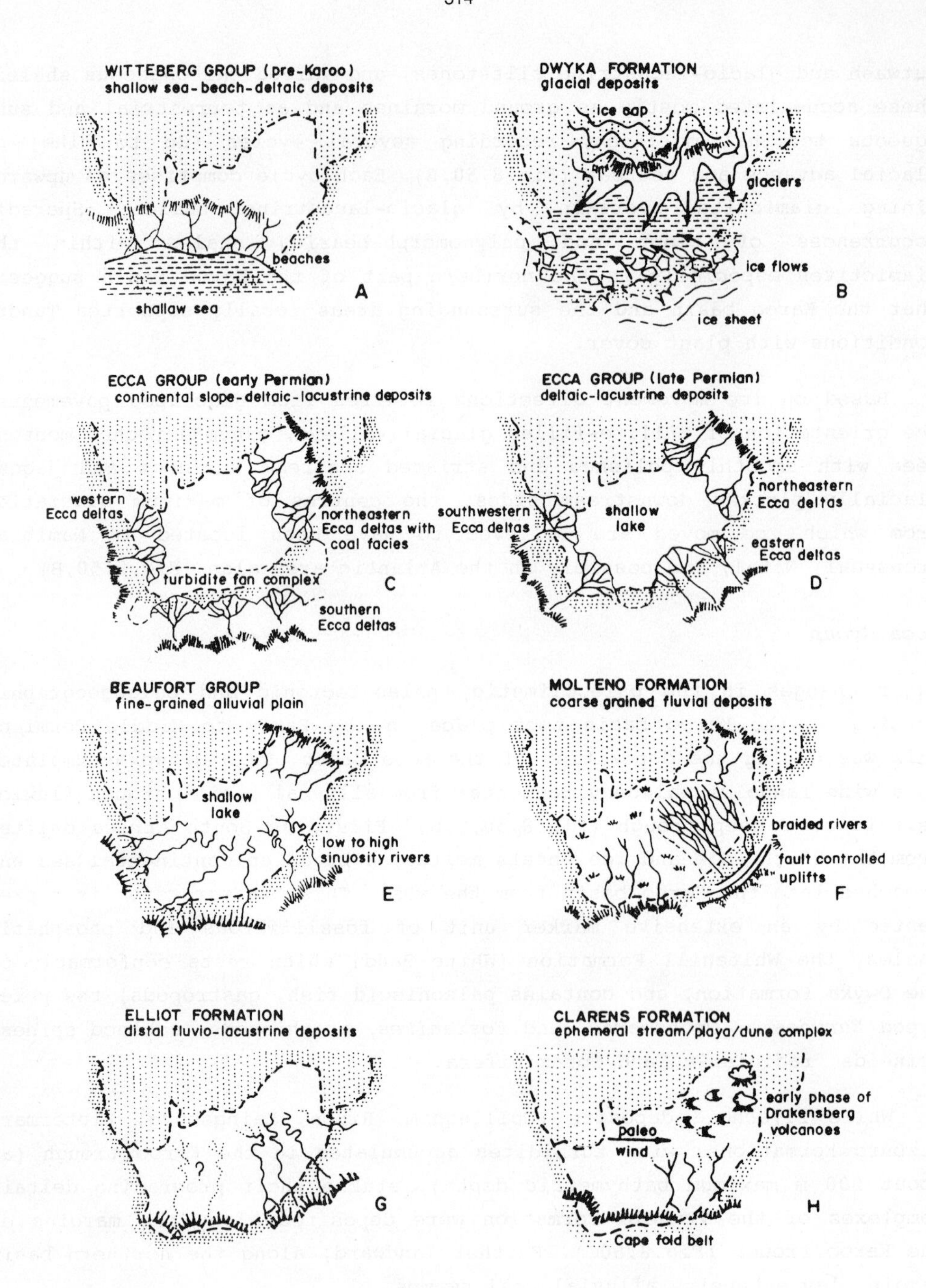

Figure 8.50: Karoo depositional models. (Redrawn from Smith, 1990.)

pansberg-Lebombo, Waterberg, Springbok Flats, and the northern Karoo basins (Witbank, eastern Transvaal, Natal) coalfields (Fig.8.49A). In these basins the distribution of coal was determined by the sedimentary facies. Coal was mostly restricted to paraglacial, upper deltaic plains, strand-plains and alluvial valleys (Caincross, 1989). Another controlling factor was the rate of subsidence, with economic coal measures occurring in stable platform areas around Witbank, northwest of Natal, whereas the less stable areas in Natal to the east, contain numerous but impersistent and thin coal seams within a considerably thickenend deltaic sequence.

Beaufort Group

This is the most extensively exposed of all Karoo strata in South Africa, covering about 200,000 km^2 (Fig.8.49A) with a maximum thickness of about 3 km in southwestern Karoo basin. The Beaufort Group of Late Permian to late Early Triassic age, is a thick sequence of fossiliferous alluvial clastic wedges that accumulated in an east-trending depocentre which had shifted northward from the Ecca depocentre (Fig.8.50,E). Uplift of the Cape fold belt to the south supplied regressive clastic wedges into the Beaufort foredeep basin (Fig.8.49B), while subordinate amounts of quartzitic clastics came from locally elevated Precambrian sourcelands which lay to the north and to the east. Consequently, the Beaufort Group thins northward. The Beaufort Group is famous for its distinctive reptilian fossils upon which its Late Permian-Early Trisassic biozonation is based (Table 8.5).

The Beaufort is made up of two subgroups. The Adelaide Subgroup comprises mostly mudstones which accumulated in distal (in relation to alluvial sediment source areas) inter-channel floodplains, bordered by natural levees and in extensive freshwater lakes (Yemane and Kelts, 1990). Its lateral equivalent, the Tarkastad Subgroup, with more sandy lithofacies, represents predominantly fluvial channel environments showing upward increase in bedload lithologies that resulted from the migration of high-energy proximal fluvial environments. The differences between Beaufort formations in the western and eastern parts of the Karoo foredeep merely reflects varying intensities of terrigenous influx into shrinking lake depositories.

Uranium Mineralization. In the highly arkosic Abrahamskraal Formation at the base of the Beaufort Group in the southwestern part of Karoo basin is a sequence of upward-fining meandering channel sandstones and inter-channel flood basin interlaminated siltstones and mudstones. Uranium mineralization occurs in the channel sandstones, especially in elongate or tabular calcareous sandstone pods which are rich in plant remains especially along the base and lateral margins of the channels. Uraninite, complex silicates of the coffinite type, pyrite, arsenopyrite, molybdenum, phosphorus and urano-organic compounds occur together. Interbedded volcaniclastics and basement granites are believed to have supplied uranium which was released through weathering, soil formation, and by subsequent flushing into the groundwater system. Mildly reducing groundwater transported the resulting uranyl carbonate complexes preferentially into the more permeable coarse-grained basal portions of the channel sandstones. Local controls over uranium mineralization were exercised by reducing conditions created by concentrations of carbonaceous debris and pyrite, and by permeability barriers provided by adjacent or interlayered argillaceous units.

Uranium mineralizations in the Karoo are of the sandstone-hosted, pelite-hosted, as well as coal-hosted, and vein types (Roux and Toens, 1987). These are also found in Karoo deposits in Zimbabwe, Zambia, Madagascar, Angola, and in Gabon where uraniferous black shale occurs.

Upper Karoo Formations

Previously termed the Stormberg Group, the upper Karoo formations comprise the basal Late Triassic Molteno Formation, a coarse fluvial molasse wedge that was derived from the uplifted Cape fold belt. The Molteno Formation contains the richest and best-known Triassic floras, the animal fossils being represented by abundant fossil insects. The Molteno represents deposits which are typical of a braided river system of higher energy than the underlying upper Beaufort, that were laid down in a series of alluvial fans under cool, braided conditions with permanent ice in the mountain heartland (Cruickshank, 1978).

The conformably overlying Elliot Formation is a sequence of redbeds which contain diamond placers in Swaziland. At the top of the Karoo sedimentary sequence is the Early Jurassic Clarens Formation which consists of eolian sandstones and associated fossiliferous playa lake sheet flood, and ephemeral stream deposits. Large-scale cross-bedded sandstones in the middle of the Clarens Formation has been interpreted as migrating transverse dune deposits with barchan dunes represented by large trough sets which accumulated under desert conditions with paleowinds blowing from the west (Fig.8.50,H).

However, less severe climatic conditions are represented in the basal part of the Clarens Formation which contains numerous lake deposits locally carrying freshwater fish, crustaceans, dinosaurs and dinosaur footprints.

Karoo sedimentation ended with the outpouring of vast basaltic lavas of the Drakensberg plateau, the Stormberg range, and of the Zimbabwe and Lebombo-Nuanetsi volcanic provinces. These lavas represent a remarkably consistent magmatic event in the Gondwana continents that heralded the break-up of the supercontinent.

8.7.4 Other Karoo Basins

Outside South Africa continental deposits of the Karoo cycle are widespread in intracratonic basins south of the Sahara. Descriptions of a few of these basins, especially the rifts, are presented below as well as their regional correlations.

Ruhuhu Basin

The Ruhuhu rift basin of southwestern Tanzania (Fig.8.48) shows the typical Karoo succession in East Africa. It is complete, with Permo-Carboniferous glacial and periglacial deposits, thick coal-bearing fluvial facies, followed by a succession of braided stream deposits, playa redbeds, and fluvio-lacustrine strata. Kreuser and Markwort (1989) described the stratigraphy of the Ruhuhu basin in detail. As already stated (Ch. 8.7.2), the Ruhuhu basin is one of the rift basins which opened as a result of strike-slip movement along the Mwembeshi shear zone (Fig.8.41). The basin evolved through three phases (Kreuser and Markwort, 1989).

Deposition began in the Early Permian when a basal diamictite (Fig.8.51) with up to three tillite horizons was deposited. These have been attributed to mountain glaciers which issued into lowland glacial lakes. The overlying lower and upper coal measures consist of cyclical fluvial sequences. Each sequence begins with an erosive base followed by strongly cross-stratified coarse clastics with coal and silicified wood fragments. The cycle fines upward into sandstone-siltstone, and ends with a mudstone/coal facies. This facies sequence has been interpreted as a succession of braided fluvial channels which switched to meandering river point-bars with crevasse splay, levee, overbank deposits, ending with swampy vegetated ox-bow lakes and floodplains (Fig.8.52A).

Then followed a tectono-sedimentary cycle which was caused by moderate uplift of the basin margins during which arkosic clastic wedges and fine clastic redbeds accumulated (Fig.8.52B). Monotonous alternations of fluvial sandstones and lacustrine mudstones and oolitic limestone and marl lenses resulted from the fluctuations of the shorelines of giant lakes (Fig.8.52C). Up to K6 beds (Fig 8.51) were deposited during this phase of fluctuating lake levels. Similar and very thick sequences of mainly shallow lacustrine marls and fine siliciclastics occur in neighbouring Zimbabwe where they are known as the Madumabisa Mudstone (Table 8.6) in Zambia, Kenya ("Maji ya Chumvi Beds") and in Malawi, suggesting the existence of vast inter-connected shallow lakes under humid warm climate. These lacustrine deposits are potential petroleum source beds in the Ruhuhu basin (Kreuser et al., 1988). They also provided reptilian habitats.

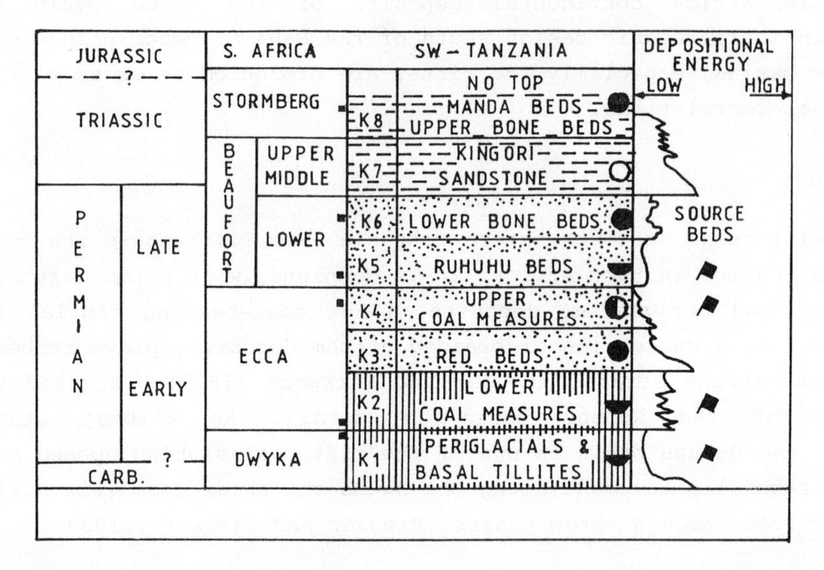

Figure 8.51: The Karoo of the Ruhuhu basin in Tanzania. Black squares, intervals dated with palynomorphs and vertebrates; black circle, red beds, half moon, gray/black colours; open circle, light colours. (Redrawn from Kreuser and Markwort, 1989.)

The Lower Bone Beds (K6) in the Ruhuhu basin mark a change in depositional pattern. In this unit nodular and thin limestone horizons and bone beds are found in a predominantly mudstone lithology of playa origin with increasing desiccation and occasional floods which supplied coarse clastic lenses (Fig.8.52C). The bone beds contain in situ skeletons and skulls including those of the sauropod *Dicynodon lacerticeps* which has been correlated with the zone that bears this taxonomic name in the uppermost Permian of the Beaufort Group in South Africa.

The terminal Karoo depositional cycle in the Ruhuhu basin, represented by the markedly disconformable Kingori Sandstone and Manda Beds,

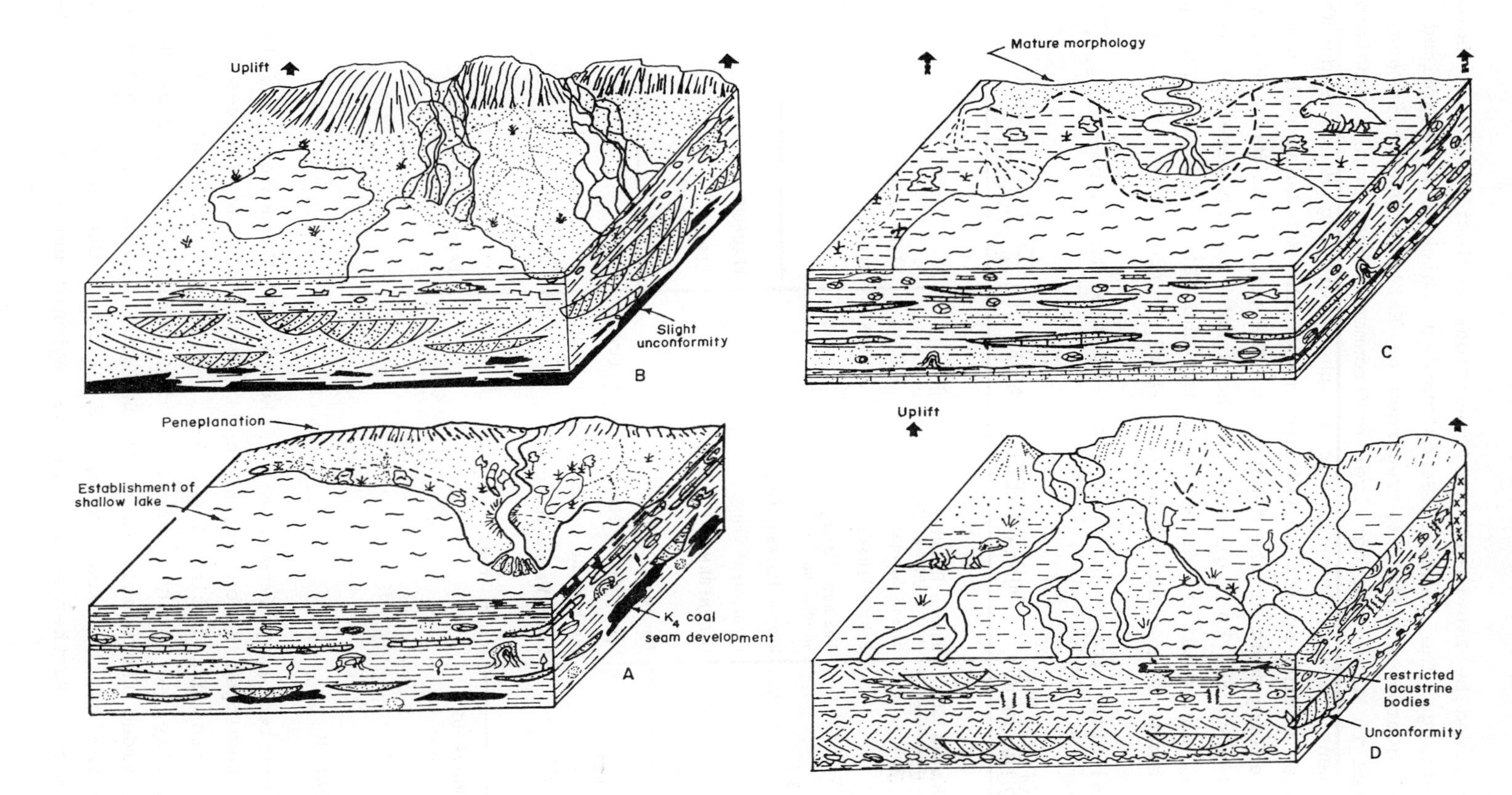

Figure 8.52: Karoo depositional models for the Ruhuhu basin. A, lacustrine model /K4/K5 sequence); B red beds (k3); C, model for K5/k6; D, model for k7/k8 unit. (Redrawn from Kreuser and Markwort, 1989.)

was characterized by fluvio-deltaic conditions reflecting a new tectono-climatic setting. The Kingori Sandstone, a thick coarse-grained and cross-bedded quartz arenite, is regarded as Early Triassic in age, and was deposited in a meandering river environment under humid conditions. Numerous extensive cyclothems occur in the overlying Manda Beds which were deposited during uplift of the margins of the Ruhuhu trough and subsidence and enlargement of the depocentre. The top of the Manda Beds were later eroded when the half grabens in the Ruhuhu trough were uplifted to the southeast and faulted, probably during the Late Jurassic. The upper part of the Manda Beds contains Early Triassic vertebrates.

Table 8.6: Stratigraphy of the Karoo of the Mid-Zambezi basin. (Redrawn from Orpen et al., 1989.)

EUROPEAN TIME EQUIVALENTS	MID-ZAMBEZI BASIN		SOUTH AFRICAN EQUIVALENTS
LOWER JURASSIC	Batoka Basalts	UPPER KAROO	
	MINOR UNCONFORMITY		
TRIASSIC	Forest Sandstone		STORMBERG SERIES
	Pebbly Arkose		
	Fine Red Marley Sandstone		
	Ripple Marked Flags		
	Escarpment Grit		BEAUFORT SERIES
	UNCONFORMITY		
PERMIAN	Madumabisa Mudstone	LOWER KAROO	
	Upper Wankie Sandstone		ECCA SERIES
	Black Shale and Coal Group		
	Lower Wankie Sandstone		
CARBONIFEROUS	Glacial Beds		DWYKA SERIES

Morondava Basin

Before it drifted away from Somalia, Kenya and Tanzania the island of Madagascar developed Karoo rifts on its western side (Reeves et al., 1987). The Morondava basin (Fig 8.53) resulted from crustal extension which eventually led to the separation of Madagascar from Africa. Basaire (1972) subdivided the nearly 12-km thick Karoo sequence in the Morondava basin into the basal Sakoa Group, the Sakamena Group, with the Isalo Group at the top. Further details of the tectonic and stratigraphic evolution of the above sequence were furnished by Nichols and Daly (1989), and by Wescott (1988). During sedimentation in the Morondava basin the regional paleoslope was to the west and southwest towards a lake basin which lay to the southwest and was intermittently inundated by the sea which lay further south.

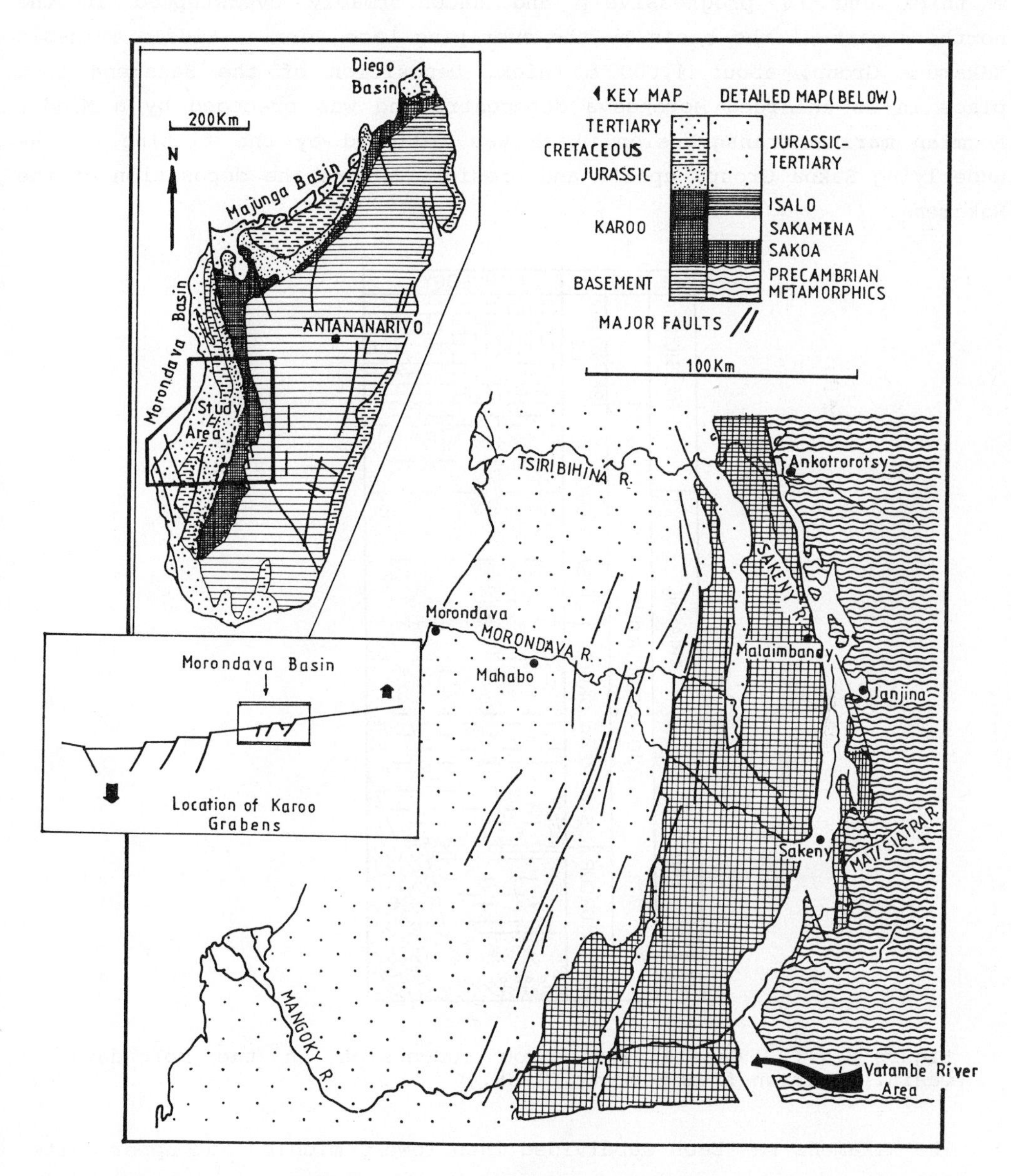

Figure 8.53: Geology and stratigraphy of the Morondava basin in Madagascar. (Redrawn from Nichols and Daly, 1989.)

Although tillites, coals and limestones occur in the Sakoa Group in the southern part of the Morondava basin, this is a predominantly fluviatile sequence (Fig.8.54) deposited when the basin was primarily one of internal drainage broken by series of horsts into sub-basins. The Sakoa Group which is of Late Carboniferous to Early Permian age is about 2,000 m thick and is progressively and unconformably overstepped in the northern part of the basin by the overlying Late Permian-Middle Triassic Sakamena Group, about 4,000 m thick. Deposition of the Sakamena took place in an enlarged Morondava depocentre and was preceded by a Middle Permian marine transgression which was followed by the tilting of the underlying Sakoa Group, uplift and erosion, before the deposition of the Sakamena.

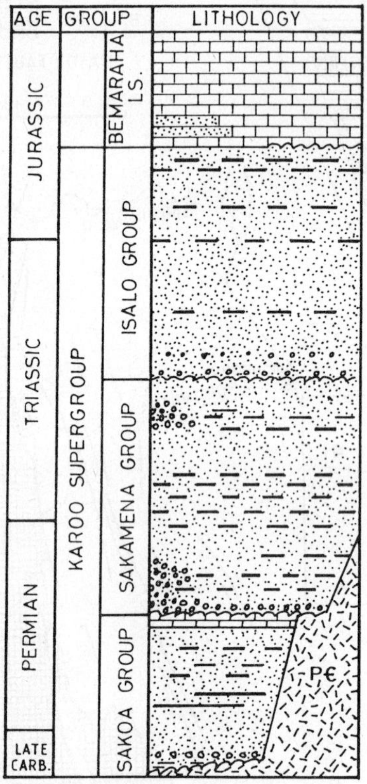

Figure 8.54: Generalized Karoo succession in the Morondava basin. (Redrawn from Wescott, 1988.)

The Sakamena has been subdivided into lower, middle, and upper units. According to Wescott (1988) the Lower Sakamena Group is characterized by very coarse-grained facies reflecting renewed tectonic activity along border faults. Braided-stream sandstones, littoral clastics and stromatolitic limestones of lagoonal, algal reef and shallow shelf origin occur in the Lower Sakamena. The Middle Sakamena, predominantly fossiliferous mudstones, siltstones and fine sandstones, is shallow marine in the basal part as a result of an Early Triassic marine transgression (Wright and Askin, 1987). Fluvial systems discharged from the east into the shallow sea. Wright and Askin placed the Permian-Triassic boundary approximately at the Lower-Middle Sakanema transition. A return to predominantly fluvial sedimentation took place in the Upper Sakamena which contains alternations of cross-bedded gravelly sandstones and sandy mudstones. The overlying Triassic to Middle Jurassic Isalo Group, about 5-6 km thick, consists mostly of alternations of cross-bedded, gravelly sandstones and sandy mudstones, all of which are of fluvial origin.

Mid-Zambezi Basin

Returning to the African mainland, another representative Karoo basin is the Mid-Zambezi basin (Fig.8.48) of Zimbabwe. Here the stratigraphic succession estabished by Bond (1967), also reflects the dominant control of changing climate and vertical tectonic movements. Table 8.6 shows a typical Karoo successon starting from the Lower Karoo with diamictites and varved mudrocks; followed by the pereglacial Lower Wankie Sandstone which grade upward into argillites containing thick economic coal seams of the Black Shale and Coal Group. The *Glossopteris*-bearing Upper Wankie Sandstone was deposited during renewed coarse clastic influx, followed by the Upper Permian Madumabisa Mudstone. After a hiatus, the Upper Karoo begins with conglomerate beds known as the Escarpment Grit following which sedimentation appears to have proceeded with uninterrupted deposition of sandstones, conglomerates and mudstones of the Ripple-Marked Flags, and the overlying clastic beds, including the eolian Forest Standstones.

In addition to *Glossopteris* and related flora, the Madumabisa Mudstone contains ostracodes, freshwater pelecypods, and reptilian bones similar to those of the Adelaide Subgroup of the Beaufort Group.

Regional Karoo Correlations

Karoo-type nonmarine beds lie along the coast of East Africa in grabens (Kent et al., 1971) where Permian marine sequences also appear, for example in the Kidodi basin of Tanzania, the Lamu embayment in Kenya; and in the Mandawa basin in Tanzania Permo-Triassic evaporites indicate restricted marine incursions (Nairn, 1978). In the coast of Kenya over 6.8 km of Karoo deposits lie in the subsurface (Mbede, 1984).

In southern Africa the Karoo succession in the Passarge basin of Botswana (Fig.8.48) and those in Namibia are direct stratigraphical extensions of the main Karoo basin in South Africa, both lithologically and paleontologically (Haughton, 1963; Nairn, 1978; Tankard et al., 1982). Strata equivalent to those described above in the Ruhuhu basin are believed to occur extensively in eastern and central Africa at least in the basal part of other Karoo rift basins in Zambia, Malawi, Tanzania, Uganda, Mozambique, Kenya and in Somalia. They also underlie the Zaire basin where they are referred to as the Lukuga Series. Table 8.7 summarizes palynological correlations of Karoo strata in various parts of sub-Saharan Africa (Kreuser, 1984).

Table 8.7: Palynological correlation of Karoo sequences in southern and central Africa. (Redrawn from Dingle, 1978.)

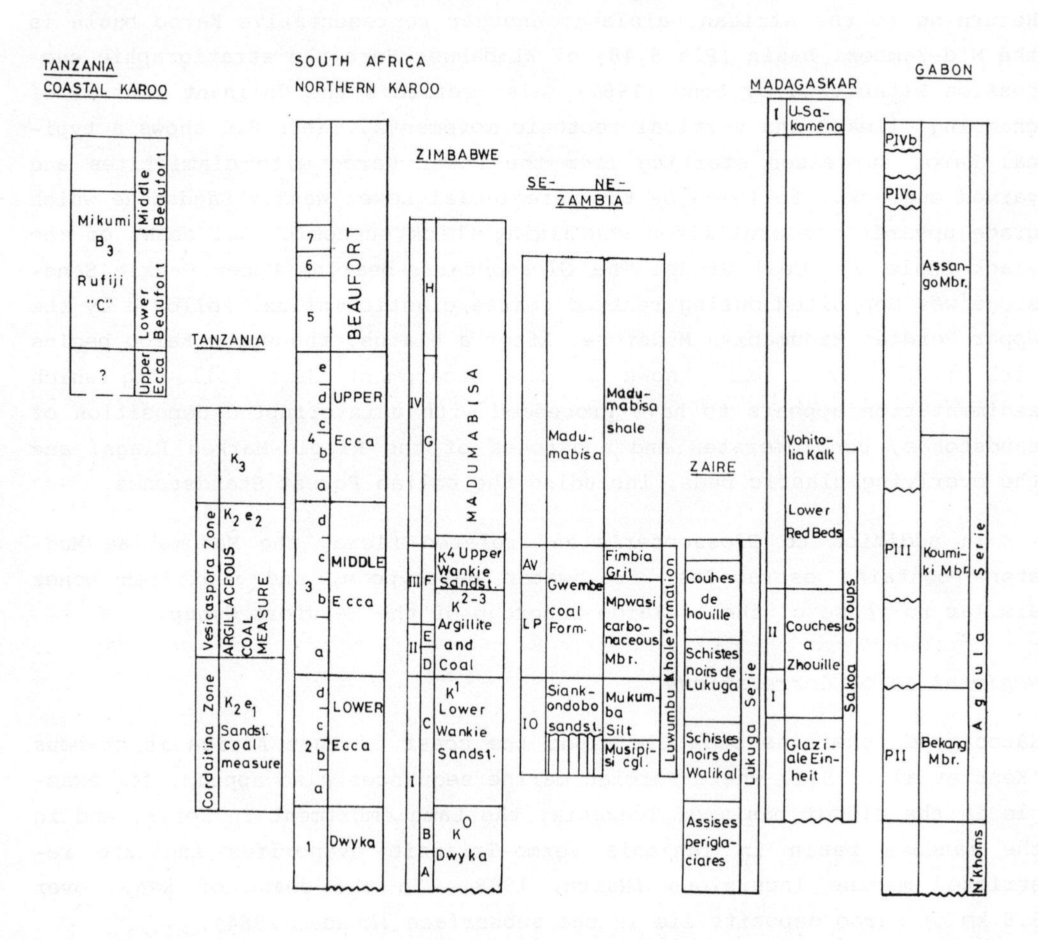

8.7.5 Aspects of Karoo Life

Plant and animal life flourished during Karoo times and left a very impressive record in South Africa that constitutes the most diverse and best known of the Gondwana flora and faunas. Since these have greatly enriched our knowledge of early terrestrial plant life and vertebrate evolution, the highlights of the Karoo paleobotanical and vertebrate record (Fig.8.55) deserve at least passing mention at this point.

Subclass	Orders and suborders	Upper Permian	Scythian: Lystrosaurus zone	Scythian: Cynognathus zone	Middle and Upper Triassic
ANAPSIDA					
SYNAPSIDA	Therapsida				
	Dinocephalia				
	Anomodontia			dominant herbivors	
	Theriodonta (mammal-like reptiles)		Lystrosaurus		
	Gorgonopsia	common, least mammal-like			
	Therocephalia				
	Cynodontia	dominant mammal-like reptiles		Cynognathus	advanced types
DIAPSIDA	(gave rise to dinosaurs, lizards, birds, crocodiles)				earliest dinosaurs ORNITHISCHIA

Figure 8.55: Stratigraphic ranges of some South African Karoo vertebrates. (Redrawn from Dingle, 1978.)

Just as the major Carboniferous marine transgression (Fig.8.4) created widespread epicontinental seas in most parts of the world and led to the formation of coals swamps and coals measures on the Saharan platform and in other low latitude regions of the world, a high latitude coal measures flora was flourishing in Karoo basins (Fig.8.56) and throughout southern Gondwana. The well-known *Glossopteris* plants which appeared during the later phase of the Permo-Carboniferous glaciation, formed the coal measures vegetation (Plumstead, 1969) which continued right into the Triassic. Global vegetation at that time consisted of four major floristic provinces (Fig.8.6), viz; the Euramerican flora, and the Cathaysian (or *Gigantopteris*) flora, both of which existed at the paleo-equatorial latitudes; the northern Angara or Kuznetsk flora; and the southern Gondwana or *Glossopteris* flora (Chaloner and Creber, 1988). Glacial conditions seemed to favour the evolution of the *Glossopteris* flora (Plumstead, 1973), which expanded with the retreating glacier so that the Early to Middle Permian Ecca coal swamps were thickly vegetated with a mixed *Glossopteris* flora. This mixed flora contained many species and genera of leaves and fructifications of *Gangamopteris* and *Glossopteris*, lycopods, a few ferns and pteridosperms, and ginkos. By the late Permian and the early Triassic when the Beaufort Group was deposited, the first signs of a *Dicroidium* flora, a lowland temperate vegetation, had appeared in the Early Triassic. *Dicroidium* flora took over in the Late Triassic Molteno Formation where it beceame the dominant and most diverse element. Interestingly the animal fossils in the Molteno Formation are represented mainly by insects with 25 genera and 32 species (Anderson and Anderson, 1984).

Figure 8.56: *Glossopteris* - *Gangamopteris* forest of Karoo times, South Africa (Redrawn from Plumstead, 1969.)

The Gondwana paleontological record was best summed up by Kummel (1970) as follows: "The most abundant and diversified reptilian fauna is in South Africa, where the Upper Dwyka shales contain the swimming reptile *Mesosaurus*. The Ecca Group bears a *Glossopteris* flora but has yielded no reptilian remains. The overlying Beaufort Group, however, is a veritable graveyard of ancient reptiles in which no less than 600 species have so far been discovered. Within the Beaufort, in fact, six distinct reptilian zones have been recognized, and they greatly aid in correlation. The overlying Stormberg Group also contains a rich reptilian fauna, sharply differentiated from that of the Beaufort. Reptiles are apparently not common in the Gondwana strata of Madagascar. The middle part of the lower group (corresponding to Ecca) has some genera also present in Tanzania. The fauna of the upper part of the series has affinities with that of the Upper Gondwana of India". The above appraisal of the Karoo fauna is even more valid today as more reptilian fossils are known and the evolutionary link between these early reptiles and dinosaurs and mammals are better understood (Halstead and Halstead, 1981).

We also know that *Mesosaurus*, a Late Carboniferous-Permian amphibious reptile that inhabited Karoo basins and similar basins in South America, also lived in the Ecca coal swamps (Plumstead, 1969), as shown in Fig.8.56; and that another reptile, *Hovasaurus boulei* (Fig.8.57A), was the most common vertebrate that lived in the Late Permian of Madagascar (Currie, 1981). During the Late Permian and Early Triassic, Madagascar had no vertebrate faunal similarities with South Africa since it was separated from the African continent by a marine barrier (Battail et al., 1987).

But by far the most fascinating aspects of the Karoo reptilian faunas are their evolutionary trends. The reptiles of the Beaufort Group (Fig.8.55) represent one of the best-preserved ecological assemblages of primitive mammal-like reptiles or paramammals in the world, which show evolutionary transitions to dinosaurs and to mammals. Paleoecologically, the Beaufort accumulated in large inland basins with terrestrial habitats comprising extensive meandering river floodplains and lakes, and heavily dissected lowlands where plant and animal life were probably concentrated around water courses (Smith, 1990). It is believed that heavy rainfall occasionally led to the overflowing of rivers into their floodplains and that the climate became hot or semi-arid with highly seasonal rainfall (King, 1990; Tankard et al., 1982). Plant-eating parammamals (dicynodonts), other reptiles, amphibians, freshwater fish and the *Glossopteris* flora were part of the complex Beaufort terrestrial, aquatic and semi-aquatic ecosystems.

Using the paramammals, the Beaufort Group has been subdivided into biozones (eg. Tankard et al., 1982). The basal Dinocephalian Zone in the western Karoo basin (Table 8.5) contained the large herbivorous and carnivorous *Dinocephalia*. The *Pristerognathus-Diictodon* Zone above contained large numbers of Early Permian dicynodonts, while the overlying Late Permian *Cistecephalus* Zone had large herbivores such as *Diictodon*

Figure 8.57: Some Karoo vertebrates from southern Africa. A, *Hovasaurus boulei*; B, *Mesosaurus*; C, *Euparkeria capensis*; D, *Lesothosaurus austrialis*; E, *Heterodontosaurus*; F, *Syntarsus*; G, *Brachiosaurus branchi*; H, *Spinosaurus*; I, *Allosaurus*; J, *Kentrosaurus*. (Redrawn mostly from Halstead and Halstead 1981; Curie, 1981.)

(Fig.8.58) and many other dicynodonts. However, by the time of the overlying Early Triassic *Lystrosaurus-Thrinaxodon* Zone, the dicynodonts had declined from about 24 genera and 200 species in the Late Permian to two genera and six to nine species in the *Lystrosaurus* Zone (Colbert, 1965; Cruickshank, 1978).

Figure 8.58: Reconstruction of *Diictodon* from the Late Permian (*Cistecephalus* zone) of South Africa. (Redrawn from King, 1990.)

There was a corresponding floral change from a Glossopteris-dominated flora to a *Dicroidium*-dominated one, which allowed many vacant niches to be filled by herbivorous representatives of other reptilian groups (Cruickshank, 1978). Thus, the *Lystrosaurus* Zone was characterized by an impoverished Early Triassic plains community dominated by the single herbivorous dicynodont genus *Lystrosaurus* (Fig.8.59). This genus took to an aquatic way of life and is believed to have been a small reptilian hippopotamus that was completely at home in rivers and lakes, perhaps feeding on aquatic vegetation (Colbert, 1965; Halstead, 1975). *Thrinaxodon*, a smaller fierce flesh eater, and *Prolacerta*, the ancestor of the lizards, both lived on land nearby (Fig.8.60).

Figure 8.59: *Lystrosaurus* from the Early Triassic of South Africa. (Redrawn from Halstead, 1975.)

Colbert (1965) pointed out that the essential features of reptilian and amphibian evolution that had been established at the beginning of Triassic times continued throughout this period with the dominance in the overlying *Kannemeyeria-Diademodon* Zone, of paramammals typical of the *Lystrosaurus* Zone. The *Kannemeyaria-Diademodon* Zone represents the land-dwelling vertebrates that lived at the close of the early Triassic. This zone contained theriodonts or mammal-like reptiles (Fig.8.55) that had approached the threshold of mammalian anatomy and physiology by the Late Triassic. An example was *Cynognathus* (Fig.8.60), a flesh-eating, medium to large, wolf-like paramammal with dagger-like canines, that was an active predator. Living at the same time and place with *Cynognathus* was the small two-legged reptile named *Euparkeria* (Fig.8.57C), an archeosaur and the immediate ancestor of the dinosaurs (Halstead and Halstead, 1981) which first appeared in the Late Triassic (Fig.8.55). Some of the most primitive dinosaurs ever known are actually from Karoo beds in Lesotho in southern Africa. These belong to the early dinosaur group known as *Ornithischia* (Fig.8.55). The ornithischians were lightly built bipedal and herbivorous dinosaurs, the oldest of which are *Lesothosaurus australis* and *Heterodontosaurus consors* (Fig.8.57D,E) from Lesotho (Halstead and Halstead, 1981). Another Karoo dinosaur was *Syntarsus* (Fig.8.57F) from Zimbabwe.

Figure 8.60: *Cynognathus* and *Prolacerta* from the Triassic of South Africa. (Redrawn from Halstead, 1975.)

The appearance of the better adapted archeosaurs in the Late Triassic brought about the demise of the paramammals, a dwindling stock which had not recovered from massive end-Permian extinction. The archeosaurs invaded the niches of the parammamals. The smallest paramammal evolved into the first true mammals that survived by resorting to a nocturnal way of life, while the rest of the paramamals disappeared completely (Halstead, 1975).

Chapter 9 Mesozoic-Cenozoic Basins in Africa

9.1 Formation of the African Plate

Rifting and break-up of Gondwana dominated the Mesozoic-Cenozoic history of Africa. The Hercynian orogeny had by Permian times welded Gondwana and Laurasia into Pangea (Fig.8.6). In northwest Africa rifting in the Triassic heralded the fragmentation of Pangea. The Paleo-Tethys Ocean which occupied the eastern part of Pangea began to encroach westward into the rifts, thus creating embayments along the Saharan platform and an Atlas gulf which extended from Tunisia to Morocco. From the Middle-Late Jurassic onward, a large flysch basin lay along the transcurrent fault zone which formed the boundary between the African and the European plates. Late Cretaceous to Late Tertiary Alpine orogenesis caused thrusting and folding across this boundary zone in Africa and southern Europe.

Triassic rifting in northwest Africa (Fig.8.2) was propagated southward, and from southern Gondwana another rift system which began in Early Jurassic times was extending northward. Towards the end of the Middle Jurassic, both rift systems had reached and cut through the equatorial region. Sea-floor spreading and the opening of the North Atlantic Ocean started during the Middle Jurassic and progressed southward. The South Atlantic began to open later in the Early Cretaceous (Late Aptian). The opening of the Equatorial Atlantic started in the Albian. During the Turonian the North and South Atlantic merged as Africa and South America finally separated.

Similarly, the eastern margin of Africa formed by the fragmentation and drowning of eastern Gondwana, but this initially involved large-scale transcurrent movements along transform faults, a process that began earlier and created aborted rifts during Karoo times. Transcurrent motion which occurred between the Middle Jurassic and the Early Cretaceous along the Falkland-Agulhas fracture zone (Fig.8.2) in the south leading to the opening of the South Atlantic, concomittantly initiated the southern Indian Ocean margin of Africa. Motion along the Davie fracture zone (Fig.8.2) to the north caused the southward drift of Madagascar, thus opening the Somali basin in northwestern Indian Ocean. Distensive tectonics subsequently widened this ocean.

Another tectonic imprint of transcurrent motion along oceanic fracture zones was that where the latter are propagated into the continent along pre-existing basement lineaments, strike-slip motion created rifts or interior fracture basins (Fig.8.2). West African interior basins such as the Benue trough and its bifurcations, and the Chad basin rift system, the Sudanese interior basins, and even the Sirte basin in Libya, are believed to be the products of shear motion along a Central African lineament, which was the landward continuation of the Gulf of Guinea transform fault system (Fig.8.2).

While the classical processes of doming, rifting and sea-floor spreading led to the break-up of the Arabian-Nubian Shield, thus creating the Red Sea and the Gulf of Aden during the Late Eocene-Early Miocene, transcurrent movement along major Precambrian basement lineaments seemed to determine the origin and alignment of the East African Rift System.

Various rift scenarios will be examined in this chapter together with the evolution of their structures and stratigraphic fill.

9.2 The Atlas Belt: An Alpine Orogen in Northwest Africa

9.2.1 Tectonic Domains

The Atlas fold and thrust belt includes the Atlas foreland belt and the Maghrebide internal and external domains. The Atlas and the Magheribes (Fig.9.1) constitute the southern chains of the Alpine orogen which lies mostly in southern Europe and beneath the Mediterranean, and evolved during Mesozoic-Cenozoic times. The Moroccan Atlas, the Saharan Atlas in Algeria, and the Tunisian Atlas, like the Jura Mountains of Europe, constitute the foreland fold belt of the Alpine chain. In North Africa the internal zones of the Alpine orogen, starting from the western Mediterranean Sea, includes the Betic-Rifian loop bordering the Alboran Sea (Fig.9.2D), and the Tell Atlas of Algeria. A major paleo-tectonic element in North Africa during the Alpine orogeny was the African promontory which occupied the northernmost part of the Saharan platform in Tunisia. It was against this feature that the Apennines and the Dinarides were overthrust (Caire, 1978).

9.2.2 Synoptic Tectonic History

The tectonic and stratigraphic evolution of the Alpine orogen in northwest Africa display the various stages of the Wilson Cycle. Caire (1978), Jacobshagen (1988) and Wildi and Favre (1989) outlined the tectonic history of this region. During the Triassic, rifts (Fig.9.2A) formed from the relative eastward movement of the African plate (Fig.9.2B); the rifts were northeast-trending pull-apart basins in the Atlas region. Dolerite lavas were extruded in the rifts and red beds were deposited. The Atlas rift system failed because there was no sea-floor spreading. Mesozoic seaways encroached westward (Fig.9.3) so that by the Late Triassic evaporitic facies and open marine sedimentation took place in the seaways. Open marine Middle and Late Liassic faunas in the Rif basin indicate the creation of an Atlantic seaway between Europe and the Portuguese and northwest African basins.

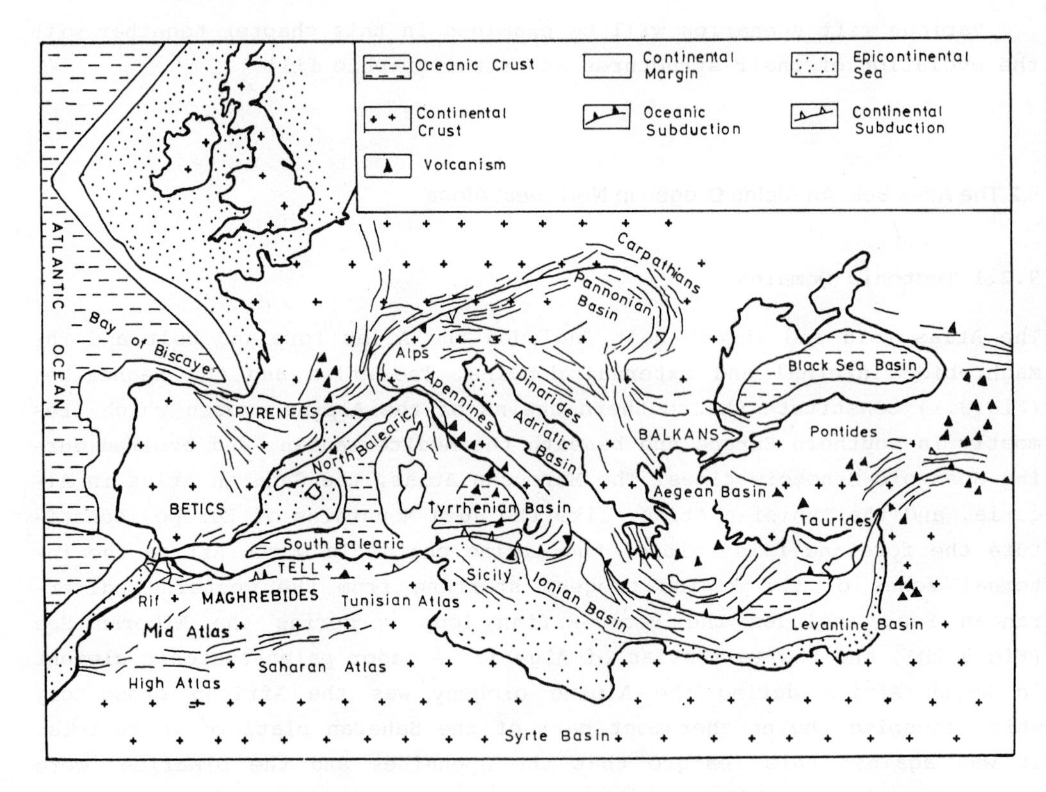

Figure 9.1: Tectonic map of the Mediterranean region. (Redrawn from Biju-Duval et al., 1977.)

During the Middle and Late Jurassic the external Betics in Spain, the Rif, the Tell, and the Alboran continental margins subsided rapidly while

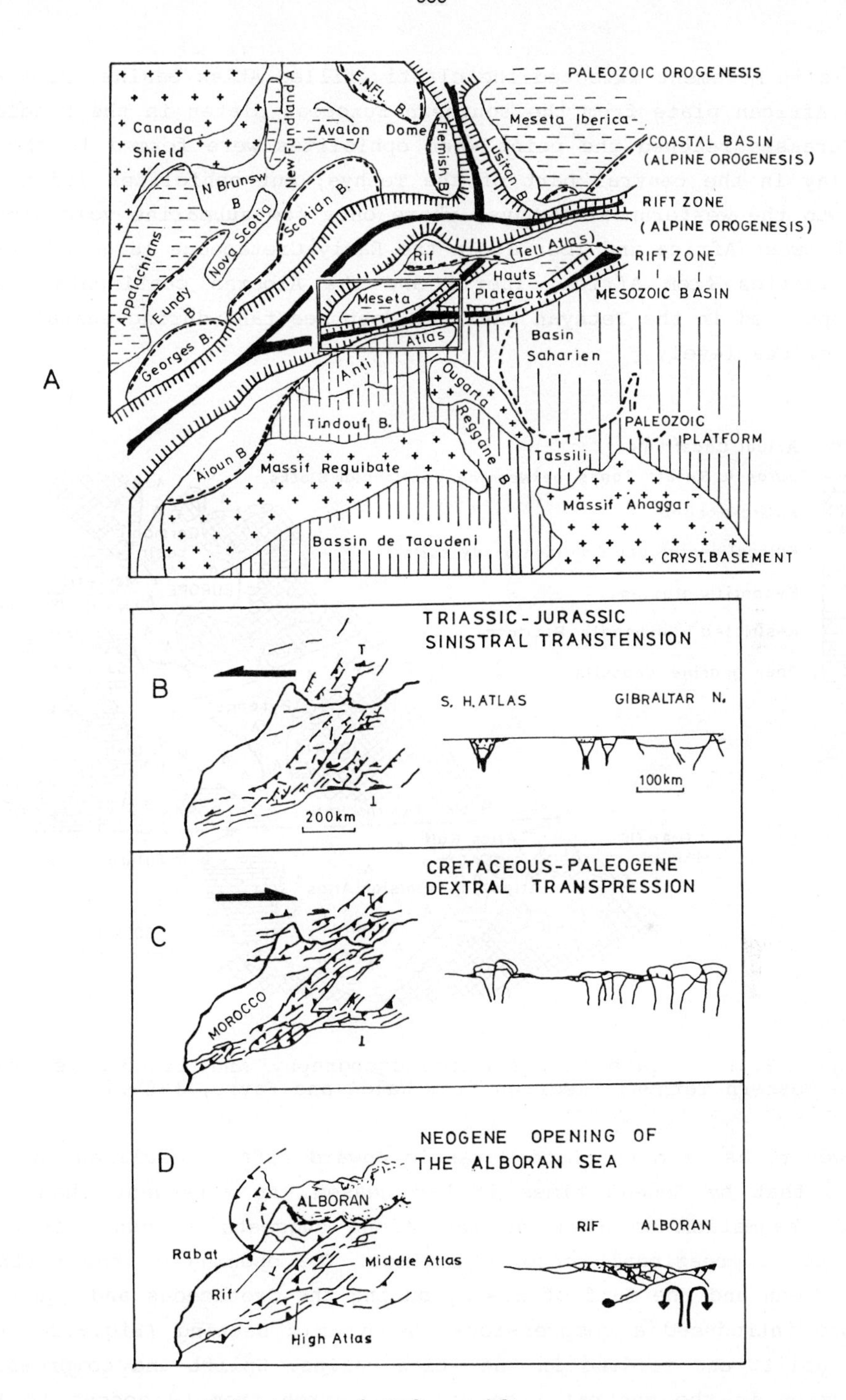

Figure 9.2: Atlantic and Atlas rift systems and their structural development. (Redrawn partly from Stets and Wurster, 1982.)

strike-slip movement affected the clastic-filled Atlas basins. Separation of the African plate from the southern European plates in the Middle and Late Jurassic reached the point when ophiolites were formed in the Alps which lay in the central part of the Tethys; but ophiolites did not extend into the westernmost Tethys, where only few submarine volcanics are known between Africa and Iberia. In the Early Cretaceous large amounts of siliciclastics from Africa, Iberia and the Alboran continental blocks were deposited in the Tethyan basin as deep sea fans during eustatic lowstands of sea level.

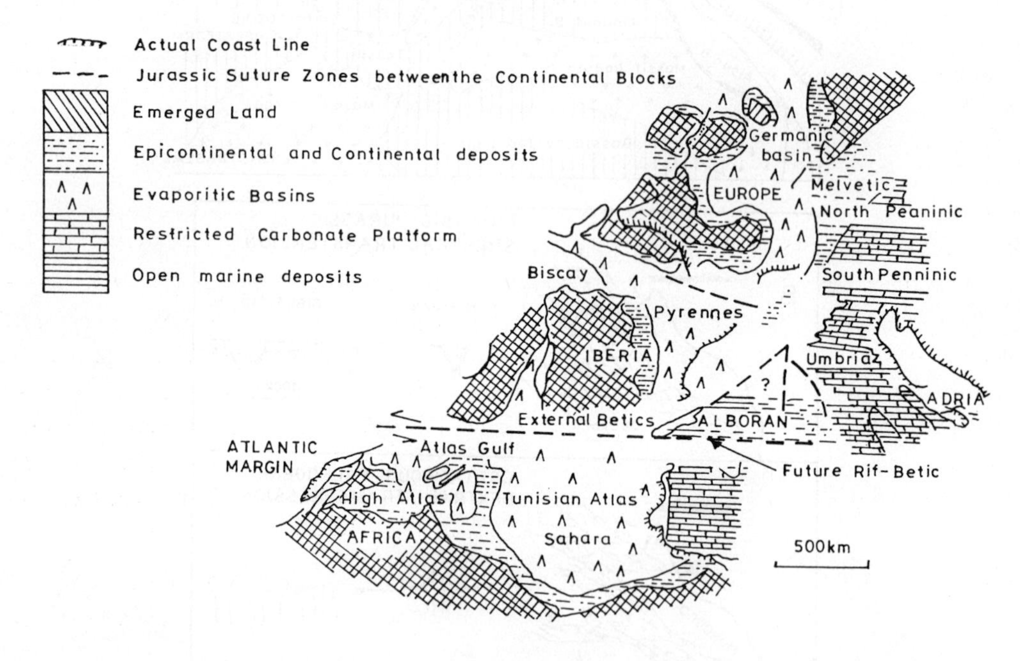

Figure 9.3: Late Triassic paleogeography and lithofacies of the Western Tethys. (Redrawn from Waldi and Favre, 1989.)

However, as from the Late Jurassic onward rifting declined considerably so that by Eocene times it had ceased. An attendant change from basaltic magmatism to alkaline intrusions suggests a change from tensional to compressional tectonic regimes. The opening of the North Atlantic Ocean and the Gulf of Biscay during the Cretaceous and the Paleogene had introduced a compressional geodynamic setting (Fig.9.2C) which caused uplift and folding in the Atlas basins. Uplift and compressional deformation in the central High Atlas occurred from Oligocene to Early Miocene times, concomittantly with the main phase of deformation in the internal zones of the Rif and the Betic chains. The last thrusting in the Rif during the Pliocene coincided with uplift in the central High Atlas.

Here uplift persisted down to the Quaternary. The terminal phase of thrusting in the Rif has been linked to crustal extension in the Alboran Sea, caused by Neogene upwelling of the asthenosphere (Fig.9.2D) along the Gibraltar fracture zone. Neogene uplift in the central High Atlas has been accompanied by subsidence of narrow and elongate foredeep basins which are parallel to the Atlas chain (Fig.9.4A) and contain continental strata.

9.2.3 The Moroccan or High Atlas

About 800 km long and 40 to 100 km wide, the Moroccan or High Atlas (Fig.9.4) consists of four main parts. There are, from west to east, the Western High Atlas; the Paleozoic High Atlas, a horst of Paleozoic terranes intruded by Carboniferous granites and thinly covered by Mesozoic strata; the Precambrian High Atlas, a horst of slightly deformed Infra-Cambrian and Precambrian rocks with very thin Mesozoic strata; and the Central and Eastern High Atlas (Schaer, 1987). In its tectonic and stratigraphic evolution the Western High Atlas was more related to the development of the Atlantic margin of northwest Africa than to the rest of the High Atlas. From the Atlantic coast of Morocco it extends eastward for about 50 to 70 km to the Argana basin (Fig.9.4B) where the Triassic is up to 5 km thick and comprises essentially detrital and locally conglomeratic facies with evaporitic sequences appearing in the west. Mesozoic strata in the Western High Atlas thicken to the west and contain Jurassic and Cretaceous calcareous marly and locally evaporitic facies, whereas continental strata are predominant in the east. The Western High Atlas was only affected by differential subsidence.

The Central and Eastern High Atlas, also known as the Calcareous High Atlas because of thick Mesozoic carbonates (Fig.9.4B), extends as a deep rift trough all the way to the Algerian border. Its stratigraphic and structural evolution, summarized by Stets and Wurster (1982) and by Warme (1988), show the following major phases. Continental rifting in the Late Triassic created the Atlas rift in which broad alluvial fans (Fig.9.5A) prograded towards the centre of the grabens (Lorenz, 1988; Manspeizer, 1982; Mattis, 1977) and deposited brick-red fluvial sandstones, conglomerates and mudstones. These mudstones are intercalated with evaporitic horizons of dolomite, gypsum and halite, and with tholeiitic dolerites at the top of the Triassic sequence.

While the supply of terrigenous clastics continued into the Jurassic, marine invasions coming along the Atlas gulf, flooded the Moroccan meseta. Another transgression proceeded eastward from the nascent

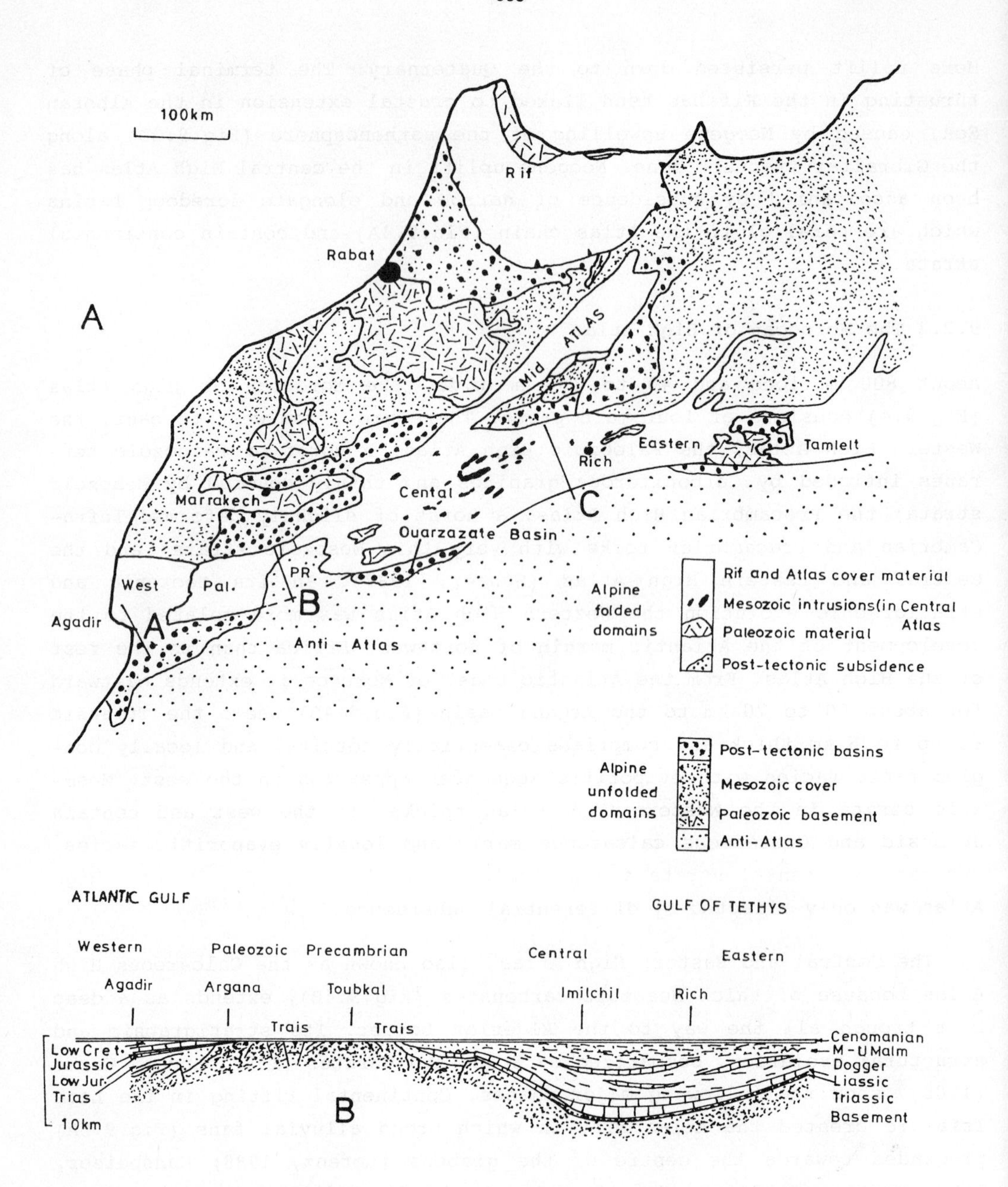

Figure 9.4: A, structural divisions of the High Atlas; B, schematic cross-section through the High Atlas from Agadir to Tamlet in Morocco. (Redrawn from Schaer, 1987.)

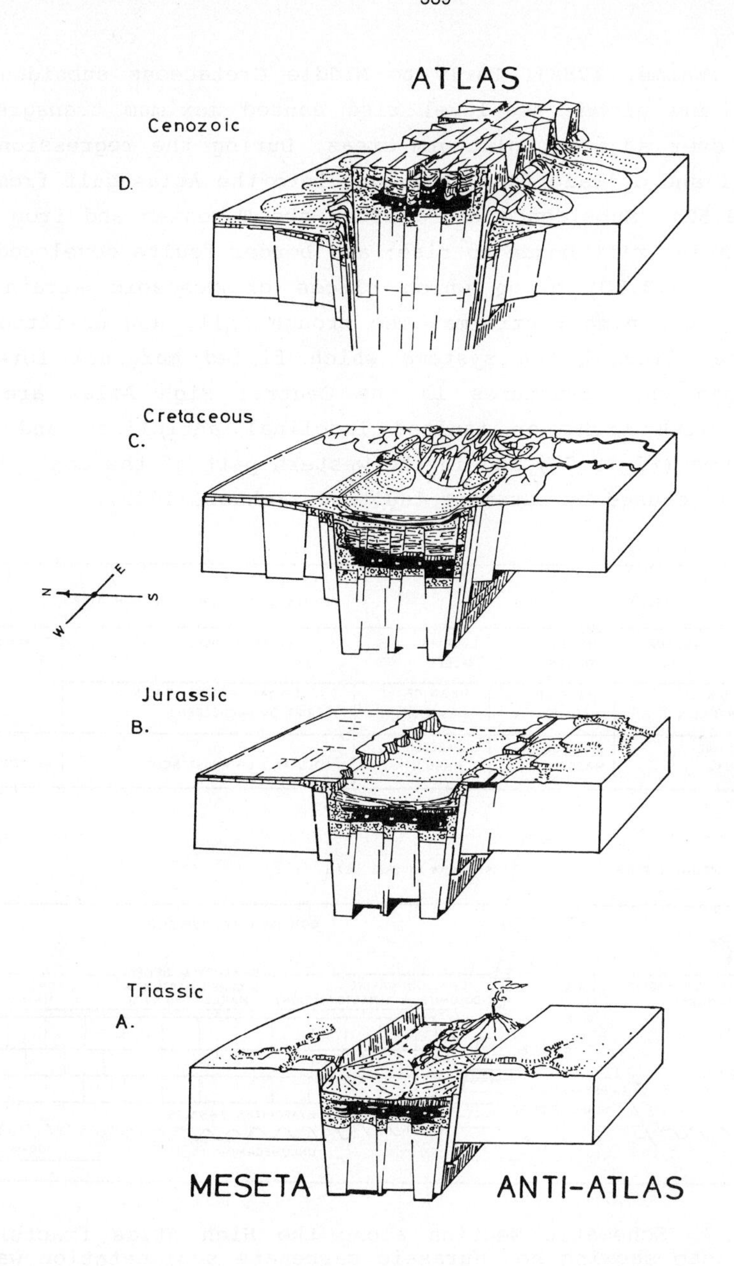

Figure 9.5: Tectonic evolution of the High Atlas. (Redrawn from Stets and Wurster, 1982.)

Atlantic Ocean to the west. From the Early to Middle Jurassic major carbonate build-ups including epicontinental limestones and reefs (Fig.9.6) were established on fault blocks which were shoal areas, whereas gravity-generated limestones and olistostromes accumulated in adjacent deeps (Warme, 1988). Early to Middle Cretaceous subsidence in the Atlas rift and global sea-level rise caused maximum transgression which extended over adjacent platform areas. During the regression that followed fluvial and deltaic fans prograded into the Atlas gulf from east and west (Fig.9.5C). Subsidence ended after the Turonian and from later Cretaceous the Atlas rift began to rise; and border faults developed into thrust faults (Fig.9.5D) along which slices of Mesozoic strata were thrust onto the adjoining platforms. The trough fill, now uplifted, was eroded into new alluvial fan systems which filled marginal foredeeps (Fig.9.5D). Among the structures in the Central High Atlas are ENE-trending folds with tight or faulted isoclinal anticlines and flat-bottomed synclines (Fig.9.7); and in the western part of the Central High Atlas folding was caused by basement faulting (Schaer, 1988).

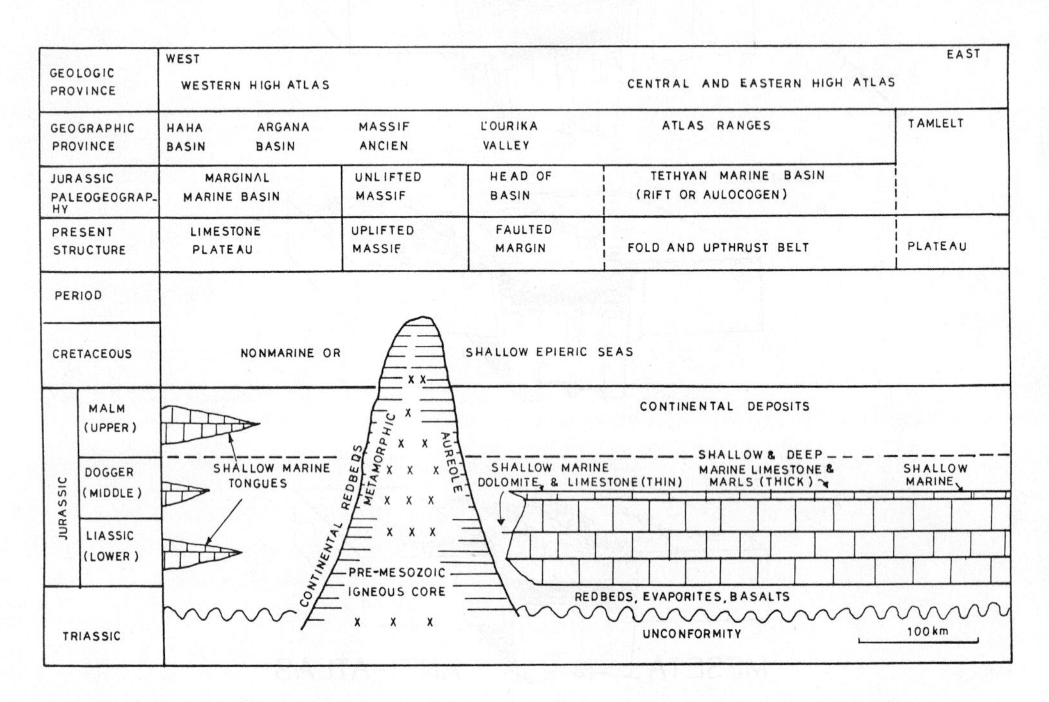

Figure 9.6: Schematic section along the High Atlas fracture zone of Morocco showing how Jurassic carbonate sedimentation was influenced by their geologic and paleogeographic settings. (Redrawn from Warme, 1988.)

9.2.4 The Saharan Atlas

The Moroccan Middle and High Atlas continue eastward into Algeria as the Saharan Atlas, where Mesozoic deposition took place in a more marine milieu (Fig.9.3). Here the Mesozoic-Cenozoic sequence is probably up to

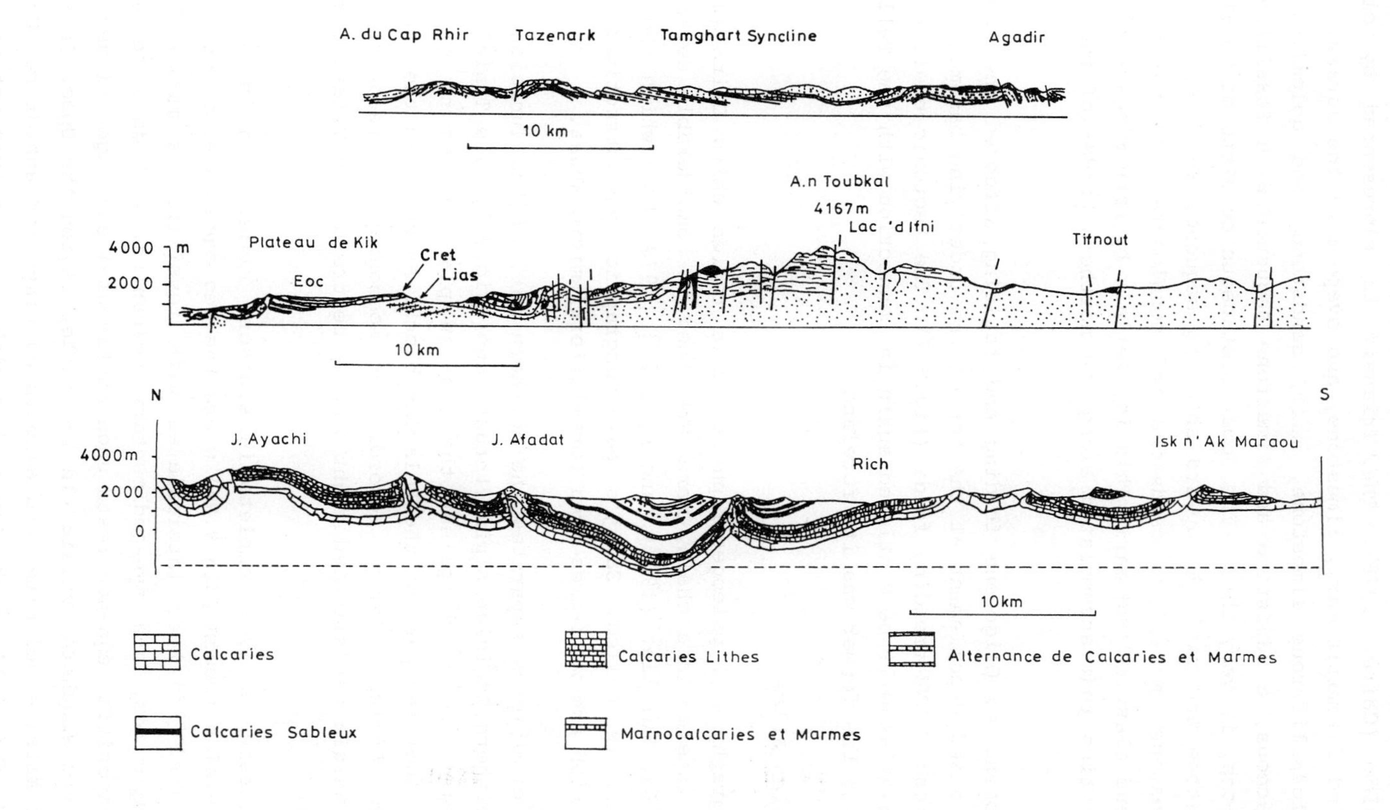

Figure 9.7: Cross-sections through the Moroccan High Atlas. (Redrawn from Schaer, 1987.)

10.7 km thick (Caire, 1978). The Triassic is represented by clays, dolomite, and dolomitic marl, limestone, and evaporite; the Jurassic consists of fossiliferous limestone, marl, sandstone, and dolomite. The Lower Cretaceous, a sandstone and limestone sequence with fossiliferous intercalations, is very thick. The Upper Cretaceous consists of fossiliferous limestone and marl. An upward shoaling sequence which starts with marine limestone, marl, and sandstone and conglomerate is followed by Lower Miocene clastics and marl. This is overlain by Late Miocene to Quaternary clastics that accumulated along the northern border of the Saharan Atlas.

Late Eocene to Oligocene folding and faulting affected the Saharan Atlas and caused décollement at the top of the underlying basement which produced broad E-NNE-trending folds (Fig.9.7). The tectonic evolution of the Saharan Atlas will be mentioned again in connection with the Tell Atlas, of which the former was its foreland.

9.2.5 Tunisian Atlas

The stratigraphy and paleogeography of this region which Caire (1978) termed the intermediate chain, were described in considerable detail by Bishop (1976), Burollet (1967), and by Salaj (1978) from which the present outline is culled. Salaj's paleogeographic reconstructions were based upon rich and well-preserved foraminiferal microfaunas.

A zone of diapirs separates the northern part of the Tunisian Atlas from the northern Tunisian Alpine thrust zone (Fig.9.8). The Tunisian Atlas terminates southward against the Saharan platform and to the east against the Tunisian part of the Pelagian block. As shown in the generalized North African Early Mesozoic paleogeographic reconstruction (Fig.9.3) Tunisia was situated right in the centre of the Triassic evaporite basin.

The Triassic is more complete in southern Tunisia and in the NNE-trending Tunisian trough (Fig.9.9A) where the sequence consists of continental detrital red beds intercalated with fossiliferous marine limestones, evaporites, and syn-sedimentary volcanics. A Middle to Upper Triassic evaporitic sequence rests upon continental and lagoonal deposits in central and southern Tunisia. In the latter region the Upper Triassic consists of massive dolomites and associated clays and anhydrite. During the Jurassic (Fig.9.9A) a shallow to brackish water environment occupied the edge of the Saharan platform. Zones of subsidence with pelagic sedimentation were located to the north in the Tunisian trough. Carbonate reefs developed on the Pelagian platform and in the Middle to Late Jurassic shallow-water algal mats became ubiquitous in the central region and on the western edge of the Pelagian platform.

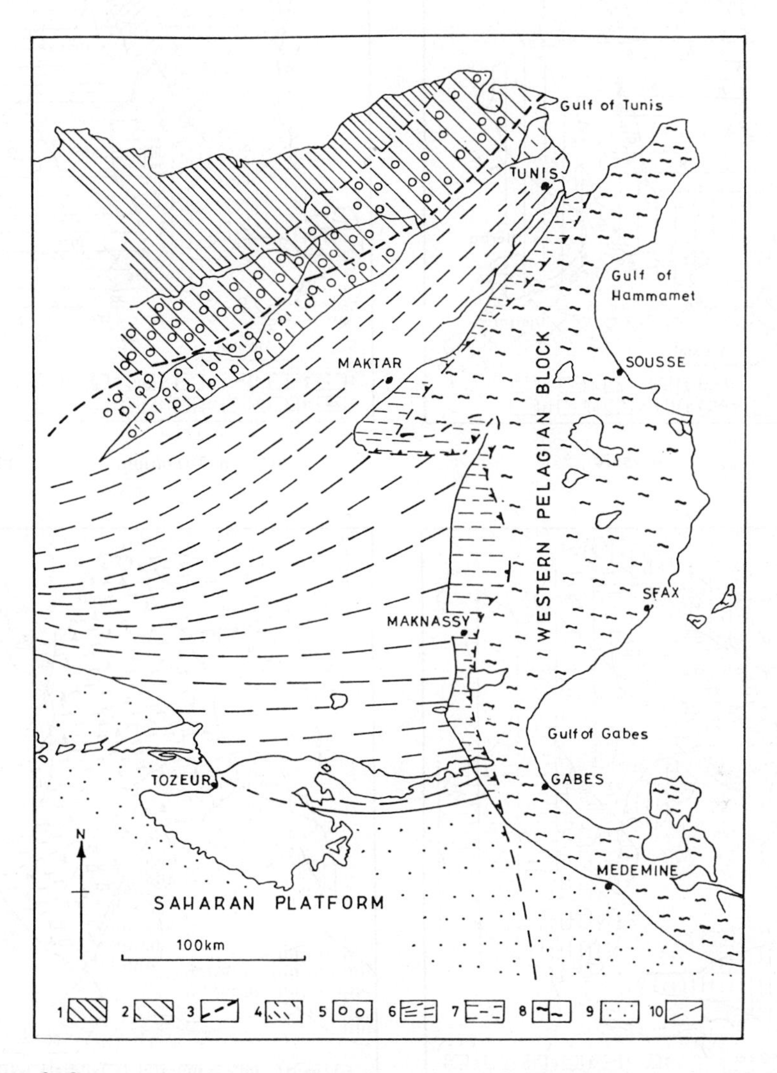

Figure 9.8: Tectonic sketch map of Tunisia. 1, Numidian nappes, Tellian units, and para-autochthons of Heldi; 2, Medjerda para-autochton and autochthon; 3, thrust zone of Teboursouk; 4, Miocene foredeep; 5, diapiric zone of Tunisian Atlas; 6, central and southern zone of Tunisian Atlas; 8, eastern platform (Western part of Pelagian block) 9, Saharan platform; 10, thrusts. (Redrawn from Salaj, 1978.)

A complete and well-exposed Cretaceous sequence occurs in Tunisia. The Lower Cretaceous is represented along the margin of the Saharan plat-

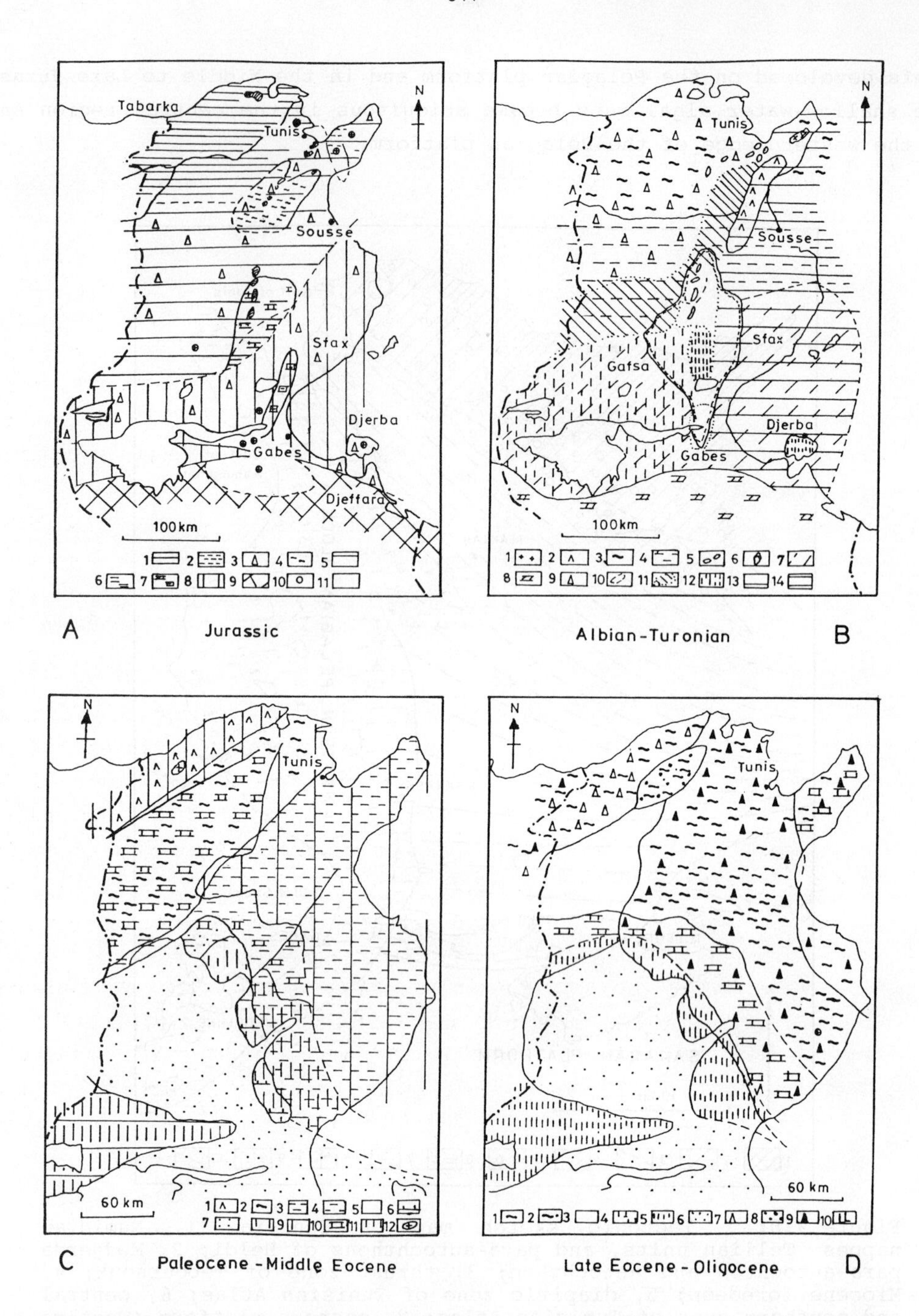

Figure 9.9: Paleogeographic maps of Tunisia. A: 1-2, 5, pelagic facies; 3, 6, 9, littoral facies. B: 1, rudistid reef; 3, 4, 9, pelagic; 7, evaporitic laguno-neritic. C: 1-4, El Haria Fm; 5-6 Metlaoui fm. D: 1,2 Souar Fm., 4, marly limestone; 5, gypsiferous strata; 7, Numidian Sst; 9, Nummulitic limestone; 10, limestone with *Lepidocyclina*.

form by neritic sandy oolitic limestones overlain by lagoonal gypsiferous shales, and gypsum with sandstone intercalations containing lagoonal ostracodes. Coeval strata in northern Tunisia consist of pelagic sublithographic limestone with marl. The Pelagian platform in Tunisia was emergent in the Early Cretaceous since it consists of nonmarine deposits. The Late Aptian in Tunisia was marked by a marine transgression in which coral-bearing and orbitolinoid limestones accumulated. This transgression climaxed in the Late Cenomanian-Turonian (Wiedmann, 1987) and deposited widespread rudistid limestone and dolomite along the margin of the Saharan platform (Fig.9.9B), and pelagic marls and limestone in the Tunisian trough. Carbonate sedimentation prevailed throughout the Late Cretaceous to Paleogene in Tunisia. Northwestern Tunisia contains classic exposures where the Cretaceous-Tertiary boundary lies within a unit (El Haria Formation) in which there was continuous deposition across this boundary (Fig.9.9C). A change from Paleocene-Early Eocene globigerine limestone and marls to Late Eocene-Oligocene nummulitic limestones (Fig.9.9C) suggests progressive shoaling and emergence during the Late Tertiary in which there was a significant transgression in the early Middle Miocene.

As in other parts of the Atlas foreland, the stratigraphic evolution of Tunisia and the creation of depocentres were directly related to the tectonic evolution of this region. However, as shown by Clifford (1986) the Pelagian basin, the main part of which lay offshore of present-day Tunisia, changed from a marginal sag basin into a wrench-modified basin (Fig.8.2) on account of transcurrent motion between North Africa and the Western Mediterranean. Consequently, the Pelagian basin was uplifted as from the Late Cretaceous. Transcurrent movement caused the reactivation of old faults; it reversed their throws; and triggered salt intrusions. The net effect was basin inversion and the creation of paleo-highs on which reefal build-ups developed. The Late Eocene-Oligocene nummulitic banks, already mentioned, flourished after the Lower Eocene deep-water globigerinid limestone had shoaled. These nummulitic banks are petroleum reservoirs in Tunisia; they are sourced by the globigerinid facies.

9.2.6 The Moroccan Rif

Palinspastic Reconstruction

The Rif Mountains of northern Morocco, up to 1,500 m high, and the Tell Atlas of Algeria, constitute the southernmost segments of the Alpine orogenic belt. Together, they form the African part of the Maghrebides, an Alpine orogenic chain which actually extends from the Betic cordillera of southeastern Spain and continues beyond Algeria into southern Italy

(Fig.9.1). The Betic and the Rif constitute the Alboran margin or the Arc of Gibraltar which rims the Alboran Sea in the Western Mediteranean (Fig.9.10A).

The sedimentary sequence in the Alboran (Betic-Rifian) margin are believed to have initially accumulated from Triassic times at a more easterly location northeast of present-day Tunisia. This was along the continental margin of an ancient microplate in southern Europe, probably the Alboran block (Fig.9.3). The sequence was tectonically transported to its present western location by progressive WSW movement of the Alboran block (Durand-Delga and Olivier, 1988). The microcontinent collided with the African plate in Oligocene-Miocene times and produced the complicated Rif overthrust (Fig.9.10B). Because of the large separation between the Rif and the High Atlas Durand-Delga and Olivier concluded that it is impossible to trace direct paleogeographic links between the Rif and the Moroccan Atlas adjacent to it, moreso as their contacts are tectonic (Fig.9.4A). These authors presented stratigraphic and paleogeographic interpretations for the various structural units in the Rif based on this palinspatic reconstruction.

Stratigraphy of the Main Structural Units in the Rif

The internal zone of the Rif (Fig.9.10B) consists of the Sebtides, the Ghomarides and the "Dorsale calcaire" (Durand-Delga and Olivier, 1988). Choubert and Faure-Muret (1973) referred to the Sebtides and the Ghomarides as the Rifides and the "Dorsale calcaire" as the Ultrarifaine. In the Sebtides mantle peridotites of uncertain age are overlain by a thick probably Precambrian to Paleozoic sequence which passes upward into Permo-Triassic strata at the greenschist facies. The latter occur as alpine nappes. Generally thrust over the Sebtides, the Ghomarides, which are structurally more complex towards the southwestern and southern borders of the internal Rifian zones, are essentially Paleozoic slates. These slates are overlain disconformably by Triassic red sandstones and by a thin and discontinuous Late Triassic-Early Jurassic carbonate cover. The youngest formation in the Ghomarides are of Eocene in age. The "Dorsale calcaire", comprising largely Mesozoic carbonates locally overlain by Paleogene detrital formations, occurs as folded and thrust sheets (Fig.9.10B). From the more internal to external parts, the "Dorsale calcaire" shows marked facies changes from shallow depositional environments ("Chaine calcaire interne") through intermediate depth facies, to deeper and more pelagic conditions ("Chaine calcaire externe").

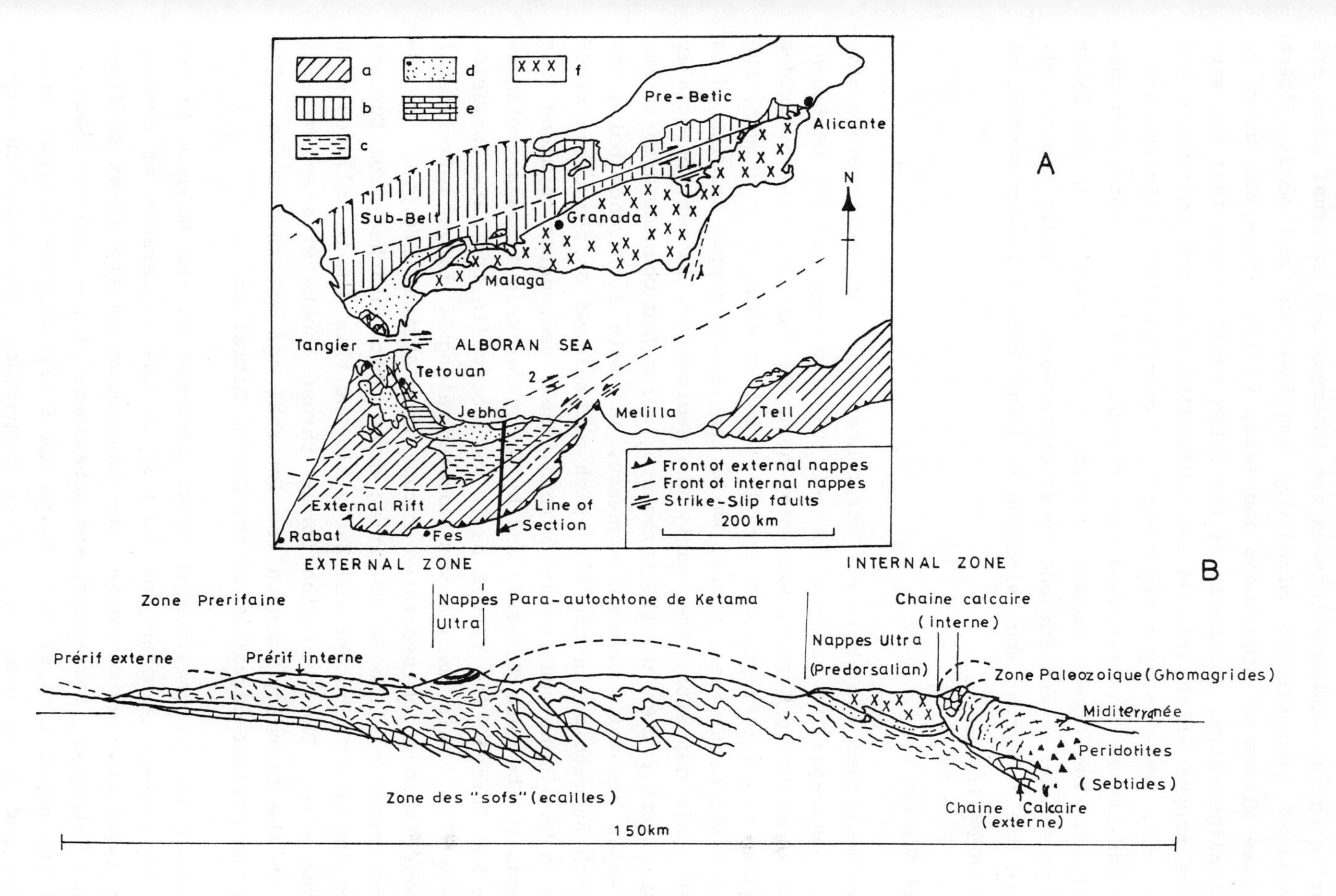

Figure 9.10: A, the Betic-Rifian orogen; a, Maghrebide external non-schist zones; b, Betic external non-schist zones; c, Maghrebide external schist zones; d, flysch and Predorsalian area; e, "Dorsale calcaire"; f, Ghomaride-Malaguides/Sebtides-Alpujarrides/Nevado-Filabride; 1, North Betic fault; 2, Jebha fault. B, schematic cross-section through the Rif. (Redrawn from Durand-Delga and Olivier, 1988; Guiraud et al., 1987.)

Occupying thrust contacts between the internal and external zones are highly deformed tectonically mixed and argillaceous and marly flysch which enclose slices of limestones and sandy flysch. These are known as the Predorsalian units. Southeast of the Jebha fault (Fig.9.10A) are several flysch nappes which include Barremian-Albian flysch sandstones and pelites locally overlain by a Mid-Cretaceous organic-rich siliceous unit, and by Aptian-Albian quartzitic-pelitic flysch, and Upper Cretaceous calcareous flysch. The flysch nappes located to the northwest of the Jebha fault are more complex and include Upper Cretaceous to Middle Eocene calcareous and marly flysch; and Paleocene to Lower Eocene flysch sandstones (Numidian nappes).

Geological History

As already mentioned the above Rif structural units are believed to have initially been deposited along a Triassic to Oligocene passive continental margin in southern Europe, somewhere north of Tunisia. The Ghomarides and Dorsalian terranes represent the shelf facies along this margin, the Predorsalian the continental slope, while the flysch nappes originated as deep ocean basin deposits. Presumably the sedimentary facies, especially the Triassic rift facies and the carbonate facies had counterparts in the Tunisian Atlas Mesozoic-Cenozoic sequence, the main difference being the thick flysch succession in the Rif which accumulated on the ocean floor along a more subsident continental margin. Shelf sedimentation along this margin began in the Late Triassic and by Early Jurassic times a continental slope had formed. The flysch basin was active in the Late Jurassic-Early Cretaceous. This margin was located along a large left-lateral transcurrent fault which separated Europe from Africa (Fig.9.3). By the end of the Cretaceous the Betic-Rifian domain had detached from Europe; and during the Late Oligocene and Early Miocene this domain collided with the African plate. The collision formed thrust sheets with an external vergence at the boundary between the internal and external zones, thus producing the present curvature of the Arc of Gibraltar.

Stretching of the continental crust occurred in the Neogene in the area of the present Alboran Sea leading to the appearance of oceanic crust and subsidence which marked the initiation of the first Mediterranean Sea. Collision, thrusting, and subsidence of the Mediterranean Sea triggered the deposition of Late Oligocene-Early Miocene conglomerates, sandstones, and marls over the Rifian Ghomarides. The Numidian flysch comprising very mature quartzose clastics was also deposited throughout the Early Miocene. After coarse clastic sedimentation ceased because of reduced subsidence in the late Early Miocene, siliceous clays, marls and

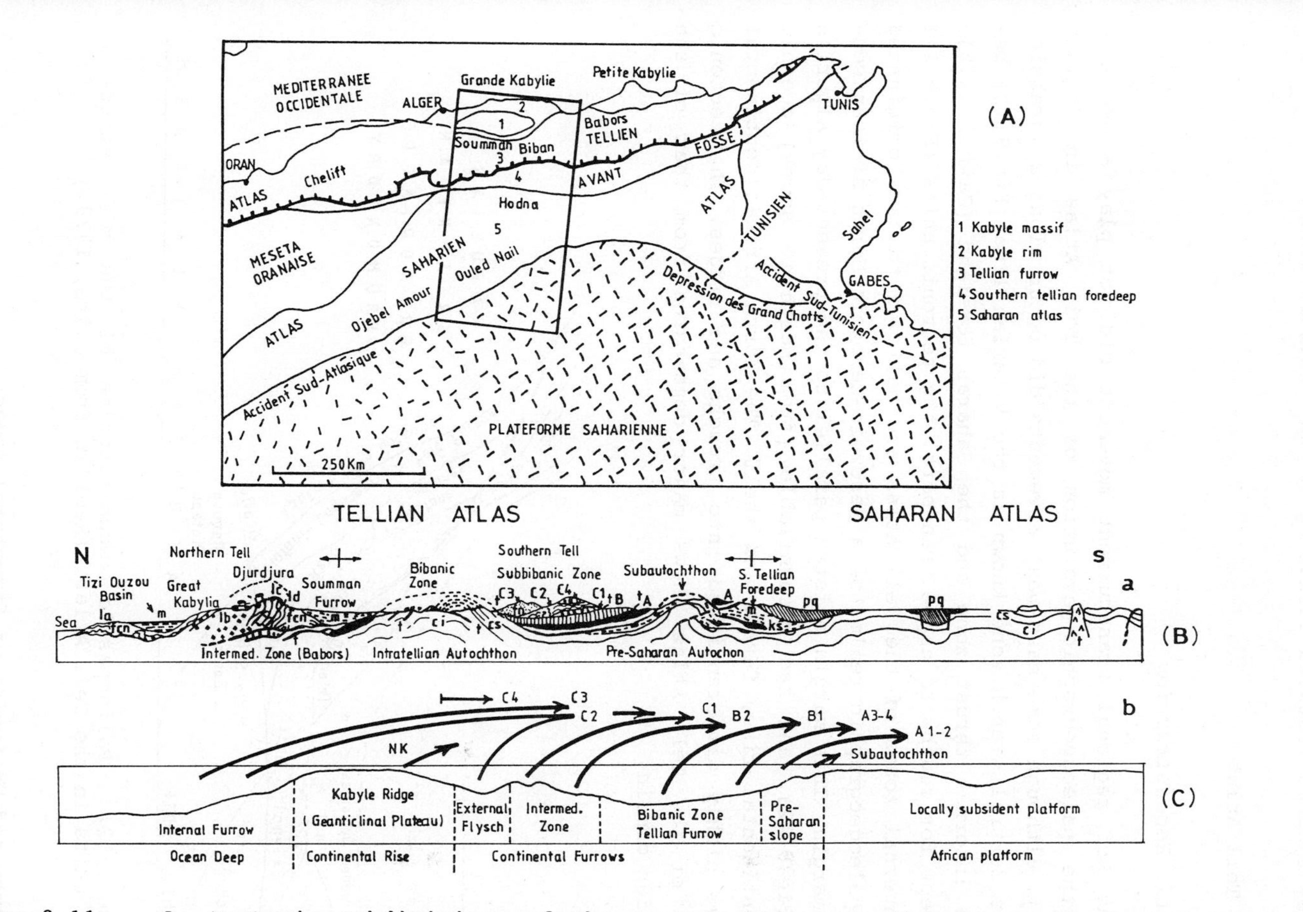

Figure 9.11: A, tectonic subdivisions of the eastern Atlas: 1, Kabyle massif; 2, Kabyle rim; 3, Tellian furrow; 4, Southern Tellian foredeep; 5, Saharan Atlas. Composite schematic N-S sections (B) and paleogeography (C) across the Tellian and Saharan Atlas of Western Algeria. (Redrawn from Cairé, 1973.)

radiolarian shales were deposited in the Rif over the Numidian flysch, followed by renewed westward thrusting of the internal zones.

9.2.7 The Tell Atlas

Palinspastic Reconstruction

Unlike the Rif regional transcurrent movement did not play a decisive role in the paleogeographic evolution of the Tell Atlas in Tunisia (Fig.9.11), although some east-west movements did occur along a Jurassic-Cretaceous dextral wrench zone known as the Vicarian line (Fig.9.12) between the internal thrust zone and the Saharan foreland (Caire, 1978). Thus, except for southward nappe transport, the tectonic units (Fig.9.11) of the internal zones of the Tell Atlas are believed to have originated from the paleogeographic realms of a Mesozoic-Cenozoic North African continental margin. This margin existed parallel to the present-day Algerian margin (Fig.9.12), and had recognizable (from north to south) oceanic basin, continental rise, furrow and shelf. The North African continental margin was highly differentiated into troughs and ridges. The tectonic evolution of the Tell Atlas shows major departures from that of the Betic-Rifian domain.

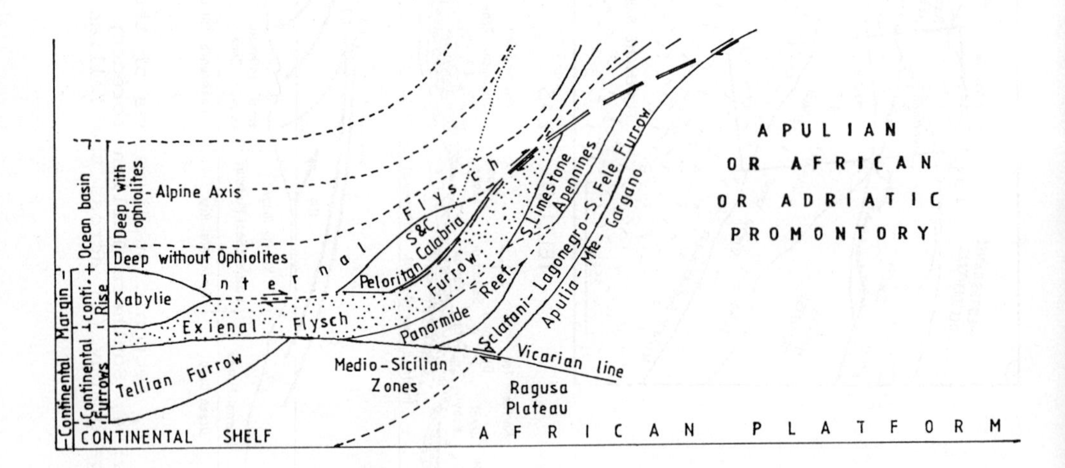

Figure 9.12: Palinspastic reconstruction of the main structural units in the Tell Atlas. (Redrawn from Cairé, 1978.)

Stratigraphy and Tectonics of Structural Zones

This will be examined from north to south along the transect (Fig.9.11A) described by Caire (1973, 1978). The Kabyle massif is a crystalline basement overlain by oceanic Paleozoic and Mesozoic limestone chain which formed atop oceanic ridges (geanticlinal plateau). Such features existed in the Cretaceous because the continental margin was broken into ridges, and troughs in which thick terrigenous clastics accumulated. Oligocene-Miocene molasse occurs in the upper sequence in the Kabyle massif. The massif was deformed by southward thrusting and sliding over the adjoining structural zone, the Kabyle rim.

The Kabyle rim is a Mesozoic geosynclinal flysch with radiolarites, underlain by Triassic sandstones and limestones, and overlain by detrital Paleogene to Miocene strata including the Oligocene-Miocene Numidian flysch sandstones and shales. The dominant thrust direction in the Senonian was also towards the south. There are, however, some northward thrusts in the Kabyle rim.

South of the above internal flysch zone starts the external zone which contains continental furrows or troughs. The continental furrows (Fig.9.11B) consist of an external flysch zone, the intermediate zone, and the Bibanic zone or Tellian furrow. With these furrows the Algerian shelf was like a continental borderland in which the furrows served as terrigenous depocentres since Early Cretaceous times with the Tellian furrow remaining active up to the Oligocene-Miocene despite frequent deformation along its margins. The Tellian furrow separated the Saharan Atlas platform which lay to the south; it received continental sands. Neritic deposits also accumulated in the oceanic Kabyle region to the north (Fig.9.12). The Intermediate zone, a tectonically active area was deformed, metamorphosed, and intruded with ultrabasic rocks as from the end of the Jurassic until the Albian when it was folded and overturned to the north at the time the ocean basin closed in the north. During the Senonian the Intermediate zone was again folded with the re-deposition of Triassic beds as olistostromes. In the Tellian furrow Cenozoic strata consist of conformable Lower Eocene with limestones, marls, phosphate beds, shell beds and sandstones. After Late Eocene folding and thrusting Oligocene-Miocene detrital sedimentation ensued including deposition of the Numidian flysch sandstones and shales, discordantly upon the Cretaceous-Eocene flysch.

South of the Tellian furrow there is an Early Miocene molasse basin, the Tellian foredeep, which is filled with fossiliferous marls and sandstones. This basin received sedimentary klippe and nappes from the north in the Early Miocene during the main Alpine orogenic phase. Figure 9.11B shows the sources and directions of slide nappes, most of which originated from the Tellian furrow (A nappes), while those that came from the internal zone (C nappes) contained Cretaceous and Numidian flysch. From the Middle Miocene onwards the Tell Atlas has been experiencing uplift.

9.3 Stratigraphic Evolution of the Eastern Saharan Platform

9.3.1 Structural Framework

Post-Hercynian structural realignments and resultant east-west structures in the Saharan platform determined the major depocentres until the disintegration of Pangea in Jurassic-Cretaceous times created new structural trends (Klitzsch, 1986). During the Mesozoic Egypt consisted of two main structural provinces, an unstable shelf to the north with complex NE-SW transcurrent faults and fault-bounded basins and horsts, and a stable shelf to the south. The stable shelf was covered by mostly Paleozoic-Mesozoic continental deposits and was intermittently overlapped by shallow seas which encroached from the north (Fig.9.13A) especially during periods of eustatic sea level rise (Kerdany and Cherif, 1990). Major intracratonic basins lie in Egypt's unstable shelf, among which are the Dakhla basin, the Misaha trough, and the Abyad basin (Fig.9.14A).

Clifford (1986) classified the Mesozoic-Cenozoic sedimentary basins of Tunisia, Libya, and Egypt into wrench modified basins and interior fracture basins (Fig.8.2). The structural inversion of the Early Mesozoic Pelagian platform in Tunisia has already been mentioned. Other major North African basins such as the Libya's Sirte and Egypt's Gulf of Suez basins are major interior fracture basins, whereas the Dakhla basin was later modified by wrenching during the Late Cretaceous dextral movement and deformation in the Atlas Mountains.

9.3.2 Paleogeographic Development

Triassic

From Sinai through the Gulf of Suez to Egypt thin Triassic clastic beds can be found with Middle Triassic limestones (Fig.9.15A), suggesting limited marine encroachment over the Arabian-Nubian Shield (Kuss, 1989). Triassic transgression was, however, more pronounced in Libya.

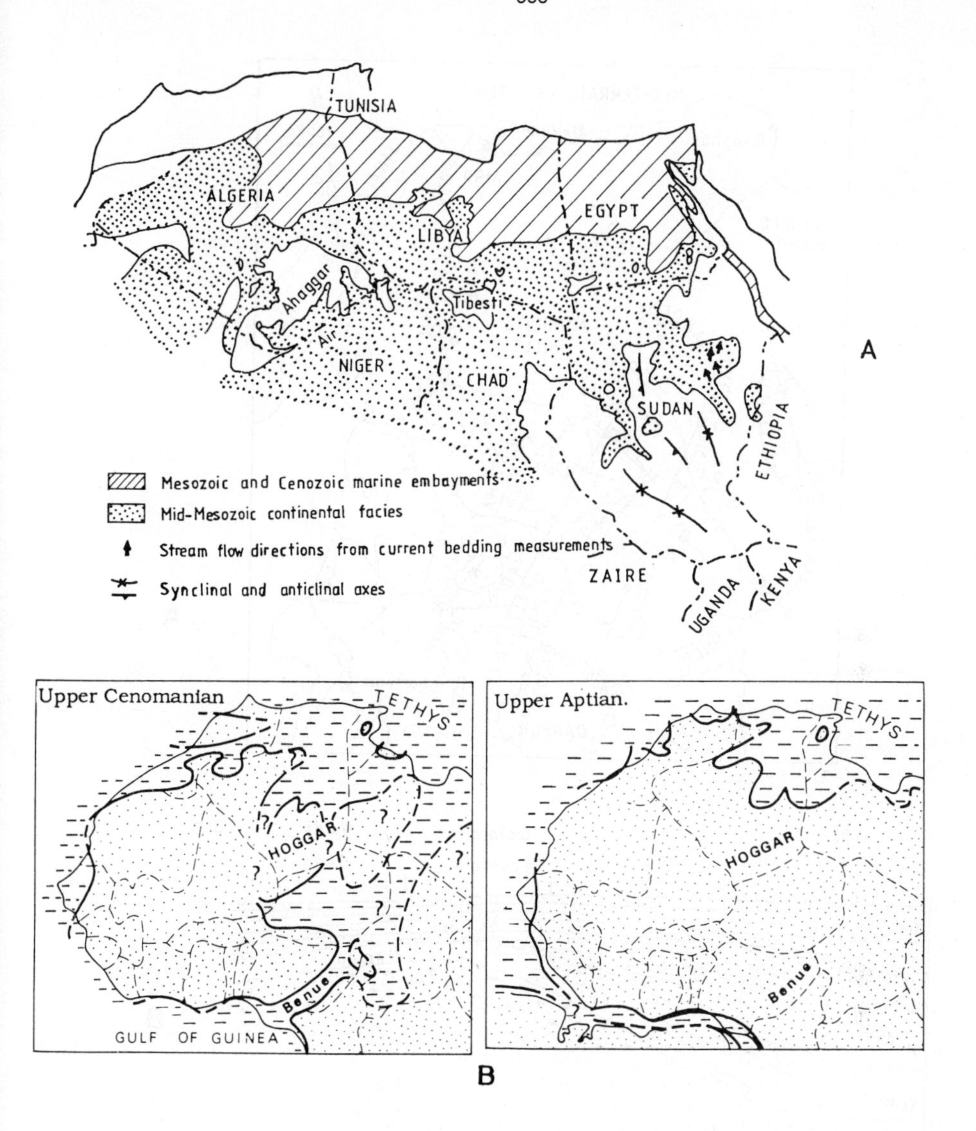

Figure 9.13: A, distribution of mid-Mesozoic environments of the Saharan platform; B, paleogeography of the Late Cretaceous of northern Africa. (Redrawn from Nairn, 1978; Clifford, 1986.)

Jurassic

This is the thickest and most complete interval in the northern part of Sinai (Fig.9.14B) with mostly lagoonal facies and Middle-Late Jurassic marine algal carbonates. The Jurassic accumulated mostly in a depocentre which extended from Sinai through the present Nile delta to the Dakhla

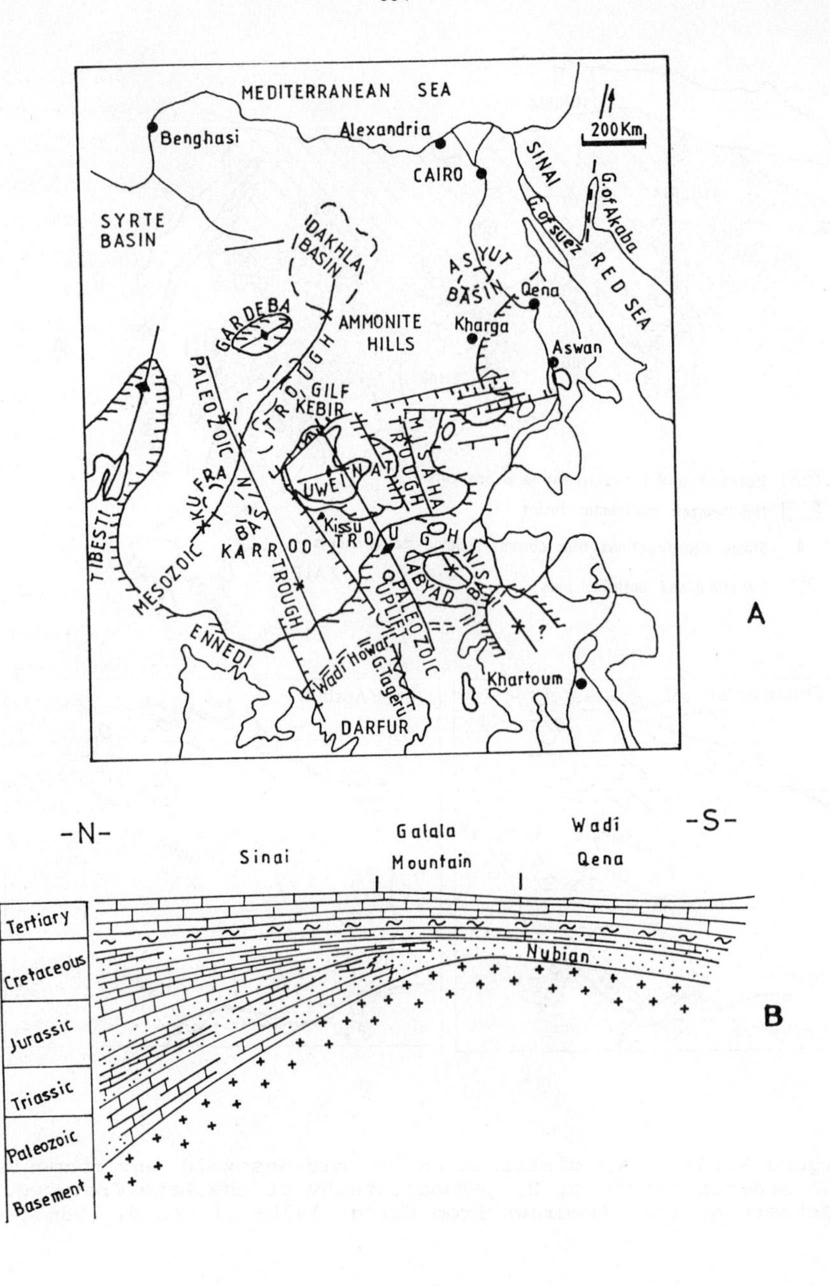

Figure 9.14: Structural elements of NE Africa (A), and section across Sinai. (Redrawn from Klitzsch, 1986; Kuss, 1989.)

basin in the western desert. Several transgressive regressive cycles occurred in the Jurassic of Egypt (Fig.8.4), the most extensive being that of the Middle Jurassic which reached Libya, Tunisia, and Algeria where carbonates and fine clastics of similar age are known. The Late Jurassic-

Early Cretaceous was regressive with extensive alluvial sedimentation (Nubian sandstone) in Egypt (Fig.9.15B) and most of North Africa.

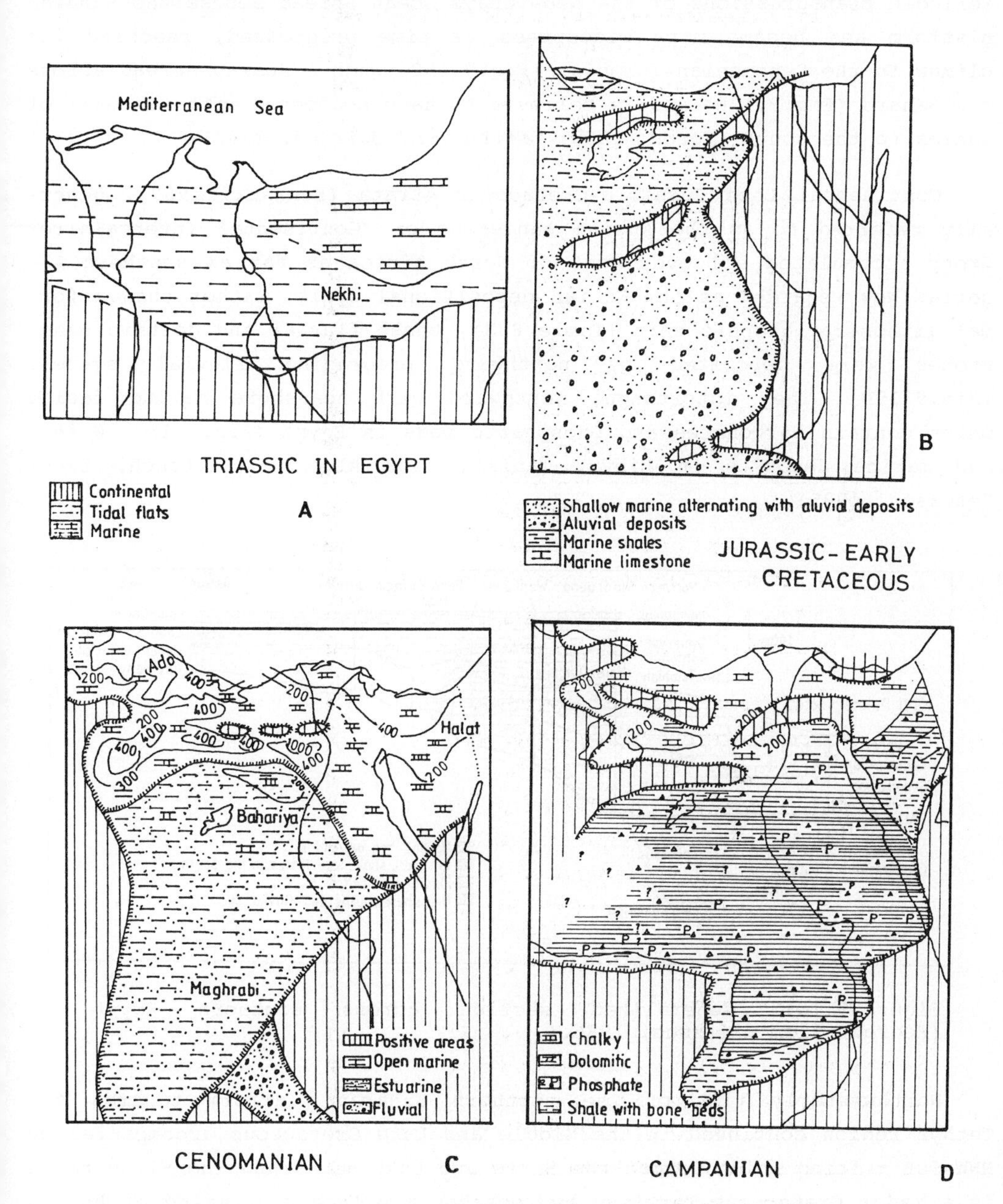

Figure 9.15: Mesozoic paleogeography of Egypt, (Redrawn from Said, 1990a.)

Cretaceous

Periodic transgressions of the Neo-Tethys Ocean spread across the Saharan platform and became more widespread as time progressed, reaching its climax in the Cenomanian-Turonian (Fig.9.15C) when a seaway spread across the Sahara (Fig.9.13B) and is believed to have connected with the Gulf of Guniea to the south (Furon, 1963; Lefranc and Guiraud, 1990).

Continental Late Jurassic-Cretaceous strata (Schrank, 1987), generally referred to as the Nubian sandstone or "Continental Intercalaire" Group accumulated all over most of North Africa as far as southern Algeria. Now regarded as the Nubian depositional cycle, rather than a formal lithostratigraphic unit (Table 8.3), these fluvial and deltaic sandstones were deposited by northward prograding fluvial systems (Fig.9.15B). They intertongue northward with nearshore marine facies which contain carbonates and phosphatic beds in Egypt (Fig.9.15D; 9.16), and marine deposits in Libya, Tunisia, and Algeria (Klitzsch, 1986; Peterson, 1985).

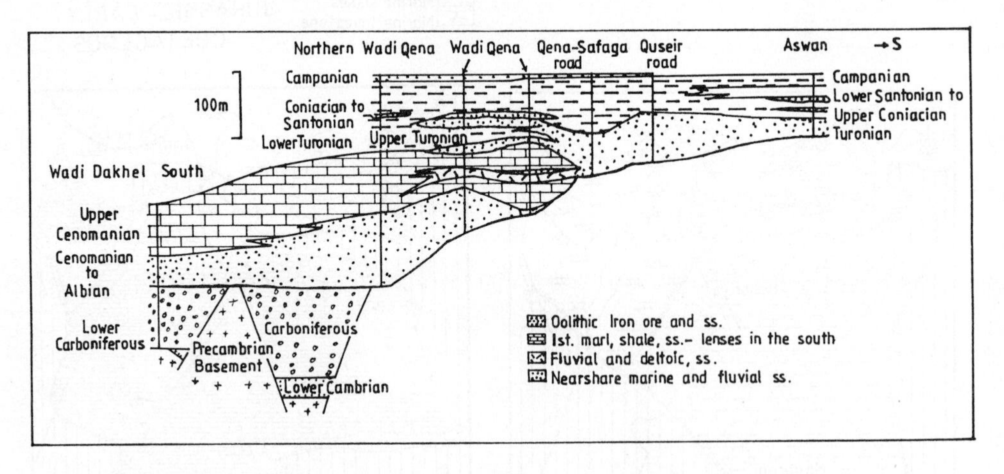

Figure 9.16: Generalized section across southern Egypt. (Redrawn from Klitzsch, 1986.)

Northwesterly tilt of the northern Saharan platform towards the Tethys region continued in the Middle and Late Cretaceous accompanied by NNW-SSE rifting which formed the Sirte and Dakhla basins. The Sirte basin is a major Cretaceous-Tertiary hydrocarbon province consisting of horsts and grabens that had a major influence on sedimentation, especially reefal build-ups which constitute the principal reservoirs (Fig.8.29). Thick, continental Lower Cretaceous sands overlie Cambrian-Ordovician quartzitic sand. Above the Lower Cretaceous sands are thick Upper Cretaceous shales with thick micritic carbonates marking the top of the Cretaceous (Clifford, 1986). There are carbonate build-ups in the Paleocene within a carbonate-shale succession; and Lower Eocene shales in the basin centre pass upward into evaporites, carbonates, and finally into marl and shale. Oil is trapped mainly in Paleocene reefs which developed on the crest of deeper horst blocks.

Paleogene

Maximum Tertiary marine transgression occurred during the Paleocene and extended as far south as the Sudan (Fig.9.17A). In Egypt where the unstable shelf had developed folds in the Late Cretaceous and Early Tertiary, there was a thinner sedimentary cover in that region than in the stable shelf where epeirogenic downwarps had created deeper depocentres (Said, 1990b). The Paleocene in Egypt (Table 8.3) is represented by the Dakhla Shale, the Tarawan, and the lower part of the Esna Shale, while Eocene strata (Upper Esna, Thebes, Mokattam and Maadi Groups) are mostly carbonates with larger foraminifera such as *Nummulites*, and *Alveolina*. The Oligocene in Egypt (Fig.9.17B) was deposited under predominantly continental conditions with fluviatite facies occuring in the south; a northern shelf facies existed in which clays and minor carbonates accumulated.

Of considerable paleoclimatic significance in the Oligocene of Egypt are the petrified forests in which silicified tree trunks suggest tropical climate and vegetation. Also, along the escarpments in the Fayum depression a unique mammalian fauna (Simons and Rasmussen, 1990) is associated with silicified logs in fluvial point bar and floodplain deposits of the Gebel Qatrani Formation (Fig.9.17B), in a paleoenvironmental setting quite reminiscent of Karoo vertebrate localities in South Africa.

Neogene

As from the Late Cretaceous pronounced structural changes took place in Egypt which culminated in the break-up of the Arabian-Nubian Shield and the formation of the Gulf of Aden, an event that is considered later in this chapter. Suffice it to state at this juncture that this tectonic event dramatically affected the structural and paleogeographical framework of Egypt in the Neogene (Said, 1990b). After the Gulf of Suez and the Red Sea grabens opened in the Early Miocene, a marine transgression spread over large areas of northern Egypt. Marine deltaic clays and fluvio-marine deposits accumulated in northern Egypt, and in the newly formed Gulf of Suez fluvial sedimentation also took place (Fig.9.17C). During a late Early Miocene regression the Gulf of Suez was isolated from the Mediterranean Sea, and evaporites formed in the Gulf of Suez, later extending into the Red Sea. Arid conditions began in the Late Miocene, in the course of which thick evaporite sequences accumulated in the Red Sea-Gulf of Suez grabens, and the Mediterranean Sea dried up (Hsu et al., 1973).

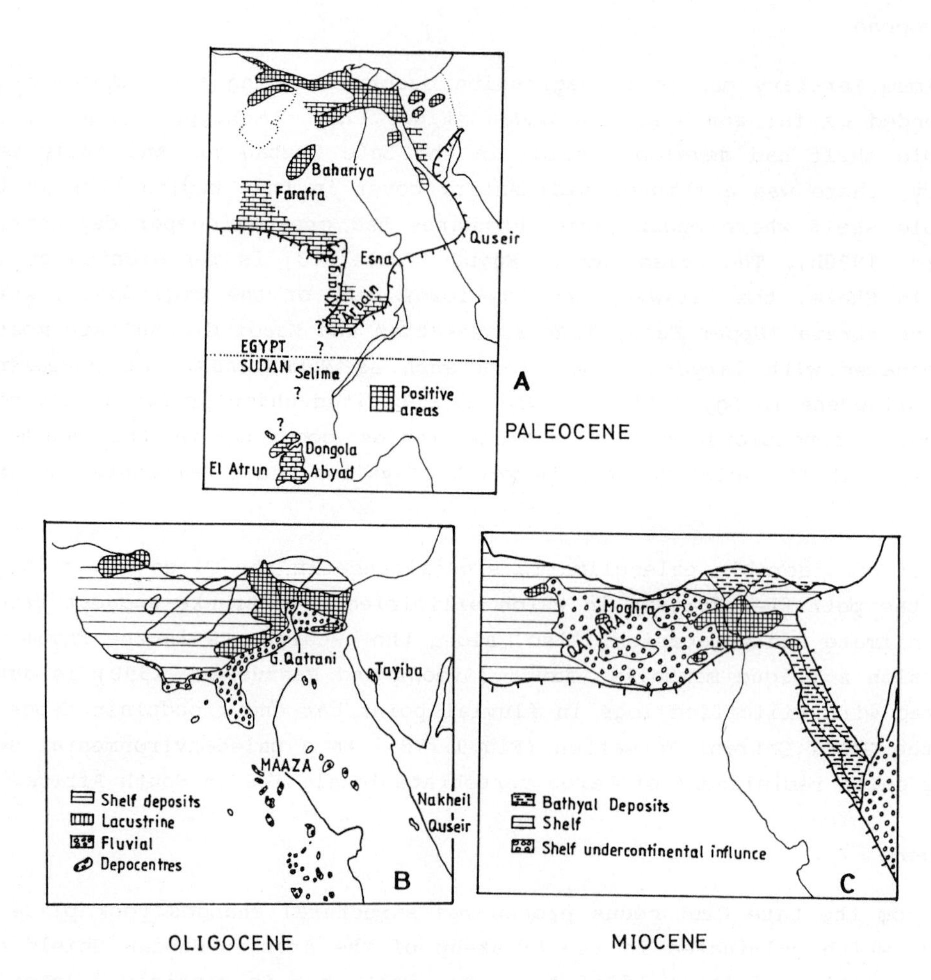

Figure 9.17: Cenozoic paleogeographic maps of Egypt. (Redrawn from Said, 1990b.)

9.4 Evolution of the Atlantic Margin of Africa

9.4.1 Origin and Structure of the African Atlantic Margin

Continental rifting in the Late Triassic in northwest Africa west of the Moroccan Atlas (Fig.9.2A) and the southward propagation of rifting produced some of the world's classic examples of marginal or divergent (Atlantic-type) basins. As shown in Fig.9.18 these basins fall into four broad categories, namely: the Northwest African coastal basins; the Equatorial Atlantic basins; the Aptian salt basins; the Southwest African coastal basins (Clifford, 1986). Each basin group has a common structural style and stratigraphic fill. Besides, there are modifications of the basic plan of the marginal sag basins of the Equatorial Atlantic by wrenching and by the construction of the Niger delta.

Uchupi (1989) observed that the African Atlantic marginal basins originated in Mesozoic rift systems that consisted of four main segments, each separated by oceanic fracture zones or transform faults. Combining the rift segments with the resultant basin types it is evident that the Northwest coastal basins correspond to Uchupi's Northern North Atlantic and Southern North Atlantic rift segments (Fig.9.19); the Equatorial Atlantic segment (Fig.9.20) contains a different group of basins that are modified by transcurrent motion, as well as the Aptian salt basins. The South Atlantic rift segment includes the Southwest African basins and the basins along the margin of Southeast Africa (Fig.9.21).

Except along the oceanic fracture shear zones which separate them, the basic basin structure consists of half-grabens (Fig.9.18). These are typical pull-apart structures. Grabens, however, characterize the fracture zones. Early Cretaceous shear motion along the Equatorial fracture zone has fragmented the basement in the Ivory coast basin, for example, into detached blocks some of which have subsided as grabens; there are steep scarps and basement ridges in the Dahomey basin (Omatsola and Adegoke, 1981) further to the east (Fig.9.20) which also resulted from wrenching.

In the South Atlantic Mesozoic rift system which starts from the Torres-Walvis ridge, a major transform margin was initiated in the Late Jurassic along the roughly east-west Falkland-Agulhas fracture zone. Motion along this fracture zone fragmented the crust into north-south-trending basement highs such as the Malvinas and Maurice Ewing banks, and

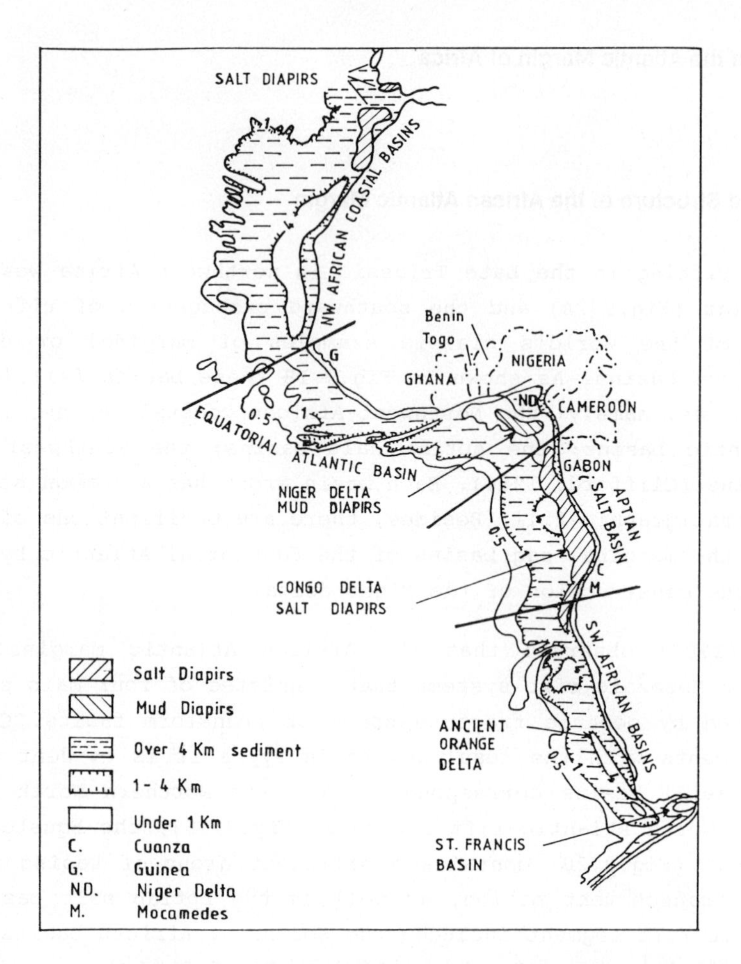

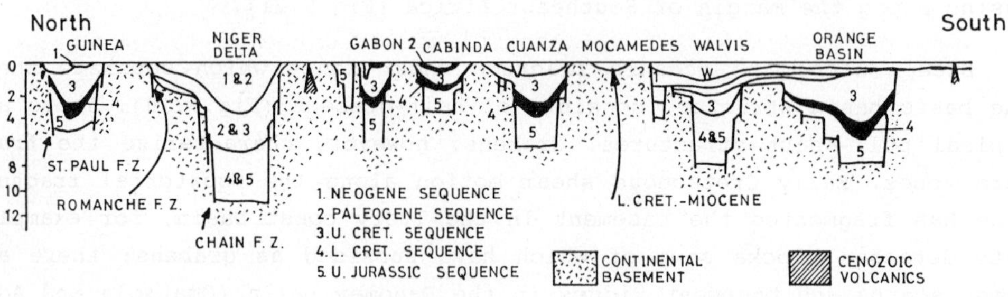

Figure 9.18: Broad subdivisions of the eastern Atlantic continental margin of Africa. (Redrawn partly from Emery et al., 1975; Dingle, 1982).

basins including the Valdez and San Jorge basins in South America, and the Outeniqua and Malvinas plateau basin of South Africa (Fig.9.21). After sea-floor spreading began in the Early Cretaceous these structures exercised much influence over sedimentation.

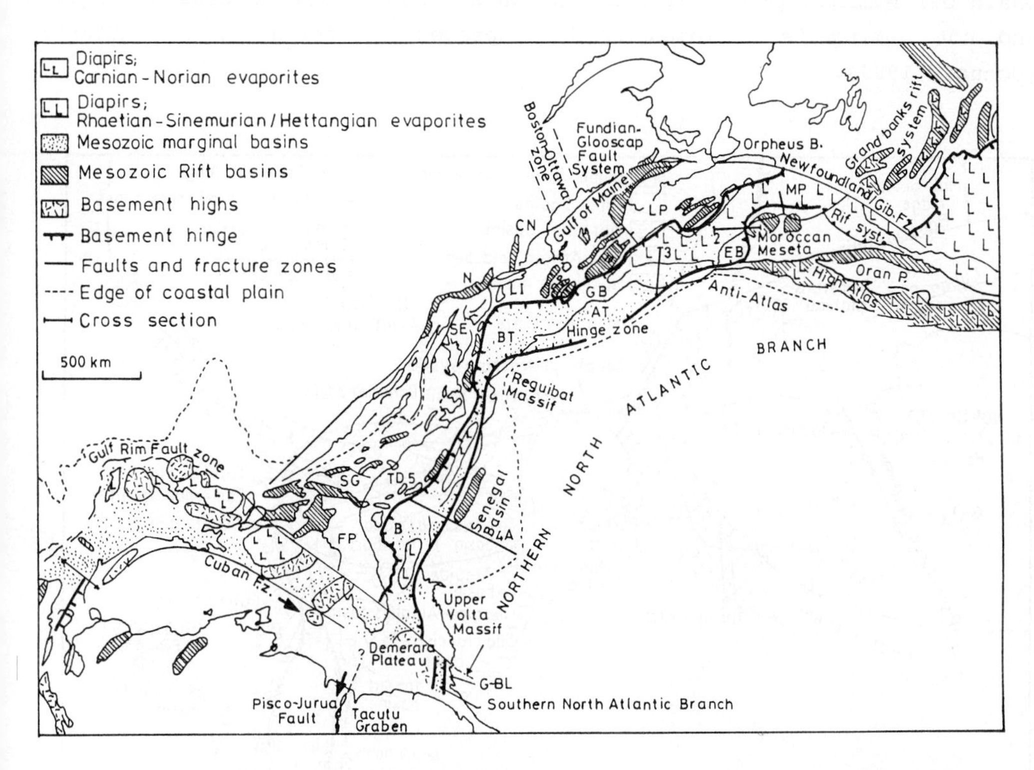

Figure 9.19: Rifts at the end of Early Jurassic before sea-floor spreading in the North Atlantic. AT, Aaiun-Tarfaya basin; B, Blake plateau basin; BSFZ, Blake spur fracture zone; BT, Baltimore canyon; CN, Clinton - Newberrry fault system; CP, Carolina platform; CT, Carolina trough; EB, Essouira basin; F, Franklin basin; FP, Florida platform; GB, Georges bank basin; G-BL, Guinea-Bisseau-Liberia plateau; LI, Long Island platform; MP, Mazagan plateau; N, Newark basin; SE, Salisbury embayment; SE, Southeast Georgia embayment. (Redrawn from Uchupi, 1989.)

The pull-apart margins between the transform faults are characterized by a steep continental basement hinge which faces seaward with a relief often exceeding 8 km, and a marginal sag basin located at the base of the hinge, and platforms and embayments on the landward side of the hinge (Fig.9.22A, B). The continental-oceanic crust boundary roughly lies at the base of the basement scarp, with an attenuated continental crust on the landward side and oceanic basement on the seaward side.

Pre-existing basement structures determined the development of the rifted zone so that where the rifts were aligned parallel to the Pan-African orogens, extension and crustal attenuation was quite pronounced, but where rifts were aligned at right angles, for example the Walvis basin off Namibia (Fig.9.21), extension has been limited. Crustal sagging and not doming is believed to have preceded rifting in the Atlantic (Uchupi, 1989).

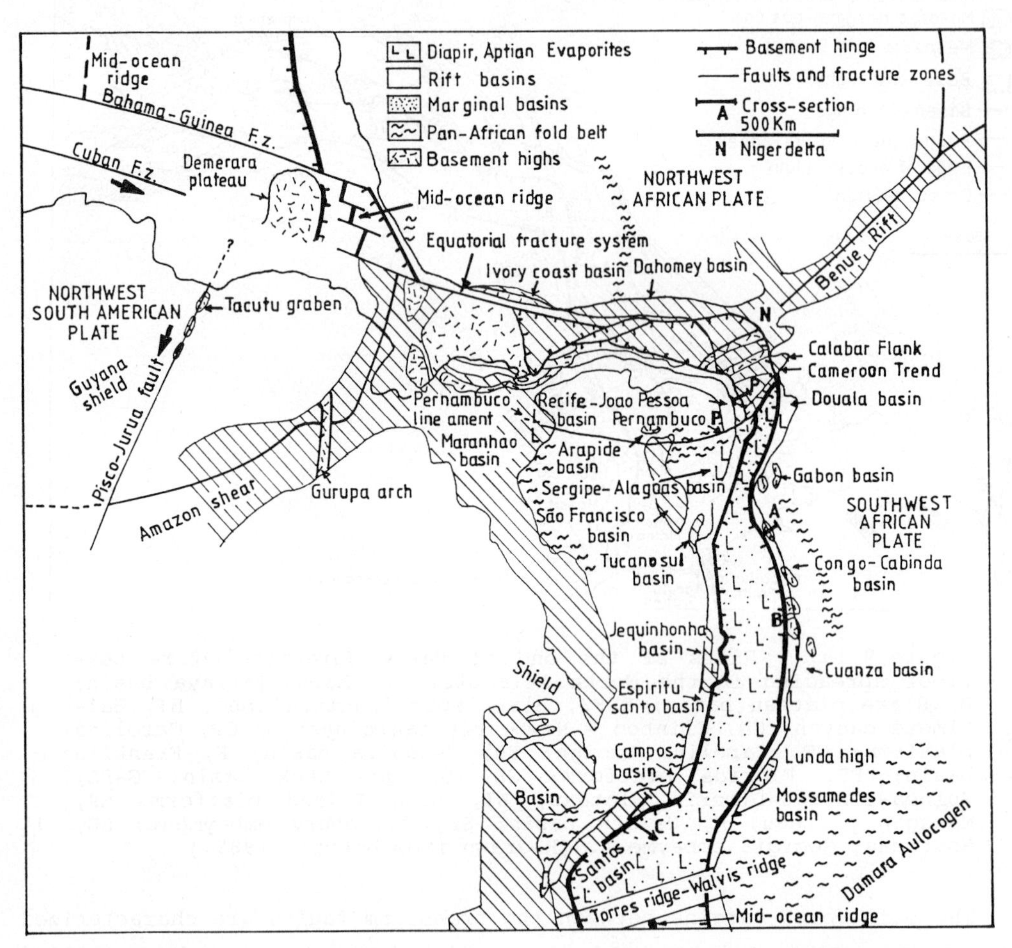

Figure 9.20: Equatorial Atlantic branch of the Atlantic Mesozoic rift system in Aptian times. (Redrawn from Uchupi, 1989.)

In their structure, lithic fill, and hydrocarbon potential the evolution of the Mesozoic Atlantic rift systems produced symmetrical and conjugate basins along the African and the North and South American margins (Figs. 9.19, 20, 21; 23).

The stratigraphic fill in these basins is a consistent record of the stages in their tectonic evolution. The sequence generally begins with syn-rift deposits such as continental red beds, volcanics, carbonates and evaporites, overlain by drift-phase deposits comprising open marine terrigenous sediments which accumulated after continental separation.

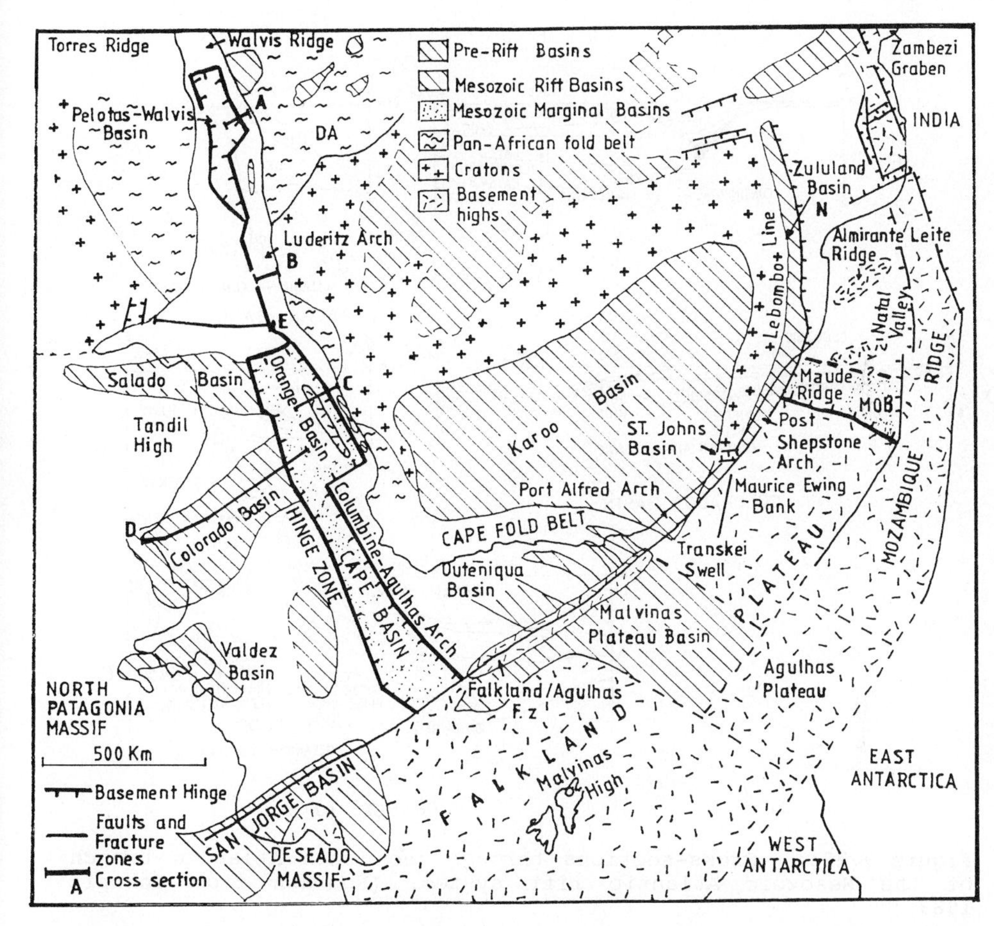

Figure 9.21: South Atlantic branch of the Mesozoic Atlantic rift system in Late Jurassic before initiation of rifting in the Luderitz arch area. (Redrawn from Uchupi, 1989.)

9.4.2 Northwest African Coastal Basins

Several basins occur along the northwestern margin; these include the Essaouira, Agadir, Aaiun-Tarfaya, and Senegal basins (Figs. 9.19; 9.23A-D). Von Rad et al.(1982) and Wiedemann (1987) furnished stratigraphical accounts of the sequences in these basins while Clifford (1986) presented a synthesis on their petroleum geology.

In the Aaiun-Tarfaya basin (Fig.9.23C) typical rift clastics, evaporites, and basalt extruded in the late tensional phase are succeeded by late rift Jurassic clastics and carbonates. These are overlain by early drift Late Jurassic clastics and marls and carbonates; and succeeded by a Cretaceous-Tertiary clastic sequence in which the Paleocene-Eocene is generally unconformable upon Cretaceous deposits.

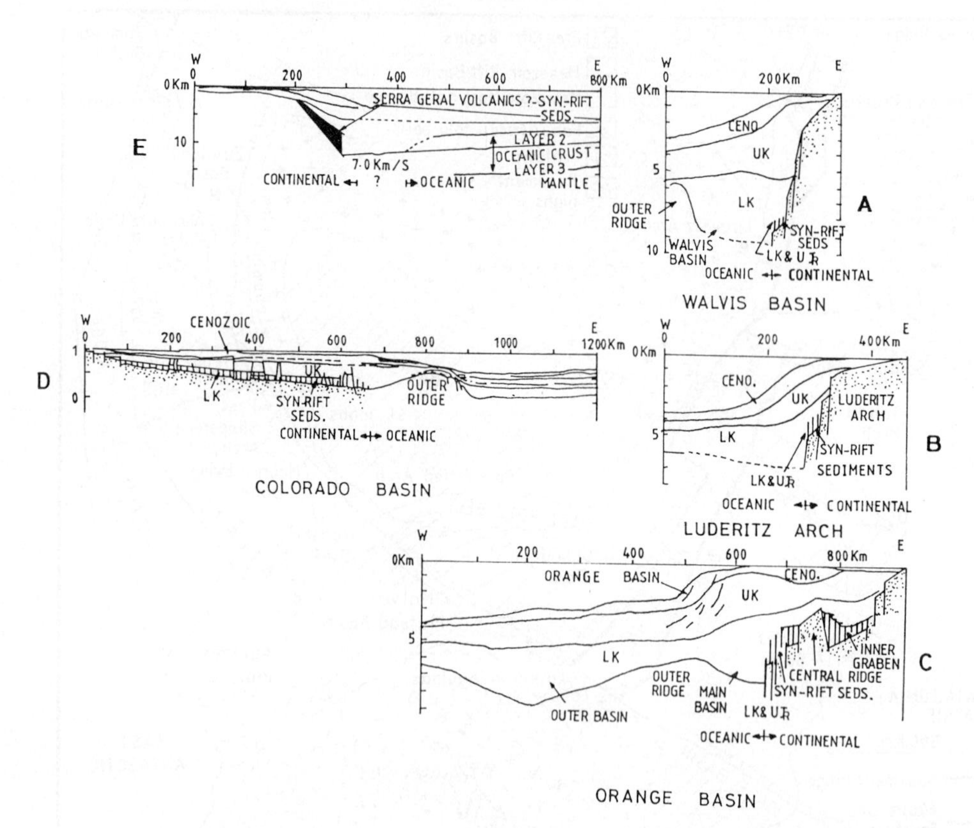

Figure 9.22: Cross-sections through the South Atlantic branch of the Mesozoic Atlantic rift system. (Redrawn from Uchupi, 1989.)

The Senegal basin, the largest onshore embayment in northwest Africa, contains over 6 km of section (Fig.9.23D) which in the onshore part starts with Late Jurassic marine dolomitic limestones. Offshore, Early Cretaceous detrital sediments overlie Early Jurassic evaporites, while more calcareous lithofacies developed on a western carbonate platform. In the overlying Late Cretaceous-Tertiary sequence continental beds in the east interfinger westward with marine terrigenous deposits. A pronounced regression occurred at the end of the Maastrichtian before a widespread Early Tertiary transgression.

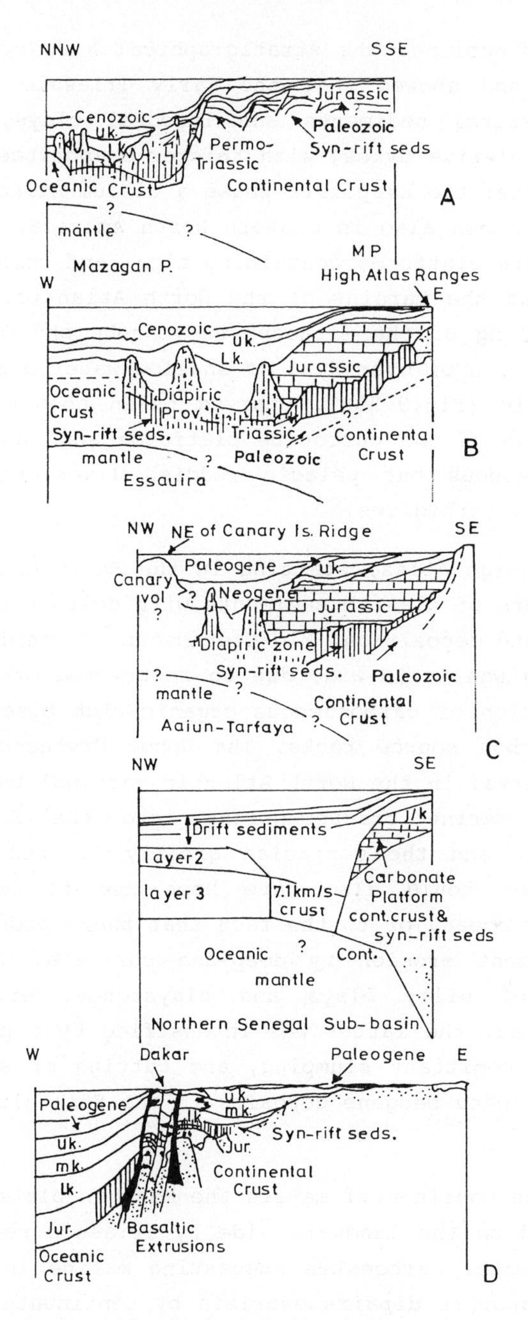

Figure 9.23: Cross-sections through the northern North Atlantic branch. (Redrawn from Uchupi, 1989.)

Seibold (1982) outlined the stratigraphical history of the Essaouira basin of Morocco and showed that the Early Triassic syn-rift deposits comprises red detrital sediments and lagoonal clays, evaporites, and basalt flows and dolerite dykes, with thicker evaporites deposited in the early Jurassic. After the evaporite phase a circum-Atlantic transgression occurred which is known also in eastern North America. This was the main phase when carbonate platforms containing algal and coral reefs were constructed all around the margins of the North Atlantic. Rapid subsidence caused by the cooling of the lithosphere (Onuoha and Ofoegbu, 1988) allowed the accumulation of over 5 km of shallow-water carbonate section in the Essaouira basin (Fig.9.23B). Starved basin conditions prevailed on the deep ocean side of the carbonate platforms and caused deposition of deep-water argillaceous but pelagic reddish limestones containing ammonites, and distal turbidites.

Uplift of the High Atlas, starting in the Early Cretaceous, supplied overwhelming amounts of clastic sediments with deltaic progradation which suppressed carbonate deposition. Mid-Cretaceous (Barremian to Cenomanian) plaeo-anoxia which was world-wide due to restricted oceanic circulation, caused the deposition of carbonaceous organic-rich black shales that are excellent hydrocarbon source rocks. The Upper Cretaceous is an anomalously reduced interval in the North Atlantic marginal basins probably because widespread marine transgressions into the Sahara during the Cenomanian-Turonian and the Coniacian greatly limited sediment supply. Thickness reduction could also have been due to enhanced carbonate dissolution in the ocean, or to the fact that there probably was penecontemporaneous sediment erosion by deep-sea currents. The Paleogene sequence consists of silty clays and claystones, with Paleocene and Oligocene turbidites; the latter was intensified by a pronounced drop of sea level with concomittant slumping, and cutting of submarine canyons. Because of arid climate Neogene deposits in the Essaouira basin form only a thin veneer.

On the Moroccan continental margin the Mazagan plateau (Fig.9.23A) is a platform mantled on the landward side by Triassic red beds with Early Jurassic shallow water carbonates suggesting marine invasion. An extensive area with evaporite dipairs overlain by continental red beds occurs seaward off the basement hinge along the western edge of this plateau. Paleo-oceanographic interpretations (Uchupi, 1989) of the Early Jurassic drift-phase black to gray carbonates and argillaceous facies on the Mazagan plateau reveal that the North Atlantic Ocean was up to 500 km wide and about 1,000 m deep with an estuarine-type vertically stratified water mass with slow bottom and intermediate circulation.

Off the coast of northwest Africa the growth and uplift of volcanic islands began in the later Paleocene and climaxed in the Miocene in the Cape Verde and Canary Island archipelago. Apart from contributing volcaniclastics to marine sediments these island chains diverted the routes of turbidity currents and also confined deep-sea currents thus intensifying submarine erosion.

9.4.3 Equatorial Atlantic Basins

The Liberian marginal sag basin to the north is the only exception in this region of wrench-modified marginal basins, which are sometimes termed the Gulf of Guinea basins. The Tano basin is an eastward extension of the Ivory Coast basin (Fig.9.20), while the Keta basin which underlies the eastern coast of Ghana is the westward continuation of the Dahomey basin. Although it originated through transcurrent movement along landward extensions of oceanic fracture zones, the Benue trough is an interior fracture basin, hence it is not included here. Rather, the Niger delta, which sits across the entrance into the Benue trough, is considered.

The southernmost embayment of the Gulf of Guinea is where the Calabar flank is located, immediately north of the Cameroon trend (Fig.9.20). This feature separates the Equatorial Atlantic basins from the Aptian salt basins. The Calabar flank constitutes the eastern hinge line of the Niger delta, and is underlain by a system of NW-SE-trending horsts and grabens which descend in a step-wise manner into the Niger delta. Highly fossiliferous marine Albian-Maastrichtian beds are exposed along the featheredge of the Calabar flank (Nyong and Akpan, 1987).

The Cenozoic Cameroon volcanic trend sits on an oceanic aseismic ridge which effectively barred Aptian evaporites of the South Atlantic from entering the Calabar flank, thus creating a major stratigraphic discontinuity between this basin and the Douala basin immediately south of the Cameroon trend.

Liberian Basin

This extends from the southern coast of Guinea Bissau through Sierra Leone to Liberia (Fig.8.2). Off the Liberian coast geophysical surveys reveal up to 8 km of sediments, but onshore only 2 km of mostly Cretaceous continental conglomerates, sandstones with plant remains belong to the Farmington Formation. Thinner Tertiary continental sandstones of the Edina Sandstone Formation are also exposed along the coast. But in

Sierra Leone the marine to estuarine Tertiary Bullom Group, at least 100 m thick, consists of sands, kaolinitic clays, and lignite beds which also bear plant and fish remains.

Ivory Coast Basin

From available subsurface information (De Klasz, 1978; Mascle et al., 1987) in offshore Ghana there is up to 2 km of basal rift continental sediments (Fig.9.24B) of Late Jurassic to Early Cretaceous age. A pronounced unconformity of Middle Cenomanian age represents the separation of Africa from South America and the end of rapid syn-rift sedimentation in a basin that was surrounded by Early Cretaceous landmasses and probably characterized by fast sedimentation and rapid subsidence. Drifting of South America from Africa caused the deposition of the overlying Upper Cretaceous-Tertiary sequence of the marginal sag basin phase, with turbidites that were proximal and distal to prograding deltaic wedges. Oil and gas occur (Clifford, 1986) in the turbidites (Belier field) and in shallow marine sands (Espoir field) as shown in Fig.9.24B).

In the Tano basin the basin fill is referred to as the Appollonian "system" in which there is a basal nonmarine Lower Cretaceous overlain by mostly marine limestones, sandstones and shales ranging in age from Albian-Cenomanian to Eocene. A Santonian regression is marked by conglomerates and bituminous sandstones, locally overlain by the Maastrichtian Nauli Limestone (Wright et al., 1985).

The effects of transcurrent deformation are very noticeable in the Ivory Coast basin (Mascle et al., 1987). Seismic lines across the offshore part of the basin show tilted sequences which fill huge asymmetric grabens with possible local syntectonic folding in the basal sequence. Other structures include faults, exposed continental basement; and a deformed sedimentary and basement wedge known as the Ivory Coast ridge (Fig.9.24A). This ridge extends eastward along the continental slope as far as Ghana where continental basement is exposed, and southward it becomes part of exposed oceanic basement. The Ivory Coast-Ghana ridge is regarded as the landward continuation of the Romanche fracture zone; this ridge is located along the continent oceanic crust boundary. It caused the landward ponding of sediments along the continental margin. According to Mascle et al. (1987) the ridge shows evidence of both tensional faults (asymmetric grabens) and compressional deformation which includes shearing folds that die out near the continent. Deformation appears to have been contemporaneous with progressive shearing when a deep basin was developing between the African and South American plates. It ended abruptly

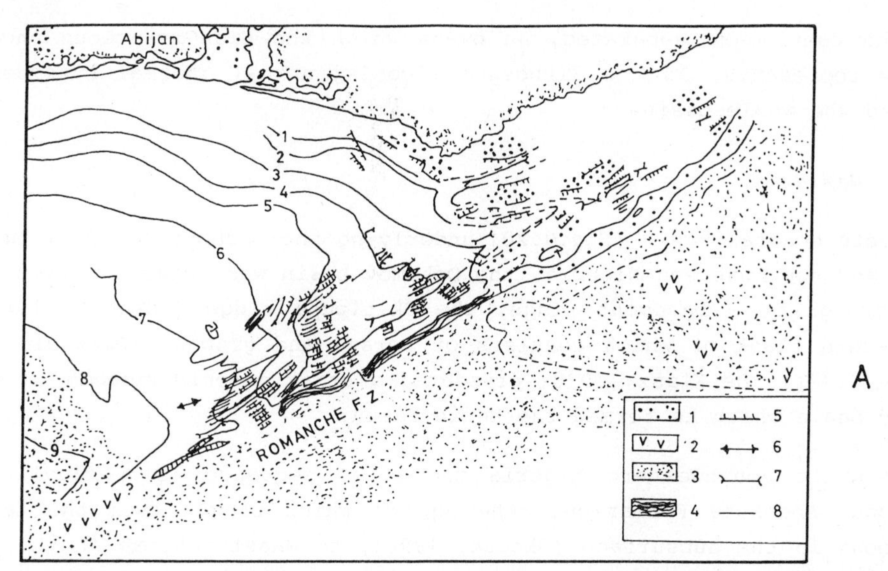

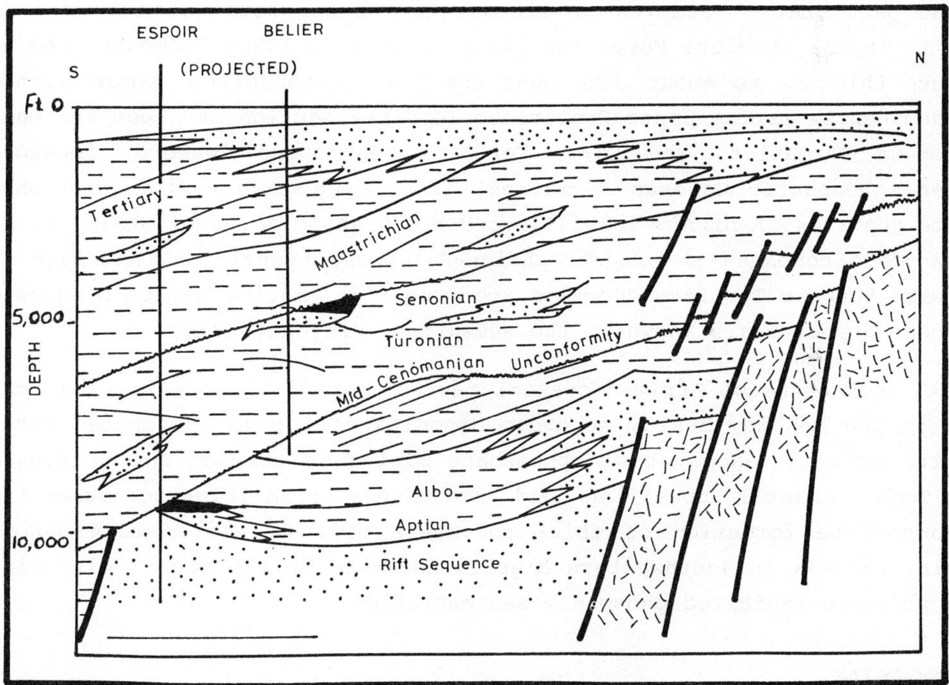

Figure 9.24: A, structural sketch map of the eastern Ivory Coast basin; 1, sub-cropping basement; 2, oceanic basement ridges; 3, oceanic basement; 4, Ivory Coast ridge (deformed sedimentary and basement wedge); 5, main distensive faults; 6, main anticline axis; 7, main synclines and basin axis; 8, isochrons of the Lower Cretaceous-Upper Cretaceous unconformity (Albian?). B, cross-section across Ivory Coast basin. (Redrawn from Mascle et al., 1987; Clifford, 1986.)

when the continents separated, an event which the Mid-Cretaceous unconformity represents. As the lithosphere cooled normal thermal subsidence affected the whole basin.

Dahomey Basin

An arcuate coastal basin (Fig.9.20) underlying the onshore parts of Togo, Benin, and southwestern Nigeria, the Dahomey basin was separated from the Benue trough by a basement ridge, the Okitipupa ridge (Adegoke, 1969), the western flank of which consists of horsts and grabens (Omatsola and Adegoke, 1981). De Klasz (1978) presented a stratigraphic summary of the Dahomey basin which he termed the Benin basin.

From Togo to southwestern Nigeria the basal sequence is a fluviatile sandstone (Abeokuta Formation), the age of which ranges from the Early Cretaceous in the subsurface (Okosun, 1990), to Maastrichtian in the exposed part. Large reserves of bitumen have been proven in tar sand deposits in the Abeokuta Formation (Adegoke et al., 1980). Downdip, the sequence thickens to about 3 km near the present shoreline, where a thick predominently marine Upper Cretaceous-Tertiary section succeeds the basal Abeokuta Formation. The Lower Tertiary (Paleocene Ewekoro Limestone, Eocene phosphatic Oshosun Formation) is exposed in limestone and phosphate quarries. A highly fossiliferous black shale (Araromi Shale) in the subsurface contains the Maastrichtian-Paleocene transition. Although the exposed younger Tertiary deposits are nonmarine, marine latest Oligocene-Miocene deposits are known in the subsurface (Fayose, 1970).

As in the northwest African coastal basins, especially the Senegal basin and in the Dahomey basin, economic deposits of Eocene phosphates occur. In the western Dahomey basin phosphate mining is carried out at Hahatoe in Togo. Coastal upwelling and remoteness from clastic deposition favoured the formation of thick phosphate beds in the western Dahomey basin, whereas in southwestern Nigeria nearness to the Niger delta clastic province inhibited phosphate sedimentation.

Niger Delta

The Niger Delta is among the world's largest petroleum provinces, and has been rated as the sixth largest oil producer and twelfth giant hydrocarbon province (Ivanhoe, 1980).

The southernmost and last of the major deltaic complexes constructed in the Benue trough, the Niger delta began to form in the Eocene and now contains over 12 km of clastics which fill the entrance into the Benue trough (Fig.9.18). The Niger delta complex is a regressive offlap sequence which prograded across the southern Benue trough (Fig. 9.25) and spread out onto cooling and subsiding oceanic crust, which had formed as Africa and South America separated. Because of its petroleum resources detailed information is available on the geophysics (eg. Hospers, 1965, Onuoha, 1981), petroleum geology (Evamy et al., 1978; Orife and Avbovbo, 1982; Short and Stauble, 1967; Whiteman, 1982) of the Niger delta. The Recent sedimentary environments (Allen, 1965) serves as a classic model for high energy, wave-dominated, constructional, arcuate-lobate tropical deltas.

The oldest formations (Paleocene-Eocene) in the Niger delta form an arcuate exposure belt along the delta frame (Fig.9.25A). These are the Paleocene Imo Shale (fossiliferous blue-gray shales with thin sandstones, marls and limestones, and locally thick nearshore sandstones); the Eocene Ameke Formation (fossiliferous calcareous clays, coastal sandstones); the Late Eocene-Early Oligocene lignitic clays and sandstones of the Ogwashi-Asaba Formation, and the Miocene-Recent Benin Formation (Coastal Plain Sands). These formations are highly diachronous and extended into the subsurface where they have been assigned different formation names (Fig.9.26). The Akata, Agbada, and Benin formations are interfingering facies equivalents representing pro-delta, delta-front and delta-top environments respectively (Fig.9.27). Figure 9.27 depicts the delta sub-environments and their sedimentary characteristics (Allen, 1965; Weber and Daukoru, 1975). Unconformities, large clay fills of ancient submarine canyons, and deep-sea fans occur in the eastern and western delta (Burke, 1972; Orife and Avbovbo, 1982). These formed mainly during Early Oligocene and Tertiary lowstands of sea-level.

Whiteman (1982) gave the following outline of the geological history of the Niger delta. By the Middle Eocene the major depocentres initiated in the Paleocene-Eocene (in the Anambra basin, Afikpo syncline, Ikang trough) were the sites of deltaic outbuilding with the Niger-Benue and the Cross River drainage systems (Fig.9.25B) accounting for the bulk of the sediment supply. Both drainage systems merged at the end of the Oligocene and formed the present Niger delta. Simple growth faults were initiated in the Oligocene. During the Miocene uplift of the Cameroon Mountains provided a new and dominant sediment supply through the Cross River, thus constructing of the Cross River delta. As shown in Fig.9.26C the shoreline progressively migrated seaward during deltaic progradation. This was greatly accelerated in Miocene-Pliocene times with attendant increase in growth faulting and large-scale diapiric movement of the Akata Shale. This involved deep mass movement of the undercompacted and over-

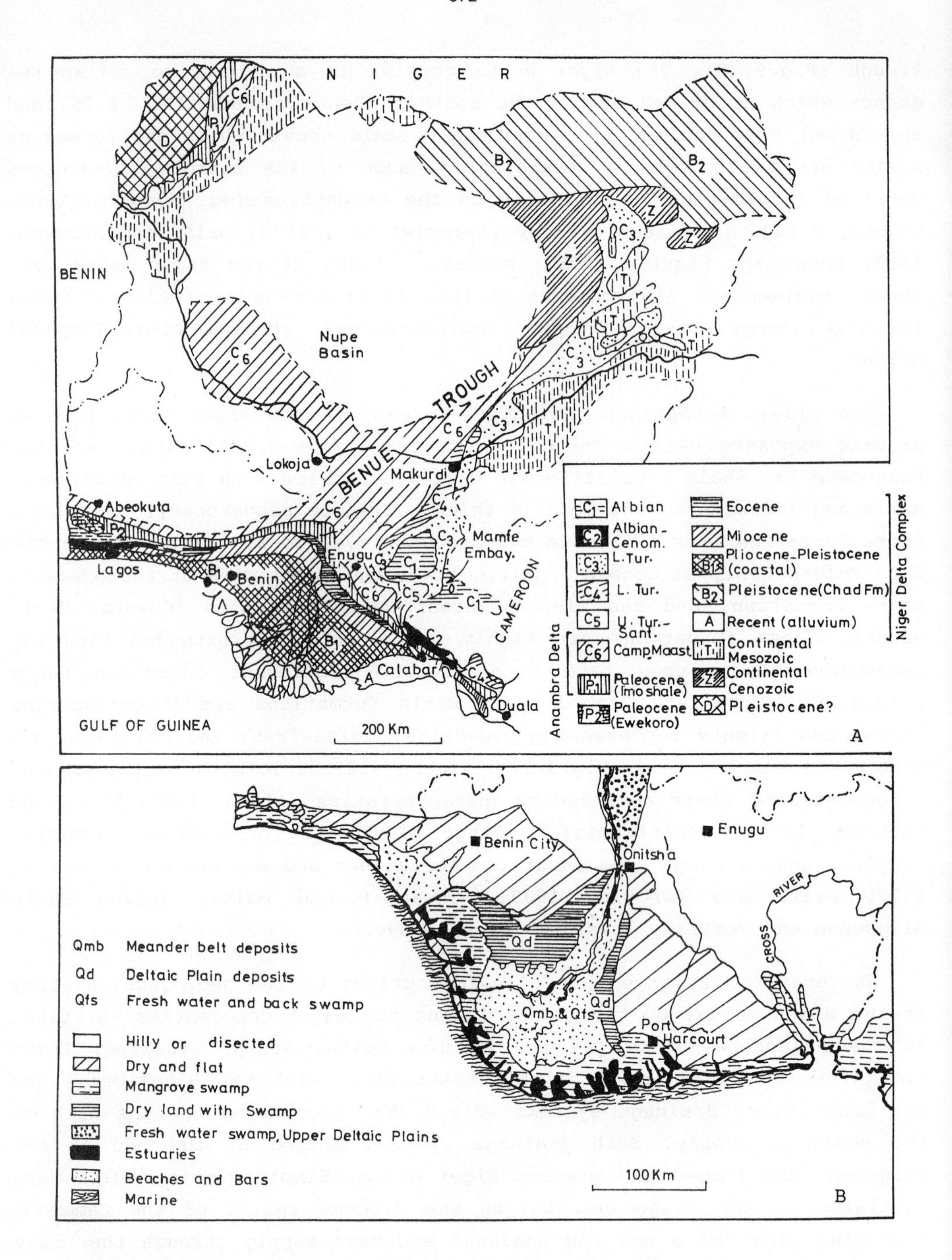

Figure 9.25: Outline geological map of Nigeria (A); B, Recent environments of the Niger delta. (Redrawn from De Klasz, 1978; Allen, 1965.)

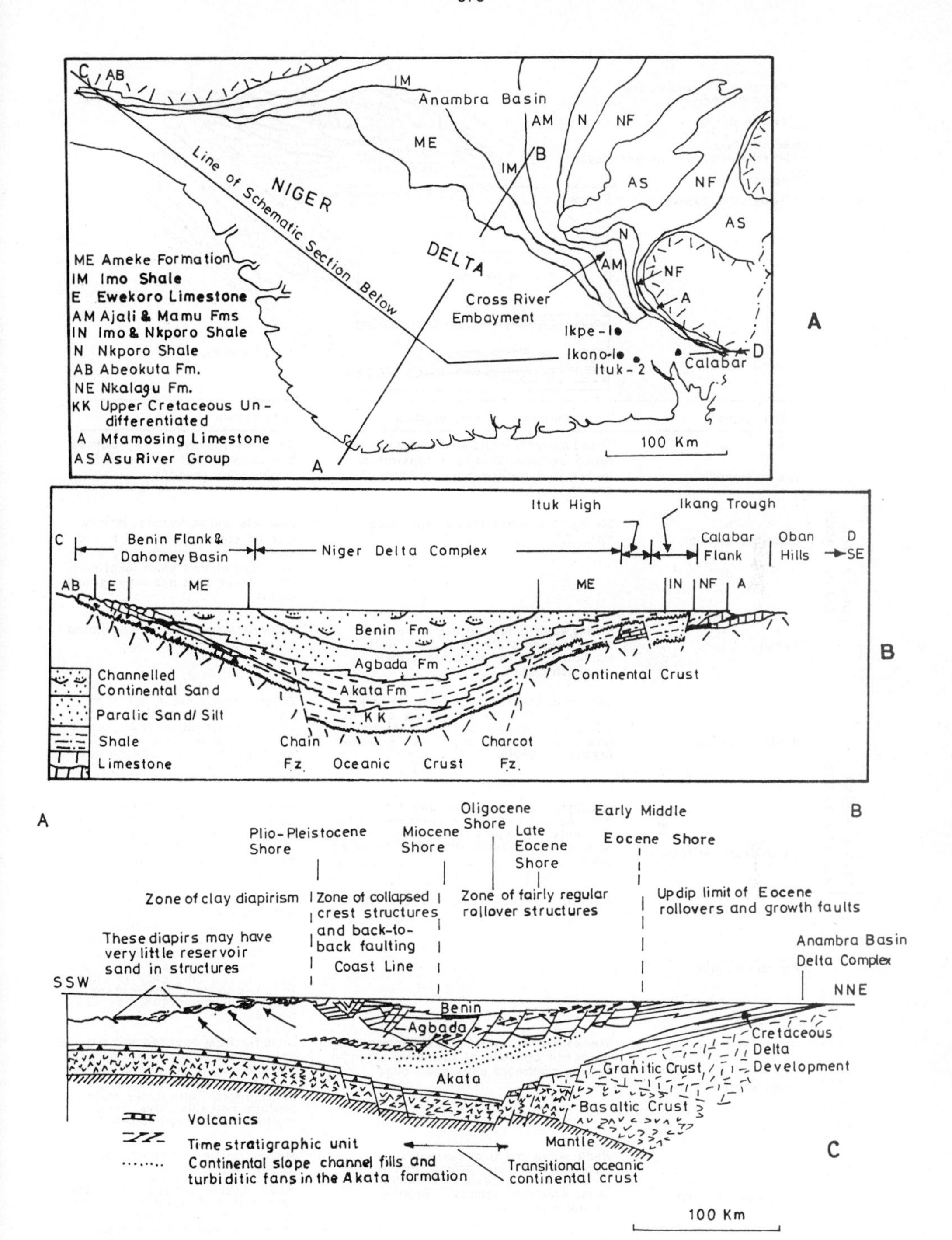

Figure 9.26: Geologic sketch map of southern Nigeria (A), and cross-sections (B,C). (Redrawn from Reijers and Petters, 1987; Whiteman, 1982.)

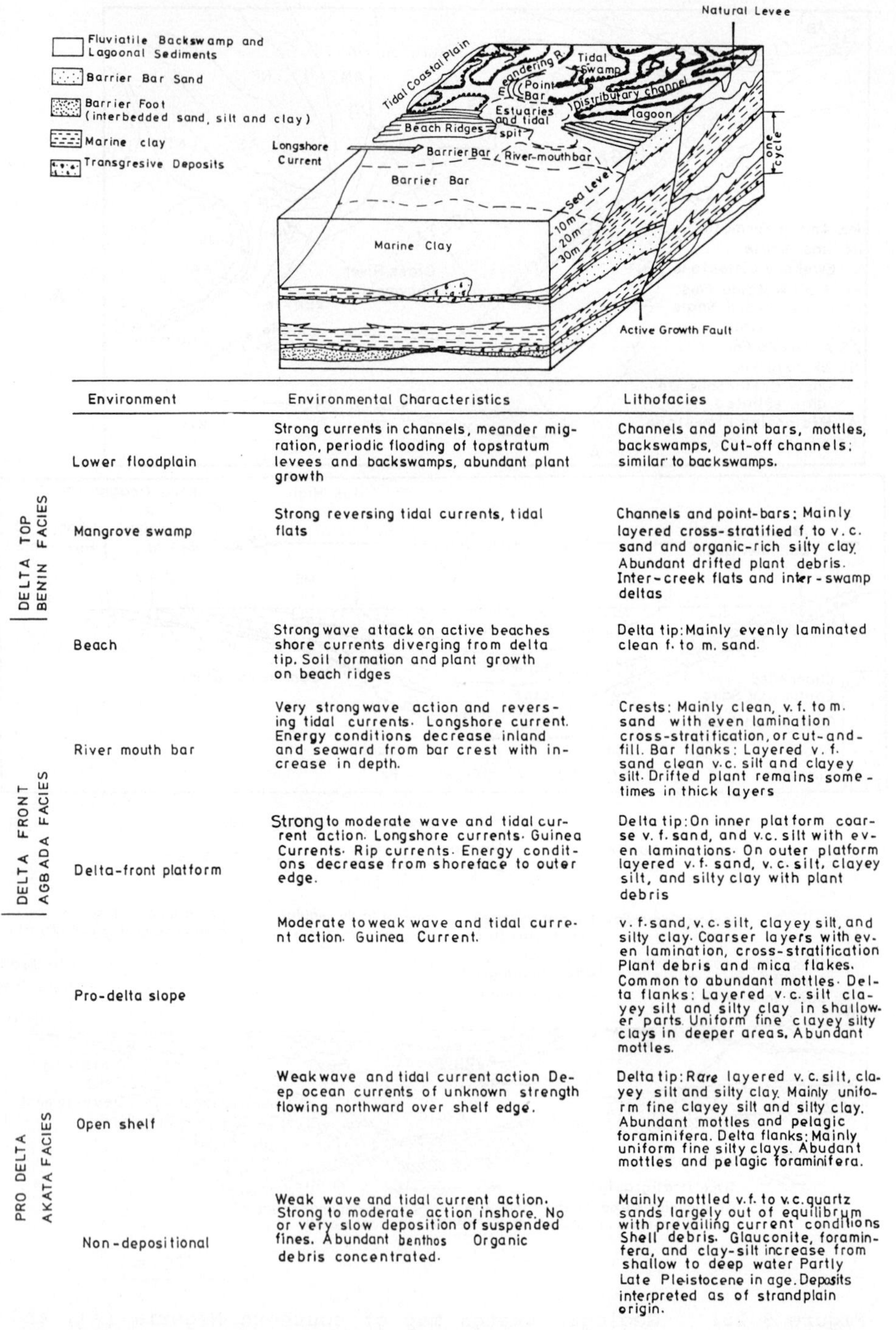

	Environment	Environmental Characteristics	Lithofacies
DELTA TOP BENIN FACIES	Lower floodplain	Strong currents in channels, meander migration, periodic flooding of topstratum levees and backswamps, abundant plant growth	Channels and point bars, mottles, backswamps, Cut-off channels; similar to backswamps.
	Mangrove swamp	Strong reversing tidal currents, tidal flats	Channels and point-bars; Mainly layered cross-stratified f. to v.c. sand and organic-rich silty clay. Abundant drifted plant debris. Inter-creek flats and inter-swamp deltas
DELTA FRONT AGBADA FACIES	Beach	Strong wave attack on active beaches shore currents diverging from delta tip. Soil formation and plant growth on beach ridges	Delta tip: Mainly evenly laminated clean f. to m. sand.
	River mouth bar	Very strong wave action and reversing tidal currents. Longshore current. Energy conditions decrease inland and seaward from bar crest with increase in depth.	Crests: Mainly clean, v.f. to m. sand with even lamination cross-stratification, or cut-and-fill. Bar flanks: Layered v.f. sand clean v.c. silt and clayey silt. Drifted plant remains sometimes in thick layers
	Delta-front platform	Strong to moderate wave and tidal current action. Longshore currents. Guinea Currents. Rip currents. Energy conditons decrease from shoreface to outer edge.	Delta tip: On inner platform coarse v.f. sand, and v.c. silt with even laminations. On outer platform layered v.f. sand, v.c. silt, clayey silt, and silty clay with plant debris
PRO DELTA AKATA FACIES	Pro-delta slope	Moderate to weak wave and tidal current action. Guinea Current.	v.f. sand, v.c. silt, clayey silt, and silty clay. Coarser layers with even lamination, cross-stratification Plant debris and mica flakes. Common to abundant mottles. Delta flanks: Layered v.c. silt clayey silt and silty clay in shallower parts. Uniform fine clayey silty clays in deeper areas. Abundant mottles.
	Open shelf	Weak wave and tidal current action Deep ocean currents of unknown strength flowing northward over shelf edge.	Delta tip: Rare layered v.c. silt, clayey silt and silty clay. Mainly uniform fine clayey silt and silty clay. Abundant mottles and pelagic foraminifera. Delta flanks: Mainly uniform fine silty clays. Abudant mottles and pelagic foraminifera.
	Non-depositional	Weak wave and tidal current action. Strong to moderate action inshore. No or very slow deposition of suspended fines. Abundant benthos Organic debris concentrated.	Mainly mottled v.f. to v.c. quartz sands largely out of equilibrum with prevailing current conditions Shell debris. Glauconite, foraminfera, and clay-silt increase from shallow to deep water Partly Late Pleistocene in age. Deposits interpreted as of strandplain origin.

Figure 9.27: Sedimentary environments of the Recent Niger delta and lithofacies characteristics. (Redrawn from Weber and Daukoru, 1975; Whiteman, 1982.)

pressured shale towards the continental slope. Deltaic growth declined in the Late Pliocene-Pleistocene during a major drop in sea-level, with sediment by-passing into deep-sea fans. A Late Pleistocene transgression flooded the Plio-Pleistocene offlap upper and lower deltaic plains, and as sea-level stabilized, a new regressive offlap sequence developed.

Deltaic-front sands account for the bulk of the Niger delta hydrocarbon production. These interfinger with pro-delta source beds. Turbidite sands are also significant reservoirs. Growth faults and shale diapirs provide the traps. Growth faults are very common in the Niger delta with primary and secondary synthetic and antithetic faults (Fig.9.26C). In the distal offshore part of the delta, petroleum accumulations are also located at the top and flanks of shale diapirs.

9.4.4 Aptian Salt Basins

This group of basins extends from Cameroon to Angola and includes the Douala, Gabon, Congo, Cuanza, and Mossamedes basins (Fig.9.20). These basins which are narrow embayments onshore, with the Gabon basin (50,000 km^2) and the Cuanza basin (22,000 km^2) having the largest land surfaces, are more extensive offshore because structurally, the basins consist of half-grabens which are mostly down-dropped seaward. The Cuanza basin (Fig.9.28), is however better exposed, with the Late Jurassic-Early Cretaceous to Quaternary sedimentary fill outcropping in a belt about 150 km wide. Otherwise, the subaerial parts of these basins range from low vegetated coastal plains in the northern part to low-latitude desert in Angola.

Considerable stratigraphic uniformity exists in the Aptian Salt basin (Fig.9.29). Above the faulted crystalline basement or the Karoo Supergroup, as the case may be in some basins (eg. Gabon basin), the Mesozoic-Cenozoic succession comprises three distinct units: a basal nonmarine sequence, overlain by evaporites, followed by normal marine strata (De Klasz, 1978; Franks and Nairn, 1973). The basal Jurassic to Early Cretaceous fluviatile and lacustrine rift sequence are variously termed "Grès de base" (Douala basin), M'Vone Sandstone and Marl, N'Dombo Sandstone and Cocobeach Series (Gabon), and the Cuvo Group (Cuanza). According to Uchupi (1989) deposition of clastic sediments in the Jurassic was followed by rifting and subsidence in the Early Cretaceous, during which a deep anoxic lake formed which extended from Angola to Gabon, and was filled with organic-rich dolomites, green shales and carbonates. During the second rift pulse which began in the Barremian the initial rift, now broken into a series of secondary basins, was filled with non-marine carbonates and shales. The region was invaded by the sea from the south as the basin subsided, and evaporites were deposited which grade upward into dolomites. A transgressive sheet sand (Gamba, Chela and Upper Cuvo Formations) accumulated before the formation of Aptian salts.

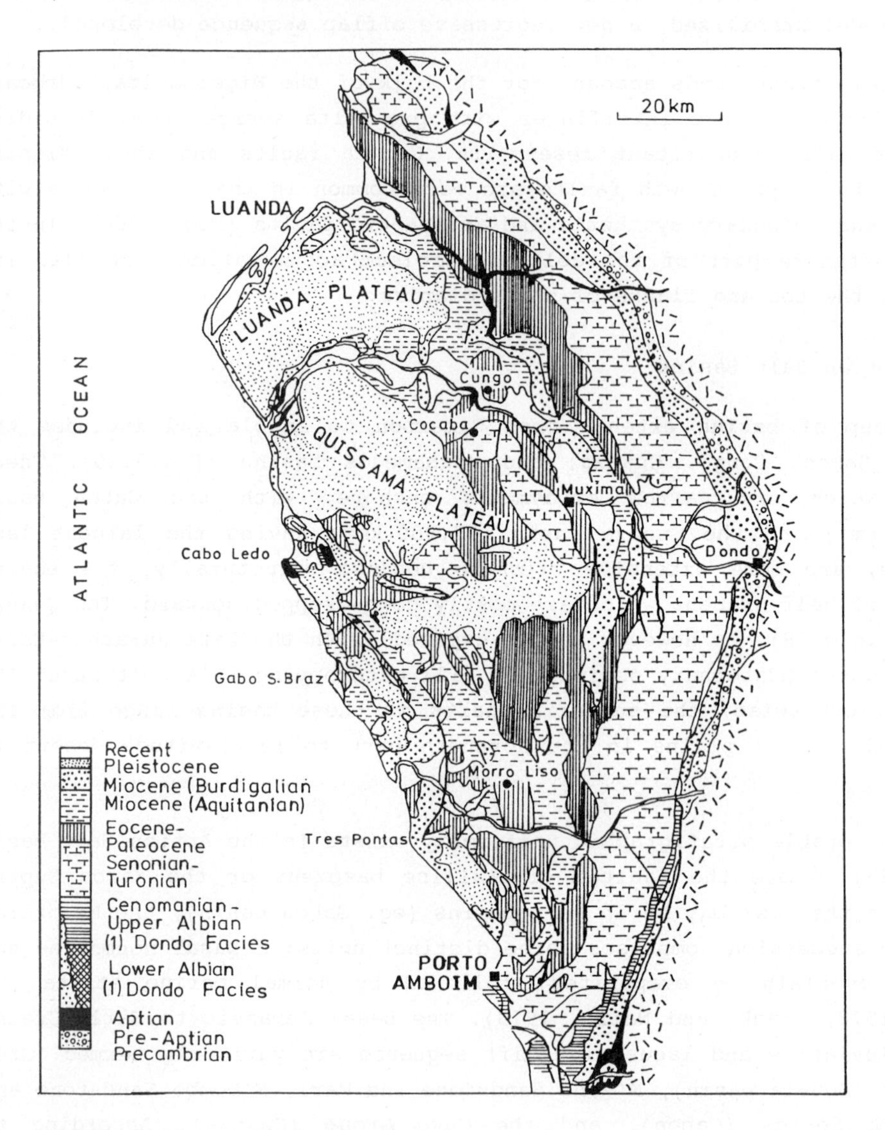

Figure 9.28: Geologic sketch map of the Cuanza basin. (Redrawn from De Klasz, 1978.)

The evaporites, mostly halite and carnallite, were deposited in cycles and show great horizontal uniformity over distances of 100 km because of deposition in a large basin in which the inflow of marine water was restricted by local barriers. The evaporites in the northern basins contain a higher proportion of potash salts and a lower proportion of sulphates than those in Angola to the south. This was probably due to a much more direct access to sea-water which came from the south, whereas the northern part of the evaporite basin was more confined and concentrated. Another difference between the southern (Angola) and northern parts of the African salt basins was that in the former area, the main evaporite is overlain by a marine sequence containing limestone and dolomite with anhydrite horizons in which there was renewed formation of halite and anhydrite.

Anoxic conditions prevailed in the Late Aptian, after the formation of the evaporites. This was because the entire basin was closed to the north and isolated from the ocean to the south by the Torres-Walvis ridge (Fig.9.20). This led to the deposition of varved clays. Massive reefal carbonate platforms were built atop basement highs in the upper Aptian-Lower Albian, above the anoxic sediments, during the transition from anoxic to oxic conditions, as a result of the breaching of the Torres-Walvis ridge barrier. The Upper Cretaceous-Tertiary sequence is mostly fossiliferous marine clastics and marl with a few carbonate cycles, and unconformities which were caused by subsidence and sea-level changes. From this stratigraphic outline it is apparent that the Aptian Salt basins originated from one rift setting, the western parts of which now constitute the Brazilian coastal basins.

The structural and stratigraphic similarities between the Mesozoic sequence in the Aptian salt basins and the Brazilian coastal basins have attracted considerable attention over the years. Strong faunal similarities exist among nonmarine and marine ostracodes (Kromelbein and Wenger, 1966), ammonites (Reyment and Tait, 1972) and vertebrates (Buffetaut and Taquet, 1979). The Brazilian coastal basins (Recife-Joao Pessoa, Sergipe-Alagoas, Reconcavo, Camamu-Almada, Jequitinhonha-Espiritu Santo, Campos, Santos) share the same depositional history with the African Aptian salt basins (Ponte and Asmus, 1973; Uchupi, 1989). The Late Jurassic infra-rift sequence in Brazil consists of alluvial sands, lacustrine red shales with sand interbeds and sandstones; overlain by Neocomian rift-phase fluvio-lacustrine sandstones, shales and some limestones, with the Serra Geral Early Cretaceous volcanics (9.22E) constituting the youngest syn-rift unit. These are overlain by Aptian euxinic shales, evaporites, and carbonates, as in the African basins. Whereas the southern Brazilian basins all subsided when sea-floor spreading began in the Albian, there was lesser subsidence in the Sergipe-Alagoas and Recife Joao Pessoa basins at the northern end. This paleogeographic situation hindered free communication between the North Atlantic and the South Atlantic. The connection was, however, established in the Turonian when the Equatorial Atlantic ocean opened.

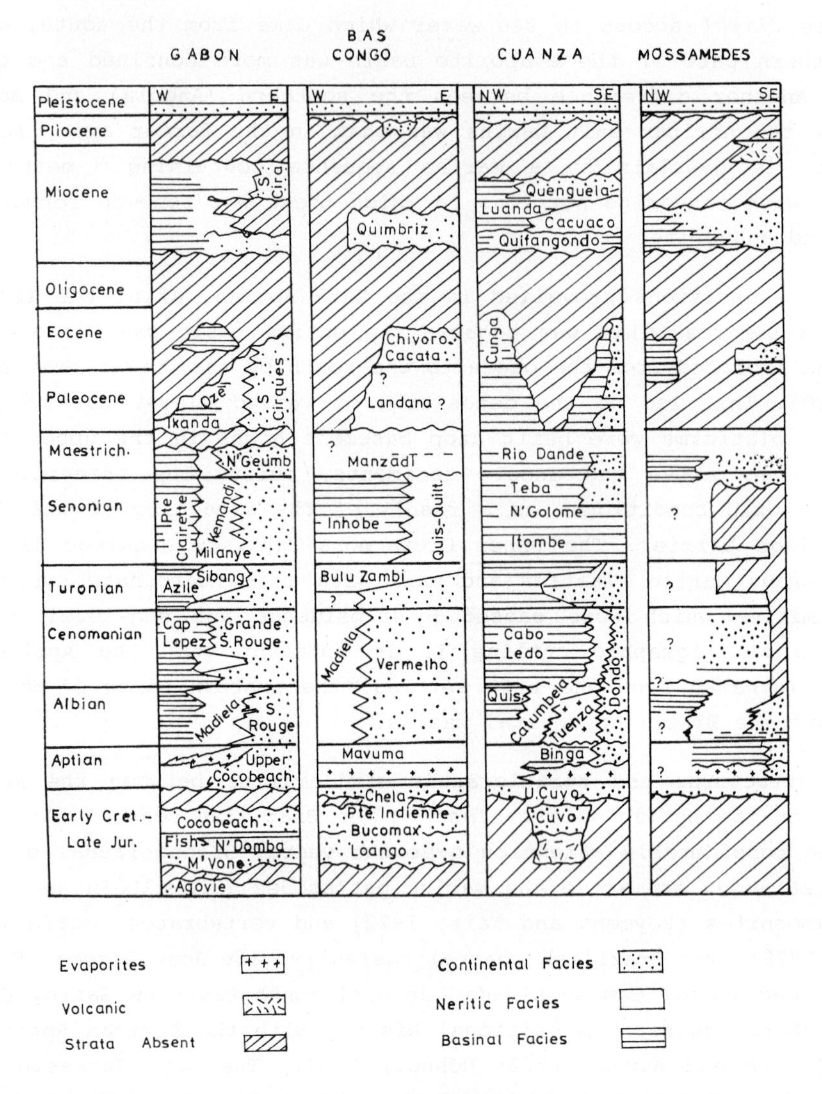

Figure 9.29: Stratigraphic correlations of South Atlantic marginal basins. (Redrawn from Franks and Nairn, 1973.)

Clifford (1986) showed that the Aptian salt basins which are major petroleum producers have as hydrocarbon reservoirs pre-salt fluviatile and turbidite (lacustrine) sands, pre-salt freshwater carbonate reefs, post-salt marine carbonates and intertidal channel or estuarine sands (Fig.9.30). The transgressive sheet sand, the Gamba, Chela, and Upper Cuvo also provide excellent reservoirs. Salt movement has played a major role in providing structural traps (Fig.9.30), especially where the salt

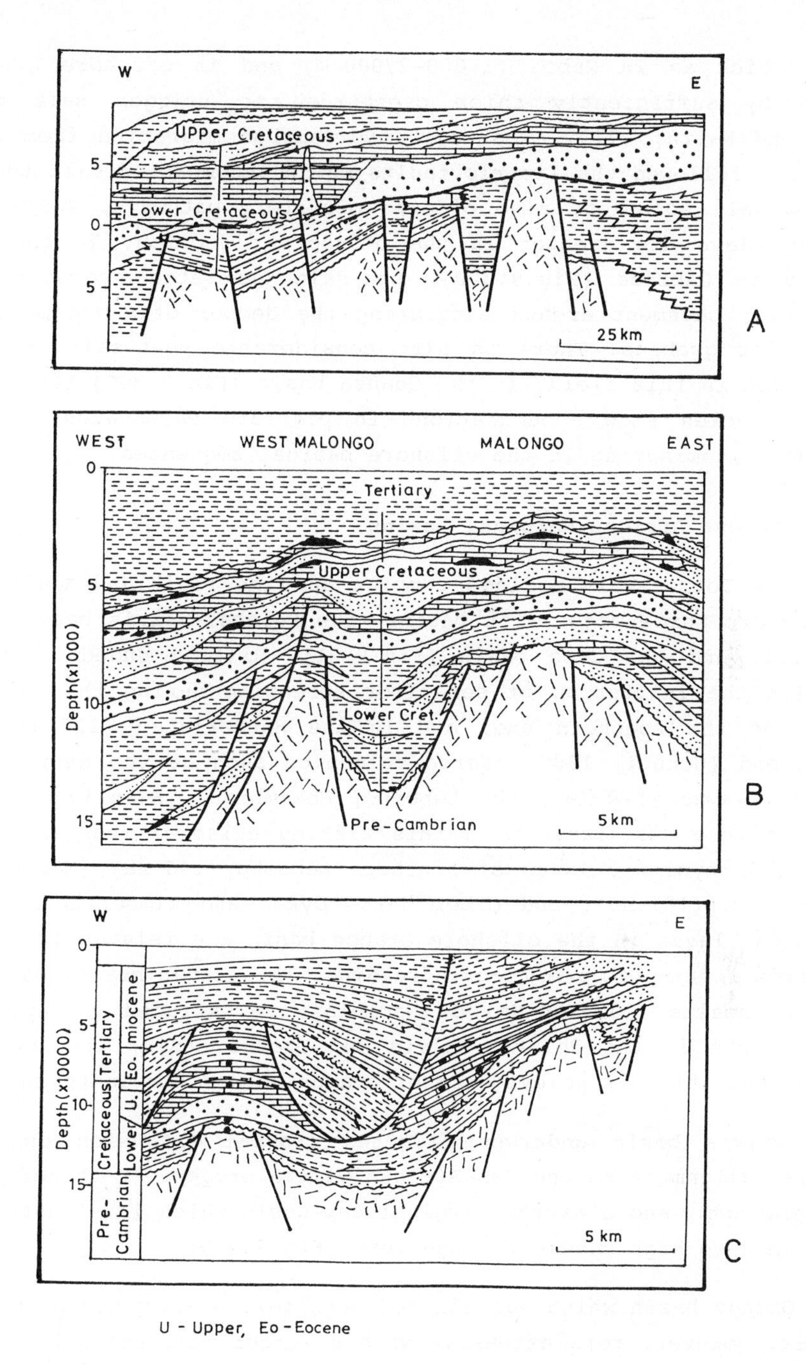

Figure 9.30: Cross-sections of A, offshore Congo basin; B, Cabinda basin; C, Cuanza basin. (Redrawn from Clifford, 1986.)

unit is thick as in Gabon (1,000-2,000 m) and in offshore Congo, and overlain by sufficiently thick overburden to trigger salt movement (Franks and Nairn, 1973). Post-salt hydrocarbon migration from pre-salt source beds followed the listric faults which accompanied salt tectonics; localized salt solutions also offered migratory pathways. Post-salt hydrocarbon plays are greater in the offshore direction. In the Malongo oil-field in Cabinda (Fig.9.30B), pre-salt generated hydrocarbons are trapped over basement arches separating the deeper offshore margin from the interior grabens. There is also considerable post-salt hydrocarbon accumulation in this field. In the Cuanza basin (Fig.9.30C) the post-salt carbonate cycles show accumulations in platform carbonates, with turbidite sands dominating in the offshore basinal sequences.

9.4.5 Southwest African Marginal Basins

Between the Torres-Walvis ridge complex and the Faulkland-Agulhas fracture zone are three major rift basins--the Pelotas-Walvis basin, the Orange basin, and the Cape basin (Fig.9.21). Rifting ended in the Cape basin with the initiation of sea-floor spreading during the Valanginian (133 Ma) at the southern end, and in the Hauterivian (126 Ma) at the northern end (Uchupi, 1989). Karoo volcanics such as the Stormberg diatremes and lavas (210 Ma), the Lebombo, Mozambique, Swaziland, and Zambezi lavas were associated with this rifting episode. Younger volcanics such as the rhyolitic lavas of southern Lebombo (146 Ma), Moveme basalt of Mozambique (137 Ma), and other Karoo dykes and kimberlite plugs and amygdaloidal lavas in the offshore Orange basin are related to the transition from rifting to sea-floor spreading. The latter stage was when the Kaokovelt basalts (132-108 Ma) of Namibia were extruded (Uchupi, 1989). Syn-rift rocks in the Pelotas-Walvis and Orange basins include clastic deposits, but there is no evidence of evaporite and carbonate deposition.

The Walvis basin underwent limited crustal extension due to its transverse alignment to the Pan-African Damara orogen. It therefore has a narrow platform, and a narrow marginal sub-basin which is situated at the base of an 8-km high basement hinge zone (Fig.9.22A).

The Orange basin which was aligned parallel to Precambrian structural trends is, however, more extensive with a larger marginal sub-basin that is divided into two segments by a central basement ridge (Fig.9.22C). The stratigraphic succession in the Orange basin described by Tankard et al. (1982) shows continuous deposition from Early Aptian through Maastrichtian times, resulting in a seaward thickening clastic wedge. At the DSDP site 361 (Fig.9.31) the sequence in the distal part (continental

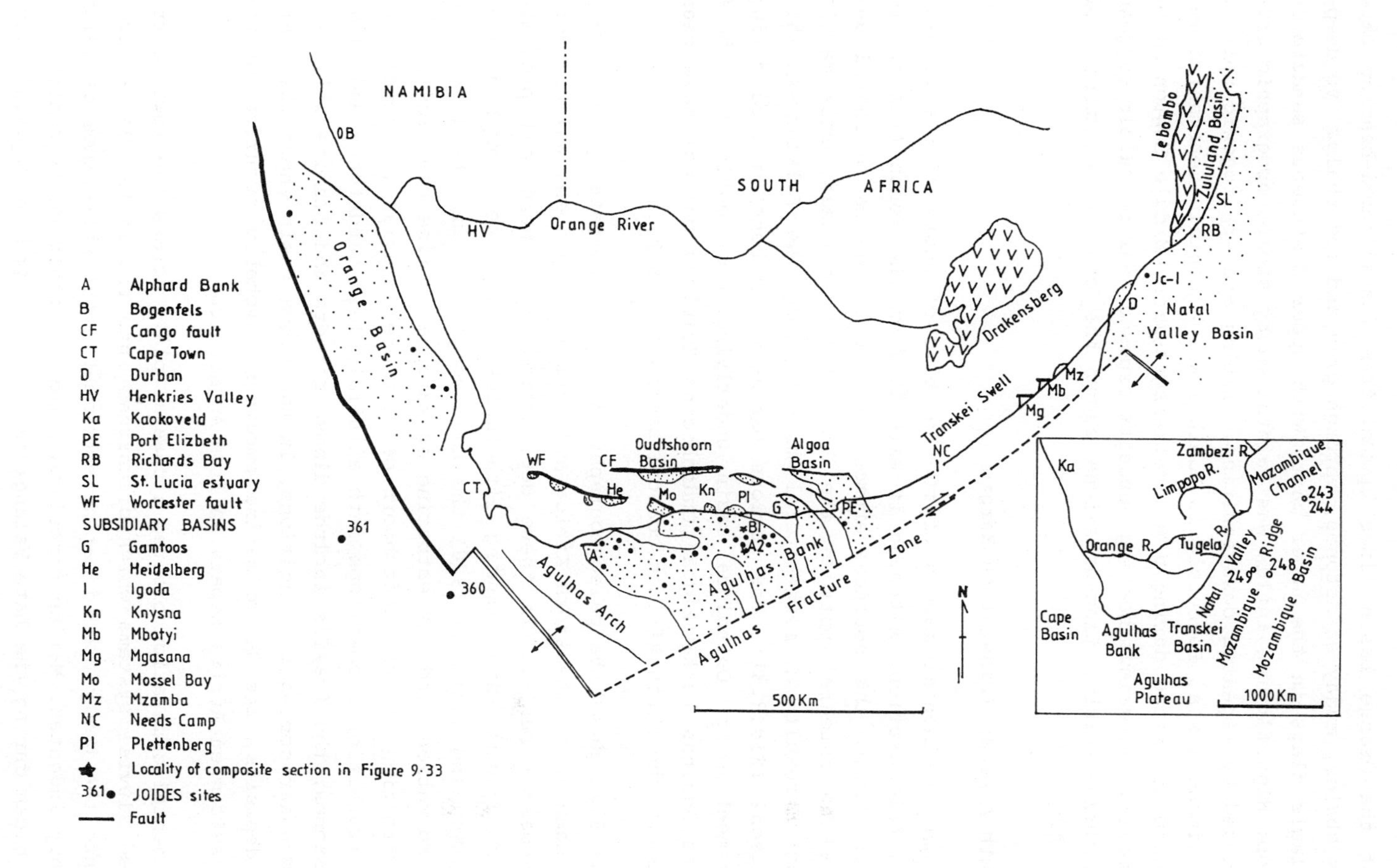

Figure 9.31: Mesozoic sedimentary basins of South Africa. (Redrawn from Tankard et al., 1982.)

slope) of the Orange basin fines upward from fossil wood-bearing carbonaceous shales, muddy sandstones, through gray and red shales, to deep-water pelagic clays in the upper part, with Upper Cretaceous sandstones at the top. High fluvial influx was maintained by steep topographic gradients caused by repeated down-faulting, tilting and epeirogenic uplift. Onshore, limestones at Bogenfels in Namibia which contain Cenomanian ammonites were deposited during a major transgression. Middle Santonian to Late Campanian inoceramus-bearing deposits exposed near Bogenfels suggest other extensive marine transgressions along the continental margin of southwest Africa.

9.4.6 South African Translation Margin

The eastern continental margin of the Republic of South Africa is discussed in this section. Although it extends from the South Atlantic to the Indian Ocean, the southeasternmost margin of Africa orginated and evolved as an integral part of the South Atlantic Ocean. This margin opened as a result of left-lateral motion along the Falkland-Agulhas fracture zone (Fig.9.31). The various basins or grabens created during this movement (e.g. Outeniqua basin underlying the continental shelf termed the Agulhas bank; Algoa; Oudtshoorn; Zululand basins) have been grouped under the South African coastal basins (Fig.8.2).

Figure 9.21 shows the paleogeography of the Middle-Late Jurassic when rifting began in southeastern Africa and formed the grabens which underlie the coastal basins. In these depocentres three principal predominantly continental intertonguing lithofacies (Uitenhage Group) accumulated, comprising (Dingle, 1978) the Enon conglomerates, the fluviatile Kirkwood Formation, and the estuarine or marginal marine clastics of the Infanta Formation. In the Oudtshoorn basin where alluvial fan conglomerates are transitional downslope into alluvial plain sandstones and playa lake mudstones, the fossils include dinosaur teeth and lignite which accumulated under semi-arid conditions. In the Algoa basin where there are 4 km of deposits, the first marine embayment probably occurred in the Late Jurassic when Africa separated from Antarctica.

Continental separation created a coastal plain across the lower parts of braided fluvial systems which discharged into the rifts. Deltas prograded into the newly opened marine embayments, the shorelines of which were highly indented. Marine circulation did not start until continental break-up began during the Late Valanginian as indicated in Fig.9.32. Earlier, in the Late Jurassic the temporary marine incursion was anoxic in the Outeniqua basin and in the Natal Valley basin (Fig.9.31). Barremian-

Aptian paleo-anoxia entended all the way from the offshore region of southwest Africa to the Falkland plateau, Maurice Ewing bank, and into the Outeniqua basin (Fig.9.32). It was not until Maurice Ewing bank had cleared South Africa at the end of the Albian that deep-water conditions entered the South Atlantic and the Indian Ocean, thus ending anoxic sedimentation.

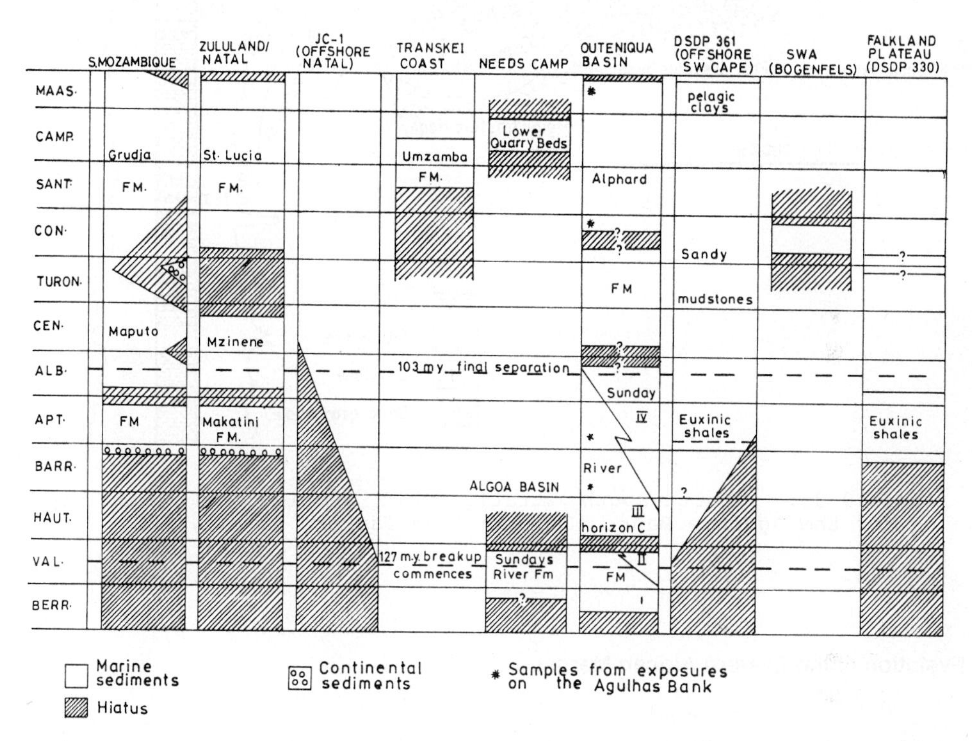

Figure 9.32: Stratigraphic subjdivisions of Cretaceous sediments in South Africa. (Redrawn from Dingle, 1978.)

Cretaceous-Tertiary deposition along the continental margin of southeastern Africa involved deltaic progradation (Fig.9.33), and regional unconformities in the Late Cenomanian-Coniacian (Fig.9.32). These regressions have been attributed to regional uplift due to kimberlite intrusions. The terminal Cretaceous regression was probably due to subsidence of oceanic crust caused by the reduced rates of sea-floor spreading.

Fragmentation of southern Gondwana induced a major paleoclimatic shift from the hyper-continentality of Karoo times with attendant desertification (e.g. Stormberg dune fields, playa lakes, seasonal streams) (Fig.8.50,F-H) to more humid conditions, with moderately high run-off

(Tankard et al., 1982). Once the continental mass had reduced in size and shallow seas expanded, continental climate ameliorated.

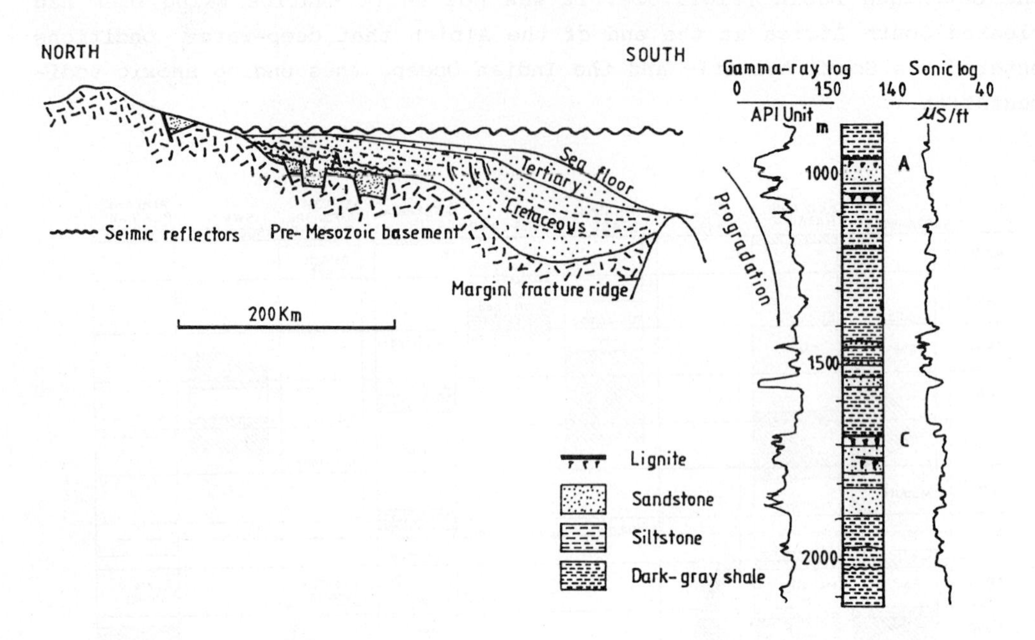

Figure 9.33: North-south stratigraphic section and gamma-ray log for the Agulhas Bank. (Redrawn from Tankard et al.,1982.)

9.5 Evolution of the Eastern African Margin

9.5.1 Plate Tectonic History

Rifting and strike-slip tectonics were involved in the break-up of eastern Gondwana, and in the opening of the Indian Ocean which began in the Middle Jurassic with the drifting of Madagascar from Kenya and Somalia. Both processes led to the creation of the continental margin of eastern Africa, a highly subsident margin (Fig.9.34) with extensive sedimentary wedges stretching from Mozambique to Somalia, over a distance of nearly 6,500 km. The coastal basins in East Africa sometimes contain over 6 km of deposits (Kamen-Kaye, 1978).

Based on interpretations of marine and coastal geological and geophysical data (Dualeh et al., 1990; Coffin and Rabinowitz, 1988; Mascle et al., 1987; Tarling, 1988) the following phases have been recognized in the evolution of the eastern African margin. First there was southeasterly transform motion of Madagascar and India (eastern Gondwana) along the Davie fracture zone (Fig.9.34) between mid-Jurassic and Early Cretaceous. This created rift basins in the Horn of Africa and the Somali basin offshore. Madagascar reached its present location off the coasts of Mozambique and Tanzania during the Early Cretaceous. Secondly, there was the NE-SW opening of the Mascarene basin in the Late Cretaceous, separating Madagascar from the Indian subcontinent. Thirdly, the separation of the Seychelles micro-continent from India took place in the Paleocene-Eocene followed by the northward drift of India. Lastly, the East Africa Rift System was initiated in the mid-Miocene.

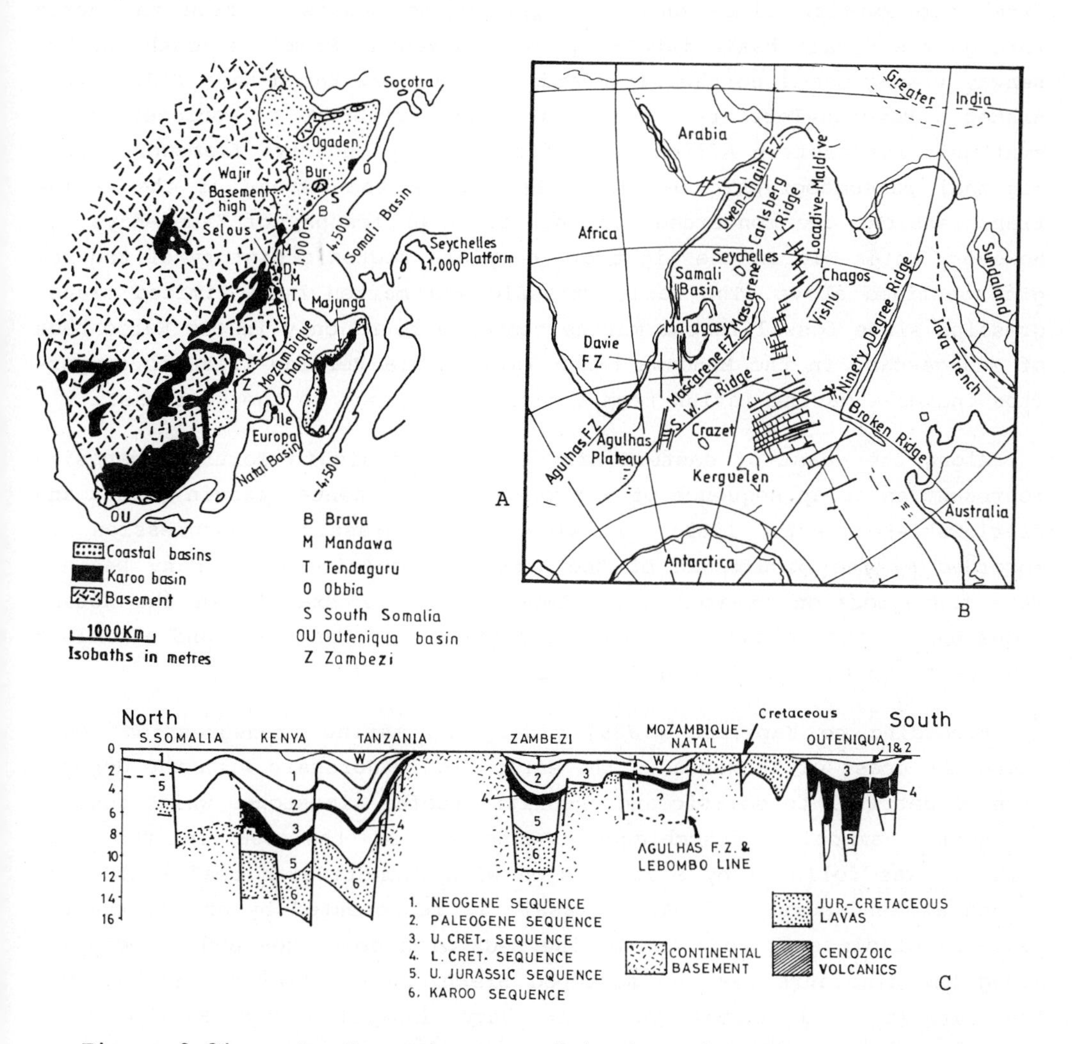

Figure 9.34: A, Jurassic coastal basins of eastern Africa; L, Lamu; M, Mombasa; D, Dar es Salaam. B, tectonic features of the Indian Ocean. (Redrawn from Nairn, 1978; Tarling, 1988.). C, North-South cross-section (Redrawn from Dingle, 1982).

Whereas virtually the entire western Somali basin is underlain by oceanic crust which extends landward to about 1,500 m bathymetric depth, the continental-oceanic crust boundary lies east of Madagascar in the Mascarene basin, with the Mozambique channel underlain by continental crust. The Somali basin contains mostly Cretaceous and younger deposits.

9.5.2 Paleogeography

Table 9.1 shows that the eastern African coastal basins are all underlain by Karoo deposits. Over the Karoo basins the sea progressively encroached from Late Permian times onwards, approaching southward from the north through the Somali basin into Kenya and Tanzania. From the south another seaway transgressed northward towards Mozambique. Kamen-Kaye (1978) furnished a Permian-Tertiary invertebrate and vertebrate paleobiogeographic synthesis for eastern Africa that offers a comprehensive framework for a regional paleogeographic overview, as far as the extent of the marine transgressions are concerned. The non-drift philosophy of Kamen-Kaye is, however, untenable in the light of the plate tectonic history of the region outlined above. The Early Triassic was marked by a limited transgression since fossiliferous marine beds are found only in the Karoo beds of Madagascar. In the Mandawa basin in Tanzania Permian-Triassic evaporites indicate marine influx from a marine embayment to the southeast.

Along the coast of eastern Africa the Jurassic to Tertiary deposits represent an onlap sequence of a regressive continental margin (Ernst and Gierlowski-Kordesch, 1989). A widespread Jurassic transgression is recorded by the occurrence of the ammonite *Bouleiceras* in many basins. This transgression created major embayments, for example in the Mozambique basin, the Selous basin and at Tendaguru in Tanzania, and all along the coast, up to Somalia and Ethiopia.

According to Kapilima (1989) widespread marine transgression began along the coast of Tanzania in the early Middle Jurassic with the deposition of detrital to oolitic and oncolitic reefal limestones which contain diagnostic ammonite assemblages. Deepening of the sea in the Late Jurassic was followed by a regression with Kimmeridgian reef build-ups. Inland at Tendaguru, the Late Jurassic is represented by an alternating sequence of dinosaur beds and marine beds with ammonites and pelecypods. Among the dinosaurs was the sauropod *Brachiosaurus branchi* (Fig.8.57G), the largest land animal known to have inhabited the Earth. Other dinosaurs at Tendaguru include *Allosaurus* (Fig.8.57I), and *Kentrosaurus* (Fig.8.57J).

Table 9.1: Correlations of the Mesozoic-Cenozoic sequences in eastern Africa. (Redrawn from Kamen-Kaye, 1978.)

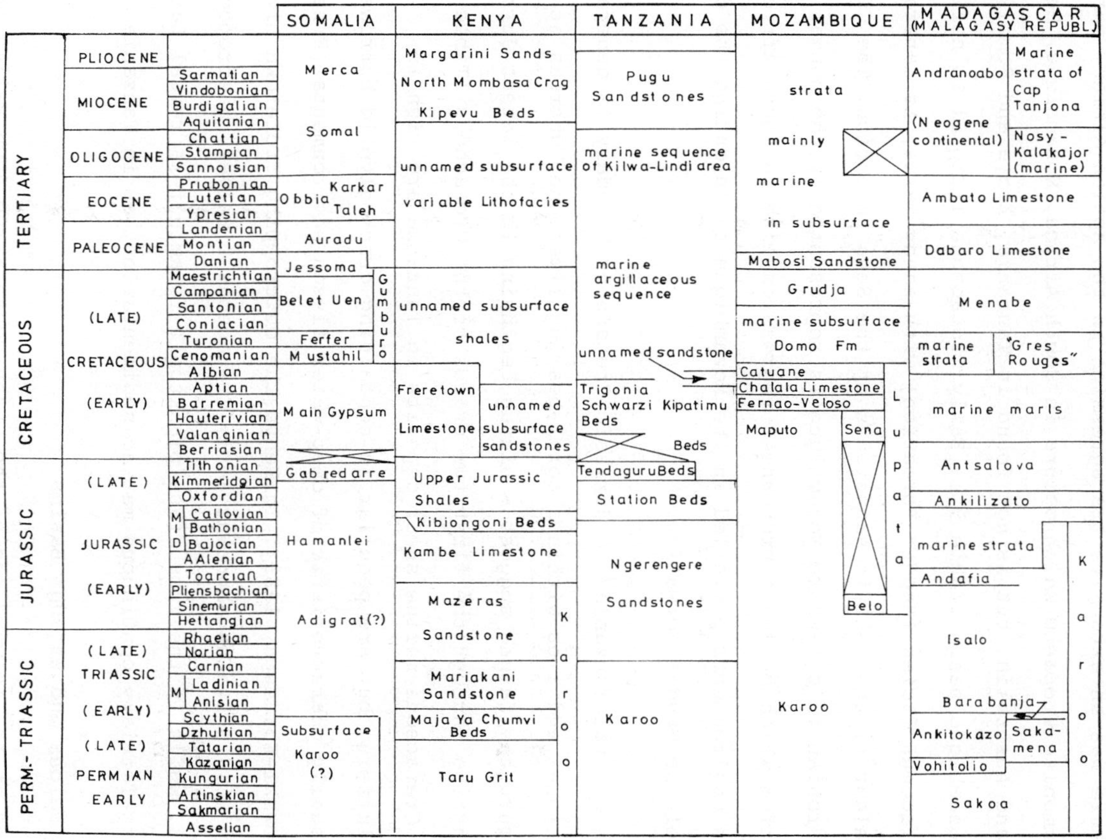

			SOMALIA	KENYA	TANZANIA	MOZAMBIQUE	MADAGASCAR (MALAGASY REPUBL)
TERTIARY	PLIOCENE			Margarini Sands			Marine strata of Cap Tanjona
	MIOCENE	Sarmatian	Merca	North Mombasa Crag	Pugu Sandstones	strata	Andranoabo
		Vindobonian					
		Burdigalian		Kipevu Beds			
		Aquitanian					(Neogene continental)
	OLIGOCENE	Chattian	Somal			mainly	Nosy-Kalakajor (marine)
		Stampian			marine sequence of Kilwa-Lindi area		
		Sannoisian		unnamed subsurface		marine	
	EOCENE	Priabonian	Obbia Karkar				
		Lutetian		variable Lithofacies			Ambato Limestone
		Ypresian	Obbia Taleh			in subsurface	
	PALEOCENE	Landenian					
		Montian	Auradu				Dabaro Limestone
		Danian	Jessoma			Mabosi Sandstone	
CRETACEOUS	(LATE)	Maestrichtian	Gumburo		marine argillaceous sequence		
		Campanian				Grudja	
		Santonian	Belet Uen	unnamed subsurface shales			Menabe
		Coniacian				marine subsurface	
		Turonian	Ferfer				
	CRETACEOUS	Cenomanian	Mustahil		unnamed sandstone	Domo Fm	marine strata / "Gres Rouges"
	(EARLY)	Albian				Catuane	
		Aptian		Freretown Limestone	Trigonia Schwarzi Beds	Chalala Limestone	
		Barremian	Main Gypsum	unnamed subsurface sandstones	Kipatimu Beds	Fernao-Veloso	
		Hauterivian				Lupata	marine marls
		Valanginian				Maputo / Sena	
		Berriasian					
JURASSIC	(LATE)	Tithonian			Tendaguru Beds		Antsalova
		Kimmeridgian	Gabredarre	Upper Jurassic Shales			
		Oxfordian			Station Beds		Ankilizato
	JURASSIC	M Callovian					
		I Bathonian		Kibiongoni Beds			
		D Bajocian	Hamanlei	Kambe Limestone			marine strata
		Aalenian			Ngerengere Sandstones		
	(EARLY)	Toarcian					Andafia
		Pliensbachian		Mazeras Sandstone			
		Sinemurian				Belo	
		Hettangian	Adigrat(?)	Karoo			Karoo
PERM.-TRIASSIC	(LATE) TRIASSIC	Rhaetian					Isalo
		Norian					
		Carnian					
		M Ladinian		Mariakani Sandstone			
	(EARLY)	Anisian					Barabanja
		Scythian		Maja Ya Chumvi Beds	Karoo	Karoo	
	(LATE) PERMIAN	Dzhulfian	Subsurface Karoo (?)				Ankitokazo / Sakamena
		Tatarian					
		Kazanian					Vohitolio
	EARLY	Kungurian		Taru Grit			
		Artinskian					
		Sakmarian					Sakoa
		Asselian					

There was an extensive transgression in the Aptian-Albian, marked by the ammonite *Tropaeum* which occurs in South Africa, Mozambique, Madagascar, and Australia, but is not found in Tanzania, Kenya and Somolia. In the Mozambique basin (Table 9.1) Early Cretaceous littoral to neritic argillaceous facies (Maputo Formation) with sandy marls, is laterally equivalent to the deltaic sandstones of the Sena Formation. The Cenomania-Turonian in off-shore Mozambique is represented by the nonmarine Domo Formation which marked a pronounced regression like in South Africa. Open marine environments prevailed in the Mozambique basin during the deposition of the Grudja Formation.

In coastal Tanzania the Jurassic-Cretaceous boundary is generally gradational and lies within a coral-bearing limestone; whereas at Tendaguru the faunas suggest a hiatus beginning in the latest Jurassic and lasting through the first two stages of the Cretaceous (Table 9.1). Upper Cretaceous marine shelf deposits in Tanzania are rich in foraminifera, but are peculiar in containing graphoglyptid burrows which are usually characteristic of deep-water flysch environments (Ernst and Gierlowski-Kordesch, 1989). In Kenya the outcropping Freretown Limestone (Table 9.1) contains abundant Early Cretaceous bivalves, gastropods, and corals, while in the subsurface planktonic foraminifera reveal Late Cretaceous-Tertiary continuous marine sedimentation, as in Tanzania.

9.5.3 Selous and Majunga Basins

Located at the seaward end of the north-south Karoo Ruvu Valley in northeastern Tanzania, the Selous basin (Figs. 8.2; 9.35A) is a Mesozoic sag basin overlying a pre-drift Karoo interior fracture basin. Transcurrent motion along the Davie fracture zone separated the eastern half of the Selous basin which now constitutes the Majunga basin in northern Madagascar (Fig.8.2). In both basins (Fig.9.35) a Middle Jurassic break-up unconformity separates nonmarine Karoo clastics and lacustrine shales from overlying marine carbonates, marls, shales, with marginal-to-marine clastics (Clifford, 1986). Poorly developed latest Triassic or earliest Jurassic evaporites mark the transition from nonmarine to marine sedimentation as the break-up of Gondwana progressed.

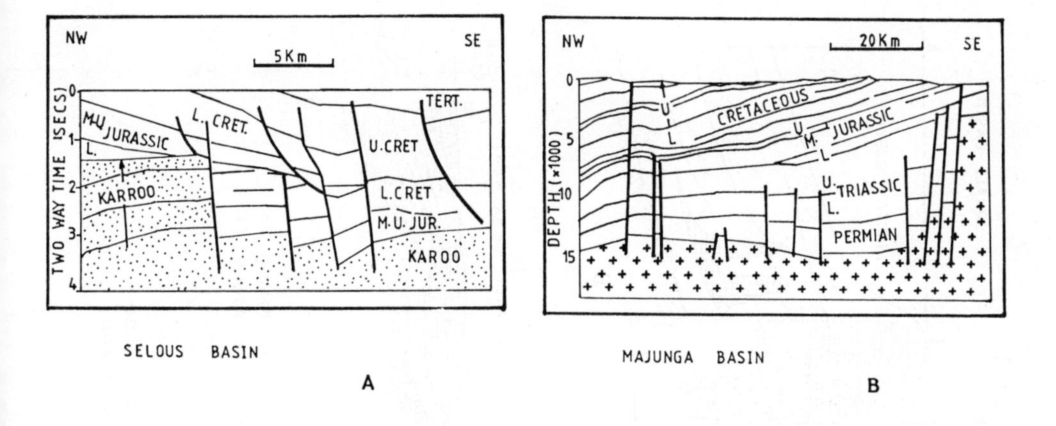

Figure 9.35: Cross-sections for the Selous and Majunga basins. (Redrawn from Clifford, 1986.)

9.5.4 Mesozoic Rift Basins in the Horn of Africa

This is a group of related basins in Kenya, Somalia, and Ethiopia which are separated by various basement uplifts (Fig.9.36). They occupy the passive Indian Ocean margin and comprise the Lamu embayment in southern Somalia and adjoining parts of Kenya; and in Somalia there are the Lugh-Mandera basin, the Somali coastal basin, the Somali embayment, the Al-Mado Darro basin, and the Barbera-Borama basin. The Lamu embayment is believed to be the southern extension of the Ogaden and the Lugh-Mandera basins, although in the subsurface an east-west basement ridge seems to separate the Lamu embayment (Whiteman, 1981). Although the structural pattern of the Lamu embayment and those along the Somali coast are generally obscured by flat-lying coastal plain deposits, in the subsurface geophysical surveys have delineated NE-SW-trending faults and pre-Tertiary highs in the Lamu embayment (Peterson, 1985; Walters and Linton, 1973; Whiteman, 1981). Uplift of the Bur Aqabar and Nogal basement highs were accompanied by downwarping in Ogaden, Somali coastal basin, Somali embayment, and in the Lugh-Mandera basin. Although separated by a north-

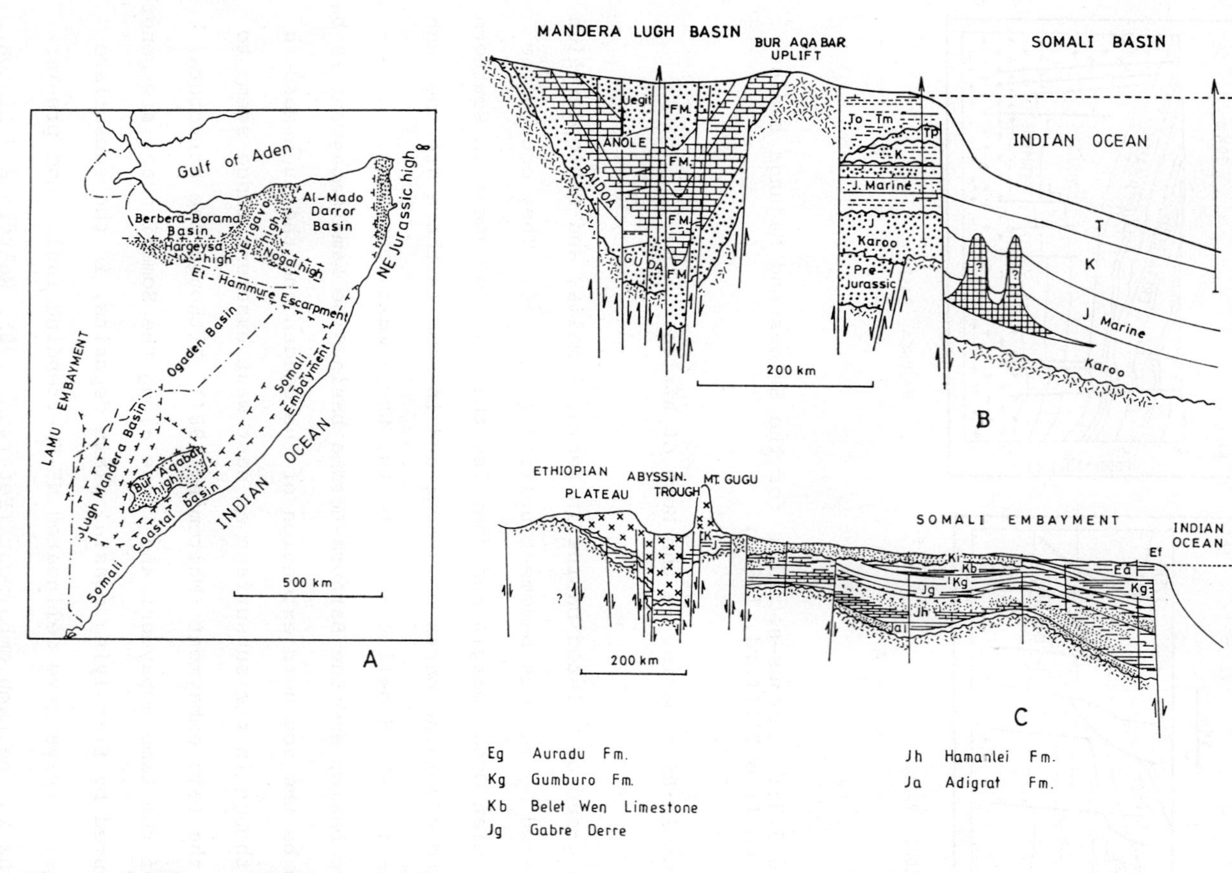

Figure 9.36: Tectonic elements and geologic sections for the Horn of Africa. (Redrawn from Dualeh et al., 1990; Peterson, 1985.)

south trend of sedimentary thinning, the Ogaden basin seems to merge eastward with the Somali embayment.

Offshore, lies the Somali basin, the thickest part of which is located along the coastline of Kenya and Somalia (Fig.9.36B) on the eastern side of a major continental margin fault system which downdropped on the ocean side, with a throw of at least 4 km (Peterson, 1985). Over 9 km of marine and continental sequence ranging from Karoo to Tertiary deposits is preserved on the downthrown side (continental slope). Peterson considered the Somali (Fig.9.36C) and Lamu embayments, and the Somali coastal basin to be the shoreward manifestations of the regional subsidence associated with the growth of the oceanic Somali basin.

Dualeh et al. (1990) reconstructed the structural and depositional histories of the basins in the Horn of Africa. Late Carboniferous-Triassic continental rifting created Karoo troughs where continental beds and shallow marine clastics and Permo-Triassic evaporites accumulated. A new phase of basin formation which marked the end of regional peneplanation of the eastern African shield was initiated by the deposition of the widespread Adigrat sandstone (Fig.9.37). The Adigrat Sandstone is a diachronous regional latest Triassic-Middle Jurassic unit, the base of which contains rift basalts, for example in the Berbera basin on the coast of the Gulf of Aden.

The Middle Jurassic recorded the first major transgression in which the Hamanlei Formation and its equivalents, were deposited in all the basins. These are sequences of Tethyan marine shelf carbonates of Middle to Late Jurassic age. They comprise oolitic and coralline limestone beds in southern Somalia and in parts of Ethiopia, with interbedded anhydrite and dark gray marine shale in northern Somalia. Shaly carbonates occur in the Somali embayment, Lugh-Mandera basin, and in the Lamu embayment. The Hamanlei Formation ranges in thickness from less than 300 m to over 2,000 m in Somalia and eastern Ethiopia, and thins and becomes sandy to the west (Peterson, 1985).

The Late Jurassic was regressive with evaporite deposition which continued into the Early Cretaceous in eastern Ogaden basin, and in the Lugh-Mandera basin (Fig.9.38A) which was partially isolated by the Bur uplift. After this the Lugh-Mandera basin (Fig.9.36B) was emergent throughout the Cretaceous-Tertiary. But marine sedimentation prevailed offshore. Following a complete retreat of the sea from Somalia in the Neocomian, there was a renewed transgression in the Aptian which did not extend very far inland. Late Cretaceous marine sediments are represented

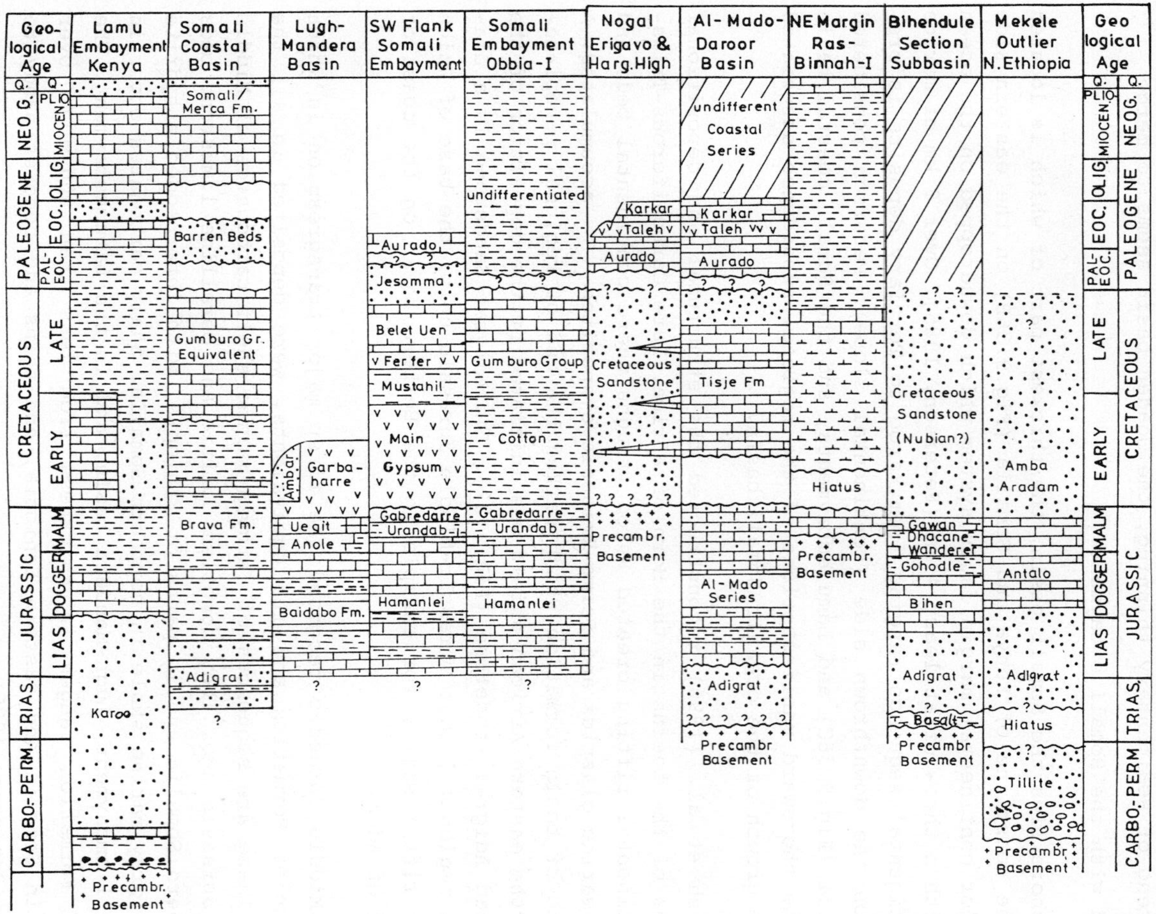

Geological Age: Q.; NEO G. (PLIO, MIOCEN); PALEOGENE (OLIG., EOC., PAL-EOC.); CRETACEOUS (LATE, EARLY); JURASSIC (MALM, DOGGER, LIAS); TRIAS.; CARBO-PERM.

Lamu Embayment Kenya: Karoo; Precambr. Basement

Somali Coastal Basin: Somali/Merca Fm.; Barren Beds; Gumburo Gr. Equivalent; Brava Fm.; Adigrat; ?

Lugh-Mandera Basin: Ambar; Garba-harre; Uegit; Anole; Baidabo Fm.; ?

SW Flank Somali Embayment: Aurado; Jesomma; Belet Uen; Ferfer; Mustahil; Main Gypsum; Gabredarre; Urandab; Hamanlei; ?

Somali Embayment Obbia-I: undifferentiated; Gumburo Group; Cotton; Gabredarre; Urandab; Hamanlei; ?

Nogal Erigavo & Harg. High: Karkar; Taleh; Aurado; Cretaceous Sandstone; Precambr. Basement

Al-Mado-Daroor Basin: undifferent Coastal Series; Karkar; Taleh; Aurado; Tisje Fm; Al-Mado Series; Adigrat; Precambr Basement

NE Margin Ras-Binnah-I: Hiatus; Precambr. Basement

Bihendule Section Subbasin: Cretaceous Sandstone (Nubian?); Gawan; Dhacane; Wanderer; Gohodle; Bihen; Adigrat; Basalt; Precambr Basement

Mekele Outlier N. Ethiopia: Amba Aradam; Antalo; Adigrat; Hiatus; Tillite; Precambr. Basement

Figure 9.37: Stratigraphic correlation chart for the Horn of Africa. (Redrawn from Dualeh et al., 1990.)

by the Gumburo Group which contains rudistid limestones at the base, with locally developed evaporites in Somalia; and in the Lamu embayment by marine shales, sandstone, and minor limestone. Cretaceous rocks in the Horn of Africa thin westward and northward to less than 500 m in much of eastern Ethiopia (Fig.9.38B).

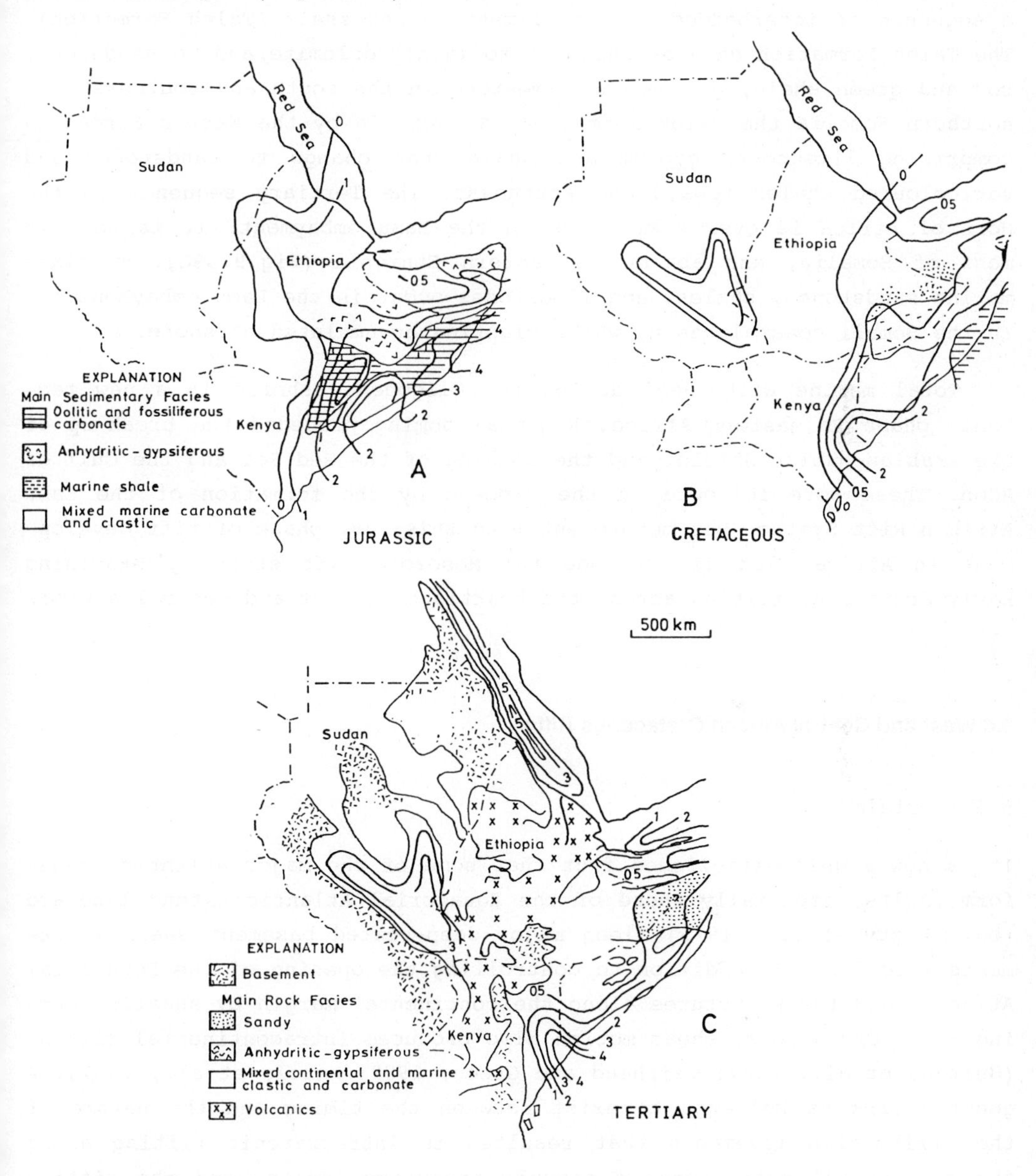

Figure 9.38: Mesozoic - Cenozoic isopach maps for the Horn of Africa showing the main lithofacies. (Redrawn from Peterson, 1985.)

There was an extensive marine transgression in the Early Tertiary during which the Paleocene-Eocene Aurado marine fossiliferous limestone was deposited in northern Somalia over end-Cretaceous regressive sandstones. The Aurado Formation changes into deep-water marine shale and shaly limestone to the east and southeast in Somalia. It is succeeded by a sequence of interbedded gypsum, limestone and shale (Taleh Formation). The Taleh Formation changes eastward to mainly dolomite, and to sandstone, red and green shale, gypsum and limestone in the south and southeast. In northern Somalia the Taleh Formation is overlain by the Karkar Formation comprising limestone, gypsum and shale that change to sandstones and varicoloured shales toward the southeast. The Tertiary sequence in the Horn of Africa is over 4 km thick in the Lamu embayment. It is thin in most of Somalia, and absent in eastern Ethiopia (Fig.9.38C). Tertiary marine sandstones, shales, and limestones occur in the Lamu embayment and in the Somali coastal basin, while clastics accumulated offshore.

Total marine withdrawal during the Oligocene ushered in a new tectonic phase in eastern Africa. Regional doming preceded the break-up of the Arabian-Nubian Shield, and the opening of the Red Sea and the Gulf of Aden. These were followed in the Miocene by the formation of the East African Rift System. But before entering this last phase of rift development in Africa, let us conclude the Mesozoic rift story by examining Early Cretaceous rifting across the heartland of west and central Africa.

9.6 West and Central African Cretaceous Rifts

9.6.1 Origin

It is now a well established fact that some of the major Atlantic transform faults, especially those of the Equatorial Atlantic extend landward (eg., Emery et al., 1975) along major deep-seated basement shear lineaments (Fig.8.2). In addition to controlling the opening of the Equatorial Atlantic and the structures along the continental margin by shearing during Early Cretaceous, shear motion also produced intracontinental rifting (Guiraud et al., 1987; Fairhead and Green, 1989; Mascle et al., 1987). A genetic link is believed to exist between the timing and the nature of the strike-slip movements that resulted in intracratonic rifting along the continental extensions of oceanic transform faults, and the rifting and break-up stages during the evolution of African Mesozoic-Cenozoic continental margins. The Gulf of Guinea oceanic transform faults extend landward as a complex system of shear lineaments, collectively

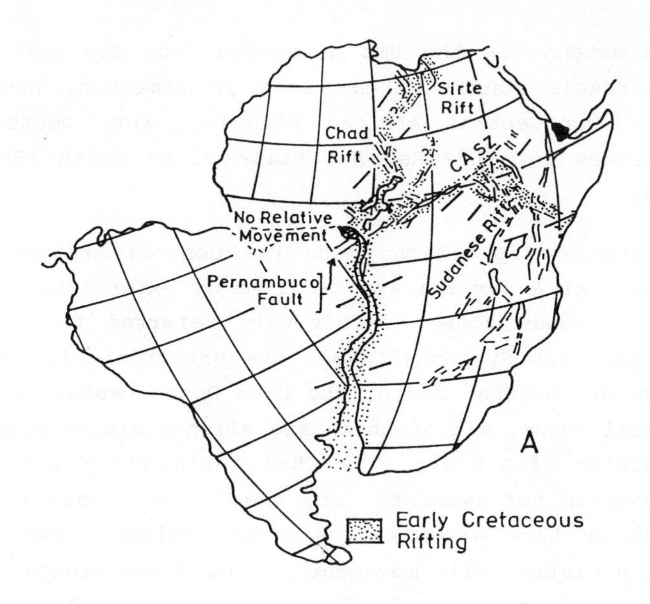

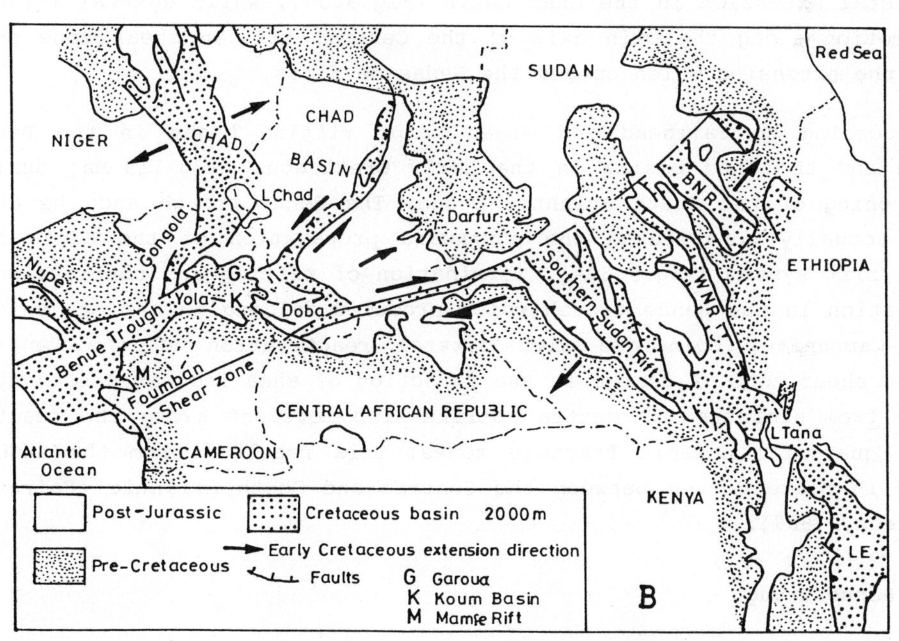

Figure 9.39: Geodynamic model and rifts systems for West Africa and the Sudan. (Redrawn from Fairhead and Green, 1989.)

termed the Central African shear zone (Fig.9.39). This shear zone, according to Fairhead and Green (1989), is ideally orientated (approximating a small circle) to transmit shear motion from the mid-oceanic ridge, deep into the African continent. The Central African shear zone extends in the ENE direction from the Gulf of Guinea through the Proterozoic Foumban shear zone in Cameroon, and runs across southern Chad, the Central African Republic, into western Sudan and probably continues into the Red Sea Hills of NE Sudan (Schandelmeier and Pudlo, 1990).

Strike-slip motion along the Central African shear zone caused the opening of a complex system of major extensional basins along both sides of the shear zone. Collectively referred to as the West and Central African Rift System (WCARS), the principal rift basins are grouped into those of the Chad basin, the Doba Doseo basin, and those of southern and central Sudan, all of which are aligned almost perpendicular to the shear direction (Fig.9.39). The Chad basin rifts are believed to have been propagated northward to form the intracratonic rifts in the Sirte basin which we have already considered. Fairhead and Green (1989) estimated that sinistral slip movement in the Benue trough also caused about 58 km of crustal extension in the Chad basin (Fig.9.39), while dextral strike-slip motion along the main axis of the Central African shear zone produced the extension which opened the Sudanese rifts.

According to Fairhead and Green major rifting began in the Benue trough and the Chad basins in the Early Cretaceous (130-119 Ma) during the opening of the South Atlantic Ocean. The Benue trough and the Chad basin actually constituted the northward propagation of the South Atlantic rift system (Fig.9.39A). Termination of rifting and compressional deformation in the Benue trough which produced folding in the Santonian-Early Campanian, coincided with dextral reactivation of the Central African shear zone. A change in the direction of shearing was caused by a change from sinistral to dextral strike-slip movement along the reactivated Equatorial oceanic fracture zones. This resulted from the sudden change in plate motion between the Central and South Atlantic (Fairhead and Green, 1989).

9.6.2 Benue Trough

This is a Cretaceous folded rift basin (Fig.9.39) which lies across Nigeria (Fig.9.25), and extends from the Niger delta and links northward with the Chad basin, through the Gongola rift. It branches into northern Cameroon through the Yola rift. Other bifurcations of the Benue trough include the Mamfe rift (Fig.9.39) in southeastern Nigeria and southwestern Cameroon, and the Nupe basin in central Nigeria. Geophysical surveys (Ajakaiye, 1981; Adighije, 1979; Ofoegbu, 1985; Ofrey, 1982; Okereke, 1984) reveal crustal thinning beneath the Benue trough, flanked on both sides by linear sub-basins (Agagu and Adighije, 1983) that were the thickest depocentres (Fig.9.39), sometimes with over 6 km of strata.

In its lithic fill, magmatism and deformation, the Benue trough has evolved in direct response to Cretaceous plate tectonic processes in the Atlantic Ocean, especially the South Atlantic (Benkhelil, 1989; Popoff, 1988). At its inception in the Early Cretaceous the Benue trough consisted of a series of isolated depocentres or sub-basins (Fig.9.40A) where there was mostly alluvial fan, braided river and lacustrine sedimentation (Bima Sandstone) in the north, with the Asu River Group in the middle and southern parts of the Benue trough.

The initial marine transgression entered the Benue trough in the Middle Albian (Reyment, 1965) from the Gulf of Guinea (Fig.9.40) and established a stratified epeiric sea where anoxic bottom conditions prevailed from Cenomanian to Coniacian times (Petters and Ekweozor, 1982), and again in the Late Campanian. While thick organic-rich muds with abundant pelagic organisms (ammonites, inoceramus, foraminifera) accumulated extensively from Turonian to Coniacian times, carbonate turbidites (Oti, 1990) were deposited in the deeper parts of the Abakaliki trough (Fig, 9.40 B) which had earlier been the site of clastic turbidites (Ojoh, 1990) during the Middle Albian transgression. Shoal water carbonates are pervasively intercalated within the Late Cenomanian-Coniacian dark gray shales, the southern parts of which have been grouped into the Nkalagu Formation (Petters and Ekweozor, 1982). The Nkalagu Formation (Fig.9.41) includes the Turonian Eze-Aku Shale, and the Conaician Awgu Shales (Fig.9.40B).

The Late Cenomanian-Coniacian witnessed the most extensive marine transgression in the Benue trough which coming from the Gulf of Guinea, extended marine influence into the Chad basin (Avbovbo et al., 1986; Oti, 1990; Reyment and Tait, 1972), and into the Yola and Mamfe rifts. Figure 9.13B shows the probable extent of the Saharan seaway during the Late Cretaceous transgression. Deposits of the epeiric sea (Figs.9.40B; 9.42A) in the northern Benue and Chad basin include highly fossiliferous (eg., ammonites, foraminifera) marine shales (Pindiga and Fika Formations), limestone (Gongila Formation), and marginal marine lithologies (Yolde, Jessu, Sukuliye, Numanha, Lamja Formations (Fig.9.41)) at various localities (Odebode and Enu, 1986). Deltaic sandstones prograded into the middle and southern parts of the Benue trough (Amajor, 1990; Nwajide, 1990) as a result of regressive pulses during the Late Cenomanian-Coniacian transgression.

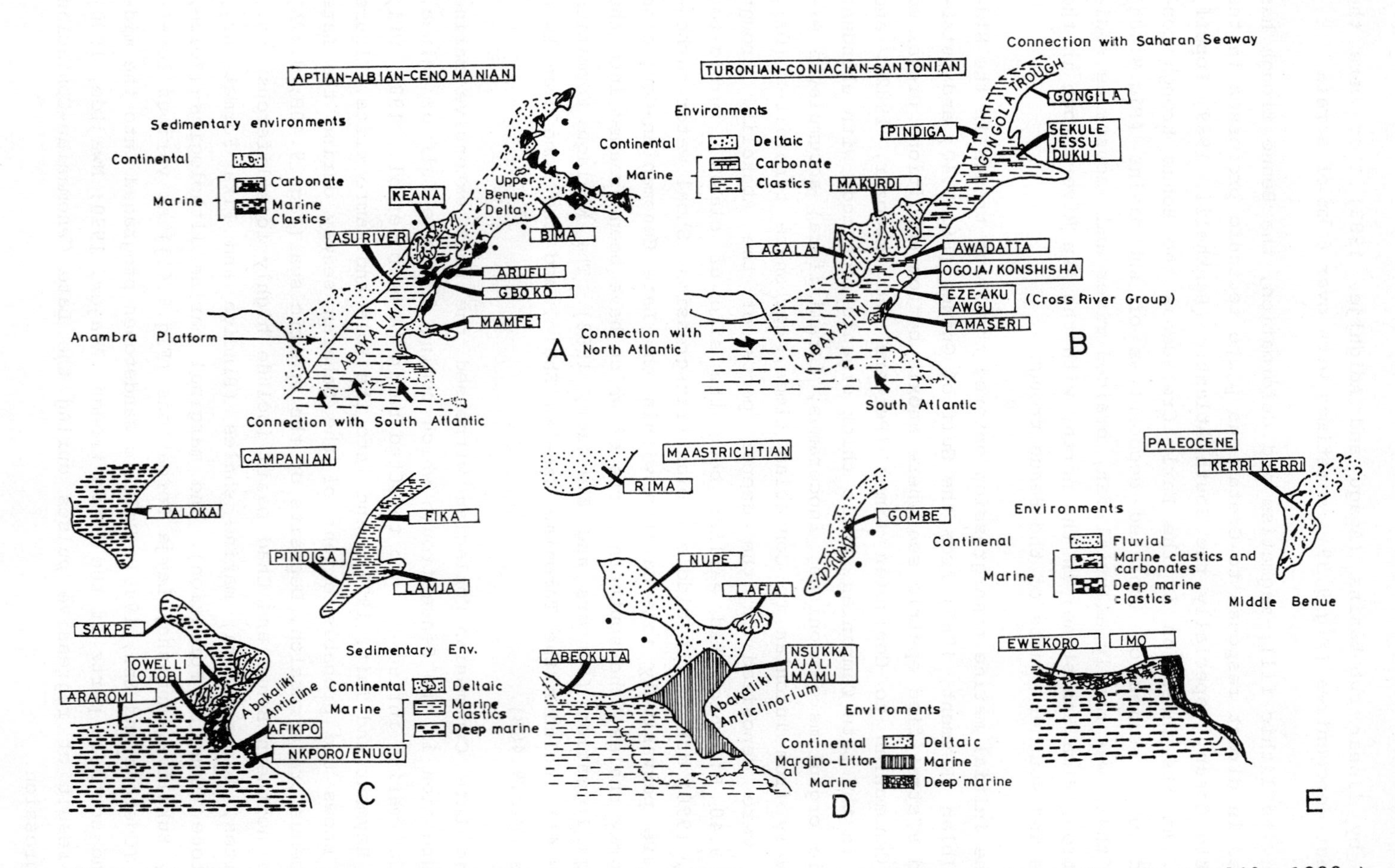

Figure 9.40: Paleogeographical development of the Benue Trough. (Redrawn from Benkhelil, 1989.)

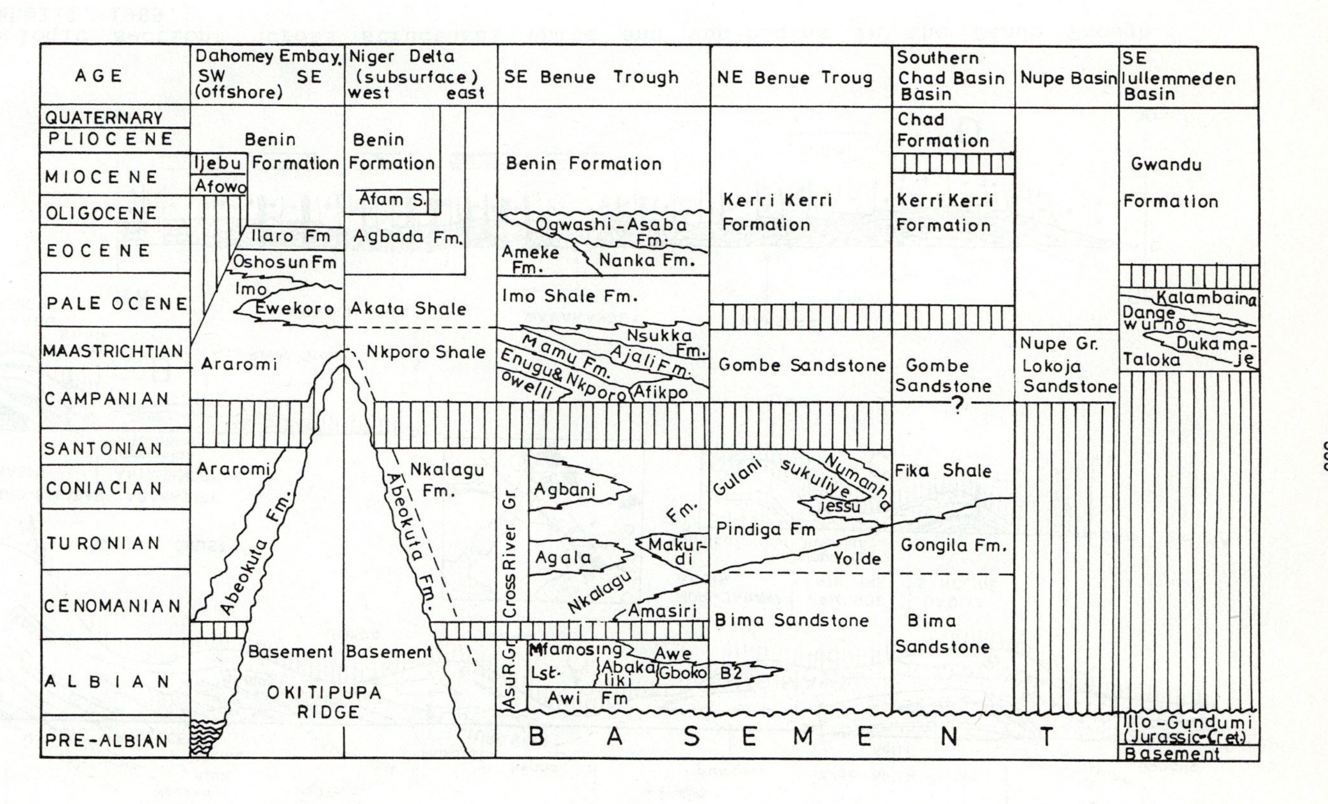

Figure 9.41: Stratigraphic correlation chart for Nigerian sedimentary basins. (Redrawn from Petters, 1982.)

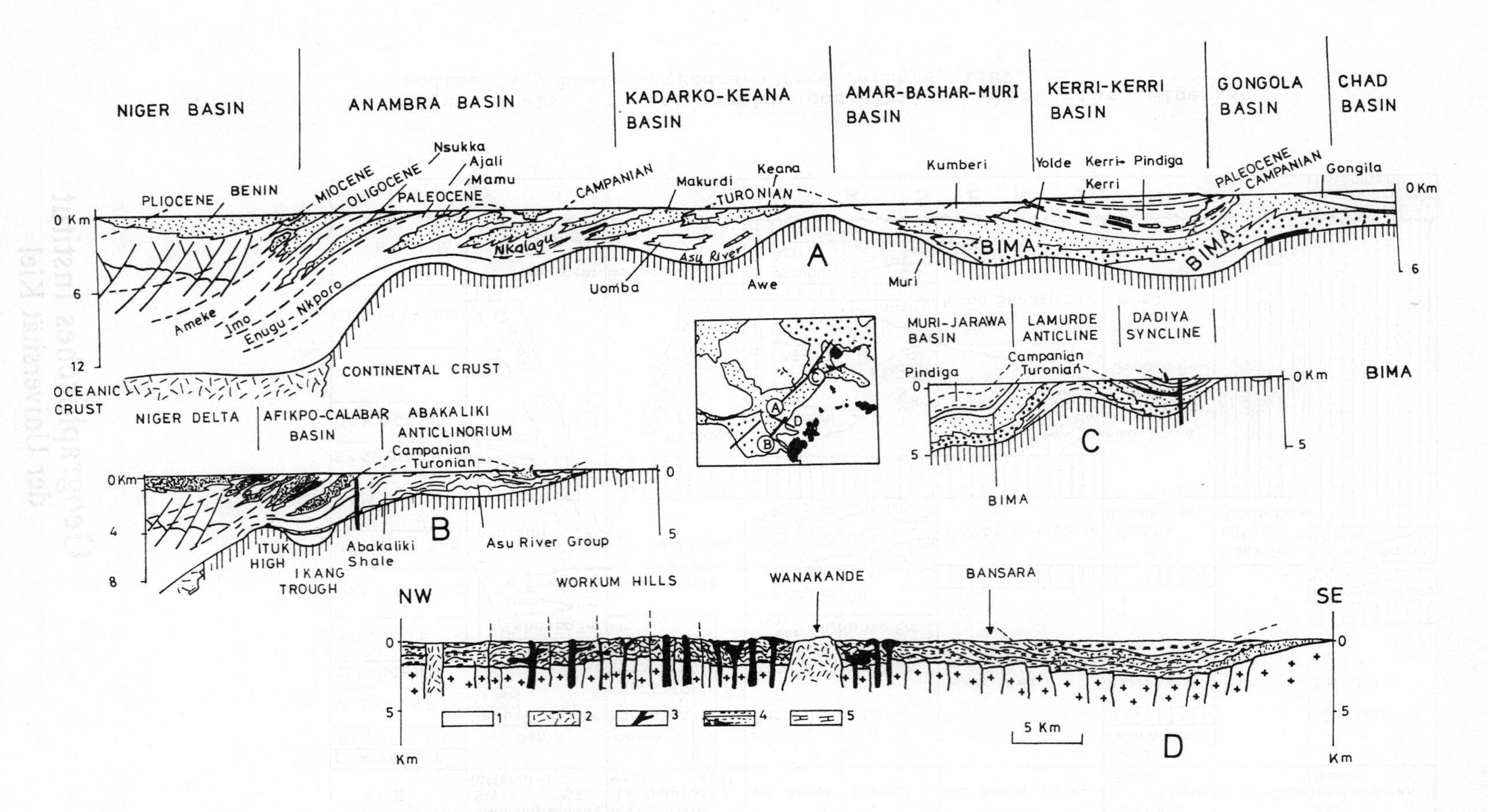

Figure 9.42: Geologic sections across structural units and sub-basins in the Benue Trough. (Redrawn from Benkhelil, 1989.)

Compressional deformation occurred in the Santonian-Early Campanian causing severe folding in the Abakaliki rift (Fig.9.42B, D). In the northeastern part of the Benue trough where deformation intensity was less, there was only moderate folding and fracturing (Benkhelil, 1989).

Alkaline magmatism accompanied rifting and initial sedimentation in the Benue trough (Okeke et al., 1988; Olade, 1978) and was particularly intense in the Abakaliki sub-basin, where in the Wanakande area of Ogoja (Fig.9.42D), Turonian alkaline intrusives were accompanied by contact metamorphism. Low-grade metamorphism also occurred in this area. Lead-zinc-copper mineralization took place in the Abakaliki sub-basin and in the middle Benue, due to the mobilization of basinal brines in the Asu River Group (Akande and Mucke, 1989; Ukpong and Olade, 1979).

After deformation the southern depocentres were displaced from the Abakaliki axis to the Anambra and the Afikpo-Calabar areas. Late Campanian-Maastrichtian deltaic coal measures accumulated in the Anambra sub-basin, while shallow marine carbonaceous shales (Nkporo Shale) were deposited southward (Fig.9.41C) under pronounced paleo-anoxia during the Late Campanian renewed transgression. Westward in the Nupe basin (Fig.9.40C) deltaic sedimentation also occurred but without the coal swamps of the adjoining Anambra basin. Rather, conditions favoured the genesis of oolitic ironstones (Sakpe and Batiti Ironstones) during this transgression (Adeleye, 1976). The Nupe basin was filled with predominantly fluviatile strata, the Nupe Group (Fig.9.41).

In the northeastern Benue trough there was deformation and uplift which affected the deltaic Gombe sandstone (Fig.9.42D), and caused the northwestward displacement of the depocentre. The post-deformation continental sequence, the Kerri-Kerri Formation thus occupies a structural position similar to the deltaic coal measures of the Anambra basin in the south (Benkhelil, 1989).

9.6.3 Chad Basin

The Chad basin (East Niger basin) refers to a group of NW-SE-trending buried rifts in west-central Chad and southeastern Niger. Gravity surveys (Louis, 1970) and stratigraphic studies (eg., Avbovbo et al., 1986; Peterson, 1985; Petters, 1981) have confirmed these burried rifts beneath a mantle of Quaternary desert sands. These rifts formed on the northern side of the Central African shear zone, already mentioned. The sequence, up to 4 km thick, begins with Permo-Jurassic to Early Cretaceous nonmarine strata of fluviatile and lacustrine origin, which belong to the

"Continental Intercalaire" Group (Fig.9.43). The dinosaurs *Ouranosaurus nigeriensis*, and *Spinosaurus* (Fig.8.57H), along with other vertebrates, and invertebrates and Early Cretaceous flora, occur in Cretaceous beds in Niger Republic (Lefranc and Guiraud, 1990). The "Continental Intercalaire" Group is overlain by Cenomanian-Coniacian marine shales and carbonates. The Upper Cretaceous is a clastic sequence with gypsiferous, glauconitic, and fossiliferous shales that record marine influence down to the Santonian-Campanian. Tertiary nonmarine beds resting unconformably on Maastrichtian-Paleocene continental sandstones with oolitic ironstones, belong to the "Continental Terminal" Group (Lang et al., 1990).

9.6.4 Cameroon Cretaceous Rifts

Poor exposures greatly limit our knowledge of the successions in African rift basins especially where drill hole data are not available. This is the case regarding the rift basins in Cameroon Republic, where extensions of the Benue trough are found in the northern and southwestern parts of that country. Small basins in northern Cameroon (Fig.9.39B) such as the Koum, Babouri-Figuil, Mayo Rey, and Hama Koussou contain nonmarine Aptian to Albian beds similar to the Bima Sandstone in the northern Benue trough. Basaltic dykes, sills and flows, as well as dolerites, also occur in these basins (Maurin and Guiraud, 1990). The Hama Koussou basin contains about 800 m of arkosic, conglomeratic beds, overlain by sandstones, and fossiliferous clays. Dinosaur footprints, turtles, crocodiles, ostracodes, and silicified wood are common in these beds (Brunet et al., 1988; Flynn et al., 1989; Jacobs et al., 1988, 1989).

The Mamfe rift starts from southeastern Nigeria and extends into Cameroon, and is filled with about 4 km of mostly Lower Cretaceous fluviatile sandstones (Mamfe Formation). Cenomanian-Turonian marginal marine shales are preserved at the top of the sequence near the entrance into the Benue trough (Fig.9.41A).

9.6.5 Sudanese Rift Basins

Geophysical investigations and petroleum exploration have uncovered 25 fault-bounded intracratonic rift basins in southern and central Sudan (Fig.9.44A), at the eastern end of the Central African shear zone. Figure 9.44A shows that these basins are generaly orientated NW-SE. They consist of variably linked half-grabens and full grabens which resulted from low-angle listric normal faulting in the continental crust. Estimated sedimentary and magmatic fill are up to 15 km in the Muglad-Sudd rift segment, over 7 km in the Melut basin, and less than 5 km in the

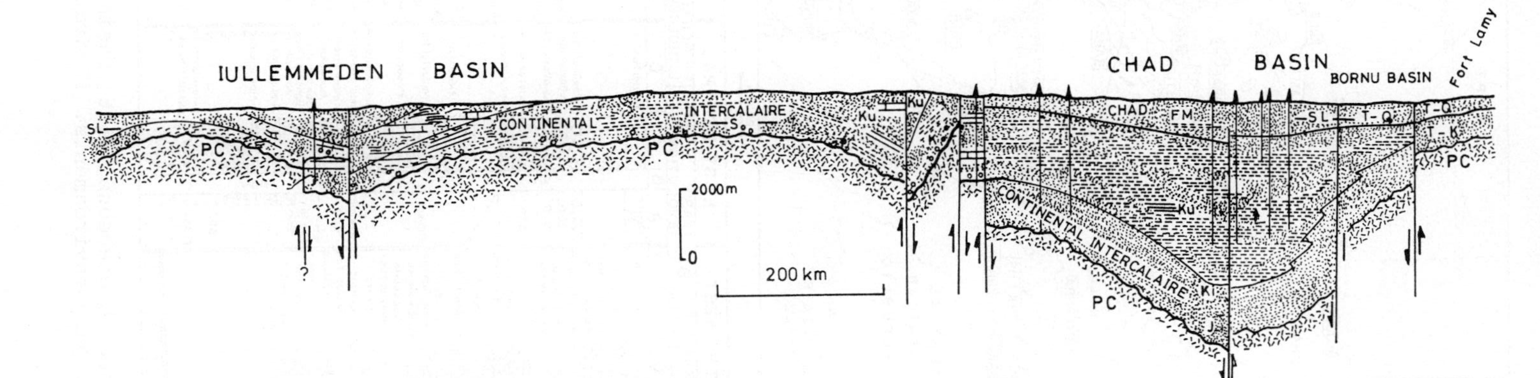

Figure 9.43: East-West cross-section through the Iullemmeden and Chad basins. (Redrawn from Peterson, 1985.)

A

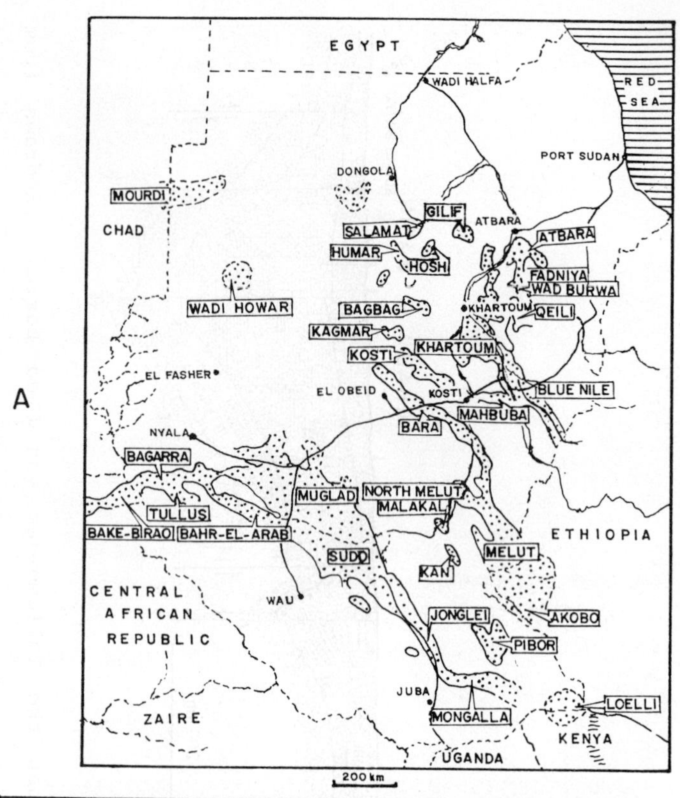

B

ERA	PERIOD	AGE (Ma)	CHRONO-STRATIGRAPHY	SOUTH-WEST A (Formation, max. thick (m))	SOUTH-WEST D	EAST-CENTRAL A	EAST-CENTRAL D	NORTH-WEST A	NORTH-WEST D
CENOZOIC	TERTIARY	2.0	Pleistocene	Zeraf 600	alluvium	Gezira 50	fluvial flood-plain		
		5.0	Pliocene		alluvial fan				
			Miocene	Adok 900	shallow lacust. braided stream				
		24.5	Oligocene	Tendi 2300	fluvial floodplain	Khaseeb 1500	proximal-mid alluvial fan	Hudi Chert <10	shallow lacust.
		38.0	Eocene	Nayil 1000	shallow lacust.-fluvial-flood-plain	1000	mid-alluvial fan	G. Abyad 50	carbonate tidal flat
		55.0	Paleocene	Amal 750	alluvial plain	350	distal alluvial fan		
MESOZOIC	CRETACEOUS	65.0	Maastrich.		braided stream	60	marginal playa	Kababish 100	fluvial lacustrine
		73.0	Campanian	Baraka 1200	alluvial fan				
		83.0	Santonian	Ghazal 300	braided stream				
		87.5	Coniacian	Zarga 300	floodplain		floodplain	W. Howar 150	fluvial
		88.5	Turonian	Aradeiba 600	fluvial/deltaic; shallow lacust.	Mansur 1150	mid-alluvial fan		
		91.0	Cenomanian		floodplain		floodplain		
		97.5	Albian	Bentiu 1500	alluvial fan; fluvial		mid-alluvial fan	W. Milk 150	fluvial
		113.0	Aptian	Abu Gabra 1800	deltaic; sub-lacust. fan; open lacustrine	Sawagir 800	distal alluvial fan; braided stream		
		119.0	Barremian				floodplain	Gilf Kebir Equiv.?	fluvial
		125.0	Hauterivian	Sharaf >370	alluvial fan; lacustrine	Abu Gin 700	open lacustrine; basaltic flows	?	
		131.0	Valanginian		fluvial flood-plain		marginal lacustrine		
		138.0	Berriasian				floodplain		
	JURASSIC	144.0	Tithonian	?					
		150.0	Kimmeridgian			500			
		156.0				El Azaza 60	distal alluvial fan; basaltic flows		
PRECAMBRIAN			Basement						

Figure 9.44: Sudanese Cretaceous rift basins, stratigraphic fill, lithofacies and paleoenvironments. (Redrawn from Wycisk et al., 1990.)

Bara, Kosti, Khartoum, Atbara, and Humar basins. The stratigraphic succession (Wycisk et al., 1990) consists of unconformity-bounded rift-related depositional cycles, which accumulated during the Early Cretaceous, Late Cretaceous and Tertiary. Vertebrates, including the remains of sauropod dinosaurs, occur in exposed continental Late Cretaceous beds in the Humar basin (Buffetaut et al., 1990).

Wycisk et al. grouped the Sudanese rift basins into those in the southwest, east-central, and northwestern parts of the country (Fig.9.44B). A probable pre-drift sequence of uncertain age, about 200 m thick, consists of strongly lithified quartz arenites. This unit overlies Pan-African basement in the Muglad-Sudd basin and is unconformably overlain by 3-5 km of syn-rift Neocomian-Barremian strata, the Sharaf Formation. The Sharaf Formation comprises cyclically bedded organic-rich lacustrine and fluvial-floodplain mudstones and fine sandstones which interfinger with coarser alluvial fan and braided stream deposits along the rift border faults. The Aptian-Albian sequence (dark, organic-rich claystones, shales with intercalated fine-grained sandstones and siltstones) represents the deposits of a deep stratified, sub-anoxic to anoxic lakes into which lacustrine deltaic sands and sub-lacustrine turbidites were discharged in a setting quite similar to the modern East African lakes.

A second rifting with an accompanying depositional phase began in the southwestern basins in the Late Cenomanian, and terminated in the Maastrichtian. The Muglad-Sudd basin was filled with a generally upward-coarsening sequence of floodplain, shallow lacustrine, braided stream, and alluvial fan clastics. The third and final rifting episode began in the late Eocene during which the basin filled up with lacustrine and floodplain clastics.

In the east-central Sudanese rift basins, a Late Jurassic-earliest Cretaceous marine transgression which came through the Horn of Africa deposited a fossiliferous and glauconitic claystone and siltstone sequence, which in the Khartoum basin, is underlain by coarse clastics and extrusive basalts. The rapidly subsiding Khartoum basin was filled in the Early to Late Cretaceous with a sequence which changes (from base upward) from floodplain through lacustrine to alluvial fan deposits. This sequence is unconformably overlain by Maastrichtian marginal playa carbonates, above which are alluvial fan deposits, and the unconformable Quaternary fluviatile sediments of the Gezira Formation, at the top.

9.7 Interior Sag Basins

Mesozoic-Cenozoic interior sag basins are mostly located in sub-Saharan Africa. Among these are the Iullemeden, Zaire, Okwango, Etosha, and Kalahari basins (Fig.8.2). Sedimentation in these basins, for the most part, was under continental conditions, and was not profoundly influenced by rifting either. However the basins in central and southern parts of western Africa are underlain by Karoo rifts.

9.7.1 Iullemmeden Basin

This basin has a sub-rectangular outline; its northern part is actually a separate Paleozoic basin as we saw earlier (Ch. 8.5.3). The Iullemmeden basin extends from Algeria, through eastern Mali, western Niger, into northwestern Nigeria where it is known as the Sokoto embayment. Basal nonmarine strata of Late Jurassic to Early Cretaceous age (Kogbe, 1973; Kogbe and Lemoigne, 1976) which belong to the Gundumi and Illo Formations are exposed in the Sokoto embayment. These are part of the "Continental Intercalaire" Group in the rest of the Iullemmeden basin (Fig.9.43).

Mesozoic marine sedimentation in the Iullemmeden basin began with the Cenomanian transgression (Reyment and Schöbel, 1983). Upper Cretaceous fossiliferous shales and limestones therefore overlie nonmarine Lower Cretaceous strata. Due to greater basin subsidence during the Maastrichtian-Paleocene, the sea encroached into the Sokoto embayment. This produced an overlap sequence in which Maastrichtian paralic beds (Taloka Formation) and gypsiferous marine and marginal marine marls and shales (Dukamaje Formation) are succeeded by the strata of a Paleocene marine cycle. Deposits of the Paleocene cycle include paralic facies (Wurno and Dange Formations), and marine marls (Kalambaina Formation). The Maastrichtian-Paleocene deposits in the Sokoto embayment contain rich vertebrate faunas (Halstead, 1974). Eocene and younger ferruginous clastics belonging to the Gwandu Formation, are part of the regional Tertiary continental deposits, the "Continental Terminal".

9.7.2 Zaire Basin

This is a broad downwarp centred on the Zaire craton, and extending into the Central African and Congo Republics (Fig.8.2.). Previously occupied by an epeiric basin in the Late Proterozoic when epicontinental stromatolitic carbonates and clastics accumulated (Fig.9.45A), the Zaire basin contains mostly Karoo, and Late Jurassic to Early Cretaceous fluviatile and lacustrine deposits (Cahen, 1983,a,b; Mateer, 1989). The Lualaba and

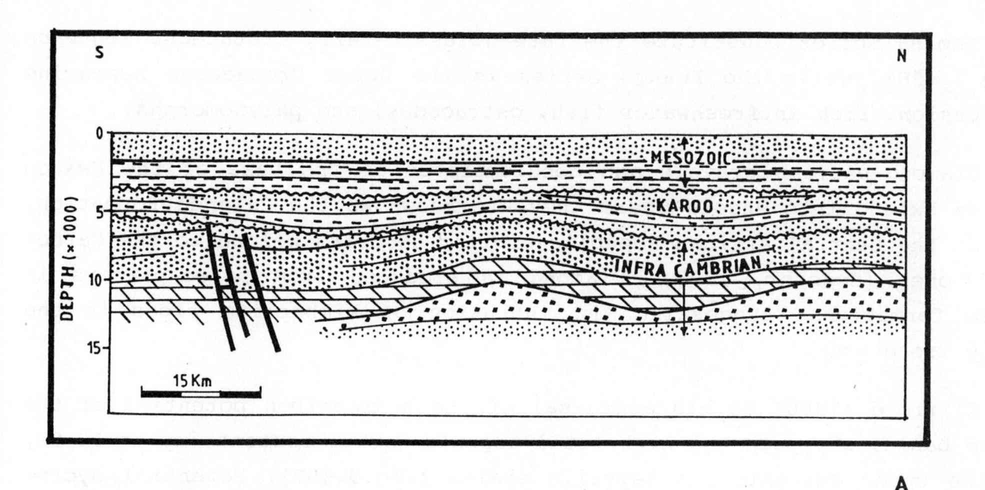

Summary of the Mesozoic Lithologies in the Zaire basin

B

Age	Unit	Lithology
UPPER CRETACEOUS	Kwango Series: Nsele Stage, max. 110 m	soft fine-grained sandstone, local lens of argillite near the base.
	Kwango Series: Inzia Stage, max. thickness observed 290 m	Kipola-Kimbau-Schwetz complex: reddish or greenish fossiliferous argillites, intercalations of sandstones and conglomerate beds, this complex contains a marine fauna. whitish, reddish coherent sandstones with occassional pebble horizons Kinko-Luzubi complex of argillites and soft sandstones, to north includes greenish shales, and grey limestones and whitish fossiliferous marls and sandstones, soft sandstones ranging from fine to coarse. Lutshima argillite: an alternation of red, calcareous, sandy micaceous shales with fine- to medium-grained sandstone.
	Kamina Series	Series of cross-bedded sands, Lignite traces, pre-end-Cretaceous
UPPER JURASSIC	Lualaba Series: Loia Stage (U.Jur.-L.Cr.)	defined in Lower Lomami it consists of: cross-bedded sandstones, soft shaly sandstones, some argillites and bituminous shales; the Litoko sandstones assigned to the Loia are now known to belong to the Stanleyville Stage
	Lualaba Series: Stanleyville Stage, max. thickness 850m, U. Jur.	(4) red and mottled marls and argillites, some bituminous horizons (3) green shales and bituminous beds (2) argillaceous and calcareous sandstones and bituminous beds (1) sandstones and conglomerates (Falls Conglomerate)
TRIASSIC		no undisputed Trias known, but red beds in various localities assigned to the Triassic, e.g., red beds of Lukuga, an alternation of red, green or gray shales with red, green, brown, or white sandstone with coarser sometimes conglomeratic base; red beds at Makunga (north 5°) contain a bivalve which resembles a Late Triassic form, fish remains and an Estheria sp. which elsewhere occur dated as Late Triassic-Rhaetic

Figure 9.45: Schematic stratigraphic section across the Zaire basin (A) and Mesozoic lithostratigraphy. (Redrawn from Clifford, 1986; Nairn, 1978.)

the Kamina Series constitute the Late Jurassic-Early Cretaceous sequence (Fig.9.45B), while the Kwango Series is the Upper Cretaceous nonmarine succession, rich in freshwater fish, ostracodes, and palynomorphs.

Diamond-bearing gravels and conglomerates at the base of the Kwango Series extend into Angola where they are known as the Calonda Formation; and in the Central African Republic similar deposits belong to the Carnot Sandstones (Censier, 1990). Diamond is produced from paleo-placers in these formations, suggesting the intrusion of kimberlite pipes in the Early Cretaceous.

Clifford (1986) in his appraisal of the hydrocarbon potential of the Zaire basin, stressed the high total organic carbon content, up to 18.5%, in the Upper Jurassic Stanleyville shales (Fig.9.45B). Potential hydrocarbon source beds and reservoirs also exist in the underlying Karoo sequence.

9.8. Tertiary Rifts and Ocean Basins

9.8.1 The Red Sea and the Gulf of Aden

Tectonic History

The Red Sea and the Gulf of Aden basins (Fig.9.46) are classic examples of the transition from Tertiary rifts to Atlantic-type marginal sags that developed along the trailing edges of young oceans. In his review of the tectonic evolution of the Red Sea and Gulf of Aden depression Beydoun (1989) stressed the following. Initial arching and uplift of the Arabian-Nubian Shield in the Late Eocene was followed by crustal extension, rifting and crustal attenuation in the Oligocene and Early Miocene. There was sea-floor spreading during which Arabia separated from Africa and rotated counter-clockwise with horizontal motion along the Gulf of Aqaba transform fault (Fig.9.46) leading to the opening of the Gulf of Suez, and the collision and suturing of Arabia with Eurasia, and the rise of the Taurus-Zagros fold and thrust belt.

The initiation of the Gulf of Aden has been attributed to the propagation westward of ocean-floor spreading along the Carlsberg ridge in the Indian Ocean (Fig.9.34), during the Miocene. The spreading ridge in the Gulf of Aden (Fig.9.46) has a deep median valley which is the site of high heat flow in contrast with the Red Sea that has no central ridge, but rather, an axial trough mostly over its southern and central parts. The Red Sea axial trough is also the site of high heat flow.

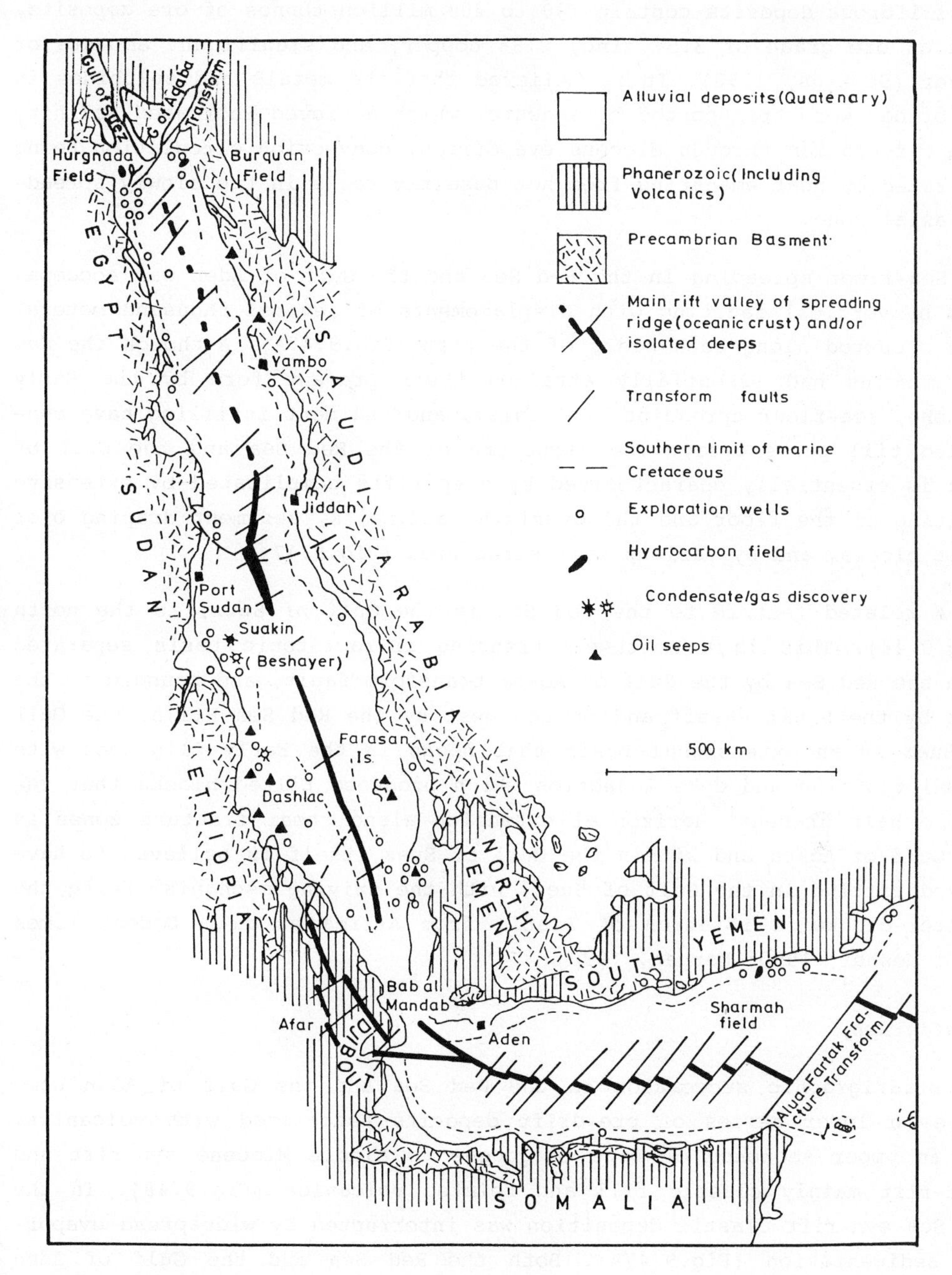

Figure 9.46: Outline geologic map of the Red Sea and Gulf of Aden. (Redrawn from Beydoun, 1989.)

A notable feature of the Red Sea median valley is the occurrence of hot brine pools below which metalliferous sediments are found. These metalliferous deposits contain 100 to 200 million tonnes of ore deposits, with an ore grade of 3.5% zinc, 0.8% copper, and significant amounts of silver (Sawkins, 1990). It is believed that the metals and sulphides in the brines were transported by seawater which achieved increased salinity from circulation through Miocene evaporites, convection circulation being generated by heat emanating from hot basaltic rocks in the slowly spreading axial zone.

Sea-floor spreading in the Red Sea and the Gulf of Aden was accompanied by vertical faulting with displacements of several thousand meters. This occurred along both sides of the rift (Fig.9.47A). Although the Red Sea margins had essentially attained their present form by the Early Miocene, sea-floor spreading, volcanism, and sediment infilling have continued till present-day. The structure of the Red Sea and the Gulf of Aden is essentially characterized by deep rifts complicated by extensive faulting of the floor and the overlying sediments, sediment draping over fault blocks, and by various salt structures (Fig.9.47).

A related feature to the Red Sea is the Gulf of Suez, to the north (Fig.9.46). This is a northwest-trending intracratonic basin separated from the Red Sea by the Gulf of Aqaba transform fault, and bounded to the east by the Sinai massif, and on the west by the Red Sea Hills. The Gulf of Suez is an extensional basin that began in the Early Oligocene with normal faulting and dyke injection which produced tilted blocks that resemble half grabens. Horizontal extension along shear fracture zones in the Gulf of Aqaba and within the Gulf of Suez itself is believed to have caused rifting in the Gulf of Suez, with the driving mechanism being the counter-clockwise rotation of Arabia from Africa as from Eocene times (eg., Meshef, 1990; Morgan, 1990).

Stratigraphy

The stratigraphic succession in the Red Sea and the Gulf of Aden comprises a lower series of pre-drift deposits associated with volcanics, and an upper sequence of Late Oligocene to Middle Miocene syn-rift and post-rift mainly clastic fill and related volcanics (Fig.9.48). In the Red Sea syn-rift clastic deposition was interrupted by widespread evaporite sedimentation (Fig.9.47A). Both the Red Sea and the Gulf of Aden basins contain a Late Miocene to Pliocene clastic fill and a more open-marine Late Pliocene to Recent post-rift sea-floor spreading sequence, with coarser clastics which are restricted to the margins.

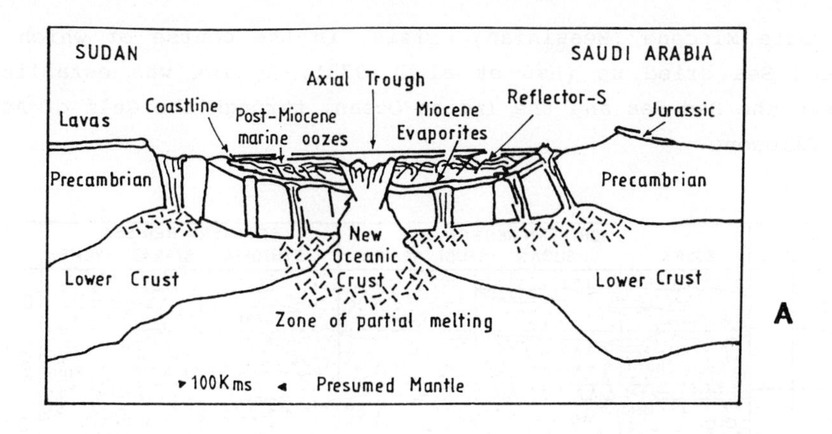

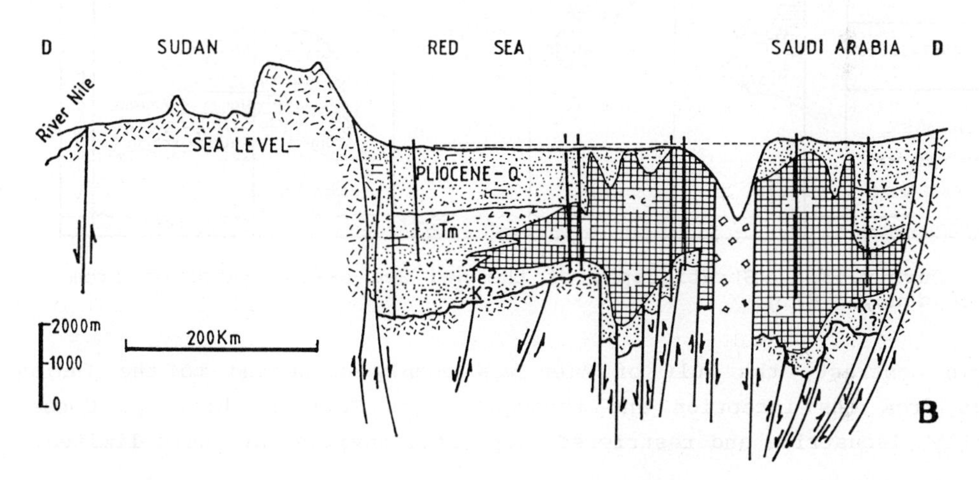

Figure 9.47: East-West structural (A) and stratigraphic (B) sections across the Red Sea. (Redrawn from Braithwaite 1987., Peterson, 1985.)

Regarding their paleogeography, it is believed that by the Late Oligocene-Early Miocene the Red Sea had become a major depression and an extensive continental rift which was occupied by a lake that received sediments. Marine incursion took place in the Early Miocene coming from the Mediterranean Sea (Fig.9.17C), but was not connected with the Gulf of Aden by the Middle Miocene, although some tenuous link may have existed through the Afar triangle. Subsequently, links with the Mediterranean also became restricted, giving rise to thick Middle to Late Miocene evaporites throughout the Red Sea basin. Evaporite formation climaxed during the Late Miocene (Messinian) crisis, in the course of which the Mediterranean Sea dried up (Hsu et al., 1973). A link was established later between the Red Sea and the Indian Ocean, through the Gulf of Aden, during the Pliocene.

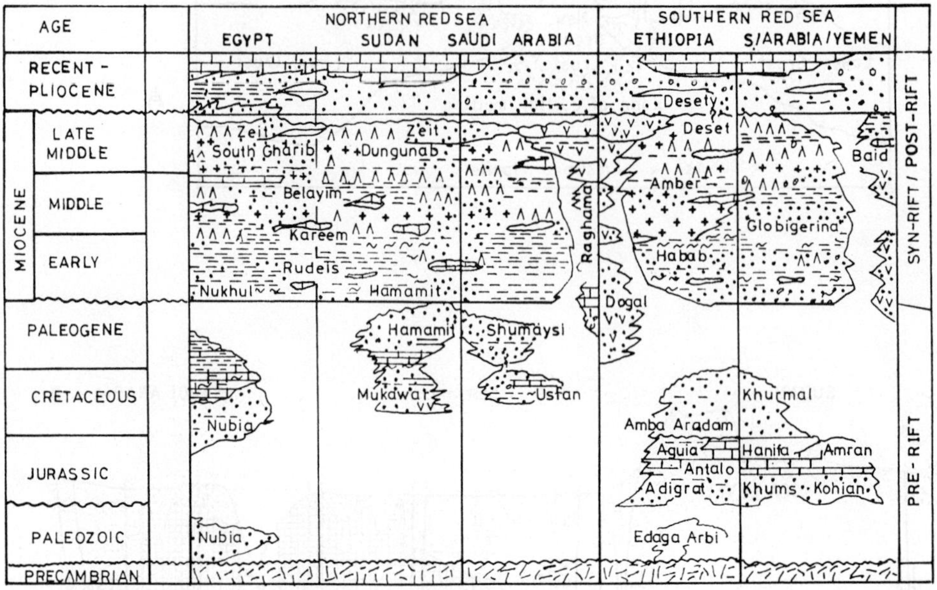

Figure 9.48: Stratigraphy of the Red Sea basin. (Redrawn from Beydoun, 1989.)

In contrast, the Gulf of Aden was openly connected to the Indian Ocean from its inception and throughout its Tertiary history. Consequently, lacustrine and restricted evaporitic environments were limited.

Although oil seeps occur along the Red Sea margin (Fig.9.46), the only major petroleum production that is directly from the Red Sea basin is from a gas/condensate field, the Suakin field, located in offshore Sudan. In this field traping is related to salt tectonics in which growth faulting is listric into Miocene evaporites, with the Miocene Globigerine Marl serving as the source rocks. There are Miocene sand reservoirs which are sealed by an upper salt layer. The Gulf of Suez also produces petroleum, but from faulted sandstones of the Nubian depositional cycle.

9.8.2 The East African Rift System

Introduction

Our survey of the development of Mesozoic-Cenozoic basins in Africa began with Triassic rifting in northwestern Africa. Having examined the propagation of rift systems all around the continent, leading to the formation of the Atlantic and Indian Ocean basins and their African continental margins from Jurassic to Early Cretaceous times, up to the birth of new oceans, the Red Sea and the Gulf of Aden, we can now conclude the rift story with a glimpse at the grand theatre of contemporary rifting--the East African Rift System (Fig.9.49). The tectonic significance of the East African Rift System lies in the enormous scale of continental breakup, involving the generation of oceanic crust--the initiation of the Wilson Cycle par excellence.

But before considering the global tectonic significance of the Great Rift Valley as it is sometimes called, it is pertinent to mention the universally acknowledged aesthetic quality of its great lakes, snow-clad mountains, and wildlife. Apart from its lure to both colonial and modern travellers, the impact of the Rift Valley on the lives of East Africans was best summed up as follows by Rogers and Rosendahl (1989): "The system of valleys has profoundly affected the social, political, and economic affairs of East Africans, from the time of *Homo erectus* to modern day--witness the abundance of archaeological finds associated with the rift terrane, the political boundaries that follow the Western Branch of the rift, and the agricultural import of rift valleys. The East African rift will continue to play an economic role in Africa, perhaps a crucial one if the boom in petroleum exploration in East Africa proves fruitful".

Since Chapter 10 is devoted to Phanerozoic magmatism, of which the volcanics of the East African Rift System form a major component, the igneous activity that was associated with this and earlier rifting will not be considered in detail in this section. Similarly, the Quaternary stratigraphy of the rift system will be treated in Chapter 11. This is because the Quaternary of the Rift Valley is the backbone of the African Quaternary. As aforementioned, the Rift Valley Quaternary contains a unique record of vertebrates, including the stages of homonid evolution, and one or the most comprehensive and best documented paleoclimatic records for the last 2.5 million years. However, in view of its importance to the understanding of the stratigraphy of the rift system in general, depositional models will be discussed in this section. What follows is mostly devoted to the geomorphology, structure, general stratigraphy, and the origin of the Rift Valley, a term commonly used for the East African Rift System.

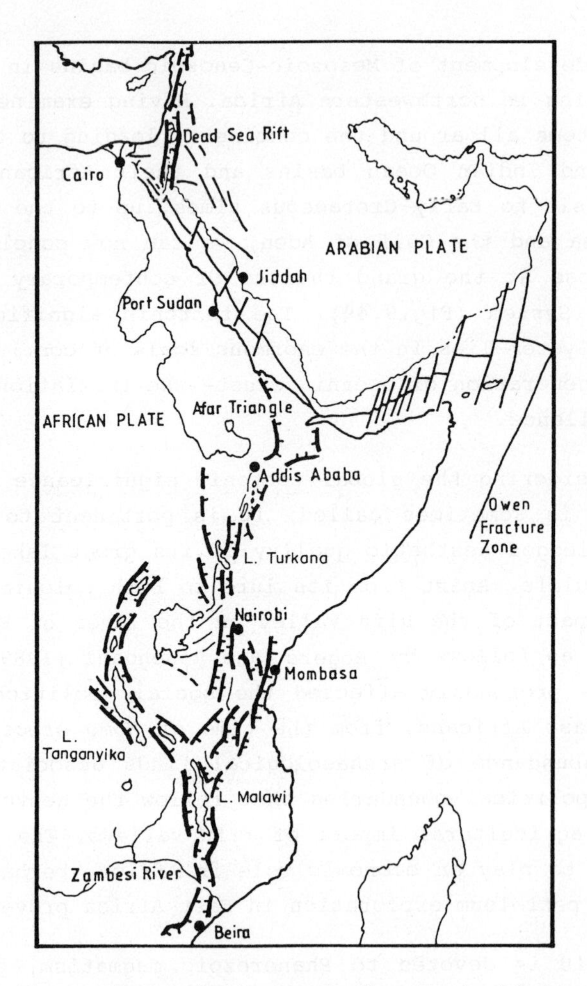

Figure 9.49: The Red Sea and the East African Rift system. (Redrawn from Braithwaite, 1989.)

Geomorphology and Structure

The East African Rift System is one of the world's most spectacular faulted land-scapes. It is part of the world's oceanic rift system. Regionally, it belongs to a system of rifts that cuts through northeastern Africa and the Middle East before entering East Africa (Fig.9.49). Along its course this rift system includes major rifts, such as the Dead Sea graben in Jordan, the Gulfs of Aqaba and Suez, the Red Sea, and the Gulf of Aden. At Afar the main rift systems veers from its NW-SE course and enters East Africa where it runs through Ethiopia to Mozambique (Fig.9.49).

The East African Rift System has been divided into the Eastern branch (Gregory Rift) and the Western branch. The Eastern branch, generally about 50-80 km wide runs from the Afar triangle southwestward through Ethiopia and Kenya before entering a diffuse region of graben and splay faults in northern Tanzania. It crosses two areas of basement uplifts, the Ethiopia and the Kenya domes. These domes reflect broad uplifts of the lithosphere-asthenosphere boundary, and topographically they form plateaus which sometimes rise to over 3,000 m.

Prichard (1979) presented a description of the geomorphology of the Rift Valley. From the wide fault-bounded triangular Danakil depression (Fig.9.50A) the Eastern branch displays prominent fault scarps (Fig.9.50B) which record throws of up to 3,000 m or more. In southern Ethiopia the normally well-defined Rift Valley is replaced by a diffuse basin-and-range province, up to 150 km wide, where the Turkana depression lies. A major rift is believed to lie beneath Lake Turkana. Both in Ethiopia and in Kenya the Rift Valley floor is occupied by a series of small lakes which in Kenya are not deeper than 16 m, except Lake Turkana (116 m).

The Western branch extends from the coastal plain of Mozambique through Lakes Malawi, Tanganyika, to Lake Mobutu where it terminates against the Aswa shear zone (Fig.9.50A). In the northern part the terrane around Lake Kivu and the Ruzizi Mountains is rugged and faulted and partly drowned to form rias. Lake Edward occupies a well-defined rift valley which divides northward into a lowland and a trough, the Semliki trough. Towards the northern end of the Western branch Lake Mobutu lies in a fault-bounded trough, the scarps of which gradually decrease in elevation until they are replaced by the zig-zag fault zones in the Nile region. The southern segment of the Western branch is a horst-and-graben landscape with sub-parallel faults, within which lies Lakes Tanganyika and Malawi. Lake Malawi occupies a deep trough (Fig.9.50C), about 80 km wide and 650 km long.

The structure of the Western branch has been described in detail by Chorowicz et al. (1987), Rosendahl et al. (1986), and Specht and Rosendahl (1989). Generally there are arcuate border faults in the Rift Valley which define half-grabens (Fig.9.50). Smaller NW-SE transverse faults located within the grabens, link the border faults and dissect them into rhomb-shaped outlines (Fig.9.50A). These transverse or transfer

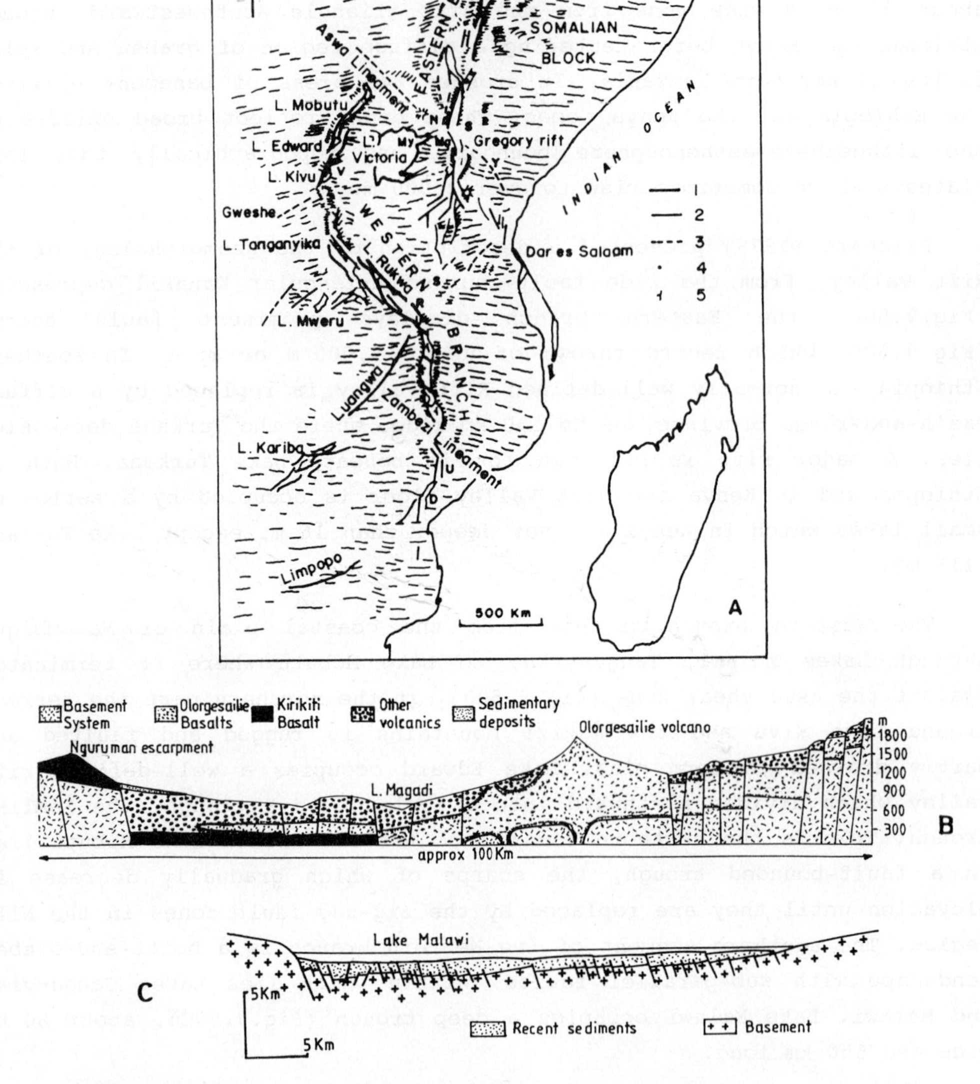

Figure 9.50: A, structural map of the East African Rift: 1, major fault zone; 2, other faults; 3, inactive fault; 4, major volcano; 5, major dip; B, Buhoro Flats; D, Dombe trough; E, Elgon volcano; G, Gilgil fault; K, Kilimanjaro; L, Livingstone fault; M. Mahali Mounts; Ma, Mau Scarp; N. Nandi fault; Ny, Nyanza rift; R, Rungwe volcano; S, Sattima scarp; U, Urema trough; V, Virunga; Y, Yatta plateau. B, cross-section of the Eastern Branch; C, cross-section of Lake Malawi basin. (Redrawn from Chorowicz, 1990; Pritchard, 1979.)

zones are characteristically defined by fault-bounded upstanding basement blocks which subdivide the half-grabens.

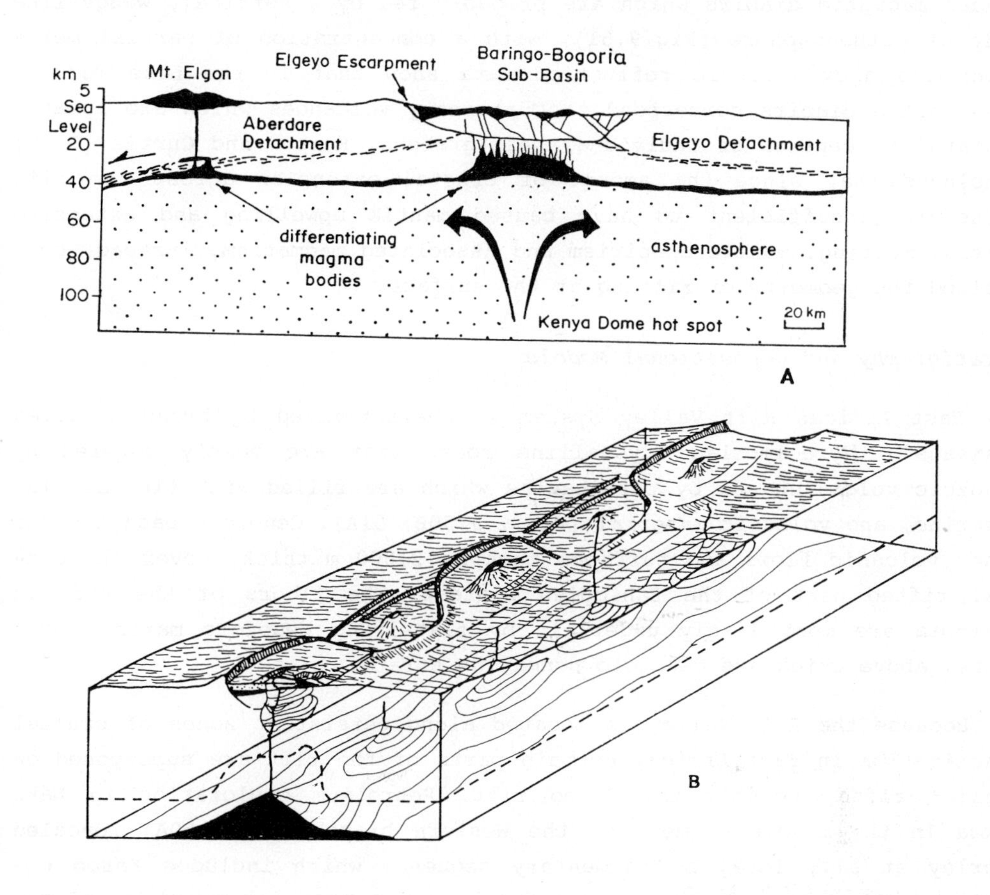

Figure 9.51: Sections showing the deep structure underneath the Eastern Branch (Gregory Rift). (Redrawn from Karson and Curtis, 1989.)

While the Eastern branch shows pronounced magmatism which varies with crustal depth (volcanic eruptions, fissuring, faulting and block rotation in the uppermost crust, with magmatic intrusions dominating in the middle and lower crustal depths), the Western branch has much less volcanism and less advanced rifting. The latter branch is therefore considered younger than the Eastern branch (Chrowicz et al., 1987). The deep structure underneath the Eastern branch revealed by geophysical surveys (eg. Karson and Curtis, 1989), is related to the magmatic processes associated with the origin of the Rift Valley. The axial positive gravity anomalies and high compressional wave velocities underneath the Rift Valley are due to basic and ultrabasic igneous bodies in the crust and upper mantle. The rift crust and upper mantle have been intruded in the Quaternary by individual magmatic diapirs which are probably fed by a vertical, wedge-like body of asthenosphere (Fig.9.51), with a concentration of partial melts along its apex. Seismic reflection data show that in the Lake Turkana area mantle diapirs correspond to Quaternary volcanoes which are located beneath the centres of individual half-grabens. Karson and Curtis (1989) concluded that since the amount of crustal extension across the rift units was insufficient to have caused mantle upwelling and extensive partial melting, mantle diapirism and associated magmatism, instead, controlled the geometry of rifting at the surface.

Stratigraphy and Depositional Models

The East African Rift Valley System is characterized by broad uplifted plateaus of Precambrian crystalline rocks that are mostly covered by Cenozoic volcanics and by rift basins which are filled with fluvial, lacustrine, and volcanic material (Figs. 9.50B; 51A). Cenozoic basaltic and other volcanic flows and tuffs as much as 2,000 m thick, cover the central rifted part of the Ethiopian plateau. The floors of the rift in Ethiopia are most likely underlain by flat-lying Mesozoic marine sediments, above which are Cenozoic nonmarine strata.

Because the Rift Valley is located along persistent zones of crustal reactivation in East Africa, certain parts of the rift are superposed on earlier rifts, notably the Karoo rift. Petroleum exploration in Lake Rukwa in the middle segment of the Western branch (Fig.9.50A) revealed (Morley et al., 1989) a sedimentary sequence which includes Karoo deposits (with coal), Jurassic or Cretaceous red beds and Tertiary to Recent deposits. Late Miocene fluvial red beds and Late Miocene-Recent lacustrine deposits have been dated palynologically underneath Lake Rukwa. The Karoo and Tertiary-Recent sequence fills a half-graben in the lake and expands greatly towards the eastern border fault where the thickness of the Late Miocene-Recent section is estimated at 6-7 km.

The enormous sedimentary thickness in Lake Rukwa may however, be atypical and quite localized, considering the Recent depositional pattern in the Rift Valley. Frostick and Reid (1989) stressed the considerable diversion of drainage and sedimentation away from the rifts because of doming and back-tilting of footwall fault blocks. Also, the segmented architecture of the rift does not permit through drainage. Consequently, parts of the Rift Valley are actually starved of sediments, while other areas were active depocentres.

Pickford (1982) suggested that in the Kenya Rift Valley sedimentation was cyclical in these depocentres and reflected initial downwarping and rapid scarp erosion and deposition of coarse alluvial fans. As the scarps are levelled by erosion finer deposits accumulate in lacustrine environments together with biogenic sediments such as diatomites, and stromatolites (Casanova, 1986). Sedimentation is often interrupted by the deposition of lava flows and pyroclastics which may bury and preserve soils that often contain rich terrestrial fossil assemblages, including footprints (Behrensmeyer and Laporte, 1981).

Cohen et al. (1989) presented two contrasting depositional models (Fig.9.52) for the Eastern branch and the Western branch. Lake Turkana which is 116 m deep, is located within a semi-arid region, in a volcanic basin. Lake Tanganyika is situated in a semi-humid belt, and is a very deep basin (1,470 m), without significant volcanism. Both basins are, however, sediment-starved. Lake Turkana sedimentation is dominated by the accumulation of oxygen-poor, terrigenous muds; rare deep-water coarse clastics, because of ponding in marginal depocentres by volcanic barriers. There are low biogenic components and low carbonate content in the sediments. Lake Tanganyika, in contrast, is dominated by organic-rich, biogenic muds. Anoxic conditions prevail below 150-200 m. Sand is deposited at great depths by subaqueous gravity flows; sediment ponding is very rare along the margin and littoral carbonates are common. The basin morphology allows only limited clastic input into the lake. Lake Tanganyika has steep relief with backsloping, limited drainage basin area, and prior sedimentation upstream in Lake Kivu, which is located to the north (Fig.9.50A). Thus, contrasting basin morphologies, volcanic activity, and climate have produced different lithofacies in the rift lakes (Fig.9.52).

Tectonic Model

The East African Rift System has generated a wealth of genetic models, including doming by asthenolith injection, mantle convection, hot spot activity, and variations in the theme of lithosperic stretching. These models were reviewed by Chorowicz et al. (1987), who also argued that the Rift Valley developed in stages (Fig.9.53), dominated by strike-slip tectonics, a model that is gaining widespread acceptance.

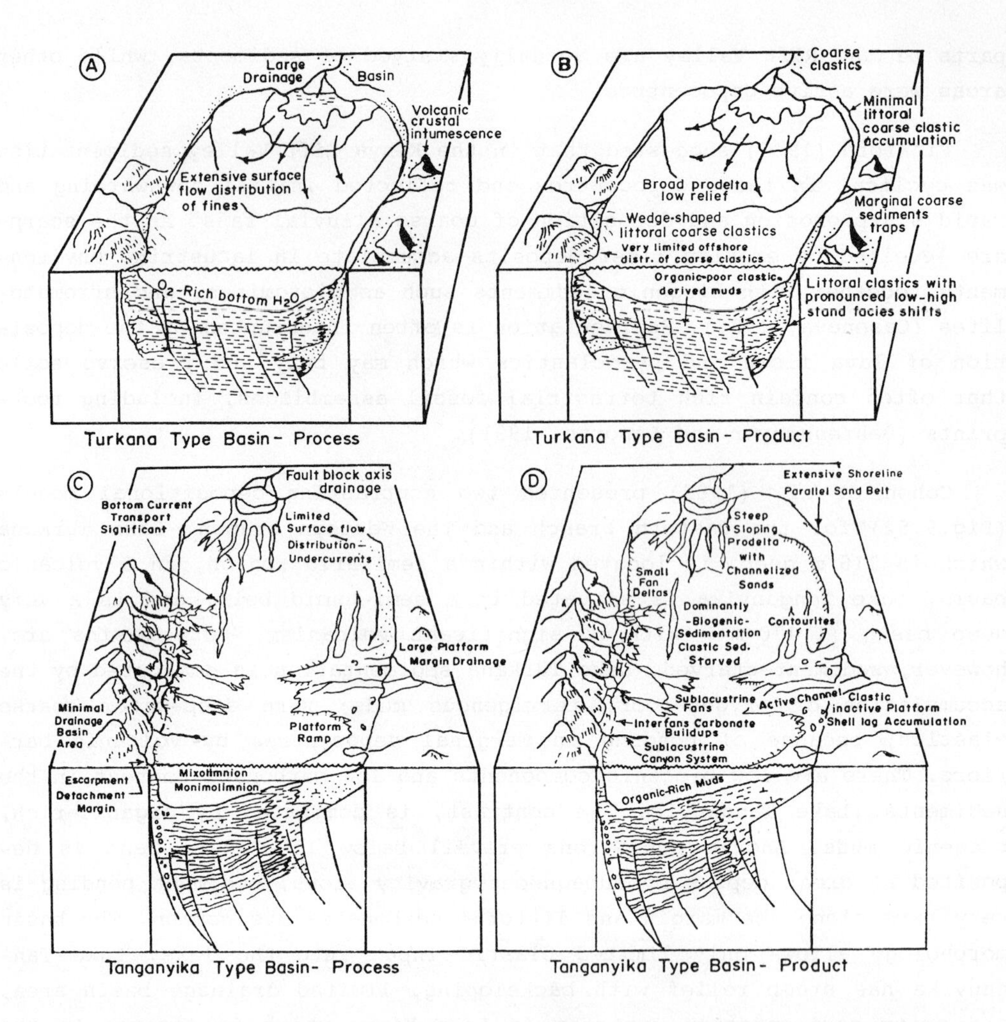

Figure 9.52: Depositional processes and resultant lithofacies in lakes Tanganyika and Turkana. (Redrawn from Cohen et al., 1989.)

At the pre-rift stage, initially horizontal motion led to dense fracturing, characterized by strike-slip faulting, while a shallow but wide topographic depression formed, and open gashes in the crust created room for tholeiitic volcanism (Fig.9.53A). Horizontal slip movement has taken place since Late Oligocene-Early Miocene times in a NW-SE direction, along fractures and lineaments such as the Aswa lineament (Fig.9.50A) and shallow lakes similar to Lake Mweru, may represent the Recent topographic manifestation of this pre-rift phase. Initial tholeiitic volcanism may have been similar to the present-day Virunga volcanic chain (Kampunzu et al., 1983).

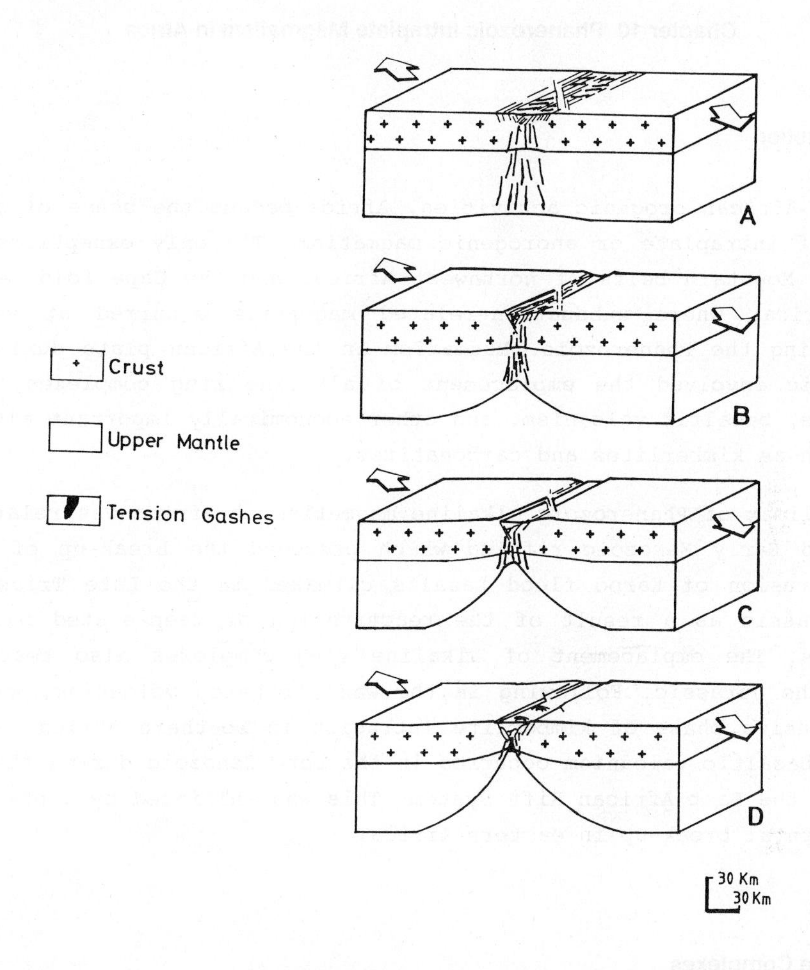

Figure 9.53: Tectonic model showing the lithosphere at different stages of rift evolution. (Redrawn from Chorowicz et al., 1987.)

After being initiated, typical rifting (Fig.9.53B, C) occured along normal faults which border the main tilted blocks, while subsidence of the rift floor and uplift along the shoulders ensued. The advanced rifting stage manifested in significant magmatic intrusions along the rift axis at which stage the initial oceanic crust had formed. In the Eastern branch the typical initial rifting stage characterized by major uplift of the rift shoulders began in the Late Miocene, while the advanced rifting stage was attained in the Pliocene. The Afar depression corresponds to the oceanic stage.

Chapter 10 Phanerozoic Intraplate Magmatism in Africa

10.1 Introduction

After Pan-African orogenic activities, Africa became the scene of a wide variety of intraplate or anorogenic magmatism. The only exceptions were the Atlas Mountain belts of northwest Africa, and the Cape fold belt of South Africa, where subduction-related magmatism occurred at various times during the Phanerozoic. Magmatism in the African plate during the Phanerozoic involved the emplacement of alkaline ring complexes, basic intrusions, basaltic volcanism, and other economically important alkaline rocks such as kimberlites and carbonatites.

The climax of Phanerozoic alkaline magmatism in Africa was related to widespread Early Mesozoic rifting which preceded the break-up of Gondwana. Extrusion of Karoo flood basalts climaxed in the Late Triassic - Early Jurassic as a result of the reactivation of deep-seated basement lineaments. The emplacement of alkaline ring complexes also reached a peak in the Jurassic. Following in the wake of Karoo volcanism, was the most intensive phase of kimberlite intrusion in southern Africa. Resurgence of basaltic volcanism occurred in the Late Cenozoic during the formation of the East African Rift System. This was initiated by a new phase of continental break-up in eastern Africa.

10.2 Alkaline Complexes

10.2.1 Types and Structure

According to Kinnaird and Bowden (1987) African Phanerozoic alkaline magmatism was characterized by small centres of subvolcanic to plutonic nature, often in the form of ring complexes. Alkaline magmatic provinces are of two types, viz. those dominated by centres with oversaturated magma, and those with undersaturated complexes and carbonatites (Fig.10.1). Prominent among the oversaturated provinces, where ring complexes are well developed, are Nigeria, Niger, Cameroon, Sudan, Egypt, Ethiopia, and Arabia (Almond, 1979; Turner, 1976; Vail, 1989b). The undersaturated provinces include Namibia, Angola, and the East African Rift System.

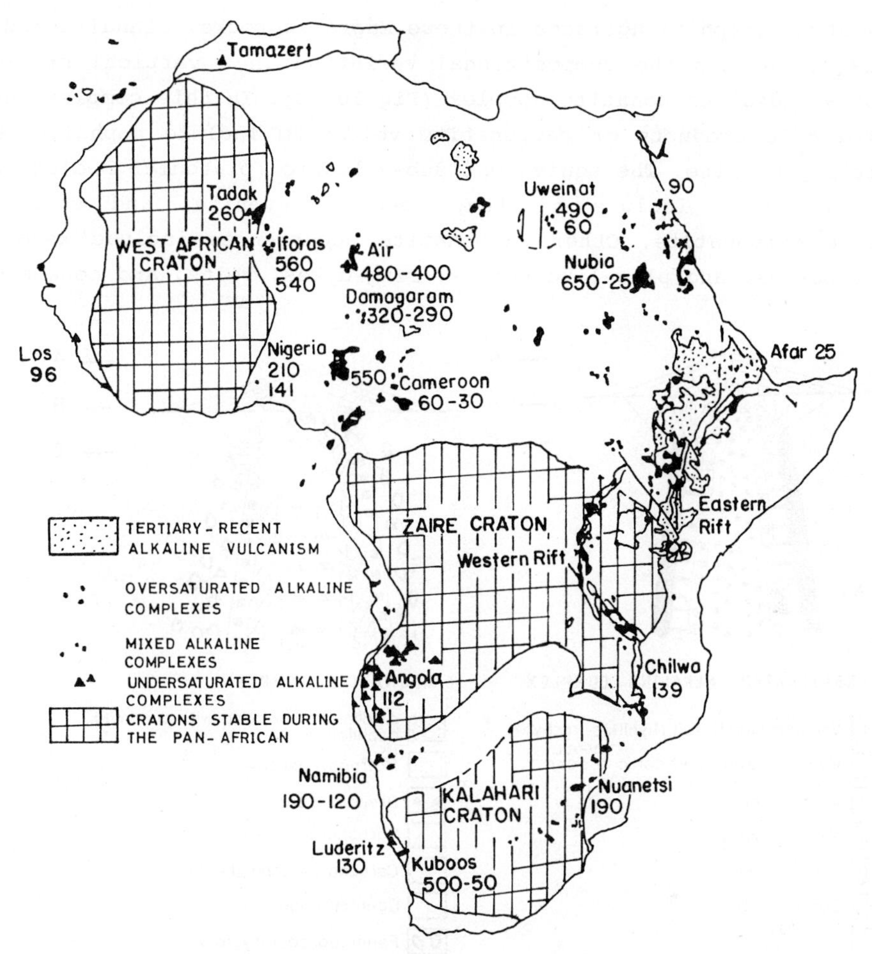

Figure 10.1: Distribution of alkaline magmatic rocks in Africa. (Redrawn from Cahen et al., 1984.)

Anorogenic alkaline centres, whether undersaturated or oversaturated, display similarities in their structural features (Fig.10.2). The centres are usually exposed at various levels, ranging from the volcanoes, to the root zones where the plutons have sharp intrusive contacts and homogeneous, unfoliated, and commonly porphyritic textures (Kinnaird and Bowden, 1987). In the oversaturated alkaline provinces the sub-volcanic intrusions are mostly syenitic to granitic in composition, with basic rocks accounting for 5% or less of the total surface area.

The classic locality for undersaturated (carbonatite) complexes is the East African Rift System. Here all levels of erosion are displayed, from alkaline volcanoes (Oldoinyo Lengai), to the partially eroded centres (e.g. Napak in Uganda), and well exposed root zones at Chilwa Island in Malawi (Fig.10.1). Combining all three different erosional (chronostratigraphic) horizons in these magmatic rocks, Kinnaird and Bowden (1987) deduced the compositional variation in a vertical cross-section of an ideal carbonatite complex (Fig.10.2B). In this complex the Recent volcanic products of carbonatite volcanism include nephelinite and agpatitic phonolite. The equivalent sub-volcanic (plutonic) bodies in the cores and roots include ijolite, nephline syenite, and sovitic or alvikitic carbonatite. Other carbonatite occurrences include xenolithic igneous bosses, and plug-like bodies, ring-dykes, veins, and cone-sheets.

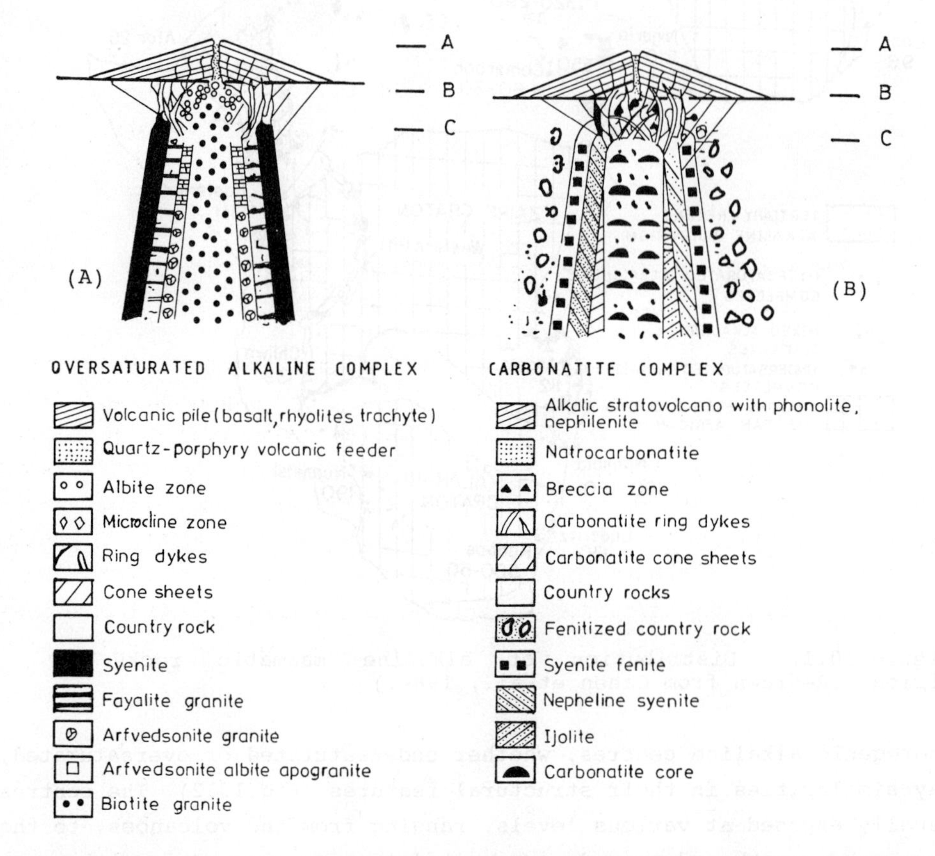

Figure 10.2: Schematic sections across oversaturated alkaline complex (A); and carbonatite complex (B) where (A) represents Oldoinyo Lengai; (B) Napak; (C) Chilwa Island. (Redrawn from Kinnaird and Bowden, 1987.)

Vail (1989b) furnished a regional account of the distribution of African ring complexes, summarized below. He observed that the African continent probably has the largest number of ring complexes in the world. Over 625 ring complexes were recognized on the continent, ranging in age from mid-Proterozoic to Tertiary . The Pan-African late or post-orogenic ring complexes, dated 720-490 Ma, were mentioned earlier in connection with the Tuareg shield (Ch. 6.4.2), and the Arabian-Nubian shield (Ch. 6.11.5). In this chapter we shall be concerned with ring complexes of Ordovician to Tertiary age.

10.2.2 The West African Younger Granite Ring Complex Province

This is one of the best known and finest alkaline ring provinces in Africa (Fig.10.3). The Early-Middle Paleozoic was a time of important anorogenic ring complex emplacement in the Sahara, from the Tuareg shield to the Nubian shield. On the Tuareg shield the Early-Middle Paleozoic complexes occur in the northernmost and oldest part of the meridional Younger Granite ring complex province which extends from northern Niger Republic through the Jos plateau in Nigeria, to the Benue valley (Fig.10.3). The northern Niger complexes consist of well developed structures with ring-dykes and cone-sheets of granite. They range in diameter from 2.5, to 65 km in a gabbroic complex. The complexes overlap and display shifting intrusive centres.

South of Aïr, in the Damagaram region of southern Niger, there is a Carboniferous-Permian alkaline ring province which extends into Nigeria (Fig.10.3). These are the best developed Late Paleozoic ring complexes in Africa. About a dozen granitic ring-dykes, approximately 2 km and more in diameter, intrude a highly deformed Proterozoic basement inlier. Their ages range from 323 Ma in southernmost Niger to 258 Ma on the Nigerian side (Rahaman et al., 1984), thus demonstrating a southward decrease in the age of the ring complexes. This suggest a southward migration of magmatic activity, a feature which characterizes this province (Turner and Webb, 1974).

The westernmost alkaline complexes in the West African province lie in Tadhak and at Timetrine nearby. These are of Permian age. A different type of alkaline magmatism prevailed here. Timetrine, located on the eastern edge of the West African craton, contains no ring structures. Rather, there are undersaturated foid syenites and carbonatites.

Like elsewhere in Africa, the Jurassic was the most intense phase of ring complex development in Africa. There are over 40 granite complexes among the Younger granites of Nigeria (Turner, 1976). They lie in a broad zone 400 km long (N-S) and 150 km wide (Fig.10.3), and display ring structures, 2 to 15 km across. They include soda pyroxene and amphibole biotite, fayalite granites, syenites, and trachytes with minor gabbros, and dolerites. There are rare volcanics such as rhyolites, tuffs, and ignimbrites (Badejoko, 1976; Ike, 1983; Jacobson et al., 1958; Turner, 1976).

The characteristic southward migration of magmatic activity already evident in the Paleozoic, was quite pronounced in the Mesozoic. The ages of the ring complexes of the Nigerian Younger Granites illustrate this trend (Fig.10.3).

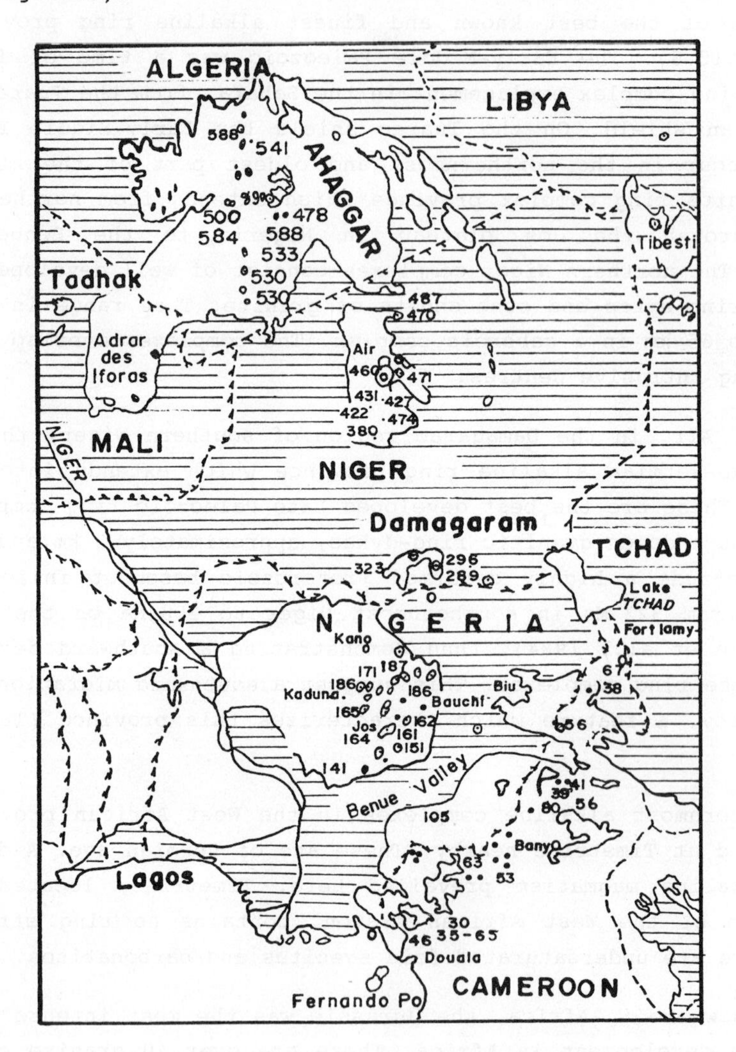

Figure 10.3: Sketch map of the Younger Granite complexes of West African. (Redrawn from Cahen et al., 1984.)

A broad regional negative gravity anomaly characterizes the Nigerian Younger Granites on the Jos plateau (Ajakaiye, 1976). Large negative anomalies over individual complexes suggest that the complexes extend to depths of 10-12 km and that they are not underlain by deeper magmatic bodies. The north-south alignment of the Younger Granites province has been attributed to their probable location along a deep-seated shear zone. Shear motion is believed to have generated frictional heating, and to have released pressure for magma to ascend through fractures. The timing and localization of shearing was probably responsible for the progressive southward age decrease.

Outside the Younger Granites province lies a NE-trending band of small Tertiary alkaline plutons which extend from the Gulf of Guinea through Cameroon into Chad Republic. Known as the "Granites ultimes" in Cameroon, they coincide with the Cameroon volcanic line. The "Granites ultimes" consist of plugs and ring complexes of granites, syenites, and micrograrites, numbering up to 38. They are obscured by younger volcanics in some places. Being petrographically and geochemically very similar to the Younger Granites, they are considered as part of the West African Younger Granite ring complex province. The "Granites ultimes" range in age from 67 Ma in the north to 38 Ma in Poli in the central part, to 46 Ma near Douala on the coast. This, however does not reflect an age migration of magmatic activity.

10.2.3 Northeast African Province

This province includes the ring complexes of the Sudan, Egypt, Ethiopia, and Uganda (Fig.10.1). In this province various anorogenic granitic, syenitic, and gabbroic plutons intrude the basement complex. Ring complexes of Early-Middle Paleozoic age occur in the Nuba Mountains of the Sudan, in the Bayuda desert, the Nile valley, and in the Red Sea Hills of northeastern Sudan. Sabaloka has the largest complex. The Paleozoic complexes are characterized by alkaline micrograrites, quartz-soda pyroxene-amphibole syenite, with the extrusive phase being trachybasalts, trachytes, alkali lavas, and pyroclastics. These are preserved in downfaulted cauldrons which are enclosed by ring-dykes.

In the Bayuda desert, Mesozoic ring complexes, sometimes indistinguishable from the earlier ones, have intrusive cores. These cores consist of soda pyroxene and amphibole granites, quartz syenite, and associated volcanics such as rhyolites and tuffs. Mesozoic alkaline ring complexes with granites and syenites also occur in the Kordofan province of central Sudan. In the Red Sea Hills of Egypt and the Sudanese ring complexes alkali granites, quartz and foid syenites, trachytes and gabbros pierce through Pan-African volcano-sedimentary-ophiolitic supracrustals. In the Sudan these complexes are meridionally aligned for over 300 km, whereas in southeastern Egypt they lie in a broad NNE-trending band 230 km long. The distribution pattern is believed to reflect the arrangement of brittle fractures in the basement.

Another group of Mesozoic alkali ring complexes occurs in a north-south trend, about 850 km long, along the Sudan-Ethiopia frontier, down to northern Uganda. A prominent group of Tertiary alkaline ring granites and foyaites, of which Jebel Uweinat is the best known example, intrudes the Uweinat Archean basement inlier (Fig.10.1).

10.2.4 Southeast African Province

Southern Africa was not exempt from alkaline intrusion during the Mesozoic. High level ring-dykes and cone sheets which extend for over 300 km along the centre of the Limpopo belt, constitute the Nuanetsi province (Fig.10.4). The Nuanetsi intrusives lie along a NNE-trending band extending from Transvaal to SE Mozambique. The intrusive phases consist of microgabbros, microgranites, and granophyres, quartz syenites, and a nepheline syenite. The known ages of these intrusives range from 186 to 173 Ma.

Another major occurrence of alkaline complexes is in the Chilwa area in southern Malawi. In this region several dozen ring complexes intrude the high-grade gneisses of the Southern Mozambique belt. These contain well preserved granite, syenite, foid syenite, and carbonatite ring-dykes, cone sheets, and radial dykes. This group of intrusives, dated between 139 Ma and 108 Ma, form prominent mountain ranges, including the Mlanje Mountain, over 3,000 m high.

10.2.5 Southwest African Province

In southern Namibia there is the Luderitz alkaline province (Fig.10.4), comprising ring complexes (nepheline syenite, foyaites, syenites), and an extensive dyke swarm, which lie in a NE-trending belt. These were intruded at about 133 Ma, probably along the continental extension of a South Atlantic oceanic fracture zone.

Northward along southwestern Africa, the Damaraland group of ring complexes lies in two ENE-trending linear belts (Fig.10.4) which extend from the coast landward for about 370 km. According to Cahen et al. (1984) the plutons in the Damaraland show well developed ring-dykes, cone sheets, and radical dyke swarms. The intrusions are predominantly granite, syenite, or gabbro; with carbonatite being present in a few of the more alkaline complexes. The Damaraland ring complexes are located along the axis of the Pan-African Damara belt. Their ages range from 194 Ma to 126 Ma. Thus, they were emplaced at about the same time as the Nuanetsi and Chilwa ring complexes.

Figure 10.4: Igneous ring complexes of southern Africa. (Redrawn from Cahen et al., 1984.)

Another NE-trending belt of arcuate alkaline complexes with carbonatites, occurs in central Angola from Mossamedes on the coast towards the kimberlite province of NE Angola. Thirty-six alkaline and carbonatite complexes are known in the Angola province. They include various plutons ranging from granites, quartz syenites, foid syenites, and trachytes to gabbros. Carbonatites are associated with several of the complexes. In age range (159-89 Ma), rock type, and structural alignment, the Angola alkaline province is similar to the Chilwa province.

10.2.6 Tectonic Controls of Ring Complex Emplacement

The origin of ring complexes belongs to the broader theme of continental magmatism, a theme that has been dominated by the hypothesis of rising mantle plumes or hot spot activity (e.g. Burke et al., 1981). Sawkins (1990) in his discussion of the link between intracontinental hot spots, anorogenic magmatism, and associated tin mineralization echoed the following widely held view regarding the likely role of hot spot activity. "Where the relative motions of hot spots and overlying continental crust are negligible or very small, mantle hot spots impinge more substantially on overlying continental areas and appear capable in specific instances of generating an array of igneous rocks. These can include basalts, peralkaline mafic rocks, and carbonatites, and peralkaline and peraluminous felsic suites". For Africa, which is characterized by crustal doming where anorogenic magmatism is concentrated, and as a continent that has been largely stationary, the hot spot mechanism for anorogenic magmatism is very plausible.

Another view (Vail, 1989) is that Phanerozoic ring complexes in Africa show a distribution pattern which suggests control by deep-seated shear zones. Control by pre-existing crustal lineaments is suggested by the occurrence of almost all complexes within definable linear groups (e.g. Fig.10.4). These usually show parallel strike with the surrounding country rocks and with associated dyke swarms and fractures. Sometimes linear trends of ring complexes are parallel (Fig.10.4). Some of these trends, such as those of Damaraland, Luderitz, Angola, and the Cameroon volcanic line, are continuous with oceanic fracture zones, thus lending credence to the hypothesis that ring complexes are the products of magma generated along the continental extensions of oceanic fracture zones (Kinnaird and Bowden, 1987). In addition to controlling the emplacement of alkaline complexes, these reactivated deep-seated lineaments are believed to have served as channelways for hydrothermal mineralizing fluids.

10.2.7 Mineralization in Alkaline Complexes

Of all the Phanerozoic alkaline complexes in Africa, the West African Younger Granites province is the most mineralized. The Younger Granites of the Jos plateau (Fig.10.3) contain some of the finest and best known examples of tin deposits which are associated with anorogenic granites

(Bowden and Kinnaird, 1978; Ekwere, 1982; Imeokparia, 1982; Olade, 1985; Sawkins, 1990). The following summary of the mineralization in the Younger Granites is drawn from Sawkins (1990) and Wright et al. (1985).

Alluvial tin mining on the Jos plateau supported Nigeria's earliest mineral export until the beginning of petroleum export in 1958. Production consequently fell from about 10,000 tonnes anually to about 1,700 tonnes in 1983, and has not recovered significantly since then. Total reserves of cassiterite and columbite are estimated at 140,000 tonnes and 70,000 tonnes respectively. As the alluvial reserves become depleted, mineralized stockworks and disseminated cassiterite in the roof portions of the biotite granites which carry the primary mineralization (Fig.10.5), hold the greatest potential for tin and niobium.

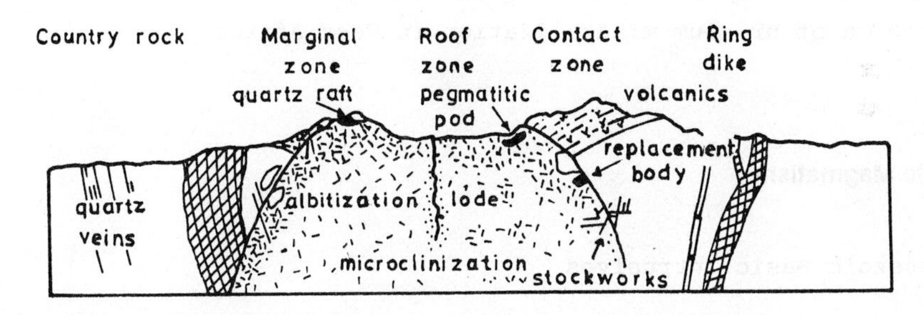

Figure 10.5: Schematic section showing the main type of bedrock tin deposits associated with the Jos Plateau granites. (Redrawn from Sawkins, 1990.)

In the biotite granites cassiterite and tantalite occur as disseminated grains, and within greisen zones and quartz veins which also contain pyrite and sulphides. The sulphides are sphalerite, chalcopyrite, and pyrite, and occasional molybdenite. Wolframite also occurs in the primary veins. Other minerals include topaz, fluorite, magnetite, ilmenite, zircon, thorite, monazite, and gem-quality beryl (aquamarines, emeralds). Placer deposits occur in gravel pockets of ancient and modern fluvial channels. Overburden depths of 30-40 m allow strip mining.

The following styles of mineralization are found among the Nigerian tin granites. Hydrothermal alterations, which are widespread, exhibit four economically important phases that are known from field , textural, and mineralogical evidence. An initial sodic metasomatism which is locally associated with the introduction of niobium (as pyrochlore or columbite), was followed by potassic metasomatism which introduced some cassiterite into the biotite granites. Next came H^+ metasomatism resulting in greisenization, with the formation of monazite, zircon, and ilmenite, followed by cassiterite and lesser amounts of wolframite and rutile. Large-scale silica metasomatism followed, with the addition of cassiterite, major amounts of sphalerite, and lesser chalcopyrite and galena. At this stage there was wall-rock alteration involving the formation of chlorite and clay minerals adjacent to the veins. The initial mineralizing fluids were highly saline with temperatures above 300°C. Kinnaird and Bowden (1987) attributed the lack of mineralization in the other African ring complexes to the lower temperatures of their residual fluids.

Because of the small total areas occupied by biotite granites (the mineralized host rocks) among the Younger Granites in Niger and Cameroon, there is lesser tin mineralization in those countries. Cameroon has small tin production coming from placers. The Younger Granites are also a potential source of uranium mineralization in West Africa.

10.3 Basaltic Magmatism

10.3.1 Mesozoic Basic Intrusives

Cahen et al. (1984) observed that a striking aspect of Phanerozoic anorogenic activity in Africa is the existence in western and southern Africa of Mesozoic dolerite dykes and associated basic-to-acid lavas which appear to be of the same age. These dykes are exposed over large areas on the continent (Fig.10.6).

In West Africa tholeiitic basic intrusives of Late Triassic to Early Jurassic age, include the dolerite dykes, sills, and irregular sheets which are extensively exposed on the West African craton (Fig.10.6). These intrusives outcrop around the Taoudeni basin, and are especially common in a belt which extends northeastwards from the Bové basin. They intrude up to the "Continental Intercalaire" Group and are not believed to be younger than Early Jurassic (Wright et al., 1985). Dyke swarms are parallel to the coast in Sierra Leone and Liberia. Sills, sometimes occurring in swarms, intrude the sedimentary sequence in the Taoudeni and Bové basins. The coastal basins, especially in the subsurface, contain sills such as the Monrovia diabase which intrudes the Paynesville Sandstone of Liberia. A layered basic body, the Freetown complex in Sierra Leone, consists of troctolitic gabbro and anorthositic rocks (Umeji, 1983; Wells, 1962). An interesting aspect of the tectonic setting of

Permo-Triassic dolerite dykes is that they are nearly all restricted to the West African craton, except in the southern part of the Mauritanides and in the Moroccan Atlas. This is in contrast to the Paleozoic-Tertiary anorogenic granites which are confined to Pan-African terranes.

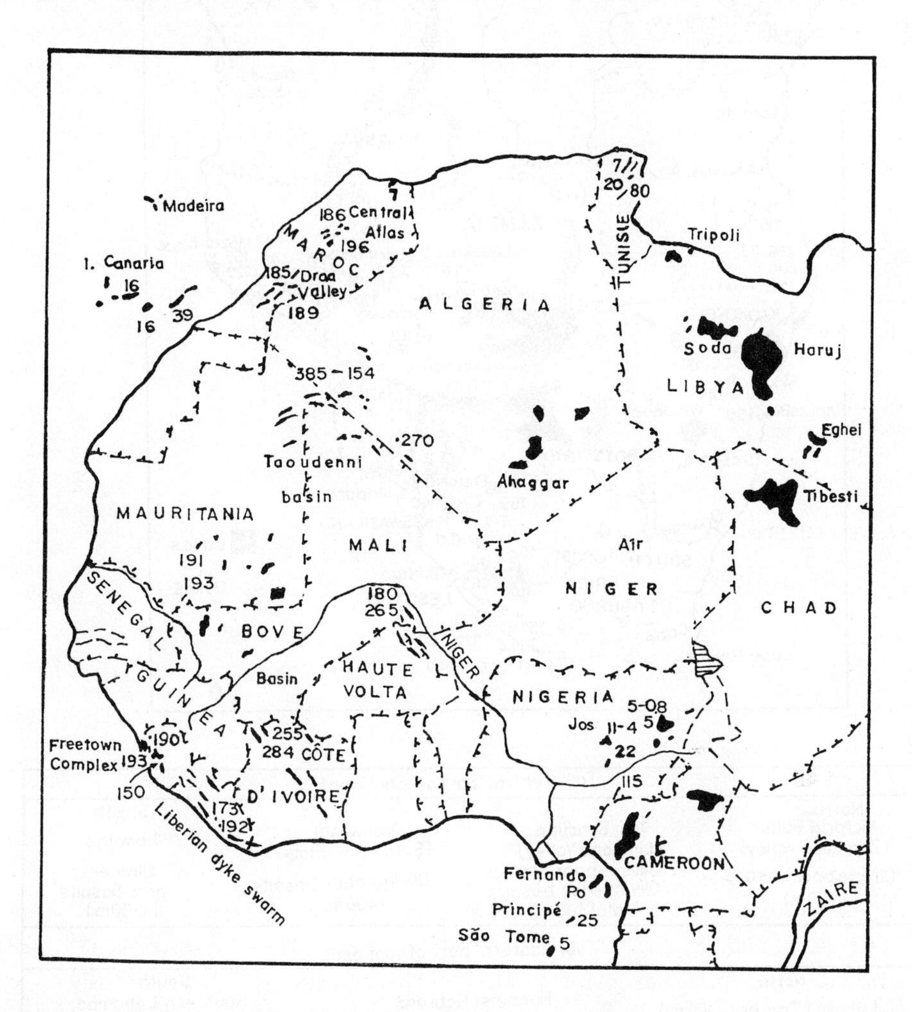

Figure 10.6: Distribution of mostly basic volcanism in West Africa. (Redrawn from Cahen et al., 1984.)

Southern Africa (Fig.10.7) shows the emplacement of similar dykes, but mostly in the Jurassic, between 195 Ma and about 155 Ma. These occur throughout the Kalahari craton, and in Karoo basins in places such as South Africa, Lesotho, and Mozambique. Major dolerite dykes provinces occur along the Lebombo monocline and the Nuanetsi syncline on the cratonic margin. They occur as fault-bounded isolated outliers in the lower Zambezi valley. Dolerite dykes also extend from the Lupata gorge to the Sabi valley.

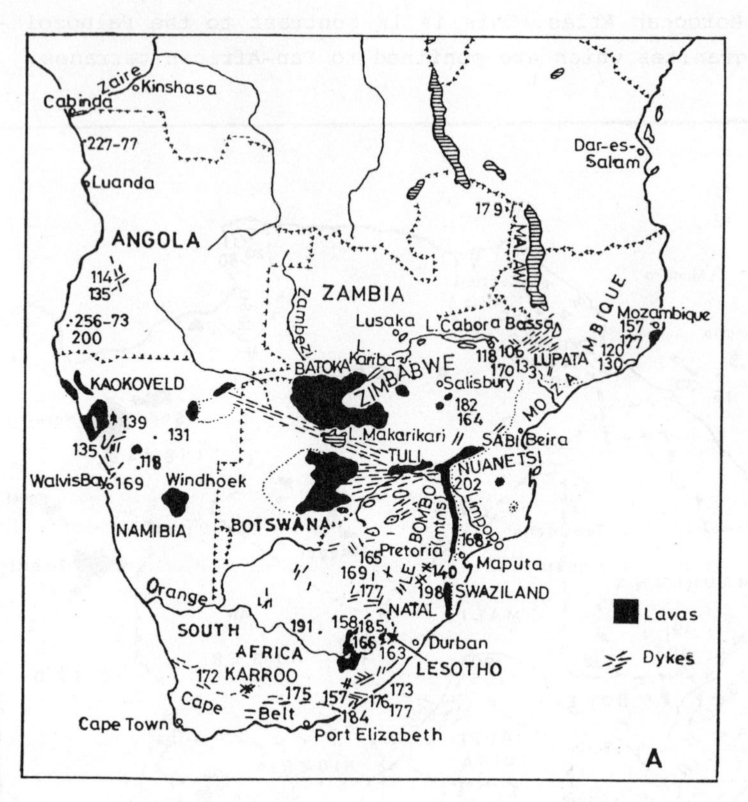

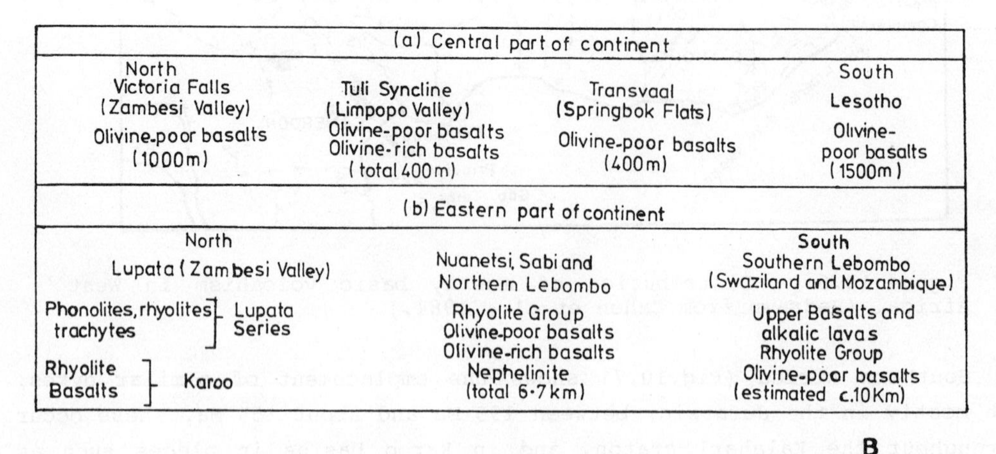

(a) Central part of continent

North Victoria Falls (Zambesi Valley)	Tuli Syncline (Limpopo Valley)	Transvaal (Springbok Flats)	South Lesotho
Olivine-poor basalts (1000m)	Olivine-poor basalts Olivine-rich basalts (total 400m)	Olivine-poor basalts (400m)	Olivine-poor basalts (1500m)

(b) Eastern part of continent

North Lupata (Zambesi Valley)	Nuanetsi, Sabi and Northern Lebombo	South Southern Lebombo (Swaziland and Mozambique)
Phonolites, rhyolites, trachytes — Lupata Series Rhyolite, Basalts — Karoo	Rhyolite Group Olivine-poor basalts Olivine-rich basalts Nephelinite (total 6·7 km)	Upper Basalts and alkalic lavas Rhyolite Group Olivine-poor basalts (estimated c.10Km)

B

Figure 10.7: A, distribution of mostly basaltic volcanic rocks of southern Africa. B, generalized stratigraphy and thicknesses of Karoo lavas. (Redrawn from Cahen et al., 1984; Burke et al., 1981)

Mesozoic basic intrusives in western and southern Africa, which are contemporaneous, were intruded during the break-up of Gondwana. This is supported by the fact that the sites of rifting and flexuring during continental separation are consistently characterized by olivine-rich basalts with rhyolites, whereas tholeiitic basalts are predominant on the cratons (Cahen et al., 1984).

In northeastern Africa dykes are unusually abundant in the Precambrian terranes of the Eastern desert of Egypt, Sinai peninsular, the Red Sea Hills, and in the Bayuda desert of the Sudan. These dykes which are narrow, steeply dipping bodies a few meters thick and several kilometres long, occur in swarms which may extend for several hundred kilometres (Cahen et al., 1984). These are doleritic, granitic, and felsitic dykes, the ages of which range from Precambrian to Cenozoic.

10.3.2 Karoo Volcanism

The Karoo Supergroup in southern Africa is overlain by very extensive plateau basalts and rhyolites of mostly Late Triassic to Early Jurassic age. Cox (1972) and Tankard et al. (1982) gave detailed accounts of Karoo volcanism, from which the following outline is drawn. Known as the Stormberg volcanics in South Africa, Karoo volcanics (Fig.10.7) cover about 140,000 km^2, a fraction of an estimated initial extent of about 2 million km^2. The average thickness is about 1,000 m, with a maximum thickness of about 10 km on the Lebombo monocline. Karoo volcanism persisted into the Cretaceous in the lower Zambezi valley where it is known as the Lupata series.

Karoo basalts occur mostly as horizontal dolerite sills of similar composition. Vertical or feeder dykes are known for the Stormberg lavas, the Drakensberg basalts, and in the Tuli syncline (Fig.10.7). The stratigraphic and compositional variations in Karoo basalts are presented in Fig.10.7. In the northern area (Lebombo, Tuli, Nuanetsi), for example, olivine basalts at the base change upward through olivine-poor basalts, to rhyolites. The breccia pipe and replacement copper deposit that is mined near Messina in South Africa formed hydrothermally by the leaching of copper from copper-rich Karoo rift basalts (Sawkins, 1990).

Together with their partial equivalents, the Late Jurassic-Early Cretaceous Parana basalts (Serra Geral Formation) of Brazil (Fig.9.22E), the outpourings of Karoo basalts resulted from the tensional regime which preceded the final separation of southwestern Africa from South America in the Early Cretaceous.

10.3.3 Kimberlites

African kimberlites cluster into two main age intervals (Cahen et al., 1984). An older Jurassic to Early Cretaceous (190-134 Ma) group of kimberlites approximately overlapped and followed closely the extrusion of Karoo volcanics. A second phase of kimberlite emplacement was during the Late Cretaceous to Eocene (93-53 Ma). In their review of the distribution of kimberlites in Africa Cahen et al. (1984) stressed the points that kimberlites seem to be confined to Africa south of the Sahara (Fig.10.8).

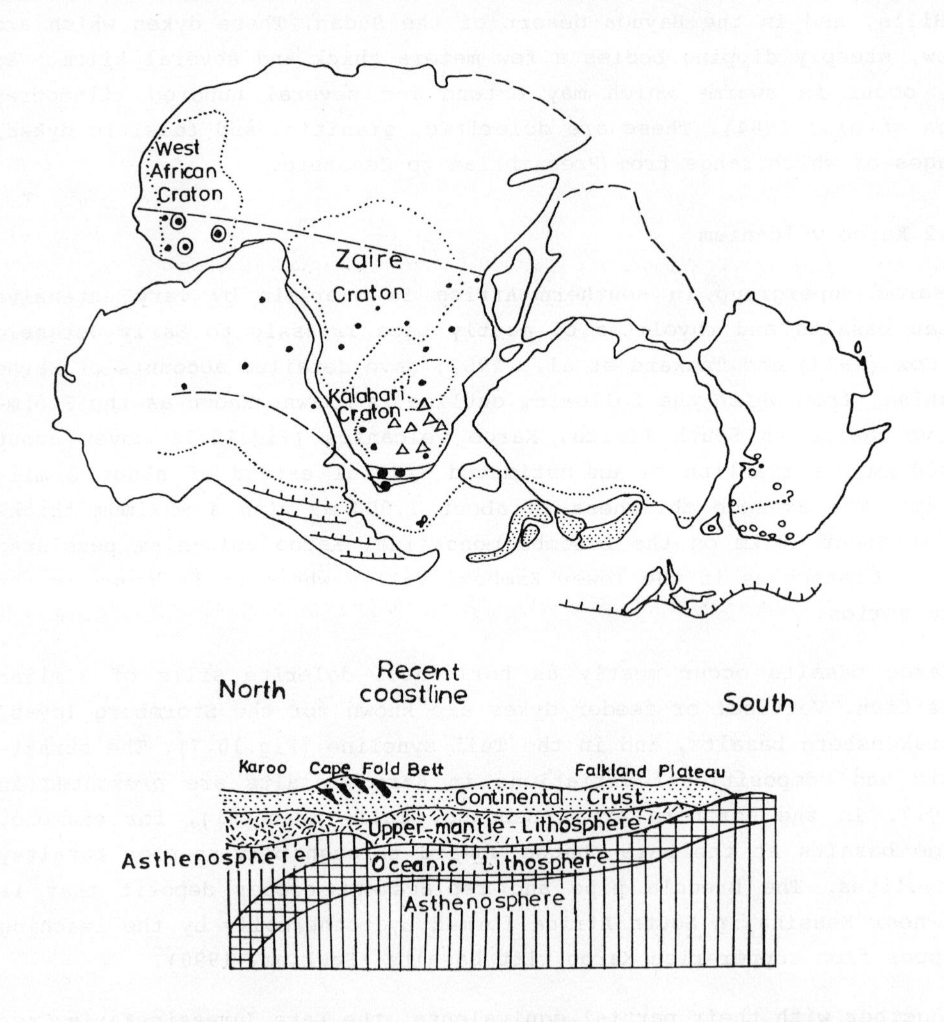

Figure 10.8: A, distribution of kimberlites in Gondwana: solid dots, kimberlites; open dots, kimberlitic rocks; triangles, diamondiferous kimberlites; dashed lines, limits of Mesozoic kimberlites; dotted lines, limits of cratons. B, flat-plate subduction model for the origin of the Cape fold belt. (Redrawn from Helmstaedt and Gurney, 1984; Tankard et al., 1982.)

They are best developed in South Africa, Tanzania, Zaire-Angola, and between Sierra Leone and Ghana where mostly the younger generation of kimberlites are found (Wright et al., 1985).

Although kimberlites constitute a vast topic on which enormous literature is available, a brief mention of their characteristics and origin (Wright et al. 1985) is adequate for our present purpose. This is necessary because the type area for kimberlites is in South Africa, and African kimberlites dominate the world's diamond production.

Kimberlites are usually porphyritic to fine-grained gray to green rocks, which comprise mainly serpentine, phlogopite, olivine, pyroxene, and carbonate, with magnesian garnet and ilmenite megacrysts. They also often carry rounded xenoliths of crustal and mantle rocks. Kimberlites are the only igneous rocks that contain diamond. They are usually emplaced as diatremes by gas-rich magmas which rise rapidly from great depths into the upper crust. Diamonds forms at great pressures which obtain at depths of 150 km and below. They survive to reach the earth's surface by rapid upward transport after their formation. At the surface since kimberlites weather rapidly, diamond is released into placer deposits which account for 90% of the world's diamond production (Guilbert and Park, 1986).

The type area for kimberlites is at Kimberley in South Africa. Here many of the kimberlite pipes intrude Karoo strata (Fig.10.9), including Karoo dolerite sills which are found in the sedimentary sequence. Since xenoliths of country rock occur in the kimberlite pipes, the pipes are therefore younger than Karoo strata.

Africa is the world's largest supplier of gem and industrial diamonds. Alluvial diamond derived from Mesozoic kimberlites occur along the Vaal and Orange valleys in South Africa. Other sources of Mesozoic alluvial diamond are Zaire, Central African Republic, and Sierra Leone in West Africa.

Helmstaedt and Gunery (1984) considered why the Jurassic to Early Cretaceous interval was the most important phase of kimberlite emplacement throughout southern Africa. They supported the view that Mesozoic kimberlites originated above a shallow subducted oceanic plate which underthrust the region during the Gondwanide orogeny (Fig.10.8) and produced the Cape fold belt in the Late Permian - Early Triassic. Large-scale shallow or flat-plate subduction under southern Gondwana, prior to continental break-up, is believed to have generated volatiles from the subducted slabs which could have caused the voluminous Karoo flood basalts. The volatiles could, additionally, have generated large-scale upper mantle metasomatism and kimberlite magmatism in the overlying sub-crustal lithosphere. The contribution of the above processes to Karoo volcanism does not, however, negate the view previously stated, that Karoo volcanism was related to Gondwana fragmentation. Rather, it suggests the fundamental processes below the lithosphere which could have accounted for the copious Karoo basaltic volcanism and attendant kimberlite intrusion which had no counterparts during the break-up of northwestern Gondwana.

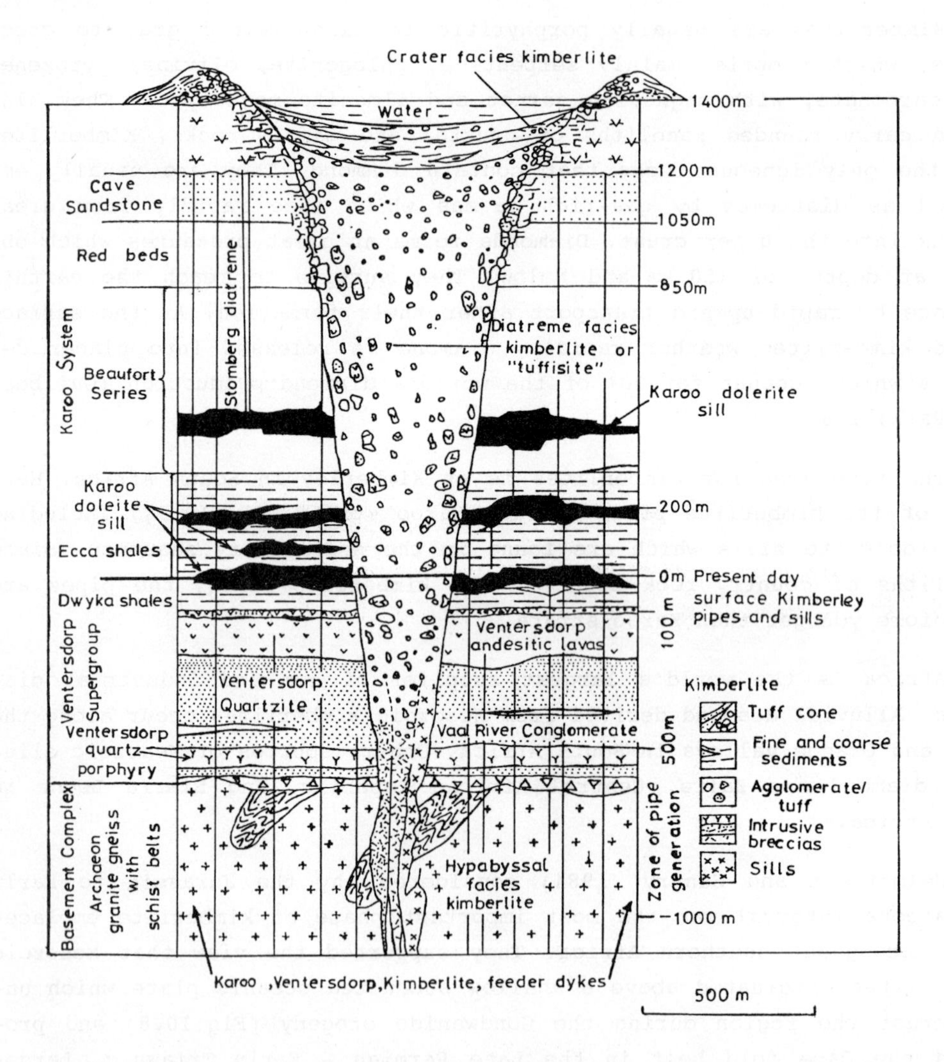

Figure 10.9: Schematic cross-section of a hypothetical kimberlite pipe shortly after emplacement; local names at the right denote outcrop erosion levels of specific pipes: (Redrawn from Guilbert and Park, 1986.)

10.3.4 Cenozoic Continental Hot Spots

East African Rift System

Burke et al. (1981) outlined the tectonic setting of African Cenozoic volcanism. Figure 10.10 shows that extensive intraplate volcanism is taking place in Africa under diverse structural settings. Most of the volcanism, is however, in the East African Rift System, where Burke et al. identified seven hot spots. The hot spots manifest as areas of crustal doming. As previously noted (Ch. 9.8.2) doming has been ascribed to mantle diapirism along the Eastern branch of the Rift Valley (Karson and Curtis, 1989). However, the domes do not carry volcanoes, and rifts are

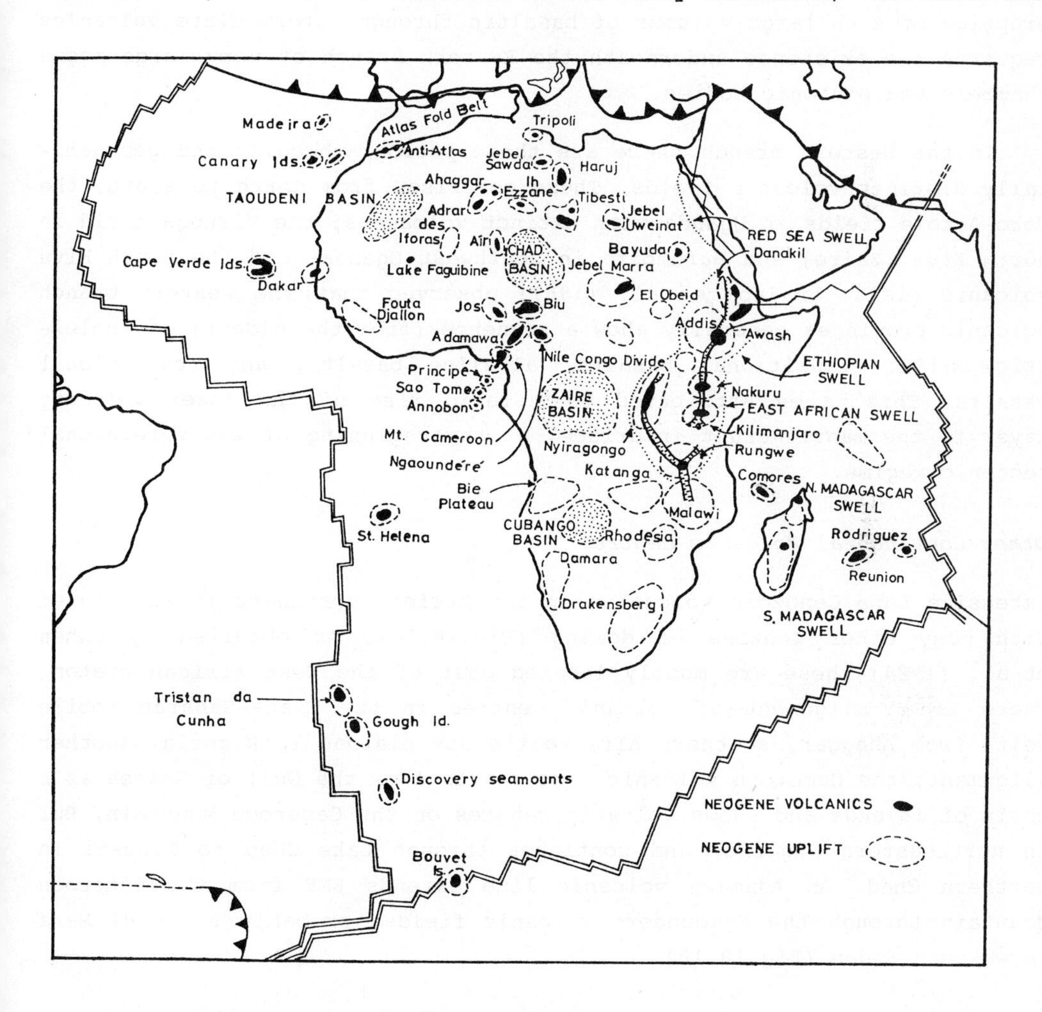

Figure 10.10: Distribution of hot spots and Neogene uplifts on the African plate. (Redrawn from Burke et al., 1981.)

mostly located not on swells but on reactivated old structures (McConnell, 1972). There is widespread occurrence of alkaline rocks in the East African Rift System, but in some areas, for example near Addis Ababa in Ethiopia, volcanism is tholeiitic. The distribution of volcanism along the rift is also very uneven.

In the Eastern branch of the Rift Valley volcanism began in the mid-Tertiary. The volcanics in Kenya range from mildly to strongly alkaline with undersaturated varieties, the commonest being phonolites, trachytes, basalts, and nephelinites. Based on the enormous volume of extrusives, Karson and Curtis (1989) estimated the volume of cumulates and other intrusive rocks underlying the Eastern branch. They concluded that the eruption of such large volumes of basaltic through intermediate volcanics requires the existence underneath the Eastern branch of very large magma chambers and plutonic bodies.

In the Western branch there are three petrographically and geochemically distinct volcanic fields. These comprise, from north to south, the Toro-Ankole fields in Uganda with extinct volcanoes; the Virunga field in north Kivu (Zaire) and Bufumbira in southwest Uganda; and the south Kivu volcanic field. Lualaba et al. (1987) observed that the Western branch volcanic provinces generally show a sequence (from the oldest) of tholeiitic and/or transitional basalts, alkaline basalts, and transitional basalts. This, they attributed to a rapid rise of the lower velocity layer to the mantle-crust interface, at the beginning of the extensional tectonic regime.

Other Continental Volcanic Centres

Extensive Late Cenozoic volcanism in the African continent is associated with many other centres of doming (Fig.10.10). As observed by Cahen et al. (1984) these are mostly located east of the West African craton. There is an alignment of volcanic centres in the Trans-Saharan mobile belt, from Ahaggar, southern Aïr, to the Jos plateau in Nigeria. Another alignment, the Cameroon volcanic line starts from the Gulf of Guinea as a chain of islands and shows volcanic centres on the Cameroon Mountain, Bui in northeastern Nigeria, and continues through Lake Chad to Tibesti in northern Chad. An Adamawa volcanic line extends ENE from the Cameroon Mountain through the Ngaoundere volcanic fields, probably to Jebel Mara in western Sudan (Fig.10.10).

In the Atlas belt of Morocco, Algeria and Tunisia there are volcanic plugs of Miocene to Quaternary age. Similar volcanics are found near Dakar in Senegal.

The above volcanics are predominantly basaltic lavas belonging to the olivine-basalt-trachyte association. Feldspathoids occur in the Cameroon Mountains. The volcanic centres show mountainous relief with large calderas. Cahen et al. (1984) attributed their distribution to pre-existing deep-seated faults. Volcanoes are sometimes situated along the faults on ring-dykes in the Younger Granite complexes. Most of these volcanics date from the mid-Tertiary, with the Miocene being the time of peak activity. Volcanism has continued down to present-day in one form or the other in some places. An example was the Lake Nyos gas eruption on August 2, 1986 on the northwestern part of the Cameroon Mountain (Freeth, 1987).

Northeast Africa also witnessed considerable Cenozoic volcanic activity. Flood basalts and shield volcanoes, and trachyte domes are exposed in Libya, where trachytic rocks and basalts of Mesozoic age have been intersepted in oil wells. In Libya basalts overlie Upper Cretaceous and Paleocene sediments at Jebel as Sawda. Approximately 40,000 km^2 of probable Oligocene olivine basalts overlie Lower Eocene and Upper Cretaceous deposits at Haruj (Fig.10.10).

In the Tibesti Mountains, the highest peak in northeast Africa, Cretaceous and Paleocene sediments are overlain by shield volcanoes and explosive volcanic centres. Here basaltic and acid ignimbritic eruptive rocks are also predominant, with hot springs indicating recent activity. Scattered basaltic and trachytic rocks occur at Jebel Uweinat; along the Red Sea coast of southern Egypt; and along the Nile Valley in the Sudan.

10.3.5 Oceanic Hot Spots

The African plate contains several volcanic archipelagoes in the Atlantic and Indian Oceans (Fig.10.10). Here we shall be concerned with the islands of the Atlantic Ocean. Some of these islands are located very close to the mid-oceanic ridges; some lie along transform faults; while others are situated quite close to the continental margin. Although the origin of volcanic islands is still being debated, we shall for our present purpose and convenience include them among hot spot activities.

Along the continental margin of northwest Africa there are two major groups of volcanic islands, the Cape Verde Archipelago, and Canary Islands. Cahen et al (1984) summarized the geology of the Cape Verde Islands. Two main groups of volcanic rocks make up these islands. A central igneous complex of probably Late Jurassic-Early Cretaceous age is overlain by limestones in Maïo, one of the islands. This complex is intruded by Late Miocene essexites, carbonatites, and nepheline syenites. These alkaline intrusions were followed by the extrusion of stratovolcanoes and phonolites in the latest Miocene and Pliocene. In general the Cape Verde Islands were constructed during an early phase of tholeiitic volcanism which formed part of the ocean floor. This was followed by strongly undersaturated alkaline magmatism.

The Canary Islands consist of seven major volcanic islands. These extend for nearly 500 km east-west. They all lie about 100 km off the northwest African coast. The largest islands include Fuerteventura, Gran Canaria, and Tenerife. Schmincke (1982) described the volcanic rocks which constitute these islands. Although volcanism began probably in Eocene times, the Canary Islands were built mainly during the last 20 Ma (Miocene).

Saturated to moderately undersaturated alkali basalt with local tholeiite was the dominant shield-building type of magma. Gran Canaria and Tenerife contain large caldera-forming ash flow eruptions, the products of more differentiated magma. Minor amounts of trachyte are found in the western islands; and phonolitic plugs occur in the central and western islands.

Schmincke (1982) linked the highly alkaline mafic undersaturated magmas in the Canaries and the Cape Verde Islands, to the presence of a thick lithosphere beneath them. This would have allowed low heat flow, and only reduced partial melting at greater depth. Schmincke (1982) found no geological or geochemical evidence for the presence of continental crust beneath any of the Canary Islands. He also discounted the location of the islands along an oceanic fracture zone. Schmincke suggested that islands are located over a zone of mantle instabilities along the boundary between oceanic and continental lithosphere. This is another variant of the hot spot hypothesis.

Chapter 11 The Quaternary in Africa

11.1 Introduction

The Quaternary is the latest geological period to which the last 2.5 to 1.8 million years of Earth's history has often been assigned. Originally referred to as the Ice Ages, the Quaternary period is now characterized as the geological interval during which the climate of the Earth witnessed spectacular alternations of cold dry phases and warm wet phases, which in Africa corresponded to the interpluvials and pluvials respectively. For this reason Quaternary geology has witnessed a recent surge in academic and public interests because of on-going global ecological and climatic changes. Quite apart from the climatic impact of industrial pollution there is a need to understand past climatic changes in order to predict what the future climate might be.

Perhaps it is in the tropical region of the world such as Africa that the impact of these climatic changes is most severe, varied, and widespread. While desert loess, laterites, soils, alluvia, terrace deposits, colluvia, lacustrine sediments, cave earth and mountain glacial deposits accumulated in different parts of the African continent, peat beds, deltaic, shelf and shoreline sands, deep-ocean fans and carbonates were laid down on the continental margin and deep-sea floor. Sea-level rose and fell as ice-caps waned and waxed; rivers rapidly filled their valleys only to resume down-cutting at the next abrupt change of base level; lake levels rose and fell and some lakes never reappeared. Forests turned into deserts. All plant and animal groups responded to these sharp climatic changes mostly by migrating back and forth. But somewhere among African vertebrates, especially the primates, a major threshold was crossed in the history of life, as Man appeared at the dawn of the Quaternary.

For the frail new creature that had suddenly appeared with an unusually prolonged infancy and childhood dependence, these rapid and hazardous ecological changes and stress contributed to the development of what turned out to be, perhaps, the most spectacular adaptive organ--the human brain. Since "man is a child of the Quaternary" he occupies a central place in the story of this period. The evolution of his adaptive strategies including his technology and culture not only ensured his survival, but using these tools, man gradually conquered and changed his environment. Man's mastery of nature in turn produced a negative feed-back on climate, vegetation, soils, and water. In this respect man appeared as a new geological agent that has greatly altered the geosphere and biosphere.

Africa occupies an important place in the current world-wide concern about global ecology. Recent ecological changes on the continent such as soil erosion, desertification and the destruction of the tropical rain forest ecosystem have attracted considerable attention.

Since the Quaternary Period is replete with these profound and on-going changes on the surface of the Earth, the study of the Quaternary is therefore a multidisciplinary science that involves, among others, the geologist, archeologist, geographer, paleoanthropologist, biologist, soil scientist, oceanographer, and engineer.

Africa is the oldest and most stable continent. It has yielded the longest and most complete record of man's origin, habitation and early technological and cultural evolution. The discoveries of homonid sites in southern Africa (Fig.11.1) since the 1920s and later at Olduvai Gorge in Tanzania, around the shores of Lake Turkana in Kenya, and in the Hadar and Awash valleys in Ethiopia unfolded an exceptionally long and unsurpassed record of homonid fossils and artefacts, thus according east and southern Africa the unrivalled position as the cradle of mankind. Unlike most parts of Africa where Quaternary deposits are unfossiliferous, incomplete and difficult to date, the East African Quaternary sequences are well preserved in the rift valley (Fig.11.1). Here earth movements have repeatedly created sedimentary basins in which excellent beds preserve fossils and artefacts. Interlayered within the sedimentary sequences are volcanic tuffs which are excellent geological marker beds that have yielded radiometric ages, by means of which homonid fossils and artefacts and the record of climatic changes have been dated and correlated with other parts of Africa, and the world.

Like other subdivisions of geologic time, the Quaternary has a dual meaning. It can be used for the last 2.5 to 1.8 million years of geologic time as well as for the rocks that formed during this time. The lower boundary of the Quaternary period or the Pliocene-Pleistocene boundary has, however, remained unresolved as some authorities have placed this boundary at 2.5 Ma (Fig.11.2) based on studies of continental glacial deposits (e.g. Boellstorff, 1978), while those working on oceanic oxygen isotope records generally favour an age of about 1.8 Ma (e.g. Williams et al., 1988).

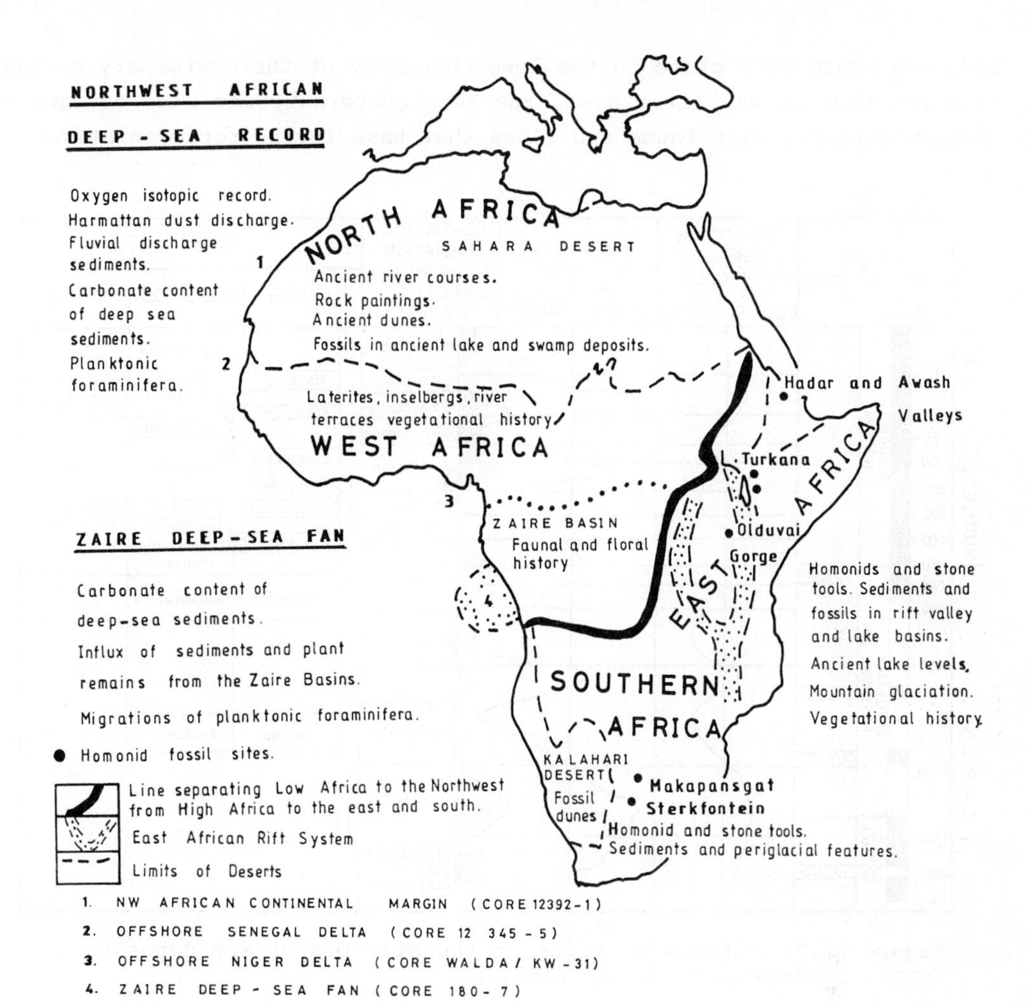

Figure 11.1: African natural regions and their Quaternary paleoclimatic indicators.

Bowen (1978) traced the historical development of the Quaternary concept in Europe where the term Quaternary was first applied in the mid-18th century to alluvial and superficial deposits. In one of the first attempts to define and standardize the use of the Quaternary in a time sense, the International Geological Congress in Great Britain in 1888 proposed that human artefacts were the characteristic element of the Quaternary age on the basis of which the Quaternary should be separated from the preceding Tertiary Period. This archeological implication in the usage of the Quaternary was later reinforced when it was discovered that the oldest human artefacts from the Olduvai Gorge was 2.5 to 2.0 m.y.

old, and hence very close to the lower boundary of the Quaternary Period. From the time it was first used, the term Quaternary was also defined to include deposits with fauna and flora that have living representatives.

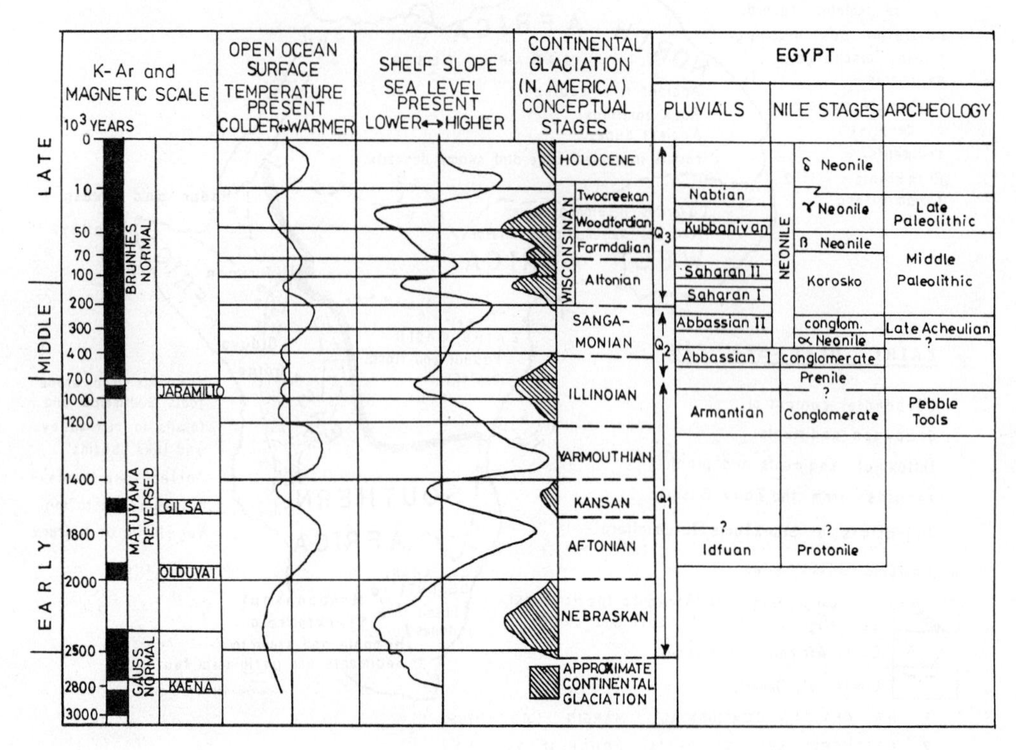

Figure 11.2: Classification of the Nile Quaternary deposits in Egypt. (Redrawn from Said, 1990c.)

The development of the glacial theory in Europe in the 19th Century also brought about the fourth and most widely adopted concept of the Quaternary. The Quaternary was accepted as the Glacial Period during which extensive and frequent continental glaciations occurred and molluscs from the cold region migrated into the lower latitudes. This redefinition of the Quaternary as "... the time distinguished by severe climatic conditions throughout the great part of the northern hemisphere" introduced a broader climatic concept, and laid the foundation for recognizing this period outside the temperate region. It called for the application of more universally acceptable stratigraphic methods, refined dating techniques, and principles, for the subdivision and correlation of Quaternary rocks. Equally important has been the need for reliable paleoclimatic indicators which can be dated and correlated. From the mid-19th century the

Pleistocene has been accepted as the lower geologic epoch of the Quaternary, and the Holocene or Recent for the present epoch.

11.2 The Quaternary Physical Geography of Africa

Before examining the Quaternary sedimentary successions in Africa it is pertinent first to consider the geomorphic and climatic factors which govern their distribution. The contemporary physical and climatic features of Africa are not only significant because "the present is the key to the past", but because the Quaternary Period is so short and the African continent has remained so stable, that at no time during this period, could the geographical framework of the continent have been radically different from what is today. Even during the peak of the pluvials ancient lakes Sudd, Araouane and Zaire existed where they are supposed to be, that is, in the great interior depressions or basins (Fig.11.3). Also the present geography and climate (Fig.11.4A) constitute the base line against which to assess and explain past paleogeographic (Fig.11.3) and paleoclimatic (Fig.11.4B) departures.

Throughout the continent and especially in West Africa the exposed basement terrains have thin Quaternary superficial deposits and alluvia. Quaternary deposits in West Africa are thickest in the Lake Chad basin (Fig.11.3) which is receiving sediments from the surrounding uplifts of the Ahaggar, Tibesti, Cameroon and Ennedi Mountains (Burke, 1976). In sharp contrast, the East African scenery is dominated by the spectacular rift valley running from Ethiopia to Mozambique and within which lie great lakes and thick piles of Quaternary fluvio-lacustrine, and volcanic debris. The Drakensberg Mountains form the backbone of southern Africa (Fig.11.3) north of which lie the Kalahari desert to the west and vast rolling plains to the east and north, where the best-known Quaternary deposits are the cave earth of Taung, Sterkfontein, Makapansgat and Kromdraai (Fig.11.1).

Climatically, three-quarters of Africa is located within the tropics and one third of the continent is affected by winds which have produced arid and semi-arid conditions (Fig.11.4). Winds from the Atlantic bring moisture to coastal West Africa in July when eastern and southern Africa are dry. While equatorial Africa remains wet in January and most of East Africa receives rainfall later in April, dry and dusty N.E. Trade winds sweep across the Sahara and most of West Africa down to the coast. Because it receives rainfall throughout the year dense rain forests used to thrive in the equatorial region before the recent deforestation. The remaining part of the continent is mostly savanna, desert, mediterranean, or montane (Fig.11.5A). Bush burning, overgrazing, and the cutting of fuel wood for rural energy have converted most of what used to be lowland forest (Fig.11.5A) to savanna, savanna to sahel, and sahel to desert. This means that the deserts are encroaching upon forested land.

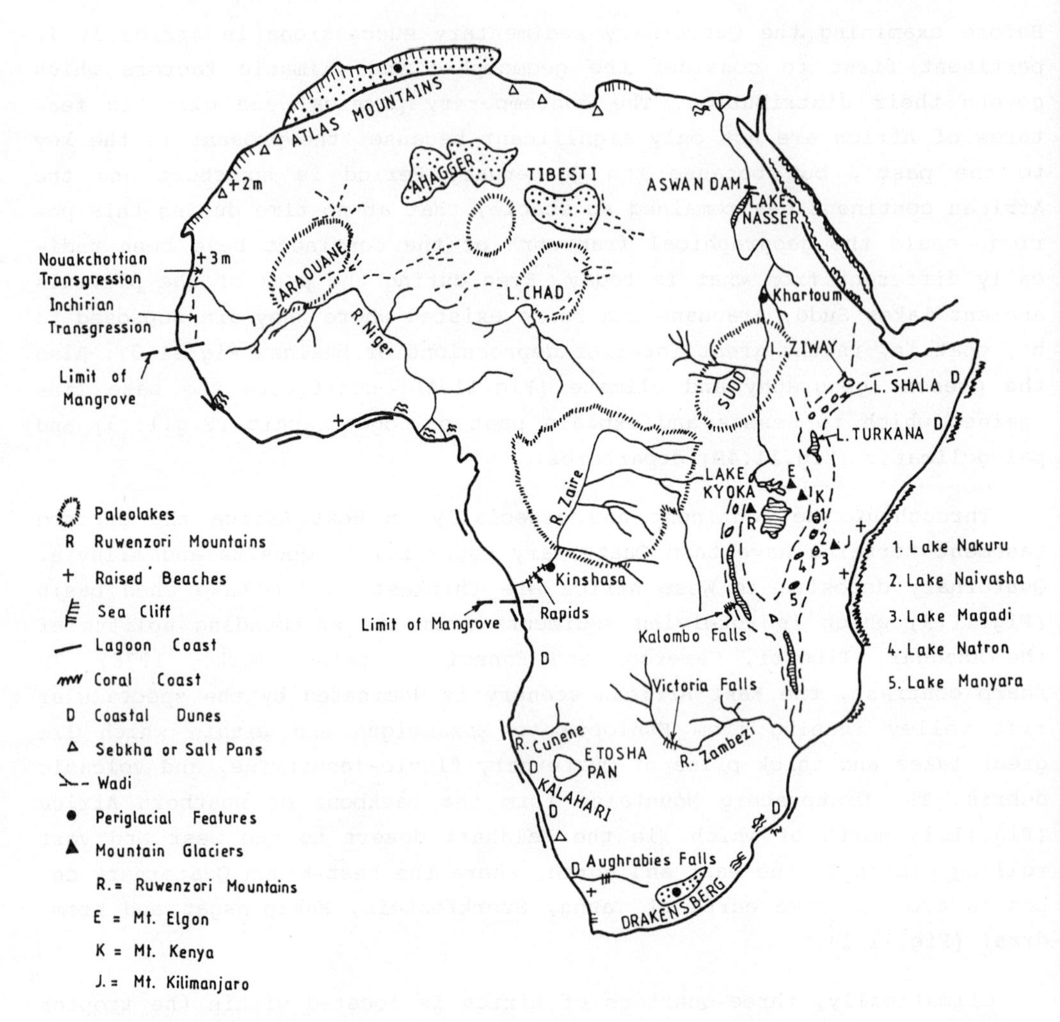

Figure 11.3: Outline geomorphological map of Africa showing major Quaternary paleo-lakes.

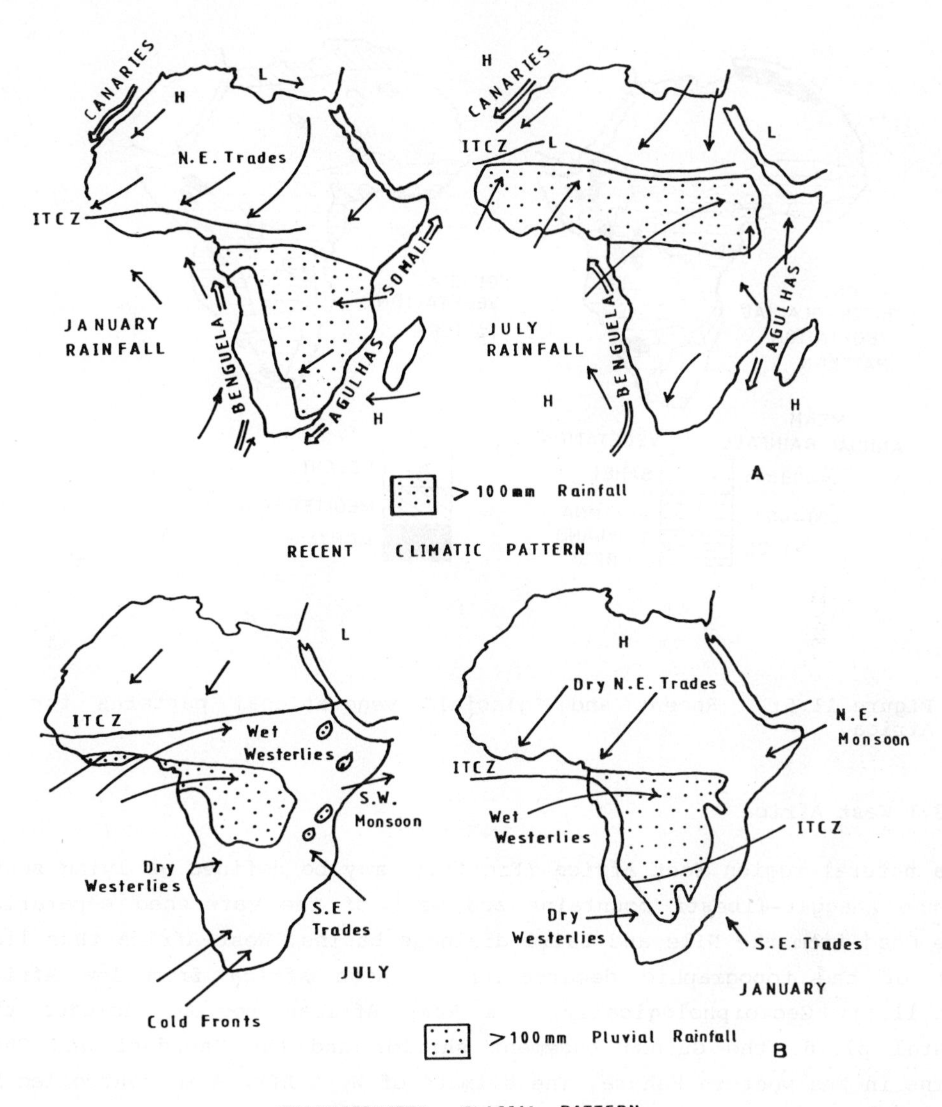

Figure 11.4: Recent and "glacial" climate regimes for Africa.

11.3 Quaternary Deposits in Africa

In this section attention will focus on representative Quaternary successions and faunas in the rather contrasting regions of west, east, north and southern Africa.

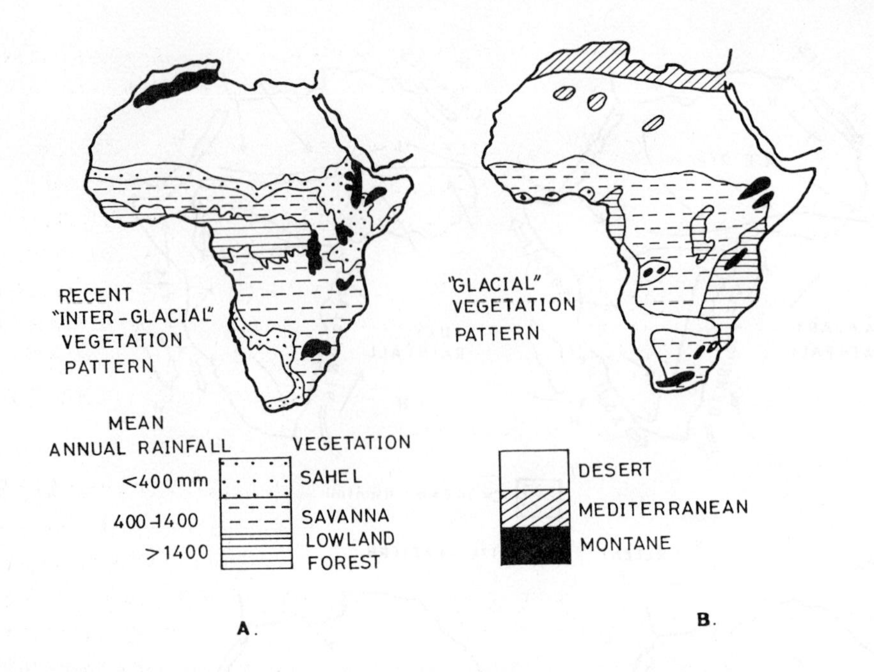

Figure 11.5: Recent and "glacial" vegetational patterns for Africa.

11.3.1 West Africa

As a natural region West Africa (Fig.11.1) may be defined as lying south of the Ahaggar-Tibesti mountains and west of the watershed separating Lake Chad from the Nile and Zaire drainage basins. West Africa thus lies west of the topographic demarcation of high Africa from low Africa (Fig.11.1). Geomorphologically, the West African region includes the coastal plain, the Guinea basement shields and the Taoudeni and Chad basins in the western Sahara. The climate of West Africa is controlled by the movement of the low pressure Inter-Tropical Convergence Zone (ITCZ) which shifts north in July and brings south-west monsoon rains over most of the southern part of the region. In January when the ITCZ is down near the coast (Fig.11.4A) the dry north-east trade winds sweep across the region. However, being very close to the equator the coastal part of west Africa is perpetually wet and humid, with bushes while most of the interior is savanna and desert (Fig.11.5A). This geomorphic and climatic setting has produced four distinct types of Quaternary sequences in West Africa, namely: (1) the coastal plain sequence, (2) the basement sequence, (3) the savanna-sahel sequence and (4) the Saharan sequence. The geomorphic-climatic control of Quaternary sequences is due to the fact that different parent or bed rocks materials undergo weathering at different rates under different climates, and different erosion and depositional processes operate under different climatic regimes.

Coastal Plain Sequences

Although known under different names in different countries (for example the Bullom Group in Sierra Leone, the Benin Formation in Nigeria), the West African coastal plain Quaternary sequence consists of predominantly cross-bedded poorly sorted fluvial sands and kaolinitic clays with lignite beds. The coastal plain sequence includes the deposits of the recent coastal depositional environments, but its base is ill-defined in the subsurface. However, the Benin Formation is about 2,000 m thick while the Bullom Group is at least 100 m thick. Andah's (1979a) review of the artefacts from the coastal region of Ghana mentioned Early and Middle Stone Age cleavers, picks and flakes from beach deposits. If explored, the West African coastal plains deposits could also yield the artefacts of prehistoric coastal fishing communities. Thus, archeological methods, including palynology, are potentially useful for dating the coastal plains sequences.

Sequences Overlying Basement in the Rain Forest and Savanna Zones

The lithological characteristics and origin of the superficial deposits in the rain forest and savanna parts of the Guinea basement arch are well known through the researches of geomorphologists (e.g. Faniran and Jeje, 1983; Thomas, 1974; Thomas and Thorpe, 1980), geologists (Burke and Durotoye, 1972), soil scientists (e.g. Smyth and Montgomery, 1962; Moss, 1968) and archaeologists (e.g. Andah, 1979a; Sowunmi, 1987). A review of basement Quaternary stratigraphic interpretations was made by Andah (1979b), who emphasized the problems of dating, correlating and interpreting the highly leached residual superficial deposits (laterites) on the West African basement complex. The Quaternary deposits on the basement of southwestern Nigeria, Asochrochona near Accra, and Sierra Leone are summarized below to show their basic lithologic successions (Fig.11.6).

The sequence usually starts with saprolite or weathered basement which may be overlain by laterite (Fig.11, 6i) or a stone layer (Fig.11, 6ii). This is overlain by sands or clays which have been washed in by rain or moved upward by termites and worms. In southwestern Nigeria this sequence, known as the Bodija Formation, is up to 8 m thick. The la-

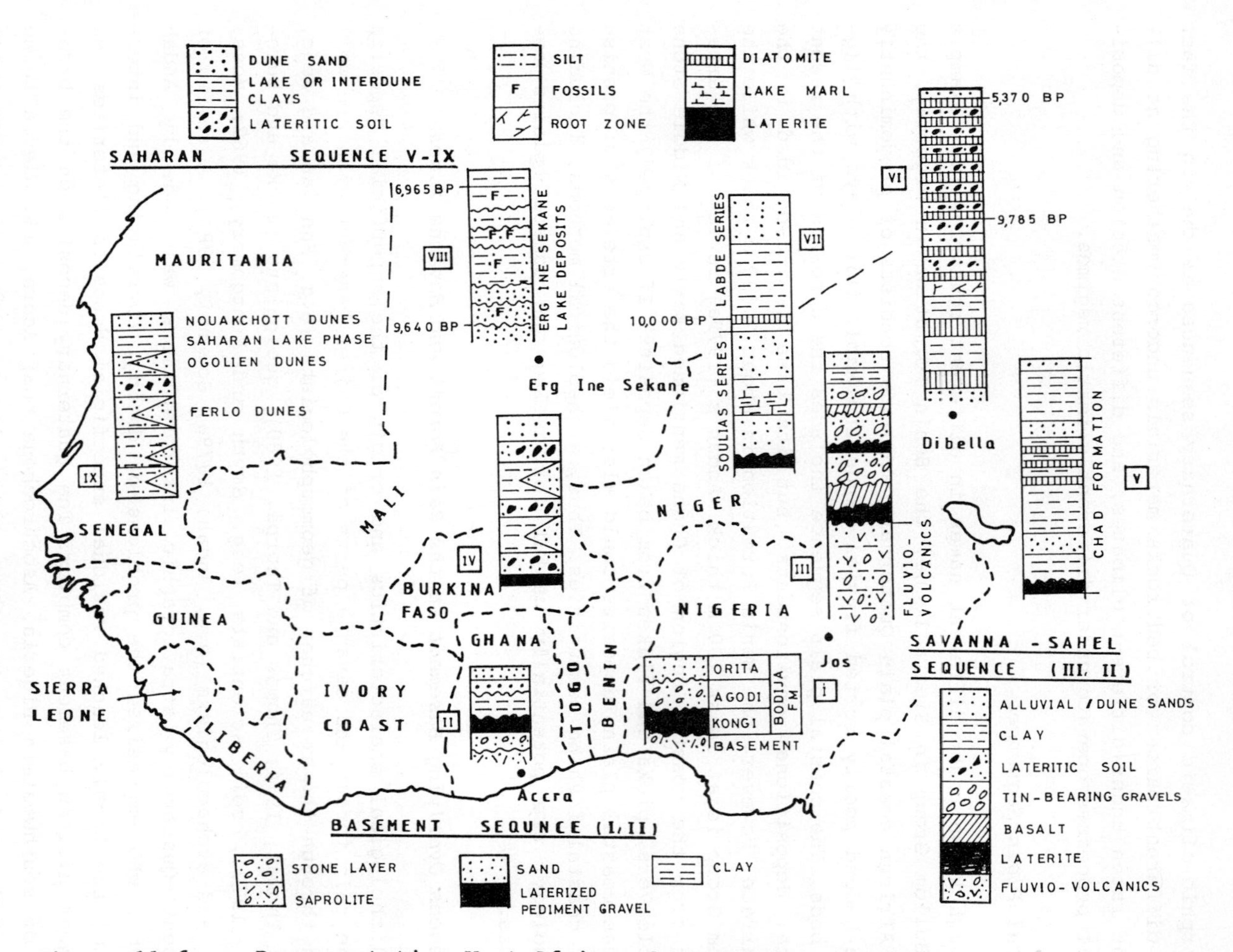

Figure 11.6: Representative West African Quaternary sequences.

terites in the sequence point to cycles of savanna wet and dry seasons; the stone layers are considered as products of mass wasting of basement and older laterites, while the sandy and clayey topsoils are forming under present-day humid climate. Because they contain little or no datable materials, the sequence of events can not be reconstructed, although there is some prospect of relative age interpretation using artefacts such as the Early Stone Age artefacts from Asochrochona (Andah, 1979a). A basement alluvial sequence is well exposed in the mining excavations in the rain forests of Sierra Leone where a basal diamond and corrundum-bearing gravel is overlain by a clayey alluvial sand layer, with loamy sandy clay at the top (Thomas and Thorpe, 1980). Radiocarbon dates on fossil wood from the alluvial deposits gave ages of 36,000-20,000 yr B.P.; 12,500-7,800 yr B.P., 3,300-1,750 yr B.P., and 1,000 yr B.P. to Recent, implying that the basement sequences are incomplete and may not represent more than the deposits of the Late Quaternary pluvials and interpluvials.

Savanna-Sahel Sequences

The sedimentary sequences on the Jos Plateau and Burkina Faso (Fig.11, 6iii, iv) exemplify the Quaternary successions of the savanna-sahel belt. In Burkina Faso, a succession (Gavaud, 1972) which is characterized by the repetition of lateritic layers and dune sands records the alternation of laterite-forming savanna climate with the southward migrations of Saharan dunes. The sequence on the Jos plateau is well exposed in mining excavations. At the base is the so-called Fluvio-volcanic Series comprising laterized, deeply weathered Plio-Pleistocene fluvio-lacustrine sands and clays. These are disconformably overlain by basalts, followed by three fining-upward fluvial cyclothems. The beginning of each fluvial cycle is marked by laterite or tin-bearing gravels which contain Neolithic artefacts of the Nok culture.

Western Saharan Successions

Late Pleistocene-Holocene lithologic sections are well exposed in pluvial lake beds in the western Sahara (Fig.11.6vi, vii, viii) and have been described at Dibella (Baumhaeur, 1987), western Niger (Gavaud, 1972) and northern Mali (Fabre and Petit-Maire, 1988; Hillaire-Marcel et al., 1983; Petit-Maire, 1987). The Quaternary sedimentary sequence on the Nigerian side of the Chad basin is up to 600 m thick and is known as the Chad Formation comprising fluvio-lacustrine clays and sand with diatomites (Fig.11.6v). The Pliocene-Pleistocene boundary in the Chad Formation is not known because this unit has not been subjected to detailed paleontological or palynological studies.

In the northern part of the Chad basin, in the great sandy plain of Bilma, deflation hollows reveal ancient Holocene lake deposits at Dibella (Fig.11.6vi), comprising a succession of diatomites which alternate with lateritic or calcrete paleosols. Radiocarbon dates and diatom species allow the recognition of small isolated saline lakes, 9,785 years old, which later merged into a large lake about 10 km^2 wide and 30 m deep. This large lake existed until 5,370 yr B.P. when it dried up. The diatom paleogeographic evidence includes saline diatom species such as *Rhopalodia gibberula*, *R. musculus* and *Campylodiscus clypeus* at the base of the Dibella sequence followed by a transgressive freshwater diatom assemblage with *Cyclotella stelligera*, *Synedra ulna* and *Melorisa granulata* var. *angustissima*. Salinity fluctuations at the terminal phase of the lake are suggested by the re-appearance of the basal saline diatom assemblage.

At the great Erg Ine Sekane in northern Mali, Hillaire-Marcel et al. (1983) uncovered and dated highly fossiliferous paleo-lake terrace sediments (Fig.11.6viii). This lake existed at the same time as the Dibella lake. The rich mollusc assemblage includes *Melanoides tuberculata*, *Bulinus truncatus*, *Biomphalaria pfeifferi*, *Aspatharia sp.*, and *Limicolaria turriformis* which is still living today in the swamps of southern Mali. Vertebrate fossils of the Nile perch, *Lates niloticus*, and bones of cows were found among the remains of Neolithic settlements which were dated 4,000 yr B.P. Artefacts such as mortars and grinders suggest sedentary agricultural life for the Neolithic people whose diet included cereals. This implies hydrological conditions which are drastically different from today's desert conditions in northern Mali. At this point it is worth mentioning that these pluvials did not only create lakes and settlements in the Sahara desert, but these were also periods when the groundwater reserves that underlie the Sahara today were replenished. This means that the aquifer systems beneath the Sahara were recharged during the pluvials.

Two groups of paleo-lakes were recently investigated in Mali by Fabre and Petit-Maire (1988), one at Taoudeni and the other in what is referred to as the Taoudeni depression (Fig.11.7). At both locations sedimentological, C^{14} dating, and paleontological criteria revealed an alternation of humid and arid phases in the hydrologically isolated Taoudeni depression which has always remained an area of inland drainage throughout its late geological history. Today this region receives only 5 mm mean annual rainfall and has not been uninhabited for over 4,000 years, thus remaining as one of the driest places on earth. The Quaternary salt at Agorgott, apart from its economic value, is very interesting paleoclimatically.

Figure 11.7: Some African Quaternary localities mentioned in the text.

Agorgott in northern Taoudeni was the site of a late Quaternary paleo-lake in which 7 million tons of mineable salt was deposited in the saline lake. Cores taken to depths just below the floor of the Agorgott salt mine revealed the following stratigraphic and paleoclimatic sequence. Sodium salt (glauberite) muds and magnesium carbonate muds with alternating layers of halite and vegetal remains occur at the bottom and have been dated 6,760 yr B.P. In the alternating dark layers pollen of

Sahelian and Sudanese elements mix with north Sahelian or Saharan "drier" species. The overlying stratigraphic unit contains five halite layers (3 of which are mined at present) separated by glauberite and clay layers which have yielded radiocarbon dates of about 6,700 to 4,000 yr B.P., for the second permanent lake episode during which salt was formed. The upper stratigraphic layers dated 3,840 yr B.P., show mud cracks suggesting that climatic deteroriation had started. The top horizon comprises a red clayey sandstone with a local brown crust indicating a shift to the present arid phase. It can therefore be surmized that while the climate was dry enough for salts to precipitate at Agorgott during the early part of the Holocene, the environment was much more humid than today so that a permanent lake could exist in the area, at least for a few thousands of years, long enough for 7 million tons of salt to accumulate! No evidence of marine waters are known at Agorgott which would have supplied the salts. Rather, the source of the salts is believed to be the late Paleozoic (mid-Carboniferous) marine and lagoonal transitional beds which underlie the Taoudeni basin.

But unlike the Taoudeni paleo-salt lake to the north, the Taoudeni depression lying to the southeast, is dissected into smaller depressions which, in the Early Quaternary, were occupied by small freshwater paleo-lakes and swamps. Fossiliferous clayey carbonate beds are now all that remain of these paleo-lakes. The first Holocene deposits of the freshwater lakes rest directly on Paleozoic formations. The oldest Holocene deposits are paleosols and muds with freshwater molluscs which date younger than 9,000 yr B.P. Further south, at Erg Ine Sakane similar deposits are dated 9,320 yr B.P., implying that large lakes existed earlier in southern Mali before conditions were humid enough for them to appear in the northern part of the country. Continuous sections of carbonate muds and clays dated 8,300 yr B.P. at the base and 4,440 yr B.P. at the top contain foraminifera, ostracodes, diatoms, charophyte oogonia and calcified vegetal stems. In these sediments epiphytic freshwater molluscs generally outnumber the more euryhaline species, thus implying low salinities and shoal water. The climatic optimum implied by this lacustrine episode persisted until 4,500 yr B.P. when the lakes dried up.

Another interesting aspect of the northern Mali paleo-lake sequence is their cyclicity. The lacustrine optimum between 8,300 and 6,700 yr B.P., includes four sequences; one section of the saline member at Agorgott contains 8 or 9 glauberitic mud/magnesitic mud/salt and clay sequences. The salt layers also exhibit microscale cyclicity which is reflected in colour variation and clay content. According to Fabre and Petit-Maire (1988) these depositional cycles, which show periodicities ranging from 400 years, 360-300 years, 250-200 years down to 50 years, are indications of more frequent climatic oscillations within the familiar long periodicity climatic events which relate to changes in ice volume, atmospheric and oceanic circulation and global eustacy. Smaller cycles could be due to a complex interaction of planetary, solar, atmospheric, hydrospheric and internal earth processes (Kutzbach and Guetter, 1984; Mörner, 1984).

11.3.2 North African Successions

Gorler et al. (1988) documented the stratigraphy and structure of the Neogene continental Ouarzazate basin (Figs. 11.7) of central Morocco, east of Marrakesh. Like the East African Rift System (Baker et al., 1988), the Ahaggar and Tibesti Mountains (Burke, 1976) and the southeastern Cape Province (Hill, 1988), the Ouarzazate basin of Morocco is a region of Quaternary tectonism involving faulting, uplift and deformation of Quaternary strata. The Oligocene-Early Pleistocene sediments of this basin are predominantly fanglomerates (Fig.11.8) which were intermittently shed during the paroxysmal uplift and deformation of the neighbouring High Atlas Mountain. These beds are tabular, cross-bedded sandstones and conglomerates which are exposed at the base of the paleo-valleys which drained the central High Atlas. The Neogene sequence terminates with an alluvial top unit of (?) Late Pliocene/Early Pleistocene sandstones with numerous conglomeratic channel fills which resulted from alluvial fan sedimentation that were triggered by renewed uplift phases of the High Atlas in Late Pliocene (?) and Early Pleistocene times. This uplift was effective throughout the Pleistocene (Stäblein, 1988) as evident from the varying levels of the indurated fluvial pediment gravel layers (glacis) (Fig.11.9), unbalanced knickpoints in the wadi gradients, and the folded and tilted gravel layers. The terrace gravel layers which occur at 5 distinct levels in the Ouarzazate basin further attest to the pronounced effects of the Quaternary climatic alternations on the geomorphological development of the region.

In neighbouring Tunisia interesting interpretations of Quaternary paleosols and desert loess deposits in the light of desert margin depositional processes, were presented by Coude-Gaussen and Rognon (1988). In southern Tunisia the calcareous Matmata (Fig.11.7) plateau which is bounded westward by the Great Eastern Sahara Erg, and eastward by the Mediterranean coastal Gulf of Gabes, is covered by deposits of peri-desert loess. Although these loessic silts originated from the Sahara and accumulated from time to time by wind deposition on the plateau, their accumulation, often mistakenly included among aeolian dunes, took place in a "pluvial" type of paleoenvironment. C^{14} dates on some of the Tunisian loess beds gave Late Pleistocene ages while "pluvial" deposition is suggested by the O^{18} and C^{13} contents of the fine fraction in the loess and in the calcareous concretions in the paleosols. The isotopic contents are similar to that of Mediterranean loesses which are accumulating today under dense steppe vegetation and soils. Steppe conditions are therefore suggested for the Tunisian paleosols which are interbedded within the loesses (Brun et al., 1988)

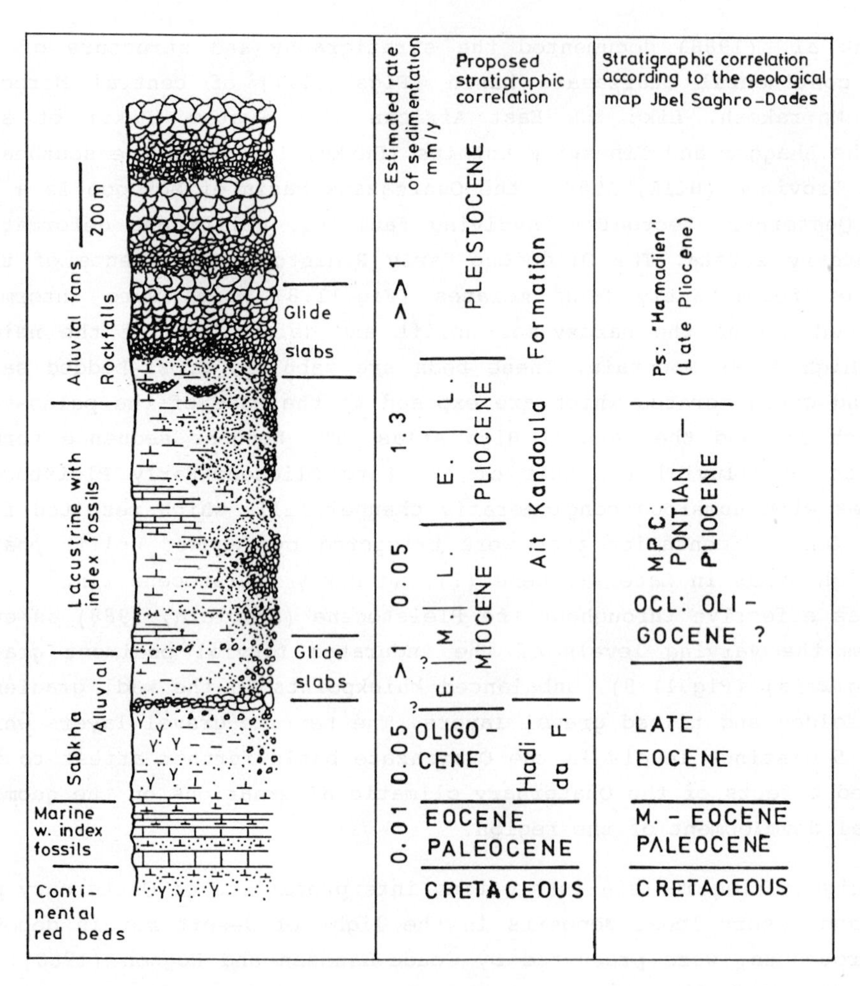

Figure 11.8: Schematic stratigraphic column for the continental Tertiary of the Ouarzazate basin in Morocco. (Redrawn from Gorler et al., 1988.)

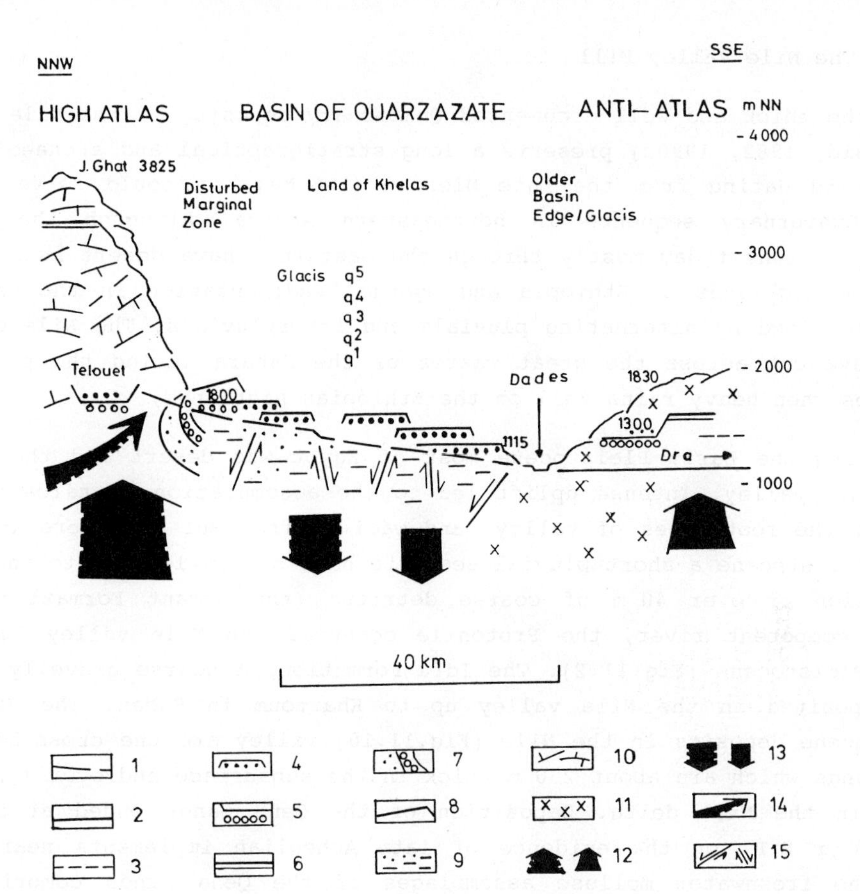

Figure 11.9: Schematic cross-section through the Ouarzazate basin. 1, wadis; 2, Dades and dra valley; 3, Pleistocene terraces of Draa valley; 4, Pleistocene glacis levels; 5, basin margin of upper Tertiary and intra-mountain basins; 6, Tertiary planation surfaces in mountains; 7, pushed and overthrust Cretaceous and Tertiary strata; 8, cuesta scarps; 9, Tertiary continental, lacustrine and marine beds; 10, locally overthrust Mesozoic; 11, basement; 12, uplifted mountain areas; 13, subsidence, compression and overthrusting; 15, sub-recent fault lines. (Redrawn from Gorler et al., 1988.)

11.3.3 The Nile Valley Fill

Since the thick and well documented Quaternary deposits of the Nile valley (Said, 1982, 1990c) preserve a long stratigraphical and archaeological record dating from the Late Miocene (5-6 Ma), it could serve as a model Quaternary sequence in northeastern Africa. Although the Nile (Fig.11.7) runs today mostly through the eastern Sahara desert being fed from the highlands of Ethiopia and Uganda, sedimentation in the valley was determined by alternating pluvials and interpluvials. The Nile could only have cut across the great wastes of the Sahara during the pluvial episodes when heavy rains fell on the Ethiopian highlands.

During the Early Pleistocene most of Egypt was desert and the Nile was a dry valley. Intense uplift led to the accumulation of talus breccias at the footslopes of valleys and wadi cliffs. But somewhere in the Early Pleistocene a short pluvial seems to have set in leading to the accumulation of over 40 m of coarse detritus (the Armant Formation). A highly competent river, the Protonile occupied the Nile valley in the Early Pleistocene (Fig.11.2). The Idfu Formation, a coarse gravelly sand was deposited in the Nile valley up to Khartoum in Sudan. The Middle Pleistocene deposits in the Nile (Fig.11.10) valley are the cross-bedded Qena Sands which are about 250 m thick in the subsurface and over 1,000 m thick in the Nile delta. Deposition of the Qena Sands ended at about 200,000 yr B.P. on the evidence of late Acheulian implements near the top. The freshwater mollusc assemblages of the Qena Sands comprise a northern assemblage with *Corbicula artinii*, *Unio gaillardoti* and *Muteline aegyptiaca* which is distinct from the southern Ethiopian fauna (with *Unio abyssinicus* and *Aspatharia calliaudi*), on the basis of which Said (1982) postulated the first connection between the northern Nile and Lake Sudd in Sudan and the Ethiopian rivers which today form the headwaters of the Nile. The Middle Pleistocene terminated with a pluvial during which the Abbassia gravel was deposited unconformably on the Qena Sands.

The overlying Late Pleistocene Neonile deposits show that the river was greatly diminished in volume. In general the deposits of this phase comprise four dry phase (aggradational) deposits known (from oldest) as the Dandara, Masma-Ballana, Sahaba Formations and Holocene silts which alternate with wet phase (recessional) depoists, the Korosko-Makhadma and the Deir el-Fakhuri Formations. The Makhadma sheet-wash deposits are thin, gravelly and sandy with Ethiopian megafauna such as horses, gazelles and hippopotamus. In the Korosko Formation the presence of wadi marls suggest a waning pluvial phase during which calcium carbonate marls were precipitated from stagnant water. The molluscan fauna of the Korosko

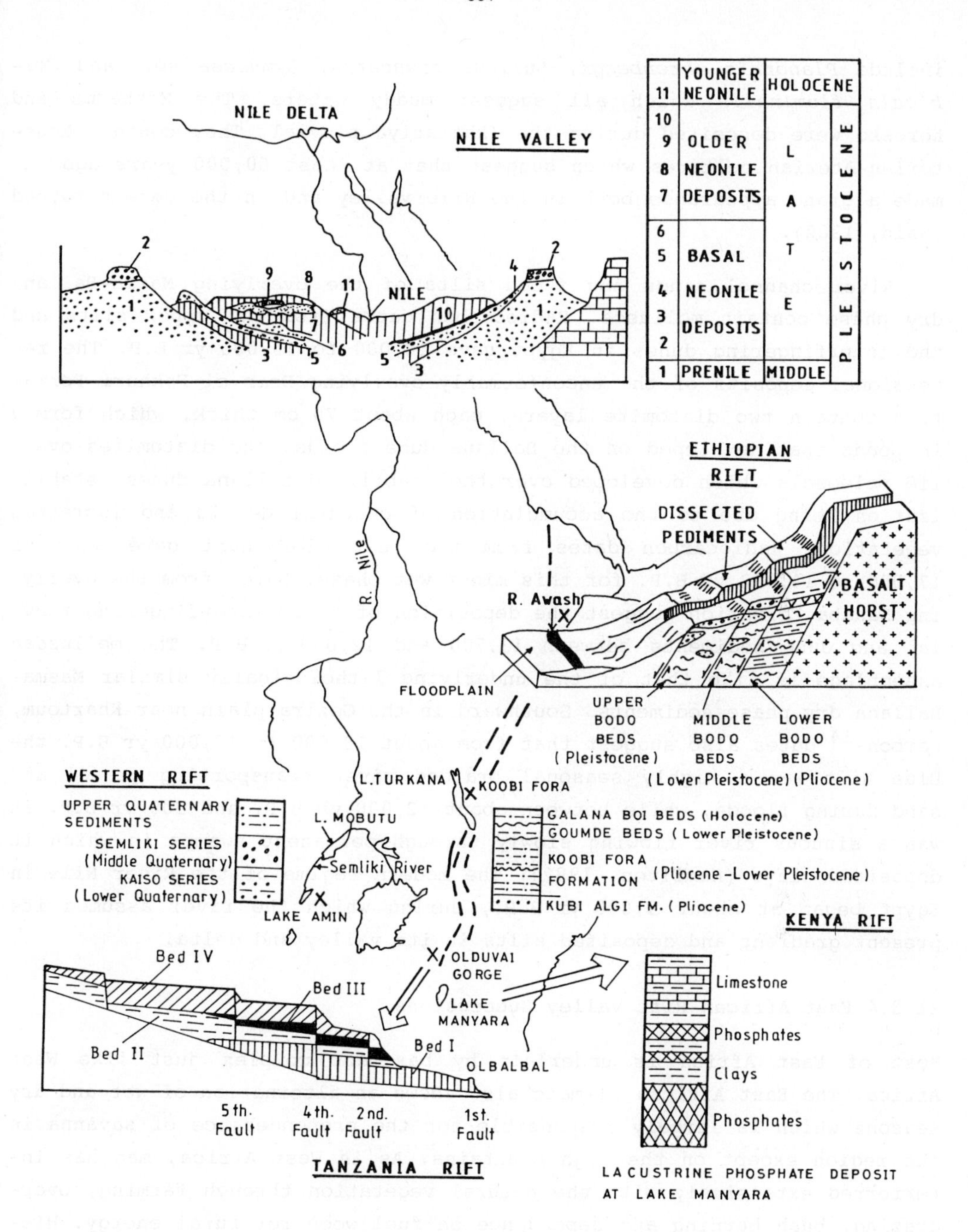

Figure 11.10: East African Quaternary successions.

include *Planorbis ehrenbergi*, *Bulinus truncatus*, *Lymnacea sp.*, and *Corbicula fluminalis* which all suggest muddy waters. The Mekhadma and Korosko were deposited during the Ikhtiariya pluvial. They contain Mousterian-Aterian cultures which suggest that at least 60,000 years ago man made a grand appearance both in the Nile valley and in the desert beyond (Said, 1982).

River channel sands and flood silts of the overlying Masma-Ballana dry phase contain molluscs. Radiocarbon dates from the upper silts and the interfingering dunes in Egypt gave 18,000 to 16,000 yr B.P. The recessional deposits of the unconformably overlying Deir el-Fakhuri Formation contain two diatomite layers, each about 70 cm thick, which formed in ponds that developed on the Ballana dune fields. The diatomites overlie paleosols which developed over the stabilized Ballana dunes, stabilization being due to the accumulation of cultural debris and increased vegetation. Radiocarbon dates from the Deir el-Fakhuri gave ages of 17,000 to 14,000 yr B.P. for this minor wet phase. Dates from the overlying Sahaba Formation suggest the deposition of these interfingering fluvial and dune sediments between 13,700 and 12,060 yr B.P. The molluscan assemblage is like that of the underlying lithologically similar Masma-Ballana dry phase sediments. Southward in the Gezira plain near Khartoum, carbon-^{14} dates also suggest that from about 18,000 to 12,000 yr B.P. the Blue Nile was a highly seasonal braided river transporting gravel and sand during floods, while between about 12,000 yr B.P. and 500 yr B.P. it was a sinuous river flowing slowly through permanent swamps in which it deposited clays (Hamilton, 1982). The modern regime of the River Nile in Egypt began at about 9,000 yr B.P., during which the river assumed its present gradient and deposited silts in its valley and delta.

11.3.4 East African Rift Valley Successions

Most of East Africa is underlain by basement complex just like West Africa. The East African climate also shows an alternation of wet and dry seasons which is largely responsible for the preponderance of savanna in the region except on the high mountains. As in West Africa, man has interferred extensively with the natural vegetation through farming, overgrazing, bush burning and dependence on fuel wood for rural energy. Historically, the concept of erosion or planation surfaces in Africa was inspired by the spectacular East African plateau sceneries. As recently summarized by Valeton and Mutakyahwa (1987) many parts of East Africa are characterized by extensively elevated plateaus of crystalline basement rocks, interrupted by isolated highlands and broad linear valleys of tectonic origin. Uplift and erosion since Jurassic times have produced different planation surfaces known (from oldest to youngest) as the Gondwana, post-Gondwana, post-African and Congo surfaces, where, as in West Africa, thick lateritic and bauxite-bearing soils varying from 5 to 40 m, are preserved on the relicts of the planation surfaces. Beyond these, there are not much similarities between the Quaternary successions of the East and West African regions. The Quaternary of East Africa is largely overshadowed by the stratigraphy of the East African Rift Valley especially of the great rift lakes, which have no counterparts elsewhere in Africa. The Quaternary of the East African Rift Valley is considered below under the four natural physiographic segments of the rift system.

Ethiopian Rift

A typical Quaternary fluvial rift valley sequence is exposed along the eastern faulted margin of the middle Awash River valley in the northern part of the Ethiopian rift valley (Fig.11.10). Informally termed the Bodo Beds (Clark et al., 1984), the sequence starts with Pliocene silts and clays (Lower Bodo Beds) which are in fault contact with the Middle Bodo Beds (brown clays with minor fossiliferous sands and gravel and volcanic ash). The Middle Bodo Beds of Early Pleistocene age are succeeded by the faulted Upper Bodo Beds with similar lithology, but of later Pleistocene age. The Bodo Beds are of great archeological interest because the lower beds contain the homonid *Australopithecus afarensis*; the middle beds contain abundant Oldowan artefacts while the upper beds carry an early Acheulian industrial complex similar to the lower Acheulian from other East African sites. At the top part of the Upper Bodo Beds upper Acheulian cleavers and hand axes are mixed with Developed Oldowan-type light duty artefacts which are associated with bones of small bovids, a canid, crocodiles and catfish (Clark et al., 1984). Sometimes the association suggests the butchery site of a hippopotamus that had been hunted and killed.

In the southern part of the Ethiopian rift valley the internal lake drainage basins such as Lake Abhé and the Ziway-Shala lakes have yielded lithologic, radiocarbon, pollen and diatom evidence of Late Quaternary lake level fluctuation (Table 11.1).

Kenya Rift

The prolific paleoanthropological and archeological sites and Quaternary stratigraphic sequences in the drainage basin of Lake Turkana have been variously described under different lithostratigraphic names (Howell and Coppens, 1976; Cohen, 1981). The Pliocene-Holocene succession at Koobi

Table 11.1: Correlation of Late Pleistocene to Recent lake levels and climatic history of Africa.

CLIMATIC PHASES	LAKE CHAD	ERG INE SEKANE (MALI)	LAKE KIVU	LAKE TURKANA	ETHIOPIAN RIFT LAKES
NIGERO - CHADIAN (13,000 BP - RECENT)	Since 9835 B P continuos lake sedimentation in southern half of lake 8,500 - 6,000 max. lake levels 11,435 - 10,265 B P diatomite marls 13,000 BP rain decreased	5400 - 4400 BP Lake episode, large fauna 6400 - 5400 BP Arid phase, eolian activity 95000 - 6400 BP High ground-water level with paleo - lake. Abundant fauna	1200 BP High level and overflow 4000 1200 BP Drop in lake level, $CaCO_3$ deposited 9500 BP Water level rose and overflowed 11,000 - 10,000 BP Major low-stand of lake level 1300 - 12500 BP Low lake phase with beach deposits, -310m below lake surface	3000 BP Fall in lake level 6000 - 4000BP Rise in lake level 7500 BP Fall in lake level 9500 BP Maximum lake level 9500 13000 BP High lake level at 60m higher than modern lake level	2700 - 1000BP Rise in lake level 6000 - 4000 BP Drastic fall in lake level to present level 9400 - 8400BP Highstand of lake level 17,000 - 10,000 B P HYPER - ARID PHASE LAKE DRIED UP
KANEMIAN (20,000 - 12,000 BP)	Sahel conditions over Sahara, lake at lowest level, Large fluviodeltaic deposits after 17,000 B P				
UPPER GHAZALIAN (29 - 20,000 BP)	First large paleo - lake, disconformable eolian and fluvial deposit over evaporites max. lake level at 20,000 B P				30,000 - 21,000 BP 3rd. lacustrine phase with tropical and temperate diatoms
LOWER GHAZALIAN (40,000 - 30,000 B P)	Groundwater outcrops in interdune areas as shallow bonds, 38 - 35,000 BP decrease in rain, sahel and evaporitic conditions			Shortly before 35,000 BP high lake stand at + 60 m lake level above present	40,000 - 30,000 BP High and deep lake with abundant diatoms
ANTE-GHAZALIAN (BEFORE 50,000 B P)	Drier climate, erg formation, large fluvio - deltaic lake				70,000 - 60,000 BP Lake Abhe´ dried up completely

Fora, about 500 m thick, has been subjected to detailed lithologic descriptions, potassium-argon dates and tuff stratigraphy, paleomagnetic studies and geochemical, paleontological and comprehensive paleoenvironmental interpretations which are summarized in Fig.11.11 based on the works of Abell (1982), Behrensmeyer (1979, 1981), Brock and Isaac (1974), McDougall (1985), Owen et al., (1982), Howell (1978), and Williamson (1982).

Essentially, the Koobi Fora sequence (Fig.11.11) comprises fluvio-lacustrine and deltaic silts, clays and sands which are abundantly interlayered by volcanic tuff. The well documented numerical time-scale based on tuff potassium-argon ages has provided an excellent numerical age framework for the sedimentary sequence and homonid and artefact successions at Koobi Fora. This should serve as the standard for the East African Quaternary, if not for the whole of Africa. The homonids *Homo habilis*, *H. erectus*, *Australopithecus boisei* and *A. robustus* co-existed between 2.0 and 1.4 m.y (Fig.11.11), thus negating the hypothesis that *Homo erectus* descended from *H. habilis*. The immediate ancestor of *H. erectus* is not known (Pilbeam, 1984). Salinity fluctuations of Lake Turkana inferred from O^{18}/O^{16} variations (Abell, 1982) suggest marked climatic alternations during this critical stage of homonid evolution. These ecological changes also had drastic effects on the Pliocene-Pleistocene molluscs (Williamson, 1982) and on diatom assemblages at Lake Turkana.

Tanzania Rift

The world-famous Quaternary locality in Tanzania is exposed in the Olduvai Gorge in the northern part of the country. Here the Olduvai Beds (Fig.11.10), as they are termed, are well exposed in a shallow lake basin. The entire sequence, about 100 m thick, comprises fluvio-lacustrine clays and sands interbedded with volcanic lavas which have yielded potassium-argon ages of 2.1 m.y to 15,000 yr B.P. (Hays, 1976). At the base is Olduvai Bed I, about 60 m thick, comprising marly lacustrine clays which interfinger eastward with alluvial fan deposits. Terrestrial gastropods such as slugs suggest damp conditions during the deposition of Bed I.

Although it overlies Bed I conformably, Bed II has an unconformity in the middle that separates a lower lacustrine sequence from an upper fluvial-lacustrine facies which implies considerable reduction in the size of the Quaternary lake. The vertebrate fauna below the middle part of Bed II is rich with mainly swamp-dwelling crocodiles and turtles (Fig.11.12),

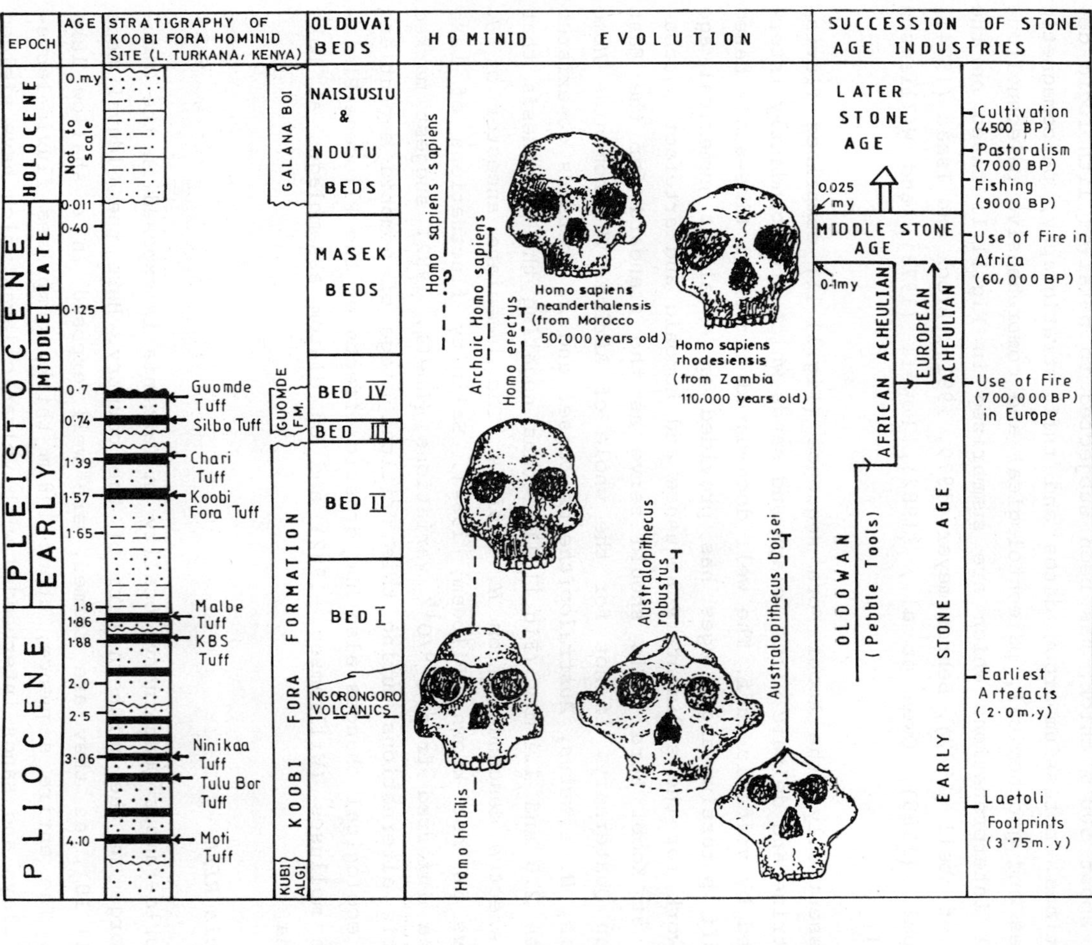

Figure 11.11: Correlation of East African (Lake Turkana, Olduvai Gorge) lithostratigraphy, hominid evolution, and cultural history.

whereas the proportion of savanna and plains dwellers (eg. lions and deer) abruptly increases at the top of Bed II. Bed II contains Oldowan artefacts in the lower part and Developed Oldowan and Acheulian tools in the upper part. *Homo habilis, H. erectus* and *Australopithecus cf. boisei* are the homonids in Bed II.

Figure 11.12: Vertebrate faunas and reconstructed scenery of Olduvai Bed I and the lower part of Bed II. The faunas include several unfamiliar animals now extinct and some modern animals. (Redrawn from display in the British Museum of Natural History, London.)

Higher in the Olduvai sequence lake deposits are replaced by predominently fluviatile strata (Bed III, IV, Masek, Ndutu and Naisiusiu Beds) which are still interbedded with volcanic tuffs. The homonid sites and vertebrates of Bed III and IV are greatly reduced (Fig.11.13). An even greater reduction of homonid occupation and activities at the time of the

Masek Beds (Middle Pleistocene) reflects the onset of climatic desiccation and large areas of dry savanna which have persisted till today.

Figure 11.13: Olduvai Bed III faunas showing an impoverished community in gravel beds. (Redrawn from display in the British Museum of Natural History, London.)

In the southern part of Tanzania different lake habitats existed. A Pliocene-Pleistocene lacustrine phosphate deposit, 10 m thick, occurs in the Lake Manyara basin and is mined at Minjingu, 5 km away from the present small alkaline Lake Manyara. Lacustrine clay beds, algal limestones and tuffs alternate with the phosphates. The rich bird, fish and mollusc faunas suggest a greatly expanded Pliocene-Pleistocene Lake Manyara which probably united with the nearby Lakes Natron and Magadi and formed a huge lake (Schluter, 1987) with abundant nutrient supply which attracted large populations of birds and fish. The faeces and bones of these organisms were the initial deposits that later underwent diagenesis into rock phosphate.

Western Rift

The Quaternary strata that underlie the Western Rift Valley are not well exposed as in the Eastern Rift. Well dated lake levels occur around Lakes Malawi, Tanganyika, and Kivu (Table 11.1). Quaternary sequences are exposed in Lakes Amin and Mobutu basins (Fig.11.10). At Lake Mobutu the sequence is estimated to be 2,500 m thick (Hamilton, 1982) and it comprises the Kaiso Beds overlain by upper Quaternary sediments. Three series of Quaternary sediments are esposed at Lake Amin, with the basal Kaiso Beds comprising fossiliferous grey to greenish clay beds with oolitic ferruginous layers, sands and volcanic tuff. The faunas of the Kaiso Beds include abundant molluscs, mammals, reptiles and fish. These beds also contain artefacts such as hammer stones, coves, flakes, cleavers and Acheulian hand axes (Bishop, 1958). The Kaiso Beds are overlain by the coarse-grained unfossiliferous Semliki Series, which are succeeded by unnamed upper Quaternary sediments (10,000-8,000 yr B.P.) with fishes and human remains and stone tools.

11.3.5 Quaternary Deposits in Southern Africa

Marine terrace deposits which accumulated at several levels during the Pleistocene sea level changes, and coastal plain eolian dune ridges occur along the South Atlantic and Indian ocean coasts of southern Africa. Inland, Quaternary deposits occur in the Kalahari Group, the Vaal-Orange drainage basin, and as the so-called Australopithecine Cave Breccias (Tankard et al., 1982). Widespread colluvial deposits, up to 10 m thick mantle east and central parts of southern Africa (Fig.11.14) and suggest the stripping of poorly vegetated pediment slopes during the dry phases between 30,000 B.P. and 12,000 yr B.P.

Kalahari Basin

Here the Quaternary is represented by thick basal fluvial conglomerates which are up to 90 m thick and are occasionally cemented by caliche; these are overlain by 180 m of red shales, marls and duricrusts. Widespread caliche crusts and dolomite in the sands of the Kalahari Group point to arid climate with saline conditions. In the Etosha pan and other paleo-lake depressions, algal stromatolites suggest the presence of temporary saline lakes in the Kalahari basin in the Late Quaternary. Although not precisely dated, three morphologically distinct groups of fixed dunes in the Kalahari basin have preserved three periods of desert-expansion in the Late Pleistocene (Lancaster, 1981). In the Namib desert the interdune areas are filled with thin calcareous lacustrine sandstones, mudstones and limestones suggesting increased availability of moisture in this normally hyperarid region during the Late Quaternary (Teller et al., 1990).

Vaal-Orange Basin and Continental Shelf

Quaternary deposits including diamond-bearing gravels, occur in a complex sequence of alluvial terraces along the Orange and Vaal rivers (Patridge and Brink, 1967). Near the confluence of both rivers (Fig.11.7) mammalian bones including the Pliocene elephant *Mammuthus subplanifrons* and artefacts suggest a Late Pliocene-Early Pleistocene age for the oldest alluvial deposits which are caliche-cemented cobble-grade conglomerates. Rare fossils and artefacts suggest a Middle Pleistocene age for a younger lithologically similar set of braided floodplain gravel and sand beds.

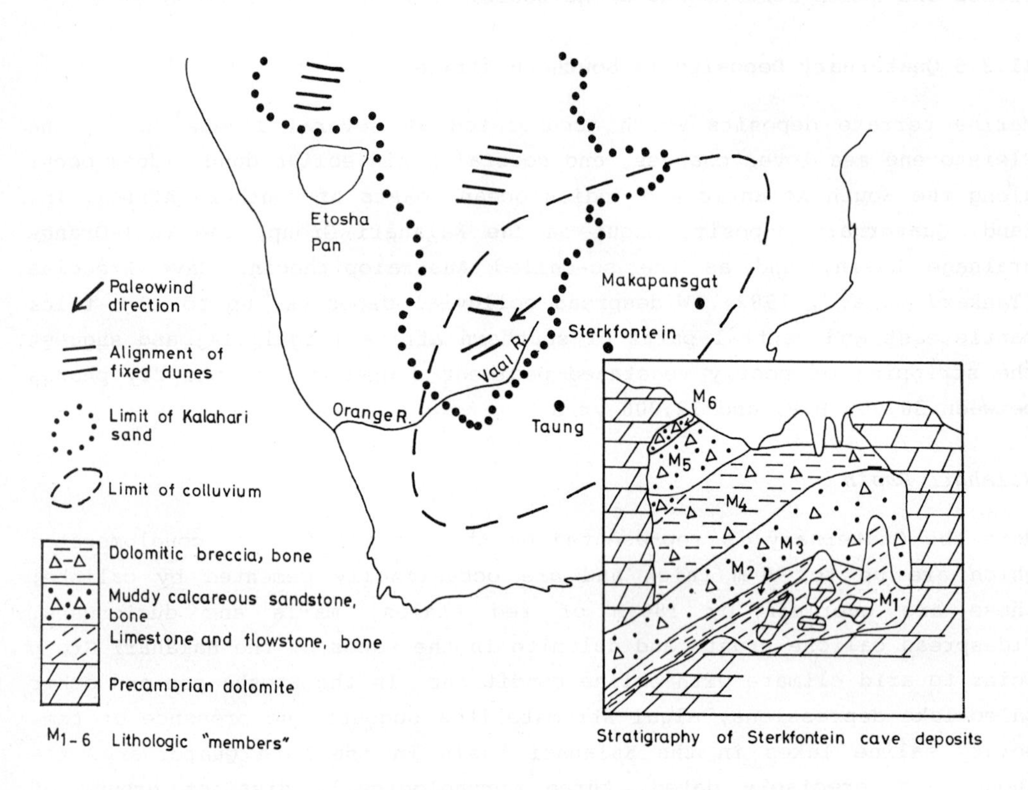

Figure 11.14: Quaternary deposits of southern Africa.

On the inner continental shelf off SW Africa between Luderitz in Namibia and St. Helena Bay in South Africa, Late Tertiary and Quaternary sea-level fluctuations have resulted in a highly unique association of the world's richest marine alluvial diamond fields with shallow shelf submarine geomorphic features which resulted from sea-level stillstands. A recent geophysical survey of the seafloor morphology of the area, including the thickness and stratigraphy of the unconsolidated phosphatic sediment cover, reveals that during the lowstands of sea-level diamonds were reworked from coastal deposits and concentrated in submarine wave-cut terraces, cliffs, paleo-channels, reefs, gullies, potholes and bedrock depressions which were eroded during the lowering of sea-level (De Decker 1989a). It was also during the last major lowering of sea-level at about 18,000 yr B.P., that the Orange River delta began to build seaward in this region. According to De Decker (1989b) sea-level fell to 120 m during the Last Glacial Maximum, and along this regressive paleo-strandline a barrier-beach and lagoonal sediment complex formed. The Orange River delta subsequently prograded over the shelf with extensive coast-parallel beach deposits mantling the shelf during the Recent transgressive phase.

Australopithecine Cave Breccias

In the Transvaal area in South Africa a system of caves in Precambrian dolomites (Fig.11.14) are known at Taung, Sterkfontein, Makapansgat, Swartkrans and Kromdraai, which contain homonid-bearing cave deposits (King, 1951 a,b; Patridge, 1978). The lithostratigraphy of a typical cave succession in the Sterkfontein cave reflects several different processes such as brecciation of cavern walls, carbonate precipitation, accumulation of insoluble cave earth and the flushing of colluvium into the caves. Homonid remains and fossils occur mostly in the carbonate-cemented colluvia and breccia. The Sterkfontein cave shows six sedimentary cycles each separated by a hiatus that may be associated with calcite or flowstone accumulation (Fig.11.14). The fourth sedimentary cycle (M4) contains *Australopithecus africanus* (synonymous with *A. afarensis*) followed by the fifth cycle (M5) with *Homo cf. habilis* (Late Pliocene-Early Pleistocene). The cave sedimentary facies suggest relatively dry climates with little variation, while the vertebrate faunas, dominated by bovids, is consistent with a savanna or wooded grassland where *Australopithecus robustus* and *A. africanus* roamed, being hunted by leopards.

11.4 Quaternary Paleoclimatic Reconstructions for Africa

Although the paleoenvironmental significance of African Quaternary successions were mentioned in the aforegoing section, there is a need to present a coherent Quaternary paleoclimatic scenario for Africa using the available evidence from some of the regions where the record is extensive and well dated. No attempt is, however made to correlate the African land record of Quaternary paleoclimatic changes with the oceanic oxygen isotope stages (eg. Williams et al., 1988). Such a correlation is, however, greatly needed.

When pieced together, the paleoclimatic information from these various sources yield a coherent picture of a fairly dry and unstable Pliocene-Early Pleistocene phase. This is evident from the lower Quaternary sequence at Olduvai Gorge in Tanzania, the Koobi Fora region of Lake Turkana in Kenya, the Australopithecus-bearing cave deposits in South Africa and the alluvial deposits in the Nile valley in Egypt. This unstable paleoclimate resulted in important speciation events amongst homonids (Fig.11.11) and amongst several groups of organisms.

While there is a Middle Pleistocene hiatus in the Lake Turkana basin with no paleoclimatic record, the Bed IV and Masek Beds at Olduvai provide the missing evidence, and reveal that climate became progressively drier but was still moistier than today. In the Nile valley in Egypt there was a wet phase, the Idfu pluvial, during the Early-Middle Pleistocene transition. From this point onward both oceanic and continental paleoclimatic data are available because deep-sea cores from the ocean bottom of northwest Africa and from the Zaire deep-sea fan have retrieved sediments as old as the Middle Pleistocene.

Calcium carbonate fluctuations in the Zaire deep-sea fan reflect cold conditions during most of the Middle Pleistocene (Jansen et al., 1984), while there is evidence of aridity in Egypt (Said, 1982) and dry conditions in East Africa. The major pluvial which terminated the Middle Pleistocene in Egypt (Fig.11.2) and caused the Nile, which up to this point was solely an Egyptian river, to capture its Ethiopian and Sudanese headwaters, was felt as a warm humid phase elsewhere in Africa. The coincidence of lake level fluctuations throughout Africa (e.g. Table 11.1), the faunal and palynological information from the Sahara and other parts of Africa (eg. Pachur et al., 1990; Voight et al., 1990),and the rich archeological record from all parts of the continent document Late Pleistocene-Holocene climatic fluctuations.

11.4.1 The Land Record

Southern and Eastern Africa

Regional overviews of the late Quaternary paleoclimatic record were recently furnished by Zinderen Bakker and Coetzee (1988) for East and southern Africa and by Deacon and Lancaster (1988), and Scott (1989) for South Africa. These workers reviewed the results of fossil pollen studies over the last 35 years. Avery's (1988) record of South African micromammals is also consistent with pollen paleoclimatic scenarios. Most of the locations mentioned are shown on Fig.11.7.

130,000 - 80,000 yr B.P. Deacon and Lancaster (1988) have shown that warm to mildly cool interglacial conditions prevailed both on the coast and in the hinterland of South Africa on the evidence of large mammals and shelfish faunas.

80,000 - 50,000 yr B.P. At the Boomplaas cave site in South Africa (Fig.11.7) where the sequence is more complete and in Zimbabwe at Redcliff, the deposits suggest cooler conditions than in the preceeding phase.

50,000 - 32,000 yr B.P. Oxygen isotopes C^{13}, and the predominance of grasses at the Wolkberg and Cango cave sites in South Africa, suggest cooling trends.

32,000 - 28,000 yr B.P. This was a warm and humid phase as evident from the upward movement of the tree line at Sacred Lake on Mount Kenya (Fig.11.7); a rapid increase in peat growth at Kamiranzovu in SW Uganda; and higher groundwater levels, stream discharges and the deposition of tufas in SW Africa (Lancaster, 1989). However, at Wolkberg cave in South Africa a temperature drop was reported. A site with warm paleoclimate is the Kashiru peatbog and valley swamps from the central African highland location in Burundi (3° 28'S, 29° 34'E). At this location, prior to 30,000 yr B.P., the occurence of a montane conifer forest, including the upper forest limit, indicates climatic conditions colder and drier than now, but more humid than in late-glacial time (Bonefille and Riollet, 1988).

28,000 - 20,000 yr B.P. During this interval the climate in East and central Africa was fairly similar to that of the Holocene moist period. The Kamiranzovou Swamp shows a slower accumulation in East Africa and slightly drier conditions. A forest period in East Africa culminated after 24,000 yr B.P., during a wet phase in which temperatures were colder and precipitation slightly greater than today. In southern Africa peat deposits at the present savanna spring-site of Wonderkrater show cool moist conditions and the existence of bushveld. The Kalahari and Namibia deserts received higher rainfall than during any period in the last 32,000 years. Towards the end of this phase, the temperature dropped and aridity spread in the Kalahari desert from south to north. The Lake Makgadikgadi area continued to be humid and in the northern Kalahari, humidity persisted at around 22,000 yr B.P. In the Namib desert the onset of arid conditions is evident in the Homeb silts which accumulated along the Kuiseb River during a drier period from 23,000 to 19,000 yr B.P.

20,000 - 16,000 yr B.P. This period represents the last glacial maximum during which aridity spread over nearly the whole of Africa, and for which there is abundant evidence. Even the tropical rainforest of the Zaire basin was considerably reduced in size and the tree line on the East African mountains was lowered by 900-1100 m, indicating a drop in mean temperature of 5-8 °C in East Africa. The Kalahari desert was arid with active dune formation (Lancaster,1988), except in the southern part and its surroundings, and the SW Cape where humidity was high. The downward expansion of the high mountain *Ericaceous* and the lower Afroalpine belts on Mount Kenya also suggest a drop in paleotemperature, while at Kamiranzovu in Uganda, Lake Mahoma on the Ruwenzori Mountains, the Kalombo Falls, Ishiba Ngandu in northern Zambia and at the Mufo archeological site in NE Angola, colder drier climates are registered on pollen evidence.

Van Zinderen Bakker and Coetzee (1988) drew comparisons between the land record and the coeval deep-sea record of the southwestern African continental margin. Fossil pollen obtained from the offshore Zaire deep-sea (Fig.11.1) from at less than 3,800 m water depth provides excellent information on the environmental conditions that prevailed in the Zaire basin on land, during the last glacial maximum around 18,600 - 15,400 yr B.P. This interval in the cores contain abundant terrigenous organic matter with a black and burnt appearance , together with grass pollen, and many spores, few forest pollen types and no mangrove pollen. A considerable change in vegetation is therefore suggested when compared with the present-day tropical rainforest in the Zaire basin with its coastal mangrove and savanna vegetation. During the glacial period savanna and grassland replaced the tropical rainforest in the Zaire basin while the coastal zone was a desert. These changes were strongly correlated with the northward penetration of the cold upwelling waters of the Benguela Current.

At Wonderkrater and Boomplaas in the southern Cape in South Africa there was a downward migration of highland vegetation suggesting a mean temperature 5-6° lower than at present. Large drops in temperature also occured in the southern Cape at Wolkberg Cave which recorded a drop of 7.5° at 18,000 yr B.P., and at Uitenhage and Cango cave sites where temperatures of about 5.5° and 5°C respectively are known. Severe frost occured between 26° and 29°S latitude as evident at Rose Cottage cave and Wonderwerk and Border caves. Periglacial landforms are known in the high mountains of Lesotho and at the southern Cape. Westward, geomorphic evidence in the Kalahari suggests less humid conditions during the last glacial maximum. Lake Paleo-Makgadikgadi periodically dried out completely between 19,000 and 12,000 yr B.P., while sub-humid to humid conditions prevailed in the northern Kalahari at 16,000 - 13,00 yr B.P. In the Malopo valley in the southern Kalahari a perennial river existed between about 17,000 and 15,000 yr B.P.

16,000 - 14,000 to 10,000 yr B.P. This transitional period witnessed considerable climatic and phytogeographic changes in which vegetation responded to a general rise in temperature and an important increase in precipitation between 12,600 and 11,000 yr B.P. On the East African mountains vegetation belts migrated to higher altitudes between 12,500 yr B.P. and 10,500 yr B.P. Trees replaced grassland at Sacred Lake, the *Ericaceous* grassbelt declined at Cherangani in Kenya, but at Lake Chesi in Zambia (Stager, 1988) dry conditions still prevailed at about 15,000-13,000 yr B.P., when the lake shrank and became chemically concentrated. The vegetation cover remained sparse on the higher and colder mountain sites of Lake Rutundu (3,140 m), Lake Mahoma (2,960 m) and at the Badda Swamps (4,040 m). Heavy rainfall caused rapid erosion at Rutundu and an influx of sand-rich loam occured in parts of Rwanda where the ground cover was sparse. In the lowlands around the northern edge of Lake Victoria vegetation changes took place which paralleled those on the high mountains. The open vegetation which existed between ca 14,500 and 14,000 yr B.P. was progressively replaced from about 12,000 yr B.P. onward, so that after about 9,500 yr. B.P., lowland forest became established. The palynological indications for higher rainfall are corroborated by the rise of the East African lake levels at about 12,000 yr B.P.

In southern Africa those parts which had been dry during the hypothermal period registered more humid climate, for example at Aliwal North, situated at the boundary between the dry semi-desert Karoo in the south and the subhumid grassland in the north. At 12,000 yr B.P. pure grassland occupied the area, suggesting colder and more humid conditions. This vegetation was replaced twice by warm dry Karoo-type vegetation which was re-established for the third time at about 9,600 yr B.P. Pollen spectra and charcoal analysis show that the previously open vegetation at Boomplaas in the south coastal region was replaced by *Olea* woodland between 14,200 and 12,000yr B.P., indicating higher rainfall and temperatures. This woodland changed into thicket at about 10,000 yr B.P. In the Kalahari humid phases started between 13,000 and 12,000 yr B.P., so that after a complete desiccation of the Makgadikgadi pan during the hypothermal period a new transgression occured here at 12,000 yr B.P. The Etosha

Pan received more rainfall from 13,000 to 12,000 yr B.P., as did Malopo valley in the south, from 13,000 to 10,000 yr B.P.

Holocene. In general humid conditions persisted in East and southern Africa during the Early Holocene until about 4,000 yr B.P. Pollen data in East Africa point to deforestation by man during the last two millennia. At Sacred Lake on Mount Kenya and on Mount Kilimanjaro the tree line moved upward during the warm and moist Early Holocene but conditions became colder and drier towards about 4,000 yr B.P., when on Mount Kenya the glacier grew and constructed a moraine at an altitude of 4,265 m. Maximum lake levels occured between 6,000 and 4,000 yr B.P. from East Africa down to Zambia and were followed by a lowstand under presumably arid conditions at about 3,500 yr B.P. (Stager, 1988).

In southern Africa two different sets of climatic change have been recognized in spite of the lack of precise data. In the northern more humid region of South Africa temperature and humidity increase was punctuated by a short colder interval between about 4,000 and 3,000 yr. B.P. However, the southern boundary of the Kalahari with its overall drier conditions exhibit two semi-arid periods separated by an arid episode around 4,000 yr. B.P., for example at Wonderwerk. This was followed by alternating wet and dry conditions in this region extending into the Kalahari and the Namib. However, radiocarbon dates obtained from fluvial deposits and calcretes in the Namib desert also suggest a humid phase between about 4,000 yr B.P. and 1,200 yr B.P. and a dry phase from A.D. 1200 to A.D. 1600 (Vogel, 1989).

Talbot and Livingstone (1989) demonstrated that organic geochemical parameters (total organic carbon; hydrogen content; and the proportion of plants using C3 photosynthesis, eg. grasses and aquatic plants, and C4 photosynthesis, eg. terrestrial plants) can be used to detect fluctuations of lake level, in addition to the usual geomorphological and micropaleontological methods. Exposure surfaces in lake sediments are marked by a decrease in total organic carbon and hydrogen due to oxidation. Using these indicators on sediment cores retrieved from the bottom of Lakes Victoria and Rukwa in East Africa, Talbot and Livingstone (1989) found that Lake Victoria fell below - 66 m some time between 17,310 and 15,120 yr B.P., while L. Rukwa in the Early Holocene was deep with poor circulation. On three occasions L. Rukwa dried up since 4,000 - 3,000 yr B.P.

The Sahara

Fabre and Petit-Maire (1988) summarized Holocene climatic evolution in the western Sahara during the Quaternary based on paleo-lake evidence and generally re-affirmed the paleoclimatic pattern (Fig.11.15) of a possible wet phase during isotopic stage 3 (ca. 40,000 - 20,000 yr B.P.), the evidence of which Fontes and Gasse (1989) and Whiteman (1982) deem to be as yet inconclusive. There was an arid phase during isotope stage 2 (ca. 20,000 - 10,000 yr B.P.). Extensive freshwater lakes and swamps occured during isotope stage 1 (10,000 - 3,000 yr B.P.) with the onset of climatic deterioriation at about 7,000 yr B.P., and the beginning of the arid phase at about 3,000 yr B.P.

In the Sudanese part of the southeastern Sahara Pachur et al. (1990) and Gabriel (1984) found Early Holocene fossiliferous lake and ancient wadi sediments (Fig.11.16), the ages of which range between 9,400 to less than 4,800 yr B.P. The lake and marsh sediments contain a diverse vertebrate fauna with fish, crocodile, hippopotamus, land turtle, and domesticated cattle (Fig.11.16). At latitude 19 °N in what is now a hyper-arid region, cattle rearing was still possible 3,500 years ago.

11.4.2 The Oceanic Record

The ocean contains both extrinsic and intrinsic Quaternary paleoclimatic records. Land-derived paleoclimatic signals constitute the extrinsic or extraneous record which comprises fluvial terrigenous sediments, desert dust fallouts, pollen and spores, other plant remains and diatoms. The intrinsic record comprises the indigenous marine sediments such as carbonates, chert and glauconite. The African continental margin from the Nile delta to western Africa will be reviewed to show the information contained in the sediments about the Quaternary paleoenvironmental changes as well as the imprints of these changes on the morphology of the continental shelf and shoreline. The paleoclimatic signals in the Zaire deep-sea fan have already been mentioned.

Foucault and Stanley (1989) inferred climatic oscillations from temporal fluctuations in amphibole-pyroxene ratios in the Late Quaternary sediments in cores from the eastern Nile delta. These oscillations are consistent with paleoclimatic interpretations based on changes of the levels of Lake Abhé and the lakes in the Ziway-Shala basin in Ethiopia. High proportions of amphibole in the cores supplied by the White Nile from Uganda and the Sudan, dated about 40,000 to 20,000 yr B.P., correspond to periods of high lake levels recorded before about 20,000-17,000 yr B.P. Low amounts of amphibole and high percentage of pyroxene in the

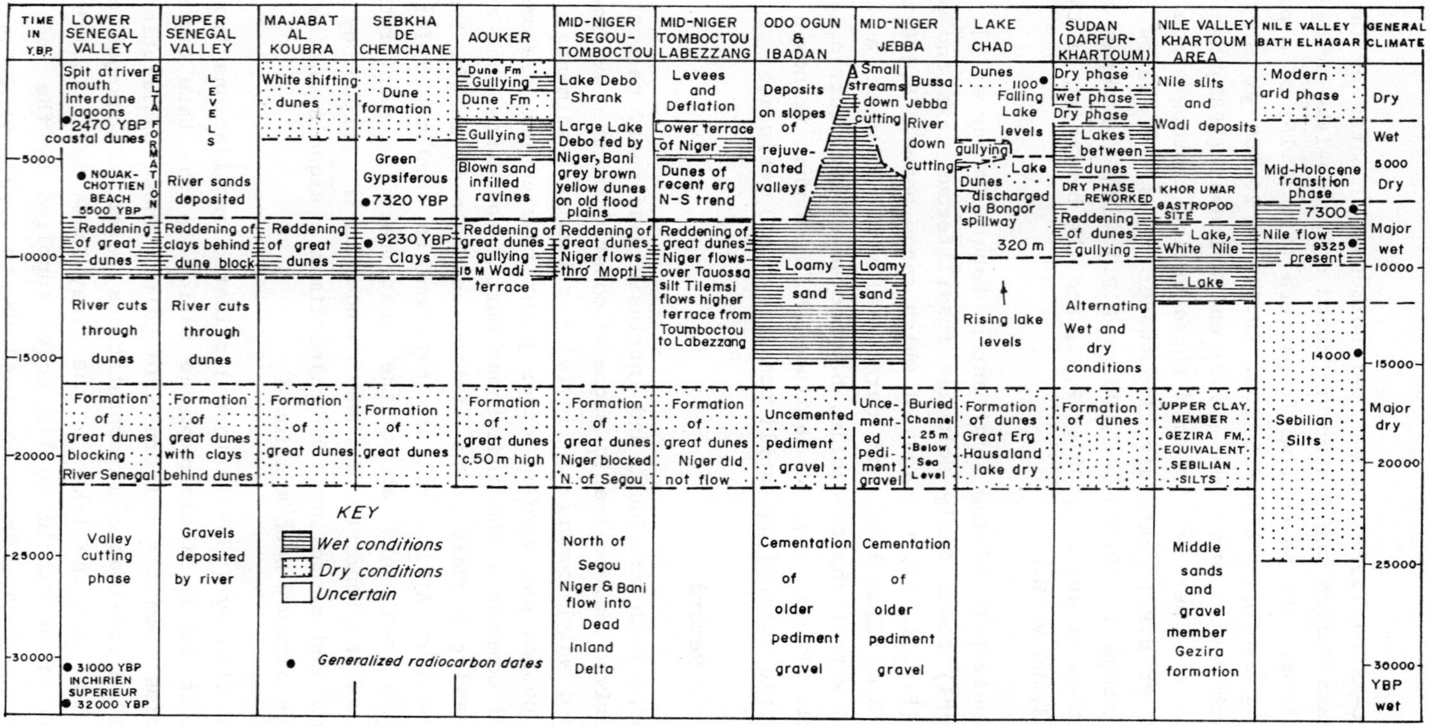

TIME IN Y.B.P.: 5000, 10000, 15000, 20000, 25000, 30000

LOWER SENEGAL VALLEY
Spit at river mouth interdune lagoons
2470 YBP coastal dunes
DELTA FORMATION
NOUAKCHOTTIEN BEACH 5500 YBP
Reddening of great dunes
River cuts through dunes
Formation of great dunes blocking River Senegal
Valley cutting phase
31000 YBP INCHIRIEN SUPERIEUR 32000 YBP

UPPER SENEGAL VALLEY
LEVELS
River sands deposited
Reddening of clays behind dune block
River cuts through dunes
Formation of great dunes with clays behind dunes
Gravels deposited by river

MAJABAT AL KOUBRA
White shifting dunes
Reddening of great dunes
Formation of great dunes

SEBKHA DE CHEMCHANE
Dune formation
Green Gypsiferous
7320 YBP
9230 YBP Clays
Formation of great dunes

AOUKER
Dune Fm
Gullying
Dune Fm
Gullying
Blown sand infilled ravines
Reddening of great dunes gullying 15 M Wadi terrace
Formation of great dunes c.50 m high

MID-NIGER SEGOU-TOMBOCTOU
Lake Debo Shrank
Large Lake Debo fed by Niger, Bani grey brown yellow dunes on old flood plains
Reddening of great dunes Niger flows thro' Mopti
Formation of great dunes Niger blocked N. of Segou
North of Segou Niger & Bani flow into Dead Inland Delta

MID-NIGER TOMBOCTOU LABEZZANG
Levees and Deflation
Lower terrace of Niger
Dunes of recent erg N-S trend
Reddening of great dunes Niger flows over Tauossa silt Tilemsi flows higher terrace from Toumboctou to Labezzang
Formation of great dunes Niger did not flow

ODO OGUN & IBADAN
Deposits on slopes of rejuvenated valleys
Loamy sand
Uncemented pediment gravel
Cementation of older pediment gravel

MID-NIGER JEBBA
Small streams down cutting
Bussa Jebba River down cutting
Loamy sand
Uncemented pediment gravel
Buried Channel 25 m Below Sea Level
Cementation of older pediment gravel

LAKE CHAD
Dunes 1100 Falling Lake levels
gullying
Lake Dunes discharged via Bongor spillway
320 m
Rising lake levels
Formation of dunes Great Erg Hausaland lake dry

SUDAN (DARFUR-KHARTOUM)
Dry phase
wet phase
Dry phase
Lakes between dunes
DRY PHASE REWORKED
Reddening of dunes gullying
Alternating Wet and dry conditions
Formation of dunes

NILE VALLEY KHARTOUM AREA
Nile silts and Wadi deposits
KHOR UMAR GASTROPOD SITE
Lake, White Nile
Lake
UPPER CLAY. MEMBER GEZIRA FM. EQUIVALENT SEBILIAN SILTS
Middle sands and gravel member Gezira formation

NILE VALLEY BATH ELHAGAR
Modern arid phase
Mid-Holocene transition phase
7300
Nile flow 9325 present
14000
Sebilian Silts

GENERAL CLIMATE
Dry
Wet
5000 Dry
Major wet
10000
15000
Major dry
20000
25000
30000 YBP wet

KEY
Wet conditions
Dry conditions
Uncertain
● Generalized radiocarbon dates

Figure 11.15: Late Pleistocene environmental correlations in West and NE Africa. (Redrawn from Whiteman, 1982.)

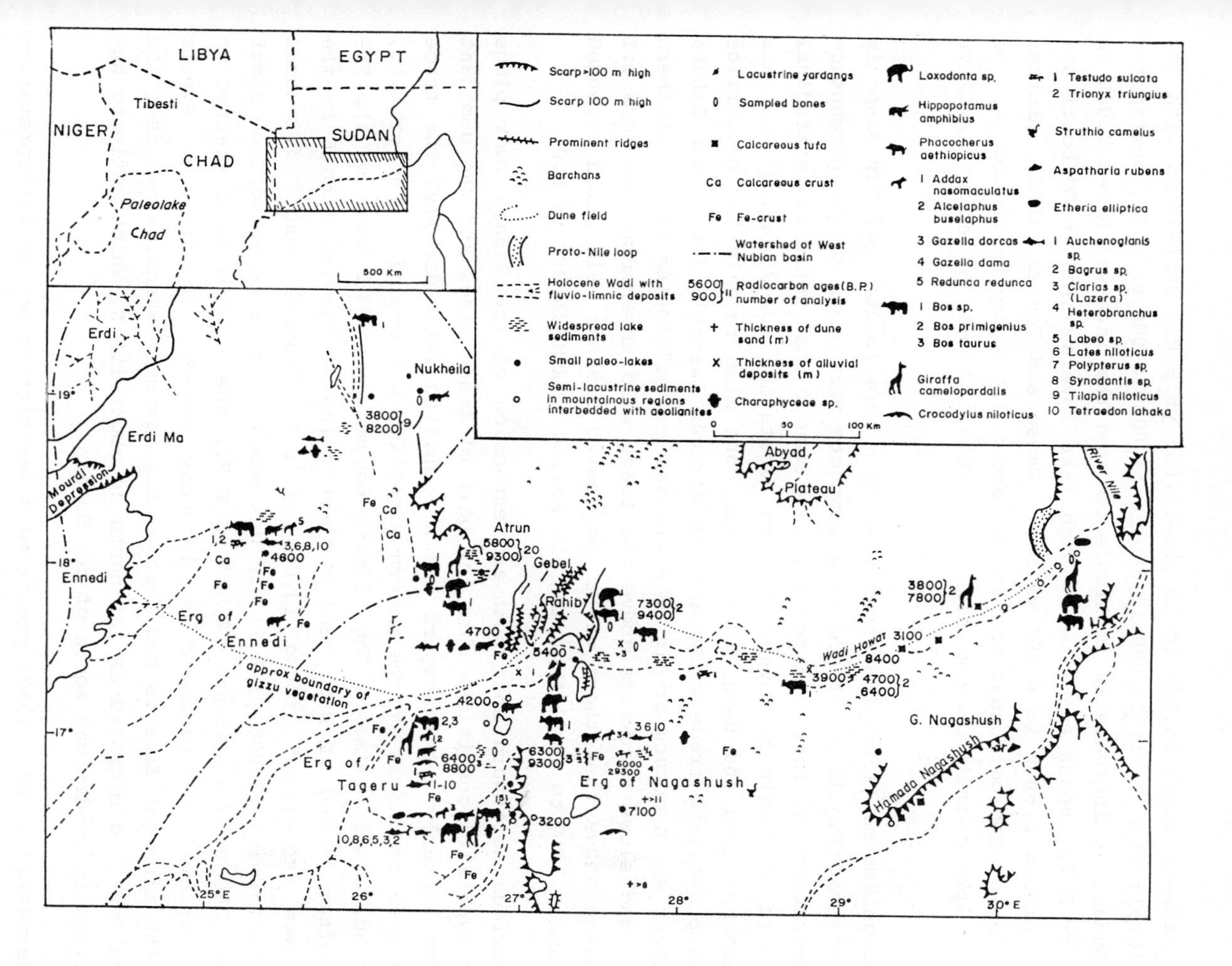

Figure 11.16: Faunas and other paleoenvironmental indicators in NW Sudan. (Redrawn from Pachur et al., 1990.)

cores between about 20,000 to 12,000-10,000 yr B.P., correspond with low lake levels from about 20,000-17,000 to 14,000-10,000 yr B.P., during the arid phase. An increase in amphibole between the interval dated 7,000 to 4,000 yr B.P., spans a period of high lake levels which in Djibouti occurred around 8,6000 to 6,000 yr B.P. (Gasse and Fontes, 1989). Another increase in amphibole near the top at about 3.4 m may be related to the climatic phase that induced high lake levels found for about 1,500 yr B.P.

Hooghiemstra (1988) has shown that in a core located off Cap Blanc in West Africa (Fig.11.7), there is a downcore variation of the percentage concentration and influx of pollen, in a manner that reflects latitudinal shifts of the main northwest African vegetation zones during the Late Quaternary (Agwu and Beug, 1982). This core also reveals the properties of the prevailing trade winds and the African Easterly Jet or the Sahara Air Layer which occurs at higher altitudes and originates in the southern Sahara and blows westwards above the trade wind inversion. The core off Cap Blanc shows that there was a compressed Savanna belt which extended from about 12°N to 14-15°N during the period 19,000-14,000 yr B.P.

Maximum northward and southward expansion of the Sahara took place under hyperarid conditions when the belt of trade winds and the dominant African Easterly Jet transport had not shifted latitudinally. The trade winds were strong as in the modern situation, but around 13,000 yr B.P., the trade winds weakened. The climate became less arid south of the Sahara after 14,000 yr B.P., and a first spike of fluvial runoff in the core was registered around 13,000 yr B.P. Fluvial runoff increased strongly around 11,000 yr B.P. and maximum runoff is recorded from about 9,000-7,000 yr B.P. Around 12,500 yr B.P., the savanna belt started to shift northward and became richer in woody species. The core also reveals that the tropical forests had reached its maximum expansion and that the Guinea zone had migrated as far north as about 15° N, which reflects humid climatic conditions south of the Sahara.

Barusseau et al. (1988) presented a description and an interpretation of Recent sedimentary environments and depositional processes on the West African continental shelves (Fig.11.7) off Congo, Côte d'Ivoire and Senegal, and demonstrated the dominant effects of paleoclimatic and paleo-physiographic controls on the evolution of the Late Quaternary paleoenvironments. Recent shelf deposition of material in the region is controlled by climatic latitudinal gradients acting on the type, volume and distribution of terrigenous and carbonate sediments. From 18,000 yr B.P. up to the present time, major climatic changes combined with eustatic sea-level rise, controlled important variations in sedimentary conditions.

During the beginning of the Wisconsinian regression fluvial sands were emplaced in paleo-valleys which were incised on the exposed continental shelf. In the tropical regions of Mauritania and Senegal, aeolian dune sands formed during the arid "glacial" period (Ogalian) on the emerged shelf, but were destroyed by the subsequent transgression. Giresse et al. (1988) up-dated the chronostratigraphy of the emerged Senegalo-Mauritanian Quaternary shoreline sequences and correlated the Tafaritian with the oldest transgressive deposits (mid-Pleistocene), the Akcharian with the next regression, the Aioujian with a transgression at 40,000-30,000 yr B.P., followed later by the Ogalian and the Nouakchottian.

Near the Equator, on the shelf of Côte d'Ivoire aeolian input was reduced, but littoral dunes of that period occurred whose remnants may be observed close to the present shoreline. Fine sediments by-passed the shelf during the low stand of sea level and were deposited in deep-sea fans on the continental rise and on the abyssal plain. Shoreline stabilization during the stillstand of the Holocene transgression allowed the accumulation of littoral deposits (fine shoreline sands, dune sands, and lagoonal deposits with mangrove peat) along paleo-shorelines which are still visible today. Offshore from the equatorial river mouths, particularly the main ones such as the Zaire, pelitic sediments settled in morphological and structural lows. In the tropical regions of Senegal and Mauritania terrigenous fluvial input is considerably reduced but, in their northernmost parts, aeolian contribution of silts and very fine material is recorded in surficial sediments. Bioclastic carbonates formed during the first stillstand of the Holocene sea-level rise (12,000 yr B.P.). This is represented by a belt of *Amphistegina* sands distributed all along the outer shelf in West Africa. This relict fauna occurs between the 80- and 120-m bathymetric depth.

Off the coast of SW Africa, climate-induced changes in the flow pattern of the Benguela ocean current (Fig.11.4,A) are reflected in the abundance and preservation of opaline or siliceous microfossils and in the diatom species composition of deep-sea sediments. There is a downcore variation of opaline silica in an oceanic core (Fig.11.7) from the continental slope in which opal-rich sediments are found in oxygen isotope stages 1 and 3 (last interglacials), whereas during the peak of the last glacial period the opal content is very much reduced. The production of opal is controlled by the Benguela Current. At the point along the coast where the Benguela Current changes its flow direction and deviates westward, upwelling of nutrient-rich coastal waters takes place and supports high productivity among siliceous plankton which contributes rich and well-preserved opal flora (diatoms) to the sediments on the seafloor. Thus, during the last interglacial the Benguela Current deviated westward at 24° S where the studied core is located (Diester-Haass et al., 1988) and generated opal-rich sediments. But during the last glacial stage the current flowed for a longer distance northward before deflecting west, in which case nutrient-rich waters were not available at the location of the core (lat. 24° S), but further to the north.

11.5 Aspects of Human Origin

As recently observed by Simons (1989) hominid evolution has remained one of the most engaging fields of Quaternary research, not only because of its direct bearing on man's ancestry but because of the many tantalizing aspects of its fragmentary record. New discoveries from African hominid sites have continued to sustain our knowledge of hominid anatomical and technological developments and of man's earliest cultural traits. However, of considerable significance to the interpretations of the course of human evolution, are the applications of improved dating techniques and genetics.

Continuing research at the famous hominid-bearing Pliocene - Pleistocene Lake Turkana basin in Kenya, this time along the western shoreline, in deposits ranging between 1.0 and 3.5 Ma, has produced the most complete specimen of *Homo erectus* known so far. This has established the presence in East Africa of a hyper-robust australopithecine at about 2.5 million years ago (Harris et al., 1988). The earliest use of fire was recently discovered in the Swartkrans cave in South Africa by Brain and Sillen (1988). Brain and Sillen found burnt bones of antelopes, zebra, warthogs and baboons in association with *Australopithecus robustus*, in deposits dated between 1.0 and 1.5 Ma. However, it is not clear who "roasted" these animals, whether *Australopithecus* or *Homo*.

Simons' (1989) review of the history of hominid discoveries in Africa and their radiation into Europe and Asia, furnishes a masterly state-of-the-art account on paleoanthropology. Starting from Raymond Dart's earliest discoveries and christening of *Australopithecus* in South Africa in the 1920's, Simons showed how our knowledge of hominids then unfolded rapidly through the discoveries of *A. africanus*, *A. afarensis* and *A.*

boisei in Tanzania and Kenya, in sediments ranging in age from 3.6 to 1.6 m.y. Although their evolutionary relationships remain illusive, *Homo habilis* (sensu lato) and *Homo erectus* are the links between the australopithecines and *Homo sapiens*, but it was not until about 1.0 million years ago that *Homo erectus* made the initial hominid migration out of Africa into Europe and Asia. How modern man subsequently originated - whether by local differentiation of *H. erectus* wherever he went, or from the African parent stock (Fig.11.11) as advocated by Stringer (1988), Stringer and Andrews (1988), and Lewin (1988) - is controversial. Stringer and Andrews (1988) summarized modern genetic data, mainly nuclear and mitochondrial DNA, and argued in favour of a sub-Saharan origin of modern man, while Simons (1989) insists that the evidence is not yet sufficiently convincing to challenge *Homo sapiens neanderthalensis* (Fig.11.11) as the ancesor of modern man in Europe and Asia. However, recent evidence which remains recalcitrant and hard for the Simon school to dismiss are the thermoluminescence age dates from a cave site in Israel which point to the fact that early modern *Homo sapiens* had appeared in Africa and in the Middle East before 92,000 yr B.P., earlier than Neandertal man, who appeared around 60,000 yr B.P. This still seems to lend credence to the "out of Africa model" of the origin of modern humans (Stringer, 1988; Stringer and Andrews, 1988; Lewin, 1988).

11.6 Reflections on Contemporary Environmental Problems

From the aforegoing, it is evident that Africa is replete with a long and well-documented record of changing climatic scenarios during the Quaternary, with concomitant changes affecting landforms and vegetation. It is now widely believed that the iterative mechanisms that perpetuated the cyclical changes which we have reviewed were global and might have been driven by cosmic or astronomical phenomena (Mörner, 1984). But within the last 2,000 years, ever since man began to deforest the woodland belts of East Africa (Hamilton, 1982), the impact of global climatic change on Africa has been seriously overprinted, exacerbated, and prolonged by local human land degradation. Human action has accelerated the rate of environmental degradation through three interrelated ecological abuses, namely: deforestation, desertification, and soil erosion. These problems also pose serious threats to African man-made lakes, especially their capacity to supply hydro-electricity, and also sustain the increasing demand for irrigated agriculture to feed Africa's teeming and at best undernourished populations.

Among the human mechanisms for global climatic change are the increasing greenhouse effect of the abnormally high amounts of carbon dioxide, the impact of other gases such as nitrous oxide, methane and ammonia, the effects of sulfur, and the depletion of the ozone layer by the addition of chlorofluorocarbon. Although these factors are threatening the global environment, their immediate and potential impact are very severe on poor African nations whose marginal agricultural productivities are further worsened by drought. Since Africa has the least ability to combat natural disasters, any global ecological disaster will bring immediate and unmitigated catastrophe to the continent. For example, an unabated increase in global temperature would mean 6-8° rise in the temperature of the polar regions of the world (Woodwell, 1984). This would melt polar ice and release enough water into the oceans to cause a rise in sea-level which will drown all the coastlands of the world, including Africa.

The annual destruction of about 7 million hectares of African forests by burning (Burngh, 1984) does not only threaten global climate through the release of copious amounts of carbon dioxide, but it also has the immediate impact on Africa of accelerating desertification and soil erosion, which have already reached crises proportions in the Sahel region of Africa and on African mountains and highlands (Grosjean and Messerli, 1988). Ironically, as evident in eastern Nigeria, soil erosion is equally severe in the sandy wet coast lands in the equatorial belts of West Africa.

The solution to African ecological problems lies partly in soil and forest conservation through communal efforts, through integrated rural development following the models which have already succeeded in Kenya and Swaziland (Hudson, 1987), and through timely governmental and international intervention. However, in planning to mitigate the impact of natural disasters in Africa, the historical and socio-political contexts of Africa's environmental crises must be borne in mind. As observed in Glantz (1987): "...drought itself is not the fundamental problem in sub-Saharan Africa: After all, drought prevails in many parts of the world and, in affluent societies, need be no more than a nuisance. The real problem in Africa is poverty -- the lack of development -- the seeds of which lie in Africa's colonial past and in unwise policy choices made in the early days of independence by national governments and external aid donors."

References

Abell, P.I. 1982. Paleoclimate at Lake Turkana, Kenya, from oxygen isotope ratios of gatropod shells. Nature 297: 321-323.

Abel, P.I., McClory, J. 1987. Stable isotopes in the carbonates and kerogen from the Archaean stromatolites in Zimbabwe. Geol. Journ. vol. 22, Thematic Issue, 45-55.

Adegoke, O.S. 1969. Eocene stratigraphy of Southern Nigeria. Coll. I'Eocene, III, Mém. Bur. Réc. Géol. Min. 69: 23-48.

Adegoke, O.S., Adeleye, D.R., Odebode, M.O., Petters, S.W., Ejeaba, D.M. 1980. Excursion to the Shagamu quarry (Paleocene-Eocene). Geological Guide to some Nigerian Cretaceous - Recent localities, Spec. Pub. 2, Nig. Min. Geosc. Soc. 1-26.

Adeleye, D.R. 1976. The geology of the middle Niger Basin. In C.A. Kogbe (ed.), Geology of Nigeria, pp. 283-287. Lagos: Elizabethan Publishing Co.

Adighije, C. 1979. Gravity field of Benue Trough, Nigeria. Nature Phys. Sci. 282: 199-201.

Affaton, P., Sougy, J., Trompette, R. 1980. The tectono-stratigraphic relationships between the upper Precambrian and Lower Paleozoic Volta basin and the Pan-African Dahomeyide orogenic belt (West Africa). Amer. Jour. Sci. 280: 224-248.

Agagu, O.K., Adighije, C.I. 1983. Tectonic and sedimentation framework of the Lower Benue Trough, S.E. Nigeria. J. Afri. Earth Sci. 1: 267-274.

Agwu, C.O.C., Beug, H.G. 1982. Palynological studies of marine sediments off West African Coast. "Meteor" Forsch. Ergebnisse 36: 1-30.

Ajakaiye, D.E. 1976. A gravity survey over the Nigerian Younger Granite Province. In C.A. Kogbe (ed.), Geology of Nigeria, pp. 207-224. Lagos: Elizabethan Publ. Co.

Ajakaiye, D.E. 1981. Geophysical investigations in the Benue Trough - a review. Earth Evolution Sci. 1: 126-136.

Ajibade, A.C., Woakes, M., Rahaman, M.A. 1987. Proterozoic crustal development in the Pan-African regime of Nigeria. In A. Kröner, (ed.), Proterozoic lithospheric evolution, pp. 259-271. Geodynamic Series 17, American Geophysical Union.

Akande, S.O., Mucke, A. 1989. Mineralogical and paragenetic studies of the Lead-Zinc-Copper mineralization in the lower Benue Trough (Nigeria) and their genetic implications. J. Afr. Earth Sci. 9: 23-29.

Ako, J.A., Wellman, P. 1985. The margin of the West African craton; the Voltaian basin. J. Geol. Soc. London 142: 625-632.

Almond, D.C. 1969. Structure and metamorphism of the basement complex of north-east Uganda. Overseas Geol. Min. Res. 10: 146-163.

Almond, D.C. 1979. Younger-Granite complexes of Sudan. In S.A. Tahoun (ed.), Evolution and mineralization of the Arabian-Nubian Shield, pp. 151-164. Oxford: Pergamon.

Almond, D.C. 1984. The concepts of "Pan-African Episode" and "Mozambique Belt" in relation to the geology of east and north-east Africa. Bull. Fac. Earth Sci., King Abdul Aziz Univ., Jeddah 6: 71-78.

Almond, D.C., Ahmed, F. 1987. Ductile Shear Zones in the Northern Red Sea Hills, Sudan and their implications for crustal collision. Geol. Journ. vol. 22, Thematic Issue, 175-184.

Amajor, L.C. 1990. Petrography and provenance characteristics of Albian and Turonian sandstones, south-eastern Nigeria. In C.O. Ofoegbu (ed.), The Benue Trough: structure and evolution, pp. 39-57. Braunschweig: Vieweg.

American Geological Institute. 1972. Glossary of geology.

Andah, B.W. 1979a. The Early Palaeolithic in West Africa: The case of Asockrochona coastal region of Accra, Ghana. W. Afr. Jour. Archaeol. 9: 47-84.

Andah, B.W. 1979b. The Quaternary of the Guinea region of West Africa: An assessment of the geomorphic evidence. W. Afr. Jour. Archaeol. 9: 9-46.

Anderson, A.M., Biljon, A.M. van (eds.). 1979. Some sedimentary basins and associated ore deposits of South Africa. Spec. Publ. Geol. Soc. S. Afr. No.6.

Anderson, J.M., Anderson, H.M. 1984. The fossil content of the Upper Triassic Molteno Formation, South Africa. Palaeont. afr. 25: 39-59.

Anderson, L.S., Unrug, R. 1984. Geodynamic evolution of the Bangweulu block, northern Zambia. Precambr. Res. 25: 187-212.

Andreoli, M.A.G. 1984. Petrochemistry, tectonic evolution and metasomatic mineralisations of Mozambique belt granulites from S. Malawi and Tete (Mozambique). Precambr. Res. 25: 161-186.

Andrews-Speed, C.P. 1989. The mid-Proterozoic Mporokoso basin, northern Zambia: sequence stratigraphy, tectonic setting and potential for gold and uranium mineralization. Precambr. Res. 44: 1-7.

Anhaeusser, C.R. 1971. The geology and geochemistry of the Archaean granites and gneisses of the Johannesburg-Pretora dome. Inf. Circ. Econ. Geol. Res. Unit, Univ. Witwatersrand, No. 62.

Anhaeusser, C.R. 1976. Archaean metallogeny in Southern Africa. Econ. Geol. 71: 16-43.

Anhaeusser, C.R. 1985. Archaean layered ultramafic complexes in the Baberton Mountain Land, South Africa. In L.D. Ayres, P.C. Thurston, K.D. Card, W. Weber (eds.), Evolution of Archaean supracrustal sequences, Geol. Assoc. Can. Spec. Pap. 28: 281-234.

Anhaeusser, C.R., Button, A. 1976. A review of southern African stratiform ore deposits - their position in time and space. In K.H. Wolf (ed.), Handbook of strata-bound and stratiform ore deposits, II Regional studies and specific deposits, pp. 257-315. Amsterdam: Elsevier.

Anhaeusser, C.R., Mason, R., Viljoen, M.J., Viljoen, R.R. 1969. A reappraisal of some aspects of Precambrian shield geology: Geol. Soc. Amer. Bull. 80: 2175-2200.

Annels, A.E. 1984. The geotectonic environment of Zambian copper-cobalt mineralization. J. Geol. Soc. London, 141: 279-289.

Annor, A.E. 1983. Metamorphism of pelitic rocks in relation to deformation episodes around Okene, Nigeria. Nig. J. Min. Geol. 20: 17-24.

Avbovbo, A.A., Ayoola, E.O., Osahon, G.A. 1986. Depositional and structural styles in Chad basin of northeastern Nigeria. AAPG Bull. 70: 1787-1798.

Avery, D.M. 1988. Micromammals and paleoenvironmental interpretation in southern Africa. Geoarchaeology 3: 45-52.

Allen, J.R.L. 1965. Late Quaternary Niger Delta and adjacent areas: Sedimentary environments and lithofacies. AAPG Bull. 48: 547-600.

Badejoko, T.A. 1976. Role of adiabatic crystallization and progressive melting in the origin of the younger granites-Sara-Fier complex. In C.A. Kogbe (ed.), Geology of Nigeria, pp. 195-206. Lagos: Elizabethan Publ. Co.

Bafor, B.E. 1982. The Zungeru metavolcanics of north-western Nigeria: Their geochemistry and tectonic setting. Nig. Jour. Min. Geol. 18: 42-52.

Baker, B.H., Mitchell, J.G., Williams, L.A.J. 1988. Stratigraphic geochronology and volcano-tectonic evolution of the Kedong - Naivasha-Kinangop region, Gregory Rift Valley, Kenya. J. Geol. Soc. London 145: 107-116.

Barr, M.W.C. 1976. Crustal shortening in the Zambezi belt. Phil. Trans. R. Soc. A 280: 555-567.

Barr, M.W.C., Brown, M.A. 1987. Precambrian gabbro-anorthosite complexes, Tete Province, Mozambique. Geol. Journ. vol. 22, Thematic Issue, 139-159.

Barton, J.M. Jr., Key, R.M. 1981. The tectonic development of the Limpopo mobile belt and the evolution of the Archaean cratons of southern Africa. In A. Kroner (ed.), Precambrian plate tectonics, pp. 185-212. Amsterdam: Elsevier.

Barusseau, J.P., Giresse, P., Faure, H., Lezine, A.M., Masse, J.P. 1988. Marine Sedimentary environments on some parts of the tropical and equatorial Atlantic margins of Africa during the late Quaternary. Continental Shelf Res. 8: 1-21.

Baudet, D. 1987. Implications of a palynological study in the Upper Precambrian from eastern Kasai and northwestern Shaba, Zaire. Geol. Journ. vol. 22, Thematic Issue, 121-137.

Baumhaeur, R. 1987. Holocene limnic accumulations in the Great Erg of Bilma, N.E. Niger. 14th Colloquium of African Geology, Technische Universität Berlin, 18-22 August 1987 (Poster session).

Battail, B., Beltan, L., Dutuit, J.-M. 1987. Africa and Madagascar during Permo-Triassic time: The evidence of vertebrate faunas. In G.D. McKenzie (ed.), Gondwana Six: Stratigraphy, sedimentology, and paleontology, pp. 147-155. Geophysical Monograph 41. Washington, D.C.: American Geophysical Union.

Behrensmeyer, A.K. 1970. Preliminary geological interpretation of a new hominid site in the Lake Rudolf basin. Nature 226(5242): 225-226.

Behrensmeyer, A.K., Laporte, L.F. 1981. Footprints of a Pleistocene hominid in northern Kenya. Nature 289(5794): 167-169.

Bell, K., Dodson, M.H. 1981. The geochronology of the Tanzanian shield. J. Geol. Chicago 89: 109-128.

Bellini, E., Massa, D. 1980. A stratigraphic contribution to the Palaeozoic of the southern basins of Libya. In M.J. Salem, M.T. Brusrewil (eds.), The geology of Libya, pp. 3-56.

Benkhelil, J. 1989. The origin and evolution of the Cretaceous Benue Trough (Nigeria). J. Afr. Earth Sci. 8: 251-282.

Bensaid, M., Kutina, J., Mahmood, A., Saadi, M. 1985. Structural evolution of Morocco and new ideas on basement controls of mineralisation. Global Tectonics and Metallogeny 3: 59-69.

Beraki, W.H., Bonavia, F.F., Getachew, T., Schmerold, R., Tarekegn, T. 1989. The Adola fold and thrust belt, Southern Ethiopia: a re-examination with implications for Pan-African evolution. Geol. Mag. 126: 647-657.

Berhe, S.M. 1990. Ophiolites in northeast and East Africa: implications for Proterozoic crustal growth. J. geol Soc. London. 147: 41-57.

Berry, W.B.N., Wilde, P. 1978. Progressive ventilation of the oceans - an explanation for the distribution of the Lower Paleozoic black shales. Am. Journ. Sci. 178: 257-275.

Bertrand, J.-M., Davison, I. 1981. Pan-African granitoid emplacement in the Adrar des Iforas mobile belt (Mali): A Rb/Sr isotope study. Precambr. Res. 14: 333-361.

Bertrand, J.M.L., Boissonnas, J., Caby, R., Gravelle, M., Lelubre, M. 1966. Existence d'une discordance dans l'antécambrien du "fossé pharusien" de l'Ahaggar occidental (Sahara central). Acad. Sci. Paris C.R. (D), 262: 2197-2200.

Bertrand, J.M.L., Lasserre, M. 1976. Pan-African and pre-Pan-African history of the Hoggar (Algerian Sahara) in the light of new geochronological data from the Aleksod area. Precambr. Res. 3: 343-362.

Bertrand-Sarfati, J. 1972. Stromatolites columnaires du Précambrian supérieur du Sahara nord-occidental. Centre Rech. Zones Arides (Paris, CNRS), Sér. géol. 14.

Bertrand-Sarfati, J., Moussine-Pouchkine, A. 1988. Is cratonic sedimentation consistent with available models? An example from the Upper Proterzoic of the West African craton. Sed. Geol. 58: 255-276.

Bertrand-Sarfati, J., Walter, M.R. 1981. Stromatolite biostratigraphy: In R. Trompette, G.M. Young (eds.), Upper Precambrian correlations, pp. 353-371. Precambr. Res., Spec. Issue.

Besaire, H. 1972. Géologie de Madagascar I. Les terrains sédimentaires. Tananarive: Malagasy Imprimerie Nationale. Ann. Géol. Madagascar. No. 35.

Bessoles, B. 1977. Géologie de l'Afrique. Le craton Ouest African. Mém. Bur. Rech. géol. Min. Paris, 88.

Beukes, N.J. 1989. Sedimentological and geochemical relationships between carbonate, iron-formation and manganese deposits in the Early Proterozoic Transvaal Supergroup, Griqualand West, South Africa. Abstr. 28th IGC, I: 143.

Beydoun, Z.R. 1989. The hydrocarbon prospects of the Red Sea - Gulf of Aden - a review. J. Petrol. Geol. 12: 125-144.

Bickle, M.J., Martin, A., Nisbet, E.G. 1975. Basaltic and peridotitic komatiites and stromatolites above a basal unconformity in the Belingwe greenstone belt, Rhodesia. Earth planet. Sci. Lett. 27: 155-162.

Biju-Duval, B., Dercourt, I., Le Pichon, X. 1977. From Tethys Ocean to Mediterranean seas. In B. Biju-Duval, L. Montadert (eds.), Structural history of the Mediterranean basins, pp. 143-164. Paris: Editions Technip.

Binda, P.L. 1972. Preliminary observations on the palynology of the Precambrian Katanga sequence, Zambia. Geol Mijnbouw 51: 315-319.

Bishop, W.F. 1976. Geology of Tunisia and adjacent parts of Algeria and Libya. AAPG Bull. 59: 413-450.

Bishop, W.W. 1958. A review of the Pleistocene stratigraphy of the Uganda protectorate. C.C.T.A., Leopodiville: 91-105.

Black, R. 1984. The Pan-African event in the geological framework of Africa. Pangea p. 6-16.

Black, R., Caby, R., Moussine-Pouchkine, A., Bayer, R., Bertrand, J.M., Boullier, A.M., Fabre, J., Lesquer, A. 1979. Evidence for late Precambrian plate tectonics in West Africa. Nature 278: 223-227.

Black, R., Fabre, J. 1983. A brief outline of the geology of West Africa. In J. Fabre (ed.), West Africa: geological introduction and stratigraphic terms, pp. 17-26, Oxford: Pergamon.

Bloomer, S.H., Karson, J., Hefferan, K., Saquaque, A., Naidoo, D. 1989. Composition and origin of the Late Precambrian Bou Azzer ophiolite, Anti-Atlas, Morocco. Abst. 28th IGC 1: 161.

Boellstorff, J. 1978. North American Pleistocene stages reconsidered in light of probable Pliocene-Pleistocene continental glaciation. Science 202: 305-307.

Bond, G. 1967. A review of the Karoo sedimentation and lithology in southern Rhodesia. First Symposium on Gondwana Stratigraphy-proceedings and papers, 173-195.

Bonhomme, M.G., Gauther-Lafaye, F., Weber, F. 1982. An example of Lower Proterozoic sediments: The Francevillian in Gabon. Precambr. Res. 18: 87-102.

Bonnefille, R., Ridlet, G. 1988. The Kashiru pollen sequence (Burundi) paleoclimatic implications for the last 40,000 yr. B.P. in tropical Africa. Quat. Res. 30: 19-35.

Boullier, A.M., Davison, I., Bertrand, J.M., Coward, M. 1978. L'unité granulitique des Iforas: une nappe de socle d'âge Pan-African précoce. Rapp. d'activité Centre Géologique et Géophysique de Montpellier, 1977.

Bowen, D.Q. 1978. Quaternary geology. London: Pergamon.

Brain, C.K., Sillen, A. 1988. Evidence from the Swartkrans cave for the earliest use for fire. Nature 336(6198): 464-466.

Briden, J.C., Whitcombe, D.N., Stuart, G.W., Fairhead, J.D., Dorbath, C., Dorbath, L. 1981. Depth of geological contrast across the West African craton margin. Nature 292: 123-128.

Brock, A., Isaac, G.Ll. 1974. Paleomagnetic stratigraphy and chronology of hominid-bearing sediments east of Lake Rudolf, Kenya. Nature 247: 344-348.

Brun, A., Guerin, C., Levy, A., Riser, J., Rognon, P. 1988. Steppic environments at the end of the upper Pleistocene in southern Tunisia (Oued el Akarit). J. Afr. Earth Sci. 7: 969-980.

Brunet, M., Dejax, J., Brillanceau, A., Congleton, J., Downs, W., Duperon-Laudoueneix, M., Eisenmann, V., Flanagan, K., Flynn, L., Heintz, E., Hell, J., Jacobs, L., Jehenne, Y., Ndjeng, E., Mouchelin, G., Pilbeam, D. 1988. Mise en evidence d'une sédimentation précoce d'âge Barrémien dans le fossé de la Bénoué en Afrique occidentale (Basin du Mayo Oulo Léré, Cameroun), en relation avec l'ouveture de l'Atlantique Sud. C.R. Acad. Sci. Paris, t. 3067 Serie II: 1125-1130.

Buffetaut, E., Bussert, R., Brinkmann, W. 1990. A new nonmarine vertebrate fauna in the upper Cretaceous of northern Sudan. Berliner Geowiss. Abh. A, 120: 183-202.

Buffetaut, E., Taquet, P. 1979. An early Cretaceous terrestrial crocodile and the opening of the South Atlantic. Nature 280: 486-487.

Buggisch, W., Flugel, E. 1988. The Precambrian/Cambrian boundary in the Anti-Atlas (Morocco) discussion and new results. In Jacobshagen, V.H. (ed.), The Atlas system of Morocco, pp. 80-90. Berlin: Springer.

Bugrov, V., Efimov, A., Lawerman, J., Mboijana, S. 1982. Mineralization in the Karagwe-Ankolean metallogenic Zone of Southwestern Uganda. In The development potential of Precambrian mineral deposits, UN Dpt. Techn. Coop. Devel. 227-250.

Burke, K., Dewey, J.F. 1972. Orogeny in Africa. In T.F.J. Dessauvagie, A.J. Whiteman (eds.), African geology, pp. 583-608. Ibadan: University Press.

Burke, K., Dewey, J.F. 1973. An outline of Precambrian plate development. In D.H. Tarling, S.K. Runcorn (eds.), Implications of Continental drift to the earth sciences, vol. 2, pp. 1033-1046. New York: Academic Press.

Burke, K., Dewey, J.F., Kidd, W.S.F. 1977. World distribution of sutures - the sites of former oceans. Tectonophysics 40: 69-99.

Burke, K., Durotoye, A.B. 1972. The Quaternary in Nigeria. In T.F.J. Dessauvagie, A.J. Whiteman (eds.), African Geology, pp. 325-347. Ibadan: University Press.

Burke, K., Freeth, S.J., Grant, N.K. 1976. The sequence of geological events in the Basement Complex of the Ibadan area, Western Nigeria. Precambrian Res. 3: 537-545.

Burke, K., Kidd, W.S.F., Kusky, T.M. 1986. Archean foreland basin tectonics in the Witwatersrand, South Africa. Tectonics 5: 439-456.

Burke, K., Sengor, C. 1986. Tectonic escape in the evolution of continental crust. In M. Barazangi, L. Brown (eds.), Reflection seismology: The continental crust, pp. 41-53. Washington, D.C.: American Geophysical Union.

Burke, K.C. 1972. Longshore drift submarine canyons and submarine fans. AAPG Bull. 56: 1975-1983.

Burke, K.C. 1976. Neogene Quaternary tectonics of Nigeria. In C.A. Kogbe (ed.), Geology of Nigeria, pp. 363-369. Lagos: Elizabethan Publ. Co.

Burke, K.C., Kidd, W.S.F., Turcotte, D.L., Dewey, J.F., Mouginis-Mark, P.J., Parmentier, E.M., Sengor, A.M., Tappioner, P.E. 1981. Tectonics of basaltic volcanism. In Basaltic volcanism on the terrestrial planets, pp. 803-898. Basaltic Volcanism Study Project. New York: Pergamon.

Burngh, P. 1984. Organic carbon in soils of the world. In G.M. Woodwell (ed.), The role of terrestrial vegetation in the global carbon cycle: Measurement by remote sensing, pp. 91-107. U.K.: Wiley.

Burrollet, P.F. 1967. General geology of Tunisia. Guidebook to the geology and history of Tunisia, Petrol. Expl. Soc. Libya, Tripoli, 51-58.

Burrollet, P.F. 1989. North African epeiric basins. Abst. 28th IGC. 1: 217.

Caby, R. 1970. La chaine pharusienne dans le Nord-Ouest de l'Ahaggar (Sahara çentral, Algérie); sa place dans l'orogénèse du Précambrien supérieur en Afrique. Thèse d'Etat, Univ. Montpellier.

Caby, R. 1987. The Pan-African belt of West African from the Sahara desert to the Gulf of Benin. In J.P. Schaer, J. Rodgers (eds.), The anatomy of mountain ranges, pp. 129-170. Princeton: University Press.

Caby, R. 1989. Precambrian terranes of Benin-Nigeria and northeast Brazil and the Late Proterozoic South Atlantic fit. Geol. Soc. Amer. Spec. Pap. 230: 145-158.

Cadoppi, P., Costa, M., Sacchi, R. 1987. A cross-section of the Namama thrust belt (Mozambique). J. Afr. Earth Sci. 6: 493-504.

Cahen, L., Lepersonne, J. 1967. The Precambrian of the Congo, Rwanda and Burundi. In K. Rankama (ed.), The Precambrian, vol. 3, pp. 145-290, New York: Interscience.

Cahen, L. 1978. La stratigraphie et la tectonique du Supergroup Ouest-congolien dans les zones médiane et externe de l'orogène ouest-congolien (pan-Africain) au Bas Zaire et dans les régions voisines. Ann. Mus. R. Afri. Centr., Tervuren, Belg., 8vo, Sci. Géol., No. 83.

Cahen, L. 1982. Geochronological correlation of the late Precambrian sequences on and around the stable zones of Equatorial Africa. Precambr. Res. 18: 73-86.

Cahen, L. 1983a. Le groupe de Stanleyville (Jurassique supérieur et Wealdien de l'intérieur de la République du Zaire). Rapp. Annu. Mus. R. Afri. Centr. Tervuren (Belg.), Dept. Géol. Min. (1981-1982).

Cahen, L. 1983b. Brèves précisions sur l'âge des groupes cretaciques post-Wealdien (Loia, Bokungu, Kwango) du Bassin intérieur du Congo (République du Zaire). Mus. R. Afr. Centr. Tervuren (Belg.), Dept. Géol. Min. (1981-1982).

Cahen, L., Snelling, N.J., Delhal, T., Vail, J.R. 1984. The geochronology and evolution of Africa. London: Oxford Science.

Caincross, B. 1989. Tectono-sedimentary settings and controls of the Karoo basin Permian coals, South Africa. Abstr. 28th IGC 1:226.

Caire, A. 1973. Les liaison alpines précoces entre Afrique du Nord et Sicile et la place de la Tunisie dans l'arc tyrrhenien. Livre Jubilaire M. Solignac, Ann. Mines. Géol. Tunis, 26: 87-110.

Caire, A. 1974. Eastern Atlas. In A.M. Spencer, (ed.), Mesozoic-Cenozoic orogenic belts, pp. 4-59. Geol. Soc. London Spec. Publ. No. 4.

Caire, A. 1978. The central Mediterranean mountain chains in the Alpine orogenic environment. In A.E.M. Nairn, W.H. Kanes (eds.), The ocean basins and margins, pp. 201-256. New York: Plenum.

Carter, G.S., Bennett, J.D. 1973. The geology and mineral resources of Malawi (2nd revised edn.), Bull. Geol. Surv. Malawi, No. 6.

Carvalho, H. 1983. Notice explicative préliminaire sur la géologie de l'Angola. Garica de Orta Sér. Géol., Lisboa, 6: 15-30.

Carvalho, H., Crasto, J.P., Silva, Z.C.G., Vialette, Y. 1987. The Kibaran cycle in Angola - a discussion. Geol. Journ. vol. 22, Thematic Issue, 85-102.

Casanova, J. 1986. East African rift stromatolites. In L.E. Frostick et al. (eds.), Sedimentation in the African Rifts, pp. 201-210. Geol. Soc. Spec. Publ.

Censier, C. 1990. Characteristics of Mesozoic fluvio-lacustrine fromations of the western Central African Republic (Carnot Sandstones) by means of mineralogical and exoscopic analyses of detrital material. J. Afr. Earth Sci. 10: 385-398.

Chaloner, W.G., Creber, G.T. 1988. Fossil plants as indicators of late Palaeozoic plate positions. In M.G. Audley-Charles, A. Hallam (eds.), Gondwana and Tethys, pp. 201-210. Oxford: University Press.

Chorowicz, J., Fournier, J.L., Vidal, G. 1987. A model for rift development in Eastern Africa. Geol. Journ. vol. 22, Thematic Issue, 495-513.

Choubert, G. 1963. Histoire géologique de l'Anti-Atlas de l'Archéen à l'aurore des temps primaires. Thèse Doct. Sci., Paris. Notes Serv. Géol., Maroc, 162.

Choubert, G., Faure-Muret, A. 1974. Moroccan Rif. In A.M. Spencer (ed.), Mesozoic - Cenozoic orogenic belts, pp. 37-46. Geol. Soc. London Spec. Publ. 4.

Chumakov, N.M. 1981. Upper Proterozoic glaciogenic rocks and their stratigraphic significance. Precambr. Res. 15: 373-395.

Clark, J.D., Asfaw, B., Assefa, G., Harris, J.W.K., Kurashina, H., Walter, R.C., White, T.D., Williams, M.A.J. (1984). Palaeoanthropological discoveries in the middle Awash Valley Ethiopia. Nature 307: 423-428.

Clemmey, H. 1976. World's oldest animal traces. Nature 261: 576-578.

Clemmey, H. 1978. A Proterozoic lacustrine interlude from the Zambian Copperbelt. Spec. Publs. int. Ass. Sediment. 2: 259-278.

Clifford, A.C. 1986. African oil - past, present, and future. In M.T. Halbouty (ed.), Future petroleum provinces of the world. AAPG. Mem. 40: 339-373.

Clifford, T.N. 1966. Tectono-thermal units and metallogenic provinces of Africa. Earth Planet Sci. Lett. 1: 421-434.

Clifford, T.N. 1970. The structural framework of Africa. In T.N. Clifford, I.G. Gass (eds.), African magmatism and tectonics, pp. 1-23. Edinburgh: Oliver and Boyd.

Cloud, P. 1968. Pre-metazoan evolution and the origins of the metazoa. In E.T. Drake (ed.), Evolution and environment, pp. 1-72. New Haven: Yale Univ.

Coffin, M.F., Rabinowitz, P.D. 1988. Evolution of the conjugate East African-Madagascar margins and the western Somali basin. Geol. Soc. Amer. Bull. 226:87.

Cohen, A. 1981. Paleolimnological research at Lake Turkana, Kenya. In J.A. Coetzee, E.M. Van Zinderen Bakker (eds.), Palaeoecology of Africa, vol. 13, pp. 61-80. Rotterdam: A.A. Balkema.

Cohen, A.S. 1989. Facies relationships and sedimentation in large rift lakes and implications for hydrocarbon exploration examples from lakes Turkana and Tanganyika. Palaeogeogr., Palaeoclimatol., Palaeoecol. 70: 65-80.

Colbert, E.H. 1965. The age of reptiles. London: Weidenfeld & Nicolson.

Condie, K.C. 1981. Archean greenstone belts. Amsterdam: Elsevier.

Condie, K.C. 1989. Plate tectonics and crustal evolution. Oxford: Pergamon.

Conrad, J. 1985. Reggan basin. In R.H. Wagner, C.F.W. Prins, L.F. Granados (eds.), The Carboniferous of the world, pp. 322-323. IUGS Publication No. 20.

Coudé-Gaussen, G., Rognon, P. 1988. The Upper Pleistocene loess of southern Tunisia: a statement. Earth Surface Processes and Landforms 13: 137-151.

Coward, M.P. 1981. The Moabi syncline: the West Congolian of Gabon and it's tectonic significance. Ann. Soc. Géol. Belg. 104: 255-259.

Coward, M.P. 1983. The tectonic history of the Damara belt. Spec. Publ. Geol. Soc. S. Afr. 11: 409-421.

Coward, M.P., Daly, M.C. 1984. Crustal lineaments and shear zones in Africa: their relationship to plate movements. Precambr. Res. 24: 27-45.

Coward, M.P., Lintern, B.C., Wright, L.I. 1976. The pre-cleavage deformation of the sediments and gneisses of the northern part of the Limpopo belt, southern Africa. In B.F. Windley (ed.), Early history of the earth, pp. 323-330, London, Wiley.

Cox, K.G. 1972. Karoo lavas and associated igneous rocks of southern Africa. Bull. volcanol. 35: 867-886.

Crowell, J.C. 1983. The recognition of ancient glaciations. Geol. Soc. Amer. Mem. 61: 289-297.

Cruickshank, A.R.I. 1978. Feeding adaptations in Triassic dicynodonts. Palaeont. afr. 21: 121-132.

Culver, S.J., Williams, H.R., Bull, P.A. 1978. Infracambrian glaciogenic sediments from Sierra Leone. Nature 274: 49-51.

Curie, P.J. 1981. *Hovasaurus boulei*, an aquatic eosuchian from the upper Permian of Madagascar. Palaeont. afr. 24: 99-167.

Dallmeyer, R.D. 1989. Contrasting accreted terranes in the southern Appalachian orogen and Atlantic-Gulf Coastal Plains and their correlations with West African sequences. Geol. Soc. Amer. Spec. Pap. 230: 247-267.

Daly, M.C. 1986a. The intracratonic Irumide belt of Zambia and its bearing on collision orogeny during the Proterozoic of Africa. In M.P. Coward, A.C. Ries (eds.), Collision tectonics, pp. 321-328. Geol. Soc. Spec. Pub. 19.

Daly, M.C. 1986b. Crustal shear zones and thrust belts: their geometry and continuity in central Africa. Phil. Trans. R. Soc. A317: 111-128.

Daly, M.C. 1988. Crustal shear zones in central Africa: a kinematic approach to Proterozoic tectonics. Episodes 11: 5-11.

Daly, M.C., Chakraborty, S.K., Kasolo, P., Musiwa, M., Mumba, P., Naidu, D., Namateba, C., Ng'ambi, O., Coward, M.P. 1984. The Lufilian arc and Irumide belt of Zambia: results of a geotraverse across their intersection. J. Afr. Earth Sci. 2: 311-318.

Daly, M.C., Unrug, R. 1982. The Muva Supergroup of northern Zambia: A craton to mobile belt sedimentary sequence. Trans. geol. Soc. S. Afr. 85: 155-165.

Darracott, B.W. 1975. The interpretation of the gravity anomaly over the Barberton Mountain Land, South Africa. Geol. Soc. S. Afr. Trans. 78: 123-128.

Davison, I. 1980. Structural geology and Rb/Sr geochronology of Precambrian rocks from eastern Iforas (Mali). Ph.D. thesis, Univ. Leeds.

Deacon, J., Lancaster, N. 1988. Late Quaternary palaeoenvironments of southern Africa. Oxford: Clarendon.

De Decker, R.H. 1989a. Geological setting of diamondiferous inner shelf of Cape west coast, South Africa. Abstr. 28th IGC 1: 374-375.

De Decker, R.H. 1989b. Model for the evolution of Orange delta. Abstr. 28th IGC 1: 375-376.

De Jager, D.H. 1988. Phosphate resources in the Palabora Igneous Complex, Transvaal, South Africa. In A.J.G. Notholt, R.P. Sheldon, D.F. Davidson (eds.), Phosphate deposits of the world, vol. 2, pp. 267-272. Cambridge: University Press.

De Klasz, I. 1978. The West African sedimentary basins. In M. Moullade, A.E.M. Nairn (eds.), The Phanerozoic geology of the world II, The Mesozoic, A. pp. 371-399. Amsterdam: Elsevier.

de Kun, N. 1987. Mineral economics of Africa. Amsterdam: Elsevier.

de la Boisse, H. 1981. Sur le métamorphisme du micaschiste eclogitique du Takamba (Mali) et ses conséquences paléogéodynamiques au Précambrien supérieur. Soc. géol. France C.R. Somm. 1981, 97-100.

Destombes, J., Hollard, H., Willefert, S. 1985. Lower Paleozoic rocks of West Africa and the western part of central Africa. In C.H. Holland (ed.), Lower Paleozoic rocks of the world, 4, pp. 91-336. Chichester: Wiley.

De Swardt., A.M.J., Drysdall, A.R. 1964. Precambrian geology and structure in northern Rhodesia. Mem. Geol. Surv. Northern Rhodesia, No. 2.

Deynoux, M. 1983. Late Precambrian and upper Ordovician glaciations in the Taoudeni basin, West Africa. In M. Deynoux (ed.), West African palaeoglaciations: characterization and evolution of glacial phenomena through space and time, pp. 43-86. Abstr. Symposium Till Mauretania 83.

Deynoux, M., Sougy, J., Trompette, R. 1985. Lower Palaeozoic rocks of west Africa and the western part of central Africa. In C.H. Holland (ed.), Lower Palaeozoic of north-western and west-central Africa, pp. 337-495. New York: Wiley.

Deynoux, M., Trompette, R. 1981. Late Precambrian tillites of the Taoudeni basin, West Africa. In M.J. Hambrey, W.B. Harland (eds.), Earth's Pre-Pleistocene glacial record, pp. 123-131. Cambridge: University Press.

Deynoux, M., Trompette, R., Clauer, N., Sougy, J. 1978. Upper Precambrian and lower most Palaeozoic correlations in West Africa and the western part of central Africa. Probable diachronism of the late Precambrian tillite. Geol. Rundsch. 67: 615-630.

Dewey, J.F., Burke, K.C.A. 1973. Tibetan Variscan and Precambrian basement reactivation: products of continental collision. J. Geol. 81: 683-692.

Diester-Hass, L., Heine, K., Rothe, P., Schrader, H. 1988. Late Quaternary history of continental climate and the Benguela current off south west Africa: Palaeogeogr. Palaeoclimatol., Palaeoecol. 65: 81-91.

Dillon, W.P., Sougy, J.M.A. 1974. Geology of West Africa and Canary and Cape Verde Islands. In A.E.M. Nairn, F.G. Stehli (eds.), The ocean basins and margins, vol. 2, The North Atlantic, pp. 315-390. New York; Plenum.

Dingle, R.V. 1978. South Africa. In M. Moullade, A.E.M. Nairn (eds.), The Phanerozoic geology of the world II, The Mesozoic, A., pp. 401-434. Amsterdam: Elsevier.

Dingle, R.V. 1982. Continental margin subsidence: A comparison between the east and west coasts of Africa. In R.A. Scruton (ed.), Dynamics of passive margins, pp. 59-71. Geodynamics Ser. 6. Washington, D.C.: American Geophysical Union/G.S.A.

Drysdall, A.R., Johnson, R.L., Moore, T.A., Thieme, J.G. 1972. Outline of the geology of Zambia. Geol. Mijnbouw 51: 265-276.

Dualeh, A.H.A., Reuther, C.-D., Scheck, P. 1990. Basement structure and sedimentary cover of Somalia. Berliner geowiss. Abh. A., 120: 505-518.

Dunelevey, J.N. 1988. Evolution of the Saldanhian orogeny and development of the Cape granite suite. South African Jour. Sci. 84: 565-568.

Durand-Delga, M., Olivier, Ph. 1988. Evolution of the Alboran block margin from early Mesozoic to Early Miocene time. In V.H. Jacobshagen (ed.), The Atlas system of Morocco, Lecture Notes in Earth Sciences, 15: 465-480 Berlin: Springer.

Du Toit, A.L. 1937. Our wandering continents. Edinburgh: Oliver and Boyd.

Ekwere, S.J. 1982. Aspects of economic geology and geochemistry of Banke ring complex. Nig. Jour. Min. Geol. 19: 62-69.

Ekwueme, B.N., Schlag, C. 1989. Compositions of monazites in pegmatites and related rocks of the Oban massif SE Nigeria: implications for economic mineral exploration. IGCP, No. 255 Newsletter/Bulletin 2: 15-20.

El Bayoumi, R.M.A. 1984. Ophiolites and melange complex of Wadi Ghadir, Eastern Desert, Egypt. Bull. Fac. Earth Sci., King Abdul Aziz Univ., Jeddah 6: 324-329.

El Gaby, S., Greiling, R.O. (eds.). 1988. The Pan-African belt of northeastern Africa and adjacent areas. Braunschweig: Vieweg.

El Gaby, S., List, F.K., Tehrani, R. 1988. Geology evolution and metallogenesis of the Pan-African belt in Egypt. In S. El Gaby, R.O. Greiling (eds.), The Pan African belt of north-east Africa and adjacent areas, pp. 17-68. Braunschweig: Vieweg.

El Ramly, M.F.R., Greiling, R.O., Kröner, A., Rashwan, A.A.A. 1984. On the tectonic evolution of the Wadi Hafafit area and environs, Eastern Desert of Egypt. Bull. Fac. Earth Sci., King Abdul Aziz Univ., Jeddah 6: 113-126.

Elueze, A.A. 1982. Mineralogy and chemical nature of meta-ultramafites in Nigerian schist belts. Nig. J. Min. Geol. 19: 21-29.

Elueze, A.A. 1986. Petrology and gold mineralization of the amphibolite belt. Ilesha area, southwestern Nigeria. Geol. Mijnbouw 65: 189-195.

Emery, K., Uchupi, O.E., Phillips, J., Bowin, C., Mascle, J. 1975. Continental margin off West Africa: Angola to Sierra Leone. AAPG Bull. 59: 2209-2265.

Eriksson, K.A., Kidd, W.S.F., Krapez, B. 1988. Basin analysis in regionally metamorphosed and deformed Early Archaean terrains: Examples from southern Africa and Western Australia. In K.L. Kleinspelin, C. Paola (eds.), New perspectives in basin analysis, pp. 371-404. New York: Springer.

Ernst, G., Gierlowski-Kordesch, E. 1989. Epeiric sea deposits and graphoglytid burrows in Upper Cretaceous of Tanzania. Abstr. 28th IGC 1: 460.

Evamy, B.D., Haremboure, J., Kammerling, R., Knaap, W.A., Molloy, F.A., Rowlands, P.H. 1978. Hydrocarbon habitat of Tertiary Niger Delta. AAPG Bull. 62: 1-39.

Fabre, J. 1982. Pan-African volcano-sedimentary formations in the Adra des Iforas (Mali). Precambr. Res. 19: 201-214.

Fabre, J., Petit-Maire, N. 1988. Holocene climatic evolution at 22-23° N from two paleolakes in the Taoudenni area (northern Mali). Palaeogeogr., Palaeoclimatol., Palaeoecol. 65: 133-148.

Fairhead, J.D., Green, C.M. 1989. Controls on rifting in Africa and the regional tectonic model for the Nigeria and East Niger rift basins. J. Afr. Earth Sci. 8: 231-249.

Fairhead, J.D., Henderson, N.B. 1977. The seismicity of southern Africa and incipient rifting. Tectonophysics 41: T19-T26.

Faniran, A., Jeje, L.K. 1983. Humid tropical geomorphology. London: Longman.

Fayose, E.A. 1970. Stratigraphical palaeontology of Afowo 1 well, south western Nigeria. Nig. Jour. Min. Geol. 5: 1-99.

Fedonkin, M.A. 1990. Precambrian metazoans. In D.E.G. Briggs, P.R. Crowther (eds.), Palaeobiology: a synthesis, pp. 17-24. Oxford: Blackwell.

Flynn, L.J., Brillanceau, A., Bruet, M., Coppens, Y., Dejax, J., Duperon-Laudoueneix, M., Ekodeck, G., Flanagan, K., Heintz, E., Hell, J., Jacobs, L.L., Pilbeam, D.R., Sen, S., Djallo, S. 1987. Vertebrate fossils from Cameroon, West Africa. Jour. Vert. Paleont. 7: 469-471.

Fontes, J.C., Gasse, F. 1989. On the ages of humid Holocene and Late Pleistocene phases in North Africa - Remarks on "Late Quaternary climatic reconstruction for the Maghreb (North Africa)" by P. Rognon. Palaeogeogr., Palaeoclimatol., Palaeoecol. 70: 393-398.

Foster, R.P., Gilligan, J.M. 1987. Archaean iron-formation and gold mineralization in Zimbabwe. In P.W.U. Appel, G.L. Laberge (eds.), Precambrian iron-formations, pp. 635-665. Athens: Theophrastus.

Foucault, A., Stanley, D.J. 1989. Late Quaternary paleoclimatic oscillations in East Africa recorded by heavy minerals in the Nile Delta. Nature 339: 44-46.

Franssen, L., André, L. 1988. The Zadinian Group (Late Proterozoic, Zaire) and it's bearing on the origin of the West-Congo orogenic belt. Precambr. Res. 38: 215-234.

Frazier, W.J., Schwimmer, D.R. 1987. Regional stratigraphy of North America. New York: Plenum.

Franks, S., Nairn, A.E.M. 1973. The equatorial marginal basins of West Africa. In A.E.M. Nairn, F.G. Stehli (eds.), The ocean basins and margins, pp. 301-350. New York: Plenum.

Freeth, S.J. 1984. The origin of the Ibadan granite gneiss: an example of large scale local segregation. In J. Klerkx, J. Michot (eds.), African geology, pp. 179-189. Tervuren.

Freeth, S.J. 1987. The Lake Nyos gas disaster - What happened, why did it happen and will it happen again? In G. Matheis, H. Schandelmeier (eds.), Current research in African earth sciences, pp. 265-269. Rotterdam: A.A. Balkema.

Fripp, R.E. 1976. Gold metallogeny in the Archaean of Rhodesia. In B.F. Windley (ed.), The early history of the earth, pp. 455-466. London: Wiley-Interscience.

Fripp, R.E.P. 1983. The Precambrian geology of the area around the Sand River near Messina, central zone, Limpopo mobile belt. In W.J. van Biljon, J.H. Legg (eds.), The Limpopo belt, pp. 89-102. Geol. Soc. S. Afr. Pub. 8.

Froitzheim, N., Stets, J., Wurster, P. 1988. Aspects of western High Atlas tectonics. In V.H. Jacobshagen (ed.), The Atlas system of Morocco, Lecture Notes in Earth Sciences, 15: 219-245. Berlin: Springer.

Frostic, L., Reid, I. 1989. Is structure the main control of river drainage and sedimentation in rifts? J. Afr. Earth Sci. 8: 165-182.

Furon, R. 1963. Geology of Africa. New York: Hafner.

Gabert, G. 1984. Structural-lithological units of Proterozoic rocks in East Africa, their base, cover and mineralisation. In J. Klerkx, J. Michot (eds.), African geology, pp. 11-21 Tervuren.

Gabert, G. 1990. Lithostratigraphic and tectonic setting of gold mineralization in the Archean cratons of Tanzania and Uganda, East Africa Precambr. Res., 46: 59-69.

Gabriel, B., Kropelin, S. 1984. Holocene lake deposits in northwest-Sudan. Palaeoecol. Afr. 16: 295-299.

Garlick, W.G., Fleischer V.D. 1972. Sedimentary environment of Zambia copper deposition. Geologie Mijnbouw 51: 277-298.

Garson, M.S., Shalaby, I.M. 1976. Precambrian-lower Palaeozoic plate tectonics and metallogenesis in the Red Sea region. Spec. Pap. geol. Ass. Can. 14: 537-596.

Gasse, F., Fontes, J.C. 1989. Palaeoenvironments and palaeohydrology of a tropical closed lake (Lake Asal, Djibouti) since 10,000 yr. B.P. Palaeogeogr., Palaeoclimatol, Palaeoecol. 69: 67-102.

Gavaud, M. 1972. Les grandes divisions du Quaternaires des regions ouest-Africaines établies sur des bases pédologiques. In T.F.J. Dessauvagie, A.J. Whiteman (eds.), African geology, pp. 395-412. Ibadan: University Press.

Ghuma, M.A., Rodgers, J.W. 1978. Geology, geochemistry and tectonic setting of the Ben Ghnema batholith, Tibesti massif, southern Libya. Bull. geol. Soc. Am. 89: 1351-1358.

Giresse, P., Mohamadoul, D., Barusseau, J.P. 1988. Litholigical mineralogical and geochemical observations of Senegalo-Mauritanian Quaternary shoreline deposits: possible chronolgical revision. Palaeogeogr., Palaeoclimatol., Palaeoecol. 68: 241-257.

Glantz, M.H. (ed.). 1987. Drought and hunger in Africa: Denying famine a future. Cambridge: University Press.

Gliksson, A.Y. 1972. Early Precambrian evidence of a primitive ocean crust and island nuclei of sodic granite. Bull. geol. Soc. Amer. 83: 3323-3344.

Goossens, P.J. 1983. Precambrian mineral deposits and their metallogeny. Ann. Mus. R. Afr. centr., Tervuren Belg., in 8 vo, Sci géol., no. 89.

Gorler, K., Helmdach, F.F., Gaemers, P., Heissig, K., Hinsch, W., Mädler, K., Schwarzhands, W., Zucht, M. 1988. The uplift of the central High Atlas as deduced from Neogene continental sediments of the Ouarzazate Province, Morocco. In V.H. Jacobshagen (ed.), The Atlas System of Morocco, Lecture Notes in Earth Sciences, 15: 361-404. Berlin: Springer.

Goudarzi, G.H. 1970. Geology and mineral resources of Libya - A reconnaissance. U.S. Geol. Surv., Prof. Pap. 660.

Grant, N.K. 1973. Orogeny and reactivation to the west and south-east of the West African craton. In A.E.M. Nairn, F.G. Stehli (eds.), The Ocean basins and margins, vol. 1, The South Atlantic, pp. 447-492. New York: Plenum.

Greiling, R.O., Kröner, A., El Ramly, M.F., Rashwan, A.A. 1988. Structural relationships between the southern and central parts of the eastern desert of Egypt: details of a fold and thrust belt. In S.El. Gaby, R.O. Greiling (eds.), The Pan-African belt of northeast Africa and adjacent areas pp. 121-146. Braunschweig: Vieweg.

Grosjean, M., Messerli, B. 1988. African mountains and highlands: potentials and constraints. Mountain Research and Development 8: 111-122.

Grotinger, J.P. 1989. Facies and evolution of Precambrian carbonate depositional systems: Emergence of the modern platform and basin development, SEPM Spec. Publ. 44: 79-103.

Guerrak, S. 1989. Time and space distribution of Palaeozoic oolitic ironstones in the Tindouf basin, Algerian Sahara. In T.P. Young, W.E.G. Taylor (eds.), Phanerozoic ironstones, pp. 197-212. Geol. Soc. Spec. Publ. no. 46.

Guiraud, R., Bellion, Y., Benkhelil, J., Moreau, C. Post-Hercynian tectonics in northern and Western Africa. Geol. Journ. vol. 22. Thematic Issue. 422-466.

Guilbert, J.M., Park, C.F. 1986. The geology of ore deposits. New York: W.H. Freeman.

Halstead, L.B. 1974. *Sokotosuchus ianwilsoni* n.g., n.sp. A new teleosaur crocodile from the Upper Cretaceous of Nigeria. Nig. Jour. Min. Geol. II: 101-103.

Halstead, L.B. 1975. The evolution and ecology of the dinosaurs. U.K.: Peter Lowe.

Halstead, L.B., Halstead, J. 1981. Dinosaurs. Poole: Blanford.

Hambrey, M.J. 1983. Correlation of late Proterozoic tillites in the North Atlantic region and Europe. Geol. Mag. 120: 209-232.

Hamilton, A.C. 1982. Environmental history of East Africa. London: Academic.

Hargraves, R.B., Onstott, T.C. 1987. Africa in the framework of Gondwana: Paleomagnetism and global tectonics. 14th Colloq. Afr. Geol. Berlin. Abstracts, CIFEG Ocassional Publication 1987/12: 35-36.

Harland, W.B. 1983. The Proterozoic glacial record. Geol. Soc. Amer. Mem. 61: 279-289.

Harland, W.P., Cox, A.V., Llewellyn, P.G., Pickton, C.A.G., Smith, A.G., Walters, R. 1982. A geologic-time scale Cambridge: University Press.

Harris, J.F. 1981. Summary of the geology of Tanganyika. Part IV: Economic Geology. Mineral Resources Division, Dodoma, Tanzania.

Harris, J.M., Brown, F.H., Leakey, M.G., Walker, A.C., Leakey, R.E. 1988. Pliocene and Pleistocene homind-bearing sites from west of Lake Turkana, Kenya. Science 239: 27-33.

Harrison, N.M. 1970. The geology of the country around Que Que, Rhodesia. Bull. Geol. Survey S. Rhodesia, 67.

Hartnady, C., Joubert, P., Sowe, C. 1985. Proterozoic crustal evolution in southwestern Africa. Episodes 8: 236-244.

Hassan, M.A., Hashad, A.H. 1990. Precambrian of Egypt. In R. Said (ed.), The geology of Egypt, pp. 201-245. Rottderdam: A.A. Balkema.

Haughton, S.H. 1963. The stratigraphic history of Africa South of the Sahara. Edinburgh: Oliver and Boyd.

Haughton, S.H. 1969. Geological history of southern Africa. Jonannesburg: Geol. Soc. S. Afr.

Hays, R.L. 1976. Geology of the Olduvai gorge. Berkeley: University of California.

Helmstaedt, H., Gurney, J.J. 1984. Kimberlites of southern Africa - Are they related to subduction processes? In J. Kornprobst (ed.), Kimberlites I: Kimberlites and related rocks, pp. 425-434. Amsterdam: Elsevier.

Hepworth, J.V. 1964. Explanation of the geology of sheets 19, 20, 28 and 29 (Southwest West Nile). Geol. Surf. Uganda, Report No. 10.

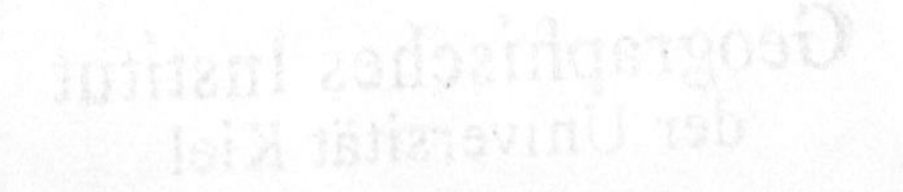

Hepworth, J.V. 1972. The Mozambique orogenic belt and its foreland in north-east Tanzania: a photogeologically-based study, J. geol. Soc. London 128: 461-500.

Hill, R.S. 1988. Quarternary faulting in the south-eastern Cape province. S. Afr. Jour. Geol. 91: 399-403.

Hillaire-Marcel, C., Riser, J., Rognon, P., Petit-Marie, N., Rosso, J.C., Soulie-Marche, I. 1983. Radiocarbon chronology of Holocene hydrologic changes in northeast Mali. Quat. Res. 20: 145-164.

Hiller, N., Theron, J.N. 1988. Benthic communities in the South African Devonian. In N.J. McMillan, A.F. Embry, D.J. Glass (eds.), Devonian of the world, vol.III, pp. 229-242. Can. Soc. Petrol. Geol.

Holmes, A. 1951. The sequence of Pre-cambrian orogenic belts in south and central Africa. 18th Int. Geol. Cong., London (1948) 14: 254-269.

Holt, R., Egbuniwe, I.G., Fitches, W.R., Wright, J.B. 1978. The relationships between low-grade metasedimentary belts, calc-alkaline volcanism and the Pan-African orogeny in N.W. Nigeria. Geol. Rdsch. 67: 631-646.

Hooghiemstra, H. 1988. Changes of major wind belts and vegetation zones in NW Africa 20,000 - 5,000 yr. B.P., as deduced from a marine pollen record near Cap Blanc. Rev. Palaeobot, Palynol. 55: 101-140.

Hospers, J. 1965. Gravity field and structure of the Niger Delta, Nigeria, West Africa. Bull. Geol. Soc. Amer. 76: 407-422.

Hottin, G. 1972. Geological map of Madagascar, scale, 1: 2,00,000 Bur. Rech. Géol. Min. Paris.

Hottin, G. 1976. Présentation et essai d'interpretation du Précambrien de Madagascar. Bull. Bur. Rech. géol. Min Paris, 2e Série. 4, No.2: 117-153.

Howell, F.C. 1978. Homidae. In V.J. Maglio, H.B.S. Cooke (eds.), Evolution of African mammals, pp. 154-234. Cambridge: Havard University.

Howell, F.C., Coppens, Y. 1976. An overview of Hominidae from the Omo succession, Ethiopia. In Y. Coppens, F.C. Howell, Isaac G. Ll., Leakey, R.E.F. (eds.), Earliest man and environments in the Lake Rudolf basin, pp. 522-532. Chicago: University Press.

Hsü, K.J., Cita, M.B., Ryan, W.B.F. 1973. The origin of the Mediterranean evaporites. In W.B.F. Ryan, K.J. Hsü (eds.), Initial Reports of the Deep Sea Drilling Project, 13/(2), pp. 1011-1099. Washington, D.C.: US Government Printing Office.

Hubbard, F.H. 1975. Precambrian crustal development in western Nigeria: indications from the Iwo region. Bull. geol. Soc. Am. 86: 548-554.

Hudson, N.W. 1987. Limiting degradation caused by soil erosion. In M.G. Wolman, F.G.A. Fournier (eds.), Land transformation in agriculture, pp. 153-169. U.K.: Wiley

Hurley, P.M., Leo, G.W., White, R.W., Fairbairn, H.W. 1971. Liberian age province (about 2700 m.y.) and adjacent provinces in Liberia and Sierra Leone. Geol. Soc. Amer. Bull. 82: 3483-3490.

Hurley, P.M., Rand, J.R. 1968. Review of age data in West Africa and South America relative to a test of continental drift. In R.A. Phinney (ed.), The history of the Earth's crust, pp. 153-160. Princeton: University Press.

Hurley, P.M., Rand, J.R. 1973. Outline of Precambrian chronology in lands bordering the South Atlantic, exclusive of Brazil. In A.E.M. Nairn, F.G. Stehli (eds.), The ocean basins and margins, vol. 1, pp. 391-410. New York: Plenum.

Hussein, A.A. 1990. Mineral deposits. In R.Said (ed.), The geology of Egypt, pp. 511-566. Rotterdam: A.A. Balkema.

Hutchison, C.S. 1983. Economic deposits and their tectonic setting. London: Macmillan.

Ike, E.C. 1983. The structural evolution of Tibchi ring complex - a case study for the Nigerian Younger Granite province. J. Geol. Soc. 140: 781-788.

Imeokparia, F.G. 1982. Geochemical relationships to mineralization of granitic rocks from the Afu granite complex, central Nigeria. Geol. Mag. 119: 39-56.

Ivanhoe, L.F. 1980. World's giant petroleum provinces. Oil Gas J. 30 June.

Jackson, N.J. 1980. Correlation of late Proterozoic stratigraphies, NE Africa and Arabia: Summary of an IGCP Project 164 Report. J. geol. Soc. London 137: 629-634.

Jackson, N.J. 1987. Precambrian evolution of northeast Africa and Arabia. 14th Colloq. Afr. Geol. Berlin, Abstracts, CIFEG Occasional Publication 1987/12; 19-23.

Jacobs, L.L., Congleton, J.D., Brunet, M., Dejax, J., Flynn, L.J., Hell, J.V., Mouchelin, G. 1988. Mammal teeth from the Cretaceous of Africa. Nature 336 (6195): 158-160.

Jacobs, L.L., Flanagan, K.M., Brunet, M., Flynn, L.J., Dejax, J., Hell, J.V. 1989. Dinosaur footprints from the Lower Cretaceous of Cameroon, West Africa. Proceedings of the First International Symposium on Dinosaur tracts and traces, pp. 349-351. Cambridge: University Press.

Jacobshagen, V.H. (ed.). 1988. The Atlas system of Morocco. Lecture Notes in Earth Sciences, 15. Berlin: Springer.

Jacobson, R.R.E., MacLeod, W.N., Black, R. 1958. Ring-complexes in the Younger Granite province of northern Nigeria. Geol. Soc. Lond. Memoir No. 1.

James, H.L. 1978. Subdivisions of the Precambrian - a brief review and a report on recent decisions by the Subcommission of Precambrian Stratigraphy. Precamb. Res. 7: 193-204.

Jansen, J.H.F., Van Weering, T.C.E., Grieles, R., Iperen, J. 1984. Middle and Late Quaternary oceanography and climatology of the Zaire-Congo fan and the adjacent eastern Angola basin. Netherlands Jour. Sea Res. 17: 201-249.

Johnson, J.G., Boucot, A.J. 1973. Devonian brachiopods. In A. Hallam (ed.), Atlas of palaeobiogeography, pp. 89-96. Amsterdam: Elsevier.

Kabengele, M., Kapenda, D., Tshimanga, K. 1989. Ubendian (Early Proterozoic) mafic rocks of Marungu plateau magmatic complex (northeastern Shaba province, Zaire): New petrological and geochemical data and geodynamical significance. Abstr. 28th 1GC 2: 143.

Kamen-Kaye, M. 1978. Permian to Tertiary Faunas and paleogeography: Somalia, Kenya, Tanzania, Mozambique, Madagascar, South Africa. Journ. Petrol. Geol. 1: 79-101

Kampunzu, A.B., Caron, J.P.-H., Kanika, M., Lubala, R.T. 1985. Evolution du rifting intracontinental et magmas tholéiitique, transitionnel et alcalin associés: cas du volcanisme phanerozoique du rift Est-african. 13th Coll. African Geol. St. Andrews-Scotland, Occasional Publication 3, CIFEG-Paris, 174-175.

Kampunzu, A.B., Vellutini, P.J., Caron, J.P.-H., Lubala, R.T., Kanika, M., Rumvegeri, B.T. 1983. Le volcanisme et l'évolution structurale du Sud-kivu (Zaire). Un modèle de interprétation géodynamique du volcanisme distensif intracontinental. Bull. Cent. Rech. Explor., Prod. Elf-Aquitaine 7: 257-271.

Kapilima, S. 1989. Depositional environment of the Ilima coal basin, Tanzania. Abstr. 28th IGC. 2: 155.

Karson, J.A., Curtis, P.C. 1989. Tectonic and magmatic processes in the Eastern Branch of the East African Rift and implications for magmatically active continental rifts. J. Afr. Earth Sci. 8: 431-453.

Kazmin, V., Shifferaw, A., Balcha, T., 1978. The Ethiopian basement: Stratigraphy and possible manner of evolution. Geol. Rdsch. 67: 531-546.

Kennedy, W.Q. 1964. The structural differentiation of Africa in the Pan-African (± 500 m.y) tectonic episode. Annu. Rep. Res. Inst. Afr. Geol., Univ. Leeds 8: 48-49.

Kent, P.E., Hunt, J.A., Johnstone, D.W. 1971. The geology and geophysics of coastal Tanzania. Geophys. paper no. 6. Inst. geol. Sci. HMSO, London.

Kerdany, M.T., Cherif, O.H. 1990. Mesozoic. In R. Said (ed.), The geology of Egypt, pp. 407-438. Rotterdam: A.A. Balkema.

Kesse, G.O. 1985. The mineral and rock resources of Ghana. Rotterdam: A.A. Balkema.

Key, R.M., Charsley, T.J., Hackman, B.D., Wilkinson, A.F., Rundle, C.C. 1989. Superimposed upper Proterozoic collision - controlled orogenies in the Mozambique orogenic belt of Kenya. Precambr. Res. 44: 197-225.

Kienast, J.R., Ouzegane, K. 1987. Polymetamorphic Al, Mg-rich granulites with orthopyroxene-sillimanite and sapphrine parageneses in Archaean rocks from the Hoggar, Algeria. Geol. Journ. vol. 22, Thematic Issue, 57-79.

Killick, M.F. 1988. Sedimentary and tectonic controls on fan-delta facies and development: an example from the Infracambrian of the High Atlas, Morocco. In W. Nemec, R.J. Steel (eds.), Fan deltas: sedimentology and tectonic settings, pp. 212-225. U.K.: Blackie and Son.

King, G. 1990. The dicynodonts - A study in palaeobiology. London: Chapman and Hall.

King, L. 1951a. The geology of the Makapan and other caves. Geol. Soc. S. Afr. Trans. 33: 121-150.

King, L. 1951b. The geology of the Cango caves, Oudtshoorn, C.P. Geol. Soc. S. Afr. Trans. 33: 457-468.

Kinnaird, J., Bowden, P. 1987. African anorogenic alkaline magmatism and mineralization - a discussion with reference to the Niger-Nigerian province. Geol. Journ. vol. 22, Thematic Issue, 297-340.

Klerkx, J., Liégeois, J.-P., Lavreau, J., Claessens, W. 1987. Crustal evolution of the northern Kibaran belt, eastern and central Africa. In A. Kröner (ed.), Proterozoic lithospheric evolution, pp. 217-233. Geodynamic Series 17, American Geophysical Union.

Klitzsch, E. 1966. South-central Libya and northern Chad. A guidebook for the geology and prehistory. Eighth Annual Field conference 1966, Petroleum Exploration Society of Libya.

Klitzsch, E. 1971. The structural development of parts of north Africa since Cambrian time. In C. Gray (ed.), Symposium on the geology of Libya, pp. 256-260. Fac. Sci., Univ. Libya.

Klitzsch, E. 1981. Lower Palaeozoic rocks of Libya, Egypt, and Sudan. In C.H. Holland (ed.), Lower Palaeozoic of the Middle East, Eastern and Southern Africa, and Antarctica, pp. 131-163 U.K.: Wiley.

Klitzsch, E. 1986. Plate tectonics and cratonal geology in northeast Africa (Egypt, Sudan). Geol. Rdsch. 75: 753-768.

Klitzsch, E. 1990. Paleozoic. In R. Said (ed.), The geology of Egypt, pp. 393-406. Rotterdam: A.A. Balkema.

Klitzsch, E., Almond, J., Barazi, N., El Hassan, A., Mansour, N., Semtner, A. 1990. Short note on recently discovered Paleozoic strata of NE Sudan (Red Sea Hills). Berliner geowiss. Abh. A. 120: 87-88.

Klitzsch, E., Lejal-Nicol, A. 1984. Flora and fauna from strata in southern and northern Sudan. Berliner geowiss. Abh., A. 50: 47-79.

Klitzsch, E., Squyres, C.H. 1990. Paleozoic and Mesozoic geological history of northeastern Africa based upon new interpretation of Nubian strata. AAPG Bull. 74: 1203-1211.

Klitzsch, E., Wycisk, P. 1987. Geology of the sedimentary basins of northern Sudan and bordering areas. Berliner geowiss., Abh., A, 75: 97-136.

Knoll, A.H. 1990. Precambrian evolution of prokaryotes and protists. In D.E.G. Griggs, P.R. Crowther (eds.), Palaeobiology: a synthesis, pp. 9-16. Oxford: Blackwell.

Kogbe, C.A. 1973. Geology of the Upper Cretaceous and Tertiary sediments of the Nigerian sector of the Iullemmeden basin (West Africa). Geol. Rdsch. 62: 197-211.

Kogbe, C.A., Lemoigne, Y. 1976. Bois de structure Gymnosphermienne provenant de la formation D'Illo (Continental Intercalaire) au nord ouest du Nigeria. Abstr. 7th African Micropal. Coll. 60.

Kreuser, Th. 1984. Karroo basins in Tanzania. In J. Klerkx, J. Michot (eds.), African geology, pp. 231-244. Tervuren.

Kreuser, T., Markwort, S. 1989. Facies evolution of a fluviolacustrine Permo-Triassic basin in Tanzania. Zbl. Geol. Paläont. H. 7/8: 821-837.

Kreuser, T., Schramedei, R., Rullkötter, J. 1988. Gas prone source rocks from cratogene Karoo basins in Tanzania. J. Petrol. Geol. 11: 169-184.

Krömmelbein, K., Wenger, R. 1966. Sur quelques analogies remarquables dans les microfaunes crétacées du Gabon et du Brésil Oriental (Bahia et Sergipe). In D. Reyre (ed.), Sedimentary basins of the African coasts, I. Atlantic coast, pp. 193-196. Assoc. Afr. Geol. Surv., Paris.

Kröner, A. 1974. The Gariep Group, Part I: Late-Precambrian formations in the western Richtersveld, northern Cape Province. Precambrian Res. Unit Univ. Cape Town Bull. No. 13.

Kröner, A. 1976. Geochronology. Precambrian Res. Unit Univ. Cape Town Ann. Rept. 13: 139-143.

Kröner, A. 1977. Precambrian mobile belts of southern and eastern Africa - Ancient sutures or sites of ensialic mobility? A case for crustal evolution towards plate tectonics. Tectonophysics 40: 101-135.

Kröner, A. 1984. Late Precambrian plate tectonics and orogeny: a need to redefine the term Pan-African. In J. Klerkx, J. Michot (eds.), African geology, pp. 23-27, Tervuren.

Kröner, A. 1989. Plate motion, crustal accretion and supercontinent assemblage since the Early Archaean. Abstr. 28th IGC, 2: 230-231.

Kröner, A., Blignault, H.J. 1976. Towards a definition of some tectonic and igneous provinces in western South Africa and southern South West Africa. Geol. Soc. S. Afr. Trans. 79: 232-238.

Kröner, A., Correia, H. 1980. Continuation of the Pan-African Damara belt into Angola: a proposed correlation of the Chela group in southern Angola with the Nosib group in Northern Namibia (S.W.A.). Trans. Geol. Soc. S. Afr. 83: 5-16.

Kröner, A., Greiling, R., Reischmann, T., Hussein, I.M., Stern, R.J., Dürr, S., Krüger, J., Zimmer, M. 1987. Pan-African crustal evolution in the Nubian segment of northeast Africa. In A. Kröner, (ed.), Proterozoic lithospheric evolution, pp. 235-257. Geodynamic Series 17, American Geophysical Union.

Kuehn, S., Ogola, J., Sango, P. 1990. Regional setting and nature of gold mineralization in Tanzania and southwest Kenya. Precambr. Res. 46: 71-82.

Küster, D., Utke, A., Leupolt, L., Lenoir, J.L., Haider, A. 1990. Pan-African granitoid magmatism in northeastern and southern Somalia. Berliner geowiss. Abh. A 120.2: 519-536.

Kummel, B. 1970. History of the earth. San Francisco: W.H. Freeman.

Kurtanjek, M.P., Tandy, B.C. 1988. The igneous phosphate deposits of Matongo-Bandaga, Burundi. In A.J.G. Notholt, R.P. Sheldon, D.F. Davidson (eds.), Phosphate depositis of the world, vol. 2, pp. 262-266. Cambridge: Cambridge University.

Kuss, J. 1989. Facies and paleogeographic importance of the pre-rift limestones from NE-Egypt/Sinai. Geol. Rdsch. 78: 487-498.

Kutzbach, J.E., Guetter, P.J. 1984. Sensitivity of late glacial and Holocene climates to the combined effects of orbital parameter changes and lower boundary condition changes: "Snapshot" simulations with a general circulation model for 18, 9, and 6 ka B.P. Annals of Glaciology 5:85-87.

Lancaster, N. 1981. Paleoenvironmental implications of fixed dune systems in southern Africa. Palaeogeogr., Palaeoclimatol., Palaeoecol. 33: 327-346.

Lancaster, N. 1988. Development of linear dunes in the southwestern Kalahari, southern Africa. Journal of Arid Environments 14: 233-244.

Lancaster, N. 1989. Late Quaternary palaeoenvironments in the southwestern Kalahari. Palaeogeogr. Palaeoclimatol., Palaeoecol. 70: 367-376.

Lancelot, J.R., Boullier, A.M., Maluski, H., Ducrot, J. 1983. Deformation and related radiochronology in a late Pan-African mylonitic shear zone, Adrar des Iforas (Mali). Contr. Min. Pet. 82: 312-326.

Lang, J., Kogbe, C., Alidou, S., Alzouma, K.A., Bellion, G., Dubois, D., Durand, A., Guiraud, R., Houessou, A., De Klasz, I., Romann, E., Salard-Cheboldaff, M., Trichet, J. 1990. The Continental Terminal in West Africa. J. Afr. Earth Sci. 10: 79-99.

Lavreau, J. 1980. Etude géologique du Haut-Zaire-Genèse et evolution dún segment lithosphérique archéen. Thèse, Univ. Brussels.

Lavreau, J. 1984. Vein and stratabound gold deposits of northern Zaire. Mineral. Deposita 19: 158-165.

Leblanc, M. 1981. The late Proterozoic ophiolites of Bou Azzer (Morocco): evidence for Pan-African plate tectonics. In A. Kröner (ed.), Precambrian plate tectonics, 435-451. Amsterdam: Elsevier.

Leblanc, M., Lancelot, J.R. 1980. Le domaine panafricain de l'Anti-Atlas (Maroc). Can. J. Earth Sci. 17:142-155.

Lécorché, J.P., Dallmeyer, R.D., Villeneuve, M. 1989. Definition of tectonostratigraphic terranes in the Mauritanide, Bassaride, and Rokelide orogens, West Africa. Geol. Soc. Amer. Spec. Pap. 230: 131-144.

Ledru, P., N'Dong, J.E., Johan, V., Prian, J.-P., Coste, B., Haccard, D. 1989. Structural and metamorphic evolution of the Gabon orogenic belt: Collision tectonics in the Lower Proterozoic. Precambr. Res. 44: 227-241.

Lefranc, J.Ph., Guiraud, R. 1990. The continental Intercalaire of northwestern Sahara and its equivalents in the neighbouring regions. J. Afr. Earth Sci. 10: 27-77.

Legrand, Ph. 1985. Lower Paleozoic rocks of Algeria. In C.H. Holland (ed.), Lower Paleozoic of north-western and west-central Africa, pp. 5-89. U.K.: Wiley.

Legrand-Blain, M. 1985. Taoudeni basin. In R.H. Wagner, C.R.W. Prins, L.F. Granados (eds.), The Carboniferous of the world, pp. 327-333, IUGS Publication No. 20.

Lemosquet, Y., Pareyn, C. 1985. Bechar basin. In R.H. Wagner, C.R.W. Prins, L.F. Granados (eds.), The Carboniferous of the world, pp. 306-315, IUGS Publication No. 20.

Leube, A., Hirdes, W., Mauer, R., Kesse, G.O. 1990. The Early Proterozoic Birimian Supergroup of Ghana and some aspects of its associated gold mineralization. Precambr. Res. 46: 139-165.

Lewin, R. 1988. Modern human origins under close scrutiny. Science 239: 1240-1241.

Leyshon, P.R., Tennick, F.P. 1988. The Proterozoic Magondi Mobile Belt in Zimbabwe - a review. S.-Afr. Tydskr. Geol. 91: 114-131.

Liégeois, J.P., Bertrand, J.M, Black, R. 1987. The subduction- and collision-related Pan-African composite batholith of the Adrar des Iforas (Mali): a review. Geol. Journ. vol. 22, Thematic Issue, 185-211.

Light, M.P.R. 1982. The Limpopo mobile belt: a result of continental collision. Tectonics 1: 325-342.

Lock, B.E. 1980. Flat-plate subduction and the Cape Fold belt of South Africa. Geology 8: 35-39.

Loock, J.C., Visser, N.J. 1985. South Africa. In R.H. Wagner, C.R.W. Prins, L.F. Granados (eds.), The Carboniferous of the world, pp. 167-175. IUGS Publication No. 20.

Lorenz, J.C. 1988. Synthesis of Late Paleozoic and Triassic redbed sedimentation in Morocco. In V.H. Jacobshagen (ed.), The Atlas System of Morocco, Lecture Notes in Earth Sciences, 15: 139-168. Berlin: Springer.

Louis, P. 1970. Contribution géophysique a la connaissance géologique du basin du Lac Tchad. Mém. Office Rech. Sci. Tech. Outre-Mer. 42.

Lowe, D.R., Knauth, L.P. 1977. Sedimentology of the Onverwacht Group (3.4 billion years), Transvaal, South Africa, and its bearing on the characteristics and evolution of the early earth. J. Geol. 85: 699-723.

Lubala, R.T., Kampunzu, A.B., Caron, J.P.-H. 1987. Petrology and geodynamic significance of the Tertiary alkaline lavas from the Kahuzi-Biega region, Western Rift, Kivu, Zaire. Geol. Journ. vol. 22, Thematic Issue, 515-535.

Lucas, J., Ilyin, A.V., Kuhn, A. 1986. Proterozoic and Cambrian phosphorites deposits: Volta basin, West Africa. In P.J. Cook, J.H. Shergold (eds.), Phosphate deposits of the world, vol. 1, Proterozoic and Cambrian phosphorites, pp. 235-243. Cambridge: University Press.

Maboko, M.A.H., Boelrijk, N.A.I.M., Priem, H.N.A., Verdurmen, E.A.Th. 1985. Zircon U-Pb and biotite Rb-Sr dating of the Wami River granulites, eastern granulites, Tanzania: evidence for approximately 715 Ma old granulite facies metamorphism and final Pan-African cooling approximately 475 Ma ago. Precambr. Res. 30: 361-378.

Macgregor, A.M. 1947. An outline of the geological history of Southern Rhodesia. Bull. Geol. Survey S. Rhodesia, 38.

Malisa, E., Muhongo, S. 1990. Tectonic setting of gemstone mineralization in the Proterozoic metamorphic terrane of the Mozambique belt in Tanzania. Precambr. Res. 46: 167-176.

Manspeizer, W. 1982. Triassic-Liassic basins and climate of the Atlantic passive margins. Geol. Rdsch. 71: 895-917.

Marsh, T.S., Bowen, M.P., Rogers, N.W., Bowen, T.B. 1989. Volcanic rocks of the Witwatersrand Triad, South Africa. II: Petrogenesis of mafic and felsic rocks of the Dominion Group. Precambr. Res. 44: 39-65.

Martin, A., Nisbet, E.G., Bickle, M.J. 1980. Archaean stromatolites of the Belingwe greenstone belt, Zimbabwe (Rhodesia). Precambr. Res. 13: 337-362.

Martin, H., 1965. The Precambrian geology of South West Africa and Namaqualand. Cape Town: Precambrian Res. Unit. Univ. Cape Town.

Martin. H. 1983. Alternative geodynamic models for the Damara orogeny. A critical discussion. In H. Martin, F.W. Eder (eds.), Intracontinental fold belts, pp. 913-945. Berlin: Springer.

Mascle, J., Mougenot, D., Blarex, E., Marinho, M., Virlogeux, P. 1987. African transform continental margins: examples from Guinea, the Ivory Coast and Mozambique. In Geol. Journ. vol. 22, Thematic Issue, 537-561.

Mason, R. 1973. The Limpopo mobile belt - Southern Africa. Philos. Trans. R. Soc. London, Ser. A, 273: 463-485.

Masters, S. 1989a. Sedimentology and copper mineralization of metamorphosed Early Proterozoic playa complex: Norah Fromation of Deweras Group. Abstr. 28th IGC. 2: 384-385.

Masters, S. 1989b. Stratabound copper-silver mineralization in red beds of an Early Proterozoic alluvial fan complex, Magunda Mine, Zimbabwe. Abstr. 28th IGC. 2: 385.

Matheis, G. 1987. Nigerian rare-metal pegmatites and their lithological framework. Geol. Journ. vol. 22, Thematic Issue, 271-291.

Mattis, A.F. 1977. Nonmarine Triassic sedimentation, central High Atlas Mountains, Morocco. Jour. Sed. Pet. 47: 107-119.

Maurin, J.C., Guiraud, R. 1990. Relationships betweeen tectonics and sedimentation in the Barremo-Aptian intracontinental basins of northern Cameroon. J. Afr. Earth Sci. 10: 331-340.

Mbede, E.I. 1984. A report on the hydrocarbon potential of Kenya. M.Sc. thesis, Univ. of London.

Meshref, W.M. 1990. Tectonic framework. In R. Said (ed.), The geology of Egypt, pp. 113-155, Roterdam: A.A. Balkema.

McConnell, R.B. 1972. Geological development of the Rift System of eastern Africa. Geol. Soc. Amer. Bull. 83: 2549-2572.

McCurry, P. 1976. The geology of the Precambrian to lower Palaeozoic rocks of northern Nigeria - a review. In C.A. Kogbe (ed.), Geology of Nigeria, pp. 15-39. Lagos: Elizabethan Pub. Co.

McDougall, I. 1985. K-Ar and $^{40}Ar/^{39}Ar$ dating of the Homonid-bearing Pliocene-Pleistocene sequence at Koobi Fora, Lake Turkana, Northern Kenya. Geol. Soc. Amer. Bull. 96: 159-175.

McElhinney, M.W., Giddings, J.W., Embleton, B.J.J. 1974. Palaeomagnetic results and late Precambrian galciations. Nature 248: 557-561.

McWilliams, O. 1981. Palaeomagnetic and Precambrian tectonic evolution of Gondwana. In A. Kröner (ed.), Precambrian plate tectonics, pp. 649-687. Amsterdam: Elsevier.

Miller, R. McG., 1983. The Pan-African Damara orogen of South West Africa/Namibia. Spec. Publ. Geol. Soc. S. Afr. 11: 431-515.

Milesi, J.P., Ledru, P., Dommanget, A., Johan, V. 1989. Lower Proterozoic succession in Senegal and Mali (West Africa): Position of sediment-hosted Au and Fe deposits of Luolo area and significance in terms of crustal evolution. Abstr. 28th IGC, 2: 433-434.

Moore, J.M., Reid, D.L., Watkeys, M.K. 1989. Evolution of the Aggeny's/Gamsberg base metal deposits, Namaqualand metamorphic complex. Abstr. 28th IGC. 456-457.

Morel, P., Irving, E. 1978. Tentative paleocontinental maps for the early Proterozoic. J. Geol. 86: 535-561.

Morgan, P. 1990. Egypt in the framework of global tectonics. In R. Said (ed.), The geology of Egypt, pp. 91-111. Rotterdam: A.A. Balkema.

Morley, C.K., Wescott, W.A., Cunningham, S.M., Harper, R.M. 1989. Recent exploration in the Lake Rukwa area, East African Rift. Abstr. 28th IGC. 2: 462-463.

Mörner, N.A. 1984. Planetary, solar, atmospheric, hydrospheric and endogene processes as origin of climatic change on the Earth. In N.A. Mörner, W. Karlen (eds.), Climatic changes on a yearly to millenial basis, pp. 483-504. Dordrecht: Reidel.

Morris, S.C. 1990. Late Precambrian-Early Cambrian metazoan diversification. In D.E.G. Briggs, P.R. Crowther (eds.), Palaeobiology: a synthesis, pp. 30-36. Oxford: Blackwell.

Moss, R.P. 1968. Soils, slopes and surfaces in tropical Africa. In R.P. Moss (eds.), The soil resources of tropical Africa, pp. 29-59. Cambridge: University Press.

Muhongo, S. 1989. Tectonic setting of the Proterozoic metamorphic terrains in eastern Tanzania and their bearing on the evolution of the Mozambique belt. Bull. IGCP 255, 2: 43-50.

Muhongo, S. (in press). Late Proterozoic collision tectonics in the Mozambique belt of East Africa. Precambr. Res.

Nairn, A.E.M. 1978. Northern and eastern Africa. In M Moullade, A.E.M Nairn (eds.), The Phanerozoic geology of the world II: The Mesozoic, A, pp. 329-370. Amsterdam: Elsevier.

Neugebauer, J. 1989. The Iapetus model: a plate tectonic concept for the Variscan belt of Europe. Tectonophysics 169: 229-256.

Nichols, G.J., Daly, M.C. 1989. Sedimentation in an intracratonic extensional basin: the Karoo of the central Morondava basin, Madagascar. Geol. Mag. 126: 339-354.

Nisbet, E.G. 1987. The young earth. Boston: Allen & Unwin.

Nyong, E.E., Akpan, E.B. 1987. Storm generated deposits in the Cretaceous of the Calabar Flank, SE Nigeria. In G. Matheis, H. Schandelmeier (eds.), Current research in African earth sciences, pp. 151-154. Rotterdam: A. A. Balkema.

Nwajide, C.S. 1990. Cretaceous sedimentation and paleogeography of the central Benue trough. In C.O. Ofoegbu (ed.), The Benue Trough: Structure and evolution, pp. 19-38. Braunschweig: Vieweg.

Odebode, M.O., Enu, E. 1986. Evidence of lateral equivalence of terminal Cretaceous formations in the Upper Benue basin, Nigeria. Newsl. Stratigr. 17:45-55.

Odeyemi, I.B. Preliminary report on the field relationships of the basement complex rocks around Igarra, Midwest Nigeria. In C.A. Kogbe (ed.), Geology of Nigeria, pp. 59-63. Lagos: Elizabethan Pub. Co.

Ofoegbu, C.O. 1985. A review of the geology of the Benue Trough of Nigeria. J. Afr. Earth Sci. 3: 283-291.

Ofrey, O. 1982. A revised Bouguer anomaly gravity map of Nigeria. Nig. Jour. min. Geol. 19: 78-79.

Ogezi, A.E.O. Geochemistry and geochronology of basement rocks from northwestern Nigeria. Ph.D. thesis, Univ. Leeds.

Ogola, J.S. 1987. Mineralization in the Migori greenstone belt, Macalder, western Kenya. Geol. Journ. vol. 22, Thematic Issue, 25-44.

Ojoh, K.A. 1990. Crataceous geodynamic evolution of the southern part of the Benue Trough (Nigeria) in the equatorial domain of the South Atlantic. Stratigraphy, basin analysis and paleo-oceanography. Bull. Centres Rech. Explor.-Prod. Elf-Aquitaine, 14: 419-442.

Okeke, P.O., Ofoegbu, C.O., Amajor, L.C. 1988. On the origin of the highly altered basalts, Benue Trough, Nigeria. Geochem. Mineral. Petrol. 24: 55-67.

Okereke, C.S. 1984. A gravity survey of the eastern sector of the Nigerian basement complex and studies of the gravity fields of the West African Rift System. Ph.D. Thesis, University of Leeds.

Okosun, E.A. 1990. A review of the Cretaceous stratigraphy of the Dahomey Embayment, West Africa. Cret. Res. 11: 17-27.

Olade, M.A. 1978. General features of a Precambrian iron ore deposit and its environment at Itakpe ridge, Okene, Nigeria. Trans. IMM B87: 1-9.

Olade, M.A. 1985. Aspects of primary tin-niobium mineralization in Nigeria's Younger Granite Province. In R.P. Taylor, D.F. Strong (eds.), Granite-related mineral deposits, pp. 200-202. Can. Inst. Min. Geol. Div.

Olade, M.A., Elueze, A.A. 1977. Petrochemistry of the Ilesha amphibolites and Precambrian crustal evolution in the Pan-African domains of SW Nigeria. Precambr. Res. 8: 303-318.

Omatsola, M.E., Adegoke, O.S. 1981. Tectonic evolution and Cretaceous stratigraphy of the Dahomey Basin. Nig. Jour. Min. Geol. 18: 130-137.

Onuoha, K.M. 1981. Sediment loading and subsidence in the Niger Delta sedimenatry basin. Nig. Jour. Min. Geol. 18: 138-140.

Orife, J.M., Avbovbo, A.A. 1982. Stratigraphic and unconformity traps in the Niger delta. AAPG. Mem. 32: 251-265.

Orpen, J.L., Swain, C.J., Nugent, C., Zhou, P.P. 1989. Wrench-fault and half-graben tectonics in the development of the Palaeozoic Zambezi Karoo basins in Zimbabwe - the "lower Zambezi" and Mid-Zambezi" basins respectively - and regional implications. J. Afr. Earth Sci. 8: 215-229.

Oti, M.N. 1990. Upper Cretaceous off-shelf carbonate sedimentation in the Benue trough: The Nkalagu Limestone. In C.O. Ofoegbu (ed.), The Benue trough: structure and evolution, pp. 321-358. Braunschweig: Vieweg.

Owen, R.B., Barthelme, J.W., Renaut, R.W., Vinces, A. 1982. Palaeolimnology and archaeology of Holocene deposits northeast of Lake Turkana, Kenya. Nature 298: 523-528.

Oyawoye, M.O. 1972. The basement complex of Nigeria. In T.F.J. Dessauvagie, A.J. Whiteman (eds.), African geology, pp. 66-102. Ibadan: University Press.

Pachur, H.-J., Kröpelin, S., Hoelzmeinn, P., Goschin, M., Altmann, N. 1990. Late Quaternary fluvio-lacustrine environments of western Nubia. Berliner geowiss. Abh. A 120: 230-260.

Papon, A. 1973. Géologie et minéralisations du Sud-ouest de la côte d'Ivoire. Mem. Bur. Rech. géol. min. Paris, 80.

Paris, I.A. 1987. The 3.5 Ga Barberton greenstone succession, South Africa: implications for modelling the evolution of the Archaean crust. Geol. Journ. vol. 22, Thematic Issue, 5-24.

Patridge, T.C. 1978. Re-appraisal of lithostratigraphy of Sterkfontein hominid site. Nature 275: 282-287.

Patridge, T.C., Brink, A.B.A. 1967. Gravels and terraces of the lower Vaal River basin. S. Afr. Geogr. Jour. 49: 21-38.

Peterson, J.A. 1985. Geology and petroleum resources of central and east-central Africa. USGS Open-File Report 85-589.

Petit-Maire, N. 1987. Local responses to recent global climatic change: Hyperarid central Sahara and coastal Sahara. In G. Matheis, H. Schandelmeier (eds.), Current research in African earth sciences, pp. 431-434. Rotterdam: A.A. Balkema.

Petters, S.W. 1979. West African cratonic stratigraphic sequences: Geology 7: 528-531.

Petters, S.W. 1981. Stratigraphy of Chad and Iullemmeden basins (West Africa). Eclogae geol. Helv. 74: 139-159.

Petters, S.W. 1982. Central West African Cretaceous-Tertiary benthic foraminifera and stratigraphy. Palaeontographica Abt. A 179: 1-104.

Petters, S.W. Ekweozor, C.R. 1982. Petroleum geology of Benue Trough and southeastern Chad Basin Nigeria. AAPG Bull. 66: 1141-1149.

Pickford, M. 1982. The tectonics, volcanics and sediments of the Nyanza Rift Valley, Kenya. Z. Geomorph. N.F. 42: 1-33.

Pilbeam, D. 1984. The descent of hominoids and hominids. Scientific American 250: 60-69.

Piper, D.P., Chikusa, C.M., Chisale, R.T.K., Kaphwiye, C.E., Malunga, G.W.P., Nkhoma, J.E.S., Klerkx, J.M. 1989. A stratigraphic and structural reappraisal of central Malawi: results of a geotraverse. J. Afr. Earth Sci. 8: 79-90.

Piper, J.D.A. 1975. Palaeomagnetic correlations of Precambrian formations of east-central Africa and their tectonic implications. Tectonophysics 26: 135-161.

Piper, J.D.A. 1980. The Precambrian paleomagnetic record: the case for the Proterozoic supercontinent. Earth Planet. Sci. Lett. 59: 61-89.

Piper, J.D.A., Briden, J.C., Lomax, K. 1973. Precambrian Africa and South America as a single continent. Nature 245: 244-248.

Pique, A., Michard, A. 1989. Moroccan Hercynides: A synopsis. The Paleozoic sedimentary and tectonic evolution at the northern margin of West Africa. Amer. J. Sci. 289: 286-330.

Plumstead, E.P. 1969. Three thousand million years of plant life in Africa. Geol. Soc. S. Afr. Trans. 72. (Annex.).

Plumstead, E.P. 1973. The late Paleozoic *Glossopteris* flora. In A. Hallam (ed.), Atlas of palaeobiogeography, pp. 187-206. Amsterdam: Elsevier.

Pohl, W. 1984. Large scale metallogenic features of the Pan African in east Africa, Nubia and Arabia. Bull. Fac. Earth Sci., King Abdul Aziz Univ., Jeddah 6: 592-601.

Pohl, W. 1987. Metallogeny of the northeastern Kibaran belt, central Africa. Geol. Journ. vol. 22, Thematic Issue, 103-119.

Pohl, W. 1988. Precambrian metallogeny of NE Africa. In S. El Gaby, R.O. Greiling (eds.), The Pan-African belt of northeast Africa and adjacent areas, pp. 319-341. Braunschweig: Vieweg.

Poidevin, J.L., Dostal, J., Dupuy, C. 1981. Archaean greenstone belt from the Central African Republic (Equatorial Africa). Precambr. Res. 16: 157-170.

Ponte, F.C., Asmus, H.E. 1978. Geological framework of the Brazilian continental margin. Geol. Rdsch. 67: 201-235.

Popoff, M. 1988. Du Gondwana à l'Atlantique sud: les connexions du fossé de la Bénoué avec les bassins du Nord-Est bresilien jusqu'à l'ouverture du golfe de Guienée au Cretacé inférieur. In J. Sougy, J. Rodgers (eds.), The West African Connection, pp. 409-431. J. Afr. Earth Sci., Special Publ. 7.

Porada, H. 1983. Geodynamic model for the geosynclinal development of the Damara orogen, Namibia/South West Africa. In H. Martin, F.W. Eder (eds.), Intracontinental fold belts, pp. 503-542. Berlin: Springer.

Porada, H. 1989. Pan-African rifting and orogenesis in southern to equatorial Africa and eastern Brazil. Precambr. Res. 44: 103-136.

Pretorius, D.A., 1975. The depositional environment of the Witwatersrand goldfields: a chronological review of speculations and observations. Minerals Sci. Eng. 7: 18-47.

Pretorius, D.A. 1979. The depositional environment of the Witwatersrand goldfields: A chronological review of speculations and observation. Geokongress 77: Geol. Soc. S. Afr. Spec. Publ. 6: 33-55.

Pritchard, J.M. 1979. Landform and landscape in Africa. London: Edward Arnold.

Prochaska, W., Pohl, W., 1984. Petrochemistry of some mafic and ultramafic rocks from the Mozambique Belt, northern Tanzania. J. Afr. Earth Sci., 1: 183-191.

Pruvost, P. 1951. L'Infracambrien. Bull. Soc. belge Géol. Palaeontol. Hydrol. 60: 43-63.

Pulfrey, W., Walsh, J. 1969. The geology and mineral resources of Kenya. Bull. Geol. Surv. Kenya, Nairobi, no. 9.

Rabinowitz, P.D., Coffin, M.F., Falvey, D. 1983. The separation of Madagascar and Africa. Science 220: 67-69.

Radulescu, J. 1982. Mineralization in the Karagwe-Ankolean System of East Africa/Burundi. In The Development Potential of Precambrian Mineral Deposits. UN Dpt. Techn. Coop. Devel. 217-225.

Rahaman, M.A., van Breeman, O., Bowden, P., Bennett, J.N. 1984. Age migrations of anorogenic ring complexes in northern Nigeria. J. Geol. 92: 173-184.

Reeves, C.V., Karanja, F.M., Macleod, I.N. 1987. Geophysical evidence for a failed Jarassic rift and triple junction in Kenya. Earth planet. Sci. Lett. 81: 299-311.

Reijers, T.J.A., Petters, S.W. 1987. Depositional environments and diagenesis of Albian carbonates on the Calabar Flank, SE Nigeria. Jour. Petrol. Geol. 10: 283-294.

Reimer, T.O. 1987. Weathering as a source of iron in iron-formations: The significance of alumina-enriched paleosols from the Proterozoic of southern Africa. In P.W.U. Appel, G.L. LaBerge (eds.), Precambrian iron-formations, pp. 601-618. Athens: Theophrastus.

Reyment, R.A. 1965. Aspects of the geology of Nigeria. Ibadan: University Press.

Reyment, R.A., Schöbel, J. 1983. Le post-Paléozoique du Mali. In J. Fabre (ed.), West Africa: Geological introduction and stratigraphic terms, pp. 140-142. Oxford: Pergamon.

Reyment, R.A., Tait, E.A. 1972. Biostratigraphical dating of the early history of the South Atlantic ocean. Phil. Trans. R. Soc. London, Ser. B. 264: 55-95.

Ries, A.C., Shackleton, R.M., Graham, R.H., Fitches, W.R. 1983. Pan-African structures, ophiolites and mélange in the Eastern Desert of Egypt, a traverse at 26°N. J. Geol. Soc. London 140: 75-95.

Ritz, M., Robineau, B., Bellion, Y. 1989. Results of geoelectric studies in southern Mauritania, West Africa. Abstr. 28th IGC, 2: 703.

Rocci, G. 1965. Essai dínterprétation de mesures géochronologique-la structure de l'Ouest African. Sci. Terre, Nancy 10: 461-478.

Rodgers, J. 1970. The tectonics of the Applachians. New York: Wiley.

Rogers, J.J., Rosendahl, B.R. 1989. Perceptions and Issues in continental rifting. J. Afr. Earth Sci. 8: 137-142.

Rollinson, H.R. 1978. Zonation of supracrustal relicts in the Archaean of Sierra Leone, Liberia, Guinea and Ivory Coast. Nature 272: 440-442.

Rosendahl, B.R., Reynolds, D.J., Lorber, P.M., Burgess, C.F., MiGill, J., Scott, D., Lambiase, J.J., Derksen, S.J. 1986. Structural expressions of rifting: lessons from Lake Tanganyika, Africa. In L. E. Frostick et al. (eds.), Sedimentation in the African Rifts, pp. 29-43. Geol. Soc. Spec. Publ. No. 25.

Roussel, J., Lécorché, J.P. 1989. Gravity study and crustal terranes relationships in West AFrica. Abstr. 28th IGC. 2: 721-722.

Roux, J.P., Toens, P.D. 1987. The Permo-Triassic Uranium deposits of Gondwana. In G.D. McKenzie (ed.), Gondwana Six: Stratigraphy, Sedimentology, and Paleontology, pp. 139-146. Geophysics Monograph 41, Washington, D.C.: American Geophysical Union.

Rust, I.C. 1975. Tectonic and sedimentary framework of Gondwana basins in southern Africa. In K.S.W. Campbell (ed.), Gondwana geology, pp. 537-564. Canberra: Australian University.

Sacchi, R., Marques, J., Costa, M., Casati, C. 1984. Kibaran events in the southernmost Mozambique belt. Precambr. Res. 25: 141-159.

Said, R. 1982. The geological evolution of the River Nile. Berlin: Springer.

Said, R. 1990a. Cretaceous paleogeographic maps. In R. Said (ed.), The geology of Egypt, pp. 439-449. Rotterdam: A.A. Balkema.

Said, R. 1990b. Cenozoic. In R. Said (ed.), The geology of Egypt, pp. 451-486. Rotterdam: A.A. Balkema.

Said, R. 1990c. Quaternary. In R. Said (ed.), The geology of Egypt, pp. 487-507. Rotterdam: A.A. Balkema.

Salaj, J. 1978. The geology of the Pelagian block: The eastern Tunisian platform. In A.E.M. Nairn, W.H. Kanes, G.G. Stehli (eds.), The ocean basins and margins, pp. 361-416. New York: Plenum.

Salop, L.J. 1983. Geological evolution of the Earth during the Precambrian. Berlin: Springer.

Sanders, L.D. 1965. Geology of the contact between the Nyanza Shield and the Mozambique belt in western Kenya. Bull. Geol. Surv. Kenya, no. 7.

Sawkins, F.J. 1990. Metal deposits in relation to plate tectonics. Berlin: Springer.

Schaer, J.-P. 1987. Evolution and structure of the High Atlas of Morocco. In J.-P. Schaer, J. Rodgers (eds.), The anatomy of mountain ranges, pp. 107-125. Princeton: University Press.

Schandelmeier, H., Darbyshire, D.P.F., Harm, U., Richter, A. 1988. The East Saharan craton: evidence for pre-Pan-African crust in northeast Africa west of the Nile. In S. El Gaby, R.O. Greiling (eds.), The Pan-African belt of northeast Arica and adjacent areas, pp. 69-94. Braunschweig: Vieweg.

Schandelmeier, H., Klitzsch, E., Henricks, F., Wycisk, P. 1987. Structural development of north-east Africa since Precambrian times. Berliner geowiss. Abh., A, 75: 5-24.

Schandelmeier, H., Pudlo, D. 1990. The central African Fault Zone (CAFZ) in Sudan - a possible continental transform fault. Berliner geowiss. Abh. A 120: 31-44.

Schandelmeier, H., Utke, A., Harms, U., Kuster, D. 1989. A review of the Pan-African evolution in NE Africa: towards a new dynamic concept for continental NE Africa. Berliner geowiss. Abh. A 120.1: 1-14.

Schluter, T. 1987. Paleoenvironment of lacustrine phosphate deposits at Minjingu, northern Tanzania, as indicated by their fossil record. In G. Matheis, H. Schandelmeier (ed.), Current research in African Earth Sciences, pp. 223-226. Rotterdam: A.A. Balkema.

Schmincke, H.-U. 1982. Volcanic and chemical evolution of the Canary Islands. In: U. von Rad., K. Hinz, M. Sarnthein, E. Seibold, (eds.), Geology of the northwest African continental margin, pp. 274-306. Berlin: Springer.

Schmitt, M. 1979. The section of Tiout (Precambriann/Cambrian boundary beds, Anti-Atlas, Morocco): stromatolites and their biostratigraphy. Thesis, Univ. Würzburg.

Schopf, J.W., Walter, M.R. 1983. Archean microfossils: new evidence of ancient microbes. In J.W. Schopf (ed.), Earth's earliest biosphere, pp. 214-239. Princeton: University Press.

Schrank, E. 1987. Palaeozoic and Mesozoic palynomorphs from Northeast Africa (Egypt and Sudan) with special reference to Late Cretaceous pollen and dinoflagellates. Berliner geowiss. Abh. A 75: 249-310.

Scotese, C.R. 1986. Phanerozoic reconstructions: A new look at the assembly of Asia. Univ. Texas Institute for Geophysics Technical Report No. 6.

Scott, L. 1989. Climatic conditions in southern Africa since the last glacial maximum, inferred from pollen analysis. Palaeogeogr., Palaeoclimatol., Palaeoecol. 70: 345-353.

Sdzuy, K. 1978. The Prcambrian-Cambrian boundary beds in Morocco (preliminary report). Geol. Mag. 115: 83-94.

Seibold, E. 1982. The northwest African continental margin - An introduction. In U. von Rad, K. Hinz, M. Sarnthein, E. Seibold (eds.), Geology of the Northwest African continental margin, pp. 3-17. Berlin: Springer.

Seilacher, A. 1984. Late Precambrian and Early Cabmrian Metazoa: preservational or real extinctions? In H.D. Holland, A.F. Trendall (eds.), Patterns of change in earth evolution, pp. 159-168. Berlin: Springer.

Seilacher, A. 1990. Paleozoic trace fossils. In R. Said (ed.), The geology of Egypt, pp. 649-670. Rotterdam: A.A. Balkema.

Shackleton, R.M. 1976. Possible late-Precambrian ophiolites in Africa and Brazil. Ann. Report Res. Inst. of Afr. Geol. 20: 3-7.

Shackleton, R.M. 1979. Precambrian tectonics of northeast Africa. In, A.M.S. Shanti (ed.), Evolution and Mineralization of the Arabian-Nubian Shield, 2, pp. 1-6, Oxford: Pergamon.

Shackleton, R.M. 1986. Precambrian collision tectonics in Africa. In M.P. Coward, A.C. Reis (eds.), Collision tectonics, Spec. Publ. Geol. Soc. London, 19: 329-349.

Short, K.C., Stauble, A.J. 1967. Outline geology of the Niger Delta. AAPG Bull. 51: 761-779.

Simons, E.L. 1989. Human origins. Science 245: 1343-1350.

Simons, E.L., Rasmussen, D.T. 1990. Vertebrate paleontology of Fayum: History of research, faunal review and future prospects. In: R. Said (ed.), The geology of Egypt, pp. 627-638. Rotterdam: A.A. Balkema.

Sims, P.K. 1980. Subdivision of the Proterozoic and Archaean eons: recommendations and suggestions by the International Subcommission on Precambrian Stratigraphy. Precambr. Res. 13: 379-380.

Skehan, J.W.S.J., Pique, A. 1989. Comparative evolution of Avalon zone of New England and of Panafrican zone of West Africa from Proterozoic through Late Paleozoic time: Abstr. 28th IGC. 3: 130-131.

Slansky, M. 1986. Proterozoic and Cambrian phosphorites-regional review: West Africa. In P.J. Cook, J.H. Shergold (eds.), Phosphate deposits of the world, vol. 1, Proterozoic and Cambrian phosphorites, pp. 108-115. Cambridge: University Press.

Smith, R.M.H. 1990. A review of Stratigraphy and sedimentary environments of the Karoo Basins of South Africa. J. Afr. Earth. Sci. 10: 117-137.

Smyth, A.J., Montgomery, R.F. 1962. Soils and land use in central western Nigeria. Ibadan: Government Printers.

Sougy, J. 1962. West African fold belt. Bull. Geol. Soc. Am. 73: 871-876.

South African Committee for Stratigraphy. 1980. Stratigraphy of South Africa. Part 1. Lithostratigraphy of South West Africa/Namibia, and the Republics of Bophuthatswana, Transkei and Venda. S. Afr. Geol. Surv. Handbook. No. 8.

Sowunmi, M.A. 1987. The environment - present and past of West Africa. West Afr. Jour. Archaeol. 17: 41-51.

Specht, T.D., Rosendahl, B.R. 1989. Architecture of the Lake Malawi rift, East Africa. J. Afr. EArth Sci. 8: 355-382.

Stager, J.C. 1988. Environmental changes at Lake Cheshi, Zambia since 40,000 yr. B.P. Quat. Res. 29: 54-65.

Stanton, W.I., Schermerhorn, L.J.G., Korpershoek, H.R. 1963. The West Congo System. Bol. Serv. Geol. Minas Angola 9: 69-78.

Stern, R.J. 1985. The Najd fault system, Saudi Arabia and Egypt: a Late Precambrian rift system? Tectonics 4:497-511.

Stets, J., Wurster, P. 1982. Atlas and Atlantic--structural relations. In U. von Rad., K. Hinz, M. Sarnthein, E. Seibold (eds.), Geology of the Northwest African continental margin, pp. 69-85. Berlin: Springer.

Stoesser, D.B., Camp, V.E. 1985. Pan-African microplate accretion of the Arabian shield. Bull. Geol. Am. 96: 817-826.

Stoesser, D.B., Stacey, J.S., Greenwood, W.R., Fischer, L.B. 1984. U/Pb Zircon geochronology of the southern portion of the Nabitah mobile belt and Pan-African continental collision in the Saudi Arabian shield. Saudi Arabia Deputy Min. Miner. Resour. Tech. Rec., USGS-TR-04-05.

Stowe, C.W. 1984. The early Archaean Selukwe Nappe, Zimbabwe. In A. Kröner, R. Greiling (eds.), Precambrian tectonics illustrated, pp. 41-56, Stuttgart: Schweitzerbart.

Stringer, C. 1988. Palaeoanthropology: the dates of Eden. Nature 331: 565-566.

Stringer, C., Andrews, P. 1988. Genetic and fossil evidence for the origin of modern humans. Science 239: 1263-1268.

Strydom, D., van der Westhuizen, W. A., Schoch, A.E. 1987. The iron-formation of Bushmanland in the north-western Cape province, South Africa. In P.W.U. Appel, G.L. LaBerge (eds.), Precambrian iron-formations, pp. 621-650. Athens: Theophrastus.

Stuart, G. W., Zengheni, T., Clark, R.A. 1986. Crustal structure of the Limpopo mobile belt, Zimbabwe. Geophysical Jour. Roy. As. Soc. 85:261.

Tagini, B. 1971. Esquisse structurale de la Côte d'Ivoire. Essai de géotechtonique regionale. Thèse Univ. Lausanne Soc. Dev. Min. Côte-d'Ivoire (Sodemi).

Talbot, M.R. 1981. Early Palaeozoic(?) diamictites of south-west Ghana. In M.J. Hambrey, W.B. Harland (eds.), Earth's pre-Pleistocene glacial record, pp. 108-111. Cambridge: University Press.

Talbot, M.R., Livingstone, D.A. 1989. Hydrogen index and carbon isotopes of lacustrine organic matter as lake level indicators. Palaeogeogr., Palaeoclimatol., Palaeoecol. 70: 121-137.

Tankard, A.J., Jackson, M.P.A., Eriksson, K.A., Hobday, D.K., Hunter, D.R., Minter, W.E.L. 1982. Crustal evolution of southern Africa. New York: Springer.

Tanner, P.W.G. 1971. Origin of the host rocks to the Kilembe orebody, Uganda. Annu. Rep. Res. Inst. Afr. Geol., Univ. Leeds 15: 28-34.

Tarling, D.H. 1988. Gondwana and the evolution of the Indian Ocean. In M.G. Audley-Charles, A. Hallam (eds.), Gondwana and Tethys, pp. 61-77. Geol. Soc. Spec. Publ. No. 37.

Tarney, J., Dalziel, I.W.D., De Wit, M.J. 1976. Marginal basin "Rocas Verdes" complex from S. Chile: a model for Archaean greenstone belt formation. In B.F. Windley (ed.), The early history of the earth, pp. 131-146. London: Wiley-Interscience.

Teller, J.T., Rutter, N., Lancaster, N. 1990. Sedimentology and paleohydrology of Late Quaternary lake deposits in the northern Namib desert sand sea, Namibia. Quat. Sci. Rev. 9.

Thomas, M. 1974. Tropical geomorphology. London: McMillan.

Thomas, M., Thorpe, M.B. 1980. Some aspects of the geomorphological interpretation of Quaternary alluvial sediments in Sierra Leone. Z. Geomorph. N.F. 36: 140-161.

Tiecerlin, J.-J., Thouin, C., Kalala, T., Mondeguer, A. 1989. Discovery of sub-lascustrine hydrothermal activity and associated massive sulfides and hydrocarbons in the north Tanganyika trough, East Africa. Geology. 17: 1053-1056.

Tissot, F., Swager, C., Berg, R., Van Straaten, P., Ingovatow, A. 1982. Mineralization in the Karagwe-Ankolean System of north-west Tanzania. In The development potential of Precambrian mineral deposits, UN Dpt. Techn. Coop. Devel. 205-215.

Torquato, J.R., Cordani, U.G. 1981. Brazil-Africa geological links. Earth Sci. Rev. 17: 155-176.

Trompette, R. 1973. Le Précambrien supérieur et Paléozoique inférieur de l'Adrar de Mauritanie (bordure occidentale du basin de Taoudeni, Afrique de l'Ouest). Un exemple de sédimentation de craton. Etude stratigraphique et sédimantoloque, Trav. Lab. Sci. Terre, Univ. Marseille St. Jerome. Serie B, No. 7.

Trompette, R. 1982. Upper Proterozoic (1800-570 Ma) stratigraphy: A survey of lithostratigraphic, paleontological, radiochronological and magnetic correlations. Precambr. Res. 18: 27-52.

Trompette, R. 1988. Phosphorites of the northern Volta basin (Burkina Faso, Niger and Benin). In A.J.G. Notholt, R.P. Sheldon, D.F. Davidson (eds.), Phosphate deposits of the world, vol. 2, phosphate rock resources, pp. 214-218. Cambridge: University Press.

Turner, D.C. 1976. Structure and petrology of the Younger Granite ring complexes. In C.A. Kogbe (ed.), Geology of Nigeria, pp. 143-158. Lagos: Elizabethan Publ. Co.

Turner, D.C., Webb, P.K. 1974. The Daura igneous complex, N. Nigeria; a link between the Younger Granite districts of Nigeria and S. Niger. J. Geol. Soc. London. 130: 71-78.

Uchupi, E. 1989. The tectonic style of the Atlantic Mesozoic rift system. J. Afr. Earth Sci. 8: 143-164.

Ukpong, E.E., Olade, M.A. 1979. Geochemical surveys for lead-zinc mineralization, southern Benue Trough, Nigeria. Trans. Inst. Min. metall. 88, B81-92.

Umeji, A.C. 1983. Geochemistry and mineralogy of the Freetown layered basic igneous complex of Sierra Leone. Chem. Geol. 39: 17-38.

Unrug, R. 1987. Geodynamic evolution of the Lufilian arc and the Kundelungu aulacogen, Angola, Zambia and Zaire, In G. Matheis, H. Schandelmeier (eds.), Current research in African earth sciences, pp. 117-120. Rotterdam: A.A. Balkema.

Unrug, R. 1988. Mineralisation controls and source of metals in the Lufilian fold belt, Shaba (Zaire), Zambia, and Angola. Econ. Geol., 83: 1247-1258.

Unrug, R. 1989. Early Proterozoic Eburnean continental crust of South-central Africa. Abstr. 28th IGC. 3: 268.

Utke, A.W. 1987. New aspects on the evolution of the Late Proterozoic crust in NW Nigeria. In G. Matheis, H. Schandelmeier (eds.), Current research in African earth Sciences, pp. 73-78. Rotterdam: A.A. Balkema.

Vachette (Caen-Vachette), M. 1979. Radiochronologie du Précambrien de Madagascar. 10 Colloq. de Géologie Africaine, Montpellier, 25-27 Avril 1979, Résumés 20-21.

Vail, J.R. 1976. Outline of the geochronology and tectonic units of the basement complex of north-east Africa. Proc. R. Soc. London, Ser. A, 350: 127-141.

Vail, J.R. 1979. Outline of the geology and mineralization of the Nubian shield east of the Nile Valley, Sudan. In S.A. Tahoun (ed.), Evolution and mineralization of the Arabian-Nubian shield, pp. 97-107. Oxford: Pergamon.

Vail, J.R. 1985. Relationship between tectonic terrains and favourable metallogenic domains in the central Arabian-Nubian shield. Trans. Instn. Min. Metall. (Sect. B: Appl. earth Sci.), 94: B1-B5.

Vail, J.R. 1987. Late Proterozoic tectonic terranes in the Arabian-Nubian shield and their characteristic mineralization. Geol. Journ. vol. 22, Thematic Issue, 161-174.

Vail, J.R. 1988a. Lexicon of geological terms for the Sudan. Rotterdam: A.A. Balkema.

Vail, J.R. 1988b. Tectonics and evolution of the Proterozoic basement of northeastern Africa. In S. El Gaby, R.O. Greiling (eds.), The Pan-African belt of northeast Africa and adjacent areas, pp. 195-226. Braunschweig: Vieweg.

Vail, J.R. 1989a. Whither the Kibaran in NE Africa. IGCP no. 255 Newsletter/Bulletin 2: 93-97.

Vail, J.R. 1989b. Ring complexes and related rocks in AFrica. J. Afr. Earth Sci. 8: 19-40.

Vail, J.R., Snelling, N.J. 1971. Isotopic age measurements for the Zambezi orogenic belt and the Urungwe Klippe, Rhodesia. Geol. Rdsch. 60: 619-630.

Valeton, I., Mutakyahwa, M. 1987. Old land surfaces and weathering on the Precambrian basement in SW and NE Tanzania, East Africa. 14th Colloquium of African Geology, Technishce Universitaet Berlin, 18-22 August 1987, (Abstract), C.I.F.E.G., Occasional Publ. 1987/12: 130-131.

Van der Merwe, S.W., Botha, B.J.V. 1989. The Groothoek thrust belt in western Namaqualand; an example of a mid-crustal structure. S. Afr. J. Geol. 92: 155-166.

Van Houten, F.B., Hargraves, R.B. 1987. Palaeozoic drift of Gondwana: palaeomagnetic and stratigraphic constraints. Geol. Journ. vol. 22, Thematic Issue, 341-359.

Van Houten, F.B., Karasek, R. 1981. Sedimentologic framework of the Late Devonian oolitic iron formation, Shatti Valley, west-central Libya. J. Sediment. Petrol. 51: 415-427.

Van Reenen, D.E., Barton, J.M. Jr., Roering, C., Smith, C.A., Van Schwalkwyk, J.F. 1987. Deep crustal response to continental collision: The Limpopo belt of southern Africa. Geology 15: 11-14.

Van Zinderen Bakker, E.M., Coetzee, J.A. 1988. A review of late Quaternary pollen studies in east, central and southern Africa. Rev. Palaeobot., Palynol. 55: 155-174.

Vaslet, D. 1989. The Paleozoic of Saudi Arabia and correlations with neighbouring regions. Abstr. 28th IGC. 3: 287.

Vearncombe, J.R. 1983. A dismembered ophiolite from the Mozambique belt, West Pokot. J. Afr. Earth Sci. 1: 133-143.

Veevers, J.J., McElhinny, M.W. 1976. The separation of Australia from other continents. Earth Sci. Rev. 12: 39-159.

Vellutini, P., Rocci, G., Vicat, J.P., Gioan, P. 1983. Mise en évidence de complexes ophiolitiques dans la chaine du Mayombe (Gabon-Angola) et nouvelles interprétations géotectoniques. Precambr. Res. 22: 1-21.

Venkatakrishnan, R., Culver, S.J. 1988. Plate boundaries in West Africa and their implications for Pangean continental fit. Geology 16: 322-325.

Verbeek, T. 1970. Géologie et lithologie du Lindien (Précambrien superieur du Nord de la République démocratique du Congo). Ann. Mus. R. Afr. cetnr., Tervuren, Belg., Sér. in 8 vo, Sci. géol., no. 66.

Vermaak, C.F. 1981. Kunene anorthosite complex, In D.R. Hunter (ed.), Precambrian of the southern hemisphere, pp. 578-598. Amsterdam: Elsevier.

Villeneuve, M. 1989. The geology of the Madina-Kouta basin (Guinea-Senegal) and its significance for the geodynamic evolution of the western part of the West African craton during the Upper Proterozoic period. Precambr. Res. 44: 305-322.

Viljoen, M.J., Viljoen, R.P. 1969. A collection of 9 papers on many aspects of the Barberton granite-greenstone belt, South Africa. Geol. Soc. S. Afr. Sp. Publ. 2.

Vogel, J.C. 1989. Evidence of past climatic change in the Namib desert. Palaeogeogr. Palaeoclimatol., Palaeoecol. 70: 355-366.

Voigt, B., Gabriel, B., Lassonczyk, B., Ghod, M.M. 1990. Quaternary events at the Horn of Africa. Berliner geowiss. Abh. A 120: 697-694.

von Knorring, O., Condliffe, E. 1987. Mineralized pegmatites in Africa. Geol. Journ. vol. 22, Thematic Issue, 253-270.

von Rad, U., Hinz, K., Sarnthein, Seibold, E. (eds.). 1982. Geology of the Northwest African continental margin. Berlin: Springer.

Wadswortzh, W.J. 1963. The Kapalugulu layered intrusion of western Tanganyika. Min. Soc. Amer. Spec. Pap. 1:108-115.

Walters, R., Linton, R.E. 1973. The sedimentary basin of coastal Kenya. In A. Blant. (ed.), Sedimentary basins of the African coast, 2, South and East coasts, p. 133-158. Assoc. Afr. Geol. Surv. Paris.

Warden, A.J., Horkel, A.D. 1984. The geological evolution of the NE-Branch of the Mozambique belt (Kenya, Somalia, Ethiopia). Mitt. Österr. geol. Ges. 77: 161-184.

Warme, J.E. 1988. Jurassic carbonate facies of the central and eastern High Atlas rift, Morocco. In V.H. Jacobshagen (ed.), The Atlas system of Morocco, Lecture Notes in Earth Sciences, 15: 169-199. Berlin: Springer.

Webber, K.J., Daukoru, E. 1975. Petroleum geology of the Niger Delta. Ninth World Petroleum Congress 2: 209-221.

Wendt, J. 1985. Disintegration of the continental margin of northwestern Gondwana: Late Devonian of the eastern Anti-Atlas (Morocco). Geology 13: 815-818.

Wendt, J. 1988. Facies pattern and paleogeography of the Middle and Late Devonian in the eastern Anti-Atlas (Morocco). In N.J. McMillan, A.F. Embry, D.J. Glass (eds.), Devonian of the world, pp. 467-481. Can. Soc. Petrol. Geol.

Wells, M.K. 1962. Structure and petrology of the Freetown Layered Basic Complex of Sierra Leone. Overseas Geol. Miner. Resour. London, Bull. Supp. no. 4.

Wescott, W.A. 1988. A late Permian fan-delta system in the southern Morondava basin, Madagascar. In W. Nemec, R.J. Steel (eds.), Fan deltas: Sedimentology and tectonic settings, pp. 226-237. U.K.: Blackie and Son.

Whiteman, A.J. 1972. The "Cambro-Ordovicien" rocks of Al Jazair (Algeria) - a review. In T.F.J. Dessauvagie, A.J. Whiteman (eds.), African geology, pp. 547-567. Ibadan: University Press.

Whiteman, A.J. 1981. East Africa basins: reserves, resources and prospects. In Petroleum exploration strategies in developing countries. United Nations meeting, Hague, Proceedings, United Nations Natural Resources and Energy Division, 51-98.

Whiteman, A.J. 1982. Nigeria: Its petroleum geology, resources and potential. London: Graham & Trotman.

Wiedman, J. 1988. Plate tectonics, sea level changes, climate--and the relationship to ammonite evolution, provincialism, and mode of life. In J. Wiedmann, J. Kullmann (eds.), Cephalopods - Present and Past, pp. 737-765. Stuttgart: Schwizerbart'sche Verlagsbuchhandlung.

Wildi, W., Favre, P. 1989. History of seaway linking western Tethys to opening Atlantic ocean. Abstr. 28th IGC. 3: 358-359.

Williams, H.R., Culver, S.J. 1982. The Rokelides of West Africa-Pan-African aulacogen or back-arc basin? Precambr. Res. 18: 262-273.

Williams, D.F., Thunnel, R.C., Tappa, E., Rio, D., Raffi, I. 1988. Chronology of the Pleistocene oxygen isotope record: 0-1.88 m.y. B.P. Palaeogeogr., Palaeoclimatol., Palaeoecol. 64: 221-240.

Williamson, P. 1982. Molluscan biostratigraphy of the Koobi Fora hominid-bearing deposits. Nature 295: 140-142.

Wilson, A.H. 1982. The geology of the Great Dyke, Zimbabwe: the ultramafic rocks. J. petrol. 23: 240-292.

Wilson, A.H. 1989. Fluid dynamic controls on formation of Cycle Unit I of Great Dyke, Zimbabwe. Abstr. 28th IGC. 3: 365.

Wilson, J.F. 1972. The Rhodesian Archaean craton--an essay on cratonic evolution. Philos. Trans. R. Soc. London A273: 389-411.

Windley, B.F. 1984. The evolving continents. Chichester: John Wiley.

Woakes, M. 1989. Mineral belts of Nigeria: A review. Global Tectonics and Metallogeny. 3: 115-123.

Woodwell, G.M. (ed.). 1984. The role of terrestrial vegetation in the global cycle: Measurement by remote sensing. U.K.: Wiley.

Wright, J.B., Hastings, D.A., Jones, W.B., Williams, H.R. 1985. Geology and mineral resources of West Africa. London: George Allen & Unwin.

Wright, R.P., Askin, R.A. 1987. The Permian-Triassic boundary in the southern Morondaya basin of Madagascar as defined by plant microfossils. In G.D. McKenzie (ed.), Gondwana Six: Stratigraphy, Sedimentology, and Paleontology, pp. 157-166. Geophysical Monograph 41, Washington D.C.: American Geophysical Union.

Wycisk, P., Klitzsch, E., Jas, C., Reynolds, O. 1990. Intracratonic sequence development and structural control of Phanerozoic strata in Sudan. Berliner geowiss. Abh. A 120: 45-86.

Yemane, K., Kelts, K. 1990. A short review of palaeoenvironments for Lower Beaufort (Upper Permain) Karoo sequences from southern to central Africa: A major Gondwana lacustrine episode. J. Afr. Earth Sci. 10: 169-185.

Lecture Notes in Earth Sciences

Vol. 1: Sedimentary and Evolutionary Cycles. Edited by U. Bayer and A. Seilacher. VI, 465 pages. 1985. (out of print).

Vol. 2: U. Bayer, Pattern Recognition Problems in Geology and Paleontology. VII, 229 pages. 1985.

Vol. 3: Th. Aigner, Storm Depositional Systems. VIII, 174 pages. 1985.

Vol. 4: Aspects of Fluvial Sedimentation in the Lower Triassic Buntsandstein of Europe. Edited by D. Mader. VIII, 626 pages. 1985.

Vol. 5: Paleogeothermics. Edited by G. Buntebarth and L. Stegena. II, 234 pages. 1986.

Vol. 6: W. Ricken, Diagenetic Bedding. X, 210 pages. 1986.

Vol. 7: Mathematical and Numerical Techniques in Physical Geodesy. Edited by H. Sünkel. IX, 548 pages. 1986.

Vol. 8: Global Bio-Events. Edited by O. H. Walliser. IX, 442 pages. 1986.

Vol. 9: G. Gerdes, W. E. Krumbein, Biolaminated Deposits. IX, 183 pages. 1987.

Vol. 10: T.M. Peryt (Ed.), The Zechstein Facies in Europe. V, 272 pages. 1987.

Vol. 11: L. Landner (Ed.), Contamination of the Environment. Proceedings, 1986. VII, 190 pages.1987.

Vol. 12: S. Turner (Ed.), Applied Geodesy. VIII, 393 pages. 1987.

Vol. 13: T. M. Peryt (Ed.), Evaporite Basins. V, 188 pages. 1987.

Vol. 14: N. Cristescu, H. I. Ene (Eds.), Rock and Soil Rheology. VIII, 289 pages. 1988.

Vol. 15: V. H. Jacobshagen (Ed.), The Atlas System of Morocco. VI, 499 pages. 1988.

Vol. 16: H. Wanner, U. Siegenthaler (Eds.), Long and Short Term Variability of Climate. VII, 175 pages. 1988.

Vol. 17: H. Bahlburg, Ch. Breitkreuz, P. Giese (Eds.), The Southern Central Andes. VIII, 261 pages. 1988.

Vol. 18: N.M.S. Rock, Numerical Geology. XI, 427 pages. 1988.

Vol. 19: E. Groten, R. Strauß (Eds.), GPS-Techniques Applied to Geodesy and Surveying. XVII, 532 pages. 1988.

Vol. 20: P. Baccini (Ed.), The Landfill. IX, 439 pages. 1989.

Vol. 21: U. Förstner, Contaminated Sediments. V, 157 pages. 1989.

Vol. 22: I. I. Mueller, S. Zerbini (Eds.), The Interdisciplinary Role of Space Geodesy. XV, 300 pages. 1989.

Vol. 23: K. B. Föllmi, Evolution of the Mid-Cretaceous Triad. VII, 153 pages. 1989.

Vol. 24: B. Knipping, Basalt Intrusions in Evaporites. VI, 132 pages. 1989.

Vol. 25: F. Sansò, R. Rummel (Eds.), Theory of Satellite Geodesy and Gravity Field Theory. XII, 491 pages. 1989.

Vol. 26: R. D. Stoll, Sediment Acoustics. V, 155 pages. 1989.

Vol. 27: G.-P. Merkler, H. Militzer, H. Hötzl, H. Armbruster, J. Brauns (Eds.), Detection of Subsurface Flow Phenomena. IX, 514 pages. 1989.

Vol. 28: V. Mosbrugger, The Tree Habit in Land Plants. V, 161 pages. 1990.

Vol. 29: F. K. Brunner, C. Rizos (Eds.), Developments in Four-Dimensional Geodesy. X, 264 pages. 1990.

Vol. 30: E. G. Kauffman, O.H. Walliser (Eds.), Extinction Events in Earth History. VI, 432 pages. 1990.

Vol. 31: K.-R. Koch, Bayesian Inference with Geodetic Applications. IX, 198 pages. 1990.

Vol. 32: B. Lehmann, Metallogeny of Tin. VIII, 211 pages. 1990.

Vol. 33: B. Allard, H. Borén, A. Grimvall (Eds.), Humic Substances in the Aquatic and Terrestrial Environment. VIII, 514 pages. 1991.

Vol. 34: R. Stein, Accumulation of Organic Carbon in Marine Sediments. XIII, 217 pages. 1991.

Vol. 35: L. Håkanson, Ecometric and Dynamic Modelling. VI, 158 pages. 1991.

Vol. 36: D. Shangguan, Cellular Growth of Crystals. XV, 209 pages. 1991.

Vol. 37: A. Armanini, G. Di Silvio (Eds.), Fluvial Hydraulics of Mountain Regions. X, 468 pages. 1991.

Vol. 38: W. Smykatz-Kloss, S. St. J. Warne, Thermal Analysis in the Geosciences. XII, 379 pages. 1991.

Vol. 39: S. E. Hjelt, Pragmatic Inversion of Geophysical Data. Approx. 240 pages (in prep.). 1991.

Vol. 40: S. W. Petters, Regional Geology of Africa. XXIII, 722 pages. 1991.